Principles of Polymer Chemistry

Principles of Polymer Chemistry

A. Ravve
Consultant in Polymer Chemistry
Lincolnwood, Illinois

PLENUM PRESS • NEW YORK AND LONDON

Library of Congress Cataloging-in-Publication Data

On file

ISBN 0-306-44873-4

A Division of Plenum Publishing Corporation
233 Spring Street, New York, N. Y. 10013

10 9 8 7 6 5 4 3 2 1

Printed in the United States of America

Preface

This book is written for graduates and advanced undergraduates in polymer chemistry and is also intended to be used as a reference by practicing chemists. It is suitable for self-education, because the topics are written without assumption of prior knowledge. The reader should possess, however, a thorough knowledge of organic and physical chemistry, at the undergraduate level.

There are eight chapters. Chapter 1 is introductory. It describes definitions, nomenclature, morphology, and relationships of chemical structures to physical properties. The mechanisms of chain-growth polymerizations, free-radical and ionic, are discussed in Chapters 2 and 3. Chapter 3 also includes coordinated anionic polymerizations. In addition, there is a section on group-transfer polymerization. It was placed there with some hesitation, because the propagation reaction takes place by a step-growth mechanism. The location appears appropriate, however, because the products of group-transfer polymerization are acrylic polymers, and these materials are mainly formed by chain-growth mechanism. Ring-opening polymerizations are presented in Chapter 4. Chapter 5 covers commercially important chain-growth polymers, their industrial preparations, properties, and performance. Step-growth polymerizations are discussed in Chapter 6 together with industrially important step-growth polymers, preparations, and properties. Naturally occurring polymers are described in Chapter 7. While cellulose and starch are presented fully, proteins and nucleic acids are described only briefly, because detailed treatment belongs in biochemistry. Chemical modifications and reactions of polymeric materials, including polymeric reagents and functional polymers, are discussed in Chapter 8.

Greater emphasis is placed on commercially important materials. Most students find employment in industry. Familiarity with materials and practices are an essential and practical part of their training. Some trade names are included for the same reason.

Two of the chapters contain appendixes with computer programs. They are offered to readers for practice. One program is there to help gain greater understanding of the meaning of molecular weights in polymers and to learn how they are calculated from gel permation chromatogram (GPC) data. The other one is to understand statistical interpretations of the propagation mechanisms from nuclear magnetic resonance (NMR) data and to gain greater insight into stereochemistry of polymers. These computer programs were not written by accomplished professional computer programmers and should not be judged as such. They are useful, however, in their present form.

There are review questions at the ends of the chapters. After reading each chapter, the students are advised to try to answer the questions first and then to compare their answers with the discussions in the book.

Silicon polymers and polyphosphazenes are discussed in Chapter 6. It can be argued that they belong in Chapter 4, because they are formed by ring-opening polymerizations. I chose Chapter 6, because silicon polymers can also be formed by reactions other than ring-opening polymerizations.

In addition, the properties and the uses of silicon polymers are of primary importance and, for this reason, I placed them with the discussions of other commercial chain-growth polymers. I also placed polyphosphazenes in this chapter, because they attain their useful properties only after replacement of the chlorine atoms with organic moieties.

This book is dedicated to all scientists whose names appear in the references.

A. Ravve

Contents

1. Introduction

2. Free-Radical Chain-Growth Polymerization

3. Ionic Chain-Growth Polymerization

4. Ring-Opening Polymerizations

5. Common Chain-Growth Polymers

7. Naturally Occurring Polymers

8. Reactions of Polymers

1

Introduction

1.1. Definitions

The word polymer is commonly understood to mean a large molecule composed of repeating units, or mers (from the Greek word *meros*—part), connected by covalent bonds. Such units may be connected in a variety of ways. The simplest is a linear polymer, or a polymer in which the units are connected to each other in a linear sequence, like beads on a string. Many examples of such linear polymers are possible, as, for instance, linear polyethylene:

$$\sim CH_2–CH_2–CH_2–CH_2–CH_2\sim \qquad -[–CH_2–CH_2–]-$$

repeat unit

The terminal units in such molecules must be different from the internal ones to satisfy valence requirements. Polyethylene, like all other polymers, can be written to show the number of repeat units, $-[–CH_2–CH_2–]_n-$, by using a number or a letter, like in this case *n*. It represents the quantity of mers present in the polymer and is called the *degree of polymerization*, or DP.

An alternative to a linear polymer is a branched one. The branches can be long or short. Low-density polyethylene, for instance, can have both short and long branches. Linear and branched molecules are illustrated in Fig. 1.1a and b. Branched polymers can also be star or comb shaped (Fig. 1.1c and d). In addition to the above, polymer molecules can also be double stranded. Such polymers are called ladder polymers (Fig. 1.1e). It is also possible for polymers to have semiladder structures (Fig. 1.1f).

When branches of different polymers become interconnected, *network* structures form. Planar networks resemble the structure of graphite. Three-dimensional networks, or space networks, however, can be compared to diamonds. A network polymer is illustrated in Fig. 1.1g.

The term polymer can apply to molecules made up from either single repeating structural units, as in the above shown polyethylene, or from different ones. If there are two or more structural units, then the term copolymer is used. An example would be a copolymer of ethyl methacrylate and styrene:

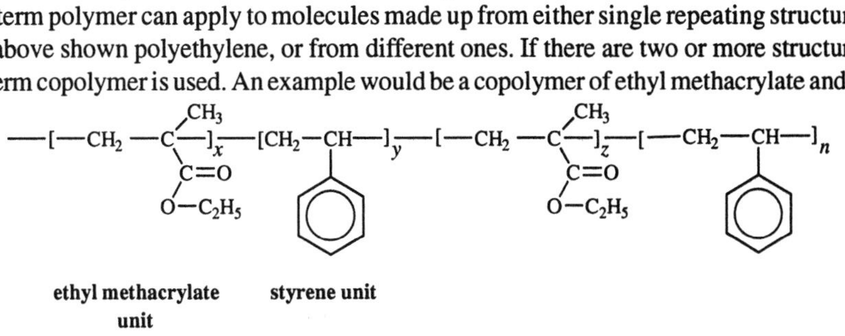

$$—[—CH_2—C(CH_3)(C{=}O\,O—C_2H_5)—]_x—[CH_2—CH(C_6H_5)—]_y—[—CH_2—C(CH_3)(C{=}O\,O—C_2H_5)—]_z—[—CH_2—CH(C_6H_5)—]_n$$

ethyl methacrylate unit **styrene unit**

A copolymer can also be linear or branched. Should there be regularity in the repetition of the structural units and should this repetition alternate, then the copolymer is called an *alternating copolymer*. An absence of such regularity would make it a *random copolymer*.

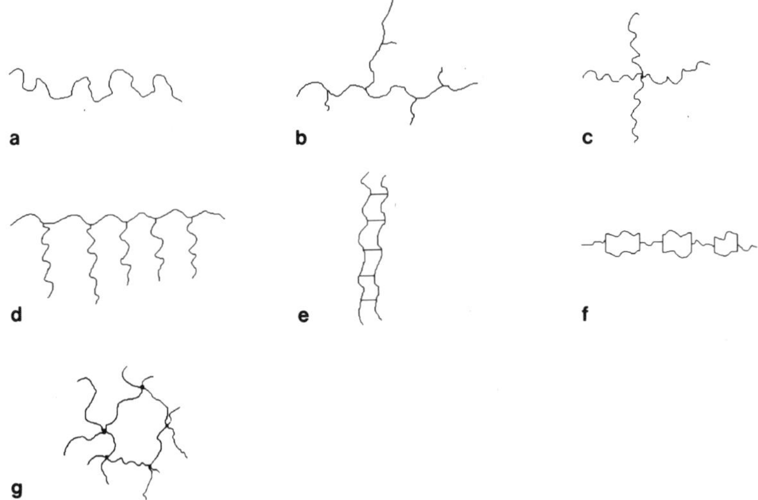

FIGURE 1.1. Shapes of polymeric molecules. (a) Linear polymer, (b) branched polymer, (c) star-shaped polymer, (d) comb-shaped polymer, (e) ladder polymer, (f) semiladder polymer, and (g) network structure.

An example of an alternating copolymer can be a copolymer of styrene with maleic anhydride:

$$-[CH_2-\underset{C_6H_5}{CH}-\underset{\underset{O}{\overset{\|}{C}}-O-\underset{O}{\overset{\|}{C}}}{CH-CH}-]_n-$$

In addition to the random sequence and an alternating one, sometimes called ordered sequence, there are also *block* copolymers. These are copolymers made up of blocks of individual polymers joined by covalent bonds. An example can be a block copolymer of styrene and isoprene:

$$-[-CH_2-\underset{C_6H_5}{CH}-]_n- \quad -[-CH_2-\overset{CH_3}{\overset{|}{C}}=CH-CH_2-]_m-$$

polystyrene block **polyisoprene block**

Still another type of copolymer is one that possesses backbones composed of one individual polymer and the branches from another one. It is called a *graft copolymer*. A graft copolymer of acrylonitrile with polyethylene can serve as an example:

$$\begin{array}{l} \qquad\qquad\qquad\qquad\quad C\equiv N \qquad C\equiv N \\ \qquad\qquad\qquad\qquad CH_2-CH-CH_2-CH\sim \\ \sim CH_2-CH-CH_2-CH_2-CH-CH_2\sim \\ \qquad\quad CH_2-CH-CH_2-CH\sim \\ \qquad\qquad\quad C\equiv N \qquad C\equiv N \end{array}$$

In both block and graft copolymers the length of the uninterrupted sequences may vary.

In 1929 Carothers[1] suggested a separation of all polymers into two classes, *condensation* and *addition* polymers. By condensation polymers he defined those polymers that lack certain atoms from the monomer units from which they were formed or to which they may be degraded by chemical means. He also defined addition polymers as polymers with identical structures of the repeat units to the monomers from which they are derived. Thus, according to the above definition, an example of a condensation polymer can be a polyamide formed by condensing a diacid chloride with a diamine and splitting off hydrochloric acid:

$$n\ \mathrm{Cl{-}\overset{O}{\overset{\|}{C}}{-}R{-}\overset{O}{\overset{\|}{C}}{-}Cl} + n\ \mathrm{H_2N{-}R'{-}NH_2} \longrightarrow \mathrm{{-}[{-}\underset{H}{\underset{|}{N}}{-}R'{-}\underset{H}{\underset{|}{N}}{-}\overset{O}{\overset{\|}{C}}{-}R{-}\overset{O}{\overset{\|}{C}}{-}]_n{-}} + n\ \mathrm{HCl}$$

Also, according to the above definition, an example of an addition polymer can be polystyrene that is formed by addition of styrene monomers:

$$n\ \mathrm{CH_2{=}CH(C_6H_5)} \longrightarrow \mathrm{{-}[{-}CH_2{-}CH(C_6H_5){-}]_n{-}}$$

monomer polymer

Note: The definition ignores loss of double bonds. The Carothers definition fails to describe all the polymers that can fit into the category of condensation polymers, yet form without an evolution of a byproduct. An example is a polyurethane that can form from a reaction of a glycol with a diisocyanate:

$$n\ \mathrm{O{=}C{=}N{-}R{-}N{-}C{=}O} + n\ \mathrm{HO{-}R'{-}OH} \longrightarrow \mathrm{{-}[{-}\overset{O}{\overset{\|}{C}}{-}\underset{H}{\underset{|}{N}}{-}R{-}\underset{H}{\underset{|}{N}}{-}\overset{O}{\overset{\|}{C}}{-}O{-}R'{-}O{-}]_n{-}}$$

Flory proposed a superior definition.[2] It is based on the reaction mechanism involved in the formation of the two classes of polymers. Into the first category (it includes all the condensation polymers) fall the macromolecules that form through reactions that occur in *discrete steps*. They are therefore called *step-growth* polymers. Such polymerizations require long periods of time for each macromolecule to form, usually measured in hours. Into the second category belong all polymers that form by *chain propagating reactions*. They are therefore called *chain-growth polymers*, as one might expect. Such reactions depend upon the presence of active centers on the ends of the growing chains. The chains grow by propagating these reactive sites through inclusion of monomers at such sites. These inclusions are very rapid and chain growth can take place in a fraction of a second, as the chains successively add monomers.

The important features of step-growth polymerizations are:

1. The monomer is consumed early in the beginning of the reaction while the increase in molecular weight occurs only slowly.
2. The growth of polymeric chains takes place by reactions between monomers, oligomers and polymers.
3. There is no termination step and the end groups of the polymers are reactive throughout the process of polymerization.
4. The same reaction mechanism functions throughout the process of polymerization.

The important features of chain-growth polymerizations are:

1. Chain-growth takes place by repeated additions of monomers to the growing chains at the reactive sites.
2. The monomer is consumed slowly and is present throughout the process of polymerization.
3. There are two distinct mechanisms during polymer formations. These are initiation and propagation.
4. In most of cases, there is also a termination step.

When the polymerization reaction takes place in three dimensions, after it has progressed to a certain point, gelation occurs. This well-defined change during polymerization is known as the *gel point*. At this point the reaction mixture changes from a viscous liquid to an elastic gel.

Before gelation, the polymer is soluble and fusible. After gelation, it is neither soluble nor fusible. This is a result of restraining effects of three-dimensional space networks. Another classification of polymers is also possible. It is based on whether the material can form crosslinked or gelled networks. The polymers that eventually reach gelation are called *thermosetting*. Such polymers are also called *crosslinkable* polymers. Once past gelation, raising the temperature will no longer attain plasticity as the molecules can no longer move past each other. For the same reason they can no longer be dissolved in any solvents.

Polymers that never gel or become crosslinked are called *thermoplastic*. Such polymers can always be reflowed upon application of heat. They can also be redissolved in appropriate solvents.

A polymer network was illustrated above by a schematic drawing. It appears desirable, however, to show what an actual thermoset polymer may look like. The following illustration of a phenol-formaldehyde thermoset resin is meant to show the phenol rings and the accompanying linkages as extending into various directions:

where ~ implies that the polymer extends further in that direction. The above illustration is one of a thermoset polymer that is formed by the step-growth mechanism. It is also possible to form crosslinked polymers by the chain-growth mechanism. This requires the presence in the polymerization mixture of a comonomer that possesses multiple functionality. Copolymerization of styrene with a comonomer like divinyl benzene can serve as an example:

An *oligomer* is a very low molecular weight polymer. It consists of only a small number of mers. The definition of a *telomer* is that of a chain-growth polymer that is composed of molecules with end groups consisting of different species from the monomer units. Telomers can form by either free-radical or by ionic chain-growth polymerization mechanism.

Telechelic polymers are macromolecules with reactive functional groups at the terminal ends of the chains. An example of telechelic polymers is polybutadiene with carboxylic acid end groups:

$$HO-\overset{O}{\overset{\|}{C}}-(-CH_2-\underset{H-C=CH_2}{\underset{|}{CH}}-)_n-\overset{O}{\overset{\|}{C}}-OH$$

1.2. Nomenclature of Polymers

The names of many polymers are based on the monomers from which they were prepared. There is, however, frequent variation in the format. A nomenclature of polymers was recommended by IUPAC[3–5] and is used in some publications. Strict adherence to the recommendation, however, is mainly found in reference works. Also, problems are often encountered with complex polymeric structures that are crosslinked or have branches. In addition some polymers derive their names from trade names. For instance, a large family of polyamides is known as nylons. When more than one functional group is present in the structure, the material may be called according to all functional groups in the structure. An example is a polyesteramide. A thermoset polymer prepared from two different materials may be called by both names. For instance, a condensation product of melamine and formaldehyde is called melamine-formaldehyde polymer.

1.2.1. Nomenclature of Chain-Growth Polymers

1. A polymer of unspecified chain length is named with a prefix *poly*. The prefix is then followed by the name of the monomer. Also, it is customary to use the common names of monomers and polymers. For instance, common names for phenylethene and polyphenylethene are styrene and polystyrene. This, however, is not an inflexible rule. When the monomer is named by a single word, then the prefix poly is simply added like polyethylene for a polymer of ethylene or polystyrene for a polymer of styrene. If, however, the monomer is named by two words or is preceded by a number, like methyl methacrylate, parentheses are used. Examples are poly(methyl methacrylate) or poly(1-hexene).

2. End groups are usually not specified in high polymers. End groups, however, can be known parts of the structure. This can be the case with telomers. Here, the end groups are named as radicals, prefixed by Greek letters, α and ω. They appear before and after the name of the polymer. The structure of a telomer, $Cl\text{–}(\text{–}CH_2\text{–})_n\text{–}CCl_3$, is therefore called α-chloro-ω-trichloromethyl poly(methylene).

3. In naming the polymer the following steps are recommended by IUPAC: (1) identify the constitutional repeating unit, (2) orient the constitutional repeating unit, and (3) name the constitutional repeating unit.

4. Random copolymers are designated by the prefix *co*, as in poly(butadiene-*co*-styrene) and poly(vinyl chloride-*co*-vinyl acetate). Alternating copolymers can be differentiated by substituting *alt* for *co*, as in poly(ethylene-*alt*-carbon monoxide).

5. The prefix *g* describes graft copolymers and the prefix *b* describes block copolymers. In this system of nomenclature, the first polymer segment corresponds to the homopolymer or copolymer that was formed during the first stage of the synthesis. Should this be a graft copolymer then this will represent the backbone polymer. For instance, if polystyrene is graft copolymerized with polyethylene, the product is called poly(ethylene-*g*-styrene). A more complex example can be poly (butadiene-*co*-styrene-*g*-acrylonitrile-*co*-vinylidine chloride). Similarly, examples of block copolymers would be poly(acrylonitrile-*b*-methyl methacrylate) or poly(methyl methacrylate-*b*-acrylonitrile).

6. Conventional prefixes indicating *cis* and *trans* isomers are placed in front of the polymer name. An example is *cis*-1,4-polybutadiene, or in *trans*-1,4-polyisoprene.

7. The nomenclature adopted by IUPAC rests upon selection of preferred *constitutional repeating* units[5] from which the polymer is a multiple. The unit is named wherever possible according to the definitive rules for nomenclature of organic chemistry.[6] For single stranded polymers this unit is a bivalent group. An example is a polymer with oxy(1-fluoroethylene) constitutional repeat unit:

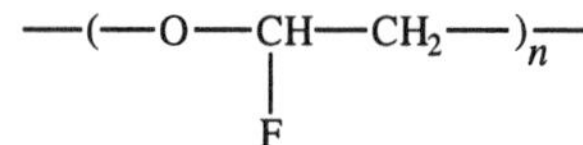

poly[oxy(1-fluoroethylene)]

The following are examples of simple constitutional repeat units:

$-(CH_2-)_n-$ **polymethylene** $-(-O-CH_2-CH_2-)_n-$ **polyoxyethylene** $-(-CH_2=CH-CH_2-CH_2-)_n-$ **poly(1-butenylene)**

8. Polymers with repeating units consisting of more than one simple bivalent radical should be named according to the order of seniority among the types of bivalent radicals: (a) heterocyclic rings, (b) chains containing heteroatoms, (c) carbocyclic rings, and (d) chains containing only carbons. This is illustrated below:

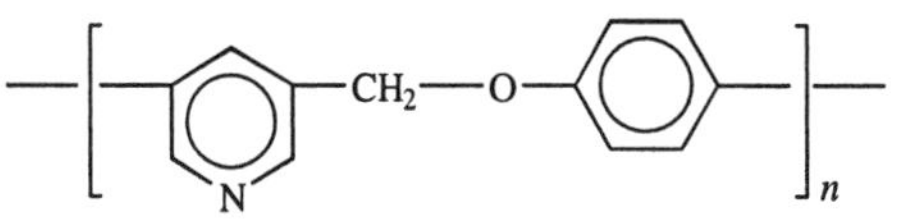

poly(3,5-pyridinediylmethyleneoxy-1,4-phenylene) 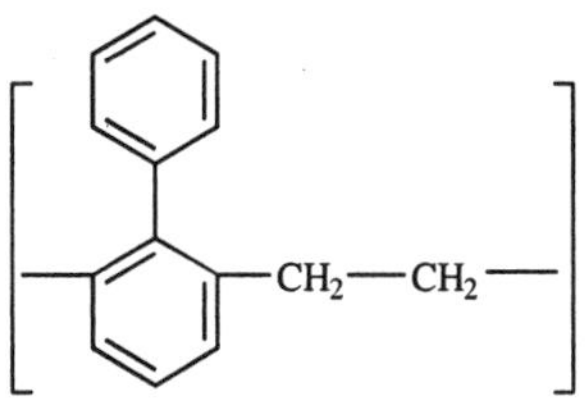 **poly(2,6-biphenyleneethylene)**

9. Double stranded or "ladder" polymers that have tetravalent repeat units are named similarly to bivalent units. The relation of the four free valences is denoted by pairs of locants separated by a chain:

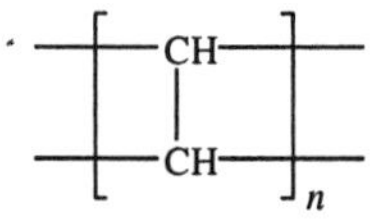

poly(1,2:1,2-ethane diylidene)

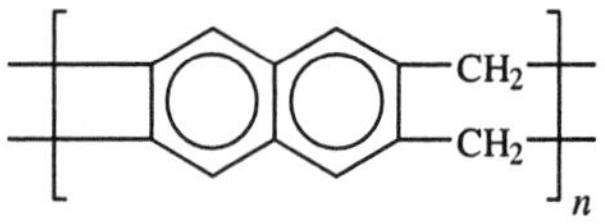

poly(2,3:6,7-naphthalenetetrayl-6,7-dimethylene)

1.2.2. Nomenclature of Step-Growth Polymers

The nomenclature for step-growth polymers is more complicated due to the possibility of having many different repeat units. Usually, the polymers are referred to according to their functional units. A polyester from ethylene glycol and terephthalic acid is called poly(ethylene terephthalate). A product of ring-opening polymerization, like, for instance, a polymer of caprolactam, might be called polycaprolactam. The repeat unit is actually not a lactam but rather an open-chain polyamide. Because it is derived from a lactam, it may still carry that name. In this particular instance, however, it is more common to call the polymer by its generic name, namely, nylon 6. The same would be true for a polymer from a lactone, like poly(β-propiolactone). The IUPAC name for this polymer, however, is:

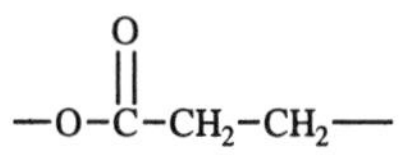

$$-O-\overset{\overset{O}{\|}}{C}-CH_2-CH_2-$$

poly[oxy(1-oxotrimethylene)]

The name is based on a presence of two subunits. Note that the carbonyl oxygen is called an *oxo* substituent. In addition, the presence of a 1-oxo substituent requires that parentheses enclose the subunit.

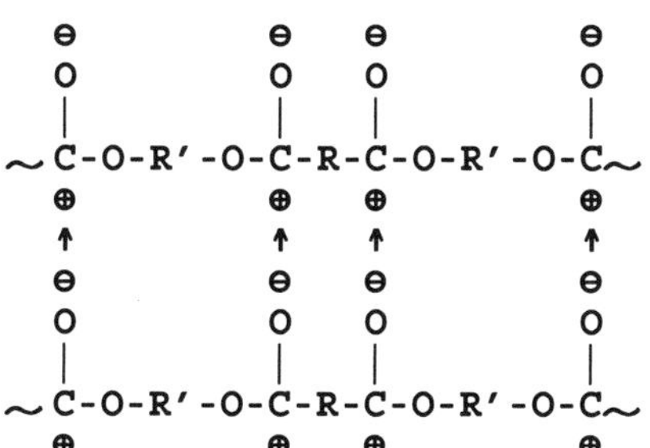

FIGURE 1.2. Intramolecular forces.

1.3. Structure and Property Relationship in Organic Polymers

Physical properties of polymers are influenced by the sizes of the molecules and by the nature of the primary and secondary bond forces. They are also influenced by the amount of symmetry or uniformity in molecular structures, and by arrangements of the macromolecules into amorphous or crystalline domains. This affects melting temperatures, solubilities, melt and solution viscosities, tensile strengths, elongation, flexibility, etc.[9]

Due to the large sizes of the polymer molecules, the secondary bond forces assume much greater roles in influencing physical properties than they do in small organic molecules. These secondary bond forces are van der Waal forces and hydrogen bonding.

The van der Waal forces can be subdivided into three types: dipole–dipole interactions, induced dipoles, and time-varying dipoles.

1.3.1. Effects of Dipole Interactions

These interactions result from molecules carrying equal and opposite electrical charges. The amount of these interactions depends upon the ability of the dipoles to align with one another. Molecular orientations of this sort are subject to thermal agitation that tends to interfere. As a result, dipole forces are strongly temperature-dependent. An illustration of dipole–dipole interaction is interaction between chains of a linear polyester. Each carbonyl group of the ester linkage sets up a weak field through polarization. The field, though weak, interacts with another field of the same type on another chain. This creates forces of cohesion. Because there are many such fields in a polyester, the net effect is strong cohesion between chains. The interaction is illustrated in Fig. 1.2.

1.3.2. Induction Forces in Polymers

These forces result from slight displacement of electrons and nuclei in covalent molecules. They take place in the proximity of electrostatic fields associated with the dipoles from other molecules. Such displacement causes interaction between the *induced dipoles* and the permanent dipoles creating induction forces. The energy of the induction forces is small and not temperature-dependent.

The third force is a result of different instantaneous configurations of the electrons and nuclei about the bonds of the polymeric chains. These are time-varying dipoles that average out to zero. They are polarizations arising from molecular motions.

The total bond energy of all the secondary bond forces combined, including hydrogen bonding, ranges between 2 and 10 kcal/mole. Of these, however, hydrogen bonding takes up the greatest share of the bond strength. Table 1.1. presents the intramolecular forces of some linear polymers.[7,13]

As can be seen in Table 1.1. polyethylene possesses much less cohesive energy than does a polyamide. This difference is primarily due to hydrogen bonding. It can be illustrated by comparing a polyamide, like nylon 11, with linear polyethylene (see Fig. 1.3.). Both have similar structures, with the exception that nylon 11 has periodic amide linkages after every tenth carbon. This is absent in polyethylene. The amide linkages participate in hydrogen bonding with neighboring chains. Due

Table 1.1. Molecular Cohesion of Some Linear Polymers[a]

Polymer	Cohesion/5 Å chain length (kcal/mole)	Repeat unit
Polyethylene	1.0	$-CH_2-CH_2-$
Polybutadiene	1.1	$-CH_2-CH=CH-CH_2-$
Polyisobutylene	1.2	$-CH_2-C(CH_3)_2-$
Polyisoprene (*cis*)	1.3	$-CH_2-C(CH_3)=CH-CH_2-$ (cis)
Poly(vinyl chloride)	2.6	$-CH_2-CH(Cl)-$
Poly(vinyl acetate)	3.2	$-CH_2-CH(O-C(=O)-CH_3)-$
Polystyrene	4.0	$-CH_2-CH(C_6H_5)-$
Poly(vinyl alcohol)	4.2	$-CH_2-CH(OH)-$
Polyamides	5.8	$-R-N(H)-C(=O)-R'-$
Cellulose	6.2	(glucose ring structure: CH_2OH, H, O, HO, OH, H)

[a]From Refs. 7 and 23, and other literature sources.

to hydrogen bonding, nylon 11 melts at 184–187 °C and is soluble only in very strong solvents. Linear polyethylene, on the other hand, melts at 130–134 °C and is soluble in hot aromatic solvents.

Intermolecular forces affect the rigidity of polymers. Should these forces be weak, because the cohesive energy is low (1 to 2 kcal per mole), the polymeric chains tend to be flexible. Such chains respond readily to applied stresses and exhibit typical properties of elastomers. High cohesive energy (5 kcal per mole or higher), on the other hand, causes the materials to be strong and tough. Such polymers exhibit resistance to applied stresses and possess good mechanical properties.

The temperatures and the flexibility of polymeric molecules govern both the sizes of molecular segments that can be in motion and the frequencies at which that can occur. This in turn determines the rate at which polymer molecules respond to molecular stresses. In flexible polymers, if the thermal energy is sufficiently high, large segments can disengage and slip past each other quite rapidly in response to applied stress. This is a property of all elastomers.

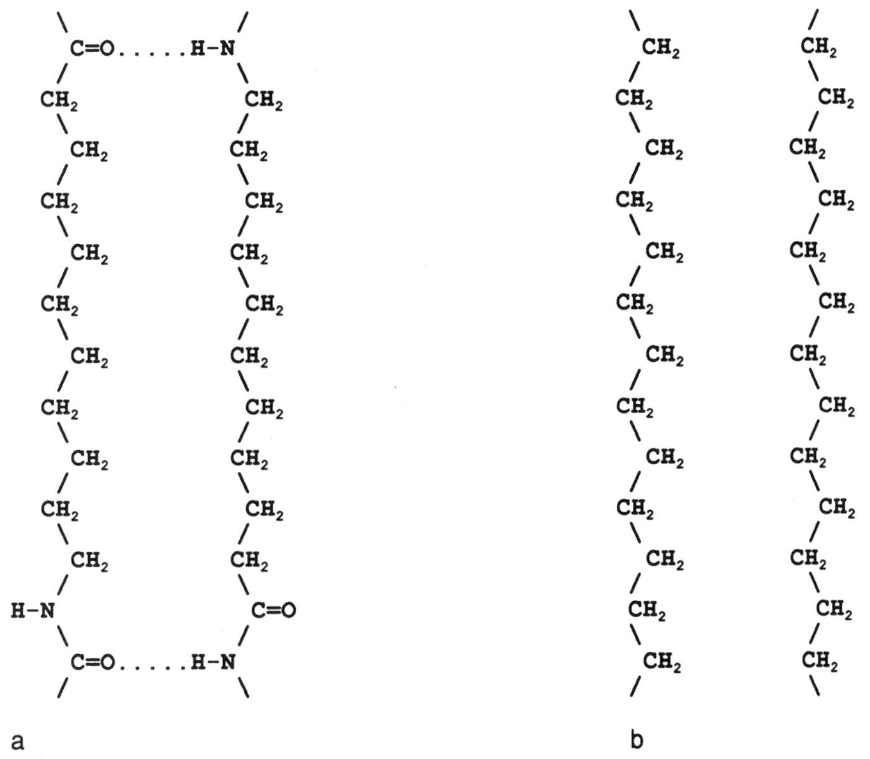

```
  \           /                      \            /
   C=O.....H-N                        CH2         CH2
  /           \                      /            \
 CH2           CH2                  CH2            CH2
  \           /                      \            /
   CH2       CH2                      CH2       CH2
  /           \                      /            \
 CH2           CH2                  CH2            CH2
  \           /                      \            /
   CH2       CH2                      CH2       CH2
  /           \                      /            \
 CH2           CH2                  CH2            CH2
  \           /                      \            /
   CH2       CH2                      CH2       CH2
  /           \                      /            \
 CH2           CH2                  CH2            CH2
  \           /                      \            /
   CH2       CH2                      CH2       CH2
  /           \                      /            \
 CH2           CH2                  CH2            CH2
  \           /                      \            /
   CH2       CH2                      CH2       CH2
  /           \                      /            \
H-N            C=O                  CH2            CH2
  \           /                      \            /
   C=O.....H-N                        CH2         CH2
  /           \                      /            \

a                                   b
```

FIGURE 1.3. Hydrogen bonding in (a) nylon 11, and (b) polyethylene.

1.4. Amorphous and Crystalline Arrangements in Thermoplastic Polymers

The *morphology* or the arrangement of the polymeric chains can be amorphous or crystalline. The term *amorphous* designates a lack of orderly arrangement. Crystalline morphology, on the other hand, means that the chains are aligned in some orderly fashion.

1.4.1. The Amorphous State

Many macromolecules show little tendency to crystallize or align the chains in some form of an order and remain disordered in the solid form. This is the condition of all molten polymers. Some of them, however, due to structural arrangement, remain amorphous upon cooling. This is called *vitrification* and in this form the polymer resembles glass. In the molten state, long-range segmented motions are possible and the molecules are free to move past each other. By comparison, in the solid state only short-range vibrational and rotational motions of the segments are possible. The amorphous state has been described fully with the aid of statistical analysis. That is beyond the scope of this book.

1.4.2. Elasticity

The phenomenon of elasticity of rubber and other elastomers is a result of a tendency of the chains of large amorphous polymers to form kinked conformations.[30] If there is also a certain amount of crosslinking, then these kinked conformations occur between the crosslinks. The distance between the ends in such polymers is much shorter than when they are fully stretched. Deformation or stretching of rubber straightens out the molecules. They tend to return to the kinked state, however, when the deformation is removed. So each segment behaves in a manner that resembles a spring. Some elastomeric materials are capable of high elongation. Some soft rubbers, for instance, can be extended by as much as 800% and even higher with full recovery.

The high degree of elasticity of rubbers is due in part to the effects of thermal motions upon the long polymeric chains. These motions tend to restrict vibrations and rotations about the single bonds in the main chain. Such restrictive forces in the lateral direction are much weaker than are the primary valence forces in the longitudinal direction. Greater amplitudes of motion can occur perpendicular to the chain rather than in the direction of the chain. The increased motion in the perpendicular direction results in a repulsive force between extended or parallel chains. This force causes them to draw together after stretching. Hence the stretched rubber molecules retract due to longitudinal tension until the irregular arrangement of molecules is achieved again. This more random conformation is actually a higher entropy state.

When unstretched rubber is heated it increases in dimension with an increase in the temperature. At higher temperatures, however, rubbers, upon elongation, have a higher tendency to contract. This higher tendency to contract is called *negative coefficient of expansion.*

In summary, polymeric materials exhibit rubber elasticity if they satisfy three requirements: (a) the polymer must be composed of long-chain molecules, (b) the secondary bond forces between molecules must be weak, and (c) there must be some occasional interlocking of the molecules along the chain lengths to form three-dimensional networks. Should the interlocking arrangements be absent, then elastomers lack memory, or have a fading memory and are not capable of completely reversible elastic deformations.

1.4.3. Crystallinity

There has been considerable controversy on the subject of polymer crystallinity.[13] Many questions still remain. When macromolecules possess a certain amount of symmetry, there is a strong accompanying tendency to form ordered domains, or crystalline regions. *Crystallinity* in polymers, however, differs in nature from that of small molecules. When the small molecules crystallize, each crystal that forms is made up of molecules that totally participate in its makeup. Crystallinity in polymers, on the other hand, means that certain elements of the polymeric system or segments of the polymeric chains have attained a form of a three-dimensional order. This order resembles orderly arrangement of small molecules in crystals. The crystalline domains, however, are much smaller than the crystals of small molecules and possess many more imperfections.

For many polymers, crystal growth can take place either from the melt or from dilute solution to yield single crystals. Crystal formations in polymers were studied intensively almost from the time of recognition of their existence in macromolecules. As a result, certain basic principles were established: (1) The melt crystallization process is a first-order phase transition[10,11] (see Section 1.4.5). (2) Crystallization from a molten polymer follows the general mathematical formulation for the kinetics of a phase change.[11,12] Equilibrium conditions, however, are seldom if ever attained[13] and complete crystallinity is not reached.

The tendency to crystallize depends upon the following[14]:

1. Structural regularity of the chains that leads to establishment of identity periods.
2. Free rotational and vibrational motions in the chains allowing different conformations to be assumed.
3. Presence of structural groups that are capable of producing lateral intermolecular bonds (van der Waal forces) and regular, periodic arrangement of such bonds.
4. Absence of bulky substituents or space irregularity, that would prevent segments of the chains from fitting into crystal lattices or prevent laterally bonding groups from approaching each other close enough for best interaction.

When polymers crystallize from the melt, X-ray diffraction studies show recognizable features in some of them. The Bragg reflections, however, appear broad and diffuse, as compared to those obtained from well developed single crystals. Such broadening could result from the crystals being

FIGURE 1.4. Fringed micelle.

small. It could also result from lattice defects. Because diffraction patterns are too weak it is impossible to reach a conclusion. The majority opinion, however, leans toward the small crystal size as being the cause of the broadening. The crystals from the melt are approximately 100×200×200 Å in size. Rough estimates from these diffraction studies indicate that the size of crystals, or crystallites, rarely exceeds a few hundred angstroms. The fact that there is a substantial background of diffuse scattering suggests that considerable amorphous areas are also present. Because the chains are very long, it has often been suggested that individual chains contribute to several different crystalline and amorphous domains. This resulted in a proposal of a composite single-phase structure, a *fringed micelle* or a *fringed crystallite* model.[11] This is illustrated in Fig. 1.4. The fringes are transition phases between the crystalline and the amorphous regions. Some analytical studies, however, failed to support this concept.[29]

Single crystals of organic polymers were observed as early as 1927 by Staudinger.[15] It became the subject of intensive investigations after an observation that linear polyethylene can crystallize into single crystals. The observations made on polyethylene were followed by observations that it is possible to grow single crystals of other polymers. Some of them are polyoxymethylene,[16] polyamides,[17] polypropylene,[18] polyoxyethylene,[19] cellulose,[20] and others.

There is good correlation between the structural regularity of a specific polymer and the appearance of its single crystals. Relatively larger single crystals with smooth, sharp edges and little random growth are formed only from polymers that are known to have high regularity in their structures.

To explain the arrangement of the chains in the single crystals, a theory of *folded chain lamella* was proposed in late 1950s. It states that the basic units in single crystals are *lamellae*, about 100 Å thick. The evidence for the existence of lamellae-like crystallites comes from microscopic observations. Direct microscopic observations, however, do not yield information about chain structure on the molecular level. The thickness of the lamellae relative to the chain length led some to postulate that the molecules are arranged perpendicular, or nearly so, to the layers. Also that sharp, folded configurations form with the fold length corresponding to the layer thickness.[15] In this view, one polymer molecule is essentially constrained to one lamellae and the interface is smooth and regularly structured with the chain folding back and forth on itself like an accordion during crystallization. Two models of chain folding are visualized. In one model, the chain folding is regular and sharp with a uniform fold period. This is called the *adjacent-reentry model*. In the other one, the molecules wander through the irregular surface of a lamella before reentering it or a neighboring lamella. This mode is called *switchboard* or *nonadjacent-reentry* model. The two are illustrated in Fig. 1.5. Some experimental evidence fails to support the lamellae concept as well.[13]

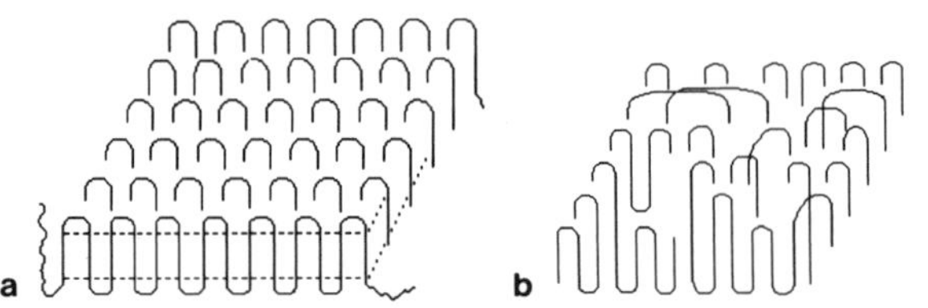

FIGURE 1.5. Models of chain folding. (a) Adjacent reentry model, and (b) nonadjacent reentry model.

Linear polyethylene single crystals often exhibit secondary structural features that include corrugations and pleats. It was suggested that the crystals actually grow in the form of pyramids, but that these pyramidal structures collapse when the solvent is removed during preparation for microscopy.[21] Various investigators described other complex structures besides pyramidal. Typical among these descriptions are sheaf-like arrays that would correspond to nuclei. Also, there were observations of dendritic growths, of clusters of hollow pyramids, of spiral growths, of epitaxial growths, of moire patterns, etc.[15]

An important parameter in the characterization of the two-phase system is the weight fraction of the crystalline regions. The degree of crystallinity that can be reached is dependent on the temperature at which crystallization takes place. At low temperatures one attains a much lower degree of crystallization than at higher temperatures. This implies that crystallization remains incomplete for kinetic reasons.

Microscopic examinations with polarized light of many polymeric materials that crystallized from the melt show the specimen packed with *spherulites*. Often these appear to be symmetrical structures with black crosses in the center.[34] It is believed[22] that these spherulites grow from individual nuclei. Ribbons of crystallites grow from one spherulitic center and fan out in all directions. Initially they are spherical but because of mutual interference irregular shapes develop. The diameters of spherulites range from 0.005 to 0.100 mm. This means that a spherulite consists of many crystalline and noncrystalline regions. The black crosses seen in the spherulites are explained[22] by assuming that the crystallites are arranged so that the chains are preferentially normal to the radii of the spherulites. Spherulitic morphology is not the universal mode of polymer crystallization.[13] Spherulitic morphology occurs usually when nucleation is started in a molten polymer or in a concentrated solution of a polymer. When the melt or the solutions are stirred, *epitaxial* crystallinity is usually observed. One crystalline growth occurs right on top of another. This arrangement is often called shish-kebab morphology. It contains lamella growth on long fibrils. Drawing of a crystalline polymer forces the spherulites to rearrange into parallel arrangements known as *drawn fibrillar morphology*.

If a polymer molecule is coiled randomly, then it cannot fit readily into a geometrically arranged, regular crystalline lattice. So the molecules must change into a uniform shape to fit into a crystal pattern. In many cases they assume either a helix or a zigzag conformation. Such arrangements are more regular than in a random coil.[40]

1.4.4. Liquid Crystal Polymers

These are macromolecules that can align into crystalline arrays while they are in solution (*lyotropic*) or while in a molten state (*thermotropic*). Such liquids exhibit *anisotropic* behavior. The regions of orderliness in such liquids are called *mesophases*. Molecular rigidity is the chief cause of liquid crystalline behavior. It excludes more than one molecule occupying a specific volume and it is not a result of intermolecular attractive forces. For a molecule to exhibit liquid crystalline behavior, it must possess molecular shape anisotropy. This is not affected by conformational changes. Some biological polymers exhibit liquid crystalline behavior, due to their rigid helical conformations. Among synthetic polymers, on the other hand, rigid rod structures are the ones that exhibit most of the liquid crystalline behavior. Based on theoretical considerations,[45] the ratio of the molecular length to the molecular diameter must be greater than six. Some other anisotropic molecular shapes, like disks, can also exhibit liquid crystalline behavior, but this is much less common.

Two types of liquid crystalline arrangement are recognized. One is *smectic* and the other one is *nematic*. Both are parallel arrangements of long molecular axes. The smectic liquid crystals are more ordered, however, than the nematic ones. This is a result of differences in the orientations of the chain ends. In smectic liquid crystals, the chain ends are lined up next to each other. In nematic ones, however, they lack any particular orientation. Also, the smectic liquid crystals are layered while the nematic ones are not. Microscopic observations can help distinguish between the two forms.

The chemical structure of the polymers determines whether the molecules can form rigid rods. If the backbone of the polymer is composed of rigid structures, then it tends to form *main chain* liquid crystals. If, however, the side chains are rigid, then the polymer will tend to form *side chain* liquid crystals.

From practical considerations, two properties are of prime interest: The effect of liquid crystalline behavior on viscosity and the ability of the polymer to retain the ordered arrangement in the solid state. Liquid crystalline behavior during the melt results in lower viscosity, because the rigid polymeric mesophases align themselves in the direction of the flow. As a result, the polymer is easier to process. Also, retention of the arrangement upon cooling yields a material with greatly improved mechanical properties. Several thermotropic liquid crystalline copolyesters are now available commercially.

A lyotropic liquid crystalline aromatic polyamide, sold under the trade name of Kevlar, is available commercially:

$$\left[-\overset{H}{\overset{|}{N}}-C_6H_4-\overset{H}{\overset{|}{N}}-\overset{O}{\overset{\|}{C}}-C_6H_4-\overset{O}{\overset{\|}{C}}- \right]_n$$

Kevlar

The polyamide forms liquid crystals in sulfuric acid solution from which it is extruded as a fiber. After the solvent is removed, the remaining fiber possesses greater uniform alignment than would be obtained by mere drawing. This gives it superior mechanical properties.

The polymers that exhibit liquid crystalline behavior often have very high melting points and are insoluble in most common organic solvents. This is a drawback, because such materials are hard to process.

1.4.5. Orientation

There is no preferred direction or arrangement in the manner in which the macromolecules align themselves in a polymeric mass during crystallization. If, however, after crystallization an external stress is applied, the crystalline material undergoes a rearrangement. From the X-ray diffraction pattern it is surmised that the chains realign themselves in the direction of the applied stress. The materials usually show considerable increase in strength in the direction of that stress.

This orientation in the direction of applied stress occurs also in amorphous materials. The amorphous polymers, like the crystalline ones, also exhibit increased strength in the direction of orientation.

1.4.6. First-Order Transition Temperature

The melting temperatures of the crystallites are designated by T_m, which is the first-order transition temperature.[23] The melting points are often not as sharp as the melting points of ordinary crystals and may melt over a range. Table 1.2 shows some typical values of T_m.

1.4.7. Second-Order Transition Temperature

When the polymer cools and the temperature drops, the mobility in the amorphous regions of the polymer decreases. The lower the temperature, the stiffer the polymer becomes until a point of transition is reached. This transition is called the second-order transition temperature[23] or the glass

Table 1.2. Melting Points, T_m, of Some Crystalline Polymers[a]

Repeat unit	Polymer	Melting point (°C)
—[—O—$(CH_2)_4$—O—C(=O)—$(CH_2)_6$C(=O)—]—	Poly(tetramethylene suberate)	45
—[—CH_2—CH_2—O—]—	Poly(ethylene oxide)	66
—[—CH_2—$CH(CH_3)$—O—]—	Poly(propylene oxide)	70
—[—CH_2—CH_2—]—	High density polyethylene	132–138
—[—CH_2—CH(CN)—]—	Polyacrylonitrile	317
—[—CH_2—CH(Cl)—]—	Poly(vinyl chloride)	212
—[—CH_2—CCl_2—]—	Poly(vinylidine chloride)	210
—[—CH_2—$CH(CH_3)$—]—	Polypropylene	168
—[—CFCl—CF_2—]—	Poly(chlorotrifluoroethylene)	210
—[—CH_2—$CH(C_6H_5)$—]—	Polystyrene	230–240
—[—$NH(CH_2)_6NH$C(=O)$(CH_2)_4$C(=O)—]—	Poly(hexamethylene adipamide)	250
—[—CF_2—CF_2—]—	Poly(tetrafluoroethylene)	327

[a]From Ref. 23, and other sources in the literature .

transition temperature, designated by T_g. Beyond stiffness, a change is manifested in specific volume, heat content, thermal conductivity, refractive index, and dielectric loss.

Above T_g, chains undergo cooperative localized motion. Below the second-order transition, however, there is insufficient energy available to enable whole segments of the polymeric chains to move. Only individual atoms can make small excursions about their equilibrium positions. The structure is now stiff and brittle and resists deformation. When, however, a sufficient amount of heat energy enters the material again and the temperature rises above T_g larger molecular motion involving coordinated movement returns and the polymer is in a rubbery or a plastic state. Above T_g large elastic deformations are possible and the polymer is actually both tougher and more pliable. Table 1.3 lists values of T_g of some common polymers.

1.5. Effect of Chemical Structure Upon Physical Properties[29]

Generally, the freedom of molecular motion along the backbones of polymeric chains contributes to the lowering of the first-order transition temperature, T_m. Substituents that interfere with this motion tend to raise the melting point. For instance, isotactic polypropylene melts at a higher temperature than linear polyethylene (see Table 1.2). If the substituent is bulky or rigid, it raises the melting point because that interferes with molecular motion. The presence of dipole forces has a similar effect. A good illustration is a comparison of poly(ethylene terephthalate) that melts at 265 °C with poly(ethylene adipate) that melts at only 50 °C. In the first polyester there is a rigid benzene ring between the ester groups, and in the second one a flexible chain of four carbons.

Table 1.3. Glass Transition Temperatures, T_g, of Some Common Polymers[a]

Repeat Unit	Polymer	T_g °C
$-[-CH_2-CH(C_6H_5)-]-$	Polystyrene	81
$-[-CH_2-CHCl-]-$	Poly(vinyl chloride)	75
$-[-CH_2-C(CH_3)(COOCH_3)-]-$	Poly(methyl methacrylate)	57–68
$-[-CH_2-C(CH_3)(COOC_2H_5)-]-$	Poly(ethyl methacrylate)	65
Cellulose ring unit with O_2N-O-, NO_2-O- and CH_2-O-NO_2 substituents	Cellulose nitrate	53
$-[-NH-C(=O)-(CH_2)_4-C(=O)-NH-(CH_2)_6-]-$	Nylon 6-6	47
$-[-CH_2-CH(O-COCH_3)-]-$	Poly(vinyl acetate)	30
$-[-CH_2-C(CH_3)(COOC_4H_7)-]-$	Poly(*n*-butyl methacrylate)	22
$-[-CH_2-CH(COOCH_3)-]-$	Poly(methyl acrylate)	5
$-[-CH_2-CH(COOC_2H_5)-]-$	Poly(ethyl acrylate)	–22
$-[-CH_2-CF_2-]-$	Poly(vinylidine fluoride)	–39
$-[-CH_2-CH(COOC_4H_7)-]-$	Poly(*n*-butyl acrylate)	–56
$-[-CH_2-C(CH_3)=CH-CH_2-]-$	Polyisoprene	–70
$-[-Si(CH_3)_2-O-]-$	Silicone rubber	–123

[a]From Ref. 23 and other sources in the literature.

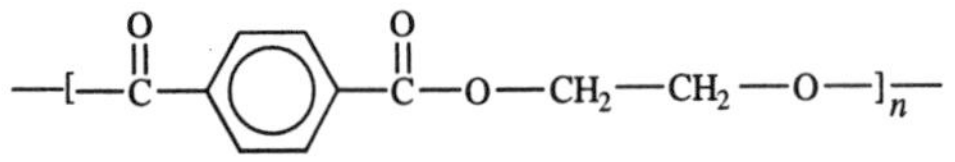

$$-[-\overset{O}{\overset{\|}{C}}-C_6H_4-\overset{O}{\overset{\|}{C}}-O-CH_2-CH_2-O-]_n-$$

poly(ethylene terephthalate)

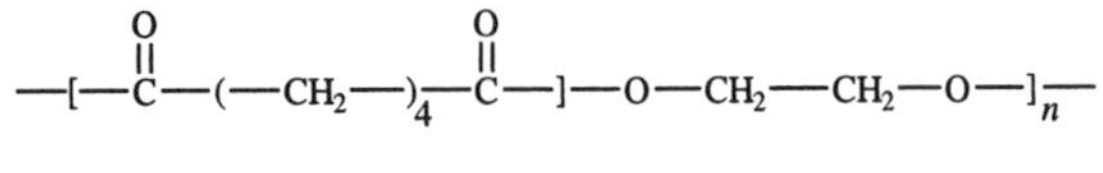

$$-[-\overset{O}{\overset{\|}{C}}-(-CH_2-)_4-\overset{O}{\overset{\|}{C}}-]-O-CH_2-CH_2-O-]_n-$$

poly(ethylene adipate)

Linear polymers that possess only single bonds in their backbones between atoms, C–C, or C–0, or C–N, can undergo rapid conformational changes.[14] Also, ether, imine, or *cis*-double bonds reduce energy barriers and, as a result, "soften" the chains, causing the polymer to become more rubbery and more soluble in various solvents. The opposite is true of cyclic structures in the backbones. Actually, cyclic structures not only inhibit conformational changes but can also make crystallization more difficult.

Among the polymers of α-olefins, the structures of the pendant groups can influence the melting point.[24] All linear polyethylene melts between 132–136 °C[35] (see Table 1.2). When the length of the side groups of isotactic poly(1-olefin)s increases, starting with polypropyline that melts at 168 °C,[26] the melting points decrease. This is accompanied by increases in flexibility[25] until polyhexene is reached. At that point, the minimum takes an upturn and there is an increase in T_m and a decrease in flexibility. This phenomenon is believed to be due to crystallization of the side chains.[28]

Alkyl substituents on the polymers of α-olefins on the α-carbon yield polymers with the highest melting points. Isomers substituted on the β-carbon, if symmetrical, yield polymers with lower melting points. Unsymmetrical substitution on the β-carbon tends to lower the melting point further. Additional drop in T_m results from substitution on the γ-carbon or further out on the side-chain. Terminal branching yields rubbery polymers.[27]

Copolymers melt at lower temperatures than do homopolymers from the individual monomers. By increasing the amount of a comonomer, the melting point decreases down to a minimum (this could perhaps be compared to a eutectic) and then rises again.

The tightest internal arrangement of macromolecules is achieved by crystallinity. As a result, the density of a polymer is directly proportional to the degree of crystallinity, which leads to high tensiles, and to stiff and hard materials that are poorly soluble in common solvents.[29] The solubility of any polymer, however, is not a function of crystallinity alone, but also of the internal structure and of the molecular weight. The solubility generally decreases with increases in the molecular weight. The fact that crystalline polymers are less soluble than amorphous ones can be attributed to the binding forces of the crystals. These binding forces must be overcome to achieve dissolution. Once in solution, however, crystalline polymers do not exhibit different properties from amorphous ones.

1.6. Molecular Weights and Molecular Weight Determinations

Among synthetic polymers, the process of polymer formation, whether by a chain propagating reaction or through a step-growth one, is governed by random events. The result is that the chains vary in length. A polymeric material cannot, therefore, be characterized by a single molecular weight, but instead it must be represented by a statistical average.[31] This average can be expressed in several ways. One way is to present the average as a *number average*. It is the sum of all the molecular weights of the individual molecules present divided by their total number. Each molecule contributes equally to the average. It can be obtained by averaging the measurements of all the colligative properties. If the total number of moles is N_i, the sum of these molecules present can be expressed as ΣN_i. The total weight ω of a sample is similarly the sum of the weights of all the molecular species present,

$$\omega = \sum \omega_i = \sum M_i N_i$$

By dividing the total weight of the molecules by their total number we have the *number average* molecular weight,

$$\overline{M}_n = \frac{\omega}{\sum N_i} = \frac{\sum M_i N_i}{\sum N_i}$$

Another way to express the molecular weight average is as a *weight average*. Each molecule in such an average contributes according to the ratio of its particular weight to that of the total,

$$\overline{M}_w = \frac{\sum M_i^2 N_i}{\sum M_i N_i}$$

The above can be illustrated quite readily by imagining that a sample consists of five molecules of molecular weights 2, 4, 6, 8, and 10, respectively. To calculate the number average molecular weight, all the weights of the individual molecules are added. The sum is then divided by the total number of molecules in the sample (in this case 5):

$$\overline{M}_n = 2/5 + 4/5 + 6/5 + 8/5 + 10/5 = 6$$

To calculate the weight average molecular weight of the above sample, the squares of each individual weight are divided by the total sum of molecular weights, in this case 30:

$$\overline{M}_w = 2^2/30 + 4^2/30 + 6^2/30 + 8^2/30 + 10^2/30 = 7.33$$

Here, $\overline{M}_w$ is more sensitive to the higher molecular weight species, while M_n is sensitive to the lower ones. This can be seen by imagining that equal weights of two different sizes of molecules are combined, $M_1 = 10{,}000$ and $M_2 = 100{,}000$. The combination would consist of ten molecules of M_1 and one molecule of M_2. The weight average molecular weight of this mixture is 55,000 while the number average molecular weight is only 18,200. If, however, the mixture consists of an equal number of these molecules, then the weight average molecular weight is 92,000 and the number average molecular weight is 55,000.

In solutions of polymers, the viscosities are more affected by the long chains than by the short ones. A correlation of the viscosity of the solution to the size of the chains or to the molecular weight of the solute allows an expression of a *viscosity average* molecular weight:

$$\overline{M}_\eta = \left(\frac{\sum M_i^{\beta+1} N_i}{\sum M_i N_i} \right)^{1/\beta}$$

where β is a constant. When $\beta = 1$, then $\overline{M}_\eta = \overline{M}_w$. In fact, the value of M_η is usually within 20% of M_w.

Solution viscosities of linear polymers relate empirically to their molecular weights. This is used in various ways to designate the size of polymers. The efflux time t of a polymer solution through a capillary is measured. This is related to the efflux time t_0 of the pure solvent. Typical viscometers, like those designed by Ubbelohde, Cannon-Fenske, and other similar ones, are used in a constant temperature bath. The following relationships are used:

Name	Symbol	Definitions
1. Relative viscosity	η_{rel}	$\eta/\eta_0 = t/t_0$
2. Specific viscosity	η_{sp}	$(\eta - \eta_0)/\eta_0 = \eta_{rel} - 1 \propto (t - t_0)/t_0$
3. Reduced viscosity	η_{red}	$\eta_{sp}/C = \eta_{rel}^{-1}/C$
4. Inherent viscosity	η_i	$\ln\eta_{rel}/C$
5. Intrinsic viscosity	$[\eta]_{C\to 0}$	$(\eta_{sp}/C)_{C=0} = (\eta_i)_{C=0}$

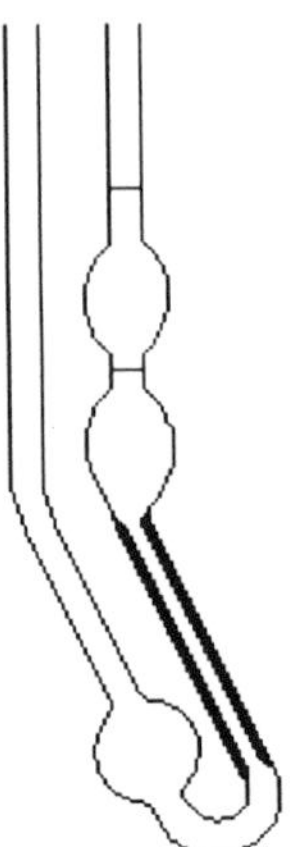

FIGURE 1.6. Cannon–Fenske capillary viscometer.

To determine the intrinsic viscosity, both inherent and reduced viscosities are plotted against concentration (C) on the same graph paper and extrapolated to zero. If the intercepts coincide, then this is taken as the intrinsic viscosity. If they do not, then the two intercepts are averaged. The relationship of intrinsic viscosity to molecular weight is expressed by the Mark–Houwink–Sakurada equation[7]:

$$[\eta]_{C=0} = K\,\overline{M}_{\eta}{}^{a}$$

where K and a are constants. Various capillary viscometers are available commercially. One common model is illustrated in Figure 1.6.

The logarithms of intrinsic viscosities of fractionated samples are plotted against log M_w or log M_n. The constants a and K of the Mark–Houwink–Sakurada equation are the intercept and the slope, respectively, of that plot. Except for the lower molecular weight samples, the plots are linear for linear polymers. Many values of K and a for different linear polymers appear in the literature.[23] When all macromolecular species are of the same size, the number average molecular weight is equal to the weight average molecular weight. On the other hand, the greater the distribution of molecular sizes, the greater is the disparity between averages. The ratio of this disparity, M_w/M_n is a measure of *polymeric dispersity*. A *monodisperse* polymer has

$$M_w/M_n = 1$$

In all synthetic polymers and in many naturally occurring ones the weight average molecular weight is greater than the number average molecular weight. Such polymers are *polydisperse*.

Two samples of the same polymer equal in weight average molecular weight may exhibit different physical properties, if they differ in the molecular weight distributions. Many effects of the molecular weight distribution on such properties as elongation, relaxation modulus, tensile strength, and tenacity were reported.[2]

The physical properties of polymers are also related to their molecular weights. Melt viscosity, hot strength, solvent resistance and overall toughness increase with molecular size. Table 1.4 illustrates the effect of molecular weights (size) upon physical properties of polyethylene.[32] The determinations of molecular weights of polymers rely, in most cases, upon physical methods. In some special cases, however, when the molecular weights are fairly low, chemical techniques can be used. Such determinations are limited to just those macromolecules that possess only one functional group that is located at the end of the chain ends. In place of the functional group, there may be a heteroatom. In that case, an elemental analysis might be sufficient to determine the molecular weight. If there is a functional group, however, a reaction of that group allows calculating the molecular weight. Molecular weights above 25,000 make chemical approaches impractical. In

Table 1.4. Properties of Low Density Polyethylene[32,33]

DP	Molecular weight	Softening temperature (°C)	Physical state at 25 °C
1	28	−169 (mp)	Gas
6	170	−12 (mp)	Liquid
35	1000	37	Grease
140	4000	93	Wax
250	7000	98	Hard wax
430	12000	104	Plastic
750	21000	110	Plastic
1350	38000	112	Plastic

chemical determination, each molecule contributes equally to the total. This is, therefore, a number average molecular weight determination.

There are various physical methods available today. The more prominent ones are ebullioscopy, cryoscopy, osmotic pressure measurements, light scattering, ultracentrifugation, and gel permeation chromatography (also called size exclusion chromatography). All these determinations are carried out on solutions of the polymers. Also, all, except gel permeation chromatography, require that the results of the measurements be extrapolated to zero concentrations to fulfil the requirements of theory. The laws that govern the various relationships used in these determinations apply only to ideal solutions. Only when there is complete absence of chain entanglement and no interaction between solute and solvent is the ideality of such solutions approached. A brief discussion of some techniques used for molecular weight determinations follows. *Ebullioscopy*, or boiling point elevation, as well as *cryoscopy*, or freezing point depression, are well known methods. They are the same as those used in determining molecular weights of small molecules. The limitation to using both methods with macromolecules is that ΔT_b and ΔT_f become increasingly smaller as the molecular sizes increase. The methods are therefore limited to the capabilities of the temperature-sensing devices to detect very small differences in temperature. This places the upper limits for such determinations to somewhere between 40,000 to 50,000. The thermodynamic relationships for these determinations are

$$[\Delta T_{b/c}]_{C\to 0} = \frac{RT^2}{\rho \Delta H_b M} \quad \text{and} \quad [\Delta T_{f/c}]_{C\to 0} = \frac{RT^2}{\rho \Delta H_f M}$$

boiling point rise **freezing point depression**

The above two determinations, because each molecule contributes equally to the properties of the solutions, yield number average molecular weights.

A method that is useful for higher molecular weight polymers is based on *osmotic pressure measurements*. It can be applied to polymers that range in molecular weight from 20,000 to 500,000 (some claim 1,000,000 and higher). The method is based on van't Hoff's law. When a pure solvent is placed on one side of a semipermeable membrane and a solution on the other, pressure develops from the pure solvent side. This pressure is due to a tendency by the liquids to equilibrate the concentrations. It is inversely proportional to the size of the solvent molecules. The relationship is as follows:

$$\overline{M}_n = RT / (\pi / C)_{C=0} + A_2 C$$

where π is the osmotic pressure, C is the concentration, T is temperature, and R is the gas constant; A_2 is a measure of interaction between the solvent and the polymer (second virial coefficient).

A *static* capillary osmometer is illustrated in Fig. 1.7. Rather than rely on the liquid to rise in the capillary on the side of the solution in response to osmotic pressure, as is done in the static method, a *dynamic equilibrium* method can be used. Here a counterpressure is applied to maintain

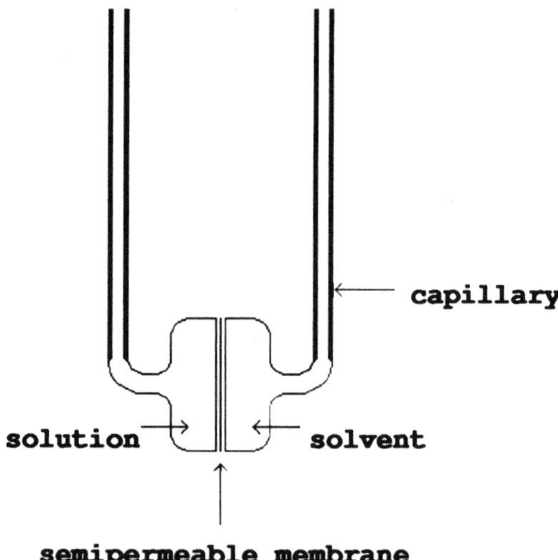

FIGURE 1.7. Membrane osometer.

equal levels of the liquid in both capillaries and prevent flow of the solvent. Different types of dynamic membrane osmometers are available commercially.

The results obtained from either method must still be extrapolated to zero concentration for van't Hoff's law to apply. Such extrapolation is illustrated in Fig. 1.8. Because all molecules contribute equally to the total pressure, osmotic pressure measurements yield the number average molecular weight.

Light scattering is a technique for determining the weight average molecular weight. When light passes through a solvent, a part of the energy of that light is lost due to absorption, conversion to heat, and scattering. The scattering in pure liquids is attributable to differences in densities that result from finite inhomogeneities in the distribution of molecules within adjacent areas. Additional scattering results from the presence of a solute in the liquid. The intensity or amplitude of that additional scattering depends upon concentration, the size, and the polarizability of the solute plus some other factors. The refractive index of pure solvent and a solution is also dependent upon the amplitude of vibration. The turbidity that arises from scattering is related to concentration:

$$\text{turbidity, } \tau = Hc\overline{M}_w$$

$$H = \frac{32\pi^3 n_0^2 (dn/dc)^2}{3\lambda^4 N_0}$$

where n_0 is the refractive index of the solvent, n the refractive index of the solution, λ the wavelength of the incident light, N_0 is Avogadro's number, and c is the concentration. The dn/dc relationship is

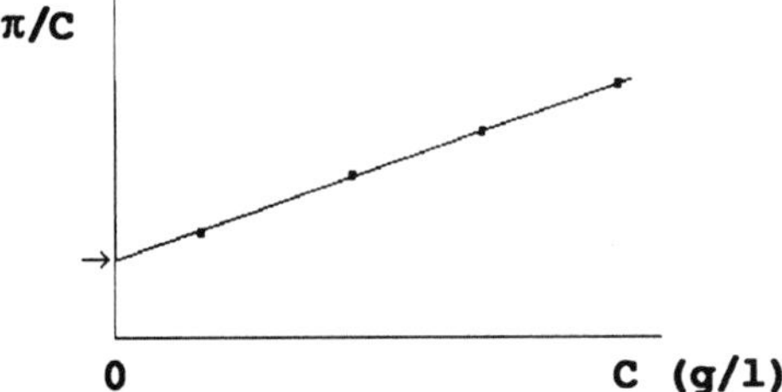

FIGURE 1.8. Extrapolation to zero concentration.

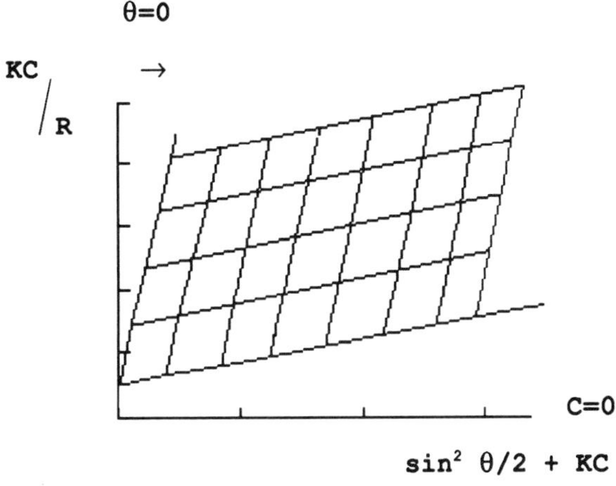

FIGURE 1.9. A typical Zimm plot.

obtained by measuring the slope of the refractive index as a function of concentration. It is constant for a given polymer, solvent, and temperature and is called the *specific refractive increment*.

Because scattering varies with different angles from the main beam of light, the results must be extrapolated to zero concentration and zero angle of scattering. This is done simultaneously by a method developed by Zimm. A typical Zimm plot is illustrated in Fig. 1.9.

A popular technique for determining molecular weights and molecular weight distributions is *gel permeation chromatography*. It is also called size exclusion chromatography.[42,43] The procedure allows one to determine M_w, M_n, M_z, and the molecular weight distribution in one operation. It is a form of HPLC that separates molecules according to their hydrodynamic volumes or their effective sizes in solutions. The separation takes place on one or more columns packed with a porous support. It results from retention of the polymer molecules by the pores of the packing as the solvent elutes the material through the columns. It was postulated in the past that the separation is due to smaller molecules diffusing into all the pores while the larger ones only diffuse into some of the pores. The largest molecules were thought to diffuse into none of the pores and pass only through the interstitial volumes. As a result, polymer molecules of different sizes travel different distances down the column. This means that the molecules of the largest size (highest molecular weight) are eluted first because they fit into the least number of pores. The smallest molecules, on the other hand, are eluted last because they enter the greatest number of pores and travel the longest path. The rest fall in between. The process, however, is more complex than the above-postulated picture. It has not yet been fully explained. It was found, for instance, that different gels display an almost identical course in the relation of dependence of V_R (retention volume) to the molecular weight. Yet study of the pores of different gels show varying cumulative distributions of the inner volumes. This means that there is no simple function correlating the volume and/or the size of the separated molecules with the size and distribution of the pores.[42] Also, the shape of the pores, that can be inferred from the ratio of the area and volume of the inner pores, is very important.[43] Different models were proposed to explain the separation phenomenon. These were reviewed thoroughly in the literature.[44] They are beyond the scope of this book.

As indicated above, the volume of the liquid that corresponds to a solute eluting from the columns is called the retention volume or elution volume (V_R). It is related to the physical parameters of the column as follows:

$$VR = V_0 + KV_1$$

where V_0 is the interstitial volume of the column(s), K the distribution coefficient, and V_1 is the internal solvent volume inside the pores.

The total volume of the columns is V_T that is, equal to the sum of V_0 and V_1. The retention volume can then be expressed by

$$V_R = V_0(1 + K) + KV_T$$

From the earlier statement it should be clear that polymer fractionation by gel permeation chromatography depends upon the spaces the polymer molecules occupy in solution. By measuring, experimentally, the molecular weights of polymer molecules as they are being eluted, one obtains the molecular weight distribution. To accomplish this, however, one must have a chromatograph equipped with dual detectors, one of which must detect the presence of polymer molecules in the effluent while the other must measure their molecular weights. Such detectors might be, for instance, a refractive index detector and a low-angle laser light scattering photogoniometer to find the absolute value of M.

Many molecular weight measurements, however, are conducted on chromatographs equipped with only one detector that monitors the presence of the solute in the effluent. The equipment must, therefore be calibrated prior to use. The relationship of the ordinate of the chromatogram, commonly represented by $F(V)$, must be related to the molecular weight. This relationship varies with the polymer type and structure. There are three methods for calibrating the chromatograph. The first, and most popular, makes use of narrow molecular weight distribution reference standards. The second is based upon a polydisperse reference material. The third method assumes that the separation is determined by molecular size. All three methods require that an experimentally established calibration curve of the relationship between the molecular size of the polymer in solution and the molecular weight be developed. A chromatogram is obtained first from every standard sample. A plot is then prepared from the logarithms of the average weights against the peak retention volumes (V_R). The values of V_R are measured from the points of injection to the appearances of the maximum values of the chromatograms. The resultant curve may appear as shown in Fig. 1.10. Above M_1 and below M_4 there is no effective fractionation, because of total exclusion in the first place and total permeation in the second. These are the limits of separation by the packing material.

To date, the standard samples of narrow molecular weight distribution polymers available commercially are mainly polystyrenes. These samples have polydispersity indexes that are close to unity and are available over a wide range of molecular weights. For determining molecular weights of polymers other than polystyrene, however, the molecular weights obtained from these samples

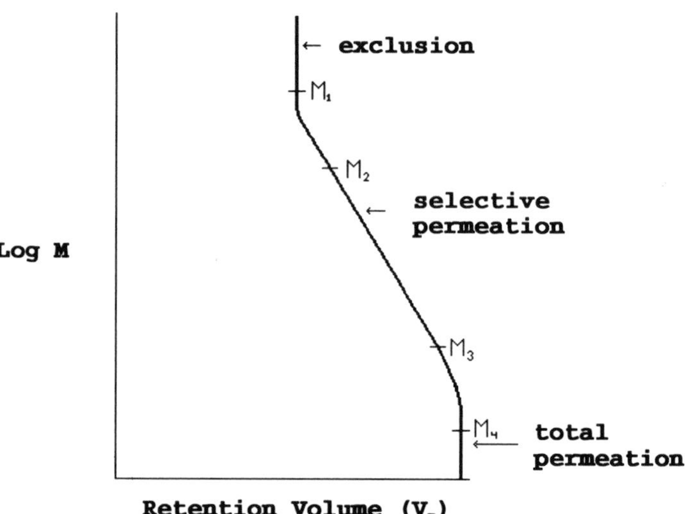

FIGURE 1.10. Molecular weight calibration curve.

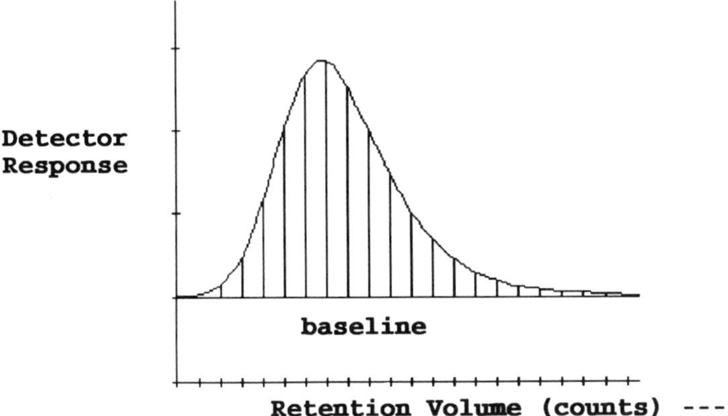

FIGURE 1.11. A typical digitized gel permeation chromatogram.

would be only approximations. Sometimes they could be in error. To overcome this difficulty a *universal calibration method* is used.

The basis for universal calibration is the observation[42] that the multiplication products of intrinsic viscosities and molecular weights are independent of the polymer types. Thus, $[\eta]M$ is the *universal calibration parameter*. As a result, a plot of log $([\eta]M)$ versus elution volume yields a curve that is applicable for many polymers. The $\log([\eta]\mathrm{M})$ *for a given column (or columns), temperature, and elution volume* may be considered a constant for all polymers.

Numerous materials have been used for packing the columns. Semirigid crosslinked polystyrene beads are available commercially. They are used quite frequently. Porous beads of glass or silica are also available. In addition, commercial gel permeation equipment is usually provided with automatic sample injection and fraction collection features. Favorites are refractive index and ultraviolet light spectroscopic detectors. Some infrared spectroscopic detectors are also in use. Commercially available instruments also contain pumps for high-pressure rapid flow and may also be equipped with a microcomputer to assist in data treatment. Also, there is usually a plotter in the equipment to plot the detector response as the samples are eluted through the column or columns. A typical chromatogram is illustrated in Figure 1.11. When polydisperse samples are analyzed, quantitative procedures consist of digitizing the chromatograms by drawing vertical lines at equally spaced retention volumes. These can be every 2.5 or 5.0 ml of volume. The resultant artificial fractions are characterized by their heights h_i, their solute concentrations C_i, and by the area they occupy within the curve A_i. The cumulative polymer weight value is calculated according to

$$I(V) = 1/A_{\mathrm{T}} \sum A_i$$

After conversion of the retention volumes V_i into molecular weights (using the primary calibration curve), the molecular weights, $\mathrm{M_w}$, M_{n}, and M_{z} can be calculated:

$$M_{\mathrm{n}} = \sum h_i/(\sum h_i/M_i); \quad M_{\mathrm{w}} = \sum h_iM_i/\sum h_i; \quad M_{\mathrm{z}} = \sum h_iM_i^2/\sum h_iM_i$$

If the chromatogram is not equipped with a microcomputer for data treatment, one can easily determine results on any available PC. Programs for data treatment have been written in various computer languages. One such program, written in Pascal language, is offered to students who may wish to practice calculations and familiarize themselves with molecular weight determinations. It can be found in Appendix A.2 at the end of this chapter.

1.7. Steric Arrangement in Macromolecules

In linear polymers, due to the polymerization process, the pendant groups can be arranged into orderly configurations or they can lack such orderliness. Propylene, for instance, can be polymerized into two types of orderly steric arrangement. It can also be polymerized into one lacking steric order. The same can be true of other monosubstituted vinyl monomers. The steric arrangement in macromolecules is called *tacticity*. Polymers can be *isotactic*, where all the chiral centers have the same configuration (see Fig. 1.12). By picturing the chain backbone as drawn in the plane of the paper and by picturing all the phenyl groups as oriented above the plane (Fig. 1.12a), isotactic polystyrene can be visualized. The orderliness can also be of the type where every other chiral center has the same configuration. Such an arrangement is called *syndiotactic* (Fig. 1.12b). A lack of orderliness or randomness in the steric arrangement is called *atactic* or *heterotactic*. Stereospecific polymers can also be prepared from 1,2-disubstituted olefins. These macromolecules can be distereoisomers, or ditactic polymers. To describe the arrangement of such polymers, a *threo–erythro* terminology is used. An erythrodiisotactic polymer is one possessing alternating substituents –(–CHR′CHR–)–*n*. If we draw the carbon chain backbone in the plane of the paper, then all the R groups would find themselves on one side of the chain and all the R′ groups on the other. They

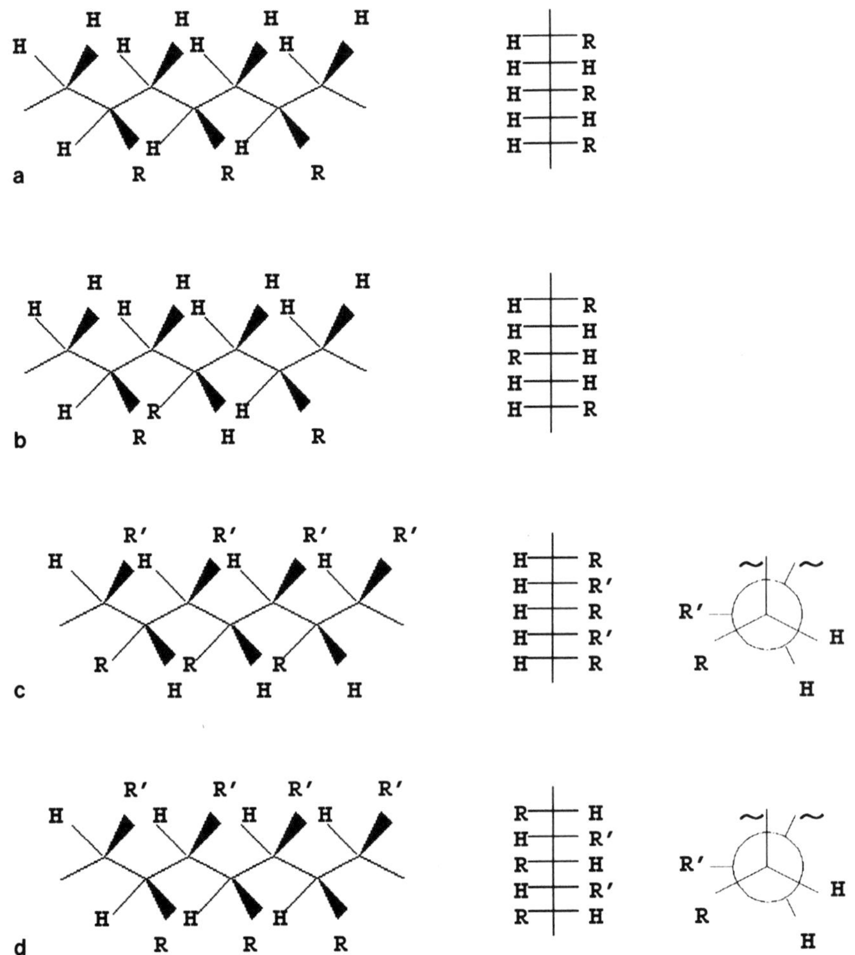

FIGURE 1.12. Steric arrangement in macromolecules (R = phenyl). (a) Isotactic polystyrene, (b) syndiotactic polymer, (c) erythrodiisotactic polymer, and (d) threodiisotactic polymer.

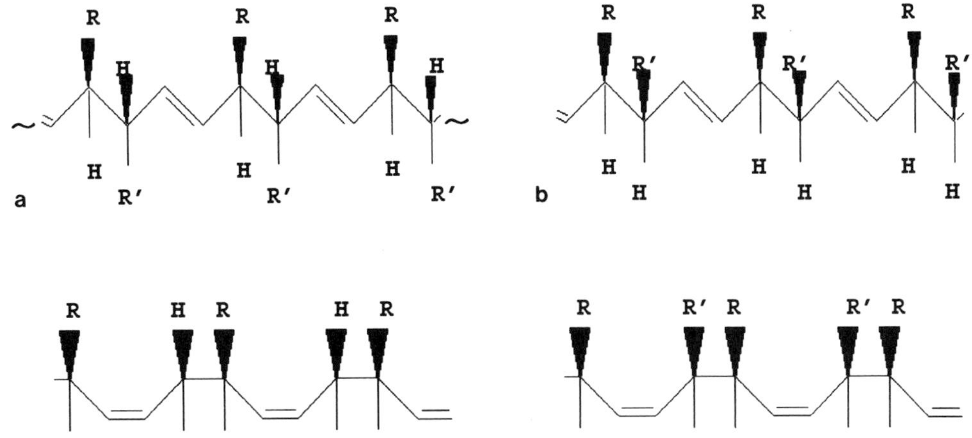

FIGURE 1.13. Tritactic polymers. (a) *Trans-erythro* tritactic polymer, (b) *trans-threo* tritactic polymer, (c) *cis-erythro* tritactic polymer, and (d) *cis-threo* tritactic polymer.

would, however, all be on the same side of the chain in a Fischer or in a Newman projection (Fig. 1.12c). A *threo* isomer or a threodiisotactic polymer can be illustrated as shown in Fig. 1.12d.

Polymerization of 1,4-disubstituted butadienes can lead to products that possess two asymmetric carbon atoms and one double bond in each repeat unit. Such *tritactic* polymers are named with prefixes of *cis* or *trans* together with *erythro* or *threo* (see Fig. 1.13).

In polymers with single carbon to carbon bonds there is free rotation, if steric hindrance is absent. This allows the molecules to assume different spacial arrangements or *conformations*. Most of the possible isomers, however, represent prohibitively high energy states. The three lowest energy states that are most probable[8] are one *trans* and two *gauche*.

Several rules that govern the configuration of repeat units along the polymeric chains were elucidated by Natta and Corradini.[41] A basic assumption is used to predict the lowest energy conformation. This assumption is known as the *equivalence postulate*. It says that all the structural units along a chain are geometrically equivalent with regard to the axis of the chain. All stereoregular polymers appear to meet this condition, with only a few exceptions.[29]

It should be possible to calculate (in principle) the lowest internal conformational energy of a given polymer molecule. To do that it is necessary to know: (1) the energy versus the bond rotation curve that relates the interactions within the structural units; and (2) the interaction energy versus the distance curve for the neighboring pendant groups of the adjacent structural units. It should be noted that the neighboring units approach each other close enough for substantial interaction. As a result, the interaction energies between the groups may distort the basic energy–rotation curves. The core electron repulsion between the units contributes most significantly to the deviations from the simple *trans* or *gauche* conformation. The specific chain conformation, therefore, is very dependent upon the exact nature of the repulsive potential. A contour plot, that represents the internal energy per mole of the polymer, can be prepared. The variables in such a plot are the relative bond rotations of the successive bonds.

1.8. Optical Activity in Polymers

Optical activity in biopolymers has been known and studied well before this phenomenon was observed in synthetic polymers. Homopolymerization of vinyl monomers does not result in structures with asymmetric centers (the role of the end groups is generally negligible). Polymers can be formed and will exhibit optical activity, however, that will contain centers of asymmetry in the backbones.[33] This can be a result of optical activity in the monomers. This activity becomes incorporated into the polymer backbone in the process of chain growth. It can also be a result of polymerization that involves asymmetric induction.[37,38] An example of inclusion of an optically active monomer into the polymer chain is the polymerization of optically active propylene oxide. (See Chapter 4 for additional discussion.) The process of chain growth is such that the monomer addition is sterically controlled by the asymmetric portion of the monomer. Several factors appear important in order to produce measurable optical activity in copolymers;[35] These are: (1) The selection of comonomer must be such that the induced asymmetric center in the polymer backbone remains a center of asymmetry. (2) The four substituents on the originally inducing center on the center portion much differ considerably in size. (3) The location of the inducing center must be close to the polymer backbone. (4) The polymerization reaction must be conducted at sufficiently low temperature to insure stereochemical selectivity. An example is a copolymerization of maleic anhydride with optically active 1-α-methylbenzyl vinyl ether. The copolymer exhibits optical activity after the removal of the original center of asymmetry.[36]

An example of an asymmetric induction from optically inactive monomers in an anionic polymerization of esters of butadiene carboxylic acids with (*R*)-2-methylbutyllithium or with butyllithium complexed with (–)-menthyl ethyl ether as the catalyst. The products, tritactic polymers, exhibit small, but measurable, optical rotations.[39] Also, when benzofuran, that exhibits no optical activity, is polymerized by cationic catalysts like aluminum chloride complexed with an optically active cocatalyst, like phenylalanine, an optically active polymer is obtained.[35]

Appendix

A.1. Additional Definitions

Term	Definition
Tensile Strength	Ability to resist stretching
Flexural Strength	Resistance to breaking or snapping
Tensile Stress (σ)	σ = force/cross-section area $\sigma = F/A$
Tensile Strain (ε)	Change in sample length (when stretched) divided by the original length $\varepsilon = \Delta l/l$
Tensile Modulus (E)	Ratio of stress to strain $E = \sigma/\varepsilon$

A.2. Program for Calculating Molecular Weights from Gel Permeation Chromatograms

```
Program GPC; uses Printer;
Const
 Max = 80;
Type
 Ary = Array[1..Max] of real;
Var
 x,x1,y,z,k,p:ary;
 I,N:integer;
 sum_k,sum_SQ,NA,WA,NA1,WA1,WZ,WZ1:real;
 W,Mw,Mn,Mw1,Mn1,Distr,Distr1,St,En,V,Me,Me1,Va,Va1,Sk,Sk1,Ku,
        Ku1:real; f,g,h,u,j,k1,l,m,o,p1,r,s,t,M1,M2,M3,M4,V2,
        V3,B,C,D,E:real;
Procedure Introduction(Var St,En,V:real;var N:integer);
Begin
writeln;
```

```
writeln;
writeln('Please enter the number of peak heights: ');
read(N);
writeln('Enter elution volumes at start and at end of the
        chromatograms:');
read(St,En);
V:=En-St;
end;

Procedure Num(var f,g,h,u,s,l,m,o,p1,r,t,M1,M2,M3,M4,V2,V3:real);

Var j,k1:real;

Begin
writeln('Enter elution volumes at two points within the
        chromatogram');
writeln('(preferably, somewhat evenly spaced betwen the start
        and the end):');
read(V2,V3);
writeln('From your calibration plot enter the molecular
        weights which');
writeln('correspond to the elution volumes at the start and at
        the end');
writeln('of the chromatogram:');
read(M1,M4);
writeln('Enter mol.wts. which correspond to the two points
        within');
writeln('the chromatogram:');
read(M2,M3);
f:=St-V2;
g:=SQR(St)-SQR(V2);
h:=St*St*St-V2*V2*V2;
u:=V2-V3;
j:=SQR(V2)-SQR(V3);
k1:=V2*V2*V2-V3*V3*V3;
l:=V3-En;
m:=SQR(V3)-SQR(En);
o:=V3*V3*V3-En*En*En;
p1:=u*g-f*j;
r:=u*h-f*k1;
s:=l*j-u*m;
t:=l*k1-u*o;
End;

Procedure Con(var B,C,D,E:real);

Var
 E1,E2:real;
Begin
 E1:=(s*u*(Ln(M1)-Ln(M2))-s*f*(Ln(M2)-Ln(M3)))/2.3026;
 E2:=(p1*l*(Ln(M2)-Ln(M3))-p1*u*(Ln(M3)-Ln(M4)))/2.3026;
 E:=(E1-E2)/(s*r*p1*t);
 D:=((u*(Ln(M1)-Ln(M2))-f*(Ln(M2)-Ln(M3)))/2.3026-r*E)/p1;
 C:=((Ln(M1)-Ln(M2))/2.3026-g*D-h*E)/f;
 B:=Ln(M1)/2.3026-St*C-D*SQR(St)-E*EXP(3*LN(St));
End;

Procedure MM(var y,z:ary);

Var
 I: integer; c1:real;z1:ary;
Begin
writeln('At what elution volume intervals were the peaks
        measured?');
read(c1);
For I:=1 to N do
```

```
 begin
  y[I]:=St;
  z1[I]:=B+C*y[I]+D*y[I]*y[I]+E*y[I]*y[I]*y[I];
  z[I]:=EXP(z1[I]*LN(10));
  St:=St+c1
 end;
End;
Procedure GtData(var x,x1:ary);

Var
 I:integer;
Begin
 writeln;
 writeln('Please enter the measured peak heights: ');
 writeln('(normalized detector response x 1000):');
 writeln;
  for I:= 1 to N do
   begin
   write(I:3,':  ':4);
   readln(x1[I]);
   x[I]:=x1[I] + 0.3;
   end;
End;

Procedure CorrectData(var k:ary);

Var
 I:integer;
 yb,sb :ary;
 D,Ds,s:real;
Begin
 writeln;
 writeln;
 D:= 0.3;
 s:=0.9;
 Ds:=0.243;
 For I:=1 to N do
  begin
   yb[I]:=V+(1/D)*LN((x[I]+Ds)/SQRT((x[I]-Ds)*(x[I]+Ds)));
   sb[I]:=SQR(s)+(1/SQR(D))*LN((x[I]-Ds)*(x[I]+Ds)/SQR(x[I]));
   k[I]:=x[I]*(s/sb[I])*EXP(-SQR(V-yb[I])/(2*SQR(sb[I])));
   end;
End;
Procedure Calculate(var NA,WA,NA1,WA1,WZ,WZ1:real);

Var
 sum_x,sum_x1,sum_k,sum_SQ,sum_k1,sum_SQ1,sum_z,sum_z1:real;
 sum_x2,sum_k2:real;
 I:integer;

Begin
 sum_x:=0; sum_k:=0; sum_x1:=0;
 sum_SQ:=0; Sum_x2:=0; sum_k2:=0;
 sum_k1:=0; sum_SQ1:=0; sum_z:=0;
 sum_z1:=0;
 For I:=1 to N do
  begin
   sum_x:=sum_x + x[I];
   sum_k:=sum_k + k[I];
   sum_x1:=sum_x1 + x[I]*z[I];
   sum_k1:=sum_k1 + k[I]*z[I];
   sum_x2:=sum_x2 + (x[I]/z[I]);
   sum_k2:=sum_k2 + (k[I]/z[I]);
   sum_SQ:=sum_SQ + SQR(x[I]*z[I]);
   sum_SQ1:=sum_SQ1 + SQR(k[I]*z[I]);
   sum_z:=sum_z + x[I]*SQR(z[I]);
```

```
   sum_z1:=sum_z1 + k[I]*SQR(z[I]);
   end;
  NA:=sum_x/sum_x2;
  NA1:=sum_k/sum_k2;
  WA:=sum_x1/sum_x;
  WA1:=sum_k1/sum_k;
  WZ:=sum_z/sum_x1;
  WZ1:=sum_z1/sum_k1;
End;

Procedure CalcP(var p:ary);

Var
 I:integer; sum_x:real;

Begin
 sum_x:=0.0;
 For I:=1 to N do
  begin
   sum_x:=sum_x + k[I];
 end;
 For I:=1 to N do
  begin
   p[I]:=(k[I]/sum_x)*100;
 end;
End;

Procedure Mean(var Me,Me1,Va,Va1:real);

Var
 su,su1,s,s1,s2,su2:real;
 I:integer;

Begin
 s:=0; s2:=0; su2:=0; s1:=0; su:=0; su1:=0;
 For I:=1 to N do
  begin
   s:=s + x[I];
   su:=su + x[I]*y[I];
   s1:=s1 + k[I];
   su1:=su1 + k[I]*y[I];
  end;
 Me:=su/s;
 Me1:=su1/s1;
 For I:=1 to N do
  begin
   s2:=s2 + x[I]*SQR(y[I]-Me);
   su2:=su2+k[I]*SQR(y[I]-Me);
  end;
 Va:=s2/s;
 Va1:=Su2/s1;
End;
Procedure Skew(var Sk,Sk1,Ku,Ku1:real);

Var
 I:integer;
 s,s1,s3,su3,s4,su4:real;

Begin
 s:=0; s1:=0; s3:=0; su3:=0; s4:=0; su4:=0;
 For I:=1 to N do
  begin
   s:=s + x[I];
   s1:=s1 + k[I];
   s3:=s3 + x[I]*(y[I]-Me)*(y[I]-Me)*(y[I]-Me);
   su3:=su3 + k[I]*(y[I]-Me1)*(y[I]-Me1)*(y[I]-Me1);
```

```
    s4:=s4 +  x[I]*(SQR(y[I]-Me)*(SQR(y[I]-Me)));
    su4:=su4 + k[I]*(SQR(y[I]-Me1)*(SQR(y[I]-Me1)));
    end;
  Sk:=(s3/s)/(EXP(1.5)*LN(Va));
  Sk1:=(su3/s1)/(EXP(1.5)*LN(Va1));
  Ku:=(s4/s)/SQR(Va);
  Ku1:=(su4/s1)/SQR(Va1)
End;

Procedure WriteData;
Var
 I:integer;
Begin
writeln(Lst,'':3,'Sample Identification:_______________');
writeln('':40,'DATA');
writeln;
writeln(Lst);
writeln(Lst,'':40,'DATA');
writeln(Lst);
writeln('':20,'El.Vol.':9,'Ex.Hts':10,'Cr.Hts':9,'%Tot.':9,'
        Ml.Wt.':12);
writeln('':9,'_____________________________________________');
writeln(Lst,'':20,'Elut.':9,'Exper.':10,'Correc.':9,'% of':9,'Ml.Wt.':12);
writeln(Lst,'':20,'Volumes':9,'Hts.':9,'Hts.':9,'Total':10);
writeln(Lst,'':14,'_____________________________________________');
For I:=1 to N do
 begin
 writeln(I:17,y[I]:10:2,x1[I]:10:2,k[I]:10:2,p[I]:10:2,z[I]:12:1);
 writeln(Lst,I:17,y[I]:10:2,x1[I]:10:2,k[I]:10:2,p[I]:10:2,z[I]:12:1);
 end;
End;

Procedure WriteResults;
Begin
writeln;
writeln(Lst);
writeln('':38,'RESULTS');
writeln;
writeln(Lst,'':38,'RESULTS');
writeln(Lst);
writeln('':30,'Experimental':20,'Corrected':20);
writeln(Lst,'':30,'Experimental':20,'Corrected':20);
writeln('':30,'____________':20,'_________':20);
writeln(Lst,'':30,'____________':20,'_________':20);
writeln('Number Ave.Mol.Wt.:':30,NA:20:2,NA1:20:2);
writeln(Lst,'Number Ave.Mol.Wt.:':30,NA:20:2,NA1:20:2);
writeln('Weight Ave.Mol.Wt.:':30,WA:20:2,WA1:20:2);
writeln(Lst,'Weight Ave.Mol.Wt.:':30,WA:20:2,WA1:20:2);
writeln('  Z-Ave.Mol.Wt.:':30,WZ:20:2,WZ1:20:2);
writeln(Lst,'  Z-Ave.Mol.Wt.:':30,WZ:20:2,WZ1:20:2);
writeln('Mol.Wt. Distr.:':30,WA/NA:20:3,WA1/NA1:20:3);
writeln(Lst,'Mol.Wt. Distr.:':30,WA/NA:20:3,WA1/NA1:20:3);
End;
Procedure WriteAnal;
Begin
writeln;
writeln(Lst);
writeln('':38,'ANALYSIS');
writeln;
writeln(Lst,'':38,'ANALYSIS');
writeln(Lst);
writeln('Mean:':30,Me:20:4,Me1:20:4);
writeln(Lst,'Mean:':30,Me:20:4,Me1:20:4);
writeln('Variance:':30,VA:20:4,VA1:20:4);
writeln(Lst,'Variance:':30,VA:20:4,VA1:20:4);
writeln('Skew:':30,SK:20:4,SK1:20:4);
```

```
writeln(Lst,'Skew:':30,SK:20:4,SK1:20:4);
writeln('Kurtosis:':30,KU:20:4,KU1:20:4);
writeln(Lst,'Kurtosis:':30,KU:20:4,KU1:20:4);
End;

Hello;
Introduction(St,En,V,N);
Num(f,g,h,u,s,l,m,o,p1,r,t,M1,M2,M3,M4,V2,V3);
Con(B,C,D,E);
MM(y,z);
GtData(x,x1);
CorrectData(k);
Calculate(NA,WA,NA1,WA1,WZ,WZ1);
CalcP(p);
Mean(Me,Me1,Va,Va1);
Skew(Sk,Sk1,Ku,Ku1);
WriteData;
WriteResults;
WriteAnal;
Plot;
End.
```

Review Questions

Section 1.1

1. Define the degree of polymerization.
2. Explain what is a linear, a branched, a star-shaped, a comb-shaped, and a ladder polymer.
3. What is a network structure?
4. What is the difference between random and alternating copolymers?
5. What is meant by graft and block copolymers? Can you illustrate?
6. What are the important features of chain-growth and step-growth polymerizations? Can you explain the difference between the two? Can you suggest an analytical procedure to determine by what mechanism a particular polymerization reaction takes place?
7. What is the DP of polystyrene with molecular weight of 104,000 and poly(vinyl chloride) with molecular weight of 630,000?
8. Explain the differences between thermosetting and thermoplastic polymers and define gel point.
9. Give the definitions of oligomer, telomer, and telechelic polymers.

Section 1.2.1

1. Can you name the following chain-growth polymers by the IUPAC system and/or by giving them trivial names:

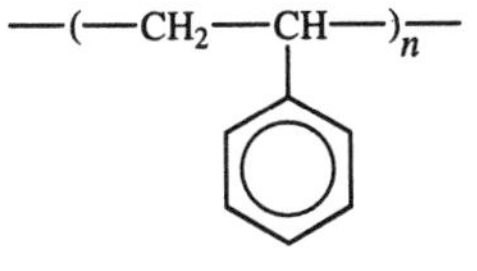

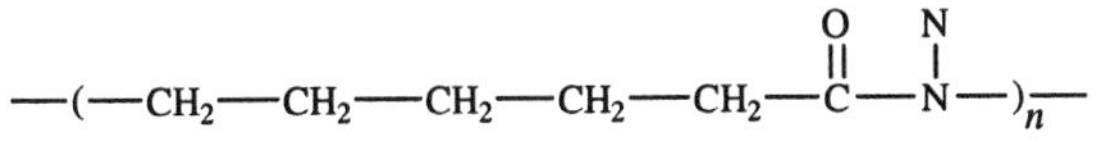

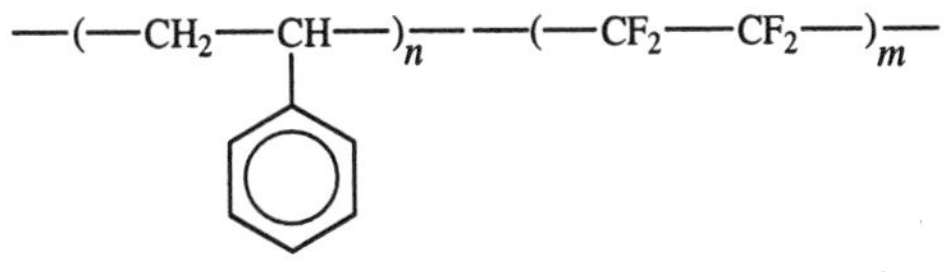

$$—(—CH_2—\underset{\displaystyle C_6H_5}{\underset{|}{CH}}—)_n—$$

$$—(—CH_2—CH_2—CH_2—CH_2—CH_2—\overset{\displaystyle O}{\overset{\|}{C}}—\overset{\displaystyle N}{\overset{|}{N}}—)_n—$$

$$—(—CH_2—\underset{\displaystyle C_6H_5}{\underset{|}{CH}}—)_n——(—CF_2—CF_2—)_m—$$

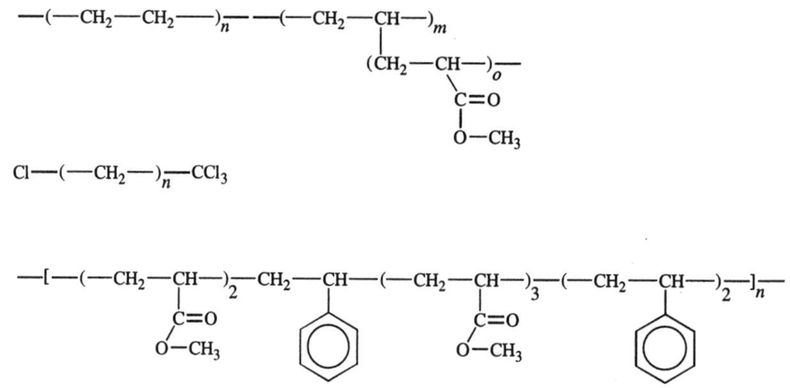

$$—(—CH_2—CH_2—)_n—(—CH_2—\underset{\displaystyle (CH_2—CH(C{=}O)(O—CH_3)—)_o—}{CH}—)_m$$

$$Cl—(—CH_2—)_n—CCl_3$$

$$—[—(—CH_2—\underset{\displaystyle \overset{|}{C}{=}O \atop O—CH_3}{CH}—)_2—CH_2—\underset{\displaystyle C_6H_5}{CH}—(—CH_2—\underset{\displaystyle \overset{|}{C}{=}O \atop O—CH_3}{CH}—)_3—(—CH_2—\underset{\displaystyle C_6H_5}{CH}—)_2—]_n—$$

Section 1.3

1. What are the secondary bond forces that influence the physical properties of macromolecules?
2. Explain dipole–dipole interactions in polymers. Can you give an example?
3. What are the induction forces?

Section 1.4

1. What is meant by the term morphology?
2. Explain the phenomenon of elasticity.
3. What is meant by the negative coefficient of expansion?
4. How does crystallinity of polymers differ from that of small molecules?
5. What are the two ways that crystal growth can take place in polymeric materials?
6. What type of polymers tend to crystallize?
7. What is the typical size of polymeric crystals formed from the melt?
8. What is a fringed micelle or a fringed crystallite model?
9. What is the folded chain lamella? An adjacent-reentry model? A switchboard or a nonadjacent-reentry model?
10. What is drawn fibrillar morphology? What is vitrification?
11. What are spherulites? How are they observed? What is the typical size of the spherulites?
12. What are liquid crystals? How do lyotropic crystalline arrays differ from thermotropic? What are mesophases? How do smectic liquid crystalline arrangements differ from nematic? What is a mesogen?
13. What is orientation and how does that benefit the properties of polymeric materials?
14. Explain what is the first-order transition temperature.
15. Explain what is the second-order transition temperature. How does it manifest itself in the physical properties of polymers and how is it detected?
16. Which polymer, A or B, would have a higher T_m and/or T_g? Explain your choice.

Section 1.5

1. Explain what type of chemical structures and chemical bonds in the backbones of the polymeric chains stiffen them and what type flexibilizes (or "softens") them?

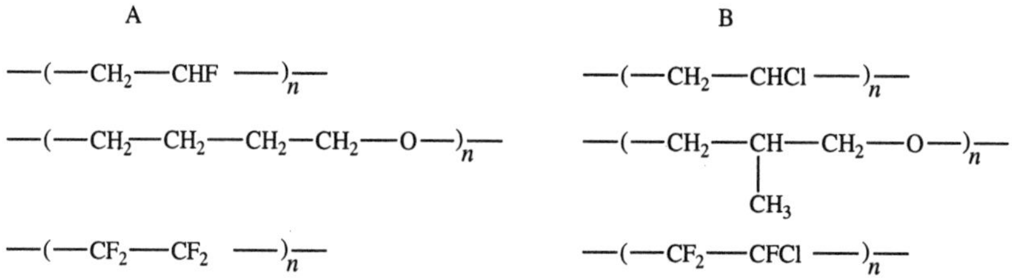

A	B
$—(—CH_2—CHF—)_n—$	$—(—CH_2—CHCl—)_n—$
$—(—CH_2—CH_2—CH_2—CH_2—O—)_n—$	$—(—CH_2—\underset{\displaystyle CH_3}{CH}—CH_2—O—)_n—$
$—(—CF_2—CF_2—)_n—$	$—(—CF_2—CFCl—)_n—$

2. How do the pendant groups affect the melting points of polymers?
3. Does copolymerization raise or lower the melting point of a polymer?

Section 1.6

1. Why must statistical averages be used to express molecular weights of polymers?
2. What is number average molecular weight? What is the equation for number average molecular weight?
3. What is a weight average molecular weight? What is the equation for the weight average molecular weight?
4. Suppose there is a mixture of two types of molecules. There are ten of each in the mixture. The molecular weight of molecules A is 10,000 and the molecular weight of molecules B is 100,000. What is the number average molecular weight of the mixture and the weight average molecular weight?
5. What is the viscosity average molecular weight and how does it differ from the weight average molecular weight?
6. What is a molecular weight distribution? What is a monodisperse polymer and a polydisperse polymer?
7. What is the Mark–Houwink–Sakurada equation? Can you suggest a way to determine constants K and a experimentally for a given polymer?
8. Give the definitions and formulas for the relative viscosity, specific viscosity, reduced viscosity, inherent viscosity, and intrinsic viscosity.
9. Why is it necessary to extrapolate to zero (explain how this is done) in order to obtain intrinsic viscosity?
10. Discuss the various methods of molecular weight determination. Explain why a particular method yields a number of a weight average molecular weight, or, as in the case of GPC, both.

Section 1.7

1. What is meant by tacticity?
2. Give a definition of and illustrate examples of isotactic, syndiotactic, and atactic arrangement of macromolecules. This should include Fischer and Newman projections.
3. Explain what is meant by erythrodiisotactic, threodiisotactic and polymers. Illustrate. Do the same for erythrodisyndiotactic and threodisyndiotactic.
4. What are tritactic polymers? Draw *cis* and *trans* tritactic polymers.

Section 1.8

1. Discuss optical activity in polymers.

References

1. W. H. Carothers, *J. Am. Chem. Soc.*, **51**, 2548 (1929); *Chem. Rev.*, **8**, 353 (1931); Collected Papers of Wallace Hume Carothers on High Polymeric Substances (H. Mark and G.S. Whitby, eds.), Wiley-Interscience, New York, 1926.
2. P. J. Flory, Principles of Polymer Chemistry, Cornell Univ. Press, Ithaca, 1951.
3. M. L. Huggins, P. Carradini, V. Desreux, O. Kratsky, and H. Mark, *J. Polym. Sci., B*, **6**, 257 (1968).
4. A. D. Jenkins and J. L. Koening, "Nomenclature," Chap. 2 in *Comprehensive Polymer Science*, Vol. I (C. Booth and C. Price, eds.), Pergamon Press, Oxford, 1989.
5. International Union of Pure and Applied Chemistry, Macromolecular Nomenclature Commission, *J. Polym. Sci., Polym. Lett.*, **11**, 389 (1973); *Macromolecules*, **6**, 149 (1973); *Pure and Appl. Chem.*, **40**, 479 (1974).
6. R. S. Cahn, *An Introduction to Chemical Nomenclature*, 3rd Ed., Butterworth, London, 1968. Also see William J. le Noble *Highlights of Organic Chemistry*, Dekker, New York, 1974.
7. H. Mark, *Ind. Eng. Chem.*, **34**, 1343 (1942); P. A. Small, *J. Appl. Chem.*, **3**, 71 (1953); E. E. Walker, *J. Appl. Chem.*, **2**, 470 (1952).
8. S. Mizushima and T. Shimanouchi, *J. Am. Chem. Soc.*, **85**, 3521 (1964).

9. P. Meares, *Polymers: Structure and Bulk Properties*, Van Nostrand, New York, 1965, p. 118; J. E. Mark, A. Eisenberg, W. W. Graessley, L. Mandelkern, and J. L. Koening, *Physical Properties of Polymers*, Am. Chem. Soc., Washington, D.C., 1984.
10. P. J. Flory, *J. Chem. Phys.*, **17**, 223 (1949).
11. L. Mandelkern, *Crystallization of Polymers*, McGraw-Hill, New York, 1964.
12. M. Avrami, *J. Chem. Phys.*, **7**, 103 (1939); *ibid.*, **8**, 212 (1940).
13. L. Mandelkern, *A.C.S. Polym. Prepr.*, **20** (1), 267 (1979); D. C. Basset, *Principles of Polymer Morphology*, Cambridge University Press, Cambridge, 1981; A. Sharples, "Crystallinity," Chap. 4 in *Polymer Science*, Vol. I (A. Jenkins, ed.), North-Holland, Amsterdam, 1972.
14. H. F. Mark, *J. Polym. Sci.*, **C9**, 1 (1965); A. Peterlin, *J. Polym. Sci.*, **C9**, 61 (1965).
15. P. H. Geil, *Polymer Single Crystals*, Wiley-Interscience, New York, 1963.
16. P. H. Geil, N. K. J. Symons, and R.G. Scott, *J. Appl. Phys.*, **30**, 1516 (1959).
17. R. H. Geil, *J. Polym. Sci.*, **44**, 449 (1960).
18. B. H. Ranby, F. F. Morehead, and N. M. Walter, *J. Polym. Sci.*, **44**, 349 (1960).
19. A. Keller, *Growth and Perfection of Crystals*, Wiley, New York, 1958; A. Keller and A. O'Connor, *Polymer*, **1**, 163 (1960).
20. R. St. J. Mauley, *Nature*, **189**, 390 (1961).
21. D. C. Basselt, F. C. Frank, and A. Keller, *Nature*, **184**, 810 (1959); W. D. Niegisch and P. R. Swan, *J. Appl Phys.*, **31**, 1906 (1960); D. H. Reneker and P. H. Geil, *J. Appl Phys.*, **31**, 1916 (1960).
22. M. G. Zachman, *Angew. Chem., Int. Ed. Engl.*, **13**, (4), 244 (1974).
23. J. Brandrup and E. H. Immergut (eds.), *Polymer Handbook*, 2nd and 3rd eds., 1989, Wiley, New York; E.A. Turi (ed.), *Thermal Characterization of Polymeric Materials*, Academic Press, New York, 1981; W. W. Wedlandt, *Thermal Analysis*, 3rd ed., Wiley, New York, 1986.
24. R. Hill and E. E. Walker, *J. Polym. Sci.*, **3**, 609 (1948).
25. L. Marker, R. Early, and S. L. Aggarwall, *J. Polym. Sci.*, **38**, 369 (1959).
26. G. Natta, F. Danusso, and G. Moraglis, *J. Polym. Sci.*, **25**, 119 (1957).
27. T. W. Campbell and A. C. Haven, Jr., *J. Appl. Polymer Sci.*, **1**, 73 (1959).
28. F. P. Redding, *J. Polym. Sci.*, **21**, 547 (1956).
29. R. D. Deanin, *Polymer Structure, Properties and Application*, Cahmers Publishing Co., Inc., Boston, Mass., 1972.
30. J. R. Collier, *Ind. Eng. Chem.*, **61**, (10), 72 (1969).
31. F. W. Billmeyer, Jr., *Textbook of Polym. Science*, Chap. 3, Wiley-Interscience, New York, 1962.
32. M. L. Miller, *Structure of Polymers*, Reinhold, New York, 1966.
33. M. Farina, M. Peraldo, and G. Natta, *Angew. Chem., Int. Ed. Engl.*, **4**, 107 (1965).
34. N. Beredjick and C. Schuerch, *J. Am. Chem. Soc.*, **78**, 2646 (1956); *ibid.*, **80**, 1933 (1958).
35. G. Natta, M. Farina, M. Peraldo, and G. Bressan, *Makromol. Chem.*, **43**, 68 (1961).
36. M. Farina and G. Bressan, *Makromol. Chem.*, **61**, 79 (1963).
37. P. Pino, F. Ciardelli, and G. P. Lorenzi, *J. Am. Chem. Soc.*, **85**, 3883 (1963).
38. P. Pino, F. Ciardelli, and G. P. Lorenzi, *Makromol. Chem.*, **70**, 182 (1964).
39. M. Goodman, A. Abe, and Y. L. Fan, *Macromol. Rev.*, **1**, 8 (1967).
40. N. G. Gaylord and H. F. Mark, *Linear and Stereoregular Addition Polymers*, Interscience, New York, 1959.
41. G. Natta and P. Corradini, *J. Polym. Sci.*, **39**, 29 (1959).
42. J. Janca (ed.), *Steric Exclusion Liquid Chromatography of Polymers*, Dekker, New York, 1984.
43. T. Provder (ed.), *Detection and Data Analysis in Size Exclusion Chromatography*, Am. Chem. Soc., Washington, 1987.
44. A. Blumstein (ed.), *Liquid Crystalline Order in Polymers*, Academic Press, New York, 1978.
45. A. Ciferri, W. R. Krigbaum, and R. B. Meyer (eds.), *Polymer Liquid Crystals*, Academic Press, New York, 1982.

2

Free-Radical Chain-Growth Polymerization

2.1. Free-Radical Chain-Growth Polymerization Process

Polymerizations by a free-radical mechanism are typical free-radical reactions. That is to say, there is an *initiation*, when the radicals are formed, a *propagation*, when the products are developed, and a *termination*, when the radical reactions end. In the polymerizations, the propagations are chain reactions. A series of very rapid repetitive steps follow each single act of initiation, leading to the addition of thousands of monomers.

This process of polymerization of vinyl monomers takes place at the expense of the double bonds:

$$n\,\rangle C{=}C\langle \longrightarrow -[-\rangle C{-}C\langle-]_n-$$

Table 2.1 illustrates the steps in free-radical polymerizations.

Formation of initiating radicals is the rate-determining step in the initiation reaction. The process can result, as shown in Table 2.1, from cleavage of compounds, like peroxides, or from other sources. Many reactions lead to formations of free radicals. The initiating radicals, however, must be energetic enough to react with the vinyl compounds. A linear correlation exists between the affinities of some radicals for vinyl monomers and the energy (calculated) required to localize a π electron at the β-carbon of the monomer.[1] By comparison to other steps in the polymerization process, initiation is slow and requires high energy of activation.

2.1.1. Kinetic Relationships in Free-Radical Polymerizations

A kinetic scheme for typical free-radical polymerizations is pictured as follows[165]:

Initiation:	$I \xrightarrow{K_d} 2\,R\bullet$	$d[R\bullet]/dt = R_x = 2\,k_d[I]$
	$R\bullet + M \xrightarrow{k_I} RM\bullet$	$-d[R\bullet]/dt = k_I[R\bullet][M]$
Propagation:	$RM\bullet + n\,M \xrightarrow{k_P} {\sim}P_{n+1}\bullet$	$R_P = k_P[M\bullet][M]$
Termination:	${\sim}P_{n+1}\bullet + {\sim}P_{m+1}\bullet \xrightarrow{k_T}$ polymer	$-d[P\bullet]/dt = 2k_T[P\bullet]^2$
Transfer:	${\sim}P_n\bullet + S \xrightarrow{k_s} {\sim}P_n + S\bullet$	
	$S\bullet + nM \longrightarrow SM_n\bullet$	

Table 2.1. Free-Radical Polymerization of Vinyl Monomers

1.	Initiation	R–R → 2 R•
		R• + CH_2=CHX → R–CH_2–CHX•
2.	Propagation	R–CH_2–CHX• + *n* CH_2=CHX → R–(–CH_2–CHX–)$_n$–CH_2–CHX•
3.	Termination	
	a. By combination	
		R–(–CH_2–CHX–)$_n$–CH_2–CHX• + •XHC–CH_2–(–XHC–CH_2–)$_n$–R → R–(–CH_2–CHX–)$_n$–CH_2–CHX–XHC–CH_2–(–XHC–CH_2–)$_n$–R
	b. By disproportionation	
		R–(–CH_2–CHX–)$_n$–CH_2–CHX• + •XHC–CH_2–(–XHC–CH_2–)$_n$–R → R–(–CH_2–CHX–)$_n$–CH_2–CH_2X + XHC=CH–(–XHC–CH_2–)$_n$–R
	c. By transfer	
		R–(–CH_2–CHX–)$_n$–CH_2–CHX• + R′H → R–(–CH_2–CHX–)$_n$–CH_2–CH_2X + R′•

In the above scheme M stands for the monomer concentration, I is the concentration of the initiator, and [R•] and [P•] mean the concentration of primary and polymer radicals, respectively; S stands for the chain transferring agent, R_I denotes the decomposition rate of the initiator, and R_P the rate of polymerization. The rate constant for initiator decomposition is k_d, for initiation is k_I, for propagation k_P, and for termination is k_T. The above is based on an assumption that k_P and k_I are independent of the sizes of the radicals. This is supported by experimental evidence, which shows that radical reactivity is not affected by the size, when the chain length exceeds dimer or trimer dimensions.[28]

The equation for the rate of propagation, shown above, contains the term [M•]. It designates radical concentration. This quantity is difficult to determine quantitatively, because it is usually very low. A *steady state* assumption is therefore made to simplify the calculations. It is assumed that while the radical concentration increases at the very start of the reaction it reaches a constant value almost instantly. This value is maintained from then on and the rate of change of free-radical concentration is assumed to quickly become and remain zero during the polymerization. At steady state the rates of initiation and termination are equal, or $R_I = R_T = 2K_I[M\bullet]^2$. It is possible to solve for [M•], which can then be expressed as:

$$[M\bullet] = (K_d[I]/K_T)^{1/2}$$

The rate of propagation is

$$R_P = K_P[M](K_d[I]/K_T)^{1/2}$$

This rate of propagation applies if the kinetic chain length is large and if the transfer to monomer is not very efficient. The rate of monomer disappearance can be expressed as

$$-d[M]/dt = R_I + R_P$$

Because many more molecules of the monomer are involved in the propagation than in the initiation step, a very close approximation is

$$-d[M]/dt = R_P$$

The average lifetime of a growing radical under steady state conditions can then be written as follows[28,54]:

$$\tau = K_P[M]/2K_T(R_P)$$

Not all primary radicals that form attack the monomer. Some are lost to side reactions. An initiator efficiency factor, f, is therefore needed. It is a fraction of all the radicals that form and can be expressed as

$$f = \text{initiating radicals/total number of radicals that form}$$

The rate of initiator decomposition and the rate equation can be expressed, respectively, as:

$$R_I = 2fK_d[I]$$

$$R_P = K_P[M](K_d[I]f/K_T)^{1/2}$$

According to the kinetic scheme, chain transfer does not affect the rate of polymerization but alters the molecular weight of the product. Also, it is important to define the average number of monomer units that are consumed per each initiation. This is the *kinetic chain length*, and it is equal to the rate of polymerization per rate of initiation:

$$\text{kinetic chain length} = \nu = K_P/K_I$$

At steady-state conditions, ν is also equal to K_P/K_T. The kinetic chain length can also be expressed as

$$\nu = K_P[M]/2K_T[M\bullet]$$

By substituting the expression for [M•] the equation becomes

$$\nu = K_P[M]/2(fK_dK_T[I])^{1/2}$$

The number average degree of polymerization, DP, is equal to 2 ν, if the termination takes place by coupling. It is equal to ν, if it takes place by disproportionation. The above kinetic relationships apply in many cases. They fail, however, to apply in all cases.[165] To account for it, several mechanisms were advanced. They involve modifications of the initiation, termination, or propagation steps. These are beyond the discussions in this book.

2.2. Reactions Leading to Formation of Initiating Free Radicals

The initiating free radical can come from many sources. Thermal decompositions of compounds with azo and peroxy groups are common sources of such radicals. The radicals can also come from "redox" reactions or through various light-induced decompositions. Ionizing radiation can also be used to form initiating radicals.

2.2.1. Thermal Decomposition of Azo Compound, and Peroxides

Azo compound and peroxides contain weak valency bonds in their structures. Heating causes these bonds to cleave and to dissociate the compounds into free radicals as follows:

$$\text{R–N=N–R} \xrightarrow{\text{heat}} 2\ \text{R}\bullet + N_2$$

For many azo compounds, such dissociations occur at convenient elevated temperatures. A commonly used azo compound is α,α'-azobisisobutyronitrile. The original synthesis of this compound was reported to be as follows[2]:

$$(CH_3)_2C{=}O + HCN \longrightarrow (CH_3)_2C(OH)C{\equiv}N$$

$$(CH_3)_2C(OH)C{\equiv}N + H_2N{-}NH_2 \longrightarrow (H_3C)_2C(C{\equiv}N){-}NH{-}NH{-}C(C{\equiv}N)(CH_3)_2 \xrightarrow{Br_2} (H_3C)_2C(C{\equiv}N){-}N{=}N{-}C(C{\equiv}N)(CH_3)_2$$

The final products of its decomposition are two cyanopropyl radicals and a molecule of nitrogen:

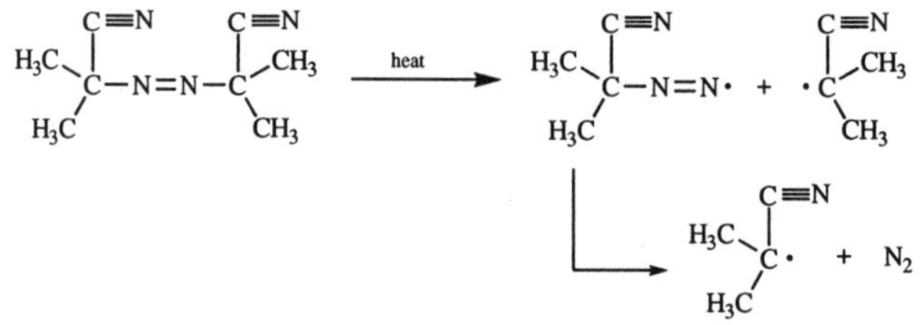

The free-radicals that form can recombine inside or outside the solvent cage, where the decompositions take place, to yield either tetramethylsuccinonitrile or a ketenimine[3]:

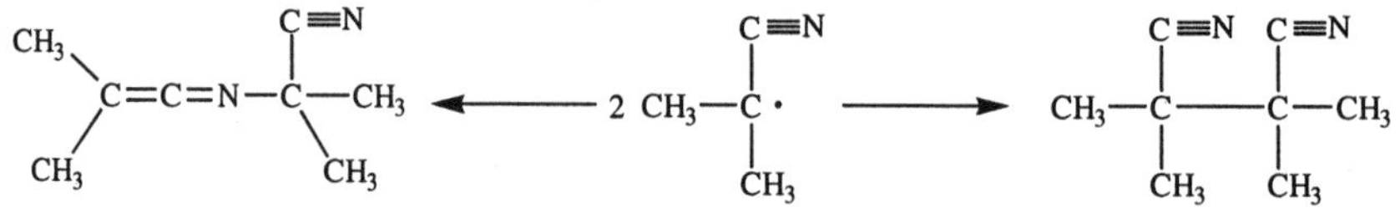

Some other, fairly efficient azo initiators include:

1. N-Nitrosoacylanilides,

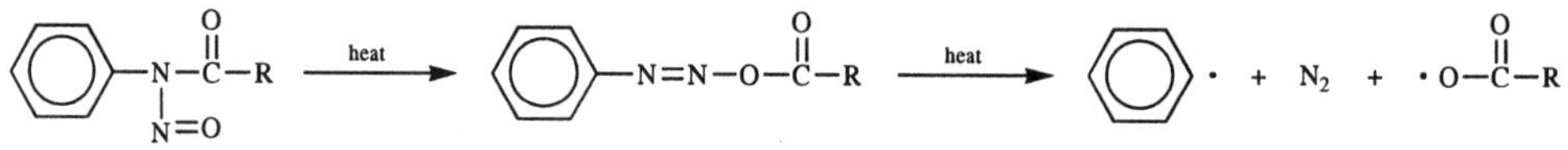

2. Bromobenzenediazohydroxide,

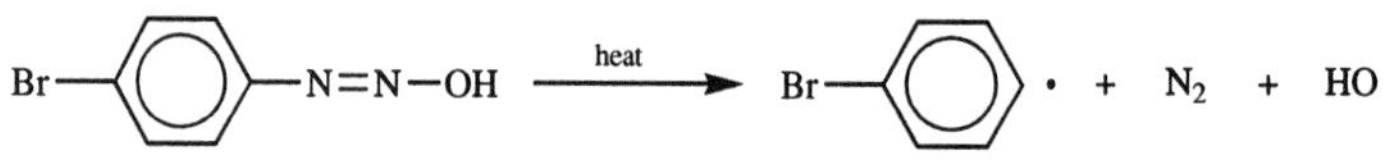

3. Triphenylazobenzene,

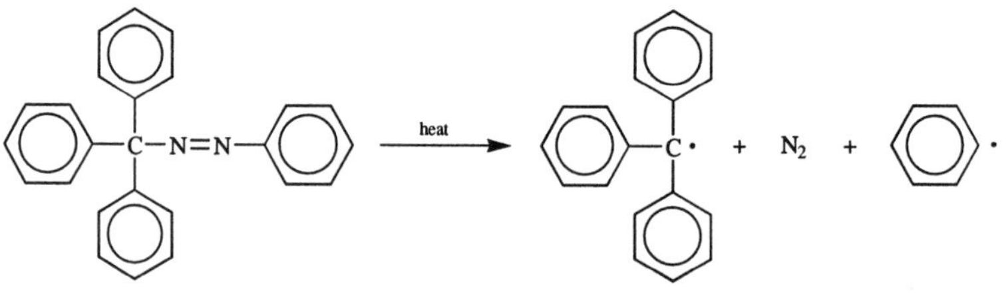

The triphenylmethyl radical shown above is resonance stabilized and unable to initiate polymerizations. The phenyl radical, on the other hand, is a hot radical. It initiates polymerizations readily. Decomposition rates of some azonitrile initiators are listed in Table 2.2. There are also many peroxides available for initiating free-radical polymerizations. These can be organic and inorganic compounds. There are, however, many more organic peroxides available commercially than the inorganic ones. The organic ones include dialkyl and diaryl peroxides, alkyl and aryl hydroperoxides, diacyl peroxides, peroxy esters and peracids. Hydrogen peroxide is the simplest inorganic peroxide.

Table 2.2. Decomposition Rates of Some Azonitrile Initiators[a]

Compound	Solvent	T (°C)	$K_d{}^b$ (s^{-1})
2,2′-azobisisobutyronitrile	benzene	78.0	8.0×10^{-5}
2,2′-azobis-2-ethylpropionitrile	nitrobenzene	100.0	1.1×10^{-3}
2,2′-azobis-2-cyclopropylpropionitrile	toluene	50.0	8.2×10^{-5}
1,1′-azobiscyclohexanenitrile	toluene	80.0	6.5×10^{-6}
2,2′-azobis-2-cyclohexylpropionitrile	toluene	80.0	8.3×10^{-6}
1,1′-azobiscyclooctanenitrile	toluene	45.0	1.5×10^{-4}

[a]From reference 4 and from other sources in the literature.
[b]$K_d = A^{(-E_a/RT)}$

Syntheses, structures, and chemistry of various peroxides were described thoroughly in the literature.[5] Here we only mention some properties of peroxides and their performance as they pertain to initiations of polymerizations. Decompositions of peroxides, like the azo compounds, are temperature dependent. This means that the rates increase with temperature. The rates are also influenced by the surrounding medium, such as the solvents that imprison or "cage" the produced pairs of free radicals. Before undergoing a net translational diffusion out of the cage, one or both of the radicals may or may not expel a small molecule like CO_2 or N_2. For instance, benzoyl peroxide can give off carbon dioxide as follows:

$$C_6H_5-\overset{O}{\overset{\|}{C}}-O-O-\overset{O}{\overset{\|}{C}}-C_6H_5 \longrightarrow 2\, C_6H_5-\overset{O}{\overset{\|}{C}}-O\bullet \longrightarrow C_6H_5\bullet + CO_2$$

The resultant radicals can still combine and yield new and completely inactive species. This can be shown on phenyl radicals,

$$2\, C_6H_5\bullet \longrightarrow C_6H_5-C_6H_5$$

It is one of the causes of inefficiency among initiators. The average time for recombination of free radicals inside a cage and the time for their diffusion out of the cage is about 10^{-10} s.[6] In addition, the efficiency of the initiator is affected by the monomer and the solvent. Viscosity of the medium is inversely proportional to the initiator efficiency, because the more viscous the solution, the greater the cage effect.[40,41]

Numerous lists are available in the literature that give the decomposition temperatures or the half-lives at certain elevated temperatures of many initiators.[5] Decompositions of peroxides may proceed via concerted mechanisms[7,8] and the rates are structure-dependent. This can be illustrated on benzoyl peroxide. The benzoyl groups, the two halves of the molecule, are dipoles. They are attached, yet they repel each other. Rupture of the peroxide link releases the electrostatic repulsion between the two dipoles. Presence of electron donating groups in the *para* position increases the repulsion, lowers the decomposition temperature, and increases the decomposition rate. The opposite can be expected from electron attracting groups in the same position.[5] The effect of substituents on the rate of spontaneous cleavage of dibenzoyl peroxide was expressed[9] in terms of the Hammett equation, $\log(K/K_0) = \rho\sigma$. This is shown in Table 2.3.

Peroxides can cleave heterolytically and homolytically. *Heterolytic* cleavage of peroxides results in formation of ions,

$$R^{\oplus} + {:}\overset{\ominus}{O}-O-R' \rightleftharpoons R-O-O-R' \rightleftharpoons R-O-\overset{\ominus}{O}{:} + R'^{\oplus}$$

Table 2.3. Effects of Substitutents on Decompostion of Dibenzoyl Peroxide in Benzene[9,10]

Substituent	$K_i \times 10^3$	$\log K/K_0$	$\sigma_1 + \sigma_2$
p,p'-Dimethoxy	7.06	0.447	−0.536
p-Methoxy	4.54	0.255	−0.268
p,p'-Dimethyl	3.68	0.164	−0.340
p,p'-Di-*t*-butyl	3.65	0.161	−0.394
Parent compound	2.52	0.000	0.000
p,p'-Dichloro	2.17	−0.065	+0.454
m,m'-Dichloro	1.58	−0.203	+0.746
m,m'-Dibromo	1.54	−0.215	+0.782
p,p'-Dicyano	1.22	−0.314	+1.30

Table 2.4. Decompostion of Benzoyl Perxoide in Various Solvents at 79.8 °C[10–13,37]

Solvent	Approximate % Decomposition in 4 hours
Anisole	43.0
Benzene	50.0
Carbon tetrachloride	40.0
Chlorobenzene	49.0
Chloroform	44.0
Cyclohexane	84.0
Cyclohexene	40.0
Ethyl acetate	85.0
Ethyl benzene	46.0
Methyl benzoate	41.0
Methylene chloride	62.0
Nitrobenzene	49.0
Tetrachloroethylene	35.0
Toluene	50.0

but *homolytic* cleavage results in formation of radicals:

$$\text{R–O–O–R}' \rightleftharpoons \text{R–O}\bullet + \bullet\text{O–R}'$$

In the gaseous phase the cleavage is usually homolytic, because it requires the least amount of energy.[11] In solution, however, the dissociation may be either one of the two, depending upon the nature of the R groups. Heterolytic cleavage may be favored, in some cases, if the two groups, R and R′, differ in electron attraction. The same is true if the reaction solvent has a high dielectric constant. Solvation of the ions that would form due to heterolytic cleavage is also a promoting influence for such a cleavage:

$$\text{A:B} + 2\text{S} \rightarrow \text{SA:}^{\ominus} + {}^{\oplus}\text{:BS}$$

where S represents the solvent.

There is much information in the literature on the rates and manner of decomposition of many peroxides in various media. Beyond that, diagnostic tests exist that can aid in determining the decomposition rates of a particular peroxide and a particular media.[12] Table 2.4 lists some examples. It is presented to show how different solvents affect the rate of decomposition of benzoyl peroxide into radicals.

2.2.2. Bimolecular Initiating Systems

Decompositions of peroxides into initiating radicals are possible through bimolecular reactions involving electron transfer mechanisms. These reactions are often called *redox* initiations and can be illustrated as follows:

$$\underset{\text{electron donor}}{\text{A}} + \underset{\text{electron acceptor}}{\text{R–O–O–R}'} \rightarrow \text{A}^{\oplus} + \text{RO}\bullet + {}^{\ominus}\text{OR}'$$

where A is the reducing agent and ROOR′ is the peroxide.

The above can be shown on a decomposition of a persulfate (an inorganic peroxide) by the ferrous ion:

$$\underset{\textbf{electron acceptor}}{{}^{\ominus}\text{O–}\overset{\text{O}}{\underset{\text{O}}{\overset{\|}{\underset{\|}{\text{S}}}}}\text{–O–O–}\overset{\text{O}}{\underset{\text{O}}{\overset{\|}{\underset{\|}{\text{S}}}}}\text{–O}^{\ominus}} + \underset{\textbf{electron donor}}{\text{Fe}^{\oplus\oplus}} \longrightarrow \underset{\textbf{initiating ion-radical}}{{}^{\ominus}\text{O–}\overset{\text{O}}{\underset{\text{O}}{\overset{\|}{\underset{\|}{\text{S}}}}}\text{–O}\cdot} + {}^{\ominus}\text{O–}\overset{\text{O}}{\underset{\text{O}}{\overset{\|}{\underset{\|}{\text{S}}}}}\text{–O}^{\ominus} + \text{Fe}^{\oplus\oplus\oplus}$$

Side reactions are possible in the presence of sufficient quantities of reducing ions:

$$^{\ominus}O-SO_2-O\cdot + Fe^{2+} \longrightarrow {}^{\ominus}O-SO_2-O^{\ominus} + Fe^{3+}$$

A redox reaction can also take place between the peroxide and an electron acceptor:

$$R{-}O{-}O{-}H + Ce^{4+} \rightarrow R{-}O{-}O\bullet + Ce^{3+} + H^{+}$$

Side reactions with an excess of the ceric ion can occur as well:

$$R{-}O{-}O\bullet + Ce^{4+} \rightarrow R^{+} + O_2 + Ce^{3+}$$

Another example is a redox reaction of *t*-butyl hydroperoxide with a cobaltous ion[14]:

$$Co^{2+} + (CH_3)_3C{-}O{-}O{-}H \longrightarrow Co^{3+} + (CH_3)_3C{-}O\cdot + OH^{\ominus}$$

The cobaltic ion that forms can act as an electron acceptor:

$$Co^{3+} + (CH_3)_3C{-}O{-}O{-}H \longrightarrow Co^{2+} + (CH_3)_3C{-}O{-}O\cdot + H^{+}$$

Side reactions can occur here too, such as:

$$(CH_3)_3C{-}O{-}O\cdot + (CH_3)_3C{-}O{-}O{-}H \longrightarrow (CH_3)_3C{-}O\cdot + O_2 + (CH_3)_3C{-}OH$$

$$(CH_3)_3C{-}O\cdot + Co^{2+} \longrightarrow (CH_3)_3C{-}O^{\ominus} + Co^{3+}$$

Nevertheless, cobaltous ions form efficient redox initiating systems with peroxydisulfate ions.[142]

Tertiary aromatic amines also participate in bimolecular reaction with organic peroxides. One of the unpaired electrons on the nitrogen atom transfers to the peroxide link, inducing decomposition. No nitrogen, however, is found in the polymer. It is therefore not a true redox type initiation and the amine acts more like a *promoter* of the decomposition.[15] Two mechanisms were proposed to explain this reaction. The first one was offered by Horner *et al.*[15]:

$$C_6H_5{-}CO{-}O{-}O{-}CO{-}C_6H_5 + :N(CH_3)_2{-}C_6H_5 \longrightarrow C_6H_5{-}CO{-}O^{\ominus} + [C_6H_5{-}N^{\cdot}(CH_3)_2]^{\oplus} + C_6H_5{-}CO{-}O\cdot$$

loss of hydrogen electron shift

$$\longrightarrow (CH_3)_2N{-}C_6H_4\cdot + C_6H_5{-}C(O){-}OH \xrightarrow{\text{dimerization}} (CH_3)_2N{-}C_6H_4{-}C_6H_4{-}N(CH_3)_2$$

$$\longrightarrow C_6H_5{-}C(O){-}OH + (CH_3)_2N{-}C_6H_4\cdot \xrightarrow{\text{dimerization}} \text{2,2'-bis(dimethylamino)biphenyl}$$

A second mechanism, proposed by Imoto and Choe,[15] shows the complex as an intermediate step:

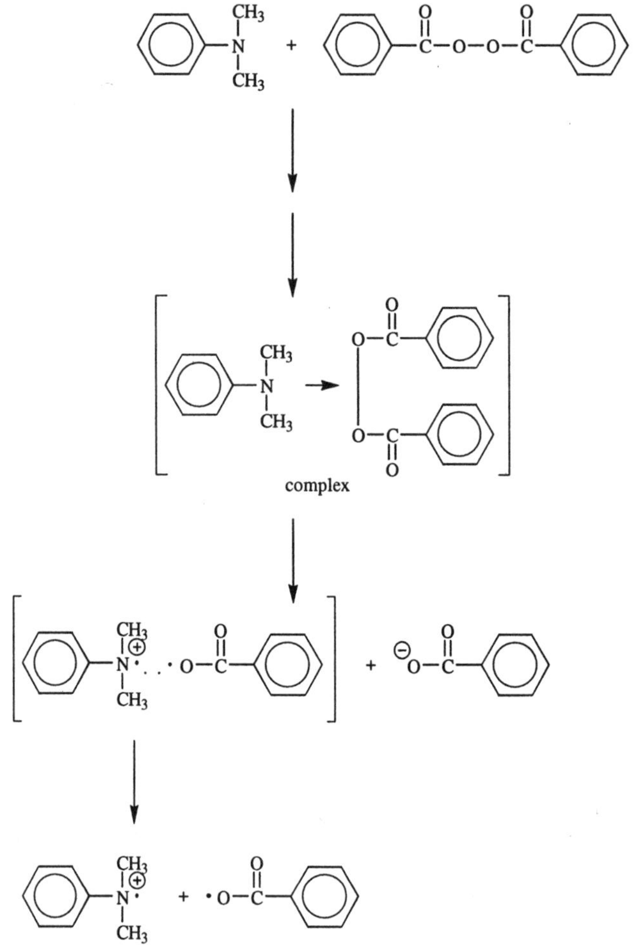

complex

The dimethyl aniline radical-cation, shown above, undergoes other reactions than addition to the monomer. The benzoyl radical is the one that initiates the polymerizations.

The presence of electron-releasing substituents on diethyl aniline increases the rate of the reaction with benzoyl peroxide.[17] This suggests that the lone-pair of electrons on nitrogen attack the positively charged oxygen of the peroxide link.[16]

By comparison to peroxides, the azo compounds are generally not susceptible to chemically induced decompositions. It was shown,[18] however, that it is possible to accelerate the decomposition of α,α'-azobisisobutyronitrile by reacting it with bis(–)-ephedrine-copper(II) chelate. The mechanism was postulated to involve reductive decyanation of azobisisobutyronitrile through coordination to the chelate.[18] Initiations of polymerizations of vinyl chloride and styrene with α,α'-azobisisobutyronitrile coupled to aluminum alkyls were investigated.[19] Gas evolution measurements indicated some accelerated decomposition. Also, additions of large amounts of tin tetrachloride to either α,α'-azobisisobutyronitrile or to dimethyl-α,α'-azobisisobutyrate increase the decomposition rates.[180] Molar ratios of $[SnCl_4]/[AIBN] = 21.65$ and $[SnCl_4]/[MAIB] = 19.53$ increase the rates by factors of 4.5 and 17, respectively. Decomposition rates are also enhanced by donor solvents, like ethyl acetate or propionitrile in the presence of tin tetrachloride.[180]

2.2.3. Boron Alkyls and Metal Alkyl Initiators of Free-Radical Polymerizations

These initiators were originally reported a long time ago.[20–22] Oxygen plays an important role in the reactions.[22,23] It reacts with the alkyls under mild conditions to form peroxides[24,25]:

$$R_3B + O\text{—}O \longrightarrow R_2B\text{—}O\text{—}O\text{—}R$$

Initiating radicals apparently come from reactions of these peroxides with other molecule of boron alkyls.[26–28] One postulated reaction mechanism can be illustrated as follows[28]:

$$\rangle B\text{—}R + R\text{—}O\text{—}O\text{—}B\langle \longrightarrow \rangle B\text{—}O\text{—}R + R\cdot + \rangle B\text{—}O\cdot$$

Another suggested reaction path is[29]

$$R_2B\text{–}O\text{–}O\text{–}R + 2\,R'_3B \rightarrow 2\,R\bullet + R'_2B\text{–}O\text{–}BR'_2 + R'_2B\text{–}O\text{–}R$$

Catalytic action of oxygen was observed with various organometallic compounds.[28] One example is dialkylzinc[30] that probably forms an active peroxide.[28] The same is also true of dialkylcadmium and of triethylaluminum.[31] Peroxide formation is believed to be an important step in all these initiations. Initiating radicals, however, do not appear to be produced from mere decompositions of these peroxides.[28]

2.2.4. Photochemical Initiators

This subject is discussed in some detail in Chapter 8, in the section on photocrosslinking reactions of coatings and films. A brief explanation is also offered here because such initiations are used, on a limited scale, in conventional polymerizations.

Many organic compounds decompose or cleave into radical upon irradiation with light of an appropriate wavelength.[33,34,143] Because the reactions are strictly light and not heat-induced, it is possible to carry out the polymerizations at low temperatures. In addition, by employing narrow wavelength bands that only excite the photoinitiators, it is possible to stop the reaction by merely blocking out the light. Among the compounds that decompose readily are peroxides, azo compounds, disulfide, ketones, and aldehydes. A photodecomposition of a disulfide can be illustrated as follows:

$$C_6H_5\text{—}S\text{—}S\text{—}C_6H_5 \xrightarrow{\text{uv light}} 2\,C_6H_5\text{—}S\cdot$$

Today, many commercially prepared photoinitiators are available. Most are aromatic ketones that cleave by the Norrish reaction or are photoreduced to form free radicals[35,149] (for further details see Chapter 8).

2.2.5. Initiation of Polymerization with Radioactive Sources and Electron Beams

Different radioactive sources can initiate free-radical polymerizations of vinyl monomers. They can be emitters of gamma rays, beta rays, or alpha particles. Most useful are strong gamma emitters, like ^{60}Co or ^{90}Sr. Electron beams from electrostatic accelerators are also efficient initiators. The products from irradiation by radioactive sources or by electron beams are similar to, but not identical to, the products of irradiation by ultraviolet light. Irradiation by ionizing radiation causes the excited monomer molecules to decompose into free radicals. Ionic species also form from initial electron captures. No sensitizers or extraneous initiating materials are required. It is commonly accepted that the free radicals and the ions are the initial products and that they act as intermediate species in these reactions. There is still insufficient information, however, on the exact nature of all of these species.[50,51] The polymerizations are predominantly by a free-radical mechanism with some monomers and by an ionic one with others.[50,51]

2.3. Capture of the Free Radical by the Monomers

Once the initiating radical is formed, there is competition between addition to the monomer and all other possible secondary reactions. A secondary reaction, such as a recombination of fragments, can be caused by the cage effect of the solvent molecules.[36,52] Other reactions can take place between a radical and a parent initiator molecule. This can lead to formation of different initiating species. It can, however, also be a dead end as far as the polymerization reaction is concerned.

After the initiating radical has diffused into the proximity of the monomer, the capture of the free radical by the monomer completes the step of initiation. This is a straightforward addition reaction:

$$R{-}\dot{C}(R_1)(R_2) + (R_3)_2C{=}C(R_3)_2 \longrightarrow R{-}C(R_1)(R_2){-}C(R_3)_2{-}\dot{C}(R_3)_2$$

The unpaired electron of the radical is believed to be in the pure *p*-orbital of a planar, sp^2, carbon atom. Occasionally, however, radicals with sp^3 configuration appear to form.[152–154]

The half-life of the initial benzoyloxy radicals from decompositions of benzoyl peroxide is estimated to be 10^{-4} to 10^{-5} seconds. Past that time, they decompose into phenyl radicals and carbon dioxide.[53] This is sufficient time for the benzoyloxy radicals to be trapped by reactive monomers. Unreactive monomers, however, are more likely to react with the phenyl radicals that form from the elimination reaction. In effect there are two competing reactions[54]:

1. Decomposition of the free radical:

$$C_6H_5{-}C({=}O){-}O\cdot \xrightarrow{K'} C_6H_5\cdot + CO_2$$

2. Addition of the free-radical to the monomer:

$$C_6H_5{-}C({=}O){-}O\cdot + {>}C{=}C{<} \xrightarrow{K''} C_6H_5{-}C({=}O){-}O{-}\overset{|}{\underset{|}{C}}{-}\dot{C}{<}$$

The ratio of the rates of the two reactions, K''/K', depends upon the reactivity of the monomers. It is shown in Table 2.5.[55,151] Benzoyloxy radical is used in this table as an illustration. Similar ratios exist for other initiating radicals. A similar comparison is possible for a redox initiating system. An initiating sulfate radical ion from a persulfate initiator can react with another reducing ion or add to the monomer:

$$SO_4^{\ominus}\cdot + Fe^{\oplus\oplus} \xrightarrow{K'} SO_4^{\ominus\ominus} + Fe^{\oplus\oplus\oplus}$$

or

$$SO_4^{\ominus}\cdot + {>}C{=}C{<} \longrightarrow SO_4^{\ominus}{-}\overset{|}{\underset{|}{C}}{-}\dot{C}{<}$$

Table 2.5. Relative Reactivities of the Benzoyloxy Radical at 60 °C: Addition to Monomer/Decomposition Rates[55,151]

Monomer	Structure	K''/K' (mol/l)
Acrylonitrile	$CH_2{=}CH{-}C{\equiv}N$	0.12
Methyl methacrylate	$CH_2{=}C(CH_3){-}COOCH_3$	0.30
Vinyl acetate	$CH_2{=}CH{-}OOC{-}CH_3$	0.91
Styrene	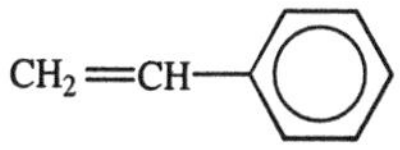 $CH_2{=}CH{-}C_6H_5$	2.50
2,5-Dimethylstyrene	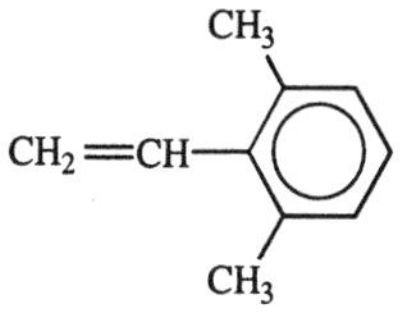 $CH_2{=}CH{-}C_6H_3(CH_3)_2$	5.0

Table 2.6 shows the relative reaction rates of $SO_4\bullet^{\ominus}$ with some monomers at 25 °C.[54,56]

The rate of addition of a radical to a double bond is affected by steric hindrance from bulky substituents. Polar effect, such as dipole interactions also influence the rate of addition. The effect of steric hindrance on the affinity of a methyl radical is illustrated in Table 2.7.[57,58]

Phenyl or methyl groups located on the carbon atom that is under a direct attack by a free radical can be expected to interfere sterically with the approach. For instance, due to steric hindrance *trans-β*-methylstylbene is more reactive toward a radical attack than is its *cis* isomer.[59] Yet, the *trans* isomer is the more stable of the two.

While 1,1-disubstituted olefins homopolymerize readily, the 1,2-disubstituted olefins are difficult to homopolymerize.[39] Some exceptions are vinyl carbonate[210,211] and maleimide derivatives.[212] Also, perfluoroethylene and chlorotrifluoroethylene polymerize readily.

Homopolymerizations of diethyl fumarate by a free-radical mechanism were reported.[213] The M_n was found to be 15,000. The same is true of homopolymerizations of several other dialkyl fumarates and also dialkyl maleates.[42,214,215] The polymerization rates and the sizes of the polymers that form decrease with increases in the lengths of linear alkyl ester groups. There is, however, an opposite correlation if the ester groups are branched. Also, the maleate esters appear to isomerize to fumarates prior to polymerization.[215]

Table 2.6. Relative Rates of Reactions of the Sulfate Ion-Radical With Some Monomers[54,56]

Monomer	Structure	Relative Rates K''/K'
Methyl methacrylate	$CH{=}C(CH_3)COOCH_3$	7.7×10^{-3}
Methyl acrylate	$CH_2{=}CH{-}COOCH_3$	1.1×10^{-3}
Acrylonitrile	$CH_2{=}CH{-}C{\equiv}N$	3.9×10^{-4}

Table 2.7. Affinity of Methyl Radical for Olefins[a,57,58]

Monomer	Structure	Methyl Affinity
Styrene	$CH_2{=}CH{-}C_6H_5$	792
α-Methyl styrene	$CH_2{=}C(CH_3){-}C_6H_5$	926
cis-β-methyl styrene	$CH_3{-}CH{=}CH{-}C_6H_5$	40
trans-β-Methyl styrene	$CH_3{-}CH{=}CH{-}C_6H_5$	92.5
α,β-Dimethyl styrene	$CH_3{-}CH{=}C(CH_3){-}C_6H_5$	66
α,β,β′-Trimethyl styrene	$(CH_3)_2C{=}C(CH_3){-}C_6H_5$	20

[a]From Carrick, Szwarc, Leavitt, Levy, and Stannett, by permission of the American Chemical Society.

2.4. Propagation

The transition state in a propagation reaction can be illustrated as follows:

$$\sim\!\sim CH_2{-}CHX^{\bullet} + CH_2{=}CHX \longrightarrow [\sim\!\sim CH_2{-}\dot{C}HX \cdots \dot{C}H_2{-}\dot{C}HX]$$

growing polymer — monomer — transition state

$$\longrightarrow \sim\!\sim CH_2{-}CHX{-}CH_2{-}CHX^{\bullet}$$

product

In the above transition state, the macroradical electron is localized on the terminal carbon. Also, the two π electrons of the double bond are localized at each olefinic carbon. Interaction takes place between *p*-orbitals of the terminal atom in the active polymer chain with associated carbon of the monomer. This results in formation of σ-bonds.[60]

The rate of the propagation reaction depends upon the reactivity of the monomer and the growing radical chain. Steric factors, polar effects, and resonance are important factors in the reaction.

2.4.1. Steric, Polar, and Resonance Effects in the Propagation Reaction

The steric effects depend upon the sizes of the substituents. The resonance stabilization of the substituents has been shown to be of the following order[61]:

$$Cl > O{-}\overset{O}{\overset{\|}{C}}CH_3 > \overset{O}{\overset{\|}{C}}{-}OCH_3 \simeq C{\equiv}N > C_6H_5 > CH{=}CH_2$$

The reactivities of the propagating polymer radicals, however, exert greater influence on the rates of propagation than do the reactivities of the monomers. Resonance stabilization of the polymer radicals is a predominant factor. This fairly common view comes from observations that a methyl radical reacts at 60 °C approximately 25 times faster with styrene than it does with vinyl acetate.[58] In homopolymerizations of the two monomers, however, the rates of propagation fall in an opposite order. Also, poly(vinyl acetate) radicals react 46 times faster with *n*-butyl mercaptan in hydrogen abstraction reactions than do the polystyrene radicals.[61] The conclusion is that the polystyrene radicals are much more resonance-stabilized than are the poly(vinyl acetate) radicals. Several structures of the polystyrene radicals are possible due to the conjugation of the unpaired electrons on the terminal carbons with the adjacent unsaturated groups. These are resonance hybrids of:

$$\sim CH_2{-}CH\cdot(C_6H_5) \longleftrightarrow \sim CH_2{-}CH{=}(C_6H_5\cdot) \longleftrightarrow \sim CH_2{-}CH{=}(C_6H_5\cdot)$$

There is no such opportunity, however, for resonance stabilization of the poly(vinyl acetate) radicals, because oxygen can accommodate only eight electrons.

In vinyl monomers both olefinic carbons are potentially subject to free-radical attack. Each would give rise to a different terminal unit:

$$R\cdot + CH_2{=}CHX \longrightarrow \begin{cases} R{-}CH_2{-}CHX\cdot \\ R{-}CXH{-}CH_2\cdot \end{cases}$$

The newly formed radicals can again potentially react with the next monomer in two ways. This means that four propagation reactions can occur:

1. $\sim CH_2{-}C^{\cdot}HX + CH_2{=}CHX \longrightarrow \sim CH_2{-}CHX{-}CH_2{-}CHX\cdot$

2. $\sim CH_2{-}C^{\cdot}HX + CH_2{=}CHX \longrightarrow \sim CH_2{-}CHX{-}CHX{-}CH_2\cdot$

3. $\sim CHX{-}C^{\cdot}H_2 + CH_2{=}CHX \longrightarrow \sim CHX{-}CH_2{-}CH_2{-}CHX\cdot$

4. $\sim CHX{-}C^{\cdot}H_2 + CH_2{=}CHX \longrightarrow \sim CHX{-}CH_2{-}CHX{-}CH_2\cdot$

Contrary to the above-shown four propagation modes, a "head-to-tail" placement strongly predominates. This is true of most free-radical vinyl polymerizations. Such placement is shown in reaction 1. It is consistent with the localized energy at the α-carbon of the monomer. Also, calculations of resonance stabilization tend to predict head-to-tail additions.[62,63]

The free-radical propagation reactions that correspond to conversions of double bonds into single bonds are strongly exothermic. In addition, the rates increase with the temperature. It is often assumed that the viscosity of the medium, or change in viscosity during the polymerization reaction, does not affect the propagation rate or the polymer growth reaction. This is because it involves diffusion of small monomer molecules to the reactive sites. Small molecules, however, can also be impeded in their process of diffusion. This can impede the growth rate.[54]

During chain growth, the radical has a great deal of freedom with little steric control over the manner of monomer placement. Decrease in the reaction temperature, however, lowers mobility of the species and increases steric control over placement. This is accompanied by an increase in stereoregularity of the product.[64,65] The preferred placement is *trans–trans*, because of lower energy required for such placement. As a result, a certain amount of syndiotactic arrangement is observed in polymerizations at lower temperatures.[66] *Trans–trans* configurations (with respect to the carbon atoms in the chains) yields zigzag backbones. This was predicted from observations of steric effects on small molecules.[80,81] It was confirmed experimentally for many polymers, like, for instance, in the formation of poly(1,2-polybutadiene)[67] and poly(vinyl chloride).[68] Also, in the free-radical polymerizations of methyl methacrylate, syndiotactic placement becomes increasingly dominant at lower temperatures. Conversely, the randomness increases at higher temperatures.[69] The same is true in the free-radical polymerization of halogenated vinyl acetate.[70]

One proposed mechanism for the above is as follows. The least amounts of steric compression within macromolecules occur during the growth reactions if the ultimate and the penultimate units are *trans* to each other. Also, if the lone electrons face the oncoming monomers during the transition states,[80–82] as shown below, syndiotactic placement should be favored:

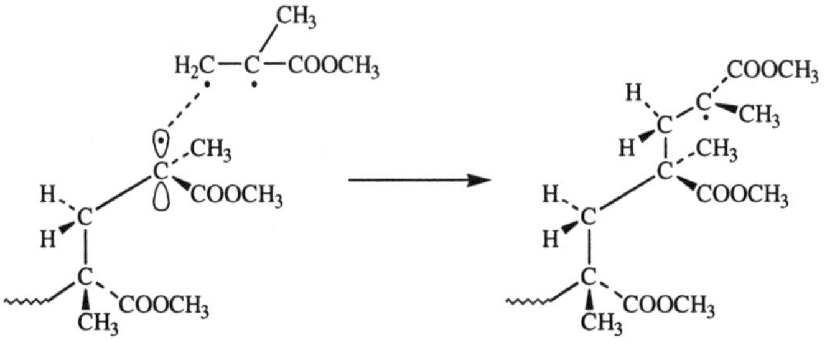

$H_2C{-}C(CH_3){-}COOCH_3$ · · · $\sim C(CH_3)(COOCH_3){-}CH_2{-}\dot{C}(CH_3)(COOCH_3)$ ⟶ $\sim C(CH_3)(COOCH_3){-}CH_2{-}C(CH_3)(COOCH_3){-}CH_2{-}\dot{C}(CH_3)(COOCH_3)$

transition state

While the above model explains the formation of syndiotactic poly(methyl methacrylate), possible interactions between the free radicals on the chain ends and the monomers are not considered. Such interactions, however, are a dominant factor in syndiotactic placement, if the terminal carbons are sp^2 planar in structures.[38]

2.4.2. Effect of Reaction Medium

The reaction medium does not affect the mode of placement in free-radical polymerizations. There were some early reports that an influence was detected in the polymerizations of vinyl chloride in aliphatic aldehydes at 50 °C.[71,72] This was not confirmed in subsequent studies.[73–75] The rate of polymerization, however, can be influenced by the pH of the reaction medium in polymerizations of monomers like methacrylic acid.[76,77] Also, the rate of polymerization and solution viscosities increase in polymerizations of acrylamide and acrylic acid with an increase in water concentration.[78] It is not quite clear whether this is due to increases in the speeds of propagations or due to decreases in the termination rates. In the free-radical polymerization of vinyl benzoate the rate of propagation varies in various solvents in the following order[79]:

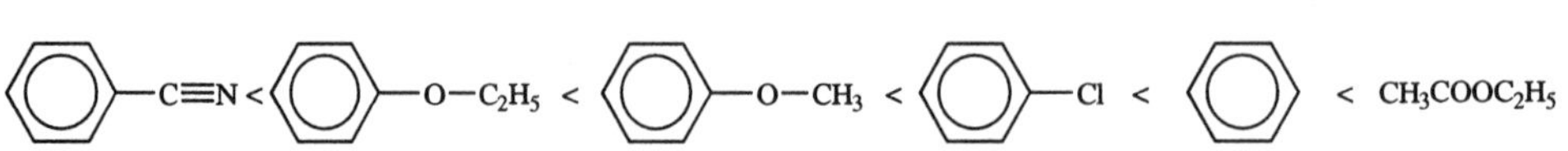

$$C_6H_5{-}C{\equiv}N < C_6H_5{-}O{-}C_2H_5 < C_6H_5{-}O{-}CH_3 < C_6H_5{-}Cl < C_6H_6 < CH_3COOC_2H_5$$

Similarly, the rate of photopolymerization of vinyl acetate is affected by solvents.[87] In most cases, however, the rate of polymerization is proportional to the square root of the initiator concentration and to the concentration of the monomer.[54]

There is a controversial suggestion that the solvent affects the propagation step in some reactions by forming "hot" radicals.[217] These radicals are supposed to possess higher amounts of energy. At the moment of their formation they obtain surplus energy from the heat of the reaction and from the activation energy of the propagation reaction. This provides the extra energy needed to activate the next chain-propagation step. The surplus energy may affect the polymerization kinetics if the average lifetime of the hot radicals is sufficient for them to react with the monomer molecules. This surplus energy is lost by the hot radicals in collisions with monomer and solvent molecules. There is a difference in the rate constants of propagation for hot and ordinary radicals, so two different reaction schemes were written[217]:

Propagation by ordinary radicals:

$$R\bullet + M \rightarrow RM\bullet \qquad (k_2)$$

Propagation by hot radicals:

$$R^* + M \rightarrow RM^* \qquad (k_2^*)$$

Energy transfer processes:

$$R^* + \text{Solvent} \rightarrow R\bullet + \text{Solvent} \qquad (k'_3{}^*)$$

where R* is the symbol for hot radicals.

The rate expression for the polymerization is then expressed as follows[217]:

$$-d[M]/dt = K_x X^{0.5}[M] \cdot [1 + 1/(\gamma + \gamma' + S/[M])$$

where $K_k = k_2(2k_1 f/k_4]^{1/2}$, $\gamma = K_1{*}/K_2{*}$ and $\gamma' = K_3{*}/K_2{*}$, S denotes solvent and [M] monomer.

In comparing free-radical polymerizations of ethyl acrylates in benzene and in dimethyl formamide at 50 °C,[218] the rates were found to be proportional to the square roots of the initiator concentration. They were not proportional, however, to the concentrations of the monomer. This was interpreted in terms of hot radicals.[218]

Similar results, however, were interpreted by others differently. For instance, butyl acrylate and butyl propionate polymerizations in benzene also fail to meet ideal kinetic models. The results, however, were explained[219] in terms of terminations of primary radicals by chain transferring.

2.4.3. Ceiling Temperature

For most free-radical polymerization reactions there are some elevated temperatures at which the chain growth process becomes reversible and depropagation takes place:

$$M_{n+1} \xrightarrow{k_{dp}} M_n\bullet + M$$

where k_{dp} is the rate constant for depropagation or depolymerization. The equilibrium for the polymerization–depolymerization reaction is temperature dependent. The reaction isotherm can be written:

$$\Delta F = \Delta F^\circ + RT \ln K$$

In the above equation, ΔF° is the free energy of polymerization of both monomer and polymer in appropriate standard states.[100] The standard state for the polymer is usually solid (amorphous or partly crystalline). It can also be a one-molar solution. The monomer is a pure liquid or a one molar solution. The relationships of monomer concentration to heat content, entropy, and free energy are shown in the following expression. This applies over a wide range of temperatures.[54]

$$\ln[M] = \Delta H^\circ_P/RT_c - \Delta S^\circ_P/R$$

In the above equation T_c is the *ceiling temperature* for the *equilibrium monomer concentration.* It is a function of the temperature. Because the heat content is a negative quantity, the concentration of the monomer (in equilibrium with polymer) increases with increasing temperatures. There are a series of ceiling temperatures that correspond to different equilibrium monomer concentrations. For any given concentration of a monomer in solution there is some upper temperature at which polymerization will not proceed. This, however, is a thermodynamic approach. When there are no active centers present in the polymer structure, the material will appear stable even above the ceiling temperature in a state of metastable equilibrium.

The magnitude of the heat of polymerization of vinyl monomers is related to two effects: (1) Steric strains that form in single bonds from interactions of the substituents. These substituents, located on the alternate carbon atoms on the polymeric backbones, interfere with the monomers entering the chains. (2) Differences in resonance stabilization of monomer double bonds by the conjugated substituents.[60]

Most 1,2-disubstituted monomers, as stated earlier, are difficult to polymerize. It is attributed to steric interactions between one of the two substituents on the vinyl monomer and the β-substituent on the ultimate unit of the polymeric chain.[100] A strain is also imposed on the bond that is being formed in the transition state.

The propagation reaction usually requires only an activation energy of about 5 kcal/mole. As a result, the rate does not vary rapidly with the temperature. On the other hand, the transfer reaction requires higher activation energies than does the chain growth reaction. This means that the average molecular weight will be more affected by the transfer reaction at higher temperature. When allowances are made for chain transferring, the molecular weight passes through a maximum as the temperature is raised. At temperatures below the maximum, the product molecular weight is lower because the kinetic chain length decreases with the temperature. Above the maximum, however, the product molecular weight is also lower with increases in the temperature. This is due to increase in the transfer reactions. The above assumes that the rate of initiation is independent of the temperature. The relationship of the kinetic chain length to the temperature can be expressed as follows[54]:

$$d\ln \nu / dT = (E_P - \frac{1}{2}E_T - \frac{1}{2}E_I)/RT^2$$

where E_P, E_T, and E_I are energies of propagation, termination, and initiation, respectively. A large E_I means that if the temperature of polymerization is raised, the kinetic chain length decreases. This is affected further by a greater frequency of chain transferring at higher temperatures. In addition, there is a possibility that disproportionation may become more significant.

2.4.4. Autoacceleration

When the concentrations of monomers are high in solution or bulk polymerizations, typical autoaccelerations of the rates can be observed. This is known as the *gel effect* or the *Trommsdorff effect*, or also as the *Norrish–Smith effect*.[61] The effect has been explained as being caused by a decrease in the rate of termination due to increased viscosity of the medium. Termination is a reaction that requires two large polymer radicals to come together and this can be impeded by viscosity. At the same time, in propagation, the small molecules of the monomer can still diffuse for some time to the radical sites and feed the chain growth.

One should not mistake an acceleration of a polymerization reaction due to a rise in the temperature under nonisothermal conditions with a true gel effect from a rise in viscosity. The gel effect can occur when the temperature of the reaction is kept constant.

A critical analysis of the gel effect suggests that the situation is complicated. In some polymerizations three different stages appear to be present when $R_P/[M][I]^{1/2}$ is plotted against conversion or against time.[229] The plot indicates that during the first stage, there is either a constant or a declining rate and during the second stage, there is autoacceleration. During the third stage there is again a constant or a declining rate.[229]

2.4.5. Polymerization of Monomers with Multiple Double Bonds

Polymerizations of monomers with multiple double bonds yield products that vary according to the locations of these bonds with respect to each other. Monomers with conjugated double bonds, like 1,3-butadiene and its derivatives, polymerize in two different ways. One way is through only one of the double bonds. Another way is through both double bonds simultaneously. Such 1,4-propagation is attributable to the effect of conjugation and hybridization of the C_2–C_3 bond that involves sp^2 hybrid orbitals.[166] All three modes of propagation are possible in one polymerization reaction so that the product can, in effect, be a copolymer. The 1,2- and 3,4-propagations can be shown as follows:

$$n\ (R_1)(R)C{=}C(R_2){-}C(R_3){=}C(R_4)(R_5) \longrightarrow -[-C(R_1)(R){-}C(R_2)(C(R_3){=}C(R_4)R_5){-}]_n- \quad \text{(1,2-placement)}$$

$$n\ (R_1)(R)C{=}C(R_2){-}C(R_3){=}C(R_4)(R_5) \longrightarrow -[-C(R_3)(C(R_2){=}C(R)R_1){-}C(R_4)(R_5){-}]_n- \quad \text{(3,4-placement)}$$

$$n\ (R_1)(R)C{=}C(R_2){-}C(R_3){=}C(R_4)(R_5) \longrightarrow -[-C(R_1)(R){-}C(R_2){=}C(R_3){-}C(R_4)(R_5){-}]_n- \quad \text{(1,4-placement)}$$

The polymerizations and copolymerizations of various conjugated dienes are discussed in Chapter 5.

2.4.5.1. Ring-Forming Polymerization

Propagation reactions of unconjugated dienes can proceed by an intra–intermolecular process. This usually results in ring formation or in *cyclopolymerization*. It can be illustrated as follows:

$$R\cdot + C{=}C{-}X{-}C{=}C \longrightarrow R{-}C{-}C\cdot{-}X{-}C{=}C \longrightarrow R{-}C{-}\overline{C{-}C{-}C\cdot{-}X}$$

$$\longrightarrow R{-}[{-}C{-}\overline{C{-}C{-}C{-}X}{-}]{-}$$

where X can designate either a carbon or a heteroatom. An example of such a polymerization is a free-radical polymerization of quaternary diethyldiallylamine[167]:

$$(H_2C{=}CH{-}CH_2)_2N^{\oplus}(C_2H_5)_2\ Br^{\ominus} \longrightarrow -[-CH_2-(\text{piperidinium ring, } N^{\oplus}(C_2H_5)_2\ Br^{\ominus})-]-$$

Another example is a polymerization of 2,6-disubstituted 1,6-heptadiene[172]:

$$I\cdot + CH_2{=}C(R){-}({-}CH_2)_3{-}C(R){=}CH_2 \longrightarrow I{-}CH_2{-}\dot{C}(R)\,[{-}CH_2{-}CH_2{-}CH_2{-}C(R){=}CH_2] \longrightarrow$$

$$\longrightarrow I{-}CH_2{-}(\text{cyclohexane ring, R, R})\cdot \longrightarrow \longrightarrow {-}[{-}CH_2{-}(\text{cyclohexane ring, R, R}){-}]_n{-}$$

where R = $COOC_2H_5$, $COOCH_3$, or COOH.

The intra-intermolecular propagations can result in ring structures of various sizes. For instance, three-membered rings can form from *transanular* polymerizations of bicycloheptadiene[168–170]:

$$n\ (\text{bicycloheptadiene}) \longrightarrow [\ \ ]_m[\ \ ]_o$$

Four-membered rings form in free-radical polymerization of perfluoro-1,4-pentadiene.[171] The size of the ring that forms depends mainly on the number of atoms between the double bonds:

$$n\ CF_2{=}CF{-}CF_2{-}CF{=}CF_2 \xrightarrow[\substack{11{,}000\text{ atm}\\110\,^{\circ}C}]{\text{gamma rays}} [{-}CF_2{-}CF{<}(CF_2)_2{>}CF{-}][{-}CF_2{-}CF{=}CF{-}CF(CF_3){-}]$$

Formation of many five-membered rings is also known. One example is a polymerization of 2,3-dicarboxymethyl-1,6 hexadiene[172]:

$$n\ CH_2{=}C(R){-}CH_2{-}CH_2{-}C(R){=}CH_2 \longrightarrow [{-}CH_2{-}(\text{cyclopentane ring, R, R}){-}]_n$$

where R= $COOCH_3$.

The polymer that forms is crosslinked, but spectroscopic analysis shows that 90% of the monomer placement is through ring formation.[173] Formation of six-membered rings is also well documented. Two examples were shown above in the polymerization of a quaternary diethyldiallylamine and in the polymerization of 2,6-disubstituted, 1,6-heptadiene. Many other 1,6-heptadienes yield linear polymers containing six-membered rings.[174]

This tendency to propagate intra–intermolecularly by the unconjugated dienes is greater than can be expected from purely statistical predictions.[175] Butler suggested that this results from interactions between the olefinic bonds.[176] Ultraviolet absorption spectra of several unconjugated diolefins do show bathochromic shifts in the absorption maxima relative to the values calculated from Woodward's rule.[177–179] This supports Butler's explanation.[176]

2.5. The Termination Reaction

The termination process in free-radical polymerization is caused, as is shown earlier, by one of three types of reactions: (1) a second-order radical–radical reaction, (2) a second-order radical–molecule reaction, and (3) a first-order loss of radical activity.

The first reaction can be either one of combination or of disproportionation. In a combination reaction, two unpaired spin electrons, each on the terminal end of a different polymer radical, unite to form a covalent bond and a large polymer molecule. In disproportionation, on the other hand, two polymer radicals react and one abstracts an atom from the other one. This results in formation of two inactive polymer molecules. The two differ from each other, however, in that one has a terminal saturated structure and the other one has a terminal double bond. Usually, the atom that is transferred is hydrogen.

It was suggested[32] that a basic rule of thumb can be applied to determine which termination reaction predominates in a typical homopolymerization. Thus, polymerizations of 1,1-disubstituted olefins are likely to terminate by disproportionation because of steric effects. Polymerization of other vinyl monomers, however, favor terminations by combination unless they contain particularly labile atoms for transferring. Higher activation energies are usually required for termination reactions by disproportionation. This means that terminations by combination should predominate at lower temperatures.

The third type of termination reaction is chain transferring. Premature termination through transferring results in a lower molecular weight polymer than can be expected from other termination reactions. The product of chain transferring is an inert polymer molecule and, often, a new free radical capable of new initiation. If, however, the new radical is not capable of starting the growth of a new chain, then this is *degenerative chain transferring*. It is also referred to as a *first-order termination reaction*. The molecules that accept the new radical sites (participate in chain transferring) can be any of those present in the reaction medium. This includes solvents, monomer molecules, inactive polymeric chains, and initiators.

The ease with which chain transferring takes place depends upon the bond strength between the labile atoms that are abstracted and the rest of the molecule to which they are attached. For instance, chain transferring in methyl methacrylate polymerization to the solvent occurs in the following order[83]:

$$\text{benzene} < \text{toluene} < \text{ethyl benzene} < \text{cumene}$$

The rate of a chain transferring reaction is

$$R_{tr} = K_{tr}\,[\mathrm{M}\bullet][\mathrm{XA}]$$

where K_{tr} is the chain transferring constant in a reaction:

$$\mathrm{M}\bullet + \mathrm{XA} \xrightarrow{K_{tr}} \mathrm{M{-}X} + \mathrm{A}\bullet$$

Examples of molecules that have particularly labile atoms and that contribute readily to chain transferring are mercaptans and halogen compounds, like chloroform, carbon tetrachloride, etc.

A polymer that was prematurely terminated in its growth by chain transferring may be a *telomer*. In most cases of telomer formation, the newly formed radical and the monomer radical are active enough to initiate new chain growth. Thus, the life of the kinetic chain is maintained.

An illustration of a telomerization reaction can be free-radical polymerization of ethylene in the presence of chloroacetyl chloride:

Initiation: $\mathrm{ROOR} \rightarrow 2\,\mathrm{RO}\bullet$

$$\mathrm{RO}\cdot + \mathrm{Cl{-}\overset{\displaystyle O}{\overset{\|}{C}}{-}CH_2Cl} \longrightarrow \mathrm{ROCl} + \cdot\mathrm{CH_2{-}\overset{\displaystyle O}{\overset{\|}{C}}{-}Cl}$$

Propagation:

$$\mathrm{Cl{-}\overset{\displaystyle O}{\overset{\|}{C}}{-}CH_2}\cdot + \mathrm{CH_2{=}CH_2} \longrightarrow \mathrm{Cl{-}\overset{\displaystyle O}{\overset{\|}{C}}{-}CH_2{-}CH_2{-}CH_2}\cdot$$

$$\xrightarrow{n\,\mathrm{CH_2{=}CH_2}} \mathrm{Cl{-}\overset{\displaystyle O}{\overset{\|}{C}}{-}CH_2{-}(\!{-}CH_2{-}CH_2{-}\!)_{n}{-}CH_2{-}CH_2}$$

Termination:

$$\mathrm{Cl{-}\overset{\overset{\displaystyle O}{\|}}{C}{-}({-}CH_2{-}CH_2{-})_n{-}CH_2{-}CH_2\cdot \; + \; ClCH_2{-}\overset{\overset{\displaystyle O}{\|}}{C}{-}Cl \longrightarrow}$$

$$\longrightarrow \mathrm{Cl{-}\overset{\overset{\displaystyle O}{\|}}{C}{-}({-}CH_2{-}CH_2{-})_n{-}CH_2{-}CH_2Cl \; + \; \cdot\, CH_2{-}\overset{\overset{\displaystyle O}{\|}}{C}{-}Cl}$$

Chain transferring is affected by temperature but not by changes in the viscosity of the reaction medium.[54] When a transfer takes place to a monomer, it is independent of the polymerization rate.[84,85] When, however, transfer takes place to the initiator, the rate increases rapidly.[86]

A chain transferring reaction to a monomer can be illustrated as follows:

1. $$\underset{\text{polymer radical}}{\mathrm{{\sim}CH_2{-}CHX\cdot}} + \mathrm{CH_2{=}CHX} \xrightarrow{\text{hydrogen abstraction}} \mathrm{{\sim}CH_2{-}CH_2X} + \underset{\text{new radical}}{\mathrm{CH_2{=}CX\cdot}}$$

or

2. $$\underset{\text{polymer radical}}{\mathrm{{\sim}CH_2{-}CHX\cdot}} + \mathrm{CH_2{=}CHX} \xrightarrow{\text{hydrogen donation}} \mathrm{{\sim}CH{=}CHX} + \underset{\text{new radical}}{\mathrm{CH_3{-}CH_2X\cdot}}$$

A transfer reaction can also occur from the terminal group of the polymer radical to a location on the polymeric backbone. This is known as *backbiting*:

$$\begin{matrix} \mathrm{{\sim}CH_2{-}CH_2} & \\ & \mathrm{CH_2} \\ \mathrm{\cdot CH_2} & | \\ & \mathrm{CH_2{-}CH_2} \end{matrix} \longrightarrow \begin{matrix} \mathrm{{\sim}\dot{C}H{-}CH_2} & \\ & \mathrm{CH_2} \\ \mathrm{CH_3} & | \\ & \mathrm{CH_2{-}CH_2} \end{matrix}$$

The new free-radical site on the polymer backbone starts chain growth that results in formation of a branch. The same reaction can take place between a polymer radical and a location on another polymer chain. In either case, fresh chain growth results in formation of a branch.

Whether chain transferring can take place to an initiator depends upon its chemical structure. It was believed in the past that chain transferring to α,α'-azobisisobutyronitrile does not occur. Later, it was shown that chain transferring to this initiator does occur as well, at least in the polymerizations of methyl methacrylate.[88–90]

The amount of chain transferring that takes place to monomers is usually low because the reaction requires breaking strong carbon–hydrogen bonds. Monomers, however, like vinyl chloride and vinyl acetate have fairly large chain transferring constants. In the case of vinyl acetate this is attributed to the presence of an acetoxy methyl group. This explanation, however, cannot be used for vinyl chloride. The chain transferring constants that are defined as:

$$C_{\mathrm{M}} = k_{\mathrm{tr,M}} / K_{\mathrm{P}} \quad \text{for monomers}$$

$$C_{\mathrm{S}} = k_{\mathrm{tr,S}} / K_{\mathrm{P}} \quad \text{for solvents}$$

$$C_{\mathrm{I}} = k_{\mathrm{tr,I}} / K_{\mathrm{P}} \quad \text{for initiators}$$

can be found in handbooks and other places in the literature. Presence of chain transferring agents in a polymerization reaction requires redefining the degree of polymerization to include the chain termination terms. The number average degree of polymerization has to be written as follows:

$$\overline{\mathrm{DP}} = \frac{R_{\mathrm{P}}}{(R_{\mathrm{T}}/2) + K_{\mathrm{tr,M}}[\mathrm{M}\bullet][\mathrm{M}] + K_{\mathrm{tr,S}}[\mathrm{M}\bullet][\mathrm{S}] + K_{\mathrm{tr,I}}[\mathrm{M}\bullet][\mathrm{I}]}$$

It can also be expressed in terms of the chain transferring constants as follows:

$$1/\overline{DP} = 2R_P/R_I + C_M + C_S[S]/[M] + C_I[I]/[M]$$

It can also be written in another form:

$$1/\overline{DP} = K_T R_P / K_P{}^2[M]^2 + C_M + C_S[S]/[M] + C_I K_T R_P / K_P 2f K_d[M]^3$$

When a polymerization reaction is conducted in a concentrated solution, or in complete absence of a solvent, the viscosity of the medium increases with time (unless the polymer precipitates out). This impedes all steps in the polymerization process, particularly the diffusions of large polymer radicals.[54] The decreased mobility of the polymer radicals affects the termination process. It appears that this is common to many, though not all, free-radical polymerizations. All molecular processes in the termination reactions are not fully understood, particularly at high conversions.[43] This is a complex process that consists of three definable steps. These can be pictured as follows. First two polymer radicals migrate together by means of translational diffusion. Second, the radical sites reorient toward each other by segmental diffusion. Third, the radicals overcome the small chemical activation barriers and react. The termination reaction is therefore diffusion-controlled. At low concentrations this will be segmental diffusion, while at medium or high concentrations it will be translational diffusion.

Present theories of terminations suggest that at intermediate conversions, terminations are dominated by interactions between short chains formed by transfer and entangled long chains.[40] When terminations are diffusion-controlled, most termination events involve two highly entangled chains whose ends move by the "reaction–diffusion" process.[45] In this process terminations occur because of the propagation-induced diffusion of the chain ends of growing macroradicals. This means that the rates of terminations depend upon the chain lengths.[45,49]

Diffusion theories have been proposed that relate the rate constant of termination to the initial viscosity of the polymerization medium. The rate determining step of termination, the segmental diffusion of the chain ends is inversely proportional to the microviscosity of the solution.[100] Yokota and Itoh[220] modified the rate equation to include the viscosity of the medium. According to that equation, the overall polymerization rate constant should be proportional to the square root of the initial viscosity of the system.

Recently, the number-average termination rate constants in a methyl methacrylate polymerization were measured with an in line ESR spectrometer. This was done by observing the radical decay rates.[43] The results are in disagreement with the concept of termination by propagation diffusion that is expected to be dominant at high conversion rates. Instead, the termination rate constants decrease dramatically in the post-effect period at high conversions. Actually, a fraction of the radicals were found trapped during the polymerization. Thus, there are two types of radicals in the reaction mixture, trapped and free radicals. In the propagations and termination reactions the two types of radical populations have very different reactivities.[43]

2.6. Copolymerization

If more than one monomer species is present in the reaction medium, a copolymer or an interpolymer can result from the polymerization reaction. Whether, however, the reaction products will consist of copolymers or just mixtures of homopolymers or of both depends largely upon the reactivity of the monomers. A useful and a simplifying assumption in kinetic analyses of free-radical copolymerizations is that the reactivity of polymer radicals is governed entirely by the terminal monomer units.[54] For instance, a growing polymer radical, containing a methyl methacrylate terminal unit, is considered, in terms of reactivity, as a poly(methyl methacrylate) radical. This assumption, not always adequate,[54] can be used to predict satisfactorily the behavior of many mixtures of monomers. Based on this assumption, the copolymerization of a pair of monomers involves four distinct growth reactions and two types of polymer radicals.

Table 2.8. Some Reactivity Ratios[4,91]

Monomer 1	Monomer 2	r_1	r_2
Styrene	Butadiene	0.78	1.39
Styrene	Methyl methacrylate	0.52	0.46
Styrene	Vinyl acetate	55.00	0.01
Vinyl acetate	Vinyl chloride	0.23	1.68
Methyl acrylate	Vinyl chloride	9.00	0.083

2.6.1. Reactivity Ratios

In a reaction of two monomers, designated as M_1 and M_2, four distinct reactions can be written as follows:

$$1.\quad \sim M_1\bullet + M_1 \xrightarrow{k_{11}} \sim M_1M_1\bullet$$

$$2.\quad \sim M_1\bullet + M_2 \xrightarrow{k_{12}} \sim M_1M_2\bullet$$

$$3.\quad \sim M_2\bullet + M_1 \xrightarrow{k_{21}} \sim M_2M_1\bullet$$

$$4.\quad \sim M_2\bullet + M_2 \xrightarrow{k_{22}} \sim M_2M_2\bullet$$

The ratios of k_{11}/k_{12} and k_{22}/k_{21} are called monomer *reactivity ratios*. They can be written as follows:

$$r_1 = k_{11}/k_{12} \qquad r_2 = k_{22}/k_{21}$$

These reactivity ratios represent the relative rates of reactions of polymer radicals with their own monomers vs. the comonomers. When $r_1 > 1$, the radical $\sim M_1\bullet$ is reacting with monomer M_1 faster than it is with the comonomer M_2. On the other hand, when $r_1 < 1$, the opposite is true. Based on the *r values*, the composition of the copolymers can be calculated from a *copolymerization equation*[54]:

$$\frac{d[M_1]}{d[M_2]} = \frac{[M_1]}{[M_2]}\left[\frac{r_1[M_1] + [M_2]}{r_2[M_2] + [M_1]}\right]$$

Table 2.8. illustrates a few typical reactivity ratios taken from the literature. Many more can be found.[91]

In an ideal copolymerization reaction,

$$k_{11}/k_{12} = k_{22}/k_{21} \qquad \text{and } r_1r_2 = 1$$

If r_1 and r_2 values are equal to or approach zero, each polymer radical reacts preferentially with the other monomer. This results in an alternating copolymer, regardless of the composition of the monomer mixture. That is, however, a limiting case. In the majority of instances $r_1 \times r_2$ is greater than zero and less than one. When the polymer radical reacts preferentially with its own monomer, and $r_1 \times r_2 > 1$, mainly a mixture of homopolymers forms and only some copolymer.

Reactivity of vinyl monomers is very often determined experimentally by studying copolymerizations. Values of many free-radical reactivity ratios have been tabulated for many different monomer pairs.[91] Also, the qualitative correlations between copolymerization data and molecular orbital calculations can be found in the literature.[92]

Some general conclusion about monomer reactivity toward attacking radicals was drawn by Mayo and Walling[93]:

1. The alpha substituents on a monomer have the effect of increasing reactivity in the following order,

$$-C_6H_5 > -CH{=}CH_2 > -COCH_3 > -C{\equiv}N >$$

$$-COOR > -Cl > -CH_2X > -O{-}COCH_3 > -OR$$

2. The effect of a second alpha substituent is roughly additive.

Giese and co-workers[221–227] developed a special technique for studying the effects of the substituents upon the relative reactivity of vinyl monomers toward free radicals. Briefly it is as follows. Free radicals are produced by reducing organomercury halides with sodium borohydride. The radicals undergo competitive additions to pairs of various substituted olefins. The adducts in turn are trapped by hydrogen transfers from the formed organomercury hydrides. Relative quantities of each product are then determined:

$$RMgX + NaBH_4 \longrightarrow R\cdot \begin{cases} \xrightarrow[K_x]{CH_2{=}CHX} RCH_2{-}CHX\cdot \xrightarrow{RHgH} RCH_2CH_2X \\ \xrightarrow[K_y]{CH_2{=}CHY} RCH_2{-}CHY\cdot \xrightarrow{RHgH} RCH_2CH_2Y \end{cases}$$

This method was used with acrylonitrile and methyl acrylate mixture.[228] It showed that the ratios of rate constants for the two monomers are independent of their concentrations.

Copolymerization reactions are affected by solvents. One example that can be cited is an effect of addition of water or glacial acetic acid to a copolymerization mixture of methyl methacrylate with acrylamide in dimethyl sulfoxide or in chloroform. This caused changes in reactivity ratios.[156] Changes in r values that result from changes in solvents in copolymerizations of styrene with methyl methacrylate is another example.[157,158] The same is true for styrene acrylonitrile copolymerization.[208] There are also some indications that the temperature may have some effect on the reactivity ratios,[159] at least in some cases.

2.6.2. *Q* and *e* Scheme

Though molecular orbital calculations allow accurate predictions of reactivity ratios,[92] many chemists rely upon the Price–Alfrey Q–e equations.[94] These are based on: (1) the polarity of the double bonds of the monomers or of the propagating chain ends, (2) mesomerism of the substituents with the double bonds or with the chain ends, and (3) the steric hindrance of the substituents. This relationship is expressed in the following equation[94]:

$$K_{12} = P_1 Q_2{}^{-e_1 e_2}$$

where K_{12} represents the rate constant for the reaction of the propagating radical ~$M_1\bullet$ with monomer M_2, P_1 represents the general reactivity of the polymer radical with the terminal unit of monomer M_1, Q_2 is the reactivity of the monomer M_2 and e_1 and e_2 are the polar characters of the radical ends and of the double bonds of the monomers.

It is possible to calculate the Q and e values from r_1 and r_2, or, conversely, r values can be obtained from the Q and e values. The relationship is as follows[94]:

$$r_1 = k_{11}/k_{12} = Q_1/Q_2 \exp\,[-e_1(e_1 - e_2)]$$

$$r_2 = k_{22}/k_{21} = Q_2/Q_1 \exp\,[-e_2(e_2 - e_1)]$$

The Q and e scheme is based on a semiempirical approach. Nevertheless, some attempts were made to develop theoretical interpretations. Thus Schwann and Price[95] developed the following relationship:

$$Q = \exp(-q/RT)$$

$$e = \varepsilon/(\gamma DRT)^{1/2}$$

In the above equation q represents the resonance of stabilization (kcal/mole), ε is the electrical charge of the transition state, and γ is the distance between the centers of the charge of the radical and the monomer, D stands for the effective dielectric constant of the reaction field. The values of q and ε are derived by calculation. In addition, more rigorous molecular orbital calculations[96] show a relationship between Q and the localized energy of a monomer, and between e and the electron affinity. Also, a scale of Q and e values was deduced from essentially molecular orbital considerations.[97] Finally, a Huckel treatment of the transition state for the monomer–radical reaction in a free-radical copolymerization was developed.[98] The resulting reactivity ratios compared well with those derived from the *Q and e* scheme. This scheme is regarded by some as a version of the molecular orbital approach.[98] Nevertheless, the scheme should only be considered as an empirical one. The precision of calculating Q and e values can be poor, because steric factors are not taken into account. It is good, however, for qualitative or semiquantitative results.

2.7. Terpolymerization

A quantitative treatment of terpolymerization, where three different monomers are interpolymerized, becomes complex. Nine growth reactions take place:

Reaction	Rate
$\sim M_1 + M_1 \xrightarrow{k_{11}} \sim M_1M_1\bullet$	$R_{11} = k_{11}[M_1\bullet][M_1]$
$\sim M_1 + M_2 \xrightarrow{k_{12}} \sim M_1M_2\bullet$	$R_{12} = k_{12}[M_1\bullet][M_2]$
$\sim M_1\bullet + M_3 \xrightarrow{k_{13}} \sim M_1M_3\bullet$	$R_{13} = k_{13}[M_1\bullet][M_3]$
$\sim M_2\bullet + M_1 \xrightarrow{k_{21}} \sim M_2M_1\bullet$	$R_{21} = k_{21}[M_2\bullet][M_1]$
$\sim M_2\bullet + M_2 \xrightarrow{k_{22}} \sim M_2M_2\bullet$	$R_{22} = k_{22}[M_2\bullet][M_2]$
$\sim M_2\bullet + M_3 \xrightarrow{k_{23}} \sim M_2M_3\bullet$	$R_{23} = k_{23}[M_2\bullet][M_3]$
$\sim M_3\bullet + M_1 \xrightarrow{k_{31}} \sim M_3M_1\bullet$	$R_{31} = k_{31}[M_3\bullet][M_1]$
$\sim M_3\bullet + M_2 \xrightarrow{k_{32}} \sim M_3M_2\bullet$	$R_{32} = k_{32}[M_3\bullet][M_2]$
$\sim M_3\bullet + M_3 \xrightarrow{k_{33}} \sim M_3M_3\bullet$	$R_{33} = k_{33}[M_3\bullet][M_3]$

There are six reactivity ratios in copolymerizations of three monomers:

$$r_{12} = k_{11}/k_{12} \quad r_{13} = k_{11}/k_{13} \quad r_{21} = k_{22}/k_{21}$$

$$r_{23} = k_{22}/k_{23} \quad r_{31} = k_{33}/k_{31} \quad r_{32} = k_{33}/k_{32}$$

The rates of disappearance of the three monomers are given by

$$-d[M_1]/dt = R_{11} + R_{21} + R_{31}$$

$$-d[M_2]/dt = R_{12} + R_{22} + R_{32}$$

$$-d[M_3]/dt = R_{13} + R_{23} + R_{33}$$

2.8. Allylic Polymerization

Compounds possessing allylic structures polymerize only to low molecular weight oligomers by a free-radical mechanism. In some cases the products consist mostly of dimers and trimers. The DP for poly(allyl acetate), for instance, is only about 14. It is a result of allylic monomer radicals being resonance-stabilized to such an extent that no extensive chain propagations occur. Instead, there is a large amount of chain transferring. Such chain transferring essentially terminates the reaction.[99] This resonance stabilization can be illustrated on an allyl alcohol radical:

$$CH_2{=}CH{-}\dot{C}H(OH) \rightleftharpoons \cdot CH_2{-}CH{=}CH(OH)$$

The transfer takes place to the allylic hydrogen, as shown on allyl acetate:

$$R\cdot + CH_2{=}CH{-}CH_2{-}O{-}\overset{O}{\overset{\|}{C}}{-}CH_3 \longrightarrow RH + CH_2{=}CH{-}\dot{C}H{-}O{-}\overset{O}{\overset{\|}{C}}{-}CH_3$$

$$\updownarrow$$

$$\cdot CH_2{-}CH{=}CH{-}O{-}\overset{O}{\overset{\|}{C}}{-}CH_3$$

In this case, transfer can also take place to the acetate moiety:

$$R\cdot + CH_2{=}CH{-}CH_2{-}O{-}\overset{O}{\overset{\|}{C}}{-}CH_3 \longrightarrow RH + CH_2{=}CH{-}CH_2{-}O{-}\overset{O}{\overset{\|}{C}}{-}CH_2\cdot$$

$$\updownarrow$$

$$CH_2{=}CH{-}CH_2{-}O{-}\overset{O\cdot}{\overset{|}{C}}{=}CH_2$$

The above-described chain transferring is called *degradative chain transferring*. Other monomers, like methyl methacrylate and methacrylonitrile, also contain allylic carbon–hydrogen bonds. They fail to undergo extensive degradative chain transferring, however, and form high molecular weight polymers. This is believed to be due to lower reactivity of the propagating radicals that form from these monomers.[54]

In spite of degradative chain transferring, polyallyl compounds can be readily polymerized by a free-radical mechanism into three-dimensional lattices. High DP is not necessary to achieve growth in three dimensions. An example of such polyallyl compounds is triallyl phosphate:

$$O{=}P(O{-}CH_2{-}CH{=}CH_2)_3$$

Many other polyallyl derivatives are offered commercially and are described in the trade literature.

2.9. Inhibition and Retardation

Free-radical polymerizations are subject to inhibition and retardation from side reactions with various molecules.[54] Such polymerization suppressors are classified according to the effect that they exert upon the reaction. *Inhibitors* are compounds that react very rapidly with every initiating free radical as it is produced. This prevents any polymerization reaction from taking place until the inhibitor is completely consumed in the process. As a result, well-defined induction periods exist.

The reaction of inhibitors with initiating radicals results in formation of new free radicals. These new free radicals, however, are too stable to initiate chain growth. After all the inhibitor is used up, polymerization proceeds at a normal rate.

Retarders are compounds that also react with initiating radicals. They do not react, however, as energetically as do the inhibitors, so some initiating radicals escape and start chain growth. This affects the general rate of the reaction and slows it down. There is no induction period and retarders are active throughout the course of the polymerization.

The efficiency of an inhibitor depends upon three factors: (1) the chain transfer constant of an inhibitor with respect to a particular monomer, (2) the reactivity of the inhibitor radical that forms, (3) the reactivity of the particular monomer.

Phenols and arylamines are the most common chain transfer inhibitors. The reaction of phenols, though not fully elucidated, is believed to be as follows[209]:

1. ~~CH_2—C(H)(X)• + HO—$C_6H_2R_3$ ⟶ ~~CH_2—CH_2X + •O—$C_6H_2R_3$ ⇌ O=$C_6H_2R_2$•(R)

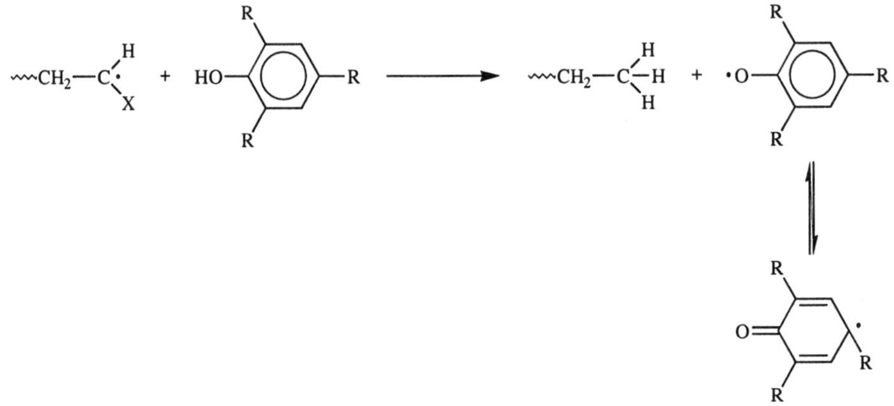

2. ~~CH_2—C(H)(X)• + •O—$C_6H_2R_3$ ⟶ ~~CH_2—CHX—O—$C_6H_2R_3$

~~CH_2—C(H)(X)• + •(R)$C_6H_2R_2$=O ⟶ ~~CH_2—CHX—(R)$C_6H_2R_2$=O

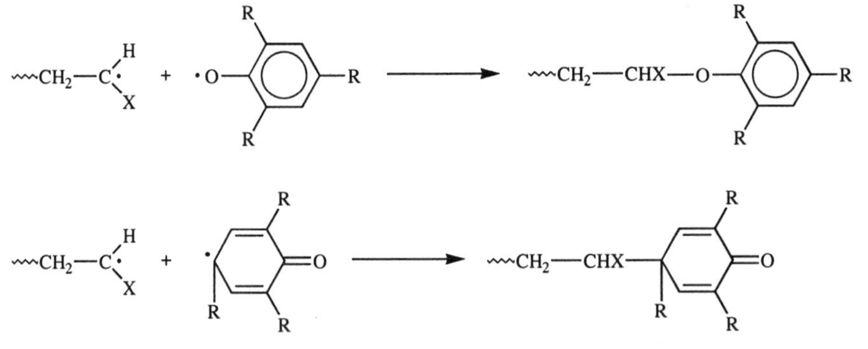

Quinones are effective inhibitors for many polymerization reactions. The reaction occurs either at an oxygen or at a ring carbon[134–136,209]:

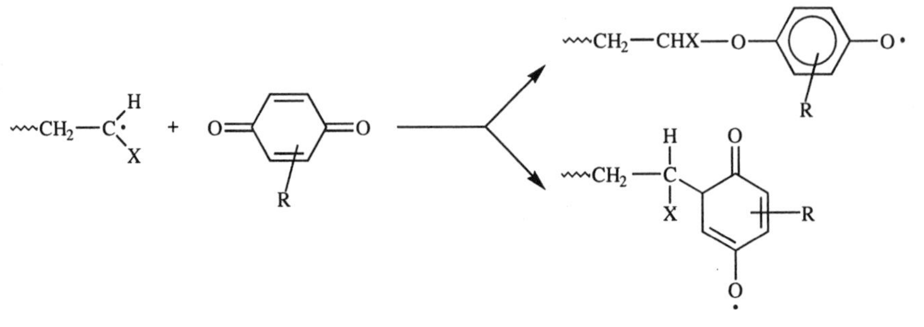

~~CH_2—C(H)(X)• + O=C_6H_3R=O ⟶ ~~CH_2—CHX—O—C_6H_3R—O• or ~~CH_2—C(H)(X)—C_6H_3R(=O)—O•

The reaction, however, is not always strict inhibition. Thus, for instance, hydroquinone acts as an efficient inhibitor for the methyl methacrylate radical but only as a retarder for the styrene radical.[136] Hydroquinone is often employed as an inhibitor, it requires, however, oxygen for activity[137,138]:

$$\sim\!CH_2-\dot{C}HX + O_2 \longrightarrow \sim\!CH_2-CHX-O-O\cdot$$

$$\sim\!CH_2-CHX-O-O\cdot + HO-C_6H_4-OH \longrightarrow \sim\!CH_2-CHX-O-O-H + HO-C_6H_4-O\cdot$$

Oxygen, however, can also act as a comonomer in a styrene polymerization:

$$\sim\!CH_2-\dot{C}H(C_6H_5) + O_2 \longrightarrow \sim\!CH_2-CH(C_6H_5)-O-O\cdot$$

$$\xrightarrow{CH_2=CH-C_6H_5} \sim\!CH_2-CH(C_6H_5)-O-O-CH_2-\dot{C}H(C_6H_5)$$

It causes marked retardation, however, in the polymerizations of methyl methacrylate.[139] The same is true of many other free-radical polymerizations.

The ability of phenols to inhibit free-radical polymerizations appears to increase with the number of hydroxyl groups on the molecule.[140] The location of these hydroxyl groups on the benzene ring in relationship to each other is important. For instance, catechol is a more efficient inhibitor than resorcinol.[140]

Aromatic nitro compounds can act as strong retarders. Their effect is proportional to the quantity of the nitro groups per molecule[134,141]:

$$\sim\!CH_2-\dot{C}HX + O_2N-C_6H_5 \longrightarrow \begin{cases} \sim\!CH_2-CHX-O-\dot{N}(\rightarrow O)-C_6H_5 \\ \sim\!CH_2-CHX-C_6H_5{=}N(O\cdot)(O) \end{cases}$$

Figure 2.1 illustrates the effect of inhibitors and retarders on free-radical polymerization.[101]

The equation that relates rate data to inhibited polymerizations is

$$2R_P{}^2K_T/K_P{}^2[M]^2 + R_P[Z]K_Z/K_P[M] - R_I = 0$$

where Z is the inhibitor or the retarder in chain growth termination:

$$\sim\!M_n\bullet + Z \xrightarrow{K_z} \sim\!M_n + Z\bullet \text{ (and/or } \sim\!M_nZ\bullet)$$

To simplify the kinetics it is assumed that Z• and $\sim M_nZ\bullet$ do not initiate new chain growth and do not regenerate Z upon termination.

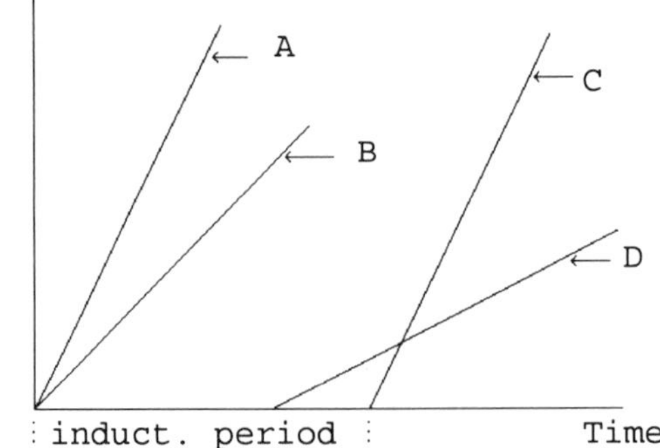

FIGURE 2.1. Effects of inhibitors and retarders. (A) Normal polymerization rate, (B) effect of a retarder, (C) effect of an ideal inhibitor, and (D) effect of a nonideal inhibitor. The time between A and C is an induction period caused by an ideal inhibitor.

2.10. Thermal Polymerization

A few monomers, like styrene and methyl methacrylate, will, after careful purification and presumably free from all impurities, polymerize at elevated temperatures. It is supposed that some ring-substituted styrenes act similarly. The rates of such thermal self-initiated polymerizations are slower than those carried out with the aid of initiators. Styrene, for instance, polymerizes only at a rate of 0.1% per hour at 60 °C, and only 14% at 127 °C. The rate of thermal polymerization of methyl methacrylate is only about 1% of the rate for styrene.[102,103] Several mechanisms of initiation were proposed earlier. The subject was reviewed critically.[104] More recently, the initiation mechanism for styrene polymerization was shown by ultraviolet spectroscopy to consist of an initial formation of a Diels-Alder dimer. The dimer is believed to subsequently transfer a hydrogen to a styrene molecule and form a free radical[46]:

2 CH_2=CH → H → styrene →

CH_3—CH• +

There are indications that photopolymerizations of pure styrene monomer may also proceed by a similar mechanism.[46]

Thermal polymerization of methyl methacrylate, on the other hand, appears to proceed through an initial dimerization into a diradical.[44] This is followed by a hydrogen abstraction from any available source in the reaction mixture.

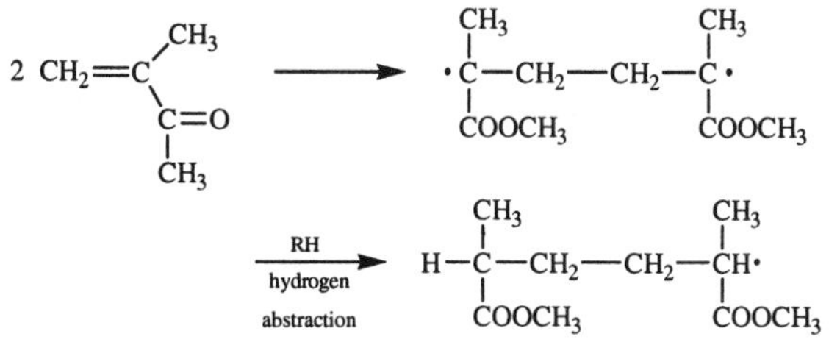

$$2\ CH_2{=}C(CH_3)(C{=}O)CH_3 \longrightarrow \cdot C(CH_3)(COOCH_3){-}CH_2{-}CH_2{-}C(CH_3)(COOCH_3)\cdot$$

$$\xrightarrow[\text{hydrogen abstraction}]{\text{RH}} H{-}C(CH_3)(COOCH_3){-}CH_2{-}CH_2{-}CH(CH_3)(COOCH_3)\cdot$$

2.11. Donor–Acceptor Complexes in Copolymerization

Polar interactions of electron donor monomers with electron acceptor monomers lead to strong tendencies toward formations of alternating copolymers. Also, some alternating copolymerizations might even result from compounds that by themselves are not capable of conventional polymerization. An example is copolymerization of dioxene and maleic anhydride. Two reaction mechanisms were proposed. One suggests that the interactions of donor monomers with acceptor radicals or acceptor monomers with donor radicals lead to decreased energies of activation for cross-propagations.[47] The transition state is stabilized by a partial electron transfer between the donor and acceptor species.[48] The second mechanism suggests that the interactions result in formations of charge-transfer complexes.[106] An electron is completely transferred from the donor monomer to the acceptor monomer. After the transfer, the complex converts to a diradical that subsequently polymerizes by intermolecular coupling. For instance, while many believe that the Diels-Alder reaction takes place by a concerted mechanism, the intermediate was postulated by some to be a charge-transfer complex. An electron is transferred from the donor to the acceptor.[107] This can be illustrated on a reaction between butadiene and maleic anhydride:

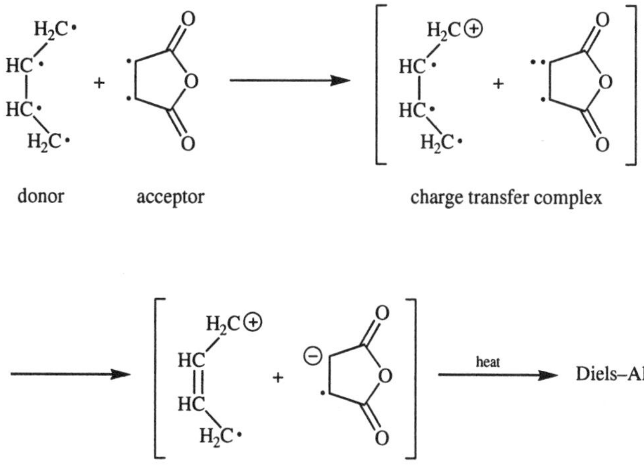

If the reaction mixture is irradiated with high energy radiation, like gamma rays, instead of being heated, an alternating copolymer forms. The complex converts to a diradical[105,107] that homopolymerizes:

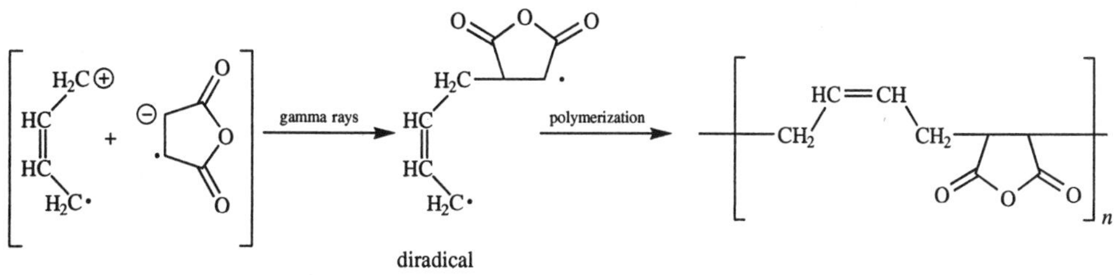

Alternating copolymerization of styrene with maleic anhydride is also explained by donor acceptor interactions.[107] A charge-transfer complex is seen as the new monomer, a diradical, that polymerizes through coupling[107,144–146]:

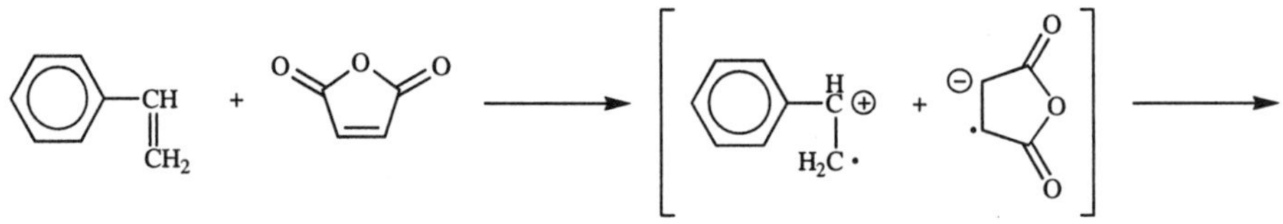

donor acceptor charge–transfer complex

diradical

Charge-transfer complexes are also claimed to be the intermediates in free-radical alternating copolymerization of dioxene or vinyl ethers with maleic anhydride[108–110]:

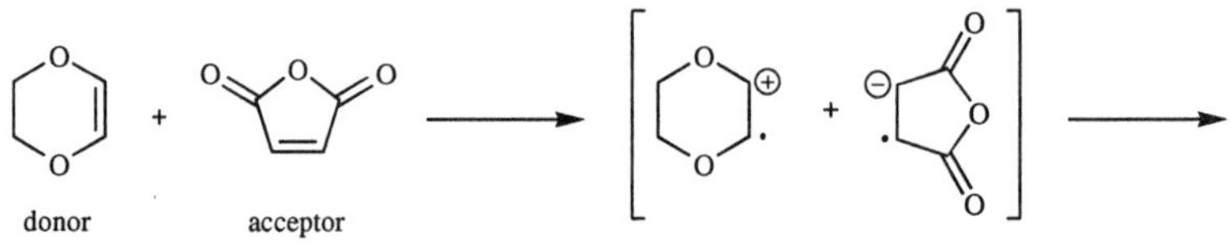

donor acceptor

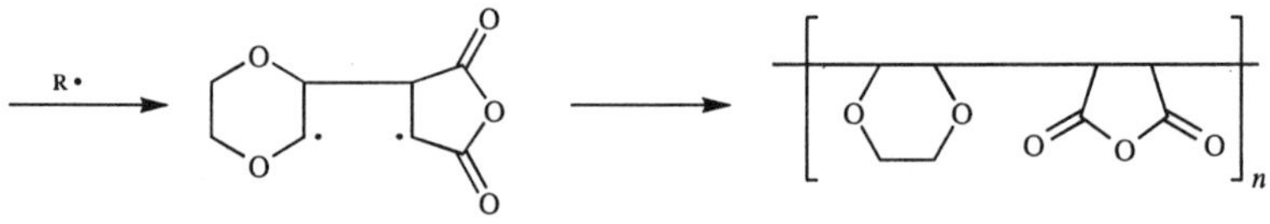

where R• is a polymerization initiating radical.

Here, a third monomer can be included to interpolymerize with the complex that acts as a unit. The product is a terpolymer.[110,111] A diradical intermediate was also postulated in sulfur dioxide copolymerizations and terpolymerizations with bicycloheptene and other third monomers.[112] These third monomers enter the copolymer chain as block segments, while the donor–acceptor pairs enter the chains in a one-to-one molar ratio. This one-to-one molar ratio of the pairs is maintained, regardless of the overall nature of the monomer mixtures.

The propagation and termination steps in the above reactions are claimed[107,144–146] to be related. As stated, an interaction and coupling between two diradicals is a propagation step. When such interactions result in disproportionations, however, they are termination steps. This means the charge-transfer mechanisms are different from conventional free-radical polymerizations. They involve not only interactions between growing polymer radicals and monomers, but also between polymer radicals and complexes. In addition, they involve interactions between the polymer radicals themselves.[107,144–146]

The stability of charge-transfer complexes depends upon internal resonance stabilization. This degree of stabilization determines how easily the diradical opens up.[107] Consequently, this stability also determines how the copolymerization occurs. It can occur spontaneously, or under the influence of light or heat, or because of an attack by an initiating free radical.

There are many examples of spontaneous reactions. When, for instance, isobutylene is added to methyl α-cyanoacrylate, a spontaneous copolymerization in a 1:1 ratio takes place at room temperature. This was explained by the following scheme[107,144–146]:

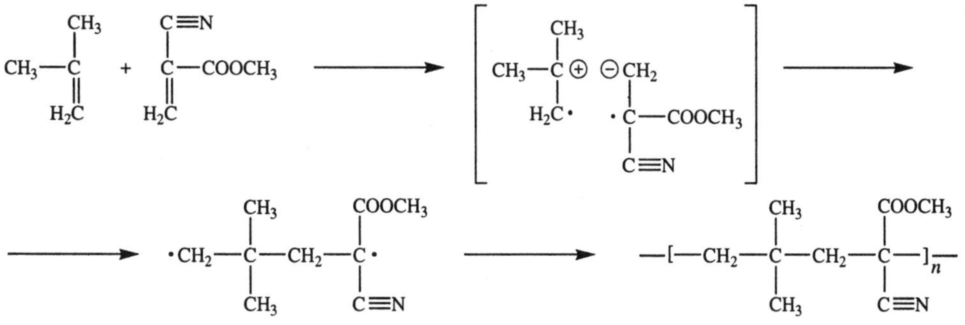

$$CH_2{=}C(CH_3)_2 + CH_2{=}C(C{\equiv}N)COOCH_3 \longrightarrow \left[CH_3\!-\!\overset{\oplus}{C}(CH_3)\!-\!H_2C\cdot \quad \overset{\ominus}{C}H_2\!-\!\cdot C(COOCH_3)(C{\equiv}N) \right] \longrightarrow$$

$$\longrightarrow \cdot CH_2\!-\!C(CH_3)_2\!-\!CH_2\!-\!C(COOCH_3)(C{\equiv}N)\cdot \longrightarrow -[-CH_2\!-\!C(CH_3)_2\!-\!CH_2\!-\!C(COOCH_3)(C{\equiv}N)-]_n-$$

The same happens when sulfur dioxide is added to bicycloheptene at –40 °C[112]:

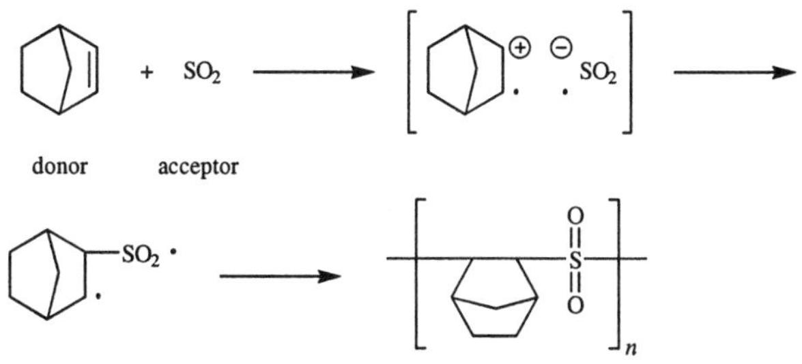

$$\text{bicycloheptene (donor)} + SO_2 \text{ (acceptor)} \longrightarrow [\text{bicycloheptene}^{\oplus}\cdot \quad {}^{\ominus}\cdot SO_2] \longrightarrow \text{bicycloheptyl}\!-\!SO_2\cdot \longrightarrow [-\text{bicycloheptylene}-S(=O)_2-]_n$$

donor acceptor

Another example is a room-temperature 1:1 copolymerization of vinylidine cyanide with styrene[114]:

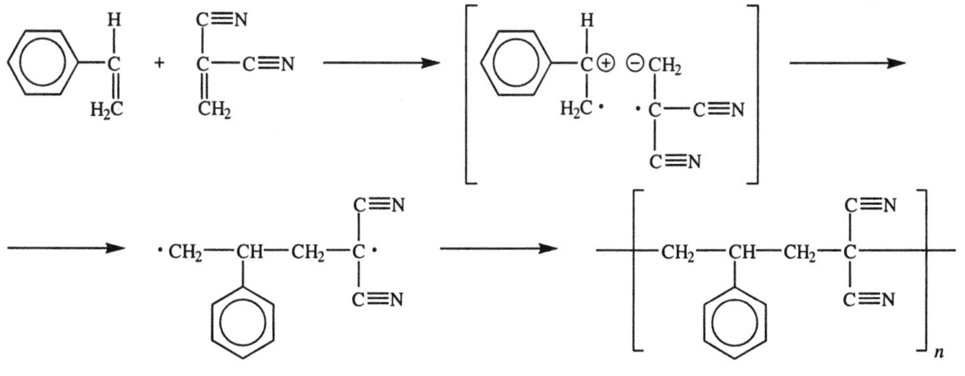

$$C_6H_5\!-\!CH{=}CH_2 + CH_2{=}C(C{\equiv}N)_2 \longrightarrow \left[C_6H_5\!-\!\overset{\oplus}{C}H\!-\!H_2C\cdot \quad \overset{\ominus}{C}H_2\!-\!\cdot C(C{\equiv}N)_2 \right] \longrightarrow$$

$$\longrightarrow \cdot CH_2\!-\!CH(C_6H_5)\!-\!CH_2\!-\!C(C{\equiv}N)_2\cdot \longrightarrow [-CH_2\!-\!CH(C_6H_5)\!-\!CH_2\!-\!C(C{\equiv}N)_2-]_n$$

Still another example is a reaction of 1,3-dioxalene with maleic anhydride.[113]

Examples of stable complexes are reactions of sulfur dioxide with styrene,[107] or vinyl ethers with maleic anhydride,[113] also α-olefins with maleic anhydride.[115–117] Also, a reaction of *trans*-stilbene with maleic anhydride.[118] In these reactions charge-transfer complexes form. They are stable and their existence can be detected by spectroscopic means. Additional energy, such as heat or a free-radical attack, converts them to diradicals and polymerizes them into alternating copolymers.[107,144–146]

Examples of intermediates between the two extremes in stability are reaction products of sulfur dioxide with conjugated dienes.[119] In this case, the reaction results in formation of a mixture of alternating copolymers and cyclic adducts:

$$O{=}S{=}O + CH_2{=}CH{-}CH{=}CH_2 \longrightarrow \left[{}^{\ominus}O{-}\dot{S}{=}O + {}^{\oplus}CH_2{-}\dot{C}H{-}\dot{C}H{-}\dot{C}H_2 \right] \longrightarrow$$

$$\longrightarrow \text{(cyclic sulfone)} + {-}[{-}CH_2{-}CH{=}CH{-}CH_2{-}SO_2{-}]_n{-}$$

The yield of the polymer increases at the expense of the cyclic structure when heat or radiation are applied. A free radical attack has the same effect.[119]

If donor–acceptor interactions and subsequent polymerizations occur upon irradiation with ultraviolet light, the reactions can be very selective. An example is a triphenylphosphine interaction with acrylic monomers[127]:

$$H_2C{=}C(CH_3){-}C({=}O{:}){-}O{-}CH_3 + P(C_6H_5)_3 \xrightarrow{h\nu} \left[H_2C{=}C(CH_3){-}\dot{C}(O^{\ominus}){-}O{-}CH_3 \quad {}^{\oplus}\dot{P}(C_6H_5)_3 \right]$$

$$\updownarrow$$

$$\left[H_2\dot{C}{-}C(CH_3){=}C(O^{\ominus}){-}O{-}CH_3 \quad {}^{\oplus}P(C_6H_5)_3 \right]$$

The reactions does not occur, however, between triphenylphosphine and styrene or vinyl acetate.[127]

The nature and the amount of solvent can influence the yield and the composition of the copolymers in these copolymerizations. Copolymerization of phenanthrene with maleic anhydride in benzene yields a 1:2 adduct. In dioxane, however, a 1:1 adduct is obtained. In dimethyl formamide no copolymer forms at all.[130] Another example is a terpolymerization of acrylonitrile with 2-chloroethyl vinyl ether and maleic anhydride or with *p*-dioxene-maleic anhydride. The amount of acrylonitrile in the terpolymer increases with an increase in the π-electron density of the solvent in the following order[131]:

xylene >> toluene >> benzene > chlorobenzene >> chloroform

The ratio of maleic anhydride to the vinyl ether in the product remains, however, equimolar.

Whether the concept of charge-transfer complexes in copolymerizations is fully accepted is not certain. Much of the accumulated evidence, to date, such as UV and NMR spectroscopy, does support it in many systems.[148] Further support comes from the strong tendencies to form alternating copolymers over a wide range of feed compositions, also, from high reaction rates at equimolar feed compositions.[107]

Some polar vinyl monomers, like methyl methacrylate or acrylonitrile interact and complex with Lewis acids. They subsequently polymerize at a faster rate and to a higher molecular weight than can be expected otherwise. The effective Lewis acids are $ZnCl_2$, $AlCl_3$, $Al(C_2H_5)_2Cl$, $AlCl_2(C_2H_5)$, $SnCl_4$, and some others.[120–122] Complexes can form on an equimolar basis and undergo homopolymerizations after application of heat. A free-radical attack, or irradiation with ultraviolet light or with gamma rays, is also effective.[123]

It was proposed[122,123,129,132,133] that the greater reactivity is due to delocalization of the electrons of the complexed molecules:

$$H_2C{=}CH{-}C{\equiv}N{\rightarrow}ZnCl_2 \rightleftharpoons [H_2C{-}CH{=}C{=}N]^{\oplus}{\rightarrow}Zn^{\ominus}Cl_2$$

Molecular orbital calculations support this.[124,125] Polymerizations can also be carried out with less than stoichiometric ratios of the Lewis acids to the monomers. As an illustration, coupling of an acrylic or a methacrylic monomer with a Lewis acid can be shown as follows[124,125]:

$$a\,M + MeX_n \rightleftharpoons (M)_a.......MeX_n$$

where MeX_n represents a Lewis acid.

Vinyl pyridine and vinyl imidazole also form complexes with zinc chloride. Here, conjugation with the metal salt results in a superdelocalizability and leads to spontaneous thermal polymerization.[124,125]

The increased reactivity toward polymerization by some monomers, even in the presence of less than equimolar amounts of Lewis acids, was suggested to be due to one of two possibilities. It may be due to greater reactivity of the portion of the monomer that complexes toward the growing chain end. It may also be due to formation of new complexes between an uncomplexed monomer and a complexed one[107]:

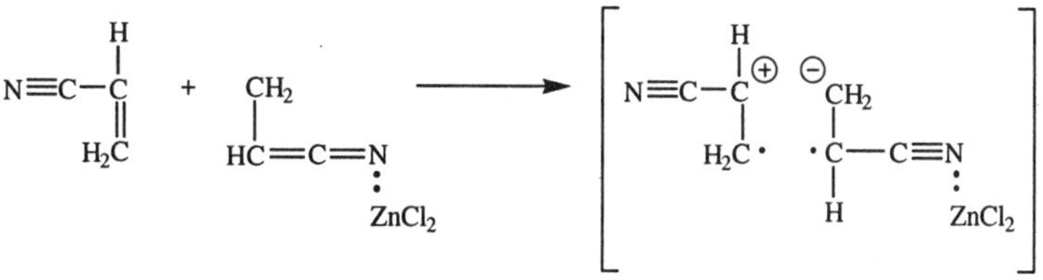

N≡C–C(H)=CH₂ + CH₂–HC=C=N:ZnCl₂ ⟶ [N≡C–C(H)$^{\oplus}$–H₂C• $^{\ominus}$CH₂–•C(H)–C≡N:ZnCl₂]

The picture is more complicated when the Lewis acids are used in combinations with donor–acceptor monomers.[155] The donor–acceptor complexes are believed to form first and then react with Lewis acids according to the following scheme[124,125]:

$$a\,M_1 + b\,M_2 + MeX_n \rightleftharpoons [(M_1)_a\ldots MeX_n] + b\,M_2 \rightleftharpoons [MeX_n..(M_2)_b]$$

A structure of methyl methacrylate and styrene complexed with $SnCl_4$ was shown to be as follows[128]:

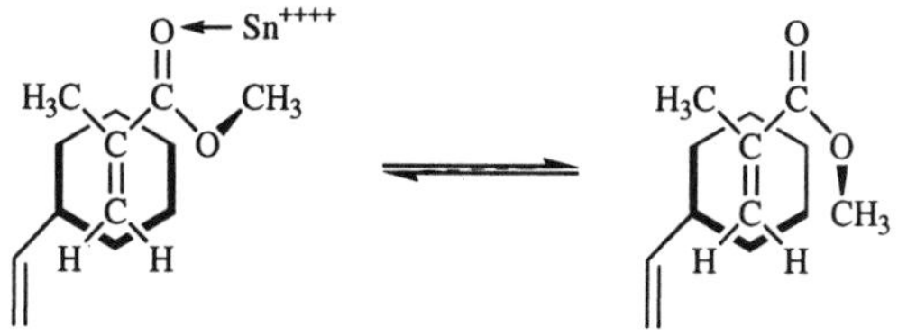

O←Sn^{++++}, H₃C, C, CH₃, O, C, C, H, H ⇌ O, H₃C, C, C, O, CH₃, C, H, H

In formations of ternary complexes, the acceptor vinyl compound must have a double bond conjugated to a cyano or to a carbonyl group. Such acceptors are acrylonitrile, methacrylonitrile, acrylic and methacrylic esters and acids, methyl vinyl ketone, acrylamide, etc. Donor monomers are styrene, α-methyl styrene, butadiene, 2-3-dimethyl butadiene, isoprene, chloroprene, etc.

One proposed mechanism[107,128] is that such charge transfer polymerizations are in effect homopolymerizations of charge transfer complexes [D⊕ A⊖. . . MeX_n]. In other words, the metal halide is complexed with the electron acceptor monomer and acts as an acceptor component.

The above opinion, however, is not universal. Others held that the increased susceptibility to ultraviolet radiation or to initiating radicals[126] is due to increased reactivity of the propagating radicals of complexed monomers toward incoming uncomplexed ones.

Arguments against the ternary complex mechanism are: (1) the physical evidence that proves the existence of the ternary molecular complexes is weak; (2) the ternary molecular complexes can have no bearing on the copolymerizations because the equilibrium concentration of the complexed monomers is low, compared to the uncomplexed ones.[160,163,164]

A third opinion is that a complex of an acceptor monomer with a Lewis acid copolymerizes alternately with the donor monomer and with an uncomplexed acceptor monomer.[161,162] This presumably takes place according to the conventional chain-growth polymerization scheme of radical copolymerization. The alternate placement of monomers is due to highly enhanced values of cross-propagation constants. It results from complexing acceptor monomers with Lewis acids. Such a mechanism fails to explain satisfactorily the completely alternating incorporation of monomers and the inefficiency of chain-transfer reagents. It also fails to explain the spontaneous initiation of alternating copolymerization.

Kabanov suggests[181,216] that during the primary free-radical formation in the Lewis acid–monomer system, both the original and complexed monomers may participate in the chain propagation. This would result in the appearance of complexed propagating radicals besides the usual ones. In the complexed ones, the last unit carrying the valence is a ligand of coordination complex:

$$\sim\overset{\bullet}{M} \ldots MeX_n$$

It excludes, however, all electron transfer reactions that may take place due to ultraviolet light irradiation.

2.13. Steric Control in Free-Radical Polymerization

In free-radical polymerization reactions, the propagating radical chain has a great amount of freedom. Atactic polymers, therefore, are usually formed. Any control that the reaction conditions exercise over the propagating species increases at lower temperatures due to lower mobilities. This leads to increased syndiotactic placement, as is discussed in the section on propagation. Special techniques, however, like the use of canal complexes, can be employed to form stereoregular polymers by a free-radical mechanism. Urea and thiourea were used originally for such purposes.[182,183] Monomers, like butadiene or others form complexes within the voids or canals of the crystal lattices of these compounds. Brief exposure to high energy radiation initiates chain growth. In the canals the monomer molecules are held in fixed positions so chain growth is restricted in one direction only. Steric control is exercised, because in these fixed positions the monomer molecules tend to be aligned uniformly. It was suggested that in the canal complexes the monomers are not just lined up end to end, but packed in an overlapping arrangement. For molecules like isobutylene or vinylidine chloride, it may be possible to lie directly on top of each other, resembling a stack of coins. Such stacking greatly facilitates reactions between guest molecules.[182,183]

Polymerizations in thiourea canal complexes yield high melting crystalline *trans*-1,4-polybutadiene, 2,3-dimethylbutadiene, 2,3-dichlorobutadiene, and 1,3-cyclohexadiene. Cyclohexadiene monoxide, vinyl chloride, and acrylonitrile also form stereoregular polymers. On the other hand, polymerizations of isobutylene and of vinylidine chloride fail to yield stereospecific polymers.

Sodium montmorilonite can also be used to polymerize polar monomers between the lamellae. Here, too, the organization of monomer molecules within the monolayers influences the structure of the resultant polymers.[184,185] Poly(methyl methacrylate) formed in sodium montmorilonite is composed of short, predominantly isotactic stereosequences.[185] The percent of isotactic component increases with an increase in the ion-exchanging population on the surface of the mineral and is independent of the temperature between 20 to 160 °C. In this way, it is possible to vary the population of isotactic triads at will up to 50% composition.[186]

Perhydrotriphenylene also forms channel-like inclusions with conjugated dienes. Polymerization of these dienes yield some steric control.[187,188]

There were attempts at controlling steric placement by a technique called *template polymerization*. An example is methyl methacrylate polymerization in the presence of isotactic poly(methyl methacrylate).[189,190] The presence of template polymers, however, only results in accelerating the rates of polymerizations.[191]

2.14. Polymer Preparation Techniques

Four general techniques are used for the preparation of polymers by a free-radical mechanism: polymerization in *bulk*, in *solution*, in *suspension*, and in *emulsion*. The *bulk* or *mass* polymerization is probably the simplest of the four methods. Only the monomer and the initiator are preset in the reaction mixture. It makes the reaction simple to carry out, though the exotherm of the reaction might be hard to control, particularly if it is done on a large scale. Also, there is a chance that local hot spots might develop. Once bulk polymerization of vinyl monomers is initiated, there can be two types of results, depending upon the solubility of the polymer. If it is soluble in the monomer, the reaction may go to completion with the polymer remaining soluble throughout all stages of conversion. As the polymerization progresses, the viscosity of the reaction mixture increases markedly. The propagation proceeds in a medium of associated polymeric chains dissolved in or swollen by the monomer until all the monomer is consumed.

If the polymer is insoluble, it precipitates out without any noticeable increase in solution viscosity. Examples of this type of reaction can be polymerizations of acrylonitrile or vinylidine chloride. The activation energy is still similar to polymerizations of soluble polymers and the initial rates are proportional to the square root of initiator concentration. Also, the molecular weights of the polymerization products are inversely proportional to the polymerization temperatures and to initiator concentrations. Furthermore, the molecular weights of the resultant polymers far exceed the solubility limits of the polymers in the monomers. The limits of acrylonitrile solubility in the monomer are at a molecular weight of 10,000. Yet, polymers with molecular weights as high as 1,000,000 are obtained by this process. This means that the polymerizations must proceed in the precipitated polymer particles, swollen and surrounded by monomer molecules.

The kinetic picture of free-radical polymerization applies best to bulk polymerizations at low points of conversion. As the conversion progresses, however, the reaction becomes complicated by chain transferring to the polymer and by gel effect. The amount of chain transferring varies, of course, with the reactivity of the polymer radical.

Bulk polymerization is employed when some special properties are required, such as high molecular weight or maximum clarity, or convenience in handling. Industrially, bulk polymerization in special equipment can have economic advantages, as with bulk polymerization of styrene. This is discussed in Chapter 5.

Solution polymerization differs from bulk polymerization, because a solvent is present in the reaction mixture. The monomer may be fully or only partially soluble in the solvent. The polymer may be (1) completely soluble in the solvent, (2) only partially soluble in the solvent, and (3) insoluble in the solvent.

When the monomer and the polymer are both soluble in the solvent, initiation and propagation occur in a homogeneous environment of the solvent. The rate of the polymerization is lower than in bulk. In addition, the higher the dilution of the reactants, the lower the rate and the lower the molecular weight of the product. This is due to chain transferring to the solvent. In addition, any solvent that can react to form telomers will also combine with the growing chains.

If the monomer is soluble in the solvent, but the polymer is only partially soluble or insoluble, the initiation still takes place in a homogeneous medium. As the chains grow, there is some increase in viscosity that is followed by precipitation. The polymer precipitates in a swollen state and remains swollen by the diffused and adsorbed monomer. Further propagation takes place in these swollen particles.

Propagation continues in the precipitated swollen polymer, so the precipitation does not exert a strong effect on the molecular weight of the product. This was demonstrated on polymerization of styrene in benzene (where the polymer is soluble) and in ethyl alcohol (where the polymer is insoluble). The average molecular weight obtained in benzene at 100 °C was 53,000 while in ethyl alcohol at the same temperature it was 51,000.[192] When the monomers are only partially soluble and the polymers insoluble in the solvents, the products might still be close in molecular weights to those obtained with soluble monomers and polymers. Polymerization of acrylonitrile in water can serve as an example. The monomer is only soluble to the extent of 5–7% and the polymer is effectively insoluble. When aqueous saturated solutions of acrylonitrile are polymerized with water-soluble initiators, the systems behaves initially as typical solution polymerizations. The polymers, however, precipitate out rather quickly as they form. Yet, molecular weights over 50,000 are readily obtainable under these conditions.

There are different techniques for carrying out solution polymerization reactions. Some can be as simple as combining the monomer and the initiator in a solvent and then applying agitation, heat, and an inert atmosphere.[193] Others may consist of feeding into a stirred and heated solvent the monomer or the initiator, or both continuously, or at given intervals. It can be done throughout the course of the reaction or through part of it.[194] Such a setup can be applied to laboratory preparations or to large-scale commercial preparations.

In both techniques the initiator concentration changes only a few percent during the early stages of the reaction, if the reaction temperature is not too high. The polymerization may therefore approach a steady-state character during these early stages. After the initial stages, however, and at higher temperatures, the square-root dependence of rates upon the initiator concentration no longer holds. This is a result of the initiator being depleted rapidly. The second technique, where the initiator, or the monomer and the initiator, are added continuously, was investigated at various temperatures and rates of addition.[195–197] If the initiator and monomer are replenished at such a rate that their ratios remains constant, steady-state conditions might be extended beyond the early stages of the reactions.[198] How long they can be maintained, however, is uncertain.

Suspension polymerization[150] can be considered as a form of mass polymerization. It is carried out in small droplets of liquid monomer dispersed in water and caused to polymerize to solid spherical particles. The process generally involves dispersing the monomer in a nonsolvent liquid into small droplets. The agitated stabilized medium usually consists of water containing small amounts of some suspending or dispersing agent. The initiator is dissolved in the monomer if it is a liquid. It is included in the reaction medium if the monomer is a gas.

To form a dispersion, the monomer must be quite insoluble in the suspension system. To decrease the solubility and to sometimes also increase the particle size of the resultant polymer bead, partially polymerized monomers or prepolymers may be used. Optimum results are obtained with initiators that are soluble in the monomer. Often, no differences in rates are observed between polymerization in bulk and suspension. Kinetic studies of styrene suspension polymerization have shown that all the reaction steps, initiation, propagation, and termination, occur inside the particles.[199]

The main difficulty in suspension polymerization is in forming and maintaining uniform suspensions. This is because the monomer droplets are slowly converted from thin immiscible liquids to sticky viscous materials and from that to rigid granules. The tendency is for the particles to stick together and to form one big mass. The suspending agent's sole function is to prevent coalescing of the sticky particles. Such agents are used in small quantities (0.01 to 0.5% by weight of the monomer). There are many different suspending agents, both organic and inorganic. The organic ones include methyl cellulose, ethyl cellulose, poly(acrylic acid), poly(methacrylic acid), salts of these acids, poly(vinyl alcohol), gelatins, starches, gums, alginates, and some proteins, like casein and zein. Among the inorganic suspending agents can be listed talc, magnesium carbonate, calcium carbonate, calcium phosphate, titanium and aluminum oxides, silicates, clays, like bentonite, and others. The diameter of the resultant beads varies from 0.1 mm to 5 mm and often depends upon the rate of agitation. It is usually inversely proportional to the particle size. Suspension polymerization is used in many commercial preparations of polymers.

Emulsion polymerization is also used widely in commercial processes.[147,148] The success of this technique is due in part to the fact that this method yields high molecular weight polymers. In addition, the polymerization rates are usually high. Water is the continuous phase and it allows efficient removal of the heat of polymerization. Also, the product from the reaction, the latex, is relatively low in viscosity, in spite of the high molecular weight of the polymer. A disadvantage of the process is that water-soluble emulsifiers are used. These are difficult to remove completely from the polymers and may leave some degree of water sensitivity.

The reaction is commonly carried out in water containing the monomer, an emulsifier, or a surface-active agent, and a water-soluble initiator. Initiation may be accomplished through thermal decomposition of the initiator or through a redox reaction. The polymer forms as a colloidal dispersion of fine particles and polymer recovery requires breaking up the emulsion.

The full mechanism of emulsion polymerization is still not completely worked out. It is not yet clear why a simultaneous increase in the polymerization rate and in the molecular weight of the product are often observed. Also, in emulsion polymerization, at the outset of the reaction the monomer is in the form of finely dispersed droplets. These droplets are about 1 μ in diameter. Yet, during the process of a typical polymerization, they are converted into polymer particles that are submicroscopic, e.g., 1000 Å in diameter.

At the start of the reaction the emulsifier exists simultaneously in three loci: (a) as a solute in water; (b) as micelles; (c) and as a stabilizing emulsifier at the interface between the monomer droplets and the water. The bulk of the emulsifier, however, is in the micelles. The monomer is also present in three loci: (a) in the monomer droplets that are emulsified and perhaps 1–10 μ in diameter; (b) solubilized in the micelles, perhaps 50 to 100 Å in diameter; (c) and as individual molecules dissolved in the water. The bulk of the monomer is in the droplets. There are on the average 10^{18}/ml of monomer swollen micelles in the reaction mixture at the outset of the reaction.[200] At the start of the reaction there are also on the average 10^{12}/ml monomer droplets that act as reservoirs. The monomer is supplied from the droplets to radical-containing micelles when the reaction progresses by a process of diffusion through the aqueous phase.

The first hypothesis of the mechanism of emulsion polymerization was formulated by Harkins.[200] According to this hypothesis, the water-soluble initiator decomposes in the aqueous phase. This results in formation of primary radicals. The primary radicals in turn react with the monomer molecules dissolved in the water (though their number may be quite small). Additional monomer molecules may add to the growing radicals in the water until the growing and propagating chains of free radicals acquire surface-active properties. At that stage, the growing radicals consist of inorganic and organic portions:

1. $$^{\ominus}O{-}\overset{O}{\overset{\|}{\underset{\downarrow}{\underset{O}{S}}}}{-}O{-}O{-}\overset{O}{\overset{\|}{\underset{\downarrow}{\underset{O}{S}}}}{-}O^{\ominus} \longrightarrow 2\;^{\ominus}O{-}\overset{O}{\overset{\|}{\underset{\downarrow}{\underset{O}{S}}}}{-}O\cdot$$

2. $$^{\ominus}O{-}\overset{O}{\overset{\|}{\underset{\downarrow}{\underset{O}{S}}}}{-}O\cdot + CH_2{=}C\begin{matrix}H\\X\end{matrix} \longrightarrow {}^{\ominus}O{-}\overset{O}{\overset{\|}{\underset{\downarrow}{\underset{O}{S}}}}{-}O{-}\overset{H}{\underset{H}{C}}{-}\dot{C}\begin{matrix}H\\X\end{matrix}$$

3. $$^{\ominus}O{-}\overset{O}{\overset{\|}{\underset{\downarrow}{\underset{O}{S}}}}{-}O{-}\overset{H}{\underset{H}{C}}{-}\dot{C}\begin{matrix}H\\X\end{matrix} + y\;CH_2{=}C\begin{matrix}H\\X\end{matrix} \longrightarrow {}^{\ominus}O{-}\overset{O}{\overset{\|}{\underset{\downarrow}{\underset{O}{S}}}}{-}O{-}({-}\overset{H}{\underset{H}{C}}{-}\overset{H}{\underset{X}{C}}{-})_y{-}\overset{H}{\underset{H}{C}}{-}\overset{H}{\underset{X}{C}}\cdot$$

growing surface–active radical

These growing radical ions tend to diffuse into the monomer–water interfaces. The probability that the diffusion takes place into monomer swollen micelles rather than into monomer droplets is backed by the considerations of the relative surface areas of the two. There are, on the average, 10^{18} micelles in each milliliter of water. These are approximately 75 Å in diameter and each swollen micelle contains on the average 30 molecules of the monomer. With the monomer droplet diameter of 1 μ and with approximately 10^{12} such droplets per milliliter of water, the micelles offer 60 times more surface for penetration than do the droplets.

Once the radicals penetrate the micelles, polymerization continues by adding monomers that are inside. The equilibrium is disturbed and the propagation process proceeds at a high rate due to the concentration and crowding of the stabilized monomers. This rapidly transforms the monomer swollen micelles into polymer particles. The changes result in disruptions of the micelles by growths from within. The amount of emulsifier present in such changing micelles is insufficient to stabilize the polymer particles. In trying to restore the equilibrium, some of the micelles, where there is no polymer growth, disintegrate and supply the growing polymer particles with emulsifier. In the process many micelles disappear per each polymer particle that forms. The final latex usually ends up containing about 10^{15} polymer particles per milliliter of water. By the time conversions reach 10–20% there are no more micelles present in the reaction mixtures. All the emulsifier is now adsorbed on the surface of the polymer particles. This means that no new polymer particles are formed. All further reactions are sustained by diffusion of monomer molecules from the monomer droplets into the growing polymer particles. The amount of monomer diffusing into the particles is always in excess of the amount that is consumed by the polymerization reaction due to osmotic forces.[202] This extra monomer supplied is sufficient for equilibrium swelling of the particles.[203] As a result, the rate of polymerization becomes zero order with respect to time.

When conversion reaches about 70%, all the remaining monomer is absorbed in the polymer particles and there are no more droplets left. At this point the reaction rate becomes first order with respect to time.

The qualitative approach of Harkins was put on a quantitative basis by Smith and Ewart.[201–204] Because 10^{13} radicals are produced per second and can enter between 10^{14} and 10^{15} particles, Smith felt that a free radical can enter a particle once every 10 to 100 seconds. It can cause the polymerization to occur for 10 to 100 seconds before another free radical would enter and terminate chain growth.[204] A period of inactivity would follow that would last 10 to 100 seconds and then the process would repeat itself. Such a "stop and go" mechanism implies that a particle contains a free radical approximately half of the time. It can also be said that the average number of radicals per particle is 0.5. This is predicted on condition that: (a) the rate of chain transfer out of the particle is negligible, and (b) the rate of termination is very rapid compared with the rate of radical entry into the particle.

The kinetic relationships derived by Smith and Ewart for the system are as follows:

$$\text{The rate of primary radical entering a particle} = r_i = R_i/N$$

$$\text{Rate of polymerization} = R_P = K_P[\text{M}]\, N/2$$

$$\text{Average degree of polymerization} = \text{DP} = NK_P[\text{M}]/R_i$$

where K_P is the constant for propagation, [M] is the concentration of monomer, N is the number of particles containing n radicals (~ 0.5), and the expression for the number of particles formed is

$$N = K(\rho/\mu)^{0.4}(A_sS)^{0.6}$$

where μ is the volume increase of the particles, A_s is the area occupied by one emulsifier molecule, S is the amount of emulsifier present, K is a constant equal to 0. 37 (based on the assumption that the micelles and polymer particles compete for the free radicals in proportion to their respective total surface areas); K can also be equal to 0.53 (based on the assumption that the primary radicals enter only micelles, as long as there remain micelles in the reaction mixture); ρ is the rate of entry into the particles.

The Smith–Ewart mechanism does not take into account any polymerization in the aqueous phase. This may be true for monomers that are quite insoluble in water, like styrene, but appears unlikely for more hydrophilic ones like methyl methacrylate or vinyl acetate. In addition, it was calculated by Flory that there is insufficient time for a typical cation radical (like a sulfate ion radical) to add to a dissolved molecule of monomer like styrene before it becomes captured by a micelle.[203] This was argued against, however, on the ground that Flory's calculations fail to consider the potential energy barrier at the micelle surfaces from the electrical double layer. This barrier would reduce the rate of diffusion of the radical ions into the micelles.[205]

A different mechanism was proposed by several groups.[206,207] It is based on a concept that most polymerizations must take place at the surface of the particles or in their outer "shell" and not within the particles. It is claimed that the interiors of the particles are too viscous for free radicals to diffuse inside at a sufficiently fast rate. Two different mechanisms were actually proposed to explain why polymerization takes place preferentially in the shell layer. One of them suggests that the monomer is distributed nonuniformly in the polymer particles. The outer shell is rich in monomer molecules, while the inside is rich in polymer molecules.[207] The other explanation is that the radical ions that form from the water soluble initiator are too hydrophilic to be able to penetrate the polymer particles.[207]

Review Questions

Section 2.1.

1. What are the three steps in free-radical polymerization? Illustrate each step.
2. What is the rate-determining step in free-radical polymerization?
3. Write the kinetic expressions for initiation, propagation, termination, and transfer.
4. What is the steady-state assumption? How is it expressed?
5. What is the expression for the rate of propagation? Rate of monomer disappearance? The average lifetime of a growing radical under steady-state conditions?
6. Why is an initiator efficiency factor needed for the rate equation? What is a kinetic chain length?

Section 2.2.

1. What sources of initiating free radicals do you know? Give the chemical equation for thermal decomposition of α,α'-azobisisobutyronitrile. Explain how free radicals can recombine to form inert compounds.
2. What kind of peroxides are available for initiations of free-radical chain-growth polymerizations? List and draw structures of various types.
3. How does the solvent "cage" affect the initiating free radicals? Explain and illustrate.
4. Explain homolytic and heterolytic cleavage of peroxides.
5. Explain and give chemical equations for *redox* initiations with Fe^{2+}, Co^{2+}, and Ce^{4+} ions in the reaction mixture.
6. How are peroxides, such as benzoyl peroxide, decomposed by aromatic tertiary amines? Show the two postulated mechanisms for the reaction of benzoyl peroxide with dimethyl aniline.

Section 2.3.

1. Describe the reaction of the free radical with the monomer. Show this reaction with equations, using a phenyl radical and styrene as an example.
2. Do the same as question 1, but with a redox mechanism, showing a sulfate ion radical adding to vinyl acetate.

Section 2.4.

1. Illustrate the transition state in the propagation reaction.

2. Explain the steric, polar, and resonance effects in the propagation reaction.
3. Explain why there is a tendency for a *trans–trans* placement in the propagation reactions.
4. How does the reaction medium affect the propagation reaction?
5. What is ceiling temperature and what is the kinetic expression for this phenomenon?
6. Explain what is meant by autoacceleration. How does it manifest itself.
7. What is cyclopolymerization? Explain and give several examples.

Section 2.5.

1. What are the three termination processes in free-radical polymerization?
2. What is meant by degenerative chain transferring?
3. What is meant by chain transferring constants?
4. Write the equation for the degree of polymerization including all the chain-transferring constants.
5. In a benzoyl peroxide initiated polymerization of 2 moles of styrene in benzene at 85 °C ($K_d = 8.94 \times 10^{-5}$ L/mole-s at 85 °C). How much benzoyl peroxide will be required in the polymerization solution to attain an average molecular weight of 250,000? Assume that termination occurs only by recombination and no chain transferring takes place.

Section 2.6.

1. Explain what is meant by reactivity ratios. How are they derived?
2. Write the copolymerization equation.
3. How do substituents on the monomer molecules affect reactivity of the monomers towards attacking radicals?
4. Explain the Q and e scheme and write the Price–Alfrey equation.
5. How can r_1 and r_2 be derived from the Q and e values? Show the relationship.
6. From chemical structures alone predict the products from free-radical copolymerizations of pairs of (1) styrene and methyl methacrylate, (2) butadiene and vinyl acetate, and (3) methyl methacrylate and butyl vinyl ether. Check your answers by consulting a polymer handbook for reactivity ratios.

Section 2.7.

1. How many reactivity ratios are there in a terpolymerization?

Section 2.8.

1. What is allylic polymerization? If allyl alcohol does not polymerize to a high molecular polymer by free-radical polymerization, why does triallyl cyanurate form a high molecular weigh network structure by the same mechanism?

Section 2.9.

1. What is inhibition and retardation? Explain.
2. Give an example of a good retarder and show by chemical equations the reaction with free radicals.
3. Show the reaction of quinone with free radicals.
4. Write the equation that relates rate data to inhibited polymerizations.

Section 2.10.

1. What is thermal polymerization? Show by chemical equations the postulated mechanism of formation of initiating radicals in styrene thermal polymerization.

Section 2.11.

1. Show the proposed charge transfer mechanism for copolymerization of styrene with maleic anhydride, dioxane with maleic anhydride.

2. What determines the stability of charge transfer complexes? Explain and give examples.

Section 2.12.

1. How do some polar monomers complex with Lewis acids? How does that affect polymerization of these monomers? Copolymerization?

Section 2.13.

1. How can canal complexes be used for steric control in free-radical polymerization? Give examples.

Section 2.14.

1. What is meant by bulk or mass polymerization? Explain and discuss.
2. What are some of the techniques for carrying out solution polymerizations?
3. Give a qualitative picture of emulsion polymerization as described by Harkins.
4. How did Smith and Ewart put the Harkins picture of emulsion polymerization on a quantitative basis? What is the equation for the rate of emulsion polymerization?

References

1. T. Fueno, T. Tsuruta, and J. Furukawa, *J. Polym. Sci.*, **40**, 487 (1959).
2. J. Thiele and K. Hauser, *Ann.*, **290**, 1 (1896).
3. G. S. Hammond, C. Wu, O. D. Trapp, J. Warkentin, and R. T. Keys, *Am. Chem. Soc. Polym. Prepr.*, **1** (1), 168 (1960).
4. J. B. Brandrup and E. H. Immergut (eds.), *Polymer Handbook*, 3rd ed., Wiley, New York, 1989; C. S. Sheppard, "Azo Compounds," in *Encyclopedia, of Polymer Science and Engineering*, Vol. 2 (H. F. Mark, N. M. Bikales, C. G. Overberger, and G. Menges, eds.), Wiley-Interscience, New York, 1985.
5. D. Swern (ed.), *Organic Peroxides*, Wiley-Interscience, New York, 1970; S. Patai, (ed.), *The Chemistry of Peroxides*, Wiley-Interscience, New York, 1983.
6. I. A. Saad and F. R. Eirich, *Am. Chem. Soc., Polym. Prepr.*, **1**, 276 (1960).
7. P. D. Bartlett and R. R. Hiatt, *J. Am. Chem. Soc.*, **80**, 1398 (1958).
8. C. G. Szwarc and L. Herk, *J. Chem. Phys.*, **29**, 438 (1958).
9. C. G. Swain, W. T. Stockmayer, and J. T. Clarke, *J. Am. Chem. Soc.*, **72**, 5726 (1950).
10. A. T. Blomquist and A. J. Buselli, *J. Am. Chem. Soc.*, **73**, 3883 (1951); C. S. Sheppard, "Peroxy Compounds," in *Encyclopedia of Polymer Science and Engineering*, Vol. 11 (H. F. Mark, N. M. Bikales, C. G. Overberger, and G. Menges, eds.), Wiley-Interscience, New York, 1988.
11. M. G. Evans, *Trans. Faraday Soc.*, **42**, 101 (1946).
12. A. V. Tobolsky and R. B. Mesrobian, *Organic Peroxides*, Wiley (Interscience), New York, 1954.
13. P. D. Bartlett and K. Nozaki, *J. Am. Chem. Soc.*, **68**, 1686 (1946); *ibid.*, **69**, 2299 (1947).
14. H. H. Dean and G. Skirrow, *Trans. Faraday Soc.*, **54**, 849 (1958).
15. L. Horner and E. Schwenk, *Angew. Chem.*, **15**, 849 (1949); M. Imoto and S. Choe, *J. Polym. Sci.*, **15**, 485 (1955).
16. W. A. Pryor and W. H. Hendrickson, Jr., *Tetrahedron Lett.*, **24**, 1459 (1983).
17. K. F. O'Driscoll, T. P. Konen, and K. M. Connolly, *J. Polym. Sci.*, **A-1,5**, 1789 (1967).
18. V. Horanska, J. Barton, and Z. Manasek, *J. Polym. Sci.*, **A-1,10**, 2701 (1972).
19. T. Hirano, T. Miki, and T. Tsuruta, *Makromol. Chem.*, **104**, 230 (1967).
20. G. S. Kolesnikov and L. S. Fedorova, *Izv. Akad. Nauk SSSR, Otd. Khim. Nauk*, 236 (1957).
21. J. Furukawa, T. Tsuruta, and S. Inoue, *J. Polym. Sci.*, **26**, 234 (1957).
22. N. Ashikari and N. Nishimura, *J. Polym. Sci.*, **28**, 250 (1958).
23. J. Furukawa and T. Tsuruta, *J. Polym. Sci.*, **28**, 227 (1958).
24. R. C. Petry and F. H. Verhoek, *J. Am. Chem. Soc.*, **78**, 6416 (1956).
25. T. D. Parsons, M. B. Silverman, and D. M. Ritter, *J. Am. Chem. Soc.*, **79**, 5091 (1957).
26. C. E. H. Bawn, D. Margerison, and N. M. Richardson, *Proc. Chem. Soc. (London)*, 397 (1959).
27. R. L. Hansen and R. R. Hammann, *J. Phys. Chem.*, **67**, 2868 (1963).
28. S. Inoue, Chapter 5 in *Structure and Mechanism in Vinyl Polymerization* (T. Tsuruta and K. F. O'Driscoll, eds.), Dekker, New York, 1969.
29. A. L. Barney, J. M. Bruce, Jr., J. N. Coker, H. W. Jackobson, and W. H. Sharkey, *J. Polym. Sci.*, **A-1,4**, 2617 (1966).

30. Y. Nakoyama, T. Tsuruta, and J. Furukawa, *Makromol. Chem.*, **40**, 79 (1960).
31. J. Furukawa, T. Tsuruta, T. Fueno, R. Sakata, and K. Ito, *Makromol. Chem.*, **30**, 109 (1959).
32. G. Oster and N. L. Yang, *Chem. Rev.*, **68** (2), 125 (1968).
33. N. J. Turo, *Molecular Photochemistry*, Benjamin, New York, 1967.
34. J. G. Calvin and J. N. Pitts, Jr., *Photochemistry*, Wiley, New York, 1967.
35. S. P. Pappas, Photopolymerization, in *Encyclopedia of Polymer Science and Engineering*, Vol. 11 (H. F. Mark, N. M. Bikalis, C. G. Overberger, and G. Menges, eds.), Wiley-Interscience, New York, 1988.
36. T. Koenig and H. Fischer, "Cage Effect," Chap. 4 in *Free-Radicals*, Vol. I (J. K. Kochi, ed.), Wiley, New York, 1973.
37. T. Koenig, "Decomposition of Peroxides and Azoalkanes," Chap. 3 in *Free-Radicals*, Vol. I (J. K. Kochi, ed.), Wiley, New York, 1973.
38. C. H. E. Bawn, W. H. Janes, and A. M. North, *J. Polym. Sci.*, **C,4**, 427 (1963).
39. M. Hagiwara, H. Okamoto, T. Kagiya, and T. Kagiya, *J. Polym. Sci.*, **A-1,8**, 3295 (1970).
40. G. T. Russell, D. H. Napper, and R. G. Gilbert, *Macromolecules*, **21**, 2133 (1988).
41. F. H. Solomon and G. Moad, *Makromol. Chem., Macromol. Symp.*, **10/11**, 109 (1987).
42. B. M. Cubertson, "Maleic and Fumaric Polymers," in *Encyclopedia of Polymer Science and Technology*, Vol. 9 (H. F. Mark, N. M. Bikalis, C. G. Overberger, and G. Menges, eds.), Wiley-Interscience, New York, 1987.
43. S. Zhu, Y. Tian, A. E. Hamielec, and E. R. Eaton, *Macromolecules*, **23**, 1144 (1990).
44. J. Lingman and G. Meyerhoff, *Macromolecules*, **17**, 941 (1984); *Makromol. Chem.*, **587** (1984).
45. G. T. Russell, R. G. Gilbert, and N. H. Napper, *Macromolecules*, **25**, 2459 (1992).
46. N. J. Barr, G. Bengough, G. Beveridge, and G. P. Park, *Eur. Polym. J.*, **14**, 245 (1978); W. B. Graham, J. G. Green, and W. A. Pryor, *J. Org. Chem.*, **44**, 907 (1979); A. Husain and A. E. Hameilec, *J. Appl. Polym. Sci.*, **22**, 1207 (1978); H. F. Kaufmann, *Makromol. Chem.*, **180**, 2649, 2665, 2681 (1979).
47. C. C. Price, *J. Polym. Sci.*, **3**, 772 (1978); C. E. Walling, E. R. Briggs, K. B. Wolfstern, and F. R. Mayo, *J. Am. Chem. Soc.*, **70**, 1537, 1544 (1948).
48. H. Hirai, *J. Polym. Sci., Macromol. Rev.*, **11**, 47 (1976); I. Kunz, N. F. Chamberlain, and F. J. Stelling, *J. Polym. Sci., Polym. Chem. Educ.*, **16**, 1747 (1978).
49. G. D. Russell, R. G. Gilbert, and N. H. Napper, *Macromolecules*, **25**, 2459 (1992).
50. A. Charlseby, Chap. 1. in *Irradiation of Polymers* (N. Platzer, ed.), Am. Chem. Soc., Advancement in Chem. Series #66, Washington, 1967.
51. J. E. Wilson, *Radiation Chemistry of Monomers, Polymers, and Plastics*, Dekker, New York, 1974.
52. J. Frank and E. Rabinowitch, *Trans. Faraday Soc.*, **30**, 120 (1934).
53. J. C. Bevington and J. Toole, *J. Polym. Sci.*, **28**, 413 (1958).
54. J. C. Bevington, *Radical Polymerization*, Academic Press, London, New York, 1961.
55. C. A. Barson, J. C. Bevington, and D. E. Eaves, *Trans. Faraday Soc.*, **54**, 1678 (1958).
56. R. J. Orr and H. L. Williams, *J. Am. Chem. Soc.*, **77**, 3715 (1955).
57. F. Carrick and M. Szwarc, *J. Am. Chem. Soc.*, **81**, 4138 (1959).
58. F. Leavitt, M. Levy, M. Szwarc, and V. Stannett, *J. Am. Chem. Soc.*, **77**, 5493 (1955).
59. A. R. Boder, R. P. Buckley, F. Leavitt, and M. Szwarc, *J. Am. Chem. Soc.*, **79**, 5621 (1957).
60. C. Walling and E. H. Huyser, *Organic Reactions*, Vol. 13 (A. C. Cope, ed.), Wiley, New York, 1963.
61. C. H. Bamford, W. G. Barb, A. D. Jenkins, and P. F. Onyon, *Kinetics of Vinyl Polymerization by Radical Mechanism*, Butterworth, London, 1958.
62. K. Hayashi, T. Yonezawa, C. Nagata, S. Okamura, and F. Fukii, *J. Polymer Sci.*, **20**, 537 (1956).
63. K. F. O'Driscoll and T. Yonezawa, *Reviews of Macromolecular Chemistry*, **1** (1), 1 (1966).
64. M. L. Huggins, *J. Am. Chem. Soc.*, **66**, 1991 (1961).
65. C. G. Fox, B. S. Garret, W. E. Goode, S. Gratch, J. F. Kinkaid, A. Spell, and J. D. Stroupe, *J. Am. Chem. Soc.*, **80**, 1968 (1958).
66. J. W. L. Fordham, *J. Polym. Sci.*, **39**, 321 (1959).
67. G. Natta, L. Porri, G. Zannini, and L. Piore, *Chim. Ind. (Milan)*, **41**, 526 (1959); *Chem. Abstr.*, **44**, 1258 (1960).
68. J. W. L. Fordham, P. H. Burleigh, and C. L. Sturm, *J. Polym. Sci.*, **42**, 73 (1959).
69. F. A. Bovey, *J. Polym. Sci.*, **46**, 59 (1960).
70. J. W. L. Fordham, G. H. McCain, and L. Alexander, *J. Polym. Sci.*, **39**, 335 (1959).
71. P. H. Burleigh, *J. Am. Chem. Soc.*, **82**, 749 (1960).
72. I. Rosen, P. H. Burleigh, and J. F. Gillespie, *J. Polym. Sci.*, **54**, 31 (1961).
73. G. M. Burnett and F. L. Ross, *J. Polym. Sci.*, **A-1, 5**, (1967).
74. G. M. Burnett, F. L. Ross, and J. N. Hay, *Makromol. Chem.*, **105**, 1 (1967).
75. T. Uryu, H. Shiroki, M. Okada, K. Hosonuma and K. Matsuzaki, *J. Polym. Sci.*, **A-2,9**, 2335 (1971).
76. S. Spinner, *J. Polym. Sci.*, **9**, 282 (1952).
77. G. Blauer, *J. Polym. Sci.*, **11**, 189 (1953).
78. F. C. De Schryver, G. Smets, and J. Van Thielen, *J. Polym. Sci., Polym. Letters*, **6**, 547 (1968).

79. M. Kamachi, J. Satoh, S. Nozakura, *J. Polym. Sci., Polym. Chem. Ed.*, **16** (8), 1789 (1978).
80. D. J. Cram and K. R. Kopecky, *J. Am. Chem. Soc.*, **81**, 2748 (1959).
81. D. J. Cram, *J. Chem. Ed.*, **37**, 317 (1960).
82. A. M. North and D. Postlethwaite, Chap. 4 in *Structure and Mechanism in Vinyl Polymerization* (T. Tsuruta and K. F. O'Driscoll, eds.), Dekker, New York, (1969).
83. S. R. Chatterjee, S. N. Kanna, and S. R. Palit, *J. Indian Chem. Soc.*, **41**, 622 (1964).
84. P. D. Bartlett and R. Altshul, *J. Am. Chem. Soc.*, **67**, 812 (1945).
85. C. H. Bamford and E. F. T. White, *Trans. Faraday Soc.*, **52**, 716 (1956).
86. C. Walling and Y. W. Chang, *J. Am. Chem. Soc.*, **76**, 4878 (1954).
87. T. Yamamoto, T. Yamamoto, A. Moto, and M. Hiroto, *Nippon Kagaku Kaishi*, (3), 408 (1979); *Chem. Abstr.*, **90** (12), Macromol. Section, 5 (June 1979).
88. D. H. Johnson, A. V. Tobolsky, *J. Am. Chem. Soc.*, **74**, 938 (1952).
89. G. H. Schulz, G. Henrici-Olive, *Z. Electrochem., Ber. Bunsenges Physik. Chem.*, **60**, 296 (1956).
90. G. Ayrey and A. C. Haynes, *Makromol. Chem.* **175**, 1463 (1974).
91. L. J. Young, *J. Polym. Sci.*, **54**, 411 (1961); *ibid.*, **62**, 515 (1962).
92. K. F. O'Driscoll and T. Yonezawa, *Reviews of Macromol. Chem.*, **1** (1), 1 (1966).
93. F. R. Mayo and C. Walling, *Chem. Rev.*, **46**, 191 (1950).
94. T. Alfrey, Jr. and C. C. Price, *J. Polym. Sci.*, **2**, 101 (1947).
95. T. C. Schwann and C. C. Price, *J. Polym. Sci.*, **40**, 457 (1959).
96. G. S. Levinson, *J. Polym. Sci.*, **60**, 43 (1962).
97. K. Hayashi, T. Yonezawa, S. Okamura, and K. Fukui, *J. Polym. Sci.*, **A-1,1**, 1405 (1963).
98. G. G. Cameron and D. A. Russell, *J. Macromol. Sci., Chem.*, **A5** (7), 1229 (1971).
99. R. C. Laible, *Chem. Rev.*, **58**, 807 (1958).
100. G. Odian, *Principles of Polymerization*, McGraw Hill, New York, 1970; 3rd ed., Wiley, New York, 1991.
101. I. M. Kolthoff and F. A. Bovey *J. Am. Chem. Soc.*, **69**, 2143 (1947).
102. F. R. Mayo, *J. Am. Chem. Soc.*, **75**, 6133 (1953).
103. G. B. Burnett and L. D. Loan, *Trans. Faraday Soc.*, **51**, 219 (1955).
104. J. R. Ebdon, *Br. Polym. J.*, **3**, 9 (1971).
105. Y. Tsuda, J. Sakai, and Y. Shinohara, IUPAC Int. Symp. on Macromol. Chem., Tokyo-Kyoto, 1966, Preprints, III, 44.
106. N. G. Gaylord, *Am. Chem. Soc., Polym. Prepr.*, **10** (1), 277 (1969).
107. N. G. Gaylord and A. Takahashi, Chap. 6 in *Addition and Condensation Polymerization Processes*, Advances in Chemistry Series #91, Am. Chem. Soc., Washington, 1969.
108. S. Iwatsuki, K. Nishio, and Y. Yamashito, *Kagyo Kagaku Zashi*, **70**, 384 (1967) (from Ref. 107).
109. S. Iwatsuki and Y. Yamashito, *J. Polym. Sci.*, **A-1,5**, 1753 (1967).
110. S. Iwatsuki and Y. Yamashito, *Makromol. Chem.*, **89**, 205 (1965).
111. S. Iwatsuki, M. Shin, and Y. Yamashita, *Makromol. Chem.*, **102**, 232 (1967).
112. N. L. Zutty, C. W. Wilson, G. H. Potter, D. C. Priest, and C. J. Whitworth, *J. Polym. Sci.*, **A-1,3**, 2781 (1965).
113. N. D. Field, *J. Am. Chem. Soc.*, **83**, 3504 (1961).
114. H. Gilbert, F. F. Miller, S. J. Averill, E. J. Carlson, V. L. Folt, H. J. Heller, F. J. Stewart, R. T. Schmidt, and H. L. Trumbull, *J. Am. Chem. Soc.*, **70**, 1669 (1956).
115. W. E. Hanford, U.S. Patent # 2,396,785 (March 19, 1946).
116. M. M. Martin and N. P. Jensen, *J. Org. Chem.*, **27**, 1201 (1962).
117. R. M. Thomas and W. J. Sparks, U.S. Patent # 2,373,067 (April 3, 1945).
118. L. K. Montgomery, K. Schueller, and P. D. Bartlett, *J. Am. Chem. Soc.*, **86**, 622 (1964).
119. E. M. Fetters and F. O. Davis, Chap. 15 in *Polyethers, Part III, Polyalkylene Sulfides and Other Polythioethers* (N. G. Gaylord, ed.), Interscience, New York, 1962.
120. C. H. Bamford, S. Brumby, and R. P. Wayne, *Nature*, **209**, 292 (1966).
121. M. Imoto, T. Otsu, and Y. Harada, *Makromol. Chem.*, **65**, 180 (1963).
122. M. Imoto, T. Otsu, and S. Shimizu, *Makromol. Chem.*, **65**, 114 (1963).
123. M. Imoto, T. Otsu, and M. Nakabayashi, *Makromol. Chem.*, **65**, 194 (1963).
124. S. Tazuke and S. Okamura, *J. Polym. Sci.*, **B5**, 95 (1967).
125. S. Tazuke, K. Tsuji, T. Yonezawa, and S. Okamura, *J. Phys. Chem.*, **71**, 2957 (1967).
126. V. P. Zubov, M. B. Lachinov, V. B. Golubov, V. P. Kulikova, V. A. Kabanov, L. S. Polak, and V. A. Kargin, IUAPC Intern. Symp. on Macromol. Chem., Tokyo-Kyoto, Japan, **2**, 56 (1966).
127. S. Tazuke, *Adv. Polym. Sci.*, **6**, 321 (1969).
128. T. Ikegami and H. Hirai, *J. Polym. Sci.*, **A-1,8**, 195 (1970).
129. N. G. Gaylord and B. Patnaik, *J. Polym. Sci., Polym. Letters*, **8**, 411 (1970).
130. Y. Nakayama, K. Hayashi, and S. Okamura, *J. Macromol Sci., Chem.*, **A2**, 701 (1968).
131. S. Iwatsuki and Y. Yamashita, *J. Polym. Sci.*, **A-1,5**, 1753 (1967).

132. N. G. Gaylord and B. K. Patnaik, *J. Macromol. Sci., Chem.*, **A5** (7), 1239 (1971).
133. N. G. Gaylord, *J. Macromol. Sci., Chem.*, **A6** (2), 259 (1972).
134. S. C. Foord, *J. Chem. Soc.*, **68**, 48 (1950).
135. B. L. Funt and F. D. Williams, *J. Polym. Sci.*, **46**, 139 (1960).
136. J. C. Bevington, N. A. Ghanem, and H. W. Melville, *J. Chem. Soc.*, 2822 (1955).
137. J. W. Breitenbach, A. Springer, and K. Hoseichy, *Ber.*, **71**, 1438 (1938).
138. R. G. Caldwell and J. L. Ihrig, *J. Polym. Sci.*, **46**, 407 (1960); *J. Am. Chem. Soc.*, **84**, 2878 (1962).
139. G. V. Schulz and G. Henrice, *Makromol. Chem.*, **18**, 437 (1956).
140. R. G. Caldwell and J. L. Ihrig, *J. Am. Chem. Soc.*, **84**, 2878 (1962).
141. E. T. Burrows, *J. Appl. Chem.*, **5**, 379 (1955).
142. S. S. Harihan and M. Maruthamuthu, *Makromol. Chem.*, **180**, 2031 (1979).
143. W. J. le Noble, *Highlights of Organic Chemistry*, Dekker, New York, 1974.
144. J. M. G. Cowie, "Alternating Copolymerization," in *Comprehensive Polymer Science*, Vol. 4 (G. Eastman, A. Ledwith, S. Russo, and P. Sigwaldt, eds.), Pergamon Press, Oxford, 1989.
145. J. Furukawa, "Alternating Copolymers" in *Encyclopedia of Polymer Science and Engineering*, Vol. 4, 2nd ed. (H. F. Mark, N. M. Bikales, C. G. Overberger, and G. Menges, eds.), Wiley-Interscience, New York, 1986.
146. H. G. Heine, H. J. Rosenkranz, and H. Rudolph, *Angew. Chem., Int. Ed. Engl.*, **11** (11), 974 (1972).
147. D. C. Blakely, *Emulsion Polymerization, Theory and Practice*, Applied Science, London, 1965.
148. D. R. Bassett and A. E. Hamielec (eds.), *Emulsion Polymers and Emulsion Polymerization*, Am. Chem. Soc. Symp. #165, Washington, 1981.
149. H. Hutchinson, M. C. Lambert, and A. Ledwith, *Polymer*, **14**, 250 (1973).
150. M. Munzer and E. Trammsdorff, "Polymerization in Suspension," Chap. 5 in *Polymerization Processes* (C. E. Schieldknecht, (with I. Skeist) ed.), Wiley-Interscience, New York, 1977.
151. J. M. Tedder, "Reactivity of Free-Radicals," Chap. 2 in *Reactivity, Mechanism, and Structure in Polymer Chemistry* (A. D. Jenkins and A. Ledwith, eds.), Wiley-Interscience, New York, 1974.
152. W. A. Pryor *Free Radicals*, McGraw-Hill, New York, 1966.
153. J. C. Martin, "Solvation and Association," Chap. 20 in *Free-Radicals*, Vol. II (J. K. Kochi, ed.), Wiley, New York, 1973.
154. R. J. Glitter and R. J. Albers, *J. Org. Chem.*, **29**, 728 (1964).
155. M. Imoto, T. Otsu, B. Yamada, and A. Shimizu, *Makromol. Chem.*, **82**, 277 (1965).
156. M. Jacob, G. Smets, and F. de Schryver. *J. Polym. Sci.*, **A-1,10**, 669 (1972).
157. G. G. Cameron and G. F. Esslemont, *Polymer*, **13**, 435 (1972).
158. G. Bonta, B. M. Gallo, and S. Russo, *Polymer*, **16**, 429 (1975).
159. A. D. Jenkins and M. G. Rayner, *Eur. Polym. J.*, **8**, 221 (1972).
161. R. E. Ushold, *Macromolecules*, **4**, 552 (1971).
162. S. Yabumoto, K. Ishii, and K. Arita, *J. Polym. Sci.*, **A-1,7**, 1577 (1969).
163. M. Hirooka, *J. Polym. Sci., Polym. Letters*, **10**, 171 (1972).
164. H. Hirai, *Macromol. Rev.*, **11**, 47 (1976).
165. G. E. Scott and E. Serrogles, *J. Macromol. Sci., Rev. of Macromol. Chem.*, **C9**(1), 49 (1973); C. H. Bamford, "Radical Polymerization," in *Encyclopedia of Polymer Science and Engineering*, Vol. 13 (H. F. Mark, N. M. Bikales, C. G. Overberger, and G. Menges. eds.), Wiley-Interscience, New York, 1988.
166. N. F. Phelan and M. Orchin, *J. Chem. Educ.*, **45**, 633 (1968).
167. G. B. Butler and R. J. Angelo, *J. Am. Chem. Soc.*, **79**, 3128 (1957).
168. P. J. Graham, E. L. Buhle, and N. J. Pappas, *J. Org. Chem.*, **26**, 4658 (1961).
169. R. H. Wiley, W. H. Rivera, T. H. Crawford, and N. F. Bray, *J. Polym. Sci.*, **61**, 538 (1962); *J. Polym. Sci.*, **A-1,2**, 5025 (1964).
170. N. L. Zutty, *J. Polym. Sci.*, **A-1,1**, 2231 (1963).
171. J. E. Fearn, D. W. Brown, and L. A. Wall, *J. Polym. Sci.*, **A-1,4**, 131 (1966).
172. C. S. Marvel and R. D. Vest, *J. Am. Chem. Soc.*, **79**, 5771 (1957).
173. C. S. Marvel and R. D. Vest, *J. Am. Chem. Soc.*, **81**, 984 (1959).
174. R. J. Cotter and M. Matzner, *Ring-Forming Polymerizations*, Academic Press, New York, 1969.
175. G. B. Butler and M. A. Raymond, *J. Polym. Sci.*, **A-1,3**, 3413 (1965).
176. G. B. Butler, *J. Polym. Sci.*, **48**, 279 (1960).
177. G. B. Butler, *Am. Chem. Soc., Polym. Prepr.*, **8** (1), 35 (1967).
178. G. B. Butler and M. A. Raynolds, *J. Org. Chem.*, **30**, 2410 (1965).
179. G. B. Butler, T. W. Brooks, *J. Org. Chem.*, **28**, 2699 (1963).
180. B. Yamada, H. Kamei, and T. Otsu, *J. Polym. Sci., Polym. Chem. Educ.*, **18**, 1917 (1980).
181. V. A. Kabanov, *J. Polym. Sci., Polym. Symposia*, **67**, 17 (1980).
182. D. M. White, *J. Am. Chem. Soc.*, **82**, 5678 (1960).
183. J. F. Brown, Jr. and D. M. White, *J. Am. Chem. Soc.*, **82**, 5671 (1960).

184. A. Blumstein, R. Blumstein, and T. H. Vanderspurt, *J. Colloid. Sci.*, **31**, 237 (1969).
185. A. Blumstein, S. L. Malhotra, and A. C. Wattson, *J. Polym. Sci.*, **A-1, 8**, 1599 (1970).
186. A. Blumstein, K. K. Parikh, and S. L. Malhotra, 23rd Int. Conf. of Pure Appl. Chem., Macromol. Prepr., **1**, 345 (1971).
187. M. Farina, U. Pedretti, M. T. Gramegna, and G. Audisio, *Macromolecules*, **3** (5), 475 (1970).
188. A. Colombo and G. Allegra, *Macromolecules*, **4** (5), 579 (1971).
189. R. Buter, Y. Y. Tan, and G. Challa, *J. Polym. Sci.*, **A-1,10**, 1031 (1972); *ibid.*, **A-1,11**, 1013, 2975 (1973).
190. J. Gons, E. J. Vorenkamp, and G. Challa, *J. Polym. Sci., Polym. Chem. Educ.*, **13**, 1699 (1975); **15**, 3031 (1977).
191. S. R. Clarke and R. A. Shanks, *J. Macromol. Sci., Chem.*, **A14** (1), 69 (1980).
192. E. Jenckel and S. Suss, *Naturwissenschaflen*, **29**, 339 (1931).
193. W. R. Sorenson and T. W. Campbell, *Preparative Methods of Polymer Chemistry*, 2nd ed., Wiley-Interscience, New York, 1968.
194. E. H. Riddle, *Monomeric Acrylic Esters*, Reinhold, New York, 1961.
195. J. Coupek, M. Kolinsky, and D. Lim, *J. Polym. Sci.*, **C4**, 1261 (1964).
196. R. F. Hoffmann, S. Schreiber, and G. Rosen, *Ind. Eng. Chem.*, **56**, 51 (1964).
197. A. Ravve, J. T. Khamis, and L. X. Mallavarapu, *J. Polym. Sci.*, **A-1,3**, 1775 (1965).
198. A. Ravve and J. T. Khamis, Chap. 4 in *Addition and Condensation Polymerization Processes*, Am. Chem. Soc. Advances in Chemistry Series # 91 (R. Gould, ed.), Washington, 1969.
199. N. N. Semenov, *J. Polym. Sci.*, **55**, 563 (1961).
200. W. D. Harkins, *J. Am. Chem. Soc.*, **69**, 1428 (1947).
201. W. V. Smith and R. H. Ewart, *J. Chem. Phys.*, **16**, 592 (1948).
202. B. M. E. Van der Hoff, Am. Chem. Soc., Advances in Chemistry Series, #34, Washington, 1962.
203. P. J. Flory, *Principles of Polymer Chemistry*, Cornell Univ. Press, 1953.
204. W. V. Smith, *J. Am. Chem. Soc.*, **70**, 3695 (1948).
205. H. J. van den Hul and J. W. Vanderhoff, Symposium Preprints, University of Manchester, Sept. 1969.
206. S. S. Medvedev, *International Symposium on Macromolecular Chemistry*, Pergamon Press, New York, 1959; S. S. Medvedev, *Ric. Sci., Suppl.*, **25**, 897 (1955).
207. M. R. Grancio and D. J. Williams, *J. Polym. Sci.*, **A-1,8**, 2617 (1970); C. S. Chern and G. W. Poehlein, *J. Polym. Sci., Polym. Chem. Educ.*, **25**, 617 (1987).
208. J. T. Asakura, *J. Macromol. Sci., Chem.*, **A15** (8), 1473 (1981).
209. T. L. Simandi, A. Rockenbauer, and F. Tudos, *Eur. Polym. J.*, **18**, 67 (1982).
210. G. Smets and K. Hayashi, *J. Polym. Sci.*, **27**, 626, (1958).
211. N. D. Field and J. R. Schaefgen, *J. Polym. Sci.*, **58**, 533, (1962).
212. M. Yamada, I. Takase, *Kobunshi Kagaku*, **22**, 626 (1965); *Chem. Abstr.*, **64**, 19803g (1966).
213. W. I. Bengough, G. B. Park, and R. Young, *Eur. Polym. J.*, **11**, 305 (1975).
214. T. Otsu, O. Ito, N. Toyoda, and S. Mori, *Macromol. Chem., Rapid Commun.*, **2**, 725 (1981).
215. T. Otsu, O. Ito, and N. Toyoda, *Makromol. Chem., Rapid Commun.*, **7**, 729 (1981).
216. V. P. Zubov, M. B. Lachinov, E. V. Ignatova, G. S. Georgiev, V. B. Golubev, and V. A. Kabanov, *J. Polym. Sci., Polym. Chem. Educ.* **20**, 619 (1982).
217. F. Tudos, *Acta Chim. Budapest*, **43**, 397; **44**, 403 (1965).
218. A. Fehervari, T. Foldes-Berzsnich, and F. Tudos, *Eur. Polym. J.*, **16**, 185 (1980).
219. W. Wunderlich, *Makromol. Chem.*, **177**, 973 (1976).
220. K. Yokota and M. Itoh, *J. Polym. Sci., Polym. Letters*, **6**, 825 (1968).
221. B. Giese, G. Kretzschmar, and J. Meixner, *Chem. Ber.*, **113**, 2787 (1980).
222. B. Giese and J. Meixner, *Angew. Chem., Int. Ed.*, **18**, 154 (1979).
223. B. Giese and K. Heuck, *Chem. Ber.*, **112**, 3759 (1979).
224. B. Giese and K. Heuck, *Tetrahedron Lett.*, **21**, 1829 (1980).
225. B. Giese and J. Meixner, *Angew. Chem., Int. Ed. Engl.*, **16**, 178 (1977).
226. B. Giese and W. Zwick, *Angew. Chem., Int. Ed. Engl.*, **17**, 66 (1978).
227. B. Giese and J. Meixner, *Chem. Ber.*, **110**, 2588 (1977).
228. S. A. Jones and D. A. Tirrell, *Am. Chem. Soc. Polym. Prepr.*, **24** (2), 24 (1983).
229. J. M. Dionisio, H. K. Mahabadi, J. H. O'Driscoll, E. Abuin, and E. A. Lissi, *J. Polym. Sci., Polym. Chem. Educ.*, **17**, 1891 (1979); J. M. Dionisio and J. H. O'Driscoll, *J. Polym. Sci., Polym. Chem. Educ.*, **18**, 241 (1980); R. Sack, G. V. Schulz, and G. Meyerhoff, *Macromolecules*, **21**, 3345 (1988).

3

Ionic Chain-Growth Polymerization

3.1. Chemistry of Ionic Chain-Growth Polymerization

Ionic polymerization can be either cationic or anionic. This difference stems from the nature of the carrier ions on the growing polymeric chains. If, in the process of growth, the chains carry positive centers, or carbon cations, the mechanism of chain growth is designated as *cationic*. On the other hand, if the growing chains carry negative ions, or carbanions, then the polymerization is designated as *anionic*.

The two different modes of polymerization can be compared further by examining the two types of initiation. The cationic ones occur by electrophilic attacks of the initiators on the monomers. Conversely, the anionic ones take place by nucleophilic attacks. The two reactions can be shown as follows:

Type of Reaction	Reagent		Monomer		Transition State
Electrophilic addition	$R^{\oplus}$	+	$CH_2{=}CHX$	$\rightarrow$	$R{-}CH_2{-}CHXR^{\oplus}$
Nucleophilic addition	$R^{\ominus}$	+	$CH_2{=}CHX$	$\rightarrow$	$R{-}CH_2{-}CHXR^{\ominus}$

Generally, the ionic polymerizations depend upon initial formations of positive and negative ions in an organic environment. Propagations in these chain reactions take place through successive additions of monomeric units to the charged or "reactive" terminal groups of the propagating chains.[1]

The transition states for the steps of propagation are formed repeatedly in liquid medium systems, containing monomer, initiator, the formed polymer, and, frequently a solvent. There are many different types of initiating reactions. These polymerization, however, never terminate by combination or by disproportionation as they do in free-radical chain-growth polymerizations. Instead, terminations of chain growths are results of unimolecular reactions, or transfers to other molecules, like monomers or solvents, or impurities, like moisture. They can also result from quenching by deliberate additions of reactive terminating species.

3.2. Kinetics of Ionic Chain-Growth Polymerization

The kinetic picture of *cationic chain polymerization* varies considerably. Much depends upon the mode of termination in any particular system. A general scheme for initiation, propagation, and termination is presented below.[96] By representing the coinitiator as A, the initiator as RH, and the monomer as M, we can write:

Cationic Initiation Process: *Rate of Initiation:*

$$A + RH \xrightleftharpoons{K} H^{\oplus}\ldots(AR)^{\ominus} \qquad R_I = KK_I[A][RH][M]$$

$$H^{\oplus}\ldots(AR)^{\ominus} + M \xrightarrow{K_i} HM^{\oplus}\ldots(AR)^{\ominus}$$

Cationic Propagation Process: *Rate of Propagation:*

$$HM_n^{\oplus}\ldots(AR)^{\ominus} + M \xrightarrow{K_p} HM_nM^{\oplus}\ldots(AR)^{\ominus} \qquad R_P = K_P[HM^{\oplus}\ldots(AR)^{\ominus}][M]$$

Cationic Termination: *Rate of Termination:*

By transfer:

$$R_T = K_T HM^{\oplus}\ldots(AR)^{\ominus}]$$

$$HM_nM^{\oplus}\ldots(AR)^{\ominus} + \xrightarrow{K_{tr}M} M_{n+1} + HM^{\oplus}\ldots(AR)^{\ominus}$$

By spontaneous termination:

$$HM_nM^{\oplus}\ldots(AR)^{\ominus} \xrightarrow{K_{ts}} M_{n+1} + H^{\oplus}\ldots(AR)^{\ominus}$$

By rearrangement of the kinetic chain:

$$HM_nM^{\oplus}\ldots(AR)^{\ominus} \longrightarrow HM_nMAR$$

In the above expressions, $[HM^{\oplus}\ldots(AR)^{\ominus}]$ represents the total concentration of all the propagating ion pairs; K is the equilibrium constant for formation of initiating cations. When a steady state exists, the rates of initiation and termination are equal to each other:

$$[HM^{\oplus}\ldots(AR)^{\ominus}] = (KK_I[A][RH][M])/K_T$$

The rate of propagation can be written as

$$R_P = KK_IK_P[A][RH][M]^2/K_T$$

The number average degree of polymerization is equal to

$$\overline{DP} = \frac{R_P}{R_T} = \frac{K_P[M]}{K_T}$$

During the reaction, terminations can occur by transferring to a monomer or to a chain transferring agent S. Termination can also occur by spontaneous termination or by combination with the counterion. If S is relatively small, the number of propagating chains will remain unchanged. The degree of polymerization will then be decreased by the chain breaking reactions. The number average degree of polymerization becomes

$$\overline{DP} = \frac{R_P}{R_T + R_{ts} + R_{tr,M} + R_{tr,S}}$$

The rates of the two transfer reactions, to a monomer or to a transferring agent (S), as well as spontaneous terminations can be written as

$$R_{tr,M} = K_{tr,M}[HM^{\oplus}\ldots(AR)^{\ominus}][M]$$

$$R_{tr,S} = K_{tr,S}[HM^{\oplus}\ldots(AR)^{\ominus}][S]$$

$$R_{ts} = K_{ts}[HM^{\oplus}\ldots(AR)^{\ominus}]$$

The number average degree of polymerization is then

$$\overline{DP} = K_P[M]/(K_T + K_{tr,M} + K_{tr,S}[S] + K_{ts})$$

The above can be rearranged to be written as follows:

$$1/\overline{DP} = K_T/K_P[M] + K_{ts}/K_P[M] + C_M + C_S[S]/[M]$$

where C_M and C_S are the chain transferring constants for the monomer and S is the chain transferring agent. Here $C_M = K_{tr,M}/K_P$ and $C_S = K_{tr,S}/K_P$. When a chain transferring agent is present and chain growth is terminated,

$$R_P = KK_IK_P[A][RH][M]^2/(K_T + K_{tr,S}[S])$$

The validity of the steady-state assumption in many cationic polymerizations may be questioned, because many reactions occur at such high rates that a steady state is not achieved. Nevertheless, the above equations were shown to be generally followed.[350,351]

The kinetic picture of *anionic chain polymerization* also depends mostly upon the specific reaction. For those that are initiated by metal amides in liquid ammonia, the rate of initiation can be shown to be as follows:

Anionic Initiation Process	*Rate of Initiation*
$KNH_2 \overset{K}{\rightleftharpoons} K^{\oplus} + NH_2^{\ominus}$	$R_I = K_I[H_2N{:}^{\ominus}][M]$
$NH_2^{\ominus} + M \xrightarrow{K_I} H_2N{-}M^{\ominus}$	$R_I = KK_I[M][KNH_2]/[K^{\oplus}]$
Anionic Propagation Process	*Rate of Propagation*
$H_2N{-}M_n^{\ominus} + M \longrightarrow H_2N{-}M_nM^{\ominus}$	$R_P = K_P[M^{\ominus}][M]$

where $[M^{\ominus}]$ represents the total concentration of propagating anions. If there is a termination reaction and it takes place by transfer to a solvent, the rate of transfer can be written

$$R_{tr} = K_{tr,S}[M^{\ominus}][NH_3]$$

When steady-state conditions exist,

$$R_P = \frac{KK_IK_P[M]^2[KNH_2]}{K_{tr,S}[K^{\oplus}][NH_3]}$$

If the quantities of potassium and amide ions are equal, which is the normal situation, then a somewhat different expression can be written:

$$R_I = K_iK^{1/2}[M][KNH_2]^{1/2}$$

$$R_P = \frac{K_IK^{1/2}[M]^2[KNH_2]^{1/2}}{K_{tr,S}[NH_3]}$$

In nonterminating or "living" polymerizations, the rate of propagation can be written

$$R_P = K_P[M^{\ominus}][M]$$

When the effect of the medium is such that the propagation takes place both with ion pairs and with free-propagating ions, the rate of propagation is written as follows[352,353]:

$$R_P = K_P{}^{\ominus}\ldots C^{\oplus}[P^{\ominus}\ldots C^{\oplus}][M] + K_P{}^{\ominus}[P^{\ominus}][M]$$

where $K_P{}^{\ominus}\ldots C^{\oplus}$ and $P^{\ominus}$ are rate constants for the propagation of the ion pairs and the free ions, respectively, $C^{\oplus}$ is the positive counterion, $P^{\ominus}\ldots C^{\oplus}$ and $P^{\ominus}$ are the concentrations of the two propagating species, and [M] is the monomer concentration.

The number average degree of polymerization of a living polymer is shown as follows:

$$\overline{DP} = [M]/[M^{\ominus}]$$

It is simply a ratio of the concentration of the monomer to the number of living ends.

A model was developed for a unified treatment of the kinetics of both cationic and anionic polymerizations.[359] It is based on a system of kinetic models that cover various initiation and propagation mechanisms and on a pseudosteady-state assumption. The treatment is beyond the scope of this book.

The principal kinetics of propylene polymerization with a magnesium chloride supported Ziegler–Natta catalyst was also developed.[362] The polymerization rate is described by a Langmuir–Hinshelwood equation showing the dependence of the rate on the concentration of the aluminum alkyl:

$$R_P = KK_A[A]/(1 + K_A[A])^2$$

where $[A] = [AlR_3]$.

Because the polymerization rate is first order with respect to the concentration of the monomer, the rate equation can be written as follows[362]:

$$R_P = K_P[M]K_A[A]/(1 + K_A[A])^2$$

3.3. Cationic Polymerization

As mentioned, in cationic polymerizations the reactive portions of the chain ends carry positive charges during the process of chain growth. These active centers can be either unpaired cations or they can be cations that are paired and associated closely with anions (counterions).

The initiations result from transpositions of electrons, either a pair or a single one. A two-electron transposition takes place when the initiating species are either protons or carbon cations. Ion generations take place through heterolytic bond cleavages or through dissociations of cationic precursors. Such initiating systems include Lewis acid/Bronsted acid combinations, Bronsted acids by themselves, stable cationic salts, some organometallic compounds, and some cation forming substances.

When, however, initiations take place by one-electron transposition, they occur as a direct result of oxidation of free radicals. They can also take place through electron transfer interactions involving electron donor monomers. The carbon cations can form from olefins in a variety of ways. One way is through direct additions of cations, like protons, or other positively charged species to the olefins. The products are secondary or tertiary carbon cations:

$$R^{\oplus} + CH_2{=}C(H)(Y) \longrightarrow R{-}CH_2{-}C^{\oplus}(H)(Y)$$

When the cations add to conjugated dienes, the charges can be distributed over several centers in the products:

$$R^{\oplus} + CH_2{=}CH{-}CH{=}CH_2 \longrightarrow \begin{cases} R{-}CH_2{-}\overbrace{CH{-}CH{-}CH_2}^{\oplus} \\ R{-}CH_2{-}CH^{\oplus}(CH{=}CH_2) \end{cases}$$

The charge may also be distributed in polar monomers like, for instance, vinyl ethers:

$$R^{\oplus} + CH_2{=}CH{-}OR' \longrightarrow R{-}CH_2{-}\overbrace{CH{-}O}^{\oplus}{-}R'$$

Chemical considerations indicate that the more diffuse the charges, the more stable are the ions.

Cationic polymerizations are not affected by common inhibitors of free-radical polymerizations. They can, however, be greatly influenced by impurities that can act as ion scavengers. These can be water, ammonia, amines, or any other compound that can be basic in character, affecting rates and molecular weights of the products. Typical cationic polymerizations proceed at high rates even at low temperatures, as low as –100 °C.[48] In the literature one can find many reports of cationic polymerizations of many different monomers with many different initiators. Often, however, such initiators are quite specific for individual monomers and their activities are strongly influenced by the solvents.

The initiation processes can be summarized as follows:

I. *Initiation by Chemical Methods*
  A. Two electron transposition (heterolytic) initiation by reactions with:
    a. Protonic acids,
    b. Lewis acids,
    c. Stable carbon cations,
    d. Certain metal alkyls,
    e. Cation forming substances.
  B. One-electron transposition (homolytic) initiations by reactions of:
    a. Direct oxidation of radicals,
    b. Cation radicals formed through charge transfers.

II. *Initiation by Physical Methods*
  A. Photochemically generated cations,
  B. High-energy irradiation,
  C. Electroinitiation.

3.3.1. Two-Electron Transposition Initiation Reactions

The step of cationic initiation can be subdivided into two separate reactions.[325] The first one consists of formation of ionic species and the second one of reactions of these ionic species with the olefins, a cationization process. This reaction, termed "priming" by Kennedy and Marechal,[325] is a process of ion formation in a non-nucleophilic media through: (1) dissociation of protonic acids to form protons and counterions, (2) reactions of Lewis acids with Bronsted acids, (3) dissociation of dimeric Lewis acids, (4) complexation of Lewis acids with water or with alkyl halides or with ethers, and so on. These reactions may take place through a series of complicated steps. The second reaction, the cationization of the olefins, may also include several intermediate steps that will eventually lead to propagating species.

3.3.1.1. Initiation by Protonic Acids

The ability of a proton from a protonic acid to initiate a polymerization depends upon the nucleophilicity of the conjugate base,[12,13] $A^{\ominus}$. If it is low, protonation or cationization of the olefin is a step of initiation[50]:

$$\mathrm{HA} \rightleftharpoons \mathrm{H}^{\oplus} + \mathrm{A}^{\ominus}$$

$$\mathrm{H}^{\oplus}\,\mathrm{A}^{\ominus} + \mathrm{CH_2{=}C(R)(R')} \longrightarrow \mathrm{H_3C{-}C^{\oplus}(R)(R')\cdots A^{\ominus}}$$

In protonic acids, where the complex anion assists ion generation, the acid may be considered as providing its own counterion, $A^{\ominus}$. When such counterions are highly nucleophilic, then complexing agents such as metal salts or metal oxides must be used to immobilize the anions. Such complexing agents are in effect coinitiators.

Thus, the differences in activities of protonic acids are due to the quality of the corresponding anion or to its tendency to form chemical bonds with the carbon cation. If the anion is unable to form such bonds without extensive regrouping or decomposition, the addition of the proton is followed by polymerization. Should the reactivity of the anion be suppressed by solvation, the tendency to polymerize is enhanced. Consequently, the efficiencies of protonic acids depend very much upon the polarities of the media and upon the reaction conditions.[30,50] Also, the stronger the protonic acid, the higher the reaction rate and the resultant degree of polymerization.[52] Generally, hydrogen halide acids do not initiate polymerizations of alkyl-substituted olefins. They may, however, initiate polymerizations of aryl-substituted olefins and vinyl ethers in polar solvents. The same is true of sulfuric acid.[51]

3.3.1.2. Lewis Acids in Cationic Initiations

Complexation of Lewis acids with water is another case of formation of electrophiles that can initiate chain growth. MeX_n represents a Lewis acid:

$$MeX_n + H_2O \rightleftharpoons H^{\oplus} \ldots MeX_n\,OH^{\ominus}$$

$$\overset{\oplus}{H}\text{---}MeX_nOH^{\ominus} + CH_2{=}C(R)(R') \longrightarrow H{-}CH_2{-}\overset{\oplus}{C}(R)(R')\text{---}MeX_nOH^{\ominus}$$

$$H{-}CH_2{-}\overset{\oplus}{C}(R)(R')\text{---}MeX_nOH^{\ominus} + n\,CH_2{=}C(R)(R') \longrightarrow \text{polymer}$$

The above can be shown on boron trifluoride:

$$BF_3 + H_2O \rightleftharpoons BF_3OH^{\ominus} \ldots H^{\oplus}$$

$$\overset{\ominus}{BF_3OH}\text{----}\overset{\oplus}{H} + CH_2{=}C(CH_3)(CH_3) \longrightarrow H{-}CH_2{-}\overset{\oplus}{C}(CH_3)(CH_3)\text{----}BF_3OH^{\ominus}$$

In the same manner, a reaction of a Lewis acid with an alkyl halide and subsequent initiation can be illustrated as follows:

$$\underset{\text{(Lewis acid)}}{MeX_n} + RX \rightleftharpoons \overset{\ominus}{MeX_{n+1}}\text{-----}\overset{\oplus}{R}$$

$$\overset{\ominus}{MeX_{n+1}}\text{-----}\overset{\oplus}{R} + CH_2{=}C(R)(R') \longrightarrow R{-}CH_2{-}\overset{\oplus}{C}(R)(R')\text{---}MeX_{n+1}^{\ominus}$$

In the above examples, the Lewis acid is actually the *coinitiator* and the water or the alkyl halide the *initiators*.

Numerous studies confirm the need by many Lewis acids for other molecules, like water or alkyl halides, to form initiating ions. For instance, pure $TiCl_4$ will not initiate the polymerization of isobutylene.[15,326] This led many to believe that none of the Lewis acids is capable of initiating cationic polymerizations of olefins by itself.[327] Subsequent investigations, however, demonstrated[2–9] that some strong Lewis acids are capable of initiating such polymerizations.

Whether a Lewis acid is capable of initiating these polymerizations by itself was tested, with the aid of an early discovery. It has been known for some time that hindered bases like crowded pyridine derivatives exhibit specificity toward reactions with protons.[360] Such bases might be used to discriminate between two types of initiating mechanisms encountered with Lewis acids.[321] The base will not interfere with direct electrophilic additions of the Lewis acids to the monomer. On the other hand, it should prevent an initiation process by protons from taking place. If both pathways are operative, then the pyridine derivatives can only quench the protonic initiation and will offer a means of assessing the relative importance of each process.

Kennedy and coworkers investigated polymerizations initiated by Lewis acids–water in the presence of one such "proton trap," 2,6-di-*t*-butylpyridine.[361] The presence of this base markedly decreased conversion from almost 100%, in some cases, to only a few percent. At the same time, there was a marked increase in the molecular weight and a narrowing of the molecular weight distribution of the products. This shows that the protonic reaction is by far the most important mode of initiation. It also suggested to Kennedy and coworkers that not all the protonic initiations occur exclusively by free protons. They might also proceed through concerted protonations. These would not be blocked by the "proton traps"[361]:

$$\underset{\substack{|\\X}}{X{-}\overset{\substack{H\diagdown\ \diagup H\\O\\\downarrow}}{Me}{-}X} + \underset{CHR}{\overset{CH_2}{\|}} \longrightarrow \left[\underset{\substack{|\\X}}{X{-}\overset{\substack{H\ \ H\\O}}{Me}{-}X}\cdots\underset{CHR}{\overset{CH_2}{\|}}\right] \longrightarrow \left[\underset{\substack{|\\X}}{X{-}\overset{\substack{H\\O}}{Me}{-}X}\right]^{\ominus} \underset{\oplus}{\overset{CH_3}{\overset{|}{CH}}}{-}R$$

where MeX_3 represents a Lewis acid. The di-*t*-butylpyridine does not block the above reaction due to steric compression. It also suggested that protonic initiations by water in the presence of Lewis acids are probably unlikely when sufficient quantities of materials that act as "proton traps" are present. This was felt to be true even in polar media.[361]

Sigwalt and coworkers, however, disputed that.[369] They based their opinions on cationic dimerizations of 1,1-diphenylethylene with $AlCl_3$–H_2O in the presence of 2,6-di-*t*-butyl-4-methylpyridine. The results suggested to them that sterically hindered amines do not inhibit cocatalytic initiation by $AlCl_3$–H_2O; also, that the free hindered amine is a powerful terminating agent and that the sterically hindered pyridine forms a strong complex with $AlCl_3$. When engaged in a complex with $AlCl_3$ the reactivity of the hindered amine in termination is much reduced.[369]

Alkyl halide solvents in the polymerization reaction mixture may cause an initiation to result from a transfer of $AlX_2^{\oplus}$ or $(RX)AlX_2^{\oplus}$ to the more basic monomer[14]:

$$2\,AlX_3 + 2\,RX \rightleftharpoons AlX_2\cdot(RX)_2^{\oplus} + AlX_4^{\ominus}$$

$$(RX)_2AlX_2^{\oplus} + \underset{\text{(monomer)}}{M} \rightarrow RX\cdot X_2AlM^{\oplus} + RX$$

where X = halogen; R = alkyl.

In summarizing the initiation mechanism by complexed Lewis acids,[331] the catalysts must be present in the reaction mixture in three forms: (1) as ionized molecules, (2) as ion pairs, and (3) as free ions. All three forms are in equilibrium. When the monomer is introduced, it complexes with some of the ionized molecules and a new equilibrium is established. Additional ion pairs form by a slow process. This depends upon a change in the monomer concentration.[12]

Many mechanisms were offered to explain the reaction paths of initiations by aluminum halides and by some other Lewis acids without any co-reactant. Some more prominent ones follow.

One mechanism[8] is based on an observation that olefins with allylic hydrogens like isoprene, methylstyrene, indene, and cyclopentadiene can be polymerized by superdry, pure Lewis acids alone. This led to a suggestion that the process may involve an allylic self-initiation[8]:

$$>C{=}\overset{|}{C}{-}\underset{|}{\overset{|}{C}}{-}H \quad + \quad MeX_n \rightleftharpoons >C{=}\overset{|}{C}{-}\underset{|}{\overset{|}{C}}\oplus\cdots\cdots MeX_nH^{\ominus}$$

The above mechanism can only apply to polymerizations of monomers with allylic hydrogens. Also, this is contradicted by the ability of 1,1-diphenylethylene to dimerize in a superdry system in the presence of aluminum halides. Yet, this compound lacks allylic hydrogens.[328]

Two other, very similar mechanisms[9] are based on a concept that initiation takes place by a process of halometalation:

$$\underset{\text{(Lewis acid)}}{MeX_n} \quad + \quad CH_2{=}C\begin{smallmatrix}R\\R'\end{smallmatrix} \longrightarrow X_{n-1}Me{-}CH_2{-}C\begin{smallmatrix}R\\ \!-X\\R'\end{smallmatrix}$$

Depending upon the substituents on the olefins, the metalorganic compound may ionize in the presence of a second molecule of a Lewis acid, or it may eliminate HX:

$$X_{n-1}Me-CH=C(R)(R') \xleftarrow{-HX} X_{n-1}Me-CH_2-C(R)(R')-X \xrightarrow{MeX_n} X_{n-1}Me-CH_2-C^{\oplus}(R)(R')\cdots\cdots MeX_{n+1}^{\ominus}$$

The above mechanism[11] is based on an observation that strong Lewis acids can form stable, molecularly bound complexes with hindered alkenes, as, for instance, with adamantyladamantane:

$$\text{adamantylideneadamantane} \xrightarrow{SbF_5} \text{adamantylideneadamantane}\cdot SbF_5\ (\pi\text{ complex})$$

The carbon cation which forms cannot initiate a cationic polymerization of excess alkene molecules that are present in the reaction mixture, due to steric hindrance.[11] The less hindered alkenes, however, like isobutylene, polymerize rapidly under the same conditions with antimony pentafluoride. It was therefore suggested[10,11] that initiation involves an opening of a previously formed π complex of SbF_5 with the double bond, through halide participation. The α,β-haloalkylantimony tetrafluoride ionizes with excess Lewis acid to the related carbon cation that is capable of initiating cationic chain growth:

$$CH_2{=}C(CH_3)_2\cdots SbF_5 \longrightarrow CH_2(SbF_4)-C(F)(CH_3)_2 \xrightarrow{SbF_5} H_2C(SbF_4)-C^{\oplus}(CH_3)_2\ SbF_6^{\ominus} \xrightarrow{n\ CH_2=C(CH_3)_2} \text{polymer}$$

Alternately, the intermediate could be shown as splitting off HF to react with another molecule of Lewis acid. The metalated alkene is a tertiary halide. It can eliminate hydrogen halide:

$$H_2C(SbF_4)-C(F)(CH_3)_2 \longrightarrow HC(SbF_4)=C(CH_3)_2 + HF \xrightarrow{SbF_5} H_2C(SbF_4)-C^{\oplus}(CH_3)_2\cdots\cdots SbF_6^{\ominus}$$

The initiation mechanism of olefins by AlB_3 was explained similarly.[11] The intermediate compound of aluminum bromide and the olefin is expected to lose HBr. Although initially the reaction mixture is free from protonic acid, it could form under conditions where initiation takes place by a conjugate acid catalyzed system. In addition, all cationic polymerizations of olefins should be considered as typical examples of general carbocationic reactivity in electrophilic reactions.[11] The separate mechanisms are to be looked upon as various examples that differ only in the nature of the initial electrophile. They always lead to the related trivalent alkyl cation when polymeric chain growth is initiated[11]:

$$\underset{\text{(electrophile)}}{E^{\oplus}} + CH_2=C(R)(R') \longrightarrow E-CH_2-C^{\oplus}(R)(R') \xrightarrow{CH_2=C(RR')} \text{polymer}$$

Much of the evidence gathered to date, however, supports the concept of initiation due to autoionization of Lewis acids. This was originally proposed in 1948.[330] Lewis acids can aggregate, generally into dimers, and then autoionize. The electrophilic portion adds to the olefins:

$$2\ AlBr_3 \rightleftharpoons AlBr_2^{\oplus}\ AlBr_4^{\ominus}$$

$$AlBr_2^{\oplus}\ AlBr_4^{\ominus} + CH_2=C(R)(R') \longrightarrow Br_2Al-CH_2-C^{\oplus}(R)(R')\cdots\cdots AlBr_4^{\ominus}$$

Later, in 1965, a separate investigation[5] also led to the conclusion that the evidence does support the above mechanism. The additional data[331] show that isobutylene can be polymerized by superdry, pure $AlBr_3$. At the same time, however, BF_3 and $TiCl_4$ require addition of water to initiate polymerizations of this monomer. Furthermore, the rate of $AlBr_3$ initiated polymerization of isobutylene is enhanced considerably by addition of other Lewis acids, like $SnCl_4$, $SbCl_5$, $TiCl_4$, and VCl_4.[332,333] Significantly, these Lewis acids by themselves are incapable of initiating polymerizations of isobutylene.[15,326] It is believed, therefore, that mixed Lewis acids interact and generate ions that lead to rapid polymerizations.

A subsequent study of isobutylene[68] polymerization yielded additional evidence. Aluminum chloride and bromide catalysts were used in alkyl halide solutions. The reaction was monitored by conductivity measurements and tritium radiotracer techniques. Conductivity changes during and after polymerization and unexchanged tritium in the polymer support the theory that initiation does take place by addition of $AlX_2^{\oplus}$ to the double bonds of the monomers. This results in formation of the carbon cations.[68] The work led to the conclusion[68] that: (1) Only a very small fraction of the aluminum halide is ionized. (2) "Active species" are formed from reactions of $AlX_2^{\oplus}$ with the double bonds and covalent bonds are formed between aluminum and carbon atoms. The cations that form start the polymerizations. Other cations may form from some impurities and may also act as initiators. The polymerizations that follow are terminated fairly rapidly because the concentration of the initiating cations is much smaller than the amount of aluminum halide. This explains the low efficiency of these particular initiators. Only a little polymer forms per mole of aluminum halide used. (3) Most of the unionized aluminum halide complexes with unreacted monomer molecules. These complexes do not act as initiators. The result is that the concentration of aluminum halide and the formation of $AlX_2^{\oplus}$ becomes very small and there is no further initiation. Such conclusions explain why yields are low when the reactions are started by additions of the initiator solutions to the monomers, but are high, however, when the monomers are added to the initiator solutions.[68]

It is not clear why other Lewis acids like BCl_3 require water or other compounds and must be added to the monomer to obtain high conversions.[16] Boron trifluoride can initiate polymerizations of styrene, presumably by itself, in a methylene chloride solvent.[17] One explanation that was offered for the different behaviors of various Lewis acids is based on differences in the electronic configurations around the metal atoms.[50]

3.3.1.3. Initiation by Stable Cations

These initiations were originally reported by Bawn and co-workers.[18] The cations must be used in low concentrations to insure complete dissociation from their respective counterions. Stable organic cations can be formed from olefins, aromatic structures, or compounds with heteroatoms possessing unshared electrons, like oxygen, nitrogen, or sulfur. Some examples of stable cations that can initiate cationic polymerizations are[19]:

Triphenylmethyl ion

Cycloheptatrienyl (tropylium) ion

Xanthylium ion

The above cations form crystalline salts with anions. These anions are $ClO_4^{\ominus}$, $SbCl_6^{\ominus}$, $BF_4^{\ominus}$, $PF_6^{\ominus}$, $SbF_6^{\ominus}$, $FeCl_4^{\ominus}$, and $AsF_6^{\ominus}$.[19]

The initiation reactions by stable cations, designated as $(X^{\oplus})$, may proceed by several possible mechanisms[94]:

(A) By direct additions to unsaturated systems:

$$X^{\oplus} + CH_2{=}CHR \rightleftharpoons X{-}CH_2{-}CHR^{\oplus}$$

(B) By hydride abstractions:

$$X^{\oplus} + CH_2{=}CHR \longrightarrow XH + CH_2 - CR^{\oplus}$$

(C) By formations of cation radicals through electron transfer or through some other mechanisms:

$$X^{\oplus} + CH_2{=}CHR \longrightarrow X\bullet + \bullet CH_2{-}CHR^{\oplus}$$

In (A), the reaction will predominate on the left-hand side. Exceptions appear to be olefins with strong electron-releasing substituents that confer thermodynamic stability to the newly formed cation, $\sim CH_2{-}CHR^{\oplus}$.[19] This either results from a suitable charge delocalization over the π-electron system or from the presence of a heteroatom. Accordingly, only those olefins that possess relatively strong nucleophilic characteristics can be polymerized by stable carbon cations. Such olefins are alkyl vinyl ethers, *N*-vinyl carbazole, *p*-methoxystyrene, indene, and vinylnaphthalenes. Styrene and α-methylstyrene, however, will not polymerize, because they are less reactive.

The initiation mechanism, as suggested by Ledwith,[19] follows the path of reaction (A) and is a result of a direct addition of the cation to the olefin. This is based on observations of the reaction of the xanthylium cation with 1,1-diphenylethylene. It is also based on the reaction of the tropillium cation with *N*-vinyl carbazole. The last may, perhaps, be influenced by steric factors. The high initiating efficiency of the tropillium ion, that has a stable six π-electron system, may be a result of formation of a charge transfer complex[18]:

$$C_7H_7^{\oplus} + CH_2{-}CHR \rightleftharpoons C_7H_7\bullet + \bullet CH_2{-}CHR^{\oplus} \rightleftharpoons C_7H_7{-}CH_2{-}CHR^{\oplus}$$

Based on the above, polymerization of *N*-vinyl carbazole can be shown as follows[19]:

3.3.1.4. Metal Alkyls in Initiations of Cationic Polymerizations

Initiations of polymerizations of vinyl and other monomers by metal alkyls generally take place by anionic mechanisms. This is discussed further in this chapter. There are, however, reports in the literature[20] of cationic polymerizations that are initiated by some metal alkyls. These are polymerizations of monomers like vinyl ethers, *o*- and *p*-methoxystyrene,[21] and isobutylene[22,23] that are initiated by compounds like dialkyl aluminum chloride.

One explanation is as follows.[24] These metals are strong electron acceptors. Their valence shells and their unfilled orbitals can accommodate electrons from donor molecules. As a result, the organometallic compounds behave like Lewis acids. This was observed in a polymerization of isobutyl vinyl ether.[224] While triethylaluminum is not by itself an initiator for the polymerizations, the reactions will take place in the presence of typical electron acceptors[24]:

$$Al(C_2H_5)_3 + ClCH_2OCH_3 \rightleftharpoons [Al(CH_2H_5)_3Cl]^{\ominus} \ldots [CH_2OCH_3]^{\oplus}$$

$$[Al(CH_2H_5)_3{}^{\ominus}Cl] \ldots [CH_2OCH_3]^{\oplus} + CH_2{=}CH{-}OC_4H_9 \rightarrow$$
$$CH_3OCH_2{-}{}^{\oplus}CH_2{-}CH(OC_4H_9) \ldots [Al(C_2H_5)_3Cl]^{\ominus}$$

The catalyst components must be combined in the presence of the monomer. This is due[24] to high instability of the carbon cation complexes $[Al(C_2H_5)_3Cl]^{\ominus}[CH_3OCH_2]^{\oplus}$. Also, polymerizations by $C_2H_5AlCl_2{-}H_2O$ show rate decreases as the reaction progresses.[25] When, however, additional water is added, rapid polymerizations start again.[25]

It is possible to modify the catalytic activity of metal alkyls by controlled additions of *modifiers*. These are water, alcohol,[24] oxygen,[27] carbon dioxide,[26] aldehydes,[27] organic peroxides,[24] and metal oxides, like V_2O_5, NiO, and HgO.[24] The exact action of these modifiers is not clear and it is not certain whether they should be regarded as coinitiators or initiators. Their addition, however, can affect catalytic activity, yield, stereospecificity, and molecular weight of the products. Oxygen, for instance, can act as a modifier for the Grignard reagent, which by itself does not initiate cationic polymerizations of vinyl ethers. Yet, introduction of oxygen to a vinyl ether–Grignard reagent system will initiate the polymerization and yield high molecular weight products.[27,28] It was suggested[24] that oxygen may cause transformation of the alkyl magnesium groups into *N*-alkoxy magnesium groups. This results in greater concentrations of magnesium dihalides that can induce polymerizations.

Substances that generate cations can vary widely. They can be molecules that dissociate into ions or react with other compounds, like solvent or monomer, to form cations. Iodine is an example of such a substance. In a system of *n*-butyl vinyl ether–iodine–diethyl ether, the iodine apparently forms an inactive π-complex with the solvent first. It subsequently dissociates and rearranges into an isomeric active ion[50]:

$$CH_2{=}CH(OR) + I_2 \rightleftharpoons CH_2{=}CH(OR){\rightarrow}I_2 \xrightarrow{I_2} I{-}CH_2{-}\overset{H}{C}{}^{\oplus}(OR)\cdots I_3^{\ominus}$$

3.3.2. One-Electron Transposition Initiation Reactions

Some *radical sources* will, in the presence of oxidizing agents,[334] or light or heat energy, initiate cationic polymerizations of monomers, like *n*-butyl vinyl ether. Those that are most readily oxidized are carbon atom centered radicals that have substituents like benzyl, allyl, alkoxy, or structures with nitrogen or sulfur. Also, radicals that are formed by addition of other radicals to alkyl vinyl ethers are particularly reactive.

Oxidants that can be used in these reactions are salts, like $(C_6H_5)_2I^{\oplus}\,PF_6^{\ominus}$. Such salts oxidize the radical and also supply the counterions, as shown below[354]:

$$(C_6H_5)_2C(OH){-}C(OH)(C_6H_5)_2 \xrightarrow[80\,^{\circ}C]{\Delta} 2\,HO{-}\dot{C}(C_6H_5)_2 \xrightarrow{2\,(toluene)_2{-}I^{\oplus}\,PF_6^{\ominus}}$$

$$\longrightarrow 2\,C_6H_5{-}CO{-}C_6H_5 + \underset{\text{initiator}}{2H^{\oplus}\,PF_6^{\ominus}} + 2CH_3{-}C_6H_4{-}I + 2CH_3{-}C_6H_4{\cdot}$$

What happens to the final radical is uncertain. High conversions were reported in polymerizations of *n*-butyl vinyl ether in the presence of azobisisobutyronitrile (as a source of radicals) with $(p\text{-tolyl})_2\text{-}I^{\oplus}\,PF_6^{\ominus}$ at 50 °C.[335]

The above is a thermal process.[334] Free radicals can also be generated with the aid of UV light. This, for instance, was done with a UV light decomposition of benzoin methyl ether in the presence of an oxidizing salt[336]:

$$C_6H_5{-}CO{-}CH(OCH_3){-}C_6H_5 \xrightarrow{h\nu} C_6H_5{-}\dot{C}{=}O + {\cdot}CH(OCH_3){-}C_6H_5$$

$$C_6H_5-\underset{H}{\overset{OCH_3}{C}}\cdot + (C_6H_5)_2-I^{\oplus}\,PF_6^{\ominus} \rightarrow C_6H_5-\underset{H}{\overset{OCH_3}{C}}{}^{\oplus}\cdots PF_6^{\ominus} + I-C_6H_5 + \cdot C_6H_5$$

3.3.2.1. Charge Transfer Complexes in Ionic Initiations

Formations of copolymers by charge transfer mechanisms in free-radical polymerizations are discussed in Chapter 2. Reactions between donor and acceptor molecules, however, can also result in some charge transfers that yield ion radicals and subsequent ionic polymerizations.

$$D + A \rightleftharpoons \underset{\text{ground state}}{[DA \leftrightarrow D\bullet^{\oplus}, A\bullet^{\ominus}]} \rightleftharpoons \underset{\text{excited state}}{[D\bullet^{\oplus} A\bullet^{\ominus} \leftrightarrow DA]}$$

The nonbonding form predominates in the ground state while the charge transfer predominates in the excited state. The energy separation between the ground state and the first excited state is small. If the ionization potential of the donor molecule is low and the acceptor molecule has a strong electron affinity, transfer of an electron can occur to a significant extent even in the ground state. Mutual oxidation–reduction takes place:

$$D + A \rightleftharpoons \text{complex} \rightleftharpoons D\bullet^{\oplus} + A\bullet^{\ominus}$$

Some vinyl compounds can function as donor molecules because they possess a low ionization potential. The acceptors can be neutral molecules, like quinones, anhydrides, nitrile compounds, etc. They can also be ionic intermediates, such as metal ions, ionized acids, and carbon cations. An interaction of an acceptor with a donor is followed by a subsequent collapse of the charge transfer complex. This can result in formation of cation radicals capable of initiating cationic polymerizations.[73] The exact mechanism of the reaction of cation radicals with olefins is still not completely determined.

One example is a combination of an alkyl vinyl ether (donor) with vinylidine cyanide (acceptor) that results in ionic polymerizations.[78] The reaction actually contains the ingredients of both cationic- and anionic-type polymerizations[77]:

$$\underset{\text{donor}}{R-O-CH=CH_2} + \underset{\text{acceptor}}{CH_2=C(C\equiv N)_2} \rightleftharpoons [CT] \longrightarrow \underset{\downarrow\ \text{cationic polymerization}}{[R-O-CH=CH_2]^{\oplus}\cdot} + \underset{\downarrow\ \text{anionic polymerization}}{[CH_2=C(C\equiv N)_2]^{\ominus}\cdot}$$

where [CT] represents a charge transfer complex.

Solvations of the charged species accelerate the transfer of electrons and the ionizations are enhanced by polar solvents.[73] Charge transfer reaction studies with tetracyanoethylene, an acceptor, and *N*-vinyl carbazole, a donor, in benzene solution demonstrated that both cation radicals and anion radicals form. This can be used in a subsequent cationic polymerizations[74,75]:

$$(N\equiv C)_2C=C(C\equiv N)_2 + CH_2=\underset{R}{CH} \rightleftharpoons [CT] \rightleftharpoons (N\equiv C)_2\overset{\ominus}{C}-\dot{C}(C\equiv N)_2 + \cdot CH_2-\underset{R}{CH}^{\oplus}$$

$$\longrightarrow (N\equiv C)_2\overset{\ominus}{C}-C(C\equiv N)_2-CH_2-\underset{R}{CH}^{\oplus} \xrightarrow{\text{monomer}} \text{polymer}$$

A similar reaction takes place between chloranil and *N*-vinyl carbazole[76]:

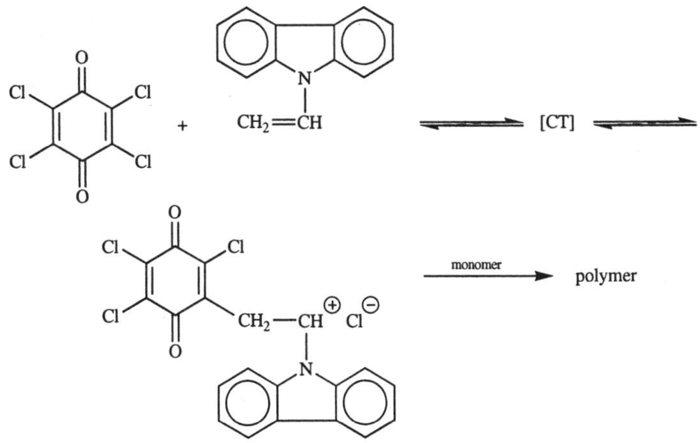

Alkyl vinyl ethers also polymerize in the presence of strong acceptors like tetracyanobezoquinone, 2,3-dichloro-5,6-dicyano-*p*- benzoquinone, and tetracyanoethylene.[74] A similar reaction mechanism was proposed.[74]

Solutions of maleic anhydride in ether will initiate cationic polymerizations of isobutyl vinyl ether or *N*-vinyl carbazole, if subjected to attacks by free radicals. The same is true if the solutions are irradiated with ultraviolet light or gamma rays.[69,70] Also, active species are generated from reactions of aldehydes or ketones with maleic anhydride when attacked by free radicals or irradiated by UV light, or gamma rays from ^{60}Co.[70] These active species are presumed to be formed through charge-transfer reactions that occur between the electron acceptors, π-acids, or electron donors, π-bases, which form cations.[71,72]

Some Lewis acids can form charge transfer complexes with monomers that yield cation radicals when irradiated with ultraviolet light.[79,80] This was shown with such Lewis acids as VCl_4, $TiCl_4$, and $TiBr_4$ in polymerizations of isobutylene. The charge transfer complexes collapse after irradiation[80]:

$$VCl_4 + M\ (\text{monomer}) \rightleftharpoons [VCl_4M]$$

$$[VCl_4M] \xrightarrow{h\nu} [VCl_4M]^* \rightarrow VCl_4^{\ominus} \ldots M\bullet^{\oplus}$$

$$2M\bullet^{\oplus} \rightarrow {}^{\oplus}M\text{–}M^{\oplus} \xrightarrow{\text{monomer}} \text{polymer}$$

3.3.2.2. Radiation-Initiated Polymerizations

Photochemical initiations of cationic polymerizations[81–86] are used commercially. This subject is discussed in Chapter 8. Irradiation by *ionizing radiation* of olefins forms several kinds of active, initiating species. Free-radical, cationic and anionic polymerizations can be initiated potentially by these active species. Generally, the characteristics of ionizing radiation-induced polymerizations can be such that free-radical and ionic polymerizations coexist.[87] Under dry conditions, polymerizations will proceed predominantly through cationic intermediates with cyclopentadiene[88] and styrene.[89,90] Also, radiation-induced polymerizations of some monomers in the liquid state are subject to retardation[337] by very low concentrations of ion scavengers such as ammonia and water.[87] The efficiency of such retardation depends upon the proton affinity of the scavengers.[87] In addition, radiation-induced cationic polymerizations of isobutylene under anhydrous conditions occur through propagation by free ions.[91]

3.3.2.3. Electroinitiation of Polymerization

These polymerizations, sometimes also called *electrolytic polymerizations*, are carried out in an electrical field. The field is applied to initiate chain growth. Passage of an electrical current through solutions of monomers in suitable solvents can produce initiating species. The majority of these species, however, are free radical in nature.[270–272]

Early studies on initiation of cationic polymerizations of styrene, isobutyl vinyl ether, and *N*-vinyl carbazole were carried out by dissolving $AgClO_4$ in pure monomers or in nitrobenzene. Electric current was then passed through them at room temperature.[338,339] Rubbery polymers formed as well as some copolymers, suggesting a cationic path of the polymerization. A mechanism was suggested, based on anodic oxidation:

$$ClO_4^{\ominus} \xrightarrow{-e} ClO_4\bullet + e$$

$$ClO_4\bullet + M \rightarrow ClO_4^{\ominus} + M^{\oplus}$$

where M is the monomer. Later, however, it was concluded that the polymerizations are due to formation of cation radicals as the initiating species.[340] Further studies of the phenomenon led to a proposed mechanism of initiation[277] of styrene polymerization in acetonitrile solution. Controlled potential electrolysis at the anode, with the aid of a salt-like tetrabutylammonium fluoroborate, is a result of direct anodic oxidation. Electrons are transferred at the anode. This is accompanied by formations of radical cations:

$$\begin{array}{c|l} + & \\ + & \cdot\, CH_2{-}CH^{\oplus} \\ + & \quad\quad\;\; | \\ + & \quad\quad\; C_6H_5 \\ + & \end{array}$$

In the above initiation, the monomer itself is the initiator and the anode can be regarded as the "co-initiator."

If a supporting electrolyte is first oxidized to a radical, indirect cationic polymerization can result. The radical in a subsequent step oxidizes the monomer to a first initiating entity, the cation radical.[276] Such an indirect initiation was also suggested for electropolymerization of isobutyl vinyl ether in the presence of $BF_4^{\ominus}$ supporting electrolyte[276]:

$$BF_4^{\ominus} \xrightarrow[\text{anode}]{e^{\ominus}} BF_4\,\cdot$$

$$BF_4\,\cdot + \underset{\substack{|\\ OR}}{CH_2{=}CH} \longrightarrow BF_4^{\ominus} + \cdot\, \underset{\substack{|\\ OR}}{CH_2{-}CH^{\oplus}}$$

3.3.3. Propagation in Cationic Polymerization

In the propagation step, the polarity of the medium affects strongly the reaction because the intimacy of the ion pair depends upon solvent polarity. The bond between the two ions can vary from a high degree of covalence to that of a pair of free, solvated ions[32]:

$$\underset{\text{covalent}}{\sim\!C{-}A} \rightleftharpoons \underset{\text{intimate ion pair}}{\sim\!\overset{\oplus}{C}\ldots\overset{\ominus}{A}} \rightleftharpoons \underset{\text{solvent separated ion pair}}{\sim\!\overset{\oplus}{C}\,/\,/\,\overset{\ominus}{A}} \rightleftharpoons \underset{\text{free, solvated ion pair}}{\sim\!\overset{\oplus}{C} + \overset{\ominus}{A}}$$

The chemical structures of the monomers also determine their reactivity toward cationic polymerizations. Electron-donating groups enhance the electron densities of the double bonds. Because the monomers must act as nucleophiles or as electron donors in the course of propagation, increased electron densities at the double bonds increase the reaction rates. It follows, therefore, that electron-withdrawing substituents on olefins will hinder cationic polymerizations. They will, instead, enhance the ability for anionic polymerization. The polarity of the substituents, however, is not the only determining factor in monomer reactivity. Steric effects can also exert considerable controls over the rates of propagation and the modes of addition to the active centers. Polymerizations

Table 3.1. Relative Rates of Reactions of Alkenes in 1,2-Dichloroethane at 24 °C with $PhCH_2^{\oplus}$ and $Ph_2CH^{\oplus}$ Cations[a,31]

Olefin	$PhCH_2^{\oplus}$	$Ph_2CH^{\oplus}$
$CH_3—CH═CH_2$	1.9×10^6	$< 10^5$
$(CH_3)_2—C═CH_2$	1.9×10^7	9.5×10^6
(cyclohexadiene)	2.7×10^7	1.5×10^7
$CH_2═CH—CH═CH_2$	8.7×10^5	$< 10^5$

[a]From Wang and Dorfmann, by permission of the American Chemical Society.

of alkyl vinyl ethers with $BF_3 \bullet (C_2H_5)_2$ in toluene or in methylene chloride at –79 °C showed that the rate of monomer consumption falls in the following order[34]:

Alkyl group = *t*-butyl > *i*-propyl > ethyl > *n*-butyl > methyl

Another example is a study of the differences in the rates of reactions of various alkenes with two cations, $PhCH_2^{\oplus}$ and $Ph_2CH^{\oplus}$, generated by electron pulses.[31] Carbon cations, free from complexities, such as ion pairing and cation aggregation that may be encountered in typical cationic polymerizations, were used. Table 3.1 shows some of the data that were reported.[31]

Chain-growth reactions with fairly tight ion pairs, that occur in medium of low polarity, require that the monomers be inserted repeatedly between the two ions. These consist of carbon cations on the terminal units paired with the counterions. The ion pairs are first loosened, or "relaxed," complexations with monomers follow, and insertions complete the process. All insertions, of course, result in formations of new carbon cations. Upon formation, they immediately pair off with the counterions, and the process continues:

$$\sim CH_2—C(X)(Y)^{\oplus}\cdots A^{\ominus} + CH_2═C(X)(Y) \longrightarrow \sim CH_2—C(X)(Y)—CH_2—C(X)(Y)^{\oplus}\cdots A^{\ominus}$$

The mechanisms of such insertions consist of repeated push–pull attacks by the ion pairs on the double bonds of the incoming monomers[29]:

$$\sim C^{\oplus}\text{------}A^{\ominus} \;\; (C{=}C) \longrightarrow \sim C—C—C^{\oplus}\cdots A^{\ominus}$$

The degree of association of the ion pairs depends also upon the nature of the counterion and on the temperature of the reaction medium. Completely dissociated ion pairs allow chain growth to take place free from the influence of counterions. The carbon cations simply add directly to the double bonds of the incoming monomers. Propagation rates for such reactions are greater than for those with tight ion pairs.[33]

The efficiency of the counterion is related to its acid strength. In isobutylene polymerization at –78 °C, the following Lewis acids were rated in the order of their efficiencies[35]:

$$BF_3 > AlBr_3 > TiBr_4 > BBr_3 > SnCl_4$$

The polymerization reactions can sometimes be complicated by two different types of propagation paths. Some chains may grow without a terminal counterion as free propagating species. Other polymeric chains, however, may be paired off with counterions. It should be noted that when references are made to free propagating ions, the ions are free from electrostatic influences of the anions. They are, however, still associated with, and interact with, polar or polarizable solvent molecules or monomers.

3.3.3.1. Steric Control in Cationic Polymerization

Ionic polymerizations yield highly stereoregular polymers when control is exercised over monomer placement. The earliest stereospecific vinyl polymerizations were observed in preparation of poly(isobutyl vinyl ether) with a BF_3–ether complex catalyst at –70 °C. An isotactic polymer formed.[36] The same catalyst was employed later to yield other stereospecific poly(vinyl ether)s.[37–39] The amount of steric placement increases with a decrease in the reaction temperature, and, conversely, decreases with an increase in the temperature.[39,40]

Various mechanisms were proposed to explain steric placement in cationic polymerization. Most of them pertain to vinyl ethers. There is no general agreement. Some of the suggested mechanisms are discussed in this section. Most were proposed for homogeneous conditions with soluble initiators like BF_3–$O(C_2H_5)_2$. There are, however, also some explanations of steric control with insoluble catalysts, like $Al(SO_4)_2{\bullet}H_2SO_4$.

3.3.3.1a. Control in Homogeneous Polymerizations. Not all explanations of steric control under homogeneous conditions give equal weight to the influence of the counterion. A Bawn and Ledwith mechanism[41] for the polymerization of vinyl ethers is based on data which suggest that only one mesomeric form of the ethers exists, presumably *trans*[42]:

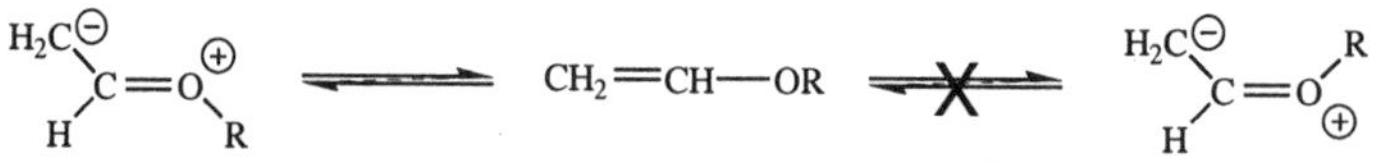

$$H_2C^{\ominus}\!-\!CH\!=\!O^{\oplus}\!-\!R \rightleftharpoons CH_2\!=\!CH\!-\!OR \not\rightleftharpoons H_2C^{\ominus}\!-\!CH\!=\!O^{\oplus}\!-\!R$$

An alkyl substituent composed of a three-carbon chain caused steric blocking of one side of the bond:

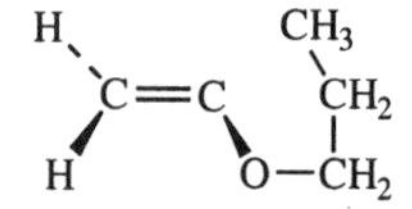

A five-carbon substituent should exhibit the highest degree of steric hindrance, which, on the other hand, should decrease with a decrease in the size of the group. There should be no blocking with an ethyl group or with an isopropyl one.[42] This was demonstrated experimentally.[41] It was suggested, therefore, that in homogeneous polymerizations of vinyl ethers, the growing cations are stabilized by a form of neighboring group interaction. This interaction (or intramolecular solvation) would be with oxygen atoms from the penultimate monomer units.[41] These are forms of "backside" stabilization of the growing chains that force reactions to occur at the opposite sides from the locations of the counterions. The mechanisms are forms of S_N2 attacks with retention of the configurations. These configurations are formed between existing and newly formed carbon cations in the transition states:

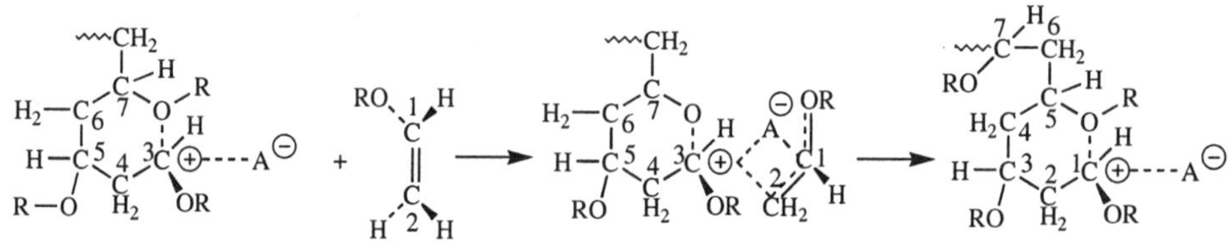

Solvations of the new cations might even occur before they are completely formed, maintaining the steric arrangement throughout, provided that the monomers enter as shown above.[41] One weakness of the above mechanism is that it fails to consider the nature of the counterions.

Another mechanism, proposed by Cram and Kopecky,[43] places emphasis on formation of six-membered rings. The growing polymeric chains in vinyl ethers occupy equatorial arrangements with the –OR groups attached to the growing ends by virtue of their size, because they are larger. In reactions between the monomers and the six-membered ring oxonium ions, the relative configurations of the two asymmetric centers that form determine total chain configurations. If the configurations are similar, the chains becomes isotactic, but if they are different they become syndiotactic. Molecular models suggest that isotactic placement should be more likely.[43]

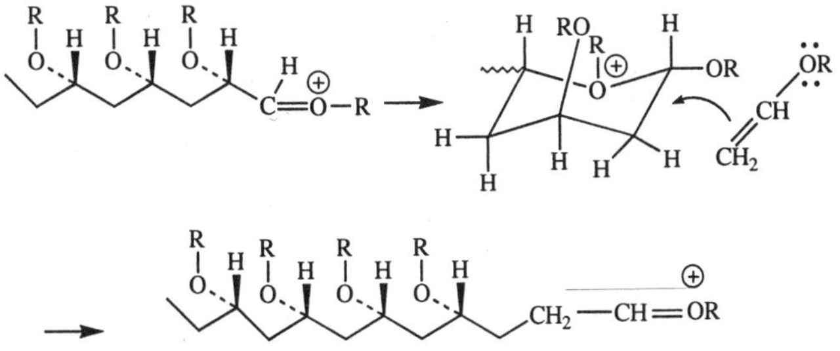

The Cram and Kopecky mechanism[43] fails to explain the influence of the various R groups upon the stereospecificity of the final product.

In a mechanism proposed by Kunitake and Aso[44] two factors were given primary importance. These are: (1) Steric repulsions determine the conformations of the propagating chains with a special arrangement of the counterions and those of the incoming monomers. (2) The directions of the monomer attacks are determined by the tightness of the growing ion pairs. It is assumed that the growing carbon cations are essentially sp^2 hybridized and that the conformations with the least steric repulsion will therefore be[44] as shown in Fig. 3.1.

The position of the counterion is assumed to be at the side of the carbon cation and away from the penultimate unit. The stability of such conformations should be very dependent on the temperature of the polymerization and on the size of the substituents. Experimental evidence confirms this. Thus, it is known that the stereoregular polymers, whether isotactic or syndiotactic, form only at low temperatures in homogeneous polymerizations (as stated earlier). This suggests that the fixation of the conformations of the growing chain ends is very important in enhancing polymer stereoregularity.

In polar solvents the counterions interact only weakly with the growing cations. The steric effects become major factors in deciding the courses of propagation. In such situations the carbon cations[44] attack the least hindered side (frontal side attacks). These give rise to syndiotactic structures. The terminal carbon cations probably can rotate freely, so the vinyl monomers should be capable of approaching from any direction. In nonpolar solvents, on the other hand, if the ion pairs are tight enough, the incoming monomers may approach the cations from the back sides only, giving rise to isotactic placements. This is illustrated in Fig. 3.2.

If there is steric hindrance to back-side approaches due to the large sizes of the penultimate substituents, front-side attacks take place. This occurs even in nonpolar medium.[44] The incoming monomers can therefore attack the cation either from the front side or from the back side. All depends upon the tightness or the Coulombic interaction of the ion pair and on the difference in the steric hindrance between the two modes of attack.

In the above reaction mechanism, the possible interactions of the counterions and the monomers are ignored. This was justified by weak interactions of electron-rich monomers, like α-methyl styrene and vinyl ethers with weak anions.[44] The nature of the counterions as such, however, is not ignored in this mechanism, because the tightness of the ion pairs is considered.

Later work[45] suggested that the sizes of the R groups of alkenyl ethers play an important role in determining the steric structures of the resultant polymers. For instance, allyl vinyl ethers can be polymerized to highly isotactic polymers with the aid of $SbCl_5$. ^1H and ^{13}C NMR data shows no evidence, however, of steric control, though, it does show a relationship between active chain ends and incoming monomers. In addition, the amounts of isotactic placement do not differ significantly at −10 °C or at −75 °C.[316] This suggests that isotactic selection is generated by orienting the

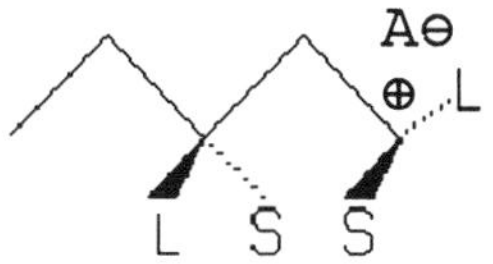

FIGURE 3.1. Steric arrangement. L = large substituent, S = small substitutent.

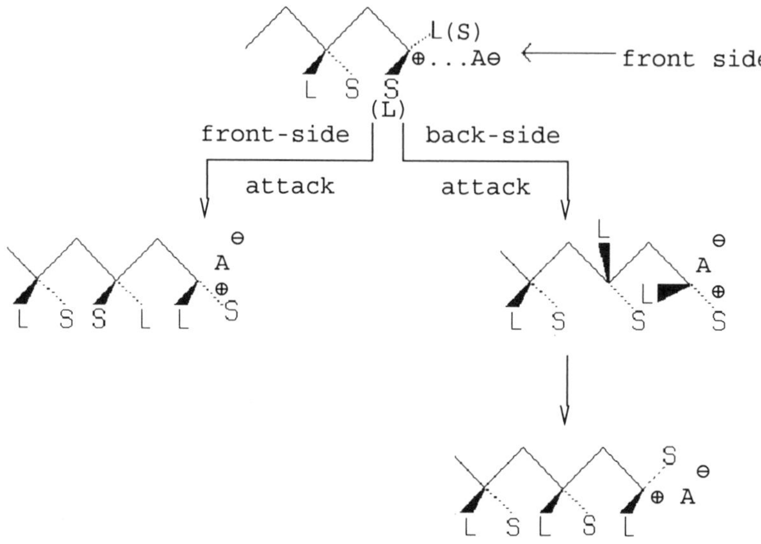

FIGURE 3.2. Propagation mechanism.

substituents in the monomer and in the chain *away* from each other. A Coulombic attraction is visualized between the counterion and the positively polarized oxygen of the monomer.

Also, in studies with optically active vinyl ethers it was observed[319] that trimethyl vinyl silane, which is bulky and nonchiral, forms highly syndiotactic polymers. Equally bulky, but chiral (–)-menthyl vinyl ether, however, produces isotactic polymers in polar solvents. This suggests that isotactic propagation is preferred in a polar medium because of helical conformation of the polymer chain and is forced by a bulky chiral substituent. Kunitake and Takarabe[320] therefore modified the original Kunitake and Aso mechanism. The growing chain ends are crowded by bulky substituents. This may result in steric interference between a bulky side groups and the counter ions. The interactions of the propagating ion pairs decrease when the sizes of the counter ions increase. A frontal attack and syndiotactic placement of the monomers results. When, however, the monomer side groups are less bulky, steric repulsion becomes insignificant. Larger counterions become responsible for retarding the front side attack and give more isotactic placement.

Studies of model reactions for cationic polymerization of alkyl propenyl ethers showed that the mode of double bond opening is independent of the geometric structure of the ether. Mainly a *threo* opening takes place, but the mode of monomer addition is dependent on the geometric structure of the monomer or on the balkiness of the substituent.[356]

Finally, in still another investigation of model systems, UV and visible spectroscopy were used together with conductivity measurements. Results showed that charge-transfer complexes do form between the counterions and π-acceptors, which can be Lewis acids or acceptor solvents.[356] This led Heublein to suggest that interactions with monomers lead to alterations of the solvation spheres of the ion pairs in the direction of the counterions. The temporary dissymmetry of the sphere of solvation affects stereoregularities of the structures of the polymers that form. As a result, the propagation reactions are seen by Hueblein as competing interaction between the chain carriers and the monomers, the counterions, and the solvents.[356]

3.3.3.1b. Control in Heterogeneous Polymerizations. Several reaction mechanisms were also proposed to explain stereospecific placement with insoluble catalysts. Furukawa[46] suggested that here the mechanism for cationic polymerization of vinyl ethers depends upon multicentered coordinations. He felt that coordinations of the polymeric chains and monomers with the catalysts are possible if the complexed counteranions have electrically positive centers. This can take place in the case of aluminum alkyl and boron fluoride:

$$R_3Al + BF_3 \rightarrow [R_2Al]^{\oplus} + [BF_3R]^{\ominus}$$

Further coordination of aluminum alkyl to the anions is possible if the coordination number of the central atoms is sufficiently large[46]:

$$[BF_3R]^{\ominus} + AlR_3 \rightarrow [R_3Al \rightarrow BF_3R]^{\ominus}$$

The products are complex counterions that enable multicentered coordination polymerization.

Thus, the mechanism of vinyl ether polymerization proposed by Furukawa[46] is as follows. Two neighboring ether oxygens that are linked to the polymer chain close to the terminal cation become coordinated to the metal center of the complexed counterion. The molecules of the monomer can then approach the growing chain only from the opposite side and isotactic placement results:

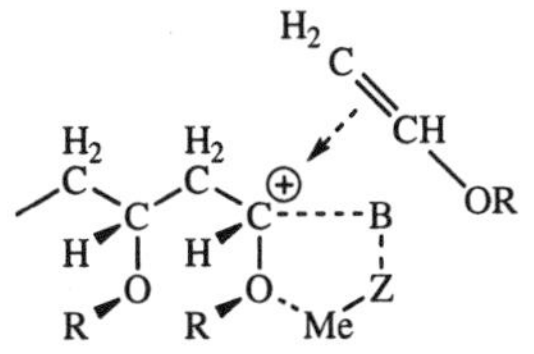

$H_2C{=}CH{-}OR$ (monomer approaching) ; $\sim C(H_2){-}C(H)(OR){-}C(H_2){-}C^{\oplus}(H)(O{-}R)\cdots B{-}Z{-}Me\cdots O$

where B represents boron and Me aluminum.

A different mechanism, however, was offered by Nakano and co-workers.[47] They felt that there must be a relationship between the crystal structures of the heterogeneous catalysts and the resultant stereoregularity of the polymers. If the crystal structures of the catalysts are tetrahedral and the crystals have active edges, stereoregular polymers should form even at room temperature. In addition, shorter active edges make the catalysts more suitable for stereospecific polymerization. The following mechanism was therefore proposed.[47]

If the terminal end of the growing chain ends

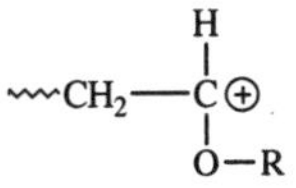

$$\sim CH_2{-}C^{\oplus}H{-}O{-}R$$

have sp^2-type configurations, vacant orbitals on the terminal carbon atoms of the growing polymeric chains are in a state of resonance with the lone pair of electrons on the adjacent oxygen atoms. This means that the positive charges are distributed to the adjacent oxygens and are not localized on the carbons:

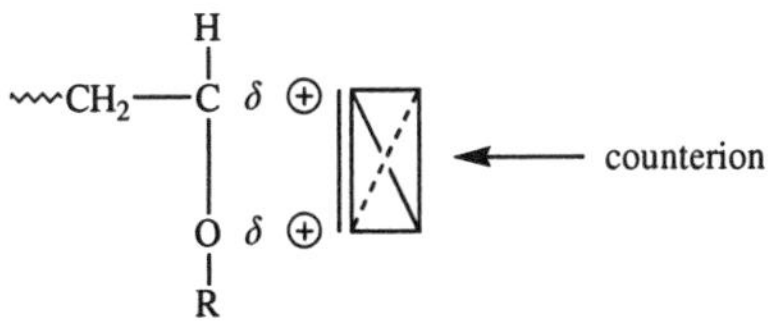

$$\sim CH_2{-}CH^{\delta\oplus}{-}O^{\delta\oplus}{-}R \quad [\text{counterion}]$$

counterion

The monomer can potentially add in four different ways:

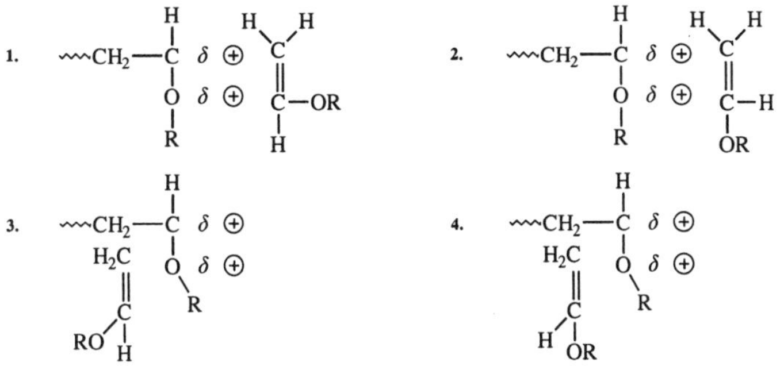

1. $\sim CH_2{-}CH^{\delta\oplus}{-}O^{\delta\oplus}{-}R \quad H_2C{=}CH{-}OR$ (C–OR end toward O)

2. $\sim CH_2{-}CH^{\delta\oplus}{-}O^{\delta\oplus}{-}R \quad H_2C{=}CH{-}OR$ (C–H end toward O)

3. $\sim CH_2{-}CH^{\delta\oplus}{-}O^{\delta\oplus}{-}R$ with $H_2C{=}C(H)(OR)$ (RO, H)

4. $\sim CH_2{-}CH^{\delta\oplus}{-}O^{\delta\oplus}{-}R$ with $H_2C{=}C(H)(OR)$ (H, OR)

Reaction 3 yields isotactic polymers and should be the mode of addition[47] when isotactic polymers forms.

3.3.3.2. Pseudocationic Polymerization

Most cationic polymerizations of olefins proceed through carbon cation carriers. There are, however, instances of cationic polymerizations where the evidence suggests that the propagating species are not carbon cations. Instead, the reactions may proceed through covalently propagating species. Such reactions are termed *pseudocationic*.[53,54] In these polymerizations the propagating species may be combinations of ionic (free ions and ion pairs) and covalently bonded species. Reaction conditions determine the relative amounts of each. Examples of such polymerizations are cationic polymerizations of styrene or acenaphthene when protonic acids like $HClO_4$ or when iodine are the initiators. The propagations[54] take place in three successive stages when the reactions are carried out in methylene chloride at –20 °C. In the first one, rapid, short-lived ionic reactions take place. In the second stage, the ions can no longer be detected by spectroscopy or conductivity measurements. In the third one, a rapid increase in the presence of ionic species can be shown (detected by conductivity measurements and spectroscopy). At temperatures between –20 °C and 30 °C, there is effectively no stage one and stage three is shorter. On the other hand, at temperatures as low as –80 °C there is only stage one. In stages one and three, propagation takes places through combinations of free ions and ion pairs. These combinations of ions result in formations of covalent perchlorate esters that are solvated and stabilized by monomer. The propagation in stage two, therefore, is pseudocationic and covalent because it consists of monomer insertion into the C–O bond of a perchlorate ester[53,54]:

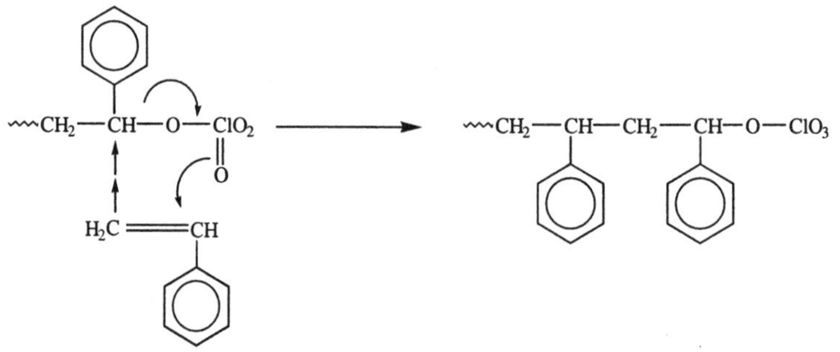

When there are insufficient amounts of monomer present to stabilize the covalent esters, such as at high conversions, ionizations take place. This leads to rapid ionic, stage three polymerizations. The propagation by free ions and ion pairs in stage three is between 10^4–10^5 liters/mole-s at low temperatures, between –60 °C to –80 °C. By comparison, polymerizations in stage two are much slower, somewhere between 0.1–20 liters/mole-s. The rates depend, of course, upon the solvent and upon the temperature. While ion–ion pair-type propagations yield polymers with molecular weights equal to 10^4, the covalent propagations only yield oligomers. Differences in molecular weights and molecular weight distributions of the products from the three stages of polymerization can be detected by size exclusion chromatography. The chromatogram shows bimodal distributions. This is supporting evidence for two modes of propagation, ionic and covalent.[54]

Similar results were reported in polymerizations of styrene with CH_3COClO_4, CF_3SO_3H, CF_3COOH, $ClSO_3H$, and FSO_3H.[53] Pseudocationic mechanism also takes place in polymerizations of some styrene derivatives, like *p*-methyl styrene, *p*-methoxystyrene, and *p*-chlorostyrene with these protonic acids.[53]

Although the pseudocationic mechanism is now fairly well accepted, it was argued against in the past and alternative mechanisms based on ion pairs were offered instead.[49,55,56]

3.3.3.3. Isomerization Polymerization

In some cationic polymerizations, the monomers may rearrange in the process of placements into the chains. The isomerizations are to energetically preferred configurations. The result is that the units in the final polymers are structurally different from the original monomers. Such rearrangements are not limited to cationic polymerizations. In this section, however, only the isomerizations in cationic polymerizations are discussed.

The isomerization polymerizations were classified by Kennedy according to the type of rearrangement that accompanies the propagation and by the particular processes[57]:

1. Propagation reactions accompanied by bond or electron rearrangement:
   A. Intra–intermolecular polymerization.
   B. Transannular polymerization.
   C. Polymerization by strain relief.
2. Propagation reactions accompanied by migration of one or more atoms.

Examples of *intra–intermolecular polymerizations* are cyclic polymerizations of nonconjugated dienes. This resembles such polymerizations by a free-radical mechanism:

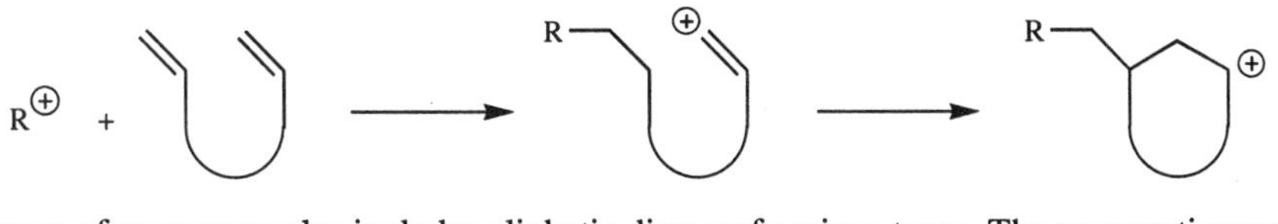

This group of monomers also includes aliphatic dienes of various types. The propagation proceeds through an internal attack by the electrophile on another part of the molecule yielding a new carbon cation. The following is an example of such a reaction[58,59]:

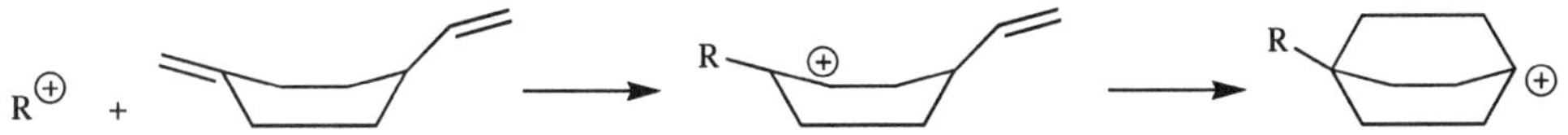

Polynortricyclene[57,60,61] forms by cationic polymerization of norbornadiene by a *transannular polymerization*. The reaction is best carried out below –100 °C to prevent crosslinking. Two propagation paths are possible:

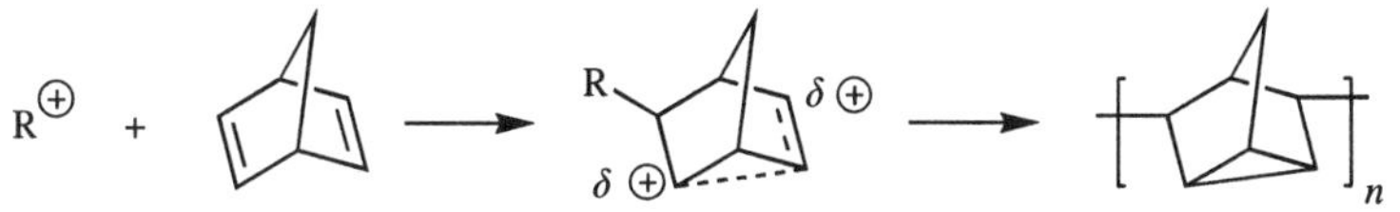

or

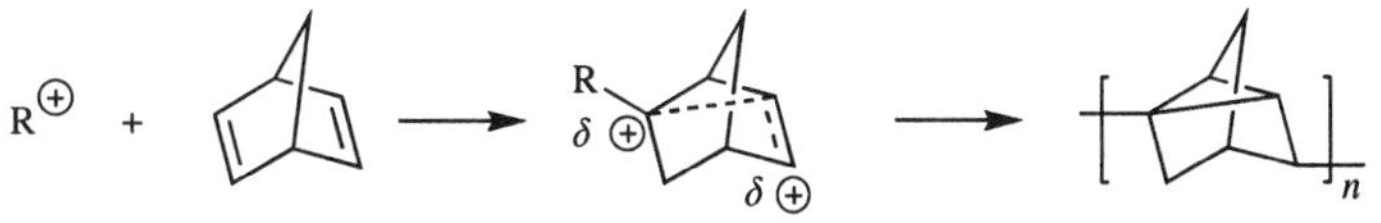

NMR and IR spectra demonstrated that both propagations shown above take place during the polymerization and the resultant polymer is in effect a copolymer of both structures.[62] Similarly, cationic polymerization of 2-methylene, 5-norbornene involves a transannular addition of the initially formed carbon cation to the strained endocyclic double bond. The product is a polymeric nortricyclene.[62]

Polymerization by strain relief occurs by opening of strained rings or by rearrangements of internal double bonds during propagation. An example is the polymerization of α-pinene. The propagation is accompanied by rearrangement of the initially formed carbon cation to a tertiary cation that relieves the ring strain[62]:

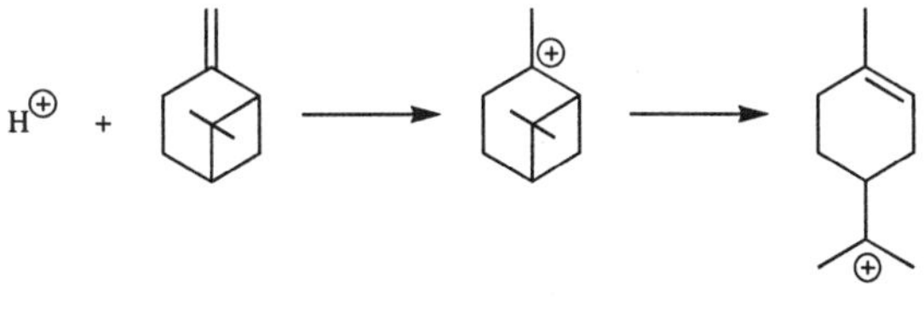

As a result, the formed polymer has the following structure:

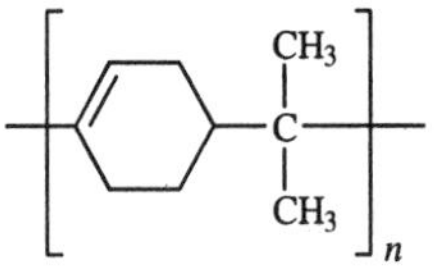

$$\left[\!-\!\text{C}_6\text{H}_8\!-\!\underset{\text{CH}_3}{\overset{\text{CH}_3}{\text{C}}}\!-\!\right]_n$$

Isomerization polymerizations by material transport include propagation reactions that are accompanied by hydride shifts. One example of this is a 1,2-hydride shift in the polymerization of 3-methylbutene-1[57]:

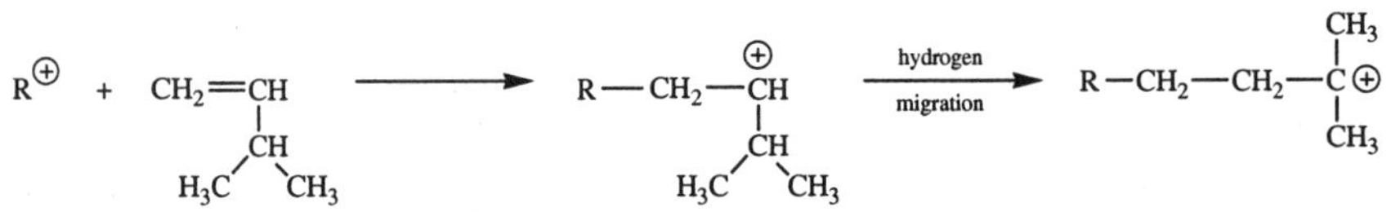

$$\text{R}^{\oplus} + \text{CH}_2\!=\!\text{CH}\!-\!\text{CH}(\text{CH}_3)_2 \longrightarrow \text{R}\!-\!\text{CH}_2\!-\!\overset{\oplus}{\text{C}}\text{H}\!-\!\text{CH}(\text{CH}_3)_2 \xrightarrow[\text{migration}]{\text{hydrogen}} \text{R}\!-\!\text{CH}_2\!-\!\text{CH}_2\!-\!\underset{\text{CH}_3}{\overset{\text{CH}_3}{\text{C}}}\oplus$$

During propagation, a tertiary hydrogen atom migrates as a hydride ion to the second carbon. It results in an energetically favored tertiary carbon cation. The final product is a 1,3-polymer[57]:

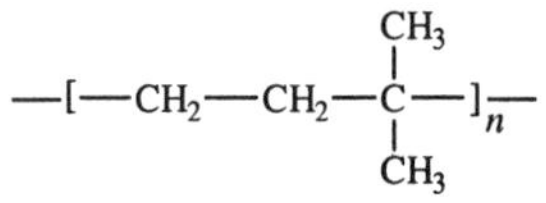

$$-[-\text{CH}_2\!-\!\text{CH}_2\!-\!\underset{\text{CH}_3}{\overset{\text{CH}_3}{\text{C}}}\!-]_n-$$

Another example of a hydride shift during propagation is polymerization of vinylcyclohexane with the aid of $AlCl_3$ at temperatures ranging from –144 °C to 70 °C[63,64]:

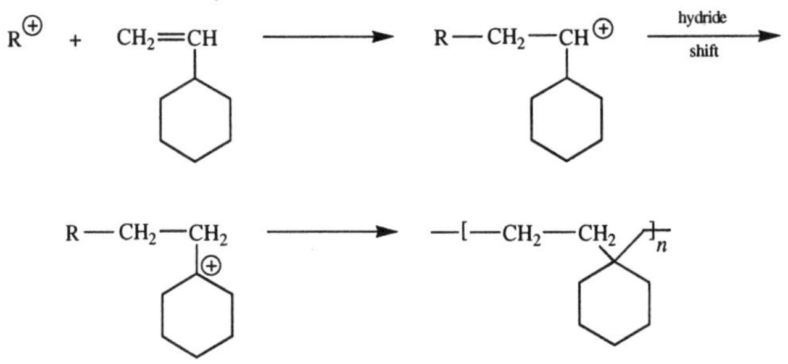

$$\text{R}^{\oplus} + \text{CH}_2\!=\!\text{CH}\!-\!\text{C}_6\text{H}_{11} \longrightarrow \text{R}\!-\!\text{CH}_2\!-\!\text{CH}^{\oplus}\!-\!\text{C}_6\text{H}_{11} \xrightarrow[\text{shift}]{\text{hydride}}$$

$$\text{R}\!-\!\text{CH}_2\!-\!\text{CH}_2\!-\!\overset{\oplus}{\text{C}}_6\text{H}_{10} \longrightarrow -[-\text{CH}_2\!-\!\text{CH}_2\!-\!\text{C}_6\text{H}_{10}\!-]_n$$

Successive hydride shifts are possible in propagation reactions where the structures of the monomers favor them[64]:

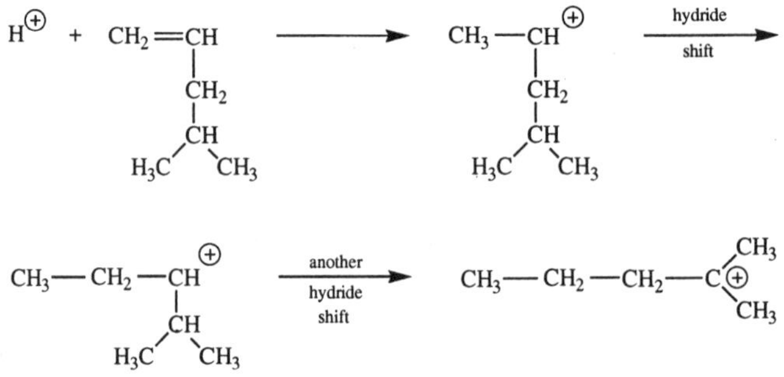

$$\text{H}^{\oplus} + \text{CH}_2\!=\!\text{CH}\!-\!\text{CH}_2\!-\!\text{CH}(\text{CH}_3)_2 \longrightarrow \text{CH}_3\!-\!\overset{\oplus}{\text{C}}\text{H}\!-\!\text{CH}_2\!-\!\text{CH}(\text{CH}_3)_2 \xrightarrow[\text{shift}]{\text{hydride}}$$

$$\text{CH}_3\!-\!\text{CH}_2\!-\!\overset{\oplus}{\text{C}}\text{H}\!-\!\text{CH}(\text{CH}_3)_2 \xrightarrow[\text{hydride shift}]{\text{another}} \text{CH}_3\!-\!\text{CH}_2\!-\!\text{CH}_2\!-\!\overset{\oplus}{\text{C}}(\text{CH}_3)_2$$

The temperature of polymerization can have some effect. This was observed in the polymerization of 3-methylbutene-1.[65] At lower temperatures the amount of 1,3 versus 1,2 placement apparently increases rapidly. At –130 °C, an essentially pure crystalline 1,3-poly(methylbutene-1) forms. The nature of the counterion also has an effect on the slope of the product composition curve. The polarity of the solvent, however, apparently does not, nor does the monomer concentration.[65]

Simultaneous migrations of hydrogen atoms and methyl groups take place in cationic polymerization of 4,4-dimethyl-1-pentene (neopentylethylene).[66] In the presence of $AlCl_3$ between −78 °C and −130 °C, the polymerization proceeds mainly as follows:

$$n\ CH_2{=}CH{-}CH_2{-}C(CH_3)_2{-}CH_3 \xrightarrow{\text{1, 4 placement}} {-}[{-}CH_2{-}CH_2{-}CH(CH_3){-}C(CH_3)_2{-}]_n{-}$$

The propagation can also involve an intramolecular halide migration. The polymerization of 3-chloro-3-methylbutene-1 proceeds at low temperatures by a chloride shift (about 50%)[67]:

$$\sim CH_2{-}CH^{\oplus}{-}C(CH_3)_2Cl \longrightarrow \sim CH_2{-}CH(CH_3){-}C^{\oplus}(CH_3)Cl \longrightarrow {-}[{-}CH_2{-}CH(CH_3){-}C(CH_3)(Cl){-}]_n{-}$$

3.3.4. Termination Reactions in Cationic Polymerizations

The termination reaction can occur by a transfer to a monomer at proper reaction conditions[92]:

$$\sim CH_2{-}C^{\oplus}(CH_3)_2\cdots A^{\ominus} + CH_2{=}C(CH_3)_2 \longrightarrow \sim CH{=}C(CH_3)_2 + H_3C{-}C^{\oplus}(CH_3)_2\cdots A^{\ominus}$$

The termination reactions in cationic polymerizations can often lead to low molecular weight products. This can be a result of various effects. Also, the counterions may be involved in the terminations.[95] Thus, with some Lewis acids the polymer cation may react with the counterion by abstracting a halogen. An example of that is polymerization of isobutylene with BCl_3. In this reaction termination by chain transferring is absent.[16] Instead, the following mechanism takes place[16]:

$$\sim CH_2{-}C^{\oplus}(CH_3)_2\cdots BCl_3OH^{\ominus} \longrightarrow \sim CH_2{-}C(CH_3)_2{-}Cl + BCl_2OH$$

Alkylaluminum-initiated reactions can terminate by a transfer of an alkyl group[93]:

$$\sim CH_2{-}C^{\oplus}(CH_3)_2\cdots (CH_3)_3AlX^{\ominus} \longrightarrow \sim CH_2{-}C(CH_3)_2{-}CH_3 + (CH_3)_2AlX$$

where X is a halogen.

When a hydrogen atom is located β to the aluminum in an alkylaluminum moiety, then termination by hydrogen transfer appears to occur preferentially[92]:

$$\sim C^{\oplus}\cdots X{-}Al^{\ominus}(C_2H_5)_2{-}CH_2{-}CH_3 \longrightarrow \sim C{-}H + (C_2H_5)_2AlX + CH_2{=}CH_2$$

Termination through a rearrangement of the propagating ion pair is also possible. Such terminations are sometimes called *spontaneous terminations* or referred to as chain transfers to counterions. The ion couple is rearranged by an expulsion of hydrogen and leaving terminal unsaturation[96]:

$$\sim CH_2{-}C^{\oplus}(CH_3)_2\cdots BF_3OH^{\ominus} \longrightarrow \sim CH{=}C(CH_3)_2 + H^{\oplus}\cdots BF_3OH^{\ominus}$$

Many substances that act as catalysts (in combination with Lewis acids), like water, are also good chain transferring agents. A solvent or an impurity can act in the same way:

$$\sim CH_2{-}C^{\oplus}(CH_3)_2\cdots A^{\ominus} + HR \longrightarrow \sim CH_2{-}C(CH_3)_2{-}H + AR$$

3.3.4.1. Living Cationic Polymerizations

In some cationic polymerizations, when conditions are carefully controlled, quasi-terminationless or terminationless systems can be achieved.[310,363–368] The "living" polymers, or "quasiliving," form as chain transfer to monomer and all other forms of termination are greatly decreased or made reversible throughout the reaction. The important aspect of living cationic polymerization is that the propagating centers are sufficiently low in reactivity. Transfer and termination reactions are suppressed. The propagation, however, is maintained. The molecular weights must increase in proportion to the cumulative amount of the monomer added. The lifetimes of the propagating species can be extended in such polymerizations by carrying out continuous slow additions of the monomer. Because chain transferring reactions are not completely eliminated in all these systems, the term "quasi" is sometimes used.

Living cationic polymerizations have been carried out with a number of monomers, such as isobutylene, styrene, *p*-methylstyrene, *p*-methoxystyrene, *N*-vinyl carbazole, and others.[370,371] To achieve living conditions, it is necessary to match the propagating carbon cation with the counterion, the solvent polarity, and the reaction temperature. Some examples are presented in Table 3.2.

By stabilizing the inherently unstable carbocationic growing species and preventing chain transfer and termination, living cationic polymerizations are achieved. This can be done by two methods: (1) use of suitable nucleophilic counterions, or (2) external additions of weak Lewis bases.[372] The ion pairs are very tight in these reactions and may border on being covalent. The propagation can be illustrated as follows:

$$\sim\!\!\sim CH(R)\text{—}I \longrightarrow \sim\!\!\sim \overset{\delta\oplus}{C}H(R)\cdots\overset{\delta\ominus}{I} + CH_2{=}CHR \longrightarrow \sim\!\!\sim CH(R)\text{—}CH_2\text{—}\overset{\delta\oplus}{C}H(R)\cdots\overset{\delta\ominus}{I}$$

$$\sim\!\!\sim CH(R)\text{—}I \xrightarrow{ZnI_2} \sim\!\!\sim \overset{\delta\oplus}{C}H(R)\cdots\overset{\delta\ominus}{I}\cdots ZnI_2 + CH_2{=}CHR \longrightarrow \sim\!\!\sim CH(R)\text{—}CH_2\text{—}\overset{\delta\oplus}{C}H(R)\cdots\overset{\delta\ominus}{I}\cdots ZnI_2$$

As shown above, the carbon–iodine bond is stretched, with or without the help of a Lewis acid. The Lewis acid assists in further stretching the carbon–iodine bond. Whether it is needed depends upon the strength of that bond. This strength, in turn, varies with the ability of the solvent, the temperature, and the substituent R to stabilize the $\delta^{\oplus}$ center. Depending upon conditions, a Lewis acid can convert the counterion, like $I^{\ominus}$, to a more stable, less nucleophilic species. Lewis bases, like dioxane or ethyl acetate, may function by reacting directly with the propagating centers.

Table 3.2. Conditions for Preparations of Some Living Polymers[a]

Monomer	Initiating system	Solvent	Temperature (°C)
Vinyl ethers	$HI + ZnI_2$	Toluene	–40 to 25
Isobutyl vinyl ether	Protonic acids[b] + $ZnCl_2$	Toluene	–40 to 0
p-Methyl styrene	CH_3COClO_4	CH_2Cl_2 toluene (1:4)	–78
Isobutylene	BCl_3 + cumyl acetate	CH_2Cl_2	–30
N-Vinylcarbazole	HI	Toluene	–40
Styrene	$CH_3CH(C_6H_5)Cl + SnCl_4 + n\text{-}C_4H_9NCl$	CH_2Cl_2	–15
Indene	Cumyl methyl ether + $TiCl_4$	Methylene chloride	–40 and –75
Indene	2-chloro-2,4,4-trimethylpentane + $TiCl_4$	Methyl chloride/ CH_3–cyclohexane	–80

[a]From various sources in the literature.

[b]Protonic acids used were CH_3SO_3H, $R_2P(O)OH$, and $R'CO_2H$, where R = OC_6H_5, C_6H_5, $n\text{-}C_4H_9$; R′ = CF_3, CCl_3, $CHCl_2$, CH_2Cl.

In summary, some typical features of living cationic polymerizations are:

1. The number average molecular weight of the polymers that form is proportional to the amount of monomer introduced into the reaction mixture. In most cases, the reactions are rapid, often to the point where it is impossible to stop them before all the monomer is consumed.
2. The concentration of the polymers formed is constant and independent of conversion. This concentration is often equal to the concentration of the initiator.
3. Addition of more monomer to a completed polymerization reaction results in further polymerization and a proportional increase in molecular weight.

3.4. Anionic Polymerization of Olefins

As is stated at the beginning of this chapter, the distinctive features of the anionic chain growth polymerizations are the negative charges at the active centers. Many different bases initiate such polymerizations, depending on the monomer. The most common ones are organometallic compounds, alkali metals, metal amides, and Grignard reagents. Monomers with carbanion stabilizing substituents, either through resonance or through induction, polymerize most readily. Such substituents, in fact, determine the base strength of the initiators needed to carry out these polymerizations by their electron-withdrawing capacity. Thus, for instance, very weak bases, such as water, initiate vinylidine cyanide polymerizations:

$$n\ CH_2{=}C(C{\equiv}N)_2 \xrightarrow{H_2O} -[-CH_2-C(C{\equiv}N)_2-]_n-$$

Potassium bicarbonate initiates polymerization of 2-nitropropene:

$$n\ CH_2{=}C(CH_3)(NO_2) \xrightarrow{KHCO_3} -[-CH_2-C(CH_3)(NO_2)-]_n-$$

On the other hand, monomers like acrylic or methacrylic esters and conjugated dienes require strong bases, like alkali metals or organometallic compounds.

3.4.1. Initiation in Anionic Chain-Growth Polymerization

Anionic initiations can take different paths depending upon the initiator, the monomer, and the solvent.[97,98] As a rough generalization, however, it is possible to separate the initiation reactions into two types: (1) those that take place through the addition of a negative ion to an olefin, and (2) those that result from an electron transfer.

The polymerization reactions can, furthermore, occur in two ways. One way is in a homogeneous environment, with both monomer and initiator soluble in the solvent. The other way is in a heterogeneous environment, where only the monomer is soluble. Organolithium compounds are examples of soluble initiators, while metal dispersions are examples of insoluble ones.

The homogeneous polymerizations can be separated further into those carried out in nonpolar solvents and those in polar ones. In nonpolar solvents they are confined mostly to organolithium initiators. Other organometallic compounds require polar solvents for solubility.

3.4.1.1. Initiation by Addition of an Anion to an Olefin

When organolithium compounds are dissolved in nonpolar solvents, there is a strong tendency of the solute molecules to associate into aggregates. For instance, butyllithium is hexameric in hexane solution. This is true of ethyllithium[101,102] as well. Addition of Lewis bases to these solutions causes formation of strong complexes between the bases and the organometallic compounds.[103,104] This causes the clusters to break up through a succession of equilibria with tetrameric and dimeric intermediates, all becoming complexed with the Lewis bases.[103] Particularly effective bases are those that allow a close approach of the lithium ion to the heteroatom. Often, the carbon–lithium bond, that normally only ranges between 20 and 40% in ionic character, becomes much more ionic.[99,100]

The breakdown of the aggregates was shown to dramatically enhance the reactivity of the organometallic compounds.[105,106] For instance, polymerizations of styrene in benzene with butyllithium are slow reactions. When, however, these polymerizations are carried out in tetrahydrofuran they are extremely rapid. Tetrahydrofuran is, of course, a Lewis base. Nevertheless, the breakdown of the aggregates even in such Lewis bases as tetrahydrofuran or diethyl ether are not complete, though the clusters are smaller and more solvated. Differences in reactivity, however, can be observed even in different nonpolar solvents.[117]

The reaction rates depend to a great extent on the nature of the organometallic compounds, such as polarity of the bonds and the degree of solvation. In polar solvents, where free solvated ion pairs predominate, the mechanism of initiation may simply consist of a direct addition of the anion to the monomer. If the solvents are nonpolar, on the other hand, the initiation is more complex. In these solvents the metal cation coordinates with the monomer first. This is followed by a rearrangement[107]:

$$\overset{\ominus}{R}\text{------}\overset{\oplus}{M} + CH_2{=}C(CH_3)X \rightleftharpoons \overset{\ominus}{R}\text{------}\overset{\oplus}{M}\text{------}[HHC{=}CXH]$$

$$\longrightarrow R{-}CH_2{-}CHX^{\ominus}\text{------}M^{\oplus}$$

In the initial coordination, the π-electron cloud of the olefin overlaps with the outer bond orbital of the metal cation. This causes stretching and eventual rupture of the R–M (metal) bond. An intramolecular rearrangement follows with the migration of the carbanion ($R^{\ominus}$) to the most electron-deficient carbon atom of the double bond. A new covalent bond and a new carbanion are formed simultaneously:

$$H_2C{=}CHX \cdots M^{\oplus} \cdots R \longrightarrow R{-}CH_2{-}\overset{\ominus}{C}HX\text{------}M^{\oplus}$$

where M represents the metal.

The basicity of the anion portion of the initiating species is also important. For instance, fluorenelithium initiates polymerizations of methyl methacrylate but fails to initiate polymerizations of styrene. A more electronegative butyl anion from butyl lithium, on the other hand, initiates polymerizations of both monomers. Yet, it was also shown that the order of reactivity is often contrary to the inherent basicity.[103] This may be due to the size of the aggregates that the particular organometallic compound forms. In addition, higher reactivity may also be due to favorable energy requirements for breaking down the aggregates.[103]

The presence of heteroatoms like oxygen or nitrogen in the monomer causes rapid complexation with the organometallic compounds. For instance, when butyllithium initiates polymerization of styrene in hydrocarbon solvents, there is an induction period and the overall reaction is slow and sluggish. When, however, it initiates polymerization of *o*-methoxystyrene under the same conditions, there is no induction period and the reaction is rapid. This is due to the initiator coordinating with the oxygen atom[107–109]:

$$C_4H_9{-}Li + CH_2{=}CH(C_6H_4OCH_3) \longrightarrow CH_2{=}CH(C_6H_4OCH_3)\cdots Li^{\delta\oplus}{-}C_4H_9^{\delta\ominus} \longrightarrow C_4H_9{-}CH_2{-}\overset{\ominus}{C}H(C_6H_4OCH_3)\cdots Li^{\oplus}$$

A similar coordination with the heteroatom takes place in polymerizations of acrylic and methacrylic esters.[114] Here, the metal coordinates with the electron cloud of the whole conjugated structure. The overlap stretches and eventually ruptures the metal alkyl bond. A new carbon-to-carbon covalent bond forms together with a new metal-to-carbon linkage:

$$\mathrm{CH_2{=}C(CH_3){-}C(=O){-}OCH_3 \cdots M^{\oplus}R'^{\ominus} \longrightarrow R{-}CH_2{-}C^{\ominus}(CH_3)(C{=}O(OCH_3)) \cdots M^{\oplus}}$$

where M represents the metal. The product might be a resonance hybrid of the two structures, where the metal is associated with the carbanion in one and with the oxygen in the other.[119]

Some abnormalities were reported in the initiations of methyl methacrylate polymerizations in toluene by butyllithium. Their nature is such that they suggest the possibility of more than one reaction taking place simultaneously.[113] One, which must be the major one, is that of the organometallic compound reacting with the carbon-to-carbon double bond as shown above. The other, minor one, may be with the carbon-to-oxygen double bond. The major reaction produces methyl methacrylate anions. The minor reaction, however, yields butyl isopropenyl ketone with an accompanying formation of lithium methoxide[113]:

$$\mathrm{C_4H_9Li + CH_2{=}C(CH_3){-}C(=O){-}OCH_3 \longrightarrow CH_2{=}C(CH_3){-}C(=O){-}C_4H_9 + LiOCH_3}$$

Lithium methoxide does not initiate polymerizations of methyl methacrylate. The ketone molecules, however, react with carbanions on the growing chain. The resultant anions are less reactive than methyl methacrylate anions and can only add new methyl methacrylate monomers slowly. Once added, however, the reaction proceeds at a normal rate.[113] Polymerizations of methyl methacrylate in polar solvents, on the other hand, proceed in what might be described as an ideal manner with formations of only one kind of ion pair.[131–134]

Other monomers can also exhibit abnormal behavior in some anionic polymerizations. Thus, for instance, organomagnesium initiation of methacrylonitrile polymerization results in formation of two types of active centers[115,116]:

$$\mathrm{RMgX + CH_2{=}C(CH_3){-}C{\equiv}N \longrightarrow \begin{cases} R{-}CH_2{-}C^{\ominus}(CH_3)(C{\equiv}N) \cdots \overset{\oplus}{M}gX \\ R{-}MgX \cdots N{\equiv}C{-}C(CH_3){=}CH_2 \end{cases}}$$

Methyl methacrylate polymerizations, initiated by organomagnesium compounds, also yield abnormal products. Here, the active centers are unusually persistent and stabile. In addition, the α-carbon atoms of the monomers were found to assume tetrahedral configurations.[110–112] This suggests that the active centers contain covalent magnesium carbon bonds. Also, gel permeation chromatography curves of the products show that more than one active center operates independently.[110–112] A "*pseudoanionic*" mechanism was therefore postulated for polymerizations of acrylic and methacrylic esters[111,112] by Grignard reagents.

Three different simultaneous reactions appear to be taking place in butyllithium initiated polymerizations of vinyl ketones in benzene[118]:

$$(C_4H_9Li)_6 \rightleftharpoons 6\ C_4H_9Li$$

$$C_4H_9Li + CH_2{=}C(CH_3){-}C({=}O){-}CH_3 \xrightarrow{1,4} C_4H_9{-}CH_2{-}C(CH_3){=}C(CH_3){-}O^{\ominus}\cdots Li^{\oplus}$$

$$\xrightarrow{1,2} C_4H_9{-}CH_2{-}C^{\ominus}(CH_3)(C({=}O)CH_3)\cdots Li^{\oplus}$$

$$\xrightarrow{3,4} CH_3{-}C({=}CH_2){-}C(CH_3)(C_4H_9){-}O^{\ominus}\cdots Li^{\oplus}$$

(monomer atoms numbered 1, 2, 3, 4: CH_2 = 1, C = 2, C = 3, O = 4)

The initiations of the polymerizations of the conjugated dienes in inert hydrocarbons are also believed to be preceded by coordination of the organometallic compounds with the π-electron clouds of the monomers:

$$\overset{\delta\ominus}{R}{-}\overset{\delta\oplus}{Li} + CH_2{=}CH{-}C(CH_3){=}CH_2 \longrightarrow \text{[π-complex of } Li^{\oplus}\text{ and } R^{\ominus}\text{ with the diene]}$$

The rearrangement that follows can result in either 1,2; 1,4; or 3,4 additions to the double bond:

$$C_4H_9Li + \underset{1}{CH_2}{=}\underset{2}{CH}{-}\underset{3}{C}(CH_3){=}\underset{4}{CH_2} \xrightarrow{1,2} C_4H_9{-}CH_2{-}CH^{\ominus}(C(CH_3){=}CH_2)\cdots Li^{\oplus}$$

$$\xrightarrow{1,4} C_4H_9{-}CH_2{-}CH{=}C(CH_3){-}CH_2^{\ominus}\cdots Li^{\oplus}$$

$$\xrightarrow{3,4} C_4H_9{-}C(CH_3)(CH{=}CH_2){-}CH_2^{\ominus}\cdots Li^{\oplus}$$

3.4.1.2. One-Electron Transfer Initiation

Alkali metals initiate anionic polymerizations of olefins in either homogeneous or heterogeneous conditions. This depends upon the metal and upon the solvent. For instance, potassium is soluble in ethers, like dimethoxyethane or tetrahydrofuran, and the initiation conditions are homogeneous. On the other hand, sodium dispersions are insoluble in hydrocarbons and the initiations are heterogeneous. Liquid ammonia is a solvent for many alkali metals, though for some, like sodium, it can be a reactant and form metal amides, if traces of iron are present. Initiation reactions in many metal solutions take place by an electron transfer from the metal to the monomer to form anion radicals. The resultant anion radicals may then undergo propagation reactions. These propagations can proceed anionically, or by a free-radical mechanism, or by both simultaneously. If the radicals are unstable, the anion radicals dimerize and the propagation proceeds by an anionic mechanism at both ends of the chain.

When liquid ammonia is employed as the solvent, as stated earlier, the particular mechanism of initiation will depend upon the metal used. Lithium metal forms solutions in liquid ammonia and initiates polymerization of monomers like methacrylonitrile by an electron transfer[136]:

$$Li + CH_2{=}C(CH_3)(C{\equiv}N) \longrightarrow \cdot CH_2{-}\overset{\ominus}{C}(CH_3)(C{\equiv}N)\cdots Li^{\oplus}$$

These radical anions couple and chain growth takes places from both ends of the chain:

$$2\ \cdot CH_2{-}\overset{\ominus}{C}(CH_3)(C{\equiv}N)\cdots Li^{\oplus} \longrightarrow Li^{\oplus}\cdots \overset{\ominus}{C}(H_3C)(N{\equiv}C){-}CH_2{-}CH_2{-}\overset{\ominus}{C}(CH_3)(C{\equiv}N)\cdots Li^{\oplus}$$

Dimerization of radical ions depends not only upon the radical's stability but also upon the π-energy changes that accompany the reaction.[124,135]

Potassium metal in ammonia, however, initiates polymerizations of monomers like methacrylonitrile or styrene differently. These reactions include additions of amide ions to the olefins and formations of amine groups at the end of the chains[136]:

$$2K + CH_2{=}CH(C_6H_5) + 2NH_3 \longrightarrow 2K^{\oplus} + CH_3{-}CH_2(C_6H_5) + 2NH_2^{\ominus}$$

$$K^{\oplus} + NH_2^{\ominus} \rightleftharpoons KNH_2$$

$$NH_2^{\ominus} + CH_2{=}CH(C_6H_5) \longrightarrow H_2N{-}CH_2{-}\overset{\ominus}{C}H(C_6H_5)$$

Solutions of potassium metal in ethers, however, form ion radicals through additions of electrons to the monomers. It should be noted that fresh solutions of potassium metal in various ethers like tetrahydrofuran or dimethoxyethane are blue in color. This blue color is attributed to the presence of spin-coupled electron pairs (e_2). The initiation of styrene polymerization that takes place between 0 and –78 °C is therefore pictured as follows[136]:

$$e_2 + CH_2{=}CH(C_6H_5) \longrightarrow e + \cdot CH_2{-}\overset{\ominus}{C}H(C_6H_5)$$

$$e + CH_2{=}CH(C_6H_5) \longrightarrow \cdot CH_2{-}\overset{\ominus}{C}H(C_6H_5)$$

When conditions are heterogeneous and sodium metal dispersions are used in polymerizations of dienes, the initiation mechanism is also by an electron transfer. It is believed[135] that initially a 1,2 anion radical forms. This is followed by a coupling reaction:

$$Na + CH_2{=}CH{-}CH{=}CH_2 \longrightarrow$$

$$\underset{\text{initial anion–radical}}{Na^{\oplus}\cdots{}^{\ominus}CH_2{-}\dot{C}H(CH{=}CH_2)} \underset{\text{rearrangement}}{\rightleftharpoons} Na^{\oplus}\cdots{}^{\ominus}CH_2{-}CH{=}CH{-}\dot{C}H_2$$

the coupling process:

$$2Na^{\oplus}\ldots{}^{\ominus}CH_2{-}CH{=}CH{-}CH_2^{\bullet} \rightarrow Na^{\oplus}\ldots{}^{\ominus}CH_2{-}CH{=}CH{-}CH_2{-}CH_2{-}CH{=}CH{-}CH_2^{\ominus}\ldots{}^{\oplus}Na$$

a different kind of coupling is also possible:

$$Na^{\oplus}\text{-----}{}^{\ominus}CH_2\text{—}\dot{C}H(CH{=}CH_2) + \dot{C}H_2\text{—}CH{=}CH\text{—}CH_2^{\ominus}\text{-----}Na^{\oplus} \longrightarrow Na^{\oplus}\text{-----}{}^{\ominus}CH_2\text{—}CH(CH{=}CH_2)\text{—}CH_2\text{—}CH{=}CH\text{—}CH_2^{\ominus}\text{-----}Na^{\oplus}$$

The following initiation mechanism was postulated[120] for heterogeneous conditions. At first, the highly polarized monomers adsorb strongly to the metal surfaces. Electron transfer takes place. The adsorbed molecules are assumed to be still sufficiently mobile to be able to rotate after adsorption. The rotation allows the radical anions that form to couple. The concentration of the anionic charges that develop on the methine carbons creates strong enough attractive forces to remove the metal cations from the surface. Solvents that strongly solvate the cations, like tetrahydrofuran, enhance the process. To fit the above mechanism,[120] the monomers must be capable of adsorbing strongly to the metal surfaces and the solvents must also be strongly solvating. When the reactions are carried out without solvents or in inert solvents like heptane, the free-energy gain due to solvation is largely lost. This increases the probability that in such instances the propagations occur on the surface of the metals until the oligomers grow to a certain length. When sufficient length is reached to create a favorable free-energy change, desorption occurs and the molecules pass into solution.[120]

Electron transfer initiations can also result from reactions of alkali metals with aromatic hydrocarbons or with aromatic ketones that result in formations of radical ions:

$$Na + \text{naphthalene} \longrightarrow [\text{naphthalene}]^{\dot{}-}\,Na^+$$

$$K + \text{anthracene} \longrightarrow \text{anthracene}^{\dot{}\ominus}\,K^{\oplus}$$

$$Li + C_6H_5\text{—}CO\text{—}C_6H_5 \longrightarrow C_6H_5\text{—}\dot{C}(O^{\ominus}\,Li^{\oplus})\text{—}C_6H_5$$

The ion radicals transfer both the electron and the ion charge to the monomers in the process of initiation, as was shown by Szwarc and co-workers[125–130]:

$$[\text{naphthalene}]^{\dot{}-}\,Na^+ + CH_2{=}CH(C_6H_5) \rightleftharpoons \text{naphthalene} + \cdot CH_2\text{—}CH^{\ominus}(C_6H_5)\text{-----}Na^{\oplus}$$

The styrene monomer radical ion is unstable and tends to dimerize:

$$2\ \cdot CH_2\text{—}CH^{\ominus}(C_6H_5)\text{-----}Na^{\oplus} \longrightarrow Na^{\oplus}\text{-----}{}^{\ominus}CH(C_6H_5)\text{—}CH_2\text{—}CH_2\text{—}CH^{\ominus}(C_6H_5)\text{-----}Na^{\oplus}$$

Metal ketyls form in reactions of alkali metals with aromatic ketones in some polar solvents, like dioxane or tetrahydrofuran. These ketyls exist in equilibrium mixtures of monomeric anion radicals and dimeric dianions.[136] Originally, there was some controversy about the mechanism of initiation of monomers like acrylonitrile or methyl methacrylate by sodium benzophenone. The following mechanism was derived from spectral evidence. The initiation is by transfer of the electron and the ion charge[137–139]:

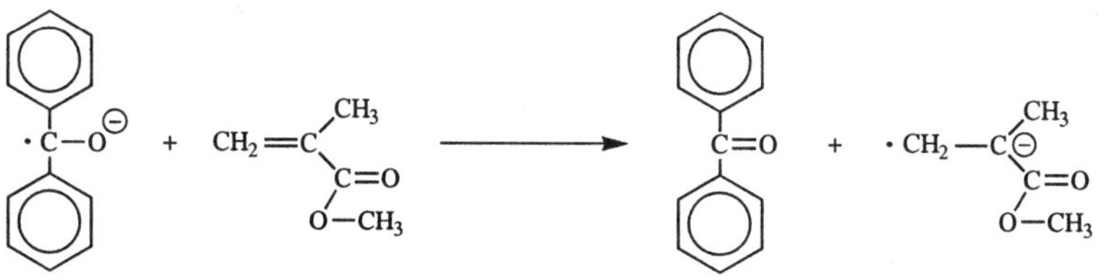

3.4.1.3. Initiations by Alfin Catalysts

These are catalysts that are formed by combining an alkyl sodium with sodium alkoxide and with an alkali metal halide.[264–269] A typical, effective catalyst for polymerization of dienes consists of allyl sodium, sodium isopropoxide, and sodium chloride. The preparation of such a catalyst is carried out by combining amyl chloride with sodium and then reacting the product with isopropyl alcohol. After that, propylene is bubbled through the reaction mixture to form allyl sodium. The reactions can be summarized as follows:

$$1.5\ C_5H_{11}Cl + 3\ Na \rightarrow 1.5\ C_5H_{11}Na + 1.5\ NaCl$$

$$C_5H_{11}Na + (CH_3)_2CHOH \rightarrow (CH_3)_2CHONa + C_5H_{12}$$

$$C_5H_{11}Na + CH_3\text{-}CH{=}CH_2 \rightarrow C_5H_{12} + CH_2{=}CH\text{-}CH_2Na$$

Diisopropyl ether may be used in place of isopropyl alcohol. In that case, the reaction does not require addition of propylene because the olefin forms in situ.[269] These catalysts are particularly effective in polymerizations of some conjugated dienes to very high molecular weight products. Styrene also polymerizes. Polymers of the dienes that form are high in *trans*-1,4 repeat units. Butadiene is polymerized by these catalysts much more rapidly than is isoprene. On the other hand, 2,3-dimethylbutadiene fails to polymerize.

There is no general agreement about the mechanism of these polymerizations. Both anionic and free-radical mechanisms were proposed[264–269] as the most probable reaction paths. The role of sodium chloride is not clear in this mixture, though it was shown that it is essential.[264–269] It may act, perhaps, as a support for the catalyst and may be a part of some sort of lattice involving both sodium alkoxide and allylsodium. The anionic mechanism is pictured as follows[264–268]:

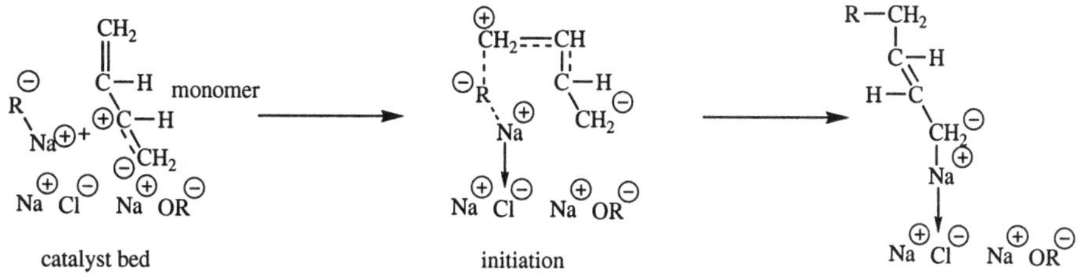

catalyst bed → initiation →

The free-radical mechanism was suggested due to the high predominance of *trans*-1,4 placement in polymerization of butadiene.[269] According to that mechanism a complex of sodium isopropoxide and allyl sodium forms first:

The monomer adsorbs on the surface and is visualized as displacing the allyl anion from the complex to form an ion pair first and then, through an electron rearrangement, a radical[269]:

$$[CH_2\text{-}CH\text{-}CH\text{-}CH_2Na]^{\oplus}\ {}^{\ominus}[CH_2 \cdots CH \cdots CH_2] \rightarrow \bullet CH_2\text{-}CH{=}CH\text{-}CH_2{}^{\ominus}Na^{\oplus} + \bullet CH_2\text{-}CH{=}CH_2$$

The polymerization is assumed to go on from this point via a free-radical mechanism until combination with an allyl radical takes place. Because the allyl radical is bound to the catalyst surface, combination does not take place readily and high molecular weights are attained.

3.4.1.4. Electroinitiation of Anionic Polymerizations

Electrolytic polymerizations were described in the section dealing with cationic polymerizations. Anionic polymerizations can also be initiated in an electric field. When $LiAlH_4$ or $NaAl(C_2H_5)_4$ are used as electrolytes in tetrahydrofuran solvent, "living" polymers can be formed from α-methylstyrene.[273] The deep red color of carbanions develops first at the cathode compartments of divided cells. By cooling the solutions to –80 °C almost quantitative conversions to high molecular weight polymers take place.[273] Similar conditions yield polystyrene.[274] 4-Vinyl pyridine polymerizes in liquid ammonia in the same manner.[274] Initiation results from reduction of the monomer by direct electron transfer at the cathode to form a red-orange vinyl pyridyl radical anion:

$$CH_2{=}CH(C_5H_4N) + e^{\ominus} \longrightarrow \cdot CH_2{-}\ddot{C}H^{\ominus}(C_5H_4N)$$

Another example is polymerization of isoprene in an electric field in tetrahydrofuran solution. Here, too, a "living" polymer forms.[275]

3.4.2. Propagation in Anionic Chain-Growth Polymerization

The propagation reaction consists of successive additions of monomer molecules to the active centers of the growing chains:

$$\sim CH_2{-}CHX^{\ominus}\text{-----}M^{\oplus} + CH_2{=}CHX \longrightarrow \sim CH_2{-}CHX{-}CH_2{-}CHX^{\ominus}\text{-----}M^{\oplus}$$

No matter what the mechanism of initiation is, the propagation reaction takes place strictly between the monomer and the growing polymeric chain with or without a counterion.

When the reaction occurs in nonpolar solvents, the propagation step is not hampered as much by a tendency of ion pairs to cluster into aggregates, as is encountered in initiation. For instance, in butyllithium-initiated polymerizations of styrene in benzene, the propagation step is much faster than the initiation.[140,141] This is probably due to an absence of aggregates. Some association between the growing polymeric chains, however, does occur.[140] It may be shown as follows:

$$2 \sim CH_2{-}CH(R)\text{-----}Li \longrightarrow \sim CH_2{-}CH(R)(\cdots Li \cdots)_2CH(R){-}CH_2\sim$$

These association equilibria, however, are mobile in character.[140] The driving force in the propagation reaction is similar to that in the initiation. In nonpolar solvents the reaction with the incoming monomers are similar to those in the initiation step. The monomers coordinate with the cations at the end of the chains first. This is followed by intramolecular rearrangements that lead to regenerations of new metal carbon linkages:

$$\sim CH_2{-}CH(C_6H_5)^{\ominus}\text{-----}Li^{\oplus} + CH_2{=}CH(C_6H_5) \longrightarrow \sim CH_2{-}CH(C_6H_5)^{\ominus}\text{-----}Li^{\oplus}\cdots(H_2C{=}CH{-}C_6H_5)$$

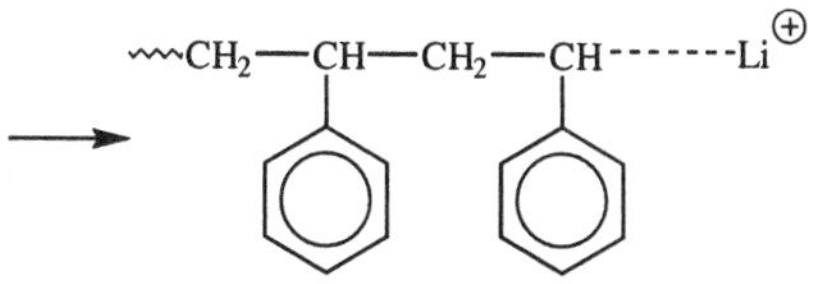

$$\longrightarrow \sim CH_2-\underset{C_6H_5}{CH}-CH_2-\underset{C_6H_5}{CH}\cdots\cdots Li^{\oplus}$$

In polar solvents, on the other hand, these reactions can go to the other extreme. The propagation can simply consist of successive additions of the monomers to the growing anions.

In homogeneous anionic polymerizations of simple vinyl monomers, steric placement is also temperature-dependent, just as it is in cationic polymerizations. Syndiotactic placement is favored in polar solvents at low temperature. In nonpolar solvents, however, isotactic placement predominates at the same temperatures. Here, too, this results mainly from the degree of association with the counterion.[130]

Much of our current knowledge of the propagation reaction is based on studies carried out in highly solvating ether solvents. Less information is available about homogeneous reactions in a nonpolar medium. Generally, though, the rate of propagation increases with solvent polarity and with the degree of ion-pair dissociation.[49] Organolithium compounds undergo the greatest degree of solvation when changed from hydrocarbons to polar ether solvents. Cesium compounds, on the other hand, are least affected by changes in solvent polarity. In addition, NMR studies of polystyryl carbanion structures associated with lithium, potassium, and cesium countercations were conducted in different solvents and at different temperatures.[315] The results show an interaction between the larger radius cations and the phenyl rings of the ultimate monomer units in the chains. The structures with potassium and cesium counterions, judging from model compounds, were found to be planar with sp^2-hybridized α-carbon. It suggests that in the presence of the larger counterions, rotation of the terminal phenyl ring in styrene polymerization is strongly hindered.[315]

Propagation rates also depend upon the structures of the monomers. For polymerizations initiated by alkali amides the following order of reactivity was observed[150]:

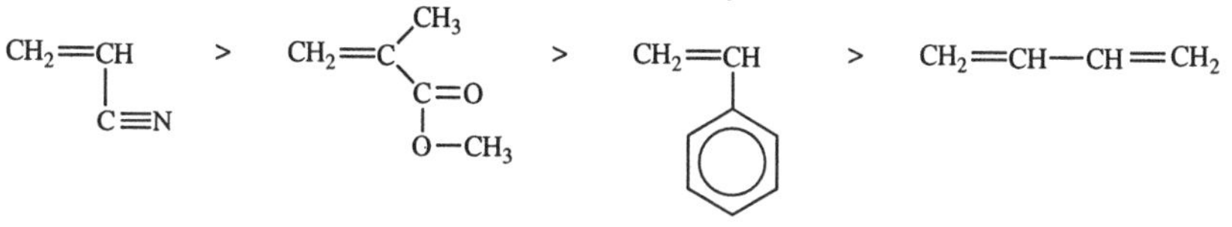

$$\underset{C\equiv N}{CH_2=CH} > CH_2=\underset{\underset{O-CH_3}{C=O}}{C(CH_3)} > \underset{C_6H_5}{CH_2=CH} > CH_2=CH-CH=CH_2$$

Also, methyl substitution on the α-carbon tends to decrease the reaction rate due to the electron-releasing effect of the alkyl group. This tends to destabilize the carbanion and also to cause steric interference with solvation of the chain end and with the addition of the monomer.[150]

3.4.2.1. Steric Control in Anionic Polymerization

Use of hydrocarbon solvents has an advantage in polymerizations of conjugated dienes, because they yield some steric control over monomer placement. This is true of both tacticity and geometric isomerism. As stated earlier, the insertions can be 1,2; 3,4; or 1,4. Furthermore, the 1,4-placements can be *cis* or *trans*. Lithium and organolithium initiators in hydrocarbon solvents can yield polyisoprene, for instance, which is 90% *cis*-1,4 in structure.[41] The same reaction in polar solvents, however, yields polymers that are mostly 1,2 and 3,4, or *trans*-1,4 in structure. There is still no mechanism that fully explains steric control in polymerization of dienes.

High *cis*-1,4 polyisoprene forms with lithium or alkyllithium initiators in nonpolar solvents, because propagation takes place through essentially covalent or intimate ion-pair lithium to carbon bonds.[41] An intermediate pseudo-six-membered ring is believed to form during the addition of the diene[41]:

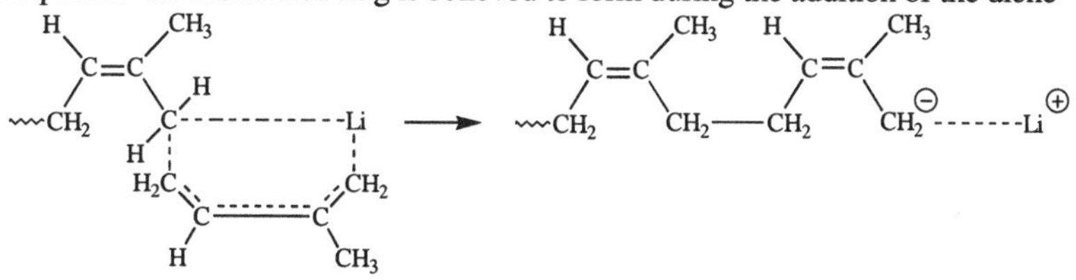

H, CH_3, C=C, $\sim CH_2$, H, C, Li, H, H_2C, CH_2, C, C, H, CH_3 $\longrightarrow$ H, CH_3, H, CH_3, C=C, C=C, $\sim CH_2$, CH_2-CH_2, $CH_2^{\ominus}\cdots\cdots Li^{\oplus}$

Formation of such intermediates is favorable for lithium, because it has a small ionic radius and is high in the proportion of *p*-character. Organometallic compounds of the other alkali metals (sodium, potassium, rubidium, and cesium) are more polar and more dissociated. They react essentially as solvated ions even in a hydrocarbon medium, yielding high 3,4 placement.

O'Driscoll, Yonezawa, and Higashimura[142] proposed a mechanism for steric control. In isoprene polymerization the terminal charges are complexed with the metal cations. These cations are close to the active centers through the occupied π-orbitals of the chain ends and the unoccupied *p*-orbitals of the lithium ions. In the transition state the monomers are complexed with the cations in the same way.[142] The lithium cations are assumed to be in hybridized tetrahedral sp^3 configurations with four vacant orbitals. The chain ends are presumed to be allylic and the diene monomers are bidentate.[142] During the propagation steps both the monomers and chain ends complex with the same counterions:

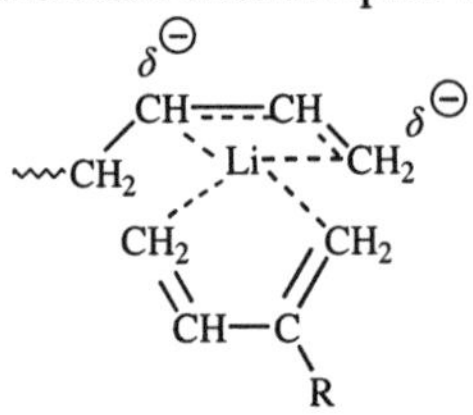

$\delta^{\ominus}$ CH=CH $\delta^{\ominus}$ ~~CH_2 Li CH_2 CH_2 CH_2 CH—C R

In hydrocarbon solvents, the complexes are tight and the rotations of the C_2–C_3 bonds are sterically hindered by the methyl groups. This constrains the 1,4-additions to *cis*-configurations. In polar solvents, however, like tetrahydrofuran, the complexes are loose and thermodynamically favored *trans* additions take place.[142]

It was observed, however, that the polymerizations of 2,3-dimethylbutadiene with organolithium initiators in nonpolar solvents result in high *trans*-1,4 structures.[143] This appears to contradict the above-proposed mechanism.

Proton NMR spectra show that solvation shifts the structures of the carbanionic chain ends from localized 1,4-species to delocalized "π-allylic" type structures[144]:

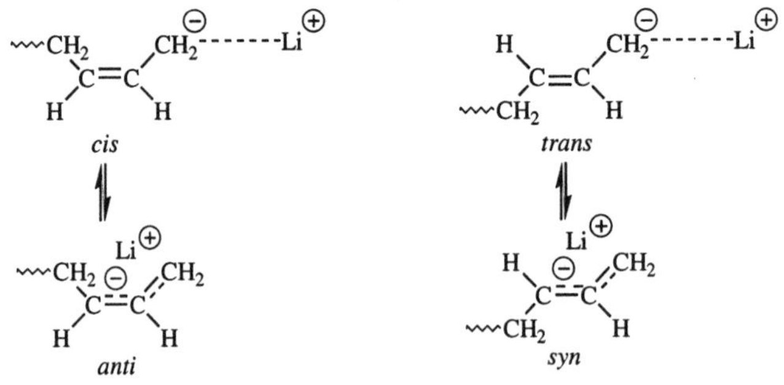

~~CH_2 $CH_2^{\ominus}$------$Li^{\oplus}$ C=C H H *cis* ⇌ ~~CH_2 $\ominus$ $Li^{\oplus}$ CH_2 C=C H H *anti*

H $CH_2^{\ominus}$------$Li^{\oplus}$ C=C ~~CH_2 H *trans* ⇌ H $\ominus$ $Li^{\oplus}$ CH_2 C=C ~~CH_2 H *syn*

The σ-bonded lithium chain can be expected to predominate. In highly solvating solvents, such as ethers, the π-allyl structure is dominant leading to high 1,2 placements. Because the 2,3-bond is maintained, the above-shown equilibria should not be expected to lead to *cis–trans* isomerization.[144] Such isomerizations do not take place for butadiene or for isoprene when they are polymerized in hydrocarbon solvents. They do occur, however, in polar solvents at high temperatures. This suggests that additional equilibria exist between the π-allylic structures and the covalent 1,2 chain ends.[144] Table 3.3 shows the manner in which different polymerization initiators and solvents affect the microstructures of polyisoprenes.

Polymerizations of polar monomers, like acrylic and methacrylic esters with alkyllithium initiators, yield the greatest amount of steric control.[151] Almost all isotactic poly(methyl methacrylate) forms at low temperatures. Addition of Lewis bases such as ethers or amines reduces the degree of isotactic placement. Depending upon the temperature, atactic or syndiotactic polymers form.[151] Also, butyllithium in heptane yields an isotactic poly(*N,N'*-dibutylacrylamide) at room temperature.[152]

The propagation rates for methyl methacrylate polymerization in polar solvents like tetrahydrofuran or dimethylformamide are lower than the rates of initiation.[145] There is no evidence,

Table 3.3. Effect of the Counterions and Solvents on the Microstructures of Polyisoprenes[a]

Initiator	Solvent[b]	Approximate % of adduct			
		cis-1,4	*trans*-1,4	1,2	3,4
Lithium dispersion	Alkane	94	—	—	6
Ethyllithium	Alkane	94	—	—	6
Butyllithium	Alkane	93	—	—	7
Sodium dispersion	Alkane	—	43	6	51
Ethylsodium	Alkane	6	42	7	45
Butylsodium	Alkane	4	35	7	54
Potassium dispersion	Alkane	—	52	8	40
Ethylpotassium	Alkane	24	39	6	31
Butylpotassium	Alkane	20	41	6	33
Rubidium dispersion	Alkane	5	47	9	39
Cesium dispersion	Alkane	4	51	8	37
Ethyllithium	Ether	6	30	5	59
Ethylsodium	Ether	—	14	10	76
Lithium dispersion	Ether	3	27	6	64

[a]From various sources in the literature.
[b]The alkane is a low boiling aliphatic hydrocarbon.

however, that more than one kind of ion pair exists.[146–148] The ion pairs that form are apparently contact-ion pairs.[145] Furthermore, based on the evidence, the counterions are more coordinated with the enolate oxygen atoms of the carbonyl groups than with the α-carbons. As a result, they exert less influence on the reactivity of the carbanions.[145] The amount of solvation by the solvents affects the reaction rates. In addition, "intramolecular solvations" from neighboring ester groups on the polymer chains also affects the rates. In solvent like dimethylformamide or tetrahydrofuran, or similar ones,[145] the propagating chain ends–ion pairs are pictured as hybrid intermediates between two extreme structures. This depends upon the counterion, the solvent, and the temperature[145]:

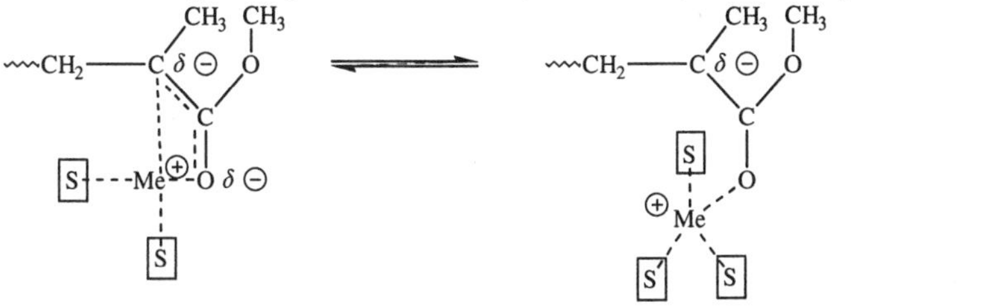

where S means a solvent molecule; Me represents a metal.

Several mechanisms were offered to explain steric control in polymerizations of polar monomers. Furukawa and co-workers[119] based their mechanism on infrared spectroscopy data of interactions between the cations and the growing polymeric chains in polymerizations of methyl methacrylate and methacrylonitrile. They observed a correlation between the tacticities of the growing molecules and the carbonyl stretching frequencies. The higher the frequency, the higher is the amount of isotactic placement in the resultant chains. The adducts, as in the initiation reactions, are resonance hybrids of two structures, **A** and **B**:

A **B**

where Me represents a metal. Furukawa and co-workers concluded that, due to the extremely low tendency of the adducts to dissociate,[119] the carbonyl absorption can only be ascribed to undissociated ion pairs. The magnitude of the carbonyl absorption and the shifts to higher frequencies show the degree of contribution of structure **B**, shown above. The absorption and the shifts were also explained by the configurations of the electrons in the antibonding orbitals of the carbonyl groups. The higher the stretching frequencies, the nearer are the positions of the counterions to the carbonyl groups of the terminal units.[149] This is accompanied by higher tacticity.[119] The carbanions on the terminal units in the transition states are located near the β-carbons of the incoming monomers. At the same time, rotations around the axis through these two carbons may be quite restricted when the cations are in the vicinity of the carbonyl groups of the terminal unit and near the incoming monomers. In this manner isotactic placement is enhanced[119]:

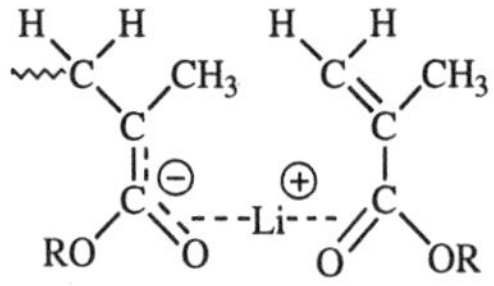

The same mechanism was proposed for the polymerization of methacrylonitrile.[119]

Cram and Kopecky[159] offered a different mechanism of steric control. According to their mechanism, during a methyl methacrylate polymerization the growing ether enolate possesses a complete alkoxide character:

Attacks by the alkoxide ion on the carbonyl groups of the penultimate units lead to formations of six-membered rings:

The six-membered ring is destroyed in the process of propagation:

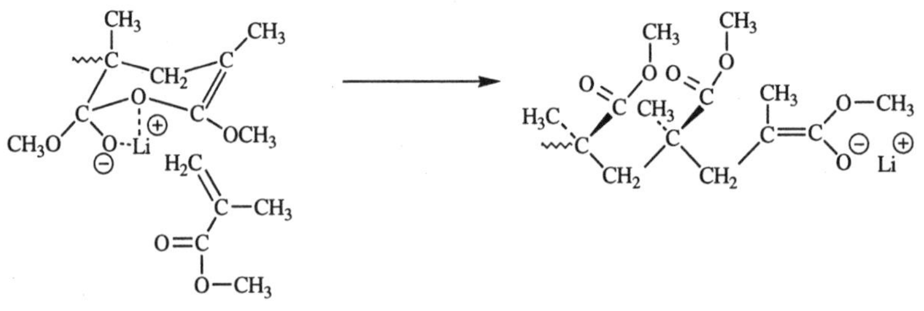

The transition state forms by a 1,4-dipolar addition to a polarized double bond. Coordination of the lithium atom to two oxygen atoms determines stereoregulation. Each new incoming monomer must approach from below the plane because the other side is blocked by an axial methyl group. This favors isotactic placement. There is doubt, however, whether it is correct to assume a rigid six-membered cyclic alkoxide structure for a propagating lithium enolate.[41]

A slightly similar model was suggested by Bawn and Ledwith.[41] It is based on the probability that a growing polymeric alkyllithium should have some enolic character, with the lithium coordinating to the carbonyl oxygen of the penultimate unit:

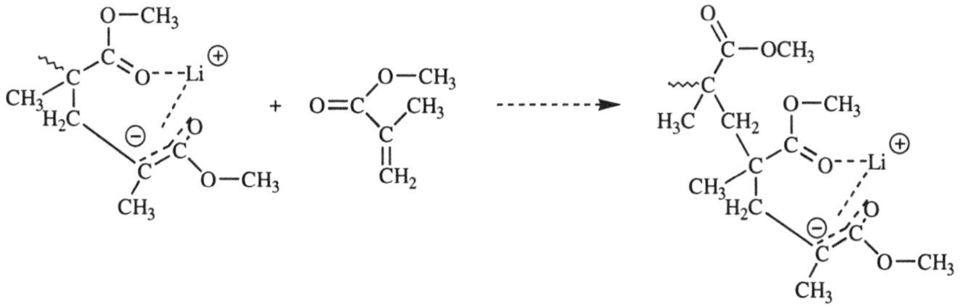

$$\text{[cyclic Li}^{\oplus}\text{ intermediate with O–CH}_3\text{, C=O, CH}_3\text{, H}_2\text{C groups]} + O{=}C(O{-}CH_3)C(CH_3){=}CH_2 \dashrightarrow \text{[chain extended by one unit, C–OCH}_3\text{, cyclic Li}^{\oplus}\text{ intermediate re-formed]}$$

The cyclic intermediate forms due to intramolecular solvation of the lithium and due to intramolecular shielding of one side of the lithium ion. The nucleophilic attack by the monomer, therefore, has to occur from the opposite side. The transition state is similar to an S*N*2 reaction. When the bond between the lithium and the incoming monomer forms, the oxygen–lithium bond ruptures. Simultaneously, the charge migrates to the methylene group of the newly added monomer. The resultant new molecule is stabilized immediately by intramolecular solvation as before. In this manner the retention of configuration is assured, if the incoming monomer always assumes the same configuration toward the lithium ion.

NMR spectra of poly(*N*,*N*-dimethylacrylamide) formed with *sec*-butyllithium in both polar and nonpolar solvents show that the penultimate unit does affect monomer placement.[153] Also, a coordination was observed with both heteroatoms,[152,153] the one on the ultimate and the one on the penultimate unit.

Many refinements were introduced into the various proposed explanations of steric control in anionic polymerizations[373] over the last twenty years. Two important features of these mechanisms are: (1) coordinations of the chain ends with the counterions, and (2) counterion solvations.

3.4.2.2. Hydrogen Transfer Polymerization

Anionic polymerization of unsubstituted acrylamide, catalyzed by strong bases, does not yield typical vinyl polymers. Instead, the product is a 1,3-adduct, poly(*β*-alanine).[154] Two alternate reaction paths were originally proposed[154]:

$$1.\quad CH_2{=}CH{-}\overset{O}{\overset{\|}{C}}{-}NH_2 + \overset{\ominus}{B}{:} \rightleftharpoons CH_2{=}CH{-}\overset{O}{\overset{\|}{C}}{-}\overset{\ominus}{NH} + BH$$

$$2.\quad CH_2{=}CH{-}\overset{O}{\overset{\|}{C}}{-}NH_2 + 2\overset{\ominus}{B}{:} \rightleftharpoons B{-}CH_2{-}\overset{\ominus}{CH}{-}\overset{O}{\overset{\|}{C}}{-}NH_2 \rightleftharpoons B{-}CH_2{-}CH_2{-}\overset{O}{\overset{\|}{C}}{-}\overset{\ominus}{NH} + BH$$

Subsequently, studies of the rate of disappearance of acrylamide in dry sulfolane or pyridine with potassium-*t*-butoxide initiator led to the following proposed mechanism[155]:

$$B{-}CH_2{-}\overset{\ominus}{CH}{-}\overset{O}{\overset{\|}{C}}{-}NH_2 + CH_2{=}CH{-}\overset{O}{\overset{\|}{C}}{-}NH_2 \longrightarrow B{-}CH_2{-}CH_2{-}\overset{O}{\overset{\|}{C}}{-}NH_2 + CH_2{=}CH{-}\overset{O}{\overset{\|}{C}}{-}\overset{\ominus}{NH}$$

Propagation proceeds in this manner[155]:

$$CH_2{=}CH{-}\overset{O}{\overset{\|}{C}}{-}\overset{\ominus}{NH} + CH_2{=}CH{-}\overset{O}{\overset{\|}{C}}{-}NH_2 \longrightarrow CH_2{=}CH{-}\overset{O}{\overset{\|}{C}}{-}\underset{H}{\underset{|}{N}}{-}CH_2{-}\overset{\ominus}{CH}{-}\overset{O}{\overset{\|}{C}}{-}NH_2$$

$$\xrightarrow{CH_2{=}CH{-}\overset{O}{\overset{\|}{C}}{-}NH_2} CH_2{=}CH{-}\overset{O}{\overset{\|}{C}}{-}NH{-}CH_2{-}CH_2{-}\overset{O}{\overset{\|}{C}}{-}NH_2 + CH_2{=}CH{-}\overset{O}{\overset{\|}{C}}{-}\overset{\ominus}{NH} \longrightarrow$$

$$CH_2{=}CH{-}\overset{O}{\overset{\|}{C}}{-}NH{-}CH_2{-}\overset{\ominus}{CH}{-}\overset{O}{\overset{\|}{C}}{-}NH{-}CH_2{-}CH_2{-}\overset{O}{\overset{\|}{C}}{-}NH_2 \longrightarrow \text{etc.}$$

The propagating center is not an ion or a radical but a carbon-to-carbon double bond at the end of the chain. The monomer anion adds to this double bond. This process is a step-growth polymerization and the monomer anion is called an *activated monomer*. Not all acrylamide polymerizations, initiated by strong bases, however, proceed by a hydrogen transfer process. Depending upon reaction conditions, such as solvent, monomer concentration, and temperature, some polymerizations can take place through the carbon-to-carbon double bonds.[156]

Cis and *trans* crotonamides can also polymerize by hydrogen transfer polymerization. Sodium *t*-butoxide in pyridine yields identical polymers from both isomers.[165] Also, hydrogen transfer polymerization of acrylamide with optically active, basic catalysts yields optically active polymers.[157] The reactions can be carried out in toluene, using optically active alcoholates of amyl alcohol. The initiating ability of the metal ions is in the following order, Na > Ba > Ca > Mg > Al.[157] Optically active polymethacrylamide forms with optically active barium and calcium alcoholates, but not with the other cations.[157] In this reaction, however, the asymmetric synthesis takes place through an intermolecular hydrogen transfer rather than through an intermolecular hydrogen migration[157]:

$$\sim CH_2-\overset{CH_3}{\underset{\underset{NH_2}{|}}{\underset{C=O}{|}}{\overset{|\ominus}{C}}}\text{------}\overset{\oplus}{Me}-OR^{*} \xrightarrow{CH_2=C(CH_3)-C(=O)-NH_2} \sim CH_2-\overset{H_3C\ \ H-N(-H)-C(=O)-C(CH_3)=CH_2}{\underset{\underset{NH_2}{|}}{\underset{C=O}{|}}{\overset{\ominus}{C}}}\text{------}\overset{\oplus}{Me}-OR^{*}$$

$$\longrightarrow \sim CH_2-\overset{*}{C}(CH_3)(H)-C(=O)-NH_2 \;+\; CH_2=C(CH_3)-C(=O)-\overset{\ominus}{N}(H)\text{------}\overset{\oplus}{Me}-OR^{*}$$

3.4.3. Termination in Anionic Polymerization

Termination reactions in anionic polymerization, particularly with nonpolar monomers and in nonpolar solvent, are not common. If carbanion quenching impurities are absent, many polymerization reactions may not terminate after the complete disappearance of the monomer. Styryl anion (one of the most stable ones), for instance, can persist for a long time, such as weeks, after the monomer is consumed. Addition of more monomer results in a continuation of the reaction and a further increase in the molecular weight. The anionic "living" polymers retain their activities for considerably longer periods of time than do the cationic "living" ones.[130]

The termination steps in anionic polymerizations can result from deliberate introductions of carbanion quenchers, such as water or acids, or from impurities. Terminations, however, can take place in some instances through chain transferring a proton from another molecule like a solvent or a monomer, or even from a molecule of another polymer. In some solvents, like liquid ammonia, transfer to solvent is extensive, as in styrene polymerization by amide ions.[159]

In addition, in some polymerizations termination might occur from the following reactions:

1. Elimination of a hydride ion to form an unsaturated end.
2. Isomerization to an inactive anion.
3. Some irreversible reaction of the active center with a molecule of a monomer or a solvent.

It was observed, for instance, that hydrogen transfer from a monomer to the growing chain can be a way of termination in polymerizations of polar monomers, like acrylonitrile[158]:

$$\sim CH_2-\overset{\ominus}{C}(C\equiv N)(H)\text{-----}\overset{\oplus}{Me} + CH_2=C(C\equiv N)(H) \longrightarrow \sim CH_2-C(C\equiv N)(H)-H + CH_2=\overset{\ominus}{C}(C\equiv N)\text{-----}\overset{\oplus}{Me}$$

where Me represents the metal cation.

As mentioned above, the polystyryl carbanions are particularly stable and persist for weeks in nonpolar solvents. Yet, even in the absence of terminating agents, the concentration of the carbanion active centers decreases with time.[159] The mechanism of decay is not fully understood. Based on spectral evidence, it is believed to consist of a hydride-ion elimination[159]:

$$\sim\!\!\sim CH_2-\underset{C_6H_5}{CH}-CH_2-\underset{C_6H_5}{\overset{H}{C}}{}^{\ominus}\cdots Na^{\oplus} \longrightarrow \sim\!\!\sim CH_2-\underset{C_6H_5}{CH}-CH=\underset{C_6H_5}{CH} + NaH$$

The above reaction can be followed by an abstraction of an allylic hydrogen from the product of elimination by another active center:

$$\sim\!\!\sim CH_2-\underset{C_6H_5}{\overset{\ominus}{CH}} + \sim\!\!\sim CH_2-\underset{C_6H_5}{CH}-CH=\underset{C_6H_5}{CH} \longrightarrow \sim\!\!\sim CH_2-\underset{C_6H_5}{CH_2} + \sim\!\!\sim CH_2-\underset{C_6H_5}{\overset{\ominus}{C}}-CH=\underset{C_6H_5}{CH}$$

Polystyryl carbanions are much less stable in polar solvents. They decay within a few days at room temperature. At lower temperatures, however, the stability is considerably better. The termination in polar solvent occurs by a mechanism of abstracting α-hydrogens and/or by a nucleophilic attack on the carbon–oxygen bonds.

Polar monomers, like methyl methacrylate, acrylonitrile, or methyl vinyl ketone, contain substituents that react with nucleophiles. This can lead to terminations and side reactions that compete with both initiation and propagation.[159] An example is a nucleophilic substitution by an intramolecular backbiting attack of a propagating carbanion:

$$\sim\!\!\sim CH_2-C(CH_3)(COOCH_3)-CH_2-C(CH_3)(COOCH_3)-CH_2-\overset{\ominus}{C}(CH_3)(COOCH_3) \longrightarrow \text{cyclic } \beta\text{-keto ester (cyclohexanone ring bearing } CH_3, COOCH_3 \text{ substituents)} + CH_3-O^{\ominus}$$

Side reactions like the one shown above can be minimized by using less nucleophilic initiators and low temperatures. This can yield "living" polymerizations of acrylic and methacrylic monomers. In addition, it is possible to add common ions like LiCl to alkyllithium to tighten the ion pairs of the propagating anion–counterion species. That also increases the tendency to form "living" polymers.[374] This approach, however, offers only limited success.

Recently, it was found that Lewis acid assisted polymerizations of methyl methacrylate with aluminum porphyrin initiators yield "living" polymers.[307] The polymerizations of methacrylate esters with alkylaluminum porphyrin initiators occur through formations of enolate aluminum porphyrin intermediate as the growing species.[307] For the sake of illustration, the methylaluminum porphyrin molecules (see Fig. 3.3) can be designated as

$$\Big\Vert\!\!\!\!\diagdown Al-CH_3 \diagup\!\!\!\!\Big\Vert$$

to demonstrate the polymerization. The reaction can be illustrated as follows[307]:

$$\Vert Al-CH_3 \Vert + m\, \underset{O}{\overset{CH_3O}{}}\!\!\!C-C\!\!\underset{CH_2}{\overset{CH_3}{}} \xrightarrow{h\nu} \Vert Al \Vert-O-\overset{CH_3O}{C}=C(CH_3)-CH_2-\Big(-\underset{CO_2CH_3}{\overset{CH_3}{C}}-CH_2-\Big)_{m-1}-CH_3 \xrightarrow{\text{monomer}} \text{polymer}$$

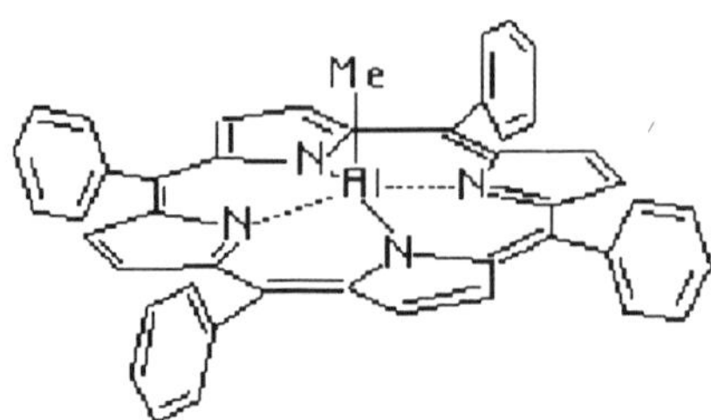

FIGURE 3.3. Mythylaluminum prophyrin molecule.

The preparatory procedure was improved[308] further by addition of sterically crowded organoaluminum phenolates to the reaction mixture:

C_4H_9 CH_3 C_4H_9

R″—⟨phenyl⟩—O——Al——O—⟨phenyl⟩—R″

R′ R′

where R′ is H and R″ is butyl, or R′ is butyl and R″ is methyl. This yields "living," very high molecular weight, monodisperse polymers of methyl methacrylate.[308] Monomer insertion might be taking place at the Al–C bond. The mechanism, however, is not known at this time.[308]

3.5. Coordinated Anionic Polymerizations

The catalysts for these polymerizations can be separated into two groups. To the first belong the so-called Ziegler–Natta catalysts, and to the second, transition metal oxides on special supports, like carbon black or silica-alumina, etc. Besides the two, there are related catalysts, like transition metal alkyls or metal halides that also catalyze some coordinated anionic polymerization. This group also includes transition metal-π-allylic compounds and transition metal hydrides.

The Ziegler–Natta catalysts received their initial attention when Ziegler showed that some transition metal halides, upon reaction with aluminum alkyls, can initiate polymerizations of ethylene.[160] Polymers that form are linear and high in molecular weight. The reactions require much lower pressures than do free-radical polymerizations of ethylene. Simultaneously, Natta demonstrated[161] that similar catalysts can polymerize various other olefins like propylene, butylene, and higher α-olefins. High molecular weight linear polymers form as well and, what is more important, highly stereospecific ones.

The two disclosures stimulated intensive research into the mechanism of catalysis. Much knowledge has been gained to date. Some uncertainties about the exact mechanisms of the reactions still persist.

Ziegler–Natta catalysts[162] are products from reactions of metal alkyls or hydrides of Groups I through III of the Periodic Table with metals, salts or complexes of Groups IV through VIII. Not all compounds, however, that fit this broad definition are actually useful catalysts. In fact, they range from very active to useless ones. Also, among them can be found materials that initiate polymerizations of some monomers by free-radical, cationic, or anionic mechanisms and not by a coordinated anionic. Nevertheless, the number of active Ziegler–Natta catalysts is still large. The catalysts that are based on metals with large numbers of d-electrons (mostly Group VIII) are effective in polymerizations of conjugated dienes. They do not appear to work too well, however, with α-olefins. On the other hand, metals of Groups IV, V, or VI with fewer d-electrons are useful in polymerizing both olefins and conjugated dienes. Also, all catalysts,[163] that polymerize propylene also polymerize ethylene. Yet, the converse is not always true.

The Ziegler–Natta catalysts can be subdivided into two groups: (1) heterogeneous insoluble catalysts, and (2) homogeneous, or soluble, ones. At times, however, it is difficult to distinguish between the two. For instance, it may be hard to determine whether a particular catalyst is truly in solution or merely in a form of a very fine colloidal suspension (and, in fact, heterogeneous).

3.5.1. Heterogeneous Ziegler–Natta Catalysts

These catalysts form when a soluble metal alkyl, like triethyl aluminum or diethyl aluminum chloride, is combined with a metal salt, like titanium chloride, in a medium of an inert hydrocarbon diluent. The transition metal is reduced during the formation of the catalyst.

The following chemical scheme illustrates the reactions that are believed to take place between aluminum alkyls and transition metal halides.[164–166] Titanium chloride is used as an example:

$$TiCl_4 + AlR_3 \rightarrow TiCl_3R + AlR_2Cl$$

$$TiCl_4 + ClAlR_2 \rightarrow TiCl_3 + Cl_2AlR + R\bullet$$

$$TiCl_4 + 2AlR_3 \rightarrow TiCl_2R_2 + 2AlR_2Cl$$

$$TiCl_2R_2 \rightarrow TiCl_2R + R\bullet$$

$$TiCl_3 + AlR_3 \rightarrow TiCl_2R + AlR_2Cl$$

$$TiCl_2R \rightarrow TiCl_2 + R\bullet$$

$$TiCl_2R + AlR_3 \rightarrow TiClR_2 + AlClR_2$$

The radicals that form in the above reactions probably undergo combinations or other reactions of radicals, or perhaps react with solvents and decay. The reduction of the tetravalent titanium is unlikely to be complete due to the heterogeneous nature of the catalyst. Better catalytic activity results when $TiCl_3$ is used directly in place of $TiCl_4$. Many catalysts, however, are prepared with $TiCl_4$. In addition, $TiCl_3$ exists in four different crystalline forms, referred to as α, β, γ, and δ. Of these, the β, γ, and δ forms yield highly stereospecific polymers from α-olefins. The α-form, however, yields polymers that are high in atactic material.

The ratio of the transition metal compounds to those of the compounds from metals in Groups I to III can affect polymerization rates. They can also affect the molecular weights, and the steric arrangement of the products. Also, additives like Lewis bases, amines, or other electron donors help increase the stereoregularity of the product. Thus, for instance, dialkylzinc plus titanium trichloride catalyst yields polypropylene that is 65% stereoregular. Addition of an amine, however, to this catalytic system raises stereoregularity to 93%.[167] Heterogeneous catalysts typically form polymers with very wide molecular weight distributions.

Many mechanisms were proposed to explain the action of heterogeneous Ziegler–Natta catalysts. All agree that the polymerizations take place at localized active sites on the catalyst surfaces. Also, it is now generally accepted that the reactions take place by coordinated anionic mechanisms. The organometallic component is generally believed to activate the site on the surface by alkylating the transition metal. Some controversy, however, still exists about the exact mechanism of catalytic action, whether it is *monometallic* or *bimetallic*. The majority opinion leans to the former. Also, it is well accepted that the monomer insertion into the polymer chain takes place between the transition metal atom and the terminal carbon of the growing polymeric chain.[168]

Only one *bimetallic mechanism* is presented here, as an example, the one originally proposed by Natta.[168] He felt that chemisorptions of the organometallic compounds to transition metal halides take place during the reactions. Partially reduced forms of the di- and tri-chlorides of strongly electropositive metals with a small ionic radius (aluminum, beryllium, or magnesium) facilitate this. These chemisorptions result in formations of electron-deficient complexes between the two metals. Such complexes contain alkyl bridges similar to those present in dimeric aluminum and beryllium alkyls.[169] The polymeric growth takes place from the aluminum–carbon bond of the bimetallic electron-deficient complexes[170,171]:

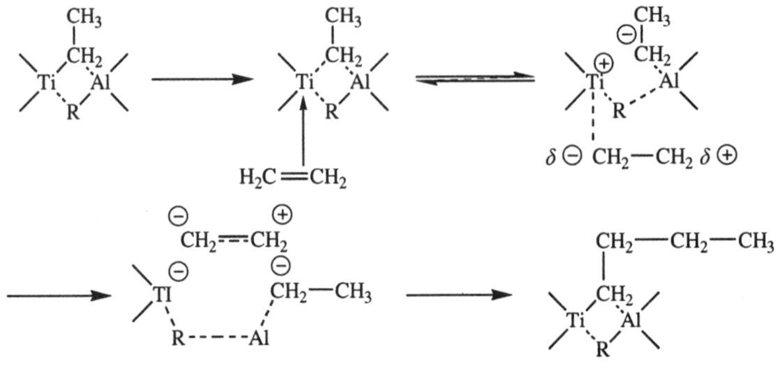

where R represents an ethyl group.

All the evidence to date, however, indicates that chain growth occurs through repeated four-center insertion reactions of the monomer into the transition metal–carbon σ-bond.[168] It is still not established whether the base metal alkyls serve only to produce the active centers and have no additional function. This would make the mechanism monometallic. If, however, the active centers must be stabilized by coordination with base metals, then the mechanism is bimetallic.[163] The two active centers are depicted as follows:

monometallic site

bimetallic site

where M_T represents the transition metal, M_B a base metal, and □ a vacant site.

Both types of active center might conceivably be present in heterogeneous Ziegler–Natta catalysts.[163] The exact locations of the sites in the solid catalyst crystals are still being debated. Some speculations center on whether they are located over the whole crystal surface or only over the edges of the crystals.[191] Most evidence points to location at the edges.

An example of the *monometallic* mechanisms is one originally proposed by Cossee and Arlman.[172,173] This mechanism assumes that the reaction occurs at a transition metal ion on the surface layer of the metal trichloride (or perhaps dichloride) lattice. Here, the halide is replaced by an alkyl group (R). The adjacent chloride site is vacant and accommodates the incoming monomer molecule. Consider titanium chloride as an illustration:

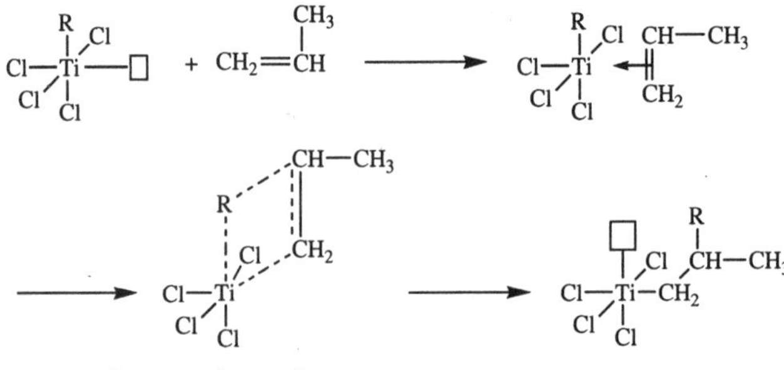

four-center intermediate

where –□ represents a vacant site in the *d*-orbital. The newly formed transition metal–alkyl bond becomes the active center and a new vacant site forms in place of the previous transition metal–alkyl bond.

The driving force for the reactions[172,173] depends on π-type olefin complexes. In these complexes, the π-electrons of the olefins overlap with the vacant *d*-orbitals of the transition metals. This results in the π-bonds being transitory. Also, the *d*-orbitals of the metals can simultaneously overlap with the vacant antibonding orbitals of the olefins. This decreases the distances between the highest filled bonding orbitals and the empty (or nearly empty) *d*-orbitals. In such situations, the carbon–metal bonds of the transition metals weaken and the alkyl groups migrate to one end of the incoming olefin.[174] The insertion process results in a *cis*-opening of the olefinic double bond.[173,182,183]

The above scheme of propagation might also be pictured for bimetallic active centers. Complexations precede monomer insertions at the vacant octahedral sites and are followed by insertion reactions at the metal–carbon bonds.[175] When the transition metals are immobilized in crystal lattices, the active centers and the ligands are expected to interchange at each propagation step.

The above model for monometallic mechanism, though now widely accepted, is still occasionally questioned. Some evidence, for instance, has been presented over the years to support a bimetallic mechanism.[176] It was shown that elimination of the organometallic portion of the complex catalyst during polymerization of propylene results in deactivation of the catalyst. By contrast, replacement of the initial organometallic compound with another one results in a change in the polymerization rate, but not in deactivation of the catalyst.

In addition, some monometallic mechanisms based on a different mode of monomer insertion were also proposed. An example is a reaction mechanism that was proposed by Ivin and co-workers.[344] This mechanism is based on an insertion mechanism involving an α-hydrogen reversible shift, carbene, and a metallocyclobutane intermediate:

$$\sim\!\!\sim\overset{\text{CH}_3}{\underset{\text{H}}{\text{C}}}-\text{M}_\text{T} \rightleftharpoons \sim\!\!\sim\overset{\text{CH}_3}{\text{C}}=\underset{\text{H}}{\text{M}_\text{T}} \rightleftharpoons \sim\!\!\sim\overset{\text{CH}_3}{\text{C}}=\underset{\text{H}}{\text{M}_\text{T}} \leftarrow \overset{\text{CH}_2}{\underset{\text{CH}_3}{\text{C}}}-\text{H}$$

$$\updownarrow$$

$$\sim\!\!\sim\overset{\text{CH}_3}{\underset{\text{H}}{\text{C}}}-\text{CH}_2-\overset{\text{CH}_3}{\underset{\text{H}}{\text{C}}}-\text{M}_\text{T} \longleftarrow \begin{matrix} \sim\!\!\sim\overset{\text{CH}_3}{\text{C}} & - & \text{CH}_2 \\ | & & | \\ \text{H}-\text{M}_\text{T} & - & \underset{\text{CH}_3}{\text{C}}-\text{H} \end{matrix}$$

where M_T means metal. The stereospecificity is dependent upon the relative configuration of the substituted carbons of the metallocyclobutane ring. Hydrogen transfer from the metal to the more substituted carbon excludes branching.[344] The following evidence supports the above mechanism.

(1) There are no unambiguous examples where a characterized metaloalkyl–olefin compound may be induced to react.

(2) There is a close identity between the catalysts that cause the Ziegler–Natta-type polymerizations and those that cause metathesis-type polymerization via a carbene mechanism (see Chapter 4).

This mechanism was argued against, however, as an oversimplification, because it ignores the experimentally observed regiospecificity of many propylene polymerizations.[345] On the other hand, it is argued that, when correct regiospecificity of the monomer is accounted for by isotactic or syndiotactic propagation, different energies of steric control can be qualitatively explained by the Ivin mechanism. This can be done through simple considerations of different distances between the substituted carbons of a four-membered ring.[345]

3.5.1.1. Steric Control with Heterogeneous Catalysts

Different independent approaches were used to investigate the mechanism of polymerization of α-olefins with heterogeneous catalysts. As a result, it was shown that isotactic polymerization of racemic mixtures of α-olefins are stereoselective.[192,193] Also, optically active polymers form with optically active catalysts.[194] Stereoelectivity and stereoselectivity are due to the intrinsic asymmetry of the catalytic centers.[197] This conclusion comes in part from knowledge that propylene coordinates asymmetrically in platinum complexes.[197] A study of chemical and configurational sequences supports this.[198] In addition, polymerization of a racemic mixture of (*R*)(*S*)-3-methyl-1-pentene with ordinary Ziegler–Natta catalysts [e.g., $Al(C_2H_5)_3 + TiCl_4$ or $Al(C_2H_5)_2Cl + TiCl_3$] yields a racemic mixture of isotactic polymers. The mixture can be separated by column chromatography into pure, optically active components. These polymeric chains are exclusively either poly[(*S*)-3-methyl-1-pentene] or poly[(*R*)-3-methyl-1-pentene].[195–197] This means that typical Ziegler–Natta catalysts have essentially two types of active site that differ only in a chiral sense. Such sites polymerize the monomers stereoselectively, i.e., either (*R*) or (*S*) with the exclusion of the other enantiomeric form:

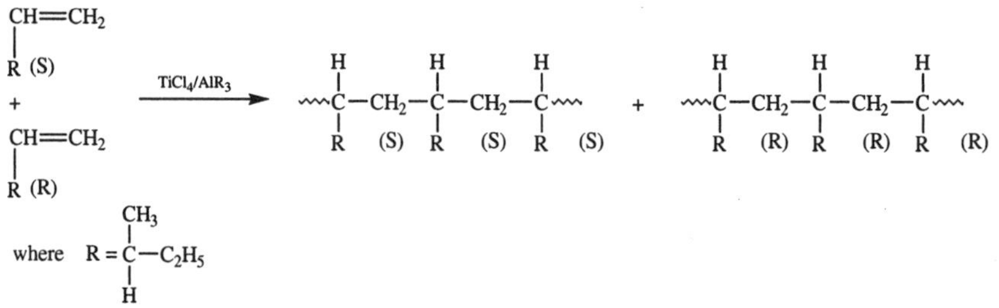

$$\begin{matrix} CH{=}CH_2 \\ | \\ R\ (S) \\ + \\ CH{=}CH_2 \\ | \\ R\ (R) \end{matrix} \xrightarrow{TiCl_4/AlR_3} \sim\!\!\underset{\underset{R}{|}}{\overset{\overset{H}{|}}{C}}-CH_2-\underset{\underset{R}{|}}{\overset{\overset{H}{|}}{C}}-CH_2-\underset{\underset{R}{|}}{\overset{\overset{H}{|}}{C}}\!\!\sim \ (S)(S)(S) \; + \; \sim\!\!\underset{\underset{R}{|}}{\overset{\overset{H}{|}}{C}}-CH_2-\underset{\underset{R}{|}}{\overset{\overset{H}{|}}{C}}-CH_2-\underset{\underset{R}{|}}{\overset{\overset{H}{|}}{C}}\!\!\sim \ (R)(R)(R)$$

$$\text{where } R = \underset{\underset{H}{|}}{\overset{\overset{CH_3}{|}}{C}}-C_2H_5$$

In the heterogeneous catalyst, like α-$TiCl_3$, the crystals are made up of elementary sheets of alternating titanium and chlorine atoms. These atoms are aligned along the principal crystal axis. The chlorines are packed hexagonally and the titanium atoms are at the octahedral interstices of the chlorine lattice. Every third titanium in the lattice is missing. There is a vacancy between pairs of titanium atoms. Many ligand vacancies are present in the crystals in order to accommodate electrical neutrality and the titanium atoms at the surface are bonded to only five chlorines instead of six. Neighboring transition metal atoms that are bridged by chlorines have opposite chiralities.[73] This means that two enantiomorphic forms exist (see Fig. 3.4). The monomers coordinate at either one of the two faces (at the vacant sites). Coordination results in formation of one of two diastereoisomeric intermediate transition states. Both result in isotactic placements, but the products are either *meso* or *racemic*.[183] In addition, enantioselectivity between the two faces of the crystals requires a minimum amount of steric bulk at the active site. This is enhanced by larger-sized monomers.[183] A schematic representation of a catalytic center, showing chirality by an asterisk, is as follows[314]:

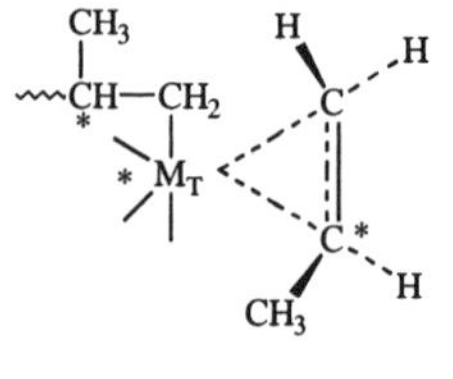

$$\sim\overset{*}{\underset{}{\overset{\overset{CH_3}{|}}{C}H}}-CH_2-\overset{*}{M_T}\cdots \begin{matrix} H\ \ C\ \ H \\ \vdots \\ CH_3\ \ \overset{*}{C}\ \ H \end{matrix}$$

The above means that the active sites act as templates or molds for successive orientations of the monomers. The monomers are forced to approach these site with the same face. This sort of monomer placement is called *enantiomorphic site control* or *catalyst site control*.

So far, the evidence for an initial complex formation between the catalyst and the olefin is not strong. The working hypothesis, commonly accepted today, is as follows. The position in the π-complex of diastereoisomeric and rotameric equilibria and/or activation energy for the insertion reaction cause large regioselectivity. That can also account for the enantioface discrimination

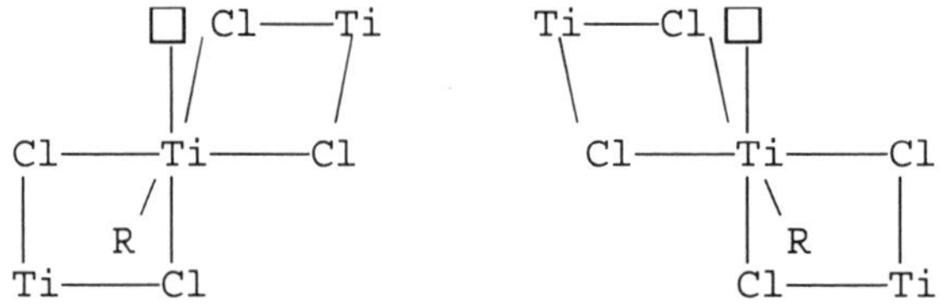

```
  □ Cl——Ti            Ti——Cl □
  |/     /             \     \|
Cl——Ti——Cl          Cl——Ti——Cl
|   /|                   |\   |
| R  |                   | R  |
Ti——Cl                   Cl——Ti
```

FIGURE 3.4. Two enantiomorphic forms of the catalyst. R indicates the ligands that were replaced by alkyl groups through alkylation by organometallic compounds from Groups I through III.

necessary for the synthesis of stereoregular poly-α-olefins.[314] ^{13}C NMR analyses of isotactic polypropylenes formed in δ-$TiCl_3$-$Al(^{13}CH_3)_2I$ catalyzed reactions showed that the enriched ^{13}C is located only in the isopropyl end group. Also, it is located predominantly at the *threo* position, relative to the methyl group on the penultimate unit.[322] This supports the concept that steric control must come from chirality of the catalytic centers.[322]

It might be interesting to note that the proponents of the carbene mechanism point out that this is also consistent with their mechanism (mentioned earlier).[323,324] The reaction can consist of (a) an insertion of a metal into an α-CH bond of a metal alkyl to form a metal–carbene hydride complex. This is followed by (b) reaction of the metal–carbene unit with an alkene to form a metal–cyclobutane–hydride intermediate. The final step, (c), is a reductive elimination of hydride and alkyl groups to produce a chain-lengthened metal alkyl. This assures that a chiral metal environment is maintained.[323] Nevertheless, it is generally believed[183] that stereospecific propagation comes from concerted, multicentered reactions, as was shown in the Cossee–Arlman mechanism earlier. The initiator is coordinated with both the propagating chain end and the incoming monomer. Coordination holds the monomer in place during the process of addition to the chain. This coordination is broken simultaneously with formation of a new coordination with the new monomer[183]:

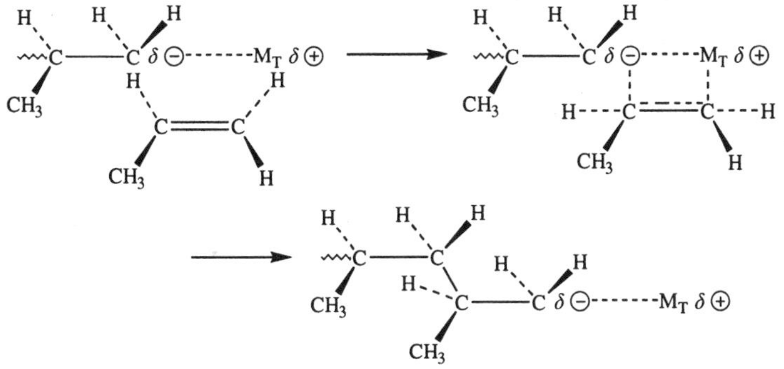

where M_T means metal.

This capability of the initiator to control the placement of the monomer overrides the common tendency for some syndiotactic placement. While syndiotactic polypropylene has been prepared with heterogeneous catalysts, the yield of syndiotactic placement is low. Soluble Ziegler–Natta catalysts, on the other hand, can yield high amounts of syndiotactic placement. This is discussed in the next section.

When 1,2-disubstituted olefins are polymerized with Ziegler–Natta catalysts, the ditacticity of the products depends on the mode of addition. It also depends on the structure of the monomer, whether it is *cis* or *trans*. A threodiisotactic structure results from a *syn* addition of a *trans* monomer. A *syn* addition of a *cis* monomer results in the formation of an erythrodiisotactic polymer. For instance, *cis* and *trans*-1-*d*-propylenes give *erythro* and *threo* diisotactic polymers, respectively.[375] To avoid 1,2-interactions in the fully eclipsed conformation, the carbon bond in the monomer units rotate after the addition of the monomer to the polymeric chain.[182]

3.5.2. Homogeneous Ziegler–Natta Catalysts

The catalysts that fit into this category are quite diverse chemically. They can be, for instance, reaction products of vanadium tetrachloride with dialkyl aluminum chloride or reaction products of bis-cyclopentadienyl titanium dichloride with aluminum alkyls. Organic titanates also form soluble catalysts upon reactions with certain aluminum alkyls.

The homogeneous catalysts can range in efficiency from very high to very low. Usually, a much narrower molecular weight distribution is obtained in polymers formed with these catalysts than with the heterogeneous ones. In addition, syndiotactic poly(α-olefins) form at low temperatures (–78 °C). The amount of stereospecific placement, however, is usually not as great as is the isotactic placement with heterogeneous catalysts.

Group I to III metal alkyl components are essential to formations of active centers.[181,182] Blue complexes of dicyclopentadienyl titanium dichloride with aluminum diethyl chloride or with aluminum ethyl dichloride were some of the early known soluble catalysts[177,178]:

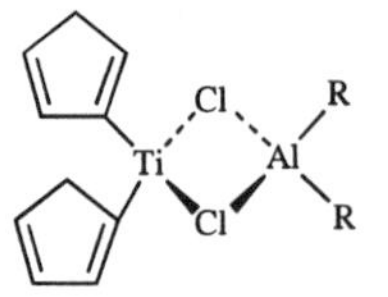

Many other soluble catalysts were developed over the last thirty years. Most of these materials must be prepared and used at low temperatures.[179–183] The polymerization mechanism is also believed to take place by an initial coordination of the monomers with the transition metals. This is followed by a four-center insertion reaction of the monomer into the transition metal–carbon bond.

3.5.2.1. Steric Control with Homogeneous Catalysts

Stereospecific placement appears to result from steric interactions between the substituents on the ultimate units of the growing chains and the incoming monomers.[181] NMR spectral evidence shows that the double bond opening is also *cis*, as with the heterogenous catalysts.[182,183]

The monomer–polymer interaction model, described above, requires sufficient steric hindrance at the active centers to prevent rotation of the penultimate units. Such rotation would cause the substituent to interfere with the incoming monomer. The fact that syndiotactic placement takes place only at low temperatures and atactic placement at elevated ones supports this concept. An illustration of the structure of the catalyst and the active center follows[182,183]:

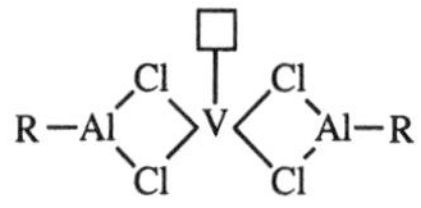

where □ denotes the vacant site.

A proposed model for syndiotactic placement can be shown as follows[163]:

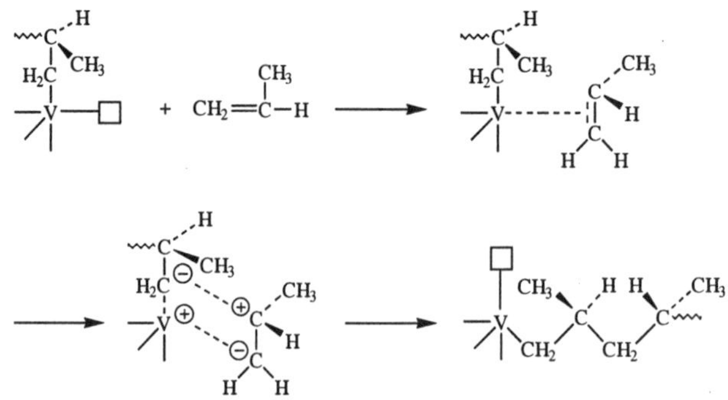

The above model shows the insertion of monomers on σ-metal bonds.[181–183] NMR spectroscopy supports the hypothesis.[345] Also, the model shows the vanadium to be pentacoordinated. This follows a Zambelli and Allegra suggestion.[345] Hexacoordinated vanadium has been pictured by others.[150] In the above model the chain ends control monomer placement. This differs significantly from the mechanism of isotactic placement by catalytic site control with the heterogeneous catalysts. The active sites of homogeneous catalyst do not discriminate between the faces of the incoming monomers in the step of coordination. Instead, steric hindrance between the substituents of the terminal units of the propagating chains and the ligands attached to the transition metals prevent rotations about the metal–carbon bonds. In addition, steric repulsion forces the monomers to coordinate at opposite faces with each successive step of propagation, resulting in syndiotactic placement.

3.5.3. Steric Control in Polymerizations of Conjugated Dienes

The subject has also received considerable attention. Nevertheless, the mechanism is still not fully understood. It is reasonable to assume that the form and structure of the catalyst and the valence of the transition metals must play a role. The conformation of the monomer (*s-cis* or *s-trans*) is probably also important.

It is known that $CoCl_2/Al(C_2H_5)Cl$ can polymerize *trans*-1,3-pentadiene but not the *cis* isomer.[357] This suggests that a two-point coordination is required. Several reaction schemes that provide for an attack at either C_1 or C_4 position were proposed over the years.[357] One mechanism for polymerization of butadiene suggests that complexes of the catalysts in solvents of low dielectric constant will either act as ion pairs or as independent solvated entities. Also, the growing chain may be bound by either a π or a σ linkage, and it is suspected that a continuous $\sigma \rightarrow \pi$ isomerism is possible[358]:

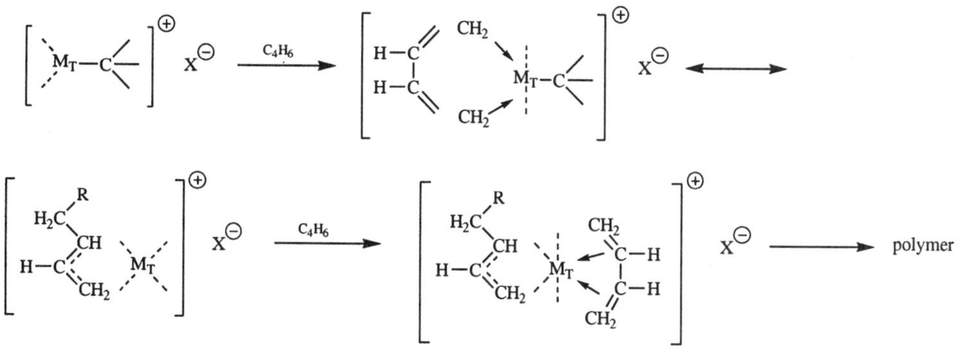

$$[\,M_T-C\,]^{\oplus}X^{\ominus} \xrightarrow{C_4H_6} [\,(CH_2=CH-CH=CH_2)\rightarrow M_T-C\,]^{\oplus}X^{\ominus} \longleftrightarrow$$

$$[\,R-CH_2-CH\cdots CH\cdots CH_2\cdots M_T\,]^{\oplus}X^{\ominus} \xrightarrow{C_4H_6} [\,R-CH_2-CH\cdots CH\cdots CH_2\cdots M_T\leftarrow (CH_2=CH-CH=CH_2)\,]^{\oplus}X^{\ominus} \longrightarrow \text{polymer}$$

where M_T means metal.

Soluble catalysts from transition metal acetoneacetonates (from nickel to titanium) combined with triethylaluminum or triethylaluminum chloride yield *cis* polymers of the dienes. The *cis* placement decreases when bases are added.[184] This decrease is proportional to the base strength. The addition is believed to decrease electron densities of the orbitals of the transition metals.[185] This suggests that electrostatic interactions between the nearly nonbonding electrons of transition metals with the dienes or with growing chain ends must play an important role. Such interactions must affect placements of the incoming monomers. Hirai and co-workers studied ESR signals obtained during polymerizations of butadiene, 1,3-pentadiene, and isoprene[186] with catalyst based on *n*-butyltitanate/triethylaluminum. They concluded that the catalyst must possess two substituted π-allyl groups and one alkoxy group during the chain-growing process[186]:

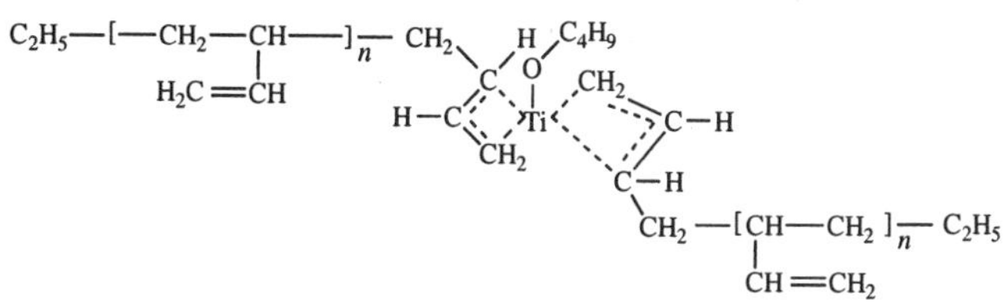

$$C_2H_5-[-CH_2-CH(CH=CH_2)-]_n-CH_2-CH\cdots CH\cdots CH_2\cdots Ti(OC_4H_9)\cdots CH_2\cdots CH\cdots CH-CH_2-[CH(CH=CH_2)-CH_2]_n-C_2H_5$$

The insertions of the monomers is believed to occur in two steps.[187,188] In the first one, the incoming monomer coordinates with the transition metal. This results in formation of a short-lived σ-allylic species. In the second one, the metal–carbon bond is transferred to the coordinated monomer with formation of a π-butenyl bond. Coordination of the diene can take place through both double bonds, depending upon the transition metal[189] and the structure of the diene. When the monomer coordinates as a monodentate ligand, then a *syn* complex forms. If, however, it coordinates as a bidentate ligand, then an *anti* complex results.[190] In the *syn* complex carbons one and four have the same chirality, while in the *anti* complex they have opposite chiralities.[187] Due to lower thermodynamic stability the *anti* complex isomerizes to a *syn* complex.[187] If the allylic system does not have a substituent at the second carbon, then the isomerization of *anti* to *syn* usually occurs spontaneously even at room temperature.[187]

Transition metal alkyls probably cannot be classified as typical Ziegler–Natta catalysts. Some of them, however, exhibit strong catalytic activity and were therefore investigated.[199–202] This also led to the conclusion that the polymerization mechanisms involve formations of π-allylic complexes as intermediates. It is similar to the mechanism visualized for the Ziegler–Natta catalysis.[203–210] The initial formation of the π-allylic ligands and the solvents used in catalyst preparations strongly influence the catalytic activity and stereospecificity of the product.[207] NMR studies of polymerizations of conjugated dienes with π-crotyl-nickel iodide[208] showed that the monomers are incorporated at the metal–carbon bonds with formations of *syn*-π-crotyl ligands. The *syn*-ligands transform to *trans*-1,4 segments next to the crotyl group and the *trans*-1,4-segments become *trans*-1,4 units in the polymers.[208] In summary, the general mechanisms of *cis* and *trans* placements by coordination catalysts were pictured as follows[206]:

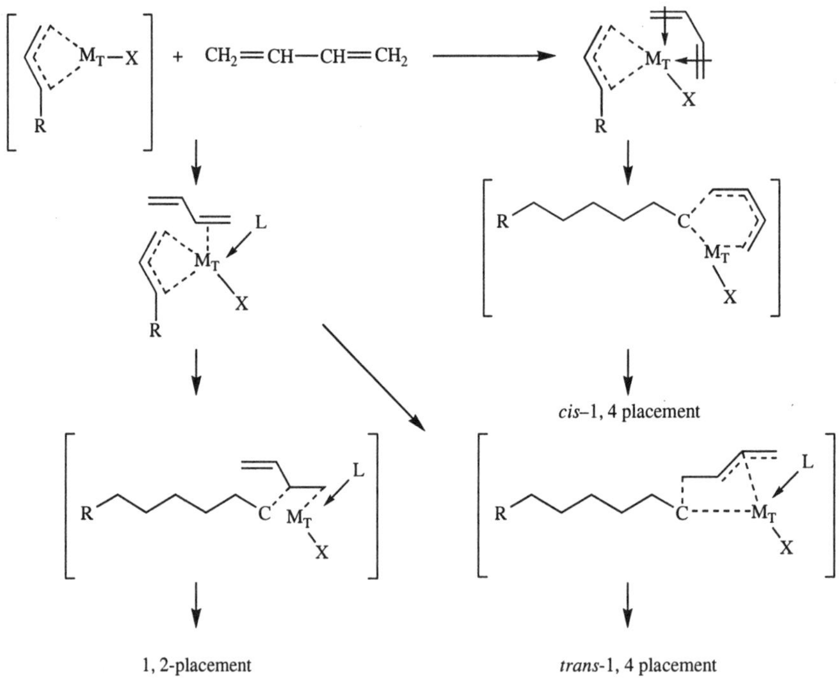

M_T—X + CH_2=CH—CH=CH_2 ⟶

cis–1, 4 placement

1, 2-placement

trans-1, 4 placement

where L means ligand and M_T means transition metal.

3.5.4. Current Industrially Developed Ziegler–Natta Catalysts

In current industrial practice, coordinated anionic catalysts differ considerably from the original ones, developments by Ziegler, Natta, and others. Using the same basic chemistry, new compounds were developed over the years that yield large quantities of polyolefins from small amounts of catalysts. In addition, catalysts can now be designed to yield products that are either wide or narrow in molecular weight distribution, as needed.[312] The new *catalysts for ethylene polymerization* can be divided into three groups[312]:

1. Products from reactions of trivalent alkoxy chlorides of transition metals with certain halogen-free organoaluminum compounds, e.g., triisobutylaluminum. Such catalysts are used without any support.

2. Products from reactions of magnesium compounds with titanium compounds. In these catalysts the transition metals are attached chemically to the surfaces of solid magnesium com-

pounds. The reactions take place between the halogen atoms from titanium compounds and the hydroxyl groups at the surfaces of magnesium hydroxide:

$$|Mg| \text{-OH} + TiX_n \rightarrow |Mg| \text{-O-}TiX_{n-1} + HX$$

Titanium compounds bonded to the surface OH groups of $Mg(OH)_2$ are mainly inactive. The active sites are the ones associated with the coordinated and unsaturated negative oxygen ions.[317] Reactions with aluminum alkyls activate the catalysts.[312] For high efficiency, special carriers must be used together with a correct balance of the reactants, and proper reaction conditions. Some choice combinations of reactants are[312]:

$$Mg(OH)Cl + TiCl_4$$

$$MgCl_2{\cdot}3Mg(OH)_2 + Ti(OR)_xCl_y$$

$$MgSO_4{\cdot}3Mg(OH)_2 + Ti(OR)_xCl_y$$

$$Mg(OH)_2 + TiCl_4$$

Highly active, unsupported catalysts form from reactions of magnesium alkoxides with tetravalent titanium chlorides. The same is true of reaction products of $MgCl_2$ or $MgCl_2$-electron donor adducts, like $MgCl_2{\cdot}6C_2H_5OH$, with tetravalent titanium compounds.

3. Products from reactions of organisilanols with chromium trioxide are also very active catalysts. The silylchromate that forms is deposited on a silica support and activated with alkyl aluminum compounds[312]:

$$2\ [C_6H_5]_3\text{—Si—OH} + CrO_3$$

$$\downarrow$$

$$[C_6H_5]_3\text{—Si—O—}\overset{\overset{O}{\|}}{\underset{\underset{O}{\|}}{Cr}}\text{—O—Si—}[C_6H_5]_3$$

$$\downarrow \ + \text{ silica}$$

supported silylchromate

$$\downarrow \ + \text{ trialkylaluminum}$$

catalyst

The activity and performance of Ziegler–Natta *catalysts for the polymerization of propylene* has been also very much improved. Titanium compounds are also supported on some carriers and then activated by reactions with aluminum alkyls.[341] The patent literature describes a variety of inorganic supports.[342] The most common ones, however, are based on $MgCl_2$ treated with various Lewis bases, like ethyl benzoate. One such catalytic system is described as being prepared by treating a complex, $TiCl_3{\cdot}3C_5H_5N$, with diethyl aluminum chloride in the presence of highly dispersed $MgCl_2$. The product of that reaction is then combined with triethylaluminum.[355] The reaction between an electron donor, a Lewis base, and $MgCl_2$ is a two-step exothermic reaction.[343] The first one is a rapid adsorption to the inorganic surface, and the second, a slower one, is formation of the complex. The most commonly used Lewis bases are ethyl benzoate, di-*n*-butyl phthalate, and methyl-*p*-toluenate. Amines, like 2,2,6,6-tetramethylpiperidine, and some phenols are also used.

A common practice is to ball mill the Lewis base with the support material first. The transition metal component is then added and the mixture is milled some more or thoroughly mixed. This

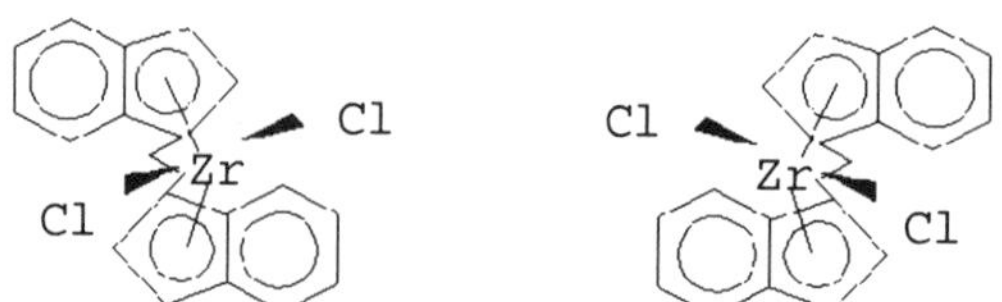

Figure 3.5. Catalyst for polypropylene polymerization

Lewis base is called the *internal* Lewis base. This is followed by addition of Group I–III metal–alkyl component with an additional Lewis base. The base that is added the second time may be the same or a different one from that used in the original milling. In either case it is called the *external* Lewis base. It is not uncommon to use an ester as the internal base and an organosilane compound, like phenyltriethoxysilane, as the external one.

Recently, a series of soluble, highly isospecific catalysts were developed for propylene polymerization.[168] These materials are zirconium, titanium, or hafnium based metallocenes, such as racemic 1,1-ethylene-di-η^5-indenylzirconium dichloride. They are rigid structures, due to an ethylene bridge between the two five-membered rings. Syntheses of these compounds yield racemic mixtures of two enantiomers. Both produce isotactic polypropylene (see Fig. 3.5)

Methylaluminoxane is required by these rigid metallocenes to form highly isospecific catalysts capable of very high isotactic placement. They are also very active, yielding very large quantities of polypropylene per each gram of zirconia.

There is little interest at present in industrial production of syndiotactic polypropylene. Nevertheless, a catalyst that yields highly syndiotactic polypropylene (86% racemic pentads) was reported recently. The material is *i*-propylene(η^5-cyclopentadienyl-η^3-fluorenyl)zirconium dichloride.[291] It may perhaps be the only catalyst that yields highly syndiotactic polypropylene that is not based on vanadium.

3.5.5. Effect of Lewis Bases

Many investigations were carried out to understand the role that Lewis bases play in affecting stereospecificity of the Ziegler–Natta catalysts. One of them, for instance, included studies with optically active bases to determine the reactions with the active sites.[314] Nevertheless, the effect of Lewis bases like tertiary amines is not fully understood. Evidence, gathered to date, suggest that when they react with aluminum alkyl halides[348,349] they increase the number of stereospecific sites by assisting in the stabilization and/or dispersal of the catalyst on the support. They appear to alter the identities of the attached ligands. This might be due to increasing the steric bulk at these sites. The Lewis bases are also given credit for reducing the reactivity of the less stereospecific sites. Because both internal and external Lewis bases are used, the enhancement of stereospecificity by the external bases is attributed to displacing the internal base and/or augmenting its effect.[376]

3.5.6. Terminations in Coordination Polymerizations

The terminations usually occur by chain transferring, either by an internal hydride transfer[183]:

$$\underset{\text{catalyst site}}{/\!/\!|}\!-CH_2-\overset{X}{\overset{|}{C}}H-(-CH_2-\overset{X}{\overset{|}{C}}H-)_n-R \longrightarrow \underset{\text{catalyst site}}{/\!/\!|}\!-H + CH_2{=}\overset{X}{\overset{|}{C}}-(-CH_2-\overset{X}{\overset{|}{C}}H-)_n-R$$

or by a transfer to a monomer:

$$/\!/\!|\!-CH_2-\overset{X}{\overset{|}{C}}H-(-CH_2-\overset{X}{\overset{|}{C}}H-)_n-R + CH_2{=}CHX \longrightarrow$$

$$/\!/\!|\!-CH_2-CH_2X + CH_2{=}\overset{X}{\overset{|}{C}}-(-CH_2-\overset{X}{\overset{|}{C}}H-)_n-R$$

Terminal unsaturation in the products was confirmed by infrared spectroscopy. Hydrogen terminates the reactions and is often used to control the molecular weights of the products. Protonic acids do the same thing:

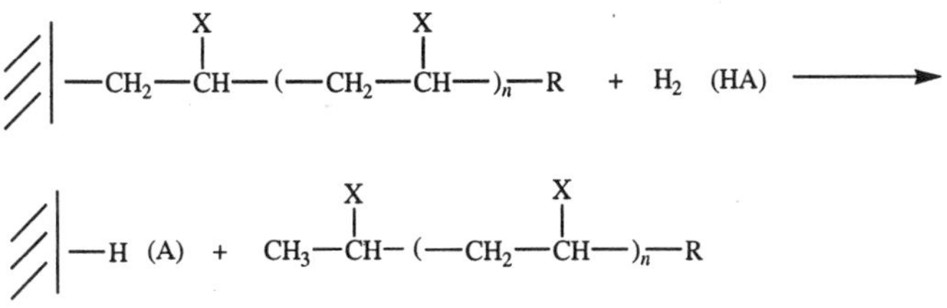

$$\text{//}|\!-\!CH_2\!-\!\underset{}{\overset{X}{\overset{|}{C}H}}\!-\!(\!-\!CH_2\!-\!\overset{X}{\overset{|}{C}H}\!-\!)_n\!-\!R \;+\; H_2\;(HA) \longrightarrow$$

$$\text{//}|\!-\!H\;(A) \;+\; CH_3\!-\!\overset{X}{\overset{|}{C}H}\!-\!(\!-\!CH_2\!-\!\overset{X}{\overset{|}{C}H}\!-\!)_n\!-\!R$$

3.5.7. Reduced Transition Metal Oxide Catalysts on Support

The catalysts that belong to this group are efficient in homopolymerizations of ethylene. They can also be used in some copolymerizations of ethylene with α-olefins. Such catalysts are prepared from TiO_2, V_2O_5, Nb_2O_5, Ta_2O_5, CrO_3, Cr_2O_3, MoO_3, WO_3, NiO, and CoO. The supports are charcoal, silica, alumina, zinc oxide, or alumina-silica. These catalysts require higher temperatures and pressures for activity than do the typical Ziegler–Natta catalysts. The temperatures range from 140 to 230 °C and pressures from 420 to 1000 psi.

One of the original disclosures of such catalyst preparation was as follows[211]: A transition metal nitrate solution is used to saturate a charcoal bed, previously leached with nitric acid. The charcoal bed is then heated to temperatures high enough to decompose the nitrate to the oxide. Following the heating, the oxide is reduced with H_2 or $NaBH_4$. Sometimes, powdered Al_2O_3 or MoO_3–Al_2O_3 are added prior to reduction.

For an Al_2O_3–SiO_2 support, a 3% metal nitrate solution is used. The saturated bed is heated to 500 °C to decompose the nitrate. This is followed by a reduction. Many modifications of the above procedure have been developed since.

Preparations based on molybdenum oxide are called Standard Oil Co. catalysts, while those based on chromium oxide are known as the Phillips Petroleum Co. catalysts. Below is a description of a preparation of a molybdenum-based catalyst. Alumina is saturated with a solution of ammonium molybdate and then subjected to heating in air at 500 to 600 °C. The oxide that forms is reduced with hydrogen at 430 to 480 °C. Reducing agents like CO, SO, or hydrocarbons are also used. Hydrogen, however, is preferred at pressures of approximately 75 psi. The catalyst may contain between 5 and 25% of the molybdenum compound dispersed on the surface.[346]

A typical chromium-based catalyst is prepared similarly. It is usually supported by a 9:1 of SiO_2:Al_2O_3[347] carrier. Either $Cr(NO_3)_3{\cdot}9H_2O$ or CrO_3 solutions in nitric acid are used to impregnate the support. The nitrates are decomposed in air at 400 to 1000 °C.[347] An optimum chromium content in these catalysts is 2–3%.

The mechanism of polymerization was initially believed to be as follows.[212] Cr–O–Si bonds are present on the catalyst surface. Contact with ethylene results in an oxidation–reduction reaction and formation of an ethylene–chromium complex. The initiations are accompanied by gains of hydride ions and formations of terminal methyl groups. Polymer growth was pictured according to the reactions shown in Fig. 3.6:

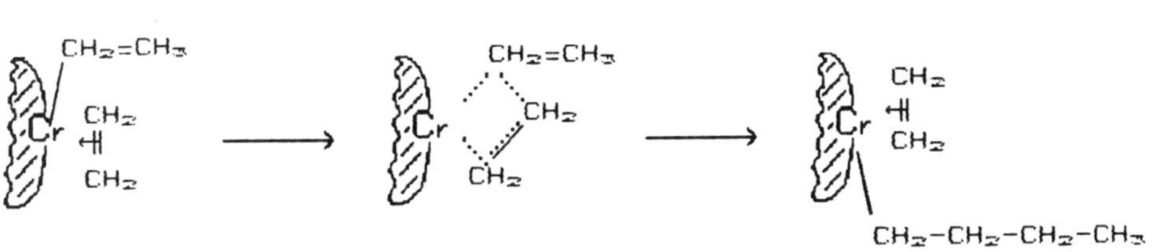

Figure 3.6. Polymer growth.

The termination reactions were believed to result from hydride-ion transfers.[150] Later, however, polyethylene, formed with these catalyst, was found to contain approximately equal numbers of saturated and unsaturated chain ends. This contradicts the above mechanism of polymerization. It may also mean that the initiation involves π-complexes that disproportionate to yield coordination of the metal to $-CH{=}CH_2$ and to $-CH_2-CH_3$ groups.[150]

Polymerization of propylene with these catalysts yields polymers with very little crystallinity. Higher α-olefins yield completely amorphous polymers. Formation of partially crystalline polymers was reported for vinylcyclohexane, allylcyclohexane, and 4-phenyl-1-butene. Styrene does not polymerize at all. Isoprene forms a *trans*-1,4 polymer.[150]

3.5.8. Isomerization Polymerizations with Coordination Catalysts

Polymerizations of many internal olefins, like 2-butene, 2-pentene, 3-heptene, 4-methyl-2-pentene, 4-phenyl-2-butene, and others, with Ziegler–Natta catalysts, are accompanied by monomer rearrangements. The isomerizations take place before insertions into the chains.[213–221] The double bonds migrate from the internal to the α-positions:

$$n\ CH_3{-}CH{=}CH{-}R \rightleftharpoons n\ CH_2{=}CH{-}CH_2R \longrightarrow -\!\!\left[-CH_2-\underset{\displaystyle CH_2R}{\underset{|}{CH}}-\right]_n\!\!-$$

The products consist exclusively of poly(1-olefin) units. Polymerizations of 2, 3 and 4 octenes with $TiCl_3/Al(C_2H_5)_3$ at 80 °C, for instance, result in the same high molecular weight homopolymer, poly(1-octene).[222]

Recently, it was reported that addition of some transition metal compounds of Group VIII to Ziegler–Natta catalysts enhances isomerization of 2-olefins.[309,378] When nickel compounds are added to $TiCl_3–(C_2H_5)_3Al$, they react and form $TiCl_3–NiX_2–(C_2H_5)_3Al$. X can be a chloride, an acetylacetonate, or a dimethyl glyoximate.[313] The product is an efficient isomerization catalyst.

3.6. Polymerization of Aldehydes

Aldehyde polymers were probably known well over 100 years ago.[223–225] In spite of that, polyoxymethylene is the only product from aldehyde polymerization that is produced in large commercial quantities. Formaldehyde polymerizes by both cationic and anionic mechanisms. An oxonium ion acts as the propagating species in cationic polymerizations.[226,227] In the anionic ones, the propagation is via an alkoxide ion.

Most polymerizations of aldehydes are conducted in aprotic anhydrous solvents because proton-yielding impurities are very efficient chain-transferring agents. Polymerization of formaldehyde can be looked upon as an exception.[227] When protonic solvents are used, however, the resultant polymer has a very wide molecular weight distribution. Usually, therefore, solvents with low dielectric constants are preferred. In all cases of aldehyde polymerizations, the polymers precipitate from solution as they form. Choice of low dielectric solvents contributes to this. In spite of that, the polymerizations often proceed as rapidly as the monomers are added to the reaction mixtures. In addition, only low molecular weight polyoxymethylene forms in some high dielectric solvents, like dimethylformamide, and the reaction is quite sluggish. Higher aldehydes fail to polymerize in dimethylformamide.

The propagation reaction continues after polymer precipitation, because the precipitated macromolecules are highly swollen by the monomers. This was shown in polymerizations of *n*-butyraldehyde in heptane.[225] The physical state of the polymers and the surface areas of their crystalline domains therefore influence the paths of the polymerizations.

The solubility of the initiators in the solvents can affect strongly the molecular weights of the resultant products and their bulk densities.[227] The whole reaction can even be affected by small changes in the chemical structures of the initiators or by changes in the counterions.[227] Chain transferring is the most important termination step in aldehyde polymerizations.

3.6.1. Cationic Polymerization of Aldehydes

The cationic polymerizations of formaldehyde can be carried out in anhydrous media with typical cationic initiators. Initiation takes place by an electrophilic addition of the initiating species to the carbonyl oxygens. This results in formations of oxonium ions[227]:

$$R^{\oplus} + O{=}CH_2 \longrightarrow R{-}\overset{\oplus}{O}{=}CH_2 \rightleftharpoons R{-}O{-}CH_2^{\oplus}$$

The oxonium ions may react as oxygen–carbon cations. The propagation steps consist of attacks by the electrophilic carbon atoms upon the carbonyl oxygens of the highly polar formaldehyde molecules:

$$R{-}O^{\oplus}{=}CH_2 + O{=}CH_2 \rightarrow R{-}OCH_2{-}O^{\oplus}{=}CH_2$$

In these polymerizations, the propagating oxonium ions are probably further solvated[226]:

$$\sim O{-}CH_2{-}\overset{\oplus}{O}(CH_2{-}O\sim)_2$$

Terminations of chain growths occur through recombinations, reactions with impurities, and by chain transfers:

$$\sim O^{\oplus}{=}CH_2 \ldots MeX_n^{\ominus} \rightarrow \sim O{-}CH_2X + MeX_{n-1}$$

$$\sim O^{\oplus}{=}CH_2 + RX \rightarrow \sim O{-}CH_2{-}X + R^{\oplus}$$

$$\sim O{-}CH_2{-}O^{\oplus}{=}CH_2 + H_2O \rightarrow \sim O{-}CH_2{-}OH + H{-}O^{\oplus}{=}CH_2$$

where MnX_n is a Lewis acid.

Formaldehyde can polymerize in an anhydrous form in all three physical states of matter: as a gas, as a liquid, or as a solid.[226] It also polymerizes in water with acid catalysts. Many Lewis acids are efficient catalysts for this reaction. In addition, some protonic acids are also effective. Among them are perchloric and sulfuric acids and monoesters of sulfuric acid.[226]

The activity of the initiators is independent of the solvents in cationic polymerizations of formaldehyde.[226] There are, however, differences in the performance of the initiators. With BF_3, the reaction reaches equilibrium at 30% conversion and stops. Much higher conversions are achieved with $SnCl_4$ and $SnBr_4$.[226] Both Lewis acids yield polyoxymethylenes that are very similar to those formed by anionic mechanism.[227–230]

Higher aldehydes can be polymerized by Lewis acids, Bronsted acids, and acidic metal salts.[226] Initiation and propagation reactions consist of repeated nucleophilic attacks by the carbonyl oxygens on the electrophiles and formations of carbon–oxonium ions[313]:

$$R'^{\oplus} + O{=}CHR \longrightarrow R'{-}\overset{\oplus}{O}{=}CHR$$

$$R'{-}\overset{\oplus}{O}{=}CHR + O{=}CHR \longrightarrow R'{-}O{-}CHR{-}\overset{\oplus}{O}{=}CHR \xrightarrow{n\ RCHO} R'{-}O{-}CHR{-}({-}O{-}CHR{-})_n{-}\overset{\oplus}{O}{=}CHR$$

The counterions are not shown in the above equations. When the initiators are protonic acids, the reactions can be illustrated as follows:

$$\overset{\oplus\ominus}{HA} + O{=}CHR \longrightarrow H{-}\overset{\oplus}{O}{=}CHR\cdots\cdots A^{\ominus} \xrightarrow{n\ RCHO} H{-}({-}O{-}CHR{-})_n{-}\overset{\oplus}{O}{=}CHR\cdots\cdots A^{\ominus}$$

Many aldehydes trimerize in acidic conditions. Paraldehyde and many other trimers fail to undergo further polymerizations to high molecular weight, linear polymers. The trimer of formaldehyde, trioxane, however, is unique. It polymerizes by a mechanism of ring-opening polymerization. This is discussed in Chapter 4. To avoid trimer formations, low dielectric solvents must be used at low temperatures. In addition, the initiator must be carefully selected.

Potentially, atactic, isotactic, and syndiotactic polymers should form from higher aldehydes. Preparation of syndiotactic polymers, however, has so far not been reported. Catalysts, like Al_2O_3, CrO_3, MoO_3 and SiO_2, form high molecular weight atactic polyacetaldehyde at –78 °C.[231] These are solid catalysts that yield polymers under heterogeneous conditions.[232] Some alumina can be converted into efficient catalysts by treating the surface with HCl, F_3CCOOH, or BF_3.[233] Acetaldehyde is polymerized by a variety of soluble initiators, such as H_3PO_4, HCl, HBO_3, F_3CCOOH, AsF_3, $AsCl_3$, $ZnCl_2$, $ZrCl_4$, and Cl_3COOH.[226] Polymerization reactions require good solvent if weak cationic initiators are used. When strong initiators are employed, the reactions are difficult to control. Low temperatures can be maintained by such techniques as boiling-off low-boiling solvents.

Formation of pure isotactic polyacetaldehyde was reported. It forms in small quantities in the presence of an atactic polymer obtained with crystalline BF_3–etherate.[234]

3.6.2. Anionic Polymerization of Aldehydes

Many patents describe polymerizations of anhydrous formaldehyde by anionic mechanism. The initiators included amines, phosphines, and metal alcoholates. Kern pictured initiations of formaldehyde polymerizations by tertiary amines as direct addition reactions[235,236]:

$$R_3N + CH_2O \rightarrow R_3N^{\oplus}\text{–}CH_2\text{–}O^{\ominus}$$

Earlier, however, Machacek suggested[237–239] that the initiations take place with the help of protonic impurities:

$$R_3N + H_2O \rightleftharpoons R_3\overset{\oplus}{N}\text{—}H\cdots\overset{\ominus}{O}H$$

$$R_3\overset{\oplus}{N}\text{—}H\cdots\overset{\ominus}{O}H + H_2C\text{=}O \longrightarrow HO\text{—}CH_2\text{—}\overset{\ominus}{O}\cdots\overset{\oplus}{N}R_3(H)$$

Much of the evidence presented since favors the Machacek mechanism of initiation.[227] By contrast, tertiary phosphenes apparently do initiate such polymerizations by a zwiterion mechanism.[240] This may, perhaps, be due to higher nucleophilicity and lower basicity than that of the tertiary amines. Phosphorus incorporates into the polymer[240] in the process.

The propagation reactions in tertiary amine initiated polymerizations can be pictured as follows[241]:

$$HO\text{—}(\text{—}CH_2\text{—}O\text{—})_n\text{—}CH_2\text{—}\overset{\ominus}{O}\cdots\overset{\oplus}{N}R_3(H) + H_2C\text{=}O \longrightarrow HO\text{—}(\text{—}CH_2\text{—}O\text{—})_{n+1}\text{—}CH_2\text{-}\overset{\ominus}{O}\cdots\overset{\oplus}{N}R_3(H)$$

The terminations probably result from chain transferring[241]:

$$HO\text{—}(\text{—}CH_2\text{—}O\text{—})_{n+1}\text{—}CH_2\text{—}\overset{\ominus}{O}\cdots\overset{\oplus}{N}R_3(H) + R'OH \longrightarrow HO\text{—}(\text{—}CH_2\text{—}O\text{—})_{n+1}\text{—}CH_2OH + R_3\overset{\oplus}{N}(H)\text{—}OR'$$

The newly formed active species can initiate new polymerizations.

Metal alkyls and metal alcoholates are very effective anionic initiators for higher aldehydes. Aldol condensation can occur, however, in the presence of strong bases. Thus, while some such initiators yield high molecular weight polymers from formaldehyde, they only yield low molecular weight polymers from higher aldehydes. Initiations by metal alkyls result from additions to the carbonyl group:

$$R{-}C(=O)H \; + \; R'Me \longrightarrow R'(R)CH{-}O^{\ominus}\cdots Me^{\oplus}$$

where Me represents a metal group.

The propagation reaction is a series of successive nucleophilic additions,[233] with the alkoxide ion as the propagating species:

$$R'(R)CH{-}O^{\ominus}\cdots Me^{\oplus} \; + \; n\,R{-}C(=O)H \longrightarrow R'{-}({-}CHR{-}O{-})_n{-}CHR{-}O^{\ominus}\cdots Me^{\oplus}$$

There are indications that many aldehyde polymerizations result in formations of "living" polymers, similarly to anionic polymerizations of vinyl compounds. Termination can occur through hydride transfer via a form of a crossed Cannizzaro reaction:

$$\sim O{-}CHR{-}O^{\ominus}\cdots Me^{\oplus} \xrightarrow{RCHO} [\text{transition state}] \longrightarrow \sim O{-}C(=O)R \; + \; R{-}CH_2{-}O^{\ominus}\cdots Me^{\oplus}$$

The alkyl substituent has a tendency to destabilize the propagating anion by increasing the charge density on the oxygen:

$$\sim CHR{-}O^{\ominus}$$

As a result, weak bases, like amines, fail to initiate polymerizations of higher aldehydes.

A *stereospecific anionic polymerization* of acetaldehyde was originally reported in 1960.[242–244] Two alkali metal compounds[244] and an organozinc[243] one were used as the initiators. Trialkylaluminum and triarylaluminum in heptane also yield crystalline, isotactic polymers from acetaldehyde, heptaldehyde, and propionaldehyde at –80 °C.[242] Aluminum oxide, activated by diethylzinc, yields stereoblock crystalline polymers from various aldehydes.[243,246] Lithium alkoxide formed polyacetaldehyde is insoluble in common solvents. It melts at 165 °C.[244]

The mechanism of stereoregulation is still being debated. Some concepts are presented in the rest of this section. Natta[242] believed that polymerizations initiated by organoaluminum compounds proceed by a coordinated anionic mechanism. The aluminum atoms were seen as forming complexes with the oxygen atoms on the penultimate units:

$$\sim CH(CH_3){-}O{-}Al< \; + \; H(CH_3)C{=}O \longrightarrow [\sim CH(CH_3){-}O\cdots CH(CH_3){=}O\cdots Al<] \longrightarrow \sim CH(CH_3){-}O{-}CH(CH_3){-}O{-}Al<$$

The activated complexes that form have steric configurations that allow minimum amounts of nonbonded interactions.

Other mechanisms have been proposed. For instance, Furukawa *et al.*[246] concluded that metal alkyl compounds must become metal alkoxides through reactions with the aldehydes:

$$R'{-}C(=O)H \; + \; MeR \longrightarrow R'(R)CH{-}O{-}Me$$

where Me means metal.

His polymerization mechanism, therefore, is based on known reactions of metal alkoxides with carbonyl compounds. Side reactions, like a Meerwin–Ponndorf or a Tischenko, should be expected and were included into the reaction scheme:

Main Reactions	Side Reactions

MeR → (+ CH_3-CHO) → $CH_3-CH(R)-O-Me$ → (+ CH_3-CHO) → [$CH_3-CH(R)-O-Me \cdots O=CH-CH_3$]

Meerwin–Ponndorf / Oppenauer (⇄): $CH_3-C(R)=O \cdots Me-O-CH_2-CH_3$

→ $CH_3-CH(R)-O-CH(CH_3)-O-Me$ → (+ CH_3-CHO) → [$CH_3-CH(R)-O-CH(CH_3)-O-Me \cdots O=CH-CH_3$]

Tischenko (⇄): $CH_3-CH(R)-O-C(CH_3)=O \cdots Me-O-CH_2-CH_3$

→ $CH_3-CH(R)-O-CH(CH_3)-O-CH(CH_3)-O-Me$ → etc.

Each monomer addition to the growing chain requires a transfer of the alkoxide anion to the carbonyl group. This results in a formation of a new alkoxide anion. (A hydride transfer from the alkoxide group to the carbon atom of the aldehyde can take place by the Meerwin–Ponndorf reduction.) Chain growth takes place by repetition of the coordination of the aldehyde, and subsequent transfer of the alkoxide anion.

In the Vogl and Bryant[247] mechanism, four oxygens are coordinated to the metal atom. These oxygens are from the penultimate and ultimate units of the growing chains and from two monomers:

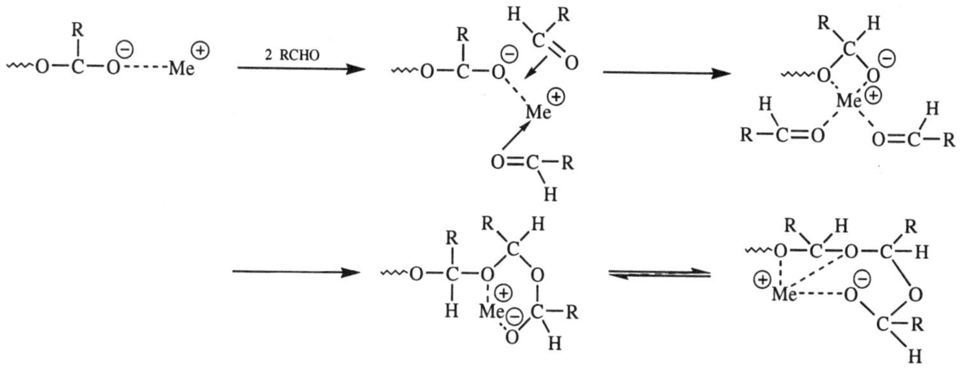

2 RCHO

A simultaneous coordination of the two aldehydes prior to addition may explain the observed sequence of isotactic dyads.

Yasuda and Tani derived their mechanism from investigations of isotactic aldehyde polymerizations by [$R_2AlOCR'NC_6H_5$] catalysts.[245,248,249] The bulkiness of the substituent group is considered to be the most important factor in steric control. The catalyst enhances the degree of stereoregulation by controlling the mode of approach through coordination. Also, the Lewis acidity of the catalyst must be confined to a narrow range for each particular aldehyde to yield isotactic polymers.[248,249] If the acidity is too strong, an amorphous polymer forms. If it is too weak, no polymerization take place.[249] In addition, special techniques, like high pressures, for instance, can result in formations of isotactic butyraldehyde and heptaldehyde.[250]

3.6.3. Polymerization of Unsaturated Aldehydes

Unsaturated aldehydes are unique, because they polymerize in special ways. Much of the effort in acrolein polymerizations is to form products with the aldehyde groups intact. This means that the reactions must be confined to the carbon-to-carbon double-bond portions of the molecules.[251–256] Organometallic catalysts like *n*-butylmagnesium bromide, however, are also initiators for polymerizations through the carbonyl groups. In fact, unsaturated aldehyde monomers can be polymerized in four ways, similar to dienes. The placement can be through carbon-to-carbon double bond, or through the carbonyl group, or it can be 1,4:

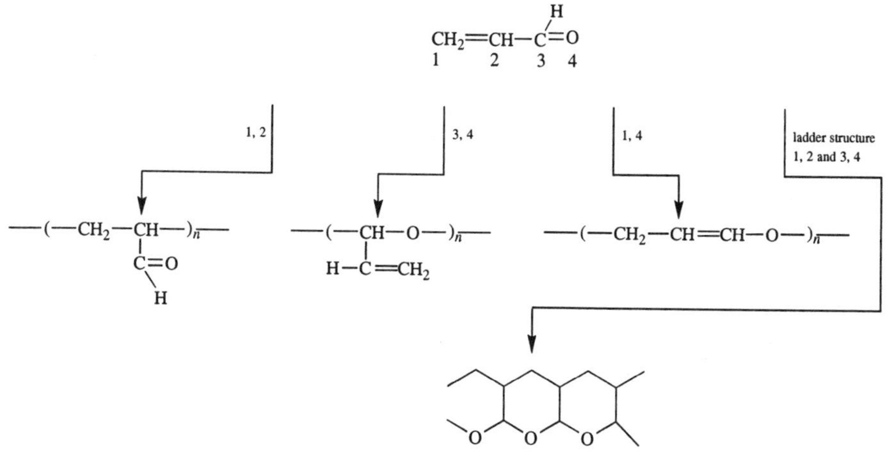

$$CH_2{=}CH{-}CH{=}O \quad (1\ 2\ 3\ 4)$$

1, 2: $-(-CH_2-CH(CHO)-)_n-$

3, 4: $-(-CH(CH{=}CH_2)-O-)_n-$

1, 4: $-(-CH_2-CH{=}CH-O-)_n-$

ladder structure 1, 2 and 3, 4

Because acrolein polymerizes by free-radical and ionic mechanisms, all of the above reactions are possible and the products are quite complex. The structures of the materials include linkages from both vinyl and carbonyl groups. In addition, tetrahydropyran rings can also form.[257]

Coordination complexes, like $CdI_2(pyridine)_2$, also initiate polymerizations of acrolein. Propagation reactions proceed through both vinyl and carbonyl groups[258]:

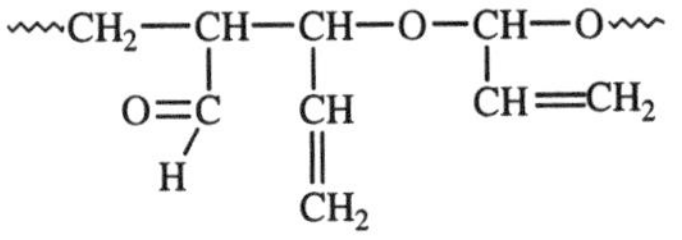

$$\sim CH_2-\underset{\underset{H}{\overset{|}{O=C}}}{CH}-\underset{\underset{CH_2}{\overset{|}{CH}}}{CH}-O-\underset{CH=CH_2}{\overset{}{CH}}-O\sim$$

The ratio of vinyl-to-carbonyl placement depends upon the nature of the complex. Polymers formed by complexes of metallic salts with triphenylphosphine contain considerably less aldehyde groups than those formed with triphenylphosphine alone.[258]

Polymerization of propynaldehyde (CH≡C–CH=O) is also unique. In dimethylformamide at O °C with sodium cyanide or with tri-*n*-butyl phosphine catalysts, the reactions yield polymers composed of two different structural units. One is a polyaldehyde and the other is a polyacetylene.[255] The reaction in tetrahydrofuran, however, at –78 °C with sodium cyanide catalyst results in a crystalline poly(ethynyl oxymethylene).[259] Radical initiated polymerizations of this monomer at 60 °C, on the other hand, proceed through the acetylenic group only.

Crotonaldehyde, like acrolein, can be expected to yield polymers with structures derived from 1,2; 3,4; or 1,4 additions. Anionic catalysts, however, yield predominantly polyacetal structures.[260]

3.6.4. Polymerizations of Dialdehydes

As one might expect, dialdehydes can be polymerized to yield polymers with cyclic ether linkages. This resembles polymerizations of nonconjugated dienes[261]:

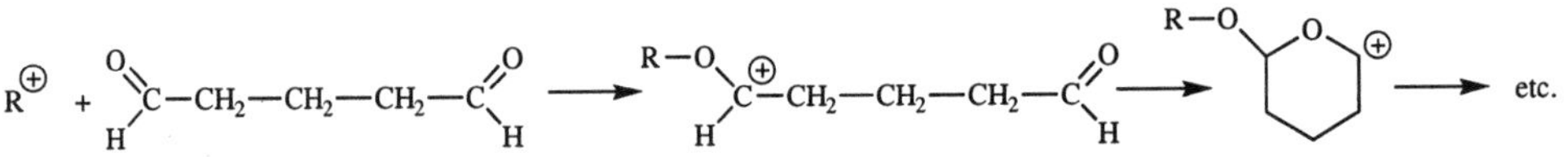

$$R^{\oplus} + OHC-CH_2-CH_2-CH_2-CHO \longrightarrow R-O-\overset{\oplus}{C}H-CH_2-CH_2-CH_2-CHO \longrightarrow \text{R–O-substituted tetrahydropyran cation} \longrightarrow \text{etc.}$$

Cationic polymerizations can take place by either one of three paths: stepwise additions, intermediate-type additions, and concerted additions. When phthalaldehyde is polymerized by cationic mechanism or by γ-irradiation at –78 °C, the products consist of dioxyphthalan units[262]:

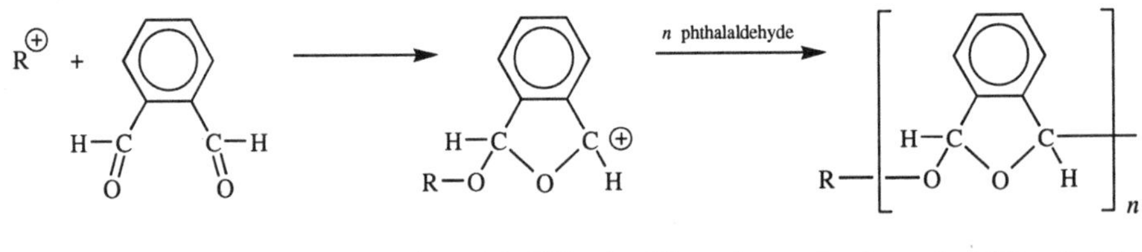

$$R^{\oplus} + \text{phthalaldehyde} \longrightarrow \text{R–O–phthalan–}\overset{\oplus}{C}H \xrightarrow{n \text{ phthalaldehyde}} R-\left[\text{O–CH(phthalan)CH–}\right]_n$$

Aromatic aldehydes usually polymerize with difficulty. The enhanced polymerizability of phthalaldehyde is due to formation of intermediates from concerted propagation reactions.[262]

In the above illustration, the anhydride rings can be either *cis* or *trans* in configuration. Cationic polymerizations yield polymers with high *cis* content. Anionic catalysts and particularly coordination catalysts, like triethylaluminum–transition metal halides, yield high *trans* polymers.[263]

3.7. Polymerization of Ketones and Isocyanates

In spite of many attempts to prepare useful polymers from ketones, this has so far not been very successful. Acetone polymerizes with the aid of magnesium.[223] The reaction requires high vacuum. Vapors of magnesium metal and dry acetone condense simultaneously in vacuum on a surface cooled by liquid nitrogen. The white elastic polymer that forms possesses a polyketal structure:

$$n\,(CH_3)_2C{=}O \longrightarrow -[-C(CH_3)_2-O-]_n-$$

Like low molecular weight ketals, the polymer is unstable and decomposes even at room temperature. Acetone also polymerizes upon irradiation in a frozen state under high vacuum. The product is a yellow, rubbery material.[223] Infrared spectra show the presence of both C–O–C linkages and carbonyl groups. Monobromoacetone polymerizes in the same manner, by irradiation in the frozen state. The hard resinous polymer that forms is more stable than polyacetone.[223]

Polymerizations of ketenes yield varieties of structures, because monomer placements are possible through either the carbon-to-carbon double bond or through the carbon-to-oxygen double bond. Dimethylketene polymerizes by anionic mechanism to a polymer with the following structures[224]:

$$n\,(CH_3)_2C{=}C{=}O \longrightarrow -(-\underset{\underset{C(CH_3)_2}{\|}}{C}-O-)_m- \quad -(-C(CH_3)_2-\overset{O}{\overset{\|}{C}}-)_o- \quad -(-C(CH_3)_2-\overset{O}{\overset{\|}{C}}-O-\overset{C(CH_3)_2}{\overset{\|}{C}}-)_p-$$

Polar solvents increase formations of ether groups. Nonpolar solvents, used with lithium, magnesium, or aluminum counterions, yield products that are high in ketones.[224] The same solvents, used with sodium or potassium counterions, form polymers with predominately polyester units.[224]

Isocyanates polymerize through the carbon-to-nitrogen double bonds by anionic mechanism. Reactions can be catalyzed by sodium or potassium cyanide at –58 °C.[278] N,N′-dimethylformamide is a good solvent for this reaction. Other anionic catalysts, ranging from alkali salts of various carboxylic acids[279] to sodium-naphthalene, are also effective.[280] In addition, polymerizations can be carried out by cationic,[281] thermal,[282] and radiation-induced[282] methods.

Anionic polymerizations yield very high molecular weight polymers. There is a tendency, however, to depolymerize at high temperatures. The products of anionic polymerization are substituted polyamides.[278] (For more information, see Chapter 6.)

3.8. Copolymerizations by Ionic Mechanism

Ionic copolymerizations are more complicated than free-radical ones. Various complicating factors arise from effects of the counterions and from influences of the solvents. These affect the reactivity ratios. In addition, monomer reactivity is affected by the substituents. They influence the electron densities of the double bonds and, in cationic polymerizations, the resonance stabilization of the resultant carbon cations. Yet, the effects of the counterions, the solvents, and even the reaction temperatures can be even greater than that of the substituents in cationic polymerizations. There are only a few studies reported in the literature, where the reactivity ratios were determined for different monomers, using the same temperature, solvent, and counterion. One such study was carried out on cationic copolymerizations of styrene with two substituted styrenes. These were α-methylstyrene, and with chlorostyrene.[283–285] The relative reactivity ratios of these substituted styrenes were correlated with Hammett $\rho\sigma$ values. The effect of the substituents on reactivity of styrene fall in the following order:

$$p\text{–}OCH_3 > p\text{–}CH_3 > p\text{–}H > p\text{–}Cl > m\text{–}Cl > m\text{–}OH$$

This information, however, is useful only for copolymerizations of substituted styrene monomers.

Copolymerization studies demonstrated that steric factors are very important in cationic copolymerizations.[286–288] For instance, β-methyl styrene is less reactive than styrene. Also, *trans*-β-methylstyrene is more reactive than the *cis* isomer.

The effect of solvents on the reactivity ratios in cationic copolymerizations can be seen from copolymerizations of isobutylene with *p*-chlorostyrene, using aluminum bromide as the initiator.[289]

The r_1 and r_2 values in hexane for isobutylene and chlorostyrene copolymerization are both equal to 1.0. In nitrobenzene, however, r_1 is equal to 14.7 and r_2 to 0.15.

In copolymerization of styrene with *p*-methylstyrene catalyzed by $SbCl_5$, AlX_3, $TiCl_4$, $SnCl_4$, $BF_3{\cdot}OEt_2$, $SbCl_3$, Cl_3COOH, and iodine, the copolymer composition depends upon the solvent and on the acid strength of the catalysts. There is no difference in copolymer composition in highly polar solvents, except for $SbCl_5$.[290] Here, the amount of styrene in the copolymer decreases as the solvent polarity increases. In solvents of low polarity, on the other hand, the amount of styrene in the copolymer decreases with a decrease in the strength of the Lewis acids. This depends upon the amount of solvation of the ion pair, and on the complexation of the solvent and monomer with the ion pair. With an increase in the tightness of the ion pairs in less polar solvents, selectivity increases. Because *p*-methylstyrene is more polar, it complexes to a greater extent than does styrene.[290]

Polymerization temperature in cationic polymerizations affects monomer reactivity strongly. This effect is considerably greater in cationic copolymerizations than in free-radical ones. No general trend, however, appears to have been established so far.

Electron-withdrawing substituents in anionic polymerizations enhance electron density at the double bonds or stabilize the carbanions by resonance. Anionic copolymerizations in many respects behave similarly to the cationic ones. For some comonomer pairs steric effects give rise to a tendency to alternate.[292] The reactivities of the monomers in copolymerizations and the compositions of the resultant copolymers are subject to solvent polarity and to the effects of the counterions. The two, just as in cationic polymerizations, cannot be considered independently from each other. This, again, is due to the tightness of the ion pairs and to the amount of solvation. Furthermore, only monomers that possess similar polarity can be copolymerized by an anionic mechanism. Thus, for instance, styrene derivatives copolymerize with each other. Styrene, however, is unable to add to a methyl methacrylate anion,[121–123] though it copolymerizes with butadiene and isoprene. In copolymerizations initiated by *n*-butyllithium in toluene and in tetrahydrofuran at –78 °C, the following order of reactivity with methyl methacrylate anions was observed.[293] In toluene the order is: diphenylmethyl methacrylate > benzyl methacrylate > methyl methacrylate > ethyl methacrylate > α-methylbenzyl methacrylate > isopropyl methacrylate > *t*-butyl methacrylate > trityl methacrylate > α,α'-dimethylbenzyl methacrylate. In tetrahydrofuran the order changes to: trityl methacrylate > benzyl methacrylate > methyl methacrylate > diphenylmethyl methacrylate > ethyl methacrylate > α-methylbenzyl methacrylate > isopropyl methacrylate > α,α'-dimethylbenzyl methacrylate > *t*-butyl methacrylate.

Copolymerizations of styrene with butadiene in hydrocarbon solvents using lithium alkyls initiators initially yield copolymers containing mainly butadiene. The amount of styrene in the copolymer increases considerably, however, in tetrahydrofuran solvent.

Anionic copolymerizations are very useful in forming block copolymers. (See Chapter 5 for discussion.) Ziegler–Natta catalysts also form block copolymers, similarly to anionic initiators. Much work on copolymerization with coordinated anionic initiators was done to develop ethylene propylene copolymers. Ethylene is considerably more reactive in these copolymerizations. To form random copolymers, soluble Ziegler–Natta catalysts are used. This is aided further by carefully controlling the monomer feed.[294]

The 1,2-disubstituted olefinic monomers will usually not homopolymerize with the Ziegler–Natta catalysts. They can, however, be copolymerized with ethylene and some α-olefins.[294] Due to poorer reactivity, the monomer feed must consist of higher ratios of the 1,2-disubstituted olefins than of the other comonomers. Copolymers of *cis*-2-butene with ethylene, where portions of the macromolecules are crystalline, form with vanadium-based catalysts. The products have alternating structures, with the pendant methyl groups in erythrodiisotactic arrangements.[150] Similarly, vanadium-based catalysts yield alternating copolymers of ethylene and butadiene, where the butadiene placement is predominantly *trans*-1,4.[150]

Copolymerizations of aldehydes take place by both anionic and cationic mechanisms.[295] An elastic copolymer of formaldehyde and acetaldehyde forms with triisobutylaluminum. The rate of copolymerization is very rapid at –78 °C. The reaction is complete within 30 minutes.[299] The product, however, is crosslinked. Aldehydes also copolymerize with some vinyl monomers.[296] An acetone block copolymer forms[223] with propylene when Ziegler–Natta catalysts are used at –78 °C. Copolymers of acetone with other olefins and with formaldehyde were also prepared.[297,298] Many initiators are effective in copolymerizations of aldehydes, ketones, and epoxies.[233,299,329]

3.9. Group Transfer Polymerization

This new technology offers considerable promise for commercial preparations of living polymers of methyl methacrylate without resorting to low-temperature anionic polymerizations. Although the mechanism or polymerization is not completely explained, the propagation is generally believed to be *covalent* in character. A silyl ketene acetal is the initiator. It forms from an ester enolate[300]:

$$(CH_3)_2{-}CH{-}C({=}O){-}OCH_3 + (H{-}C(CH_3)_2{-})_2{-}N{-}Li \longrightarrow (CH_3)_2{-}C{=}C(O{-}Li)(OCH_3)$$

$$\xrightarrow{(CH_3)_3{-}Si{-}Cl} (CH_3)_2{-}C{=}C(O{-}Si{-}(CH_3)_3)(OCH_3)$$

The initiation, which is catalyzed by either a nucleophilic or a Lewis acid catalyst, was explained as consisting of a concerted attack by the ketene acetal on the monomer[301]:

$$CH_2{=}C(CH_3){-}C(OCH_3){=}O + (CH_3)_2{-}C{=}C(OCH_3){-}O{-}Si{-}(CH_3)_3 \longrightarrow (CH_3)_2{-}C(C({=}O)OCH_3){-}CH_2{-}C(CH_3){=}C(OCH_3){-}O{-}Si{-}(CH_3)_3$$

This results in a transfer of the silyl ketene acetal center to the monomer. The process is repeated in each step of the propagation. The ketene double bond acts as the propagating center[300]:

$$CH_3{-}O{-}C({=}O){-}C(CH_3)_2{-}[{-}CH_2{-}C(CH_3)(C({=}O)OCH_3){-}]_n{-}CH_2{-}C(CH_3){=}C(OCH_3){-}O{-}Si{-}(CH_3)_3 + CH_2{=}C(CH_3){-}C({=}O){-}OCH_3 \longrightarrow$$

$$CH_3{-}O{-}C({=}O){-}C(CH_3)_2{-}[{-}CH_2{-}C(CH_3)(C({=}O)OCH_3){-}]_{n+1}{-}CH_2{-}C(CH_3){=}C(OCH_3){-}O{-}Si{-}(CH_3)_3$$

The above mechanism shows each step of chain growth involving transfer of the trialkylsilyl group from the silyl ketene acetal at the chain end to the carbonyl group of the incoming monomer. This is disputed because it excludes silyl exchange between growing chains.[302] Such an exchange, however, was observed[301] and led to a suggestion that the mechanism can involve ester enolate anion intermediates. These are reversibly complexed with silyl ketene acetal chain ends, as the propagating species.[303] It should be noted, however, that the results do not exclude the possibility of transfer reactions occurring as well.

Difunctional initiators cause chain growth to proceed from each end.[304] Because group-transfer polymerizations are "living" polymerizations, once all the monomer has been consumed a different monomer can be added and block copolymers can be formed.

The most effective nucleophilic catalysts for this reaction are bifluoride (HF_2^-) and fluoride ions. They can be generated from soluble reagents like tris(dimethylamino)sulfonium bifluoride. Other nucleophiles, like CN^- and nitrophenolate, have also been used. These nucleophilic catalysts function by assisting the displacement of the trialkylsilyl group. They are effective in concentrations below 0.1 mole percent of the initiator. Among electrophilic catalysts are Lewis acids, like zinc chloride, zinc bromide, zinc iodide, and dialkylaluminum chloride. Such catalysts probably function by coordinating with the carbonyl oxygens of the monomers and increasing the electrophilicity of the double bonds. This makes them more reactive with nucleophilic reagents. They must be used, however, in much higher concentrations.

Water and compounds with active hydrogen must be excluded from the reaction medium. Oxygen, on the other hand, does not interfere with the reaction. Tetrahydrofuran, acetonitrile, and aromatic solvents are commonly used in polymerizations catalyzed by nucleophiles. Chlorinated solvents and dimethylformamide are utilized in many reactions catalyzed by electrophiles. Living polymerizations of methacrylate esters can be carried out at 0 to 50 °C. The acrylate esters, however, require temperatures below 0 °C for living, group-transfer polymerizations, because they are more reactive and can undergo side reactions.

Weakly acidic compounds, such as methyl α-phenylacetate or α-phenylpropionitrile, are added to terminate the reaction. They are effective with anionic catalysts.[305] The trialkylsilyl group is transferred from the chain end to the transfer agent:

$$\sim CH_2-\overset{\overset{\displaystyle CH_3}{|}}{C}=C(OCH_3)-O-Si(CH_3)_3 \; + \; \underset{\text{(acid)}}{AH} \xrightarrow{HF_2^{\ominus}} \sim CH_2-\overset{\overset{\displaystyle CH_3}{|}}{\underset{\underset{\displaystyle H}{|}}{C}}-C(=O)OCH_3 \; + \; A-Si(CH_3)_3$$

Group-transfer polymerizations yield very narrow molecular weight distribution polymers. When mixtures of monomers are used, random copolymers form. The polymerization reaction is very tolerant of other functional groups in the monomer. Thus, for instance, *p*-vinylbenzyl methacrylate is converted to poly(*p*-vinylbenzyl methacrylate) without the polymerization of the vinyl group.[306] In addition, it is possible to form polymers with high syndiotactic content.

3.10. Configurational Statistics and the Propagation Mechanism in Chain-Growth Polymerization

Analyses of polymers to determine stereosequence distributions and understand the propagation mechanism can be carried out with NMR spectroscopy aided by statistical propagation models.[307,311,375] A detailed discussion of the subject is beyond this book. The following is a brief explanation of the concepts. The Bernoulli, Markov, and Colman–Fox, models describe propagation reactions with chain end control over monomer placement. The Bernoulli model assumes that the last monomer unit in the propagating chain end determines the stereochemistry of the polymer. No consideration is given to the penultimate unit or other units further back. In such an event, two modes

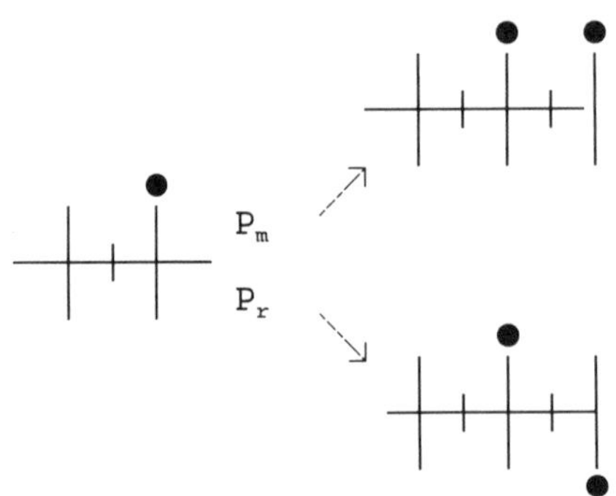

Figure 3.7. Two modes of propagation.

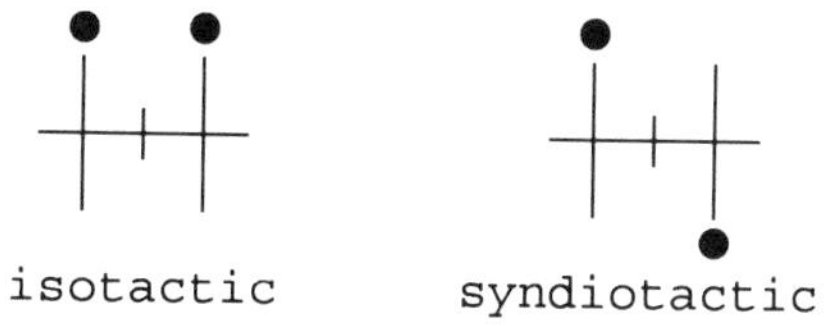

isotactic syndiotactic

Figure 3.8. Dyads.

of propagation are possible, meso and racemic (see Fig. 3.7.). The statistical probabilities, P_m and P_r, which are called *conditional probabilities* or *transitions*, of forming meso or racemic dyads are defined by the following equations:

$$P_m = R_m/(R_m + R_r)$$

$$P_r = R_r/(R_m + R_r)$$

$$P_m + P_r = 1$$

where R_m and R_r are the rates of meso and racemic dyad placement. These dyads can be isotactic or syndiotactic to one another (see Fig. 3.8). Such dyads are more frequently called meso and racemic. The probabilities represent the dyad tactic fractions m and r. The triad tacticity represent isotactic, syndiotactic, and heterotactic (or atactic) arrangements. They are designated as *mm*, *rr*, and *mr*, respectively. One way to understand these definitions is by examining a representation of a section of a polymeric chain which has a total of 9 repeating units but only 8 dyads and 7 triads (see Fig. 3.9). In the above segment there are 2 racemic dyads and 6 meso dyads. In addition, there are also 4 isotactic, 1 syndiotactic, and 2 atactic dyads.

The triads are defined as follows:

$$(mm) = P_m^2$$

$$(mr) = 2P_m(1 - P_m)$$

$$(rr) = (1 - P_m)^2$$

Note that the atactic triad can be produced in two ways, as *mr* and *rm*.

The tetrad probabilities can be defined as:

$$(mmm) = P_m^3 \qquad (mrm) = P_m^2(1 - P_m)$$

$$(mmr) = 2P_m^2(1 - P_m) \qquad (rrm) = 2P_m(1 - P_m)^2$$

$$(rmr) = P_m(1 - P_m)^2 \qquad (rrr) = (1 - P_m)^3$$

The pentad probabilities are given as:

$$(mmmm) = P_m^4 \qquad (rmrm) = 2P_m^2(1 - P_m)^2$$

$$(mmmr) = 2P_m^3(1 - P_m) \qquad (rmrr) = 2P_m(1 - P_m)^3$$

$$(rmmr) = P_m^2(1 - P_m)^2 \qquad (mrrm) = P_m^2(1 - P_m)^2$$

$$(mmrm) = 2P_m^3(1 - P_m) \qquad (rrrm) = 2P_m(1 - P_m)^3$$

$$(mmrm) = 2P_m^2(1 - P_m)^2 \qquad (rrrr) = (1 - P_m)^4$$

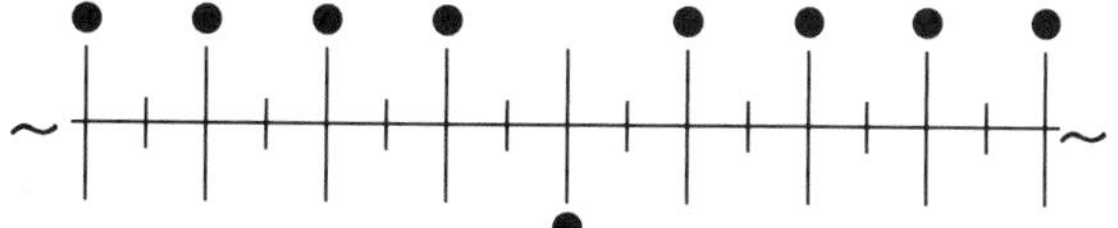

Figure 3.9. Polymeric chain.

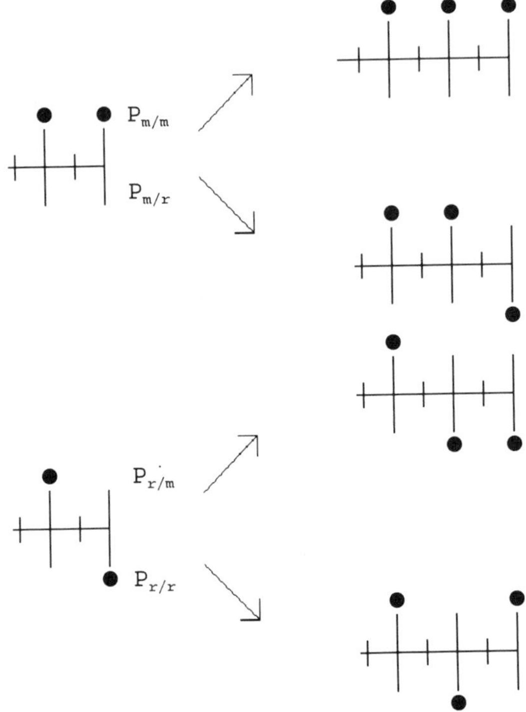

Figure 3.10. Modes of propagation.

In the Bernoulli model, there is no influence of the stereochemistry of the growing polymeric chain on the incoming monomer. The first-order Markov model, on the other hand, is based on the concept that the adding monomer is influenced by the stereochemistry of the growing end. This end may be m or r. This means that there are four probabilities for characterizing the addition process. The monomer can add in an m fashion to an m chain end, or m to an r chain end, and so on. These can be designated as $P_{m/m}$, $P_{r/m}$, $P_{m/r}$, and $P_{r/r}$. These modes of propagation are illustrated in Fig. 3.10. There are four probabilities for propagation, $P_{m/m}$, $P_{m/r}$, $P_{r/m}$, and $P_{r/r}$. With conservation of the relationship we can write:

$$P_{m/m} + P_{m/r} = 1$$

$$P_{r/m} + P_{r/r} = 1$$

We can recast the dyads, triads, tetrads, and pentads in terms of $P_{m/r}$ and $P_{r/m}$. The dyads functions are:

$$(m) = P_{r/m}/(P_{m/r} + P_{r/m})$$

$$(r) = P_{m/r}/(P_{m/r} + P_{r/m})$$

The triad fractions are:

$$(mm) = (1 - P_{m/r})P_{r/m}/(P_{m/r} + P_{r/m})$$

$$(mr) = 2P_{m/r}P_{r/m}/(P_{m/r} + P_{r/m})$$

$$(rr) = (1 - P_{r/m})P_{m/r}/(P_{m/r} = P_{r/m})$$

There are six tetrad functions,

$$(mmm) = P_{r/m}(1 - P_{m/r})^2/(P_{m/r} + P_{r/m})$$

$$(mmr) = 2P_{m/r}P_{r/m}(1 - P_{m/r})/(P_{m/r} + P_{r/m})$$

$$(mrm) = P_{m/r}P_{r/m}^2/(P_{m/r} + P_{r/m})$$
$$(mrr) = 2P_{m/r}P_{r/m}(1 - P_{r/m})/(P_{m/r} + P_{r/m})$$
$$(rmr) = P_{m/r}^2 P_{r/m}/(P_{m/r} + P_{r/m})$$
$$(rrr) = P_{m/r}(1 - P_{r/m})^2/(P_{m/r} + P_{r/m})$$

Ten pentad functions,

$$(mmmm) = P_{r/m}(1 - P_{m/r})^3/(P_{m/r} + P_{r/m})$$
$$(mmmr) = 2P_{m/r}P_{r/m}(1 - P_{m/r})^2/(P_{m/r} + P_{r/m})$$
$$(rmrm) = 2P_{m/r}^2 P_{r/m}^2/(P_{m/r} + P_{r/m})$$
$$(rmrr) = 2P_{m/r}^2P_{r/m}(1 - P_{r/m})/(P_{m/r} + P_{r/m})$$
$$(rmmr) = P_{m/r}^2P_{r/m}(1 - P_{m/r})/(P_{m/r} + P_{r/m})$$
$$(mrrm) = P_{m/r}P_{r/m}^2(1 - P_{r/m})/(P_{m/r} + P_{r/m})$$
$$(mmrm) = 2P_{m/r}P_{r/m}^2(1 - P_{m/r})/(P_{m/r} + P_{r/m})$$
$$(mrrr) = 2P_{m/r}P_{r/m}(1 - P_{r/m})^2/(P_{m/r} + P_{r/m})$$
$$(mmrr) = 2P_{m/r}P_{r/m}(1 - P_{m/r})(1 - P_{r/m})/(P_{m/r} + P_{r/m})$$
$$(rrrr) = P_{m/r}(1 - P_{r/m})^3/(P_{m/r} + P_{r/m})$$

The second-order Markov model requires the specification of eight conditional probabilities. This is due to the influence of the last three pseudoasymmetric centers of the growing chain. The details of the second-order Markov model were described by Bovey.[375] For details, the reader is advised to consult the reference. For convenience, the eight conditional probabilities are designated by Greek letters[375]:

$$P(mm/m) = \alpha \qquad P(mm/r) = \alpha'$$
$$P(mr/m) = \beta \qquad P(mr/r) = \beta'$$
$$P(rm/m) = \gamma \qquad P(rm/r) = \gamma'$$
$$P(rr/m) = \delta \qquad P(rr/r) = \delta'$$

In the above designation, $P(mm/m)$ is the probability of a monomer adding to the chain end in the m manner; $P(mm/r)$ is the probability of it adding to the chain in the r manner, and so on.

Because

$$\alpha + \alpha' = 1; \qquad \beta + \beta' = 1; \qquad \gamma + \gamma' = 1; \qquad \delta + \delta' = 1$$

there are four independent probabilities. When $\alpha = \gamma$ and $\beta = \delta$ the model reduces to the first-order Markov model.

The Coleman–Fox model attempts to explain "block-like" configurations that are exhibited in varying degrees by most propagating species that deviate from Bernoullian statistics. They proposed that "block" configurations are generated because the propagating chain ends might exist in two or more states. These states would correspond to chelation by the counterion and to interruption of chelation by solvation. Here, too, for further details the reader is advised to go to the original literature.[311]

The enantiomorphic site control is based on the probability of the monomer adding either to an R or to an S site of the catalyst. The propagation occurs through both faces of R and S monomers. The model is described in terms of a single parameter that is commonly designated as σ.[377] It is the probability of an R monomer adding at an R site and an S monomer adding at an S site.

The dyads are described as

$$(m) = \sigma^2 + (1 + \sigma)^2 \qquad (r) = 2\sigma(1 - \sigma)$$

The triad functions are given as

$$(mm) = 1 - 3\sigma(1-\sigma), \qquad (mr) = 2\sigma(1-\sigma), \qquad (rr) = \sigma(1-\sigma)$$

the tetrad fractions are

$$\begin{aligned}(mmm) &= 2\sigma^4 - 4\sigma^3 + 6\sigma^2 - 4\sigma + 1\\ (mmr) &= (mrr) = -4\sigma^4 + 8\sigma^3 - 6\sigma^2 + 2\sigma\\ (mrm) &= (rmr) = (rrr) = 2\sigma^4 - 4\sigma^3 + 2\sigma^2\end{aligned}$$

and the pentad fractions are

$$\begin{aligned}(mmmm) &= 5\sigma^4 - 10\sigma^3 + 10\sigma^2 - 5\sigma + 1\\ (mrrm) &= -3\sigma^4 + 6\sigma^3 - 4\sigma^2 + \sigma\\ (mmmr) &= (mmrr) = -6\sigma^4 + 12\sigma^3 - 8\sigma^2 + 2\sigma\\ (rmmr) &= (rrrr) = \sigma^4 - 2\sigma^3 + \sigma^2\\ (mmrm) &= (rmrr) = (rmrm) = (mrrr) = 2\sigma^4 - 4\sigma^3 + 2\sigma^2\end{aligned}$$

NMR spectroscopy allows one to test whether, in a particular polymerization, the propagation follows the Bernoulli, Markov, or enantiomorphic statistical form. Attempts are usually made to fit data for dyads, triads, tetrads, and higher-sequence fractions to the equation for the different models. Spectral intensities can be associated with theoretical expressions involving reaction probability parameters. Theoretical intensities are compared with the observed ones. This is optimized to obtain the best-fit values of reaction probability parameters and fully characterize the structure of the macromolecule. The fitting of data can be carried out with the aid of computers. Cheng published a very useful program in Basic language that applies the model-fitting approach to various polymeric systems.[318] A somewhat similar, though less sophisticated, program in Pascal language was put together by this writer. It lacks the simplex algorithm of the Cheng program to fine-tune the results. Nevertheless, an interested reader might still find it useful. It is therefore included in the appendix of this chapter. A good computer programmer should be able to introduce improvements.

By determining which statistical model is followed in a polymerization, such as Bernoullian, or Markov, or other, it should be possible to understand better the mechanism of steric control. Thus the Bernoulli model describes those reactions in which the chain ends determine the steric arrangement. These are polymerizations that are carried out under conditions that yield mostly atactic polymers. The high isotactic sequences follow the enantiomorphic site model and the high syndiotactic ones usually follow the Markov models.

Review Questions

Section 3.1

1. Explain the differences and similarities in anionic and cationic chain-growth polymerizations.

Section 3.2

1. Write chemical equations for cationic initiation, propagation, and termination steps and show what the kinetic expressions are.
2. Write the cationic kinetic expression for the degree of polymerization and the rate of propagation.
3. Repeat questions 1 and 2 for anionic polymerizations.

Section 3.3

1. Explain what is meant by the expression that initiation in cationic polymerization results from transposition of either one or two electrons. What type of initiating species are involved?
2. Summarize the initiation process in cationic chain-growth polymerization.
3. Show by chemical equations the initiation process by protonic acids, by Lewis acids complexed with water, and by alkyl halides.

4. What is the proposed mechanism of initiation of polymerization by the Lewis acid like aluminum chloride or bromide through an autoionization process?
5. What are the several proposed mechanisms of initiations by stable cations?
6. Explain how some cationic polymerizations can be initiated by metal alkyls and substances that generate cations.
7. Show the mechanism of one-electron transposition initiation with chemical equations based on a free-radical oxidation by diphenyliodonium hexafluorophosphate. (Let the source of the free-radical be photodecomposition of methyl benzoin ether.)
8. Discuss briefly the role of charge transfer complexes in initiations of cationic polymerizations.
9. What is the mechanism of electroinitiation of cationic polymerization?
10. How does the tightness of the ion pairs affect propagation in cationic polymerization?
11. What is the mechanism of chain growth with tight ion pairs in cationic polymerization?
12. Discuss the Kunitake and Takarabe mechanism for steric control in homogeneous cationic polymerizations.
13. What is the Nakano *et al.* mechanism for steric control in heterogeneous cationic polymerizations?
14. What is pseudocationic polymerization? Explain and illustrate on polymerization of styrene initiated by perchloric acid.
15. Describe and give examples of chemical reactions of isomerization polymerization where the propagation is accompanied by bond or electron rearrangement.
16. Give two examples of chemical reactions where the propagation in isomerization polymerization is accompanied by migration of one or more atoms.
17. Illustrate the termination reactions in cationic polymerizations.
18. What are the important aspects of living cationic polymerization? Can you give examples?

Section 3.4

1. What are the two types of initiation in anionic polymerization?
2. Describe and illustrate by chemical equations the initiation mechanism in a homogeneous polymerization by addition of an butyllithium to an olefin in a nonpolar solvent. What is the effect of heteroatoms, such as oxygen present in the monomers? Illustrate.
3. Describe and illustrate one-electron transfer initiations of anionic polymerizations and give several examples.
4. Describe initiations and propagation processes by Alfin catalysts.
5. Discuss electoinitiation of anionic polymerization.
6. How does the propagation reaction in anionic polymerization differ in the presence of polar and nonpolar solvents?
7. How do structures of monomers influence propagation rates in anionic polymerizations? Illustrate.
8. Discuss steric control in homogeneous anionic polymerizations of methyl methacrylate in nonpolar solvents.
9. Discuss steric control in homogeneous anionic polymerization of isoprene in nonpolar solvents.
10. Discuss steric control in homogeneous anionic polymerizations in polar solvents.
11. Discuss and illustrate hydrogen transfer polymerization of acrylamide.
12. How do termination reactions take place in anionic chain-growth polymerizations?
13. Discuss living anionic polymerizations.
14. How are living minodisperse methyl methacrylate polymers prepared with the aid of enolate aluminum porphyrin intermediates? Explain and illustrate.

Section 3.5

1. What metals form catalysts in Ziegler–Natta coordinated anionic polymerizations.
2. Illustrate the reactions that take place between aluminum alkyls and transition metals, like titanium chloride, in formations of Ziegler–Natta catalysts.

3. Describe the bimetallic polymerization mechanism of G. Natta.
4. Describe the Cossee and Arlman monometallic mechanism.
5. Describe steric control in heterogeneous coordinated anionic polymerizations.
6. What are the homogeneous Ziegler–Natta catalysts?
7. Discuss steric control with homogeneous catalysts.
8. Describe the currently used industrial Ziegler–Natta catalysts on support and soluble ones.
9. Discuss the effect of Lewis bases on Ziegler–Natta-type coordinated anionic polymerizations.
10. What is the termination reaction in coordinated anionic polymerizations? Illustrate.
11. Describe the transition metal oxide on support catalysts, how they are prepared and used.
12. How does isomerization polymerization take place by coordinated anionic mechanism?

Section 3.6

1. Describe cationic polymerization of aldehydes.
2. Describe anionic polymerizations of aldehydes.
3. What is the mechanism of control in anionic polymerizations of aldehydes?
4. Discuss polymerizations of unsaturated aldehydes, such as crotonaldehyde or acrolein.
5. Discuss polymerizations of dialdehydes.

Section 3.7

1. Discuss polymerizations of ketones and isocyanates.

Section 3.8

1. Discuss copolymerizations by ionic mechanism. What are some of the problems that are encountered?

Section 3.9

1. Discuss the proposed mechanisms of group transfer polymerizations.

Section 3.10

1. What are Bernoulli, Markov, and Colman–Fox statistical models for the propagation reactions? Can you describe each?

Appendix. Computer Program for Analyses of Polymers with Configurational Statistics and NMR Spectra

```
Program Match; uses printer;

   Label  1,2,3,4,5,6;
   Const  Max=25;  variables=10;

   Type  Ary=Array[1..Max] of real;
         Ary1=Array[1..Max,1..Max] of real;

   Var
     Shft,Int,Hi,Ge,O1,R3,U1,L1,M3,X1,Y1:ary;     S1:ary1;
     I,N,Answer,M2:integer;
     m,r,mm,rr,mr,mmm,mmr,rmr,mrm,mrr,rrr,mmmm,rmmr,mmmr,rmrr,mmrr,rmrm:real;
     mrmr,mmrm,rrrr,mrrr,mrrm,mrmm,Pmr,Prm,m1,r1:real;
     mm1,rr1,mr1,mmm1,mmr1,rmr1,mrm1,mrr1,rrr1,mmmm1,rmmr1,mmmr1,rmrr1:real;
     mmrr1,rmrm1,mmrm1,mrmr1,rrrr1,mrrr1,mrrm1,mrmm1:real;
     a,b,c,d,e,f,g,h,j,k,l,o,p,q,s,t,u,v,w,x,ov,vp,klu:real;
     q1,ts,G1,G2,G3,G4,G5,G6,G7,G8,G9,G10,tol:real;
     ch:char;
     mmmms,rmmrs,mmmrs,rmrrs,mmrrs,rmrms,mrmrs,mmrms,rrrrs,mrrrs,mrrms:real;
     mrmms:real;
```

```
Procedure Introduction(var ch:char);

Begin
 writeln;
 writeln;
 writeln('':10,'When you are ready to proceed, press the <ENTER> key');
 read(ch);
   If Eoln then
    ch:=chr(13);
 writeln;
 writeln;
 writeln;
 writeln;
 writeln;
 writeln;
 writeln;
 writeln;
 writeln;
 writeln('':10,'This software uses statistical probabilities to calculate');
 writeln('':10,'configurational sequences of polymers from NMR data.');
 writeln('':10,'First-order and Second-order Markov propagation models ');
 writeln('':10,'are developed from initial guess values of the dyads  ');
 writeln('':10,'which are fitted into the statistical equations. If the');
 writeln('':10,'intensities of the triads are known, then the dyads are
       ');
 writeln('':10,'also calculated by the computer from the observed triads');
 writeln('':10,'and then used to develop Second-order Markov values for');
 writeln('':10,'the tetrads and pentads. If you wand to avoid this, sim-
       ply');
 writeln('':10,'enter zeros when asked for the triad values');
 writeln;
 writeln('':10,'NOTE:');
 writeln('':10,'YOU MUST HAVE A PRINTER CONNECTED AND TURNED ON TO OPER-
       ATE');
 writeln('':10,'THIS PROGRAM');
 writeln;
 writeln('':10,'When ready to proceed, press the <ENTER> key');
 read(ch);
  if eoln then
  ch:=chr(13);
 writeln;
 writeln;
 writeln;
 writeln;
 writeln;
 writeln;
 writeln;
 writeln;
 writeln;
 writeln;
 writeln;
 writeln;
 writeln;
 writeln;
 writeln;
 writeln;
 writeln;
 writeln;
 writeln;
 writeln;
 writeln;
 writeln;
 writeln;
 writeln;
 writeln;
 end;
```

```
Procedure Intensities(var a,b,c,d,e,f,g,h,j,k,l,o,p,q,s,t,u,v,w,x:real);

Var I:integer;

Begin

writeln('':15,'Please enter the observed relative intensities (enter ');
writeln('':15,'zero when the peak is not present or is unresolved):');
write('':20,'mm:');
readln(a);
write('':20,'mr:');
readln(b);
write('':20,'rr:');
readln(c);
write('':20,'mmm:');
readln(d);
write('':20,'mmr:');
readln(e);
write('':20,'mrm:');
readln(f);
write('':20,'mrr:');
readln(g);
write('':20,'rmr:');
readln(h);
write('':20,'rrr:');
readln(j);
write('':20,'mmmm:');
readln(k);
write('':20,'mmmr:');
readln(l);
write('':20,'mmrm:');
readln(o);
write('':20,'mmrr:');
readln(p);
write('':20,'mrmr:');
readln(q);
write('':20,'mrrm:');
readln(s);
write('':20,'mrrr:');
readln(t);
write('':20,'rmmr:');
readln(u);
write('':20,'rmrr:');
readln(v);
write('':20,'rrrr:');
readln(w);
write('':20,'mrmm:');
readln(x);
End;

Procedure Unresolved(var ov,vp,klu,ql,ts:real);

Begin
Intensities(a,b,c,d,e,f,g,h,j,k,l,o,p,q,s,t,u,v,w,x);
writeln('':20,'For the unresolved peaks (enter zero if not present):');
write('':20,'mmrm + rmrr:');
readln(ov);
write('':20,'rmrr + mmrr:');
readln(vp);
write('':20,'mmmm + mmmr + rmmr:');
readln(klu);
write('':20,'rmrm + mmrm:');
readln(ql);
write('':20,'mrrr + mrrm:');
readln(ts);
End;
```

```
Procedure Diads(var m,r:real);

Begin
 writeln;
 writeln;
 writeln;
 writeln;
 write('':10,'Please enter your initial estimates of the dyads, m and r:');
 read(m,r);
End;

Procedure TriTet(var mm,mr,rr,Pmr,Prm,mmm,mmr,rmr,mrm,mrr,rrr:real);


Var  I:integer; F:real;

Begin
  mm:=Sqr(m);
  mr:=2*m*r;
  rr:=Sqr(r);
  Pmr:=mr/(2*mm+mr);
  Prm:=mr/(2*rr+mr);
  mmm:=(Sqr(1-Pmr)*Prm)/(Pmr+Prm);
  mmr:=(2*Pmr*(1-Pmr)*Prm)/(Pmr+Prm);
  mrm:=(Pmr*Prm*Prm)/(Pmr+Prm);
  mrr:=(2*Pmr*Prm*(1-Prm))/(Pmr+Prm);
  rmr:=(Pmr*Pmr*Prm)/(Pmr+Prm);
  rrr:=(Pmr*(1-Prm)*(1-Prm))/(Pmr+Prm);
  End;

Procedure Pent(var mmmm,mmmr,mmrm,mmrr,mrmr,mrrm,mrrr,rmmr,rmrr,rrrr,
               mrmm:real);

Var F:real;

begin
  TriTet(mm,mr,rr,Pmr,Prm,mmm,mmr,rmr,mrm,mrr,rrr);
  F:=Pmr*(1-Prm)+2*Pmr*Prm+Prm*(1-Pmr);
  mmmm:=(Prm*(1-Pmr)*(1-Pmr)*(1-Pmr))/F;
  mmmr:=(2*Pmr*Prm*Sqr(1-Pmr))/F;
  mmrm:=(2*Pmr*Sqr(Prm)*(1-Pmr))/F;
  mmrr:=(2*Pmr*Prm*(1-Pmr)*(1-Prm))/F;
  mrmr:=(2*Sqr(Pmr)*Sqr(Prm))/F;
  mrrm:=(Pmr*Sqr(Prm)*(1-Prm))/F;
  mrrr:=(2*Pmr*Prm*(1-Prm)*(1-Prm))/F;
  rrrr:=(Pmr*(1-Prm)*Sqr(1-Prm))/F;
  mrmm:=(2*Pmr*Sqr(Prm)*(1-Pmr))/F;
  rmmr:=(Prm*Sqr(Pmr)*(1-Pmr))/F;
  rmrr:=(2*Prm*Sqr(Pmr)*(1-Prm))/F;
End;

Procedure Markov(var mm1,mr1,rr1,mmm1,mmr1,mrm1,mrr1,rmr1,rrr1,mmmm1,mrmr1,
          m1,r1,mmmr1,mmrm1,mmrr1,mrrm1,mrrr1,rmmr1,rmrr1,rrrr1,mrmm1:real);

Var  alpha,beta,gamma,delta,alphap,betap,gammap,deltap,s:real;

Begin
  If (a>0) and (d>0) then
  alpha:=d/a
  else
  alpha:=mmm/mm;
  If (f>0) and (b>0) then
  beta:=(2*f)/b
  else
  beta:=(2*mrm)/mr;
  If (e>0)and (b>0) then
```

```
  gamma:=e/b
  else
  gamma:=mmr/mr;
  If (g>0) and (c>0) then
  delta:=g/(2*c)
  else
  delta:=mrr/(2*rr);
  If (e>0) and (a>0) then
  alphap:=e/(2*a)
  else
  alphap:=mmr/(2*mm);
  If (g>0) and (b>0) then
  betap:=g/b
  else
  betap:=mrr/mr;
  If (h>0) and (b>0) then
  gammap:=(2*h)/b
  else
  gammap:=(2*rmr)/mr;
  If (j>0)and (c>0) then
  deltap:=j/c
  else
  deltap:=rrr/rr;
  s:=alphap*betap + 2*alphap*delta + gamma*delta;
  m1:=((alphap + gamma)*delta)/s;
  r1:=1-m1;
  mm1:=(gamma*delta)/s;
  mr1:=(2*alphap*delta)/s;
  rr1:=(alphap*betap)/s;
  mmm1:=(alpha*gamma*delta)/s;
  mmr1:=(2*alphap*gamma*delta)/s;
  mrm1:=(alphap*beta*delta)/s;
  mrr1:=(2*alphap*betap*delta)/s;
  rmr1:=(alphap*gammap*delta)/s;
  rrr1:=(alphap*betap*deltap)/s;
  mmmm1:=(Sqr(alpha)*gamma*delta)/s;
  mmmr1:=(2*alpha*alphap*gamma*delta)/s;
  mmrm1:=(2*alphap*beta*gamma*delta)/s;
  mmrr1:=(2*alphap*betap*gamma*delta)/s;
  mrmr1:=(2*alphap*beta*gammap*delta)/s;
  mrrm1:=(alphap*betap*Sqr(delta))/s;
  mrrr1:=(2*alphap*betap*delta*deltap)/s;
  rmmr1:=(Sqr(alphap)*gamma*delta)/s;
  rmrr1:=(2*alphap*betap*gammap*delta)/s;
  rrrr1:=(alphap*betap*deltap*deltap)/s;
  mrmm1:=(2*alphap*beta*gamma*delta)/s;
End;

Procedure Bicatalytic(Var G1,G2,G3,G4,G5,G6,G7,G8,G9,G10:real);

Var A,B,C,D,E,beta,sigma,omega:real;

Begin
  writeln('':15,'Please enter your estimates of beta, sigma, and omega:');
  read(beta,sigma,omega);
  A:=beta;
  B:=Sqr(beta);
  C:=1-sigma;
  D:=1-omega;
  G1:=omega*(1-5*beta+5*B) + D*sigma*sigma*sigma*sigma;
  G2:=omega*(2*beta-6*B) + 2*D*C*sigma*sigma*sigma;
  G3:=omega*B + D*Sqr(sigma*C);
  G4:=omega*(2*beta-6*B) + 2*D*sigma*sigma*C*C;
  G5:=2*omega*B + 2*(1-omega)*sigma*sigma*sigma*C;
  G6:=2*omega*B + 2*D*sigma*C*C*C;
  G7:=omega*B + D*C*C*C*C;
```

```
  G8:=2*omega*B + 2*D*sigma*C*C*C;
  G9:=omega*(beta - 3*B) + D*sigma*sigma*C*C;
  G10:=2*omega*B+2*D*sigma*sigma*C*C;
End;

Procedure Results;

Begin
 Pent(mmmm,mmmr,mmrm,mmrr,mrmr,mrrm,mrrr,rmmr,rmrr,rrrr,mrmm);
 Markov(mm1,mr1,rr1,mmm1,mmr1,mrm1,mrr1,rmr1,rrr1,mmmm1,mrmr1,m1,r1,mmmr1,
       mmrm1,mmrr1,mrrm1,mrrr1,rmmr1,rmrr1,rrrr1,mrmm1);
 writeln;
 writeln(Lst);
 writeln(Lst);
 writeln('':30,'STATISTICAL MODEL FITTING');
 writeln(Lst,'':30,'STATISTICAL MODEL FITTING');
 writeln('':30,'________________________');
 writeln(Lst,'':30,'________________________');
 writeln('':16,'Dyad');
 writeln(Lst,'':16,'Dyad');
 writeln('':18,'(m):',m:4:2);
 writeln(Lst,'':18,'(m):',m:4:2);
 writeln('':18,'(r):',r:4:2);
 writeln(Lst,'':18,'(r):',r:4:2);
 writeln;
 writeln(Lst);
 writeln('':24,'Observed':10,'First Order':14,'Second Order':13);
 writeln('':24,'        ':10,'  Markov    ':14,'   Markov    ':13);
 writeln(Lst,'':24,'Observed':10,'First Order':14,'Second Order':13);
 writeln(Lst,'':24,'        ':10,'  Markov    ':14,'   Markov    ':13);
 writeln('':30,'_____________________________');
 writeln(Lst,'':30,'_____________________________');
 If (e>0) and (a>0) then
 writeln('':21,'Based on observed values of triads,(m) = ':5,m1:5:2);
 If (e>0) and (a>0) then
 writeln(Lst,'':21,'Based on observed values of triads, (m) = ':5,m1:5:2);
 If (e>0)and (a>0) then
 writeln('':21,'Based on observed values of triads, (r) = ':5,r1:5:2);
 If (e>0) and (a>0) then
 writeln(Lst,'':21,'Based on observed values of triads, (r) = ':5,r1:5:2);
 writeln('':16,'Triad');
 writeln(Lst,'':16,'Triad');
 writeln('':18,'(mm)',a:10:2,mm:12:2,mm1:13:2);
 writeln(Lst,'':18,'(mm)',a:10:2,mm:12:2,mm1:13:2);
 writeln('':18,'(mr)',b:10:2,mr:12:2,mr1:13:2);
 writeln(Lst,'':18,'(mr)',b:10:2,mr:12:2,mr1:13:2);
 writeln('':18,'(rr)',c:10:2,rr:12:2,rr1:13:2);
 writeln(Lst,'':18,'(rr)',c:10:2,rr:12:2,rr1:13:2);
 writeln;
 writeln(Lst);
 End;

 Procedure Tetrads;

 begin
 writeln('':16,'Tetrad');
 writeln(Lst,'':16,'Tetrad');
 writeln('':18,'(mmm)',d:9:2,mmm:12:2,mmm1:13:2);
 writeln(Lst,'':18,'(mmm)',d:9:2,mmm:12:2,mmm1:13:2);
 writeln('':18,'(mmr)',e:9:2,mmr:12:2,mmr1:13:2);
 writeln(Lst,'':18,'(mmr)',e:9:2,mmr:12:2,mmr1:13:2);
 writeln('':18,'(rmr)',h:9:2,rmr:12:2,rmr1:13:2);
 writeln(Lst,'':18,'(rmr)',h:9:2,rmr:12:2,rmr1:13:2);
 writeln('':18,'(mrm)',f:9:2,mrm:12:2,mrm1:13:2);
 writeln(Lst,'':18,'(mrm)',f:9:2,mrm:12:2,mrm1:13:2);
 writeln('':18,'(mrr)',g:9:2,mrr:12:2,mrr1:13:2);
```

```
writeln(Lst,'':18,'(mrr)',g:9:2,mrr:12:2,mrr1:13:2);
writeln('':18,'(rrr)',j:9:2,rrr:12:2,rrr1:13:2);
writeln(Lst,'':18,'(rrr)',j:9:2,rrr:12:2,rrr1:13:2);
writeln;
writeln(Lst);
End;

Procedure Pentads;

Begin
writeln('':16,'Pentad');
writeln(Lst,'':16,'Pentad');
If K > 0 then
writeln('':17,'(mmmm)',k:9:2,mmmm:12:2,mmmm1:13:2);
If K > 0 then
writeln(Lst,'':17,'(mmmm)',k:9:2,mmmm:12:2,mmmm1:13:2);
If l > 0 then
writeln('':17,'(mmmr)',l:9:2,mmmr:12:2,mmmr1:13:2);
If l > 0 then
writeln(Lst,'':17,'(mmmr)',l:9:2,mmmr:12:2,mmmr1:13:2);
If o > 0 then
writeln('':17,'(mmrm)',o:9:2,mmrm:12:2,mmrm1:13:2);
If o > 0 then
writeln(Lst,'':17,'(mmrm)',o:9:2,mmrm:12:2,mmrm1:13:2);
If p > 0 then
writeln('':17,'(mmrr)',p:9:2,mmrr:12:2,mmrr1:13:2);
If p > 0 then
writeln(Lst,'':17,'(mmrr)',p:9:2,mmrr:12:2,mmrr1:13:2);
If q > 0 then
writeln('':17,'(mrmr)',q:9:2,mrmr:12:2,mrmr1:13:2);
If q > 0 then
writeln(lst,'':17,'(mrmr)',q:9:2,mrmr:12:2,mrmr1:13:2);
If s > 0 then
writeln('':17,'(mrrm)',s:9:2,mrrm:12:2,mrrm1:13:2);
If s > 0 then
writeln(Lst,'':17,'(mrrm)',s:9:2,mrrm:12:2,mrrm1:13:2);
If t > 0 then
writeln('':17,'(mrrr)',t:9:2,mrrr:12:2,mrrr1:13:2);
If t > 0 then
writeln(Lst,'':17,'(mrrr)',t:9:2,mrrr:12:2,mrrr1:13:2);
If u > 0 then
writeln('':17,'(rmmr)',u:9:2,rmmr:12:2,rmmr1:13:2);
If u > 0 then
writeln(Lst,'':17,'(rmmr)',u:9:2,rmmr:12:2,rmmr1:13:2);
If v > 0 then
writeln('':17,'(rmrr)',v:9:2,rmrr:12:2,rmrr1:13:2);
If v > 0 then
writeln(Lst,'':17,'(rmrr)',v:9:2,rmrr:12:2,rmrr1:13:2);
If w > 0 then
writeln('':17,'(rrrr)',w:9:2,rrrr:12:2,rrrr1:13:2);
If w > 0 then
writeln(lst,'':17,'(rrrr)',w:9:2,rrrr:12:2,rrrr1:13:2);
If x > 0 then
writeln('':17,'(mrmm)',x:9:2,mrmm:12:2,mrmm1:13:2);
If x > 0 then
writeln(Lst,'':17,'(mrmm)',x:9:2,mrmm:12:2,mrmm1:13:2);
End;

Procedure Resolution;

Var  XY,AB,CD,EF,GH,IK:real;
begin
Results;
Tetrads;
Pentads;
XY:=mmm1+mmmr1+rmmr1;
```

```
AB:=mrrr1+mrrm1;
CD:=mmmm+mmmr+rmmr;
EF:=rmrr+mmrr;
GH:=mrmr+mmrm;
IK:=rmrr+mmrm;
If ov > 0 then
writeln('':14,'(mmrm+rmrr)',ov:7:2,IK:12:2,(mmrm1+rmrr1):13:2);
If ov > 0 then
writeln(Lst,'':14,'(mmrm+rmrr)',ov:7:2,IK:12:2,(mmrm1+rmrr1):13:2);
If vp > 0 then
writeln('':14,'(mmrr+rmrr)',vp:7:2,EF:12:2,(mmrr1+rmrr1):13:2);
If vp > 0 then
writeln(Lst,'':14,'(mmrr+rmrr)',vp:7:2,EF:12:2,(mmrr1+rmrr1):13:2);
If klu > 0 then
writeln('':9,'(mmmm+mmmr+rmmr)',klu:7:2,CD:12:2,XY:13:2);
If klu > 0 then
writeln(Lst,'':9,'(mmmm+mmmr+rmmr)',klu:7:2,CD:12:2,XY:13:2);
If ql > 0 then
writeln('':14,'(rmrm+mmrm)',ql:7:2,GH:12:2,(mrmr1+mmrm1):13:2);
If ql > 0 then
writeln(Lst,'':14,'(rmrm+mmrm)',ql:7:2,GH:12:2,(mrmr1+mmrm1):13:2);
If ts > 0 then
writeln('':14,'(mrrr+mrrm)',ts:7:2,(mrrr+mrrm):12:2,AB:13:2);
If ts > 0 then
writeln(Lst,'':14,'(mrrr+mrrm)',ts:7:2,(mrrr+mrrm):12:2,AB:13:2);
end;

Procedure WteBC;

Begin
Bicatalytic(G1,G2,G3,G4,G5,G6,G7,G8,G9,G10);
writeln;
writeln;
writeln(Lst);
writeln(Lst);
writeln('':30,'Bi-Catalytic Propagation Model');
writeln(Lst,'':30,'Bi-Catalytic Propagation Model');
writeln('':30,'______________________________');
writeln(Lst,'':30,'______________________________');
writeln('':24,'Observed':12,'Calculated':19);
writeln(Lst,'':24,'Observed':12,'Calculated':19);
If k > 0 then
writeln('':19,'(mmmm)',k:9:2,G1:15:2);
If k > 0 then
writeln(Lst,'':19,'(mmmm)',k:9:2,G1:15:2);
If l > 0 then
writeln('':19,'(mmmr)',l:9:2,G2:15:2);
If l > 0 then
writeln(Lst,'':19,'(mmmr)',l:9:2,G2:15:2);
If o > 0 then
writeln('':19,'(mmrm)',o:9:2,G5:15:2);
If o > 0 then
writeln(Lst,'':19,'(mmrm)',o:9:2,G5:15:2);
If p > 0 then
writeln('':19,'(mmrr)',p:9:2,G4:15:2);
If p > 0 then
writeln(Lst,'':19,'(mmrr)',p:9:2,G4:15:2);
If q > 0 then
writeln('':19,'(mrmr)',q:9:2,G10:15:2);
If q > 0 then
writeln(Lst,'':19,'(mrmr)',q:9:2,G10:15:2);
If s > 0 then
writeln('':19,'(mrrm)',s:9:2,G9:15:2);
If s > 0 then
writeln(Lst,'':19,'(mrrm)',s:9:2,G9:15:2);
If t > 0 then
```

```
writeln('':19,'(mrrr)',t:9:2,G8:15:2);
If t > 0 then
writeln(Lst, '':19,'(mrrr)',t:9:2,G8:15:2);
If u > 0 then
writeln('':19,'(rmmr)',u:9:2,G3:15:2);
If u > 0 then
writeln(Lst,'':19,'(rmmr)',u:9:2,G3:15:2);
If v > 0 then
writeln('':19,'(rmrr)',v:9:2,G6:15:2);
If v > 0 then
writeln(Lst,'':19,'(rmrr)',v:9:2,G6:15:2);
If x > 0 then
writeln('':19,'(mrmm)',x:9:2,G5:15:2);
If x > 0 then
writeln(Lst,'':19,'(mrmm)',x:9:2,G5:15:2);
If w > 0 then
writeln('':19,'(rrrr)',w:9:2,G7:15:2);
If w > 0 then
writeln(Lst,'':19,'(rrrr)',w:9:2,G7:15:2);
If ov > 0 then
writeln('':16,'(mmrm+rmrr)',ov:7:2,(G5+G6):15:2);
If ov > 0 then
writeln(Lst,'':16,'(mmrm+rmrr)',ov:7:2,(G5+G6):15:2);
If vp > 0 then
writeln('':16,'(mmrr+rmrr)',vp:7:2,(G4+G6):15:2);
If vp > 0 then
writeln(Lst,'':16,'(mmrr+rmrr)',vp:7:2,(G4+G6):15:2);
If klu  >0 then
writeln('':12,'(mmmm+mmmr+rmmr)',klu:7:2,(G1+G2+G3):15:2);
If klu > 0 then
writeln(Lst,'':12,'(mmmm+mmmr+rmmr)',klu:7:2,(G1+G2+G3):15:2);
If ql > 0 then
writeln('':16,'(rmrm+mmrm)',ql:7:2,(G10+G5):15:2);
If ql > 0 then
writeln(Lst,'':16,'(rmrm+mmrm)',ql:7:2,(G10+G5):15:2);
If ts > 0 then
writeln('':16,'(mrrr+mrrm)',ts:7:2,(G8+G9):15:2);
If ts > 0 then
writeln(Lst,'':16,'(mrrr+mrrm)',ts:7:2,(G8+G9):15:2);
end;

Procedure Conclusion;

Var  nm,nr:real;
Begin
writeln;
writeln(Lst);
writeln('':20,'Pm/r = ',Pmr:3:2,'Pr/m = ':10,Prm:3:2);
writeln(Lst,'':20,'Pm/r = ',Pmr:3:2,'Pr/m = ':10,Prm:3:2);
writeln('':20,'Based on first-order Markov statistics, the persistance');
writeln(Lst,'':20,'Based on first-order Markov statistics, the persis-
     tance');
writeln('':20,'ratio =',1/(Pmr+Prm):5:2);
writeln(Lst,'':20,'ratio =',1/(Pmr+Prm):5:2);
If (Pmr>0.5) and (Prm>0.5) then
writeln('':20,'This polymer is mainly heterotactic.');
If (Pmr>0.5) and (Prm>0.5) then
writeln(Lst,'':20,'This polymer is mainly heterotactic.');
If (Pmr<0.5) and (Prm>0.5) then
writeln('':20,'This polymer contains isotactic sequences.');
If (Pmr<0.5) and (Prm>0.5) then
writeln(Lst,'':20,'This polymer contains isotactic sequences.');
If (Pmr>0.5) and (Prm<0.5) then
writeln('':20,'This polymer contains syndiotactic sequences.');
If (Pmr>0.5) and(Prm<0.5) then
writeln(Lst,'':20,'This polymer contains syndiotactic sequences.');
```

```
 nm:=(1-mm/rr)/(1-(mm*r)/(rr*m));
 nr:=(1-mm/rr)/(m/r-mm/rr);
 If (Pmr<0.5) and (Prm>0.5) then
 writeln('':20,'Number average length of isotactic sequences = ',nm:3:2);
 If (Pmr<0.5) and (Prm>0.5) then
 writeln(Lst,'':20,'Number average length of isotactic sequences =
       ',nm:3:2);
 If (Pmr>0.5) and (Prm<0.5) then
 writeln('':20,'Average length of syndiotactic sequences = ',nr:3:2);
 If (Pmr>0.5) and (Prm<0.5) then
 writeln(Lst,'':20,'Average length of syndiotactic sequences = ',nr:3:2);
 If (mm*rr*4/Sqr(mr)>1) then
 writeln('':20,'The polymer deviates from Bernullian statistics');
 If (mm*rr*4/Sqr(mr)>1) then
 writeln(Lst,'':20,'The polymer deviates from Bernullian statistics');
 End;

Procedure Choice;

Begin
 writeln;
 writeln;
 writeln;
 writeln('':15,'In order to repeat the above calculations with a ');
 writeln('':15,'different set of values for m and r, enter 1; to fit ');
 writeln('':15,'a bi-catalytic mechanism enter 2; to complete the run.');
 writeln('':15,'enter 3.');

Procedure Choice2;

begin
   writeln('':15,'If you wish to repeat the Bi-Site calculations');
   writeln('':15,'enter 2 again; 1 to go back to beginning of program;');
   writeln('':15,'3 to complete the run.');
   read(Answer);
end;

Procedure GetData;

var   I,N:integer; H,X:ary;

begin
 writeln('Please enter the total number of peaks: ');
 read(N);
 writeln('Please enter the observed intensities: ');
 For I:=1 to N do begin
   write('':15,I:2,': ');
   readln(H[I]);
 end;
 writeln('':15,'Please enter you initial guess of the values of reaction');
 writeln('':15,'probabilities (3 values)');
 read(X[1],X[2],X[3]);
End;

Procedure Markov2;

Var I:integer;
   X,R,G,H:ary; G1,G6,G7,G8,G9:real;

Begin
  GetData;
   For I:=1 to 3 do begin
    If X[I]<-0.002 then
      R[I]:=10
    else
    If X[I]>1.0 then
```

```
        R[I]:=10;
        end;
        G1:=0.0;
        G7:=X[1] + X[2];
        G8:=X[1] + X[3];
        G9:=X[1] + X[2] + X[3];
        G[4]:=100*G7*Sqr(G8)/Sqr(G9);
        G[5]:=200*G7*X[2]*G8/Sqr(G9);
        G[7]:=100*G7*Sqr(X[2])/Sqr(G9);
        G[1]:=100*X[3];
        G[2]:=100*(X[1] + X[2]);
        G6:=100*G7;
        G[3]:=100*X[3];
        G[6]:=G6*Sqr(X[2]);
        G[8]:=2*G6*X[1]*X[2] + 2*G6*X[2]*X[3] + G6*G7*X[3];
        G[9]:=G6*Sqr(X[1]) + 2*G6*X[1]*X[3] + 2*G6*Sqr(X[3]);
        G[10]:=G6*Sqr(X[3]);
        G[11]:=100*X[3]*Sqr(X[3]);
        For I:=1 to N do begin
          G1:=G1 + ABS(H[I] - G[I]);
        end;
        R[I]:=G1/110;
   End;

    Procedure WrteMarkov2;

    Var I:integer; H,G:ary;

    Begin
      Markov2;
      writeln('':3,'Peak #':12,'Int.(obs.)':13,'Int.(calc.)':13,'Dev':12);
      writeln(Lst,'':3,'Peak #':12,'Int.(obs.)':13,'Int.(calc.)':13,'Dev':12);
    For I:=1 to N do begin
      writeln('':3,I:12,H[I]:13:2,G[I]:13:2,H[I]-G[I]:12:2);
      writeln(Lst,'':3,I:12,H[I]:13:2,G[I]:13:2,H[I]-G[I]:12:2);
      end;
    End;

  Begin
    Introduction(ch);
    Unresolved(ov,vp,klu,ql,ts);
    1:Diads(m,r);
    Resolution;
    Choice;
    Case Answer of
    1:goto 1;
    2:goto 2;
    3:goto 3;
    end;
    2:WteBC;
    Choice2;
    Case Answer of
    1:goto 1;
    2:goto 2;
    3:goto 3;
    end;
    3:Conclusion;
    4:writeMarkov2;
 End.
```

References

1. J.P. Kennedy and E. Marechal, *Cationic Polymerization*, Wiley-Interscience, New York, 1982; M. Morton, *Anionic Polymerization, Principles and Practice*, Academic Press, New York, 1983.
2. R.H. Biddulph, P.H. Plesch, and P.P. Rutheford, *J. Chem. Soc.*, 275 (1965).

3. J.P. Kennedy and R.M. Thomas, *J. Polym. Sci.*, **46**, 233 (1960).
4. J.E. Beard, P.H. Plesch, and P.P. Rutheford, *J. Chem. Soc.*, 2566 (1964).
5. M. Marek and M. Chmelir, *International Symposium of Macromolecular Chemistry*, Prague, 1965, p. 4.
6. W.H. Hunter and R.V. Yohe, *J. Am. Chem. Soc.*, **55**, 1248 (1933).
7. B.B. Fleischfresser, W. J. Chang, J.M. Pearson, and M. Szwarc, *J. Am. Chem. Soc.*, **90**, 2172 (1968).
8. J.P. Kennedy, *J. Macromol. Sci.*, **6-A**, 329 (1972).
9. P. Sigwalt, *Makromol. Chem.*, **175**, 1017 (1974).
10. G. Sauvet, J.P. Vairon, and P. Sigwalt, *Bull. Soc. Chim. France*, 4031 (1970).
11. G. A. Olah, *Makromol. Chem.*, **175**, 1039 (1974).
12. M. Chmelir, *J. Polym. Sci., Symp.*, **56**, 311 (1976).
13. D.C. Pepper, *Makromol. Chem.*, **175**, 1077 (1974).
14. P.H. Plesch, *Makromol. Chem.*, **175**, 1065 (1974).
15. N.A. Ghanem and H. Mark, *Eur. Polym. J.*, **8**, 999 (1972).
16. J.P. Kennedy, S.Y. Huang, and S.C. Feinberg, *J. Polym. Sci.*, **15**, 2801 (1977).
17. P. Giusti, F. Andrizzi, P. Cerrai, and G.L. Possanzini, *Makromol Chem.*, **136**, 97 (1970).
18. C.E.H. Bawn, C. Fitzsimmons, and A. Ledwith, *Proc. Chem. Soc.*, 391 (1964).
19. A. Ledwith, *Makromol. Chem.*, **175**, 1117 (1974).
20. G. Natta, G. Dall Asta, G. Mazzanti, U. Giannini, and S. Cesea, *Angew. Chem.*, **71**, 205 (1959).
21. G. Natta, G. Dall Asta, G. Mazzanti, and G. Casale, *Makromol. Chem.*, **58**, 217 (1962).
22. K. Ziegler, *Angew. Chem.*, **64**, 323 (1952).
23. J.P. Kennedy, *Am. Chem. Soc., Polym. Prepr.*, **7**, 485 (1966); *J. Polym. Sci., Symp.*, **56**, 1 (1976).
24. T. Saegusa, Chapt. 9 in *Structure and Mechanism in Vinyl Polymerization* (T. Tsuruta and H.S. O'Driscoll, eds.), Dekker, New York, 1969.
25. H. Sinn, H. Winter, and W. V. Tirptiz, *Makromol. Chem.*, **48**, 59 (1961).
26. T. Saeguea, H. Imai, and J. Furukawa, *Makromol. Chem.*, **65**, 60 (1963).
27. K. Iwasaki, H. Fukutani, Y. Tsuchida, and S. Nakano, *J. Polym. Sci.*, **A-1,1**, 2371 (1963).
28. J.M. Bruce and D.W. Farren, *Polymer*, **4**, 407 (1963).
29. J. P. Kennedy, Chapt. 5 in *Copolymerization* (G. B. Ham, ed.), Wiley-Interscience, New York, 1964.
30. D.D. Eley, in *The Chemistry of Cationic Polymerization* (P.H. Plesch, ed.), Macmillan, New York, 1963.
31. Y. Wang and L. M. Dorfmann, *Macromolecules*, **13**, 63 (1980).
32. S. Winstein and G.C. Robertson, *J. Am. Chem. Soc.*, **80**, 89 (1958).
33. D.C. Pepper and P.J. Reilly, *J. Polym. Sci.*, **58**, 639 (1962).
34. T. Higashimura, S. Okamura, and J. Masamoto, *Kibunski Kagaku*, **25**, (282), 702 (1968) (private translation).
35. A.G. Evans and M. Polanyi, *J. Chem. Soc.*, 252 (1947).
36. C.E. Schildknecht. A.O. Zoss, and E. McKinley, *Ind. Eng. Chem.*, **39**, 180 (1948).
37. G. Natta, I.W. Bossi, and P. Carradini, *Makromol. Chem.*, **18**, 455 (1955).
38. C. E. Schildknecht, *Ind. Eng. Chem.*, **50**, 107 (1958).
39. S. Okamura, T. Higashimura, and J. Sakurada, *J. Polym. Sci.*, **39**, 507 (1959).
40. T. Saegusa, H. Imai, and J. Furukawa, *Makromol. Chem.*, **79**, 207 (1964).
41. C. E. H. Bawn and A. Ledwith, *Quart. Rev. (London)*, **16**, 361 (1962).
42. J. Freenley, A. Ledwith, and R.H. Sutcliffe, *J. Chem. Soc.*, 2021 (1962).
43. D.J. Cram and K.R. Kopecky, *J. Am. Chem. Soc.*, **81**, 2748 (1959).
44. T. Kunitake and C. Aso, *J. Polym. Sci.*, **A-1,8**, 665 (1970).
45. Y. Hirokawa, T.R. Higashimura, K. Matsuzaki, T. Kawamura, and T. Uryu, *J. Polym. Sci., Chem. Ed.*, **17**, 3923 (1979).
46. J. Furukawa, *Polymer*, **3**, 487 (1962).
47. S. Nakano, K. Iwasaki, and H. Fukutani, *J. Polym. Sci.*, **A-1,1**, 3277 (1963).
48. J. P. Kennedy and R.M. Thomas, *Adv. Chem. Ser.*, **34**, 111 (1962).
49. S. Bywater and D.J. Worsford, *Can. J. Chem.*, **44**, 1671 (1966).
50. Z. Zlamal, Chapt. 6 in *Vinyl Polymerization*, Part II (G.E. Ham, ed.), Dekker, New York, 1969.
51. R.G. Heiligman, *J. Polym. Sci.*, **6**, 155 (1951).
52. O. Wichterle, M. Kolinsky, and M. Marek, *Chem. Listy*, **52**, 1049 (1958).
53. O. H. Plesch, *Br. Polym. J.*, 1 (1973); O.H. Plesch, *Makromol. Chem., Macromol. Symp.*, **13/14**, 375, 393 (1988); K. Matyjaszewski, *Makromol. Chem., Macromol. Symp.*, **13/14**, 389 (1988).
54. A. Gandini and P. H. Plesch, *Eur. Polym. J.*, **4**, 55 (1968); D.C. Pepper, *Makromol. Chem.*, **175**, 1077 (1974).
55. T. Masuda and T. Higashimura, *J. Polym. Sci., Polym. Lett.*, **9**, 783 (1972).
56. S.D. Hamann, A.J. Murphy, D.M. Solomon, and R.I. Willing, *J. Macromol. Sci.-Chem.*, **A6**, 771 (1972).
57. J.P. Kennedy, *Trans. N.Y. Acad. Sci.*, 1080 (1966).
58. G.B. Butler, M.L. Miles, and W.S. Brey, *J. Polym. Sci.*, **A-1,3**, 723 (1965).
59. G.B. Butler and M.L. Miles, *Polym. Eng. Sci.*, 71 (1966).

60. J.P. Kennedy and J.A. Hinlicky, *Polymer*, **6**, 133 (1965).
61. G. Santori, A. Valvassori, V. Turba, and M. P. Lachi, *Chim. Ind. (Milano)*, **45**, 1529 (1963).
62. J.P. Kennedy and H.S. Makowski, *J. Polym. Sci.*, **C(22)**, 247 (1968).
63. J.P. Kennedy, J.J. Elliott, and W. Naegele, *J. Polym. Sci.*, **A-1,2**, 5029 (1964).
64. C.G. Wanless and J.P. Kennedy, *Polymer*, **6**, 111 (1965).
65. J.P. Kennedy, W.W. Schulz, R.G. Squires, and R.M. Thomas, *Polymer*, **6**, 289 (1965).
66. G.A. Sartori, H. Lammens, J. Siffert, and A. Bernard, *J. Polym. Sci., Polym. Lett.*, **9**, 599 (1971).
67. J.P. Kennedy, P. Berzel, W. Naegele, and R.G. Squires, *J. Macromol. Sci.-Chem.*, **93**, 191 (1966).
68. D.W. Grattan and P.H. Plesch, *Makromol. Chem.*, **181**, 751 (1980).
69. K. Takakura, K. Hayashi, and S. Okamura, *J. Polym. Sci., Polym. Lett.*, **3**, 565 (1965).
70. K. Takakura, K. Hayashi, and S. Okamura, *J. Polym. Sci., Polym. Lett.*, **4**, 509 (1965).
71. M. Irie, S. Tominto, and R. Hayashi, *J. Polym. Sci., Polym. Lett.*, **8**, 809 (1970).
72. N.G. Gaylord, B. Patnaik, and A. Takahashi, *J. Polym. Sci., Polym. Lett.*, **8**, 809 (1970).
73. L.P. Ellinger, *Adv. Macromol Chem.*, **1**, 169 (1968).
74. R.F. Tarvin, S. Aoki, and J.K. Stille, *Macromolecules*, **5**, 663 (1972).
75. T. Nakamura, M. Soma, T. Onishi, and K. Tamaru, *Makromol. Chem.*, **135**, 241 (1970).
76. T. Natsuume, M. Nashiniura, M. Fujimatsu, M. Shimizu, Y. Shirota, H. Hirata, S. Kikabayashi, and H. Mikawa, *Polym. J. Japan*, **1**, 181 (1970) (From a private translation).
77. A. Ledwith, *J. Polym. Sci., Symp.*, **56**, 483 (1976).
78. H. Gilbert, F.F. Miller, S.J. Averill, B.H. Carlson, J.L. Folt, H.J. Heller, F.D. Stewart, R.F. Schmidt, and H. L. Trumbull, *J. Am. Chem. Soc.*, **78**, 1669 (1956).
79. M. Marek and L.Toman, *J. Polym. Sci., Polym. Symp.*, **42**, 339 (1973).
80. M. Marek and L.Toman, *Makromol. Chem., Rapid Commun.*, **1**, 161 (1980).
81. J.V. Crivello and J.H. Lam, *J. Polym. Sci., Symp.* **56**, 383 (1976).
82. J.V. Crivello and J.H. Lam, *J. Polym. Sci., Polym. Lett.*, **16**, 563 (1978).
83. J.V. Crivello and J.H. Lam, *J. Polym. Sci., Chem. Ed.*, **17**, 1047 (1979); *ibid.*, **18**, 2677 (1980); *ibid*, **18**, 2697 (1980).
84. J.V. Crivello and J. H. Lam, *J. Polym. Sci., Polym. Lett.*, **17**, 759 (1979).
85. J.V. Crivello and J.H. Lam, *J. Polym. Sci., Chem. Ed.*, **17**, 2877 (1979).
86. J.V. Crivello and J. H. Lam, *J. Polym. Sci., Chem. Ed.*, **18**, 1021 (1980).
87. K. Hayasai and S. Okamura, *J. Polym. Sci., Symp.* **22**, 15 (1968).
88. F. Williams *et al.*, *Discuss. Faraday Soc.*, **36**, 257 (1963).
89. B.K. Ueno, K. Hayashi, and S. Okamura, *J. Polym. Sci., Polym. Lett.*, **3**, 363 (1965); *Polymer*, **7**, 451 (1966).
90. R.C. Potter, R.M. Bretton, and D.J. Metz, *J. Polym. Sci.*, **A-1,4**, 419, 2259 (1966).
91. R.B. Taylor and F. Williams, *J. Am. Chem. Soc.*, **91**(14), 3728, (1969).
92. P.H. Plesch, in *Progress in High Polymers* (J.C. Robb, ed.), CRC Press, Cleveland, 1968.
93. J.P. Kennedy, *J. Polym. Sci., Symp.*, **56**, 1 (1976).
94. C.E.H. Bawn, *Pure Appl. Chem.*, **16** (2–3), 385 (1968).
95. J.P. Kennedy, A. Schinakawa, and F. Williams, *J. Polym. Sci.*, **A-1,9**, 1551 (1971).
96. J.P. Kennedy and E.Merechal, *Cationic Polymerization*, Wiley-Interscience, New York, 1982; T.E. Hogen-Esch and J. Smid (eds.), *Recent Advances in Anionic Polymerization*, Elsevier, New York, 1987.
97. J.E. Mulvaney, C.G. Overberger, and A.M. Schiller, *Fortschr. Hochpolymer Forsch.*, **3**, 106 (1961).
98. D.H. Richards, *Polymer*, **19**, 109 (1978).
99. H.F. Ebel, *Tetrahedron*, **21**, 699 (1965).
100. G. Fraenkel, D.G. Adams, and J. Williams, *Tetrahedron Lett.*, 767 (1963).
101. D. Margerison and J.P. Newport, *Trans. Faraday Soc.*, **59**, 2058 (1963).
102. T.L. Brown, R.L. Gerteis, D.A. Bafus, and J.A. Ladd, *J. Am. Chem. Soc.*, **86**, 2134 (1964).
103. J. Smid, Chapt. 11 in *Structure and Mechanism in Vinyl Polymerization* (T. Tsuruta and K.F. O'Driscoll, eds.), p. 350, Dekker, New York, 1969.
104. T.L. Brown, *Adv. Organometallic Chem.*, **3**, 365 (1965).
105. D.J. Worsford and S. Bywater, *Can. J. Chem.*, **40**, 1564 (1962).
106. C.G. Screttas and J.F. Eastham, *J. Am. Chem. Soc.*, **87**, 3276 (1965).
107. J.P. Kennedy and A.W. Langer, Jr., *Fortschr. Hochpolymer Forsch.*, **3**, 544 (1964).
108. J. Geerts, M. Van Beylen, and G. Smets, *J. Polym. Sci.*, **A-1,7**, 2859 (1969).
109. Ch.B. Tsvetanov, *Eur. Polym. J.*, **15**, 503 (1979).
110. P.E.M.. Allen, *J. Macromol Sci.*, *A*,**14**, 11 (1980).
111. B.L. Erussalimski, I.V. Kulevskaya, and V.V. Mazurek, *J. Polym. Sci.*, **16**, 1355 (1967).
112. B.L. Erussalimski, *Vysokomol. Soed.*, **A13**, 1293 (1971).
113. K. Hatada, T. Kitayama, K. Fujikawa, K. Ohta, and H. Yuki, *Am. Chem. Soc., Polym. Prepr.*, **21** (1), 59, 1980.
114. A.A. Korotkov, S.P. Mitzengendler, and V.N. Krasulina, *J. Polym. Sci.*, **53**, 217 (1961).

115. Y. Joh, Y. Kotake, T. Yoshihara, F. Ide, and K. Nakatsuka, *J. Polym. Sci.*, **A-1,5**, 593, 605 (1967).
116. Ch.B. Tsvetanov, *Eur. Polym. J.*, **15,** 503 (1979).
117. A. Zilkha, S. Barzakay, A. Ottolenghi, *J. Polym. Sci.*, **A-1,1**, 1813 (1963).
118. A.R. Lyons and Catterall, *Eur. Polym. J.*, **7**, 839 (1970).
119. T. Fujimoto, N. Kawabata, and J. Furukawa, *J. Polym. Sci.*, **A-1,6**, 1209 (1968).
120. D.H. Richards, *Polymer*, **19**, 109 (1972).
121. K.F. O'Driscoll, R.J. Boudreau, and A.V. Tobolsky, *J. Polym. Sci.*, **31**, 115 (1958).
122. K.F. O'Driscoll and A.V. Tobolsky, *J. Polym. Sci.*, **31**, 123 (1958); *ibid*, **32**, 363 (1959).
123. A.V. Tobolsky and D.B. Hartley, *J. Polym. Sci.*, **A-1,1**, 15 (1963).
124. C.G. Overberger and H. Yamamoto, *J. Polym. Sci.*, **B,3**, 569 (1965); *J. Polym. Sci.*, **A-1,4**, 3101 (1966).
125. M. Szwarc, *Nature*, **178**, 1168 (1956).
126. M. Brady, M. Ladaki, R. Milkovich, and M. Szwarc, *J. Polym. Sci.*, 221 (1957).
127. M. Szwarc, M. Levy, and R. Milkovich, *J. Am. Chem. Soc.*, **78**, 2656 (1956).
128. M. Szwarc, *Makromol. Chem.*, **35**, 132 (1960).
129. M. Levy and M. Szwarc, *J. Am. Chem. Soc.*, **82**, 521 (1960).
130. M. Szwarc, *Carbanions, Living Polymers, and Electron Transfer Processes*, Interscience, New York, 1968.
131. G. Lohr and G.V. Schulz, *Eur. Polym. J.*, **10**, 121 (1974).
132. V. Warzelhan, H. Hocker, and G.V. Schulz, *Makromol. Chem.*, **179**, 2221 (1978).
133. R. Kraft, A.H.E. Muller, V. Warzelhan, H. Hocker, and G.V. Schulz, *Macromolecules*, **11**, 1093 (1978).
134. R. Kraft, A.H.E. Muller, H. Hocker, and G.V, Schulz, *Makromol. Chem., Rapid Commun.*, **1**, 363 (1980).
135. K. Higashi, H. Baba, and A. Rembaum, *Quantum Organic Chemistry*, Wiley-Interscience, New York, 1965.
136. N.G. Gaylord and S.S. Dixit, *Macromol. Rev.*, **8**, 51 (1974).
137. A. Zilkha. P. Neta, and M. Frankel, *J. Chem. Soc.*, 3357 (1960).
138. S. Inoue, T. Tsuruta, and J. Furukawa, *Makromol. Chem.*, **42**, 12 (1960).
139. I.M. Panagotov, I.B. Rashkov, and I. N. Yukhnovski, *Eur. Polym. J.*, **7**, 749 (1971).
140. F.J. Welch, *J. Am. Chem. Soc.*, **81**, 1345 (1959).
141. K.F. O'Driscoll and A.V. Tobolsky, *J. Polym. Sci.*, **35**, 259 (1959).
142. K.F. O'Driscoll, T. Yonezawa, and T. Higashimura, *J. Macromol. Sci.-Chem.*, **1**, 1 (1966).
143. M. Morton, *Ind. Eng. Chem., Prod. Res. Dev.*, **11**, 106, (1972).
144. M. Morton, R. D. Sanderson, and R. Sakata, *J. Polym. Sci., Polym. Lett.*, **9**, 61 (1971).
145. R. Kraft, A.H.E. Muller, H. Hoker, and G.V. Schulz, *Makromol. Chem., Rapid Commun.*, **1**, 363 (1980).
146. G. Lohr and G.V. Schulz, *Makromol. Chem.*, **172**, 137 (1973).
147. C. Warzelhan, H. Hocker, and G.V. Schulz, *Makromol. Chem.*, **179**, 2221 (1978).
148. R. Kraft, A.H.E. Muller, V. Warzelhan, H. Hocker, and G.V. Schulz, *Macromolecules*, **11**, 1093 (1978).
149. W. Fowells, C. Schuerch, F.A. Bovey, and F.P. Hood, *J. Am. Chem. Soc.*, **89**, 1396 (1967).
150. J. Boor Jr., *Ziegler–Natta Catalysts and Polymerizations*, Academic Press, New York, 1979.
151. D.L. Glusker, E. Stiles, and B. Yoncoskie, *J. Polym. Sci.*, **49**, 297 (1961).
152. K. Butler, P.R. Thomas, and J. Tyler, *J. Polym. Sci.*, **48**, 357 (1960).
153. Huynh ba Gia and J.E. McGrath, *Am. Chem. Soc., Polym. Prepr.*, **21** (1), 74 (1980).
154. D.J. Breslow, G.E. Hulse, and A.S. Matlack, *J. Am. Chem. Soc.*, **79**, 3760 (1957).
155. L.W. Bush and D.S. Breslow, *Macromolecules*, **1**, 189 (1968).
156. Y. Kobuke, K. Hanji, J. Furukawa, and T. Fueno, *J. Polym. Sci.*, **A-1,9**, 431 (1971); T. Otsu, B. Yamada, M. Itahashi, and T. Mori, *J. Polym. Sci., Polym. Chem. Ed.*, **14**, 1347 (1976).
157. K. Yamaguchi and M. Minoura, *J. Polym. Sci.*, **A-1,10**, 1217 (1972).
158. A. Ottolenghi and A. Zilkha, *J. Polym. Sci.*, **A-1,1**, 687 (1963); Y. Imanishi, "Carbanionic Polymerization: Hydrogen Migration Polymerization," in *Comprehensive Polymer Science*, Vol. 3 (G.C. Eastmond, A. Ledwith, S.Russo, and P. Sigwalt, eds.), Pergamon Press, London, 1989.
159. D.J. Cram and K.R. Kopecky, *J. Am. Chem. Soc.*, **81**, 2748 (1959); T.E. Hogen-Esch and J. Smid (eds.), *Recent Advances in Anionic Polymerization*, Elsevier, New York, 1987; A.H.E. Muller, "Carbanionic Polymerization Kinetics and Thermodynamics," in *Comprehensive Polymer Science*, Vol. 3 (G.C. Eastmond, A. Ledwith, S. Russo, and P. Sigwalt, eds.), Pergamon Press, London, 1989.
160. K. Ziegler, E. Holzkamp, H. Breil, and H. Martin, *Angew. Chem.*, **67**, 541 (1955).
161. G. Natta, *J. Polym. Sci.*, **16**, 143 (1955).
162. W. Cooper, Chapt. I in *Vinyl and Allied Polymers*, Vol. I (P.D. Ritchie, ed.), CRC Press, Boca Raton, Florida, 1968.
163. N. C. Billingham, *Br. Polym. J.*, **6**, 299 (1974).
164. H. Bestian, K. Clauss, H. Jensen, and E. Prinz, *Angew. Chem., Intern. Ed.*, **2**, 32 (1963).
165. W.M. Saltman, W.E. Gildes, and J. Lal, *J. Am. Chem. Soc.*, **80**, 5615 (1958).
166. G. Natta, *J. Inorg. Nuc. Chem.*, **8**, 589 (1958).
167. D.O. Jordan, Chapt. I in *Stereochemistry of Macromolecules*, Vol. I (A.V. Ketley, ed.), Dekker, New York, 1967.

168. B. Rieger, X. Mu, D.T. Mallin, M.D. Raush, and J.C.W. Chen, *Macromolecules*, **23**, 383, (1990); N.H. Cheng and J.A. Ewen, *Makromol. Chem.*, **190**, 1931 (1989); A. Grassi, A. Zambelli, L. Resconi, E. Albizzati, and R. Mazzocchi, *Macromolecules*, **21**, 617 (1988); Y.V.Kissin, T.E. Nowlin, and R.I. Mink, *Macromolecules*, **26**, 2151 (1993).
169. Lewis and Rundle, *J. Chem. Phys.*, **21**, 986 (1953).
170. G. Natta, P. Pino, G. Mazzanti, V. Gianini, E. Mantica, and M. Peraldo, *J. Polym. Sci.*, **26**, 120 (1957).
171. G. Natta and G. Mazzanti, *Tetrahedron*, **8**, 86 (1960).
172. P. Cossee, *Tetrahedron Lett.*, No. 17, 12, 17 (1960); *Trans. Faraday Soc.*, **58**, 1226 (1962).
173. P. Cossee and E. J. Arlman, *J. Catalysis*, **3**, 99 (1964); P. Cossee, *J. Catalysis*, **3**, 80 (1964).
174. J. Chatt and B.L. Shaw, *J. Chem. Soc.*, 705 (1959).
175. A. Zambelli, A. L. Segre, M. Farina, and G. Natta, *Makromol. Chem.*, **110**, 1 (1967).
176. T. Kahara, M. Shinoyama, Y. Doi, and T. Keii, *Makromol. Chem.*, **180**, 2199 (1979).
177. J.W.L. Fordham, P.H. Burleigh, and C.L. Sturm, *J. Polym. Sci.*, **42**, 73 (1959).
178. F.A. Bovey, *J. Polym. Sci.*, **46**, 59 (1960).
179. G. Natta, I. Pasquon, and A. Zambelli, *J. Am. Chem. Soc.*, **84**, 1488 (1962).
180. A. Zambelli, G. Natta, and I. Pasquon, *J. Polym. Sci.*, **C4**, 411, (1963).
181. A. Zambelli, I. Pasquon, R. Signorini, and G. Natta, *Makromol. Chem.*, **112**, 160 (1968).
182. A. Zambelli, M.G. Giongo, and G. Natta, *Makromol. Chem.*, **112**, 183 (1968); M. Goodman, *Concepts of Polymer Stereochemistry, Topics in Stereochemistry* (N.G. Allinger and E.L. Eliel, eds.), Wiley-Interscience, New York, 1967.
183. A. Zambelli, *23rd. Int. Cong. of IUPAC, Macromol. Preprints*, **1**, 124 (1971); A. Zambelli and G. Gatti, *Macromolecules*, **11**, 485 (1978); A. Zambelli, P. Locatelli, M.C. Sacchi, and E. Rigamonte, *Macromolecules*, **13**, 798 (1980); P. Amandolla, T. Tancredi, and A Zambelli, *Macromolecules*, **19**, 307 (1986); I. Tritto, M.C. Zacchi, and P. Locatelli, *Makromol. Chem.*, **187**, 2145 (1986).
184. K. Matsuzaki and T. Yasukawa, *J. Polym. Sci.*, **A-1,5**, 521 (1967).
185. K. Matsuzaki and T. Yasukawa, *J. Polym. Sci.*, **A-2,5**, 511 (1967).
186. H. Hirai, K. Hiraki, I. Noguchi, T. Inoue, and S. Makishima, *J. Polym. Sci.*, **A-1,8**, 2395 (1970).
187. S. Destri, M.C. Gallazzi, A. Giarrusso, and L. Pori, *Makromol. Chem., Rapid Commun.*, **1**, 193 (1980).
188. R.P. Hughes and J. Powell, *J. Am. Chem. Soc.*, **94**, 7723 (1972).
189. H.O. Murdoch and E. Weiss, *Helv. Chim. Acta*, **45**, 1156 (1962).
190. C.A. Tolman, *J. Am. Chem. Soc.*, **92**, 6785 (1970).
191. P. Corradini, G. Guerra, R. Rusco, and V. Barone, *Eur. Polym. J.*, **16**, 835 (1980).
192. P. Pino, G.P. Lorenzi, and L. Lardicci, *Chim. Ind. (Milan)*, **42**, 712 (1960).
193. P. Pino, F. Ciardelli, and G.P. Lorenzi, *J. Am. Chem. Soc.*, **84**, 1487 (1962).
194. P. Pino, F. Ciardelli, and G.P. Lorenzi, *J. Polym. Sci.*, **C,4**, 21 (1963).
195. P. Pino, *Fortschr. Hochpolym. Forsch.*, **4**, 393 (1965).
196. G. Allegra, *Macromol. Chem.*, **145**, 235 (1971); F. Ciardelli, *Macromolecules*, **3**, 527 (1970).
197. P. Carradini, G. Pajaro, and A. Pannunzi, *J. Polym. Sci.*, **C,16**, 2905 (1967).
198. A. Zambelli, P. Locatelli, M. C. Sacchi, and E. Rigamoti, *Macromolecules*, **13**, 798 (1980).
199. A. Yammoto and T. Yamamoto, *J. Polym. Sci., Macromol. Reviews*, **13**, 161 (1978).
200. G. Wilke and B. Bogdanovic, *Angew. Chem.*, **73**, 756 (1961).
201. G. Wilke, *Angew. Chem., Intern. Ed.*, **2**, 105 (1963).
202. M.L.H. Green, *Organometallic Compounds*, Vol. 2, Methuen, London, 1968, p. 39.
203. G. Wilke, *Angew. Chem., Intern. Ed.*, **5**, 151 (1966).
204. D.G.H. Ballard, E. Jones, T. Medinger, and A.J.P. Pioli, *Makromol. Chem.*, **148**, 175 (1971).
205. D.G.H. Ballard, *Adv. Catal.*, **23**, 263 (1973).
206. F. Dawans and P. Teyssie, *Ind. Eng. Chem., Prod. Res. Develop.*, **10** (3), 261 (1971).
207. P. Bourdauducq and F. Dawans, *J. Polym. Sci.*, **A-1,10**, 2527 (1972).
208. V.I. Klepkova, G.P. Kondratenkov, V.A. Kormer, M.I. Lobach, and L.A. Churlyaeva, *J. Polym. Sci., Polym. Lett.*, **11**, 193 (1973).
209. J.P. Durand, F. Dawans, and Ph. Teyssie, *J. Polym. Sci., Polym. Lett.*, **5**, 785 (1967).
210. F. Borg-Visse, F. Dawans, and B. Marechal, *J. Polym. Chem. Polym. Chem. Ed.*, **18**, 2481 (1980).
211. N. G. Gaylord and H. F. Mark, *Linear and Stereoregular Addition Polymers*, Wiley-Interscience, New York, 1959.
212. J.P. Hogan, *J. Polym. Sci.*, **A-1,8**, 2637 (1970).
213. K. Endo and T. Otsu, *J. Polym. Sci. Polym. Chem. Ed.*, **17**, 1453 (1979).
214. A. Shimizu, T. Otsu, and M. Imoto, *J. Polym. Sci., Polym. Lett.*, **3**, 449, 1031 (1965).
215. A. Shimizu, T. Otsu, and M. Ito, *J. Polym. Sci.*, **A-1,4**, 1579 (1966).
216. J.P. Kennedy and T. Otsu, *Adv. Polym. Sci.*, **7**, 369 (1970).
217. A. Shimizu, E. Itakura, T. Otsu, and M. Imoto, *J. Polym. Sci.*, **A-1,7**, 3119 (1969).

218. T. Otsu, H. Nagahama, and K. Endo, *J. Polym. Sci., Polym. Lett.*, **10**, 601 (1972).
219. T. Otsu, H. Nagahama, and K. Endo, *J. Macromol. Sci.-Chem.*, **9**, 1249 (1975).
220. T. Otsu and K. Endo, *J. Macromol. Sci.-Chem.*, **9**, 899 (1975).
221. T. Otsu, A. Shimizu, K. Itakura, and K. Endo, *J. Polym. Sci., Polym. Chem. Ed.*, **13**, 1589 (1975).
222. K. Endo, H. Nagahama, and T. Otsu, *J. Polym. Sci., Polym. Chem. Ed.*, **17**, 3647 (1979).
223. J. Furukawa and T. Saegusa, *Polymerization of Aldehydes and Oxides*, Wiley-Interscience, New York, 1963.
224. J.C. Bevington, in *Polyethers #I* (N. Gaylord, ed.) (High Polymers, Vol.13), Wiley-Interscience, New York, 1963.
225. O. Vogl, *Polyaldehydes*, Dekker, New York, 1967.
226. O. Vogl, *Makromol. Chem.*, **175**, 1281 (1974).
227. O. Vogl, *J. Macromol. Sci., Rev. Macromol. Chem.*, **C12** (1), 109 (1975).
228. W. Kern, E. Eberius, and V. Jaacks, *Makromol. Chem.*, **141**, 63 (1971).
229. V. Jaacks, K. Boehlke, and E. Eberius, *Makromol. Chem.*, **118**, 354 (1968).
230. K. Boehlke and V. Jaacks, *Makromol. Chem.*, **142**, 189 (1971).
231. J. Furukawa, T. Saegusa, T. Tsurata, H. Fujii, A. Kawasaki, and T. Tatano, *Makromol. Chem.*, **33**, 32 (1960).
232. J. Furukawa, T. Saegusa, H. Fujii, A. Kawasaki, and T. Tatano, *J. Polym. Sci.*, **36**, 546 (1969).
233. K. Weissermel and W. Schneider, *Makromol. Chem.*, **51**, 39, (1962).
234. O. Vogl, *J. Polym. Sci.*, **46**, 261 (1960).
235. W. Kern, *Chem. Z.*, **88**, 623 (1964).
236. E. Kunzel, A. Gieffer, and W. Kern, *Makromol. Chem.*, **96**, 17 (1966).
237. Z. Machacek, J. Mejzlick, and J. Pac, *Vysokomol. Soed.*, **3**(9), 1421 (1961).
238. J. Mejzlick, J. Mencikova, and Z. Machacek, *Vysokomol. Soed.*, **4**, 769, 776 (1962).
239. K. Vesely and J. Mejzlick, *Vysokomol. Soed.*, **5**(9), 1415 (1963).
240. N. Mathes and K. Jaacks, *Makromol. Chem.*, **135**, 49 (1970).
241. C.E. Schweitzer, R.N. MacDonald, and J.O. Punderson, *J. Appl. Polym. Sci.*, **1**, 158 (1959).
242. G. Natta, G. Mazzanti, P. Corradini, and J.W. Bassi, *Makromol. Chem.*, **37**, 156 (1959).
243. J. Furukawa, T. Saegusa, H. Fujii, A. Kawasaki, H. Imai, and Y. Fujii, *Makromol. Chem.*, **37**, 149 (1959).
244. O. Vogl, *J. Polym. Sci.*, **46**, 161 (1960).
245. H. Tani, *Fortschr. Hochpolymer. Forsch.*, **11**, 57 (1973).
246. J. Furukawa, T. Saegusa, and H. Fujii, *Makromol. Chem.*, **44–46**, 398 (1961).
247. O. Vogl and W.M.D. Bryant, *J. Polym. Sci.*, **A-1,2**, 921 (1964).
248. H. Tani, T. Araki, and H. Yasuda, *J. Polym. Sci., Polym. Lett.*, **4**, 727 (1966).
249. H. Yasuda and H. Tani, *Macromolecules*, **6**, 17 (1973).
250. A. Novak and E. Whaley, *Can. J. Chem.*, **37**, 1710, 1718 (1959).
251. R.C. Schulz, H. Fauth, and W. Kern, *Makromol. Chem.*, **21**, 227 (1956).
252. R.C. Schulz, H. Cherdron, and W. Kern, *Makromol. Chem.*, **24**, 141 (1957).
253. A. Henglein, W. Schnabel, and R.C. Schulz, *Makromol. Chem.*, **31**, 131 (1959).
254. R.C. Schulz and W. Passmann, *Makromol. Chem.*, **60**, 139 (1963).
255. R.C. Schulz, G. Wegner, and W. Kern, *J. Polym. Sci.*, **C,16**, 989 (1967).
256. R.C. Schulz, S. Suzuki, H. Cherdron, and W. Kern, *Makromol. Chem.*, **53**, 145 (1962).
257. C. Aso and S. Tagami, *Macromolecules*, **2**, 414 (1969).
258. Y. Katahama and S. Ishida, *Makromol. Chem.*, **119**, 64 (1968).
259. K. Kobayashi and H. Sumimoto, *J. Polym. Chem., Polym. Lett.*, **10**, 703 (1972).
260. V.V. Amerik, B.A. Krentsel, and M.V. Shiskina, *Vysokomol. Soyed.*, **7**(10), 1713 (1965).
261. G.P. Pregaglia and M. Benaghi, Chapt. 3 in *Stereochemistry of Macromolecules* (A.D. Ketley, ed.), Vol. 2, Dekker, New York, 1967.
262. C. Aso, S. Tagami, and T. Kunitake, *J. Polym. Sci.*, **A-1,7**, 497 (1969).
263. C. Aso and S. Tagami, *Macromolecules*, **2**, 414 (1969).
264. A.A. Morton, B.B. Magat, and R.L. Letsinger, *J. Am. Chem. Soc.*, **69**, 950 (1947).
265. A.A. Morton, F.H. Bolton, F.W. Collins, and B.F. Cluff, *Ind. Eng. Chem.*, **44**, 2876 (1952).
266. A.A. Morton, *Rubber Age*, **72**, 473 (1953).
267. A.A. Morton, *Ind. Eng. Chem.*, **42**, 1488 (1950).
268. A.A. Morton and B.J. Laupher, *J. Polym. Sci.*, **44**, 233 (1960).
269. L. Reich and A. Schindler, *Polymerization by Organometallic Compounds*, Wiley-Interscience, New York, 1966.
270. J.W. Breitenback and C.H. Srna, *Pure Appl. Chem.*, **4**, 245 (1962).
271. B.L. Funt and S.W. Laurent, *J. Polym. Sci.*, **A-1,2**, 865 (1964).
272. B.L. Funt and K.C. Yu, *J. Polym. Sci.*, **62**, 359 (1962).
273. N. Yamazaki, S. Nakamura, and S. Kambara, *J. Polym. Sci., Polym. Lett.*, **3**, 57 (1965).
274. D. Laurin and G. Parravane, *J. Polym. Sci., Polym. Lett.*, **4**, 797 (1966).
275. B.L. Funt and S.N. Bhadani, *J. Polym. Sci.*, **C,17**, 1 (1968).
276. B.L. Funt and T.J. Blain, *J. Polym. Sci.*, **A-2,9**, 115 (1971).

277. U. Akbulut, J.E. Fernandez, and R.L. Birke, *J. Polym. Sci., Poly. Chem. Ed.*, **13**, 133 (1975).
278. V.E. Shashoua, W. Sweeny, and R.F. Tietz, *J. Am. Chem. Soc.*, **82**, 866 (1960).
279. Y. Iwakura, K. Uno, and N. Kobayashi, *J. Polym. Sci.*, **A-1,6**, 793 (1968).
280. G.C. East and H. Furukawa, *Polymer*, **20**, 659 (1979).
281. G. Natta, J. Di Pietro, and M. Cambini, *Makromol. Chem.*, **56**, 200 (1962).
282. H.C. Beachell and C.P. NgocSon, *J. Polym. Sci., Polym. Lett.*, **1**, 25 (1963).
283. D.C. Pepper, in *Friedel-Craft and Related Reactions*, Vol. II (G.A. Olah, ed.), Interscience, New York 1964.
284. J.F. Dunphy and C.S. Marvel, *J. Polym. Sci.*, **47**, 1 (1960).
285. R.B. Cundall, Chap. 15 in *The Chemistry of Cationic Polymerization* (P.H. Plesch, ed.), Macmillan, New York, 1963.
286. C.G. Overberger, L.H. Arnold, and J.J. Taylor, *J. Am. Chem. Soc.*, **73**, 5541 (1951).
287. C.G. Overberger, R.J. Ehrig, and D. Tanner, *J. Am. Chem. Soc.*, **76**, 772 (1954).
288. C.G. Overberger, D.H. Tanner, and E.M. Pierce, *J. Am. Chem. Soc.*, **80**, 4566 (1958).
289. C.G. Overberger and V.G. Kamath, *J. Am. Chem. Soc.*, **81**, 2910 (1959).
290. K.F. O'Driscoll, T. Yonezawa, and H. Higashimura, *J. Macromol. Sci.-Chem.*, **1**, 17 (1966).
291. T. Asanuma, Y. Nishimori, M. Ito, N. Uchicawa, and T. Shiomura, *Polym. Bull.*, **25**, 567 (1991).
292. H. Yuki, K. Kosai, S. Murahashi, and J. Hotta, *J. Polym. Sci., Polym. Lett.*, **3**, 1121 (1964).
293. H. Yuki, Y. Okamoto, K. Ohta, and K. Hatada, *J. Polym. Sci., Polym. Chem. Ed.*, **13**, 1161 (1975).
294. C.A. Lukach and H.M. Spurlin, Chapt. 4A in *Copolymerization* (G.E. Kain, ed.), Wiley-Interscience, New York, 1964.
295. H.F. Mark and N. Ogata, *J. Polym. Sci.*, **A-1,1**, 3439 (1963).
296. Y.P. Castille and V. Stannett, *J. Polym. Sci.*, **A-1,4**, 2063 (1966).
297. J. Furukawa, T. Saegusa, T. Tsuruta, S. Ohta, and G. Wasai, *Makromol. Chem.*, **52**, 230 (1962).
298. H.O. Colomb, F.E. Bailey, Jr , and R.D. Lundberg, *J. Polym. Sci., Polym. Lett.*, **16**, 507 (1978).
299. J. Furukawa and O. Vogl, *Ionic Polymerization*, Dekker, New York, 1976.
300. A.H.E. Muller, *Makromol. Chem., Macromol. Symp.*, **32**, 87 (1990); W. Schubert and F. Bandermann, *Makromol. Chem.*, **190**, 2721 (1989); W. Schubert, H.D. Sitz, and F. Bandermann, *Makromol. Chem.*, **190**, 2193 (1989); D.Y. Sogah, W.R. Hertler, Dicker, I.B., P.A. Depra, and J.R. Butera, *Makromol. Chem., Macromol. Symp.*, **32**, 75 (1990).
301. R.P. Quirk and J. Ren, *Macromolecules*, **25**, 6612 (1992).
302. W.V. Farman and D.Y. Sogah, *Am. Chem. Soc. Polym. Prepr.*, **27**, 167 (1986).
303. R.P. Quirk and G.P. Bidinger, *Polym. Bull.*, **22**, 63 (1989).
304. M.T. Reetz, R. Osterek, K.E. Piejko, D. Arlt, and D. Bomer, *Ang. Chem., Int. Ed.*, **25**, 1108 (1986).
305. W.T. Herter, *Macromolecules*, **20**, 2976 (1987).
306. C. Pugh and V. Percec, *Polym. Bull.*, **14**, 109 (1985).
307. M. Kuroki, T. Watanabe, T. Aida, and S. Inoue, *J. Am. Chem. Soc.*, **113**, 5903 (1991).
308. T. Adachi, H. Sugimoto, T. Aida, and S. Inoue, *Macromolecules*, **25**, 2280 (1992).
309. W. Endo and T. Otsu, *J. Polym. Sci., Polym. Chem. Ed.*, **29**, 847 (1991).
310. M. Kamigaito, K. Yamaoka, M. Sawamoto, and T.H. Higashimura, *Macromolecules*, **25**, 6400 (1992); J.P. Kennedy, S. Midha, and B. Keszler, *Macromolecules*, **26**, 424 (1993).
311. T. Saegusa, S. Kobayshi, and J. Furukawa, *Macromolecules*, **10**, 73 (1977).
312. R. Wessermel, H. Chedron, J. Berthod, B. Diedrich, K.D. Keil, K. Rust, H. Strametz, and T. Toth, *J. Polym. Sci., Symp.*, **51**, 187 (1975).
313. P. Kubisa, K. Neeld, J. Starr, and O. Vogl, *Polymers*, **21**, 1433 (1980).
314. P. Pino and R. Melhaupt, *Angew. Chem., Intern. Ed.*, **19**, 857 (1980).
315. K. Matsuzaki, Y. Shinobara, and T. Kanai, *Makromol Chem.*, **18**, 1923 (1980).
316. D.J. Sikkema and H. Angad-Gaur, *Makromol. Chem.*, **181**, 2259 (1980).
317. A. Simon and A. Grobler, *J. Polym. Sci., Polym. Chem. Ed.*, **18**, 3111 (1980).
318. H.N. Cheng, *J. Chem. Inf. Comput. Sci.*, **27**, 8 (1987).
319. A. Ledwith, B. Chiellini, and R. Solano, *Macromolecules*, **12**, 240 (1979).
320. T. Kunitake and K. Takarabe, *Makromol. Chem.*, **182**, 817 (1981).
321. M.J. Moulis, J. Collomb, A. Gandini, and H. Cheradame, *Polym. Bull. (Berlin)*, **3** (4), 197 (1980); *Chem. Abstr.* **94**, 31143u (1981).
322. A. Zambelli, P. Locatelli, M.C. Sacchi, and B. Rigamonti, *Macromolecules*, **13**, 798 (1980).
323. C.P. Casey, *Macromolecules*, **14**, 464 (1981).
324. M.L.H. Green, *Pure Appl. Chem.*, **50**, 27 (1978).
325. J.P. Kennedy and B. Marechal, *J. Polym. Sci., Macromol. Rev.*, **16**, 123 (1981).
326. A.G. Evans, G.W. Meadows, and M. Polanyi, *Nature*, **158**, 94 (1946); *ibid*, **160**, 869 (1947).
327. P.H. Plesch, in *The Chemistry of Cationic Polymerization* (P.H. Plesch, ed.), Macmillan, New York, 1963.
328. G. Sauvet, J.P. Vairon, and P. Sigwalt, *J. Polym. Sci., Polym. Chem. Ed.*, Chem. **16**, 3047 (1978).

329. T. Saegusa, *J. Macromol. Sci.*, **A,6**, 997 (1972).
330. V. Korshak and N.N. Lebedev, *J. Gen. Chem. U.S.S.R.*, **18**, 1766 (1948).
331. M. Chmelir, M. Marek, and O. Wichterle, *J. Polym. Sci.*, **C-16**, 833 (1967).
332. M. Chmelir and M. Marek, *J. Polym. Sci.*, **C,22**, 177 (1968).
333. P. Lopour and M. Marek, *Makromol. Chem.*, **134**, 23 (1970).
334. F.A.M. Abdul-Rasoul, A. Ledwith, and Y. Yagci, *Polymer*, **19**, 1219 (1978).
335. F.A.M. Abdul-Rasoul, A. Ledwith, and Y. Yagci, *Polym. Bull.*, **1**, 1 (1978).
336. A. Ledwith, *Polymer*, **19**, 1217 (1978).
337. F. Williams, *J. Am. Chem. Soc.*, **86**, 3954 (1964).
338. J.W. Breitenbach, Ch. Srna, and O. F. Olaj, *Macromol. Chem.*, **42**, 171 (1960).
339. J.W. Breitenbach and H. Gabler, *Monatsch. Chem.*, **91**, 202 (1960).
340. F. Sommer and J.W. Breitenbach, *IUPAC Int. Symp. Macromol. Chem.*, **1**, 257 (1969).
341. Y. Doi, E. Suzuki, and T. Keii, *Makromol. Chem., Rapid Commun.*, **2**, 293 (1981).
342. S. Sivaram, *Ind. Eng. Chem., Prod. Res. Dev.*, **16**, 121 (1977).
343. B. Kezler, A. Grobler, E. Takacs, and A. Simon, *Polymer*, **22**, 818 (1981).
344. K.J. Ivin, J.J. Rooney, C.D. Stewart, M.L.H. Green, and R. Mahtab, *J. Chem. Soc., Chem. Commun.*, 604 (1978); K.J. Ivin, *Proc. Eur. Symp. Polym. Spectrosc.*, **11**, 267 (1978).
345. A. Zambelli, M.C. Sacchi, and P. Locatelli, *Macromolecules*, **12**, 1051 (1979); A. Zambelli and G. Allegra, *Macromolecules*, **13**, 42 (1980).
346. E.F. Peters, A. Zletz, and B.L. Evering, *Ind. Eng. Chem.*, **48**, 1879 (1957).
347. A. Clark, J.P. Hogan, B.L. Banks, and W.C. Lanning, *Ind. Eng. Chem.*, **48**, 1152 (1956).
348. A.D. Caunt, *Br. Polym. J.*, **13**, 22 (1981).
349. A.D. Caunt, *J. Polym. Sci*, **C,4**, 49 (1964).
350. M.L. Burstall and F.E. Treloar, Chapt. 1 in *The Chemistry of Cationic Polymerization* (P.H. Plesch, ed.), Macmillan, New York, 1963.
351. A.B. Mathieson, Chapt. 6 in *The Chemistry of Cationic Polymerization* (P.H. Plesch, ed.), Macmillan, New York, 1963.
352. D.N. Bhattacharyya, C.L. Lee, J. Smid, and M. Szwarc, *J. Phys. Chem.*, **69**, 612 (1965).
353. D.N. Bhattacharyya, J. Smid, and M. Szwarc, *J. Phys. Chem.*, **69**, 624 (1965).
354. K. Matsuzaki, H. Morii, N. Inoue, and T. Kanai, *Makromol. Chem.*, **182**, 2421 (1981).
355. K. Soga and M. Terano, *Makromol. Chem.*, **182**, 2439 (1981).
356. G. Heublin, *J. Macromol. Sci.-Chem.*, **A,16**, 563 (1981).
357. B.G. Dolgoplosk, P.A. Vinogradov, O.P. Parenago, E.I. Tinyatova, and B.S. Turov, *IUAPC Int. Symp. Macromol. Chem. Prague Preprints*, p. 314, 1965.
358. A. Ledwith and D.C. Sherrington, "Reactivity and Mechanism In Polymerization by Complex Organometallic Derivatives," in *Reactivity, Mechanism and Structure in Polymer Chemistry* (A.D. Jenkins and A. Ledwith, ed.), Wiley-Interscience, London, 1974.
359. G.T. Chen, *J. Polym. Sci., Polym. Chem. Ed.*, **20**, 2915 (1982).
360. H.C. Brown and B.W. Kanner, *J. Am. Chem. Soc.*, **75**, 3865 (1963).
361. J.P. Kennedy, T. Kelen, S.C. Guhaniyogi, and R.T. Chou, *J. Macromol. Sci.-Chem.*, **A,18,** 129 (1982).
362. T. Keii, E. Suzuki, M. Tamura, M. Murata, and Y. Doi, *Makromol. Chem.*, **183**, 2285 (1982).
363. J.P. Kennedy, T. Kelen, and F. Tudos, *J. Macromol. Sci.- Chem.*, **A,18**, 1189 (1982–1983).
364. R. Faust, A. Fehervari, and J.P. Kennedy, *J. Macromol. Sci.- Chem.*, **A,18**, 1209 (1982–1983).
365. J. Puskas, G. Kaszas, J.P. Kennedy, T. Kelen, and F. Tudos, *J. Macromol. Sci.-Chem.*, **A,18**, 1229 (1982–1983).
366. J. Puskas, G. Kaszas, J.P. Kennedy, T. Kelen, and F. Tudos, *J. Macromol. Sci.-Chem.*, **A,18**, 1245 (1982–1983).
367. J. Puskas, G. Kaszas, J.P. Kennedy, T. Kelen, and F. Tudos, *J. Macromol. Sci.-Chem.*, **A,18**, 1263 (1982–1983).
368. M. Sawamoto and J.P. Kennedy, *J. Macromol. Sci.-Chem.*, **A,18**, 1275 (1982–1983).
369. M. Masure and P. Sigwalt, *Makromol. Chem., Rapid Commun.*, **4**, 269 (1983).
370. R. Faust and J.P. Kennedy, *J. Polym. Sci., Polym. Chem. Ed.*, **25**, 1847 (1987); T. Higashimura, S. Aoshima, and M. Sawamoto, *Makromol. Chem., Macromol. Symp., 13/14*, 457 (1988); Y. Ishihara, M. Sawamoto, and T. Higashimura, *Polym. Bull.*, **24**, 201 (1990).
371. G. Kaszas, J.E. Puskas, C.C. Chen, and J.P. Kennedy, *Macromolecules*, **23**, 3909 (1990); J.P. Kennedy, *Makromol. Chem., Macromol. Symp.*, **32**, 119 (1992); O. Nuyken and H. Kroner, *Makromol. Chem.*, **191**, 1 (1990); L. Thomas, A. Poton, M. Tardi, and P. Sigwalt, *Macromolecules*, **25**, 5886 (1992).
372. M. Sawamoto and T. Higashimura, *Am. Chem. Soc., Polym. Prepr.*, **32**, 312 (1991).
373. T.E. Hogen-Esch and J. Smid (eds.), *Recent Advances in Anionic Polymerization*, Elsevier, New York, 1987.
374. Ph. Teyssie, R. Fayt, J.P. Hautekeer, C. Jacobs, R. Jerome, L. Leemans, and S.K. Varshney, *Macromol. Chem., Macromol Symp.*, **32**, 61 (1990); S.K. Varshney, J.P. Hantekeer, R. Fayt, R. Jerome, and Ph. Teyssie, *Macromolecules*, **23**, 2618 (1990).
375. F.A. Bovey, *High Resolution NMR of Macromolecules*, Academic Press, New York, 1972.

376. J.C.W. Chien, J.C. Wu, and C.I. Kuo, *J. Polym. Sci., Polym. Chem. Ed.*, **20**, 2019 (1982); K. Soga, T. Shiono, and Y. Doi, *Makromol Chem.*, **189**, 1531 (1988); M.C. Sacchi, C. Shan, P. Locatelli, and S. Tritto, *Macromolecules*, **23**, 383 (1990).
377. Y. Doi, *Makromol. Chem., Rapid Commun.*, **3**, 635 (1982); M. Farina, "The Stereochemistry of Linear Macromolecules," in *Topics in Stereochemistry*, Vol. 17 (E.L. Eliel and S.H. Wilen, eds.), Wiley, New York, 1987; A. Le Borgne, N. Spassky, C.L. Jun, and A. Momtaz, *Makromol. Chem.*, **189**, 637 (1988).
378. W. Endo, R. Ueda, and T. Otsu, *Macromolecules*, **24**, 6849 (1991).

4

Ring-Opening Polymerizations

4.1. Chemistry of Ring-Opening Polymerizations

Formation of polymers through ring-opening reactions of cyclic compounds is an important process in polymer chemistry. In such polymerizations, chain growth takes place through successive additions of the opened structures to the polymer chain:

$$n\ \overline{\underline{\mathrm{R}\text{———}\mathrm{X}}} \longrightarrow \text{—[—R——X—]}_{n}\text{—}$$

An example of the above is a ring-opening polymerization of ethylene oxide that results in formation of poly(ethylene oxide), a polyether:

$$n\ \overset{\mathrm{O}}{\mathrm{CH_2—CH_2}} \longrightarrow \text{—[—CH}_2\text{—CH}_2\text{—O—]}_{n}\text{—}$$

The cyclic monomers that undergo ring-opening polymerizations are quite diverse. They include cyclic alkenes, lactones, lactams, and many heterocyclics with more than one heteroatom in the ring. Such polymerizations are ionic in character and may exhibit characteristics that are typical of ionic chain-growth polymerizations (e.g., effect of counterion and solvent). It would, however, be wrong to assume that these polymerizations necessarily take place by chain-propagating mechanisms. Actually, many such reactions are step-growth in nature, with the polymer size increasing slowly throughout the course of the process. There are, on the other hand, some cyclic monomers that do polymerize in a typical chain-growth manner.

4.2. Kinetics of Ring-Opening Polymerizations

There is general similarity between the kinetics of many ring-opening polymerizations and those of step-growth polymerizations that are discussed in Chapter 6. Some kinetic expressions in ring-opening polymerizations, on the other hand, resemble chain-growth ionic ones.

When there is no termination in the ring-opening polymerizations and they behave like living polymerizations, the rate of propagation can be described similarly by

$$R_P = K_P[M*][M]$$

where [M*] represent the concentration of propagating ions. Such ions could be oxonium or sulfonium, etc. When, however, there is a propagation–depropagation equilibrium, it can be expressed as follows:

$$\mathrm{M}_n{*} + \mathrm{M} \underset{K_{\mathrm{DP}}}{\overset{K_{\mathrm{P}}}{\rightleftharpoons}} \mathrm{M}_{n+1}{*}$$

The rate expression can then be written as the rate of propagation–depropagation:

$$R_{\mathrm{P}} = -[d\mathrm{M}]/dt = K_{\mathrm{P}}[\mathrm{M}*][\mathrm{M}] - K_{\mathrm{DP}}[\mathrm{M}*]$$

At conditions of equilibrium, if we designate the monomer concentration as $[\mathrm{M}]_{\mathrm{c}}$ and the polymerization rate is zero, we can write

$$K_{\mathrm{P}}[\mathrm{M}]_{\mathrm{c}} = K_{\mathrm{DP}}$$

Hirota and Fukuda[217] described the quantitative dependence of the degree of polymerization on various reaction parameters for an equilibrium polymerization. The initiation can be described as

$$\mathrm{I} + \mathrm{M} \overset{K_{\mathrm{I}}}{\rightleftharpoons} \mathrm{M}*$$

where I is the initiating species. It is assumed that the equilibrium constants for initiation and propagation are independent of the size of the propagating species. The concentration of the propagating chains $[\mathrm{M}_n*]$ of size n at an equilibrium concentration c can then be written as

$$[\mathrm{M}_n*] = K_{\mathrm{I}}[\mathrm{I}]_c[\mathrm{M}]_c(K_{\mathrm{P}}[\mathrm{M}]_c)^{n-1}$$

The total concentration of polymer molecules size N can be expressed as follows:

$$[\mathrm{N}] = \sum [\mathrm{M}_n*] = K_{\mathrm{I}}[\mathrm{I}]_c[\mathrm{M}]_c/(1 - K_{\mathrm{P}}[\mathrm{M}]_c)$$

The total concentration of monomer segments that are incorporated into the polymer can also be expressed as follows:

$$[\mathrm{W}] = \sum n[\mathrm{M}_n*] = K_{\mathrm{I}}[\mathrm{I}]_c\,[\mathrm{M}]_c/(1 - K_{\mathrm{P}}[\mathrm{M}]_c)^2$$

This allows us to express the average degree of polymerization, namely [W]/[N] as follows:

$$\overline{X}_n = 1/(1 - K_{\mathrm{P}}[\mathrm{M}]_c)$$

4.3. Polymerization of Oxiranes

Polymerizations of oxiranes or epoxides occur by one of three different mechanisms: (1) cationic, (2) anionic, and (3) coordination. In this respect, the oxiranes differ from the rest of the cyclic ethers that can only be polymerized with the help of strong cationic initiators. It appears, though, that sometimes coordination catalysis might also be effective in polymerizations of some oxetanes. The susceptibility of oxirane compounds to anionic initiation can be explained by the fact that these are strained ring compounds. Because the rings consist of only three atoms, the electrons on the oxygen are crowded and are vulnerable to attack.[1]

4.3.1. Cationic Polymerization

Various Lewis and protonic acids are capable of initiating cationic polymerization of epoxies. Among them, the following metal salts are effective in polymerizations of ethylene and propylene oxides[2,3]: $ZnCl_2$, $AlCl_3$, $SbCl_5$, BF_3, BCl_3, $BeCl_2$, $FeCl_3$, $SnCl_4$, and $TiCl_4$. Often, these polymerizations can be carried out in bulk without any solvent, particularly in the laboratory. The mechanism of these reactions can be complex, however, depending upon the particular Lewis acid used. In fact, not all of these polymerizations can even be treated in general terms as cationic. For instance, ferric chloride initiated polymerizations of epoxides initially proceed by a mechanism that has all the superficial features of cationic polymerization. After the initial stages, however, the polymerizations proceed by a coordination mechanism. This is discussed further in this section.

Stannic chloride yields only low molecular weight poly(ethylene oxide) from ethylene oxide (molecular weight below 5000) when the reaction is carried out in ethylene chloride at room temperature. Some dioxane and dioxolane also form in the process. The following reaction scheme was proposed[1–5]:

Initiation:

$$SnCl_4 + 4\ \underset{\text{(O)}}{CH_2{-}CH_2} \longrightarrow \ominus SnCl_4\ominus \left[O{-}CH_2{-}CH_2{-}\overset{\oplus}{O}(CH_2{-}CH_2)\right]_2$$

Propagation:

$$\ominus SnCl_4\ominus \left[O{-}CH_2{-}CH_2{-}\overset{\oplus}{O}(CH_2{-}CH_2)\right]_2 \xrightarrow{\text{slow}}$$

$$\ominus SnCl_4\ominus \left[O{-}CH_2{-}CH_2{-}O{-}CH_2{-}CH_2\oplus\right]_2 \underset{\underset{\text{(O)}}{CH_2{-}CH_2}}{\overset{\text{fast}}{\rightleftharpoons}} \ominus SnCl_4\ominus \begin{matrix} O{-}CH_2{-}CH_2{-}O{-}CH_2{-}CH_2\oplus \\ O{-}CH_2{-}CH_2{-}O{-}CH_2{-}CH_2{-}\overset{\oplus}{O}(CH_2{-}CH_2) \end{matrix}$$

$$\longrightarrow \ominus SnCl_4\ominus \begin{matrix} O{-}CH_2{-}CH_2{-}O{-}CH_2{-}CH_2\oplus \\ O{-}CH_2{-}CH_2{-}O{-}CH_2{-}CH_2{-}O{-}CH_2{-}CH_2\oplus \end{matrix}$$

$$\ominus SnCl_4\ominus \begin{matrix} O{-}CH_2{-}CH_2{-}O{-}CH_2{-}CH_2\oplus \\ O{-}CH_2{-}CH_2{-}\overset{\oplus}{O}(\text{dioxane ring}) \end{matrix} \longrightarrow \ominus SnCl_4\ominus \begin{matrix} O{-}CH_2{-}CH_2{-}O{-}CH_2{-}CH_2\oplus \\ O{-}CH_2{-}CH_2\oplus \end{matrix} + \text{dioxane}$$

Termination:

$$\ominus SnCl_4\ominus \begin{matrix} O{-}CH_2{-}CH_2{-}O{-}CH_2{-}CH_2\oplus \\ O{-}CH_2{-}CH_2{-}({-}O{-}CH_2{-}CH_2{-})_x{-}O{-}CH_2{-}CH_2\oplus \end{matrix} \longrightarrow$$

$$\ominus SnCl_3 \begin{matrix} O{-}CH_2{-}CH_2{-}O{-}CH_2{-}CH_2\oplus \\ O{-}CH_2{-}CH_2{-}({-}O{-}CH_2{-}CH_2{-})_x{-}O{-}CH_2{-}CH_2Cl \end{matrix}$$

The initiation step depends upon formation of oxonium ions. Because a carbon cation intermediate is indicated, it was suggested[3] that the propagation probably occurs by ether exchange which results from a nucleophilic attack by the monomer on the oxonium ion.

Boron trifluoride forms complexes with oxygen-containing compounds, like water, alcohols, and ethers. When it initiates the polymerization of epoxides, it can associate simultaneously with several different moieties. These are the monomeric cyclic ethers, as well as the open-chain polymeric ether groups, and the hydroxy groups on the chain ends. In addition, it can also associate with the hydroxy groups of water. The following illustration shows the type of equilibrium that can take place[1]:

$$R{-}\overset{BF_3}{\ddot{O}}\cdots H{-}\overset{H}{O}{-}R + R'{-}O{-}R' \rightleftharpoons R{-}\overset{BF_3}{\ddot{O}}\cdots H{-}O\begin{matrix}R'\\R'\end{matrix} + ROH$$

The alcohols and open-chain ethers have comparable basicities toward the coordinated acid ROH:BF_3. Ethylene oxide, on the other hand, is much less basic than the open-chain ethers.[6] In the initiation step, therefore, the monomer reacts with the coordinated acid[1]:

$$H{-}\ddot{O}(BF_3){-}H + CH_2{-}CH_2(O) \longrightarrow H{-}\ddot{O}(BF_3)\cdots H\cdots O(CH_2)_2 \longrightarrow H{-}\ddot{O}(BF_3){-}CH_2{-}CH_2{-}OH$$

During propagation, three different reactions can occur[1]:

1. $$R{-}\ddot{O}(BF_3)\cdots H\cdots O(CH_2)_2 \longrightarrow R{-}\ddot{O}(BF_3){-}CH_2{-}CH_2{-}OH$$

2. $$R{-}\ddot{O}(BF_3)\cdots H\cdots O(H)(R) + CH_2{-}CH_2(O) \longrightarrow R{-}\ddot{O}(BF_3){-}H + O(CH_2{-}CH_2OH)(R)$$

3. $$R{-}\ddot{O}(BF_3)\cdots H\cdots O(CH_2{-}CH_2{-}O\sim)_2 + CH_2{-}CH_2(O) \longrightarrow$$

$$HO{-}CH_2{-}CH_2{-}\overset{\oplus}{O}(CH_2{-}CH_2{-}O\sim)_2 \cdots RO{-}\overset{\ominus}{BF_3} \xrightarrow{R'{-}O{-}CH_2{-}CH_2{-}OH}$$

$$R'{-}O{-}CH_2{-}CH_2{-}O{-}CH_2{-}CH_2{-}OH + R{-}\ddot{O}(BF_3)\cdots H\cdots O(CH_2{-}CH_2{-}O\sim)_2$$

This reaction is also accompanied by formation of dioxane. It is a step of depolymerization:

$$\sim O{-}CH_2{-}CH_2{-}\overset{\oplus}{O}(CH_2{-}CH_2{-}OH)(\overset{\ominus}{ROBF_3}){-}CH_2{-}CH_2{-}O{-}CH_2{-}CH_2{-}O{-}CH_2{-}CH_2{-}O\sim \longrightarrow \sim O{-}CH_2{-}CH_2{-}\overset{\oplus}{O}(CH_2{-}CH_2{-}OH)(\overset{\ominus}{ROBF_3}){-}CH_2{-}CH_2{-}O\sim + \text{dioxane}$$

The ring-opening reaction, a nucleophilic substitution, usually takes place with an inversion of configuration at the carbon atom that undergoes the nucleophilic attack.[7,8,10] This can be illustrated as follows:

$$\sim O(\text{C(S)H(CH}_3\text{)}{-}\text{C(S)H(CH}_3\text{)}) + O(\text{C(S)H(CH}_3\text{)}{-}\text{C(S)H(CH}_3\text{)}) \longrightarrow \sim O{-}C(H)(CH_3){-}C_{(R)}(H)(CH_3){-}\overset{}{O}(C_{(S)}(H)(CH_3){-}C_{(S)}(H)(CH_3))$$

Alkyl substituents on the ethylene oxide ring enhance the process of cationic polymerization. For instance, ethylene oxide yields only low molecular weight oils with strong Lewis acids. Tetramethylethylene oxide, on the other hand, is converted readily by BF_3 into high molecular weight polymers that are insoluble in common solvents.[9]

When proton donors initiate the polymerizations of epoxides, only low molecular weight products result. The reaction is quite straightforward. Oxonium ions form during the initiation step as follows:

$$HA + CH_2{-}CH_2(O) \rightleftharpoons H{-}\overset{\oplus}{O}(CH_2{-}CH_2)\cdots \overset{\ominus}{A}$$

Propagation is the result of a ring-opening attack by a monomer:

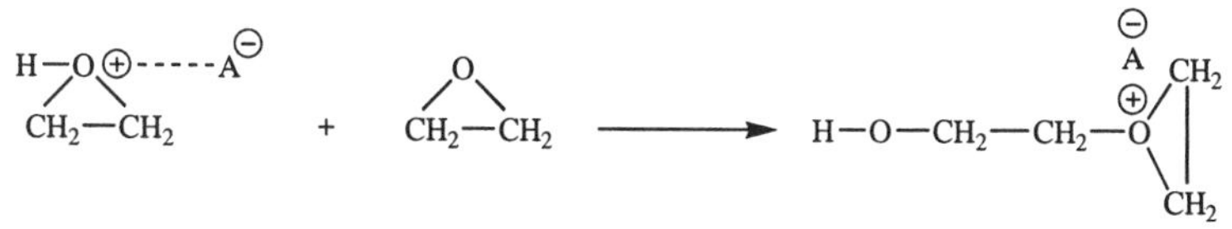

$$H{-}\overset{\oplus}{O}(CH_2{-}CH_2)\cdots A^{\ominus} + (CH_2{-}CH_2)O \longrightarrow H{-}O{-}CH_2{-}CH_2{-}\overset{\oplus}{O}(CH_2CH_2)\ A^{\ominus}$$

Chain growth can terminate by a reaction with water:

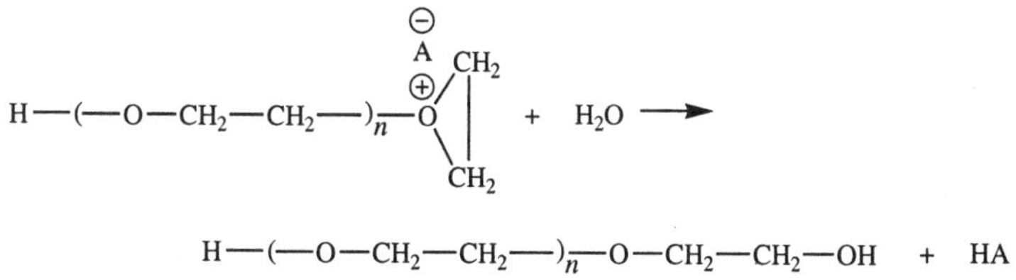

$$H{-}({-}O{-}CH_2{-}CH_2{-})_n{-}\overset{\oplus}{O}(CH_2CH_2)\ A^{\ominus} + H_2O \longrightarrow$$

$$H{-}({-}O{-}CH_2{-}CH_2{-})_n{-}O{-}CH_2{-}CH_2{-}OH + HA$$

In cationic polymerizations of propylene oxide, the ring-opening step involves a direct attack on the oxonium ion at the carbon that bears a more labile bond to the oxygen:

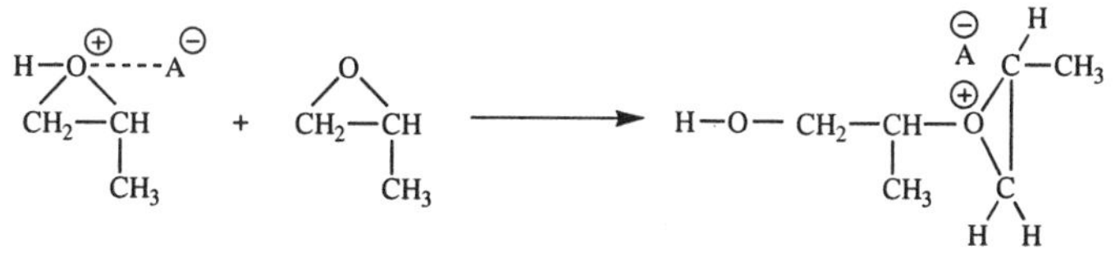

$$H{-}\overset{\oplus}{O}(CH_2{-}CH(CH_3))\cdots A^{\ominus} + (CH_2{-}CH(CH_3))O \longrightarrow H{-}O{-}CH_2{-}CH(CH_3){-}\overset{\oplus}{O}(CH(CH_3){-}CH_2)\ A^{\ominus}$$

4.3.2. Anionic Polymerization

Anionic polymerizations of ethylene oxide were originally observed as early as the end of the last century.[11] A step-growth mechanism for these polymerizations was proposed later.[12] This mechanism is now well established.[13] The conversion increases linearly with time and the molecular weight also increases with conversion. Reactions with bases like sodium or potassium hydroxides or alkoxides yield only low molecular weight polymers. The initiation, an S_N2 displacement, results in the formation of an alkoxide ion:

$$R{-}O^{\ominus} + (CH_2{-}CH_2)O \longrightarrow R{-}O{-}CH_2{-}CH_2{-}O^{\ominus}$$

Subsequent propagation may occur by nucleophilic displacement involving a new alkoxide ion:

$$R{-}O{-}CH_2{-}CH_2{-}O^{\ominus} + (CH_2{-}CH_2)O \longrightarrow R{-}O{-}CH_2{-}CH_2{-}O{-}CH_2{-}CH_2{-}O^{\ominus}$$

Termination takes place by a transfer to a hydroxyl group of another molecule. It can also be to a terminal hydroxyl group of a formed polymer. This starts new chain growth from the second unit:

$$R{-}O{-}CH_2{-}CH_2{-}O{-}CH_2{-}CH_2{-}O^{\ominus} + HO{-}CH_2{-}CH_2{-}O{-}CH_2{-}CH_2{\sim} \rightarrow$$
$$R{-}O{-}CH_2{-}CH_2{-}O{-}CH_2{-}CH_2{-}OH + {}^{\ominus}O{-}CH_2{-}CH_2{-}O{-}CH_2{-}CH_2{\sim}$$

The chain ends remain active and higher molecular weights can be obtained by further additions of the monomer.

The reaction is often more complex than is shown above, because many such polymerizations, catalyzed by alkali hydroxides or alkoxides, are carried out in the presence of alcohols. This is done to achieve a homogeneous system. Such conditions, however, lead to exchange reactions:

$$R{-}({-}O{-}CH_2{-}CH_2{-})_n{-}O{-}CH_2{-}CH_2{-}O^{\ominus}Na^{\oplus} + ROH \rightarrow R{-}({-}O{-}CH_2{-}CH_2{-})_n{-}O{-}CH_2{-}CH_2{-}OH + RO^{\ominus}Na^{\oplus}$$

Of course, the exchanges as shown above affect the kinetics of the process. The extent of these reactions is subject to the acid strength of the alcohols present, including the terminal hydroxyl groups of formed polymers. If their acidities are approximately equal, the exchange reactions take place throughout the course of the polymerization.[1]

One of the reasons for the relatively low molecular weights of the products is the low reactivity of the epoxide ring toward anionic propagation. Another reason is the tendency to chain transfer to monomers, particularly to substituted ones like propylene oxide:

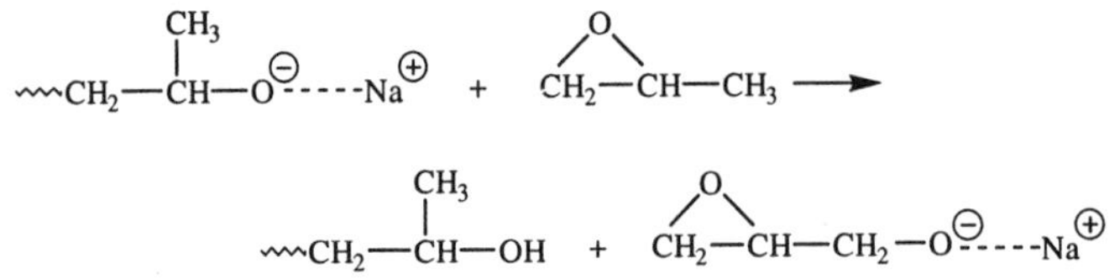

$$\sim\!CH_2-CH(CH_3)-O^{\ominus}\cdots Na^{\oplus} + CH_2-CH-CH_3\ (\text{epoxide}) \longrightarrow \sim\!CH_2-CH(CH_3)-OH + CH_2-CH-CH_2-O^{\ominus}\cdots Na^{\oplus}\ (\text{epoxide})$$

The newly formed species rearranges rapidly:

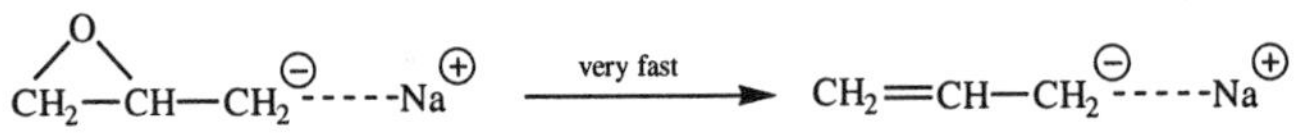

$$CH_2-CH-CH_2^{\ominus}\cdots Na^{\oplus}\ (\text{epoxide}) \xrightarrow{\text{very fast}} CH_2{=}CH-CH_2^{\ominus}\cdots Na^{\oplus}$$

Such transfer reactions are *E*-2 type eliminations. This was shown on tetramethyletheylene oxide that undergoes the reaction when treated with catalytic amounts of potassium *t*-butoxide[4]:

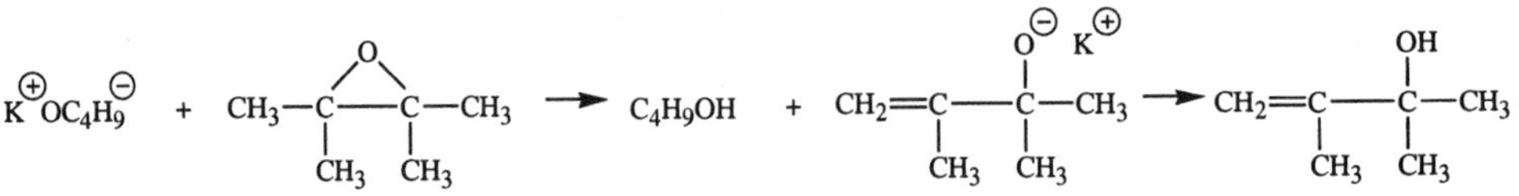

$$K^{\oplus}OC_4H_9^{\ominus} + CH_3-C(CH_3)-C(CH_3)-CH_3\ (\text{epoxide}) \longrightarrow C_4H_9OH + CH_2{=}C(CH_3)-C(CH_3)(O^{\ominus}K^{\oplus})-CH_3 \longrightarrow CH_2{=}C(CH_3)-C(CH_3)(OH)-CH_3$$

In propylene oxide polymerization, therefore, the *E*-2 type elimination reaction is in competition with propagation:

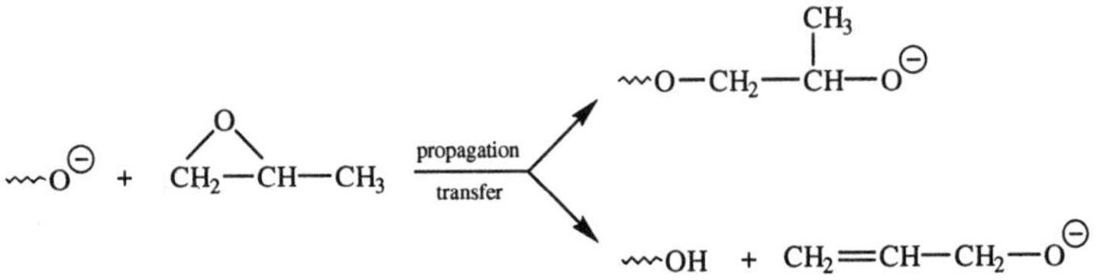

$$\sim\!O^{\ominus} + CH_2-CH-CH_3\ (\text{epoxide}) \begin{cases} \xrightarrow{\text{propagation}} \sim\!O-CH_2-CH(CH_3)-O^{\ominus} \\ \xrightarrow{\text{transfer}} \sim\!OH + CH_2{=}CH-CH_2-O^{\ominus} \end{cases}$$

There are both allyl and propenyl ether end groups in the products, according to infrared spectra.[4] This suggests that, in addition to the *E*-2 type elimination, an intramolecular transfer takes place by an allylic hydrogen.

4.3.3. Polymerization by Coordination Mechanism

The coordination catalysts for these reactions are diverse. They can be compounds of alkaline earth metals, like calcium amide, or calcium amide-alkoxide. They can also be Ziegler–Natta-type catalysts. These can be alkoxides of aluminum, magnesium, or zinc combined with ferric chloride. Others are reaction products of dialkylzinc with water or alcohol. They can also be bimetallic μ-oxoalkoxides, such as $[(RO)_2AlO_2]Zn$. Other catalysts are aluminum or zinc metalloporphyrin derivatives (see Fig. 4.1).

From propylene oxide these catalysts yield crystalline, isotactic polymers.[15,16] Living polymerizations with metalloporphyrin derivatives are difficult to terminate and are therefore called by some *immortal*.[216] Catalysts like $(C_6H_5)_3$–$SbBr_2$–$(C_2H_5)_3N$ in combination with Lewis acids also yield crystalline poly(propylene oxide). Others, like pentavalent organoantimony halides, are useful in polymerizations of ethylene oxide.[17]

Polymerizations of epoxides by a *coordinated anionic* mechanism result in high molecular weight products. The details of the reaction mechanism have not been fully resolved yet, but it is commonly believed to involve coordination of the monomers to electrophilic centers of the catalyst. This is followed by an activation for an attack by the anion.[1] The mechanism[1] can be illustrated by the following reactions:

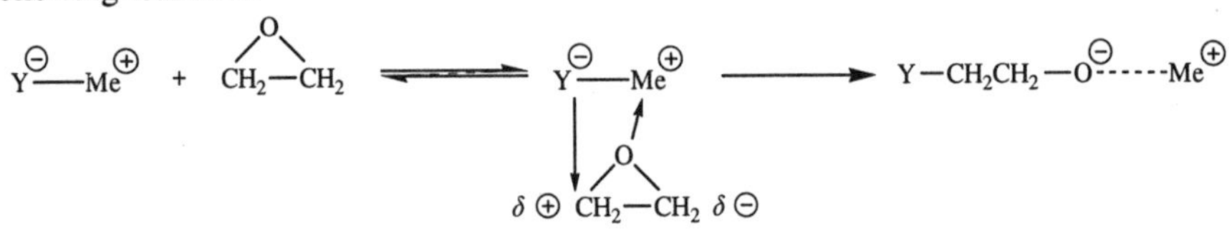

$$Y^{\ominus}-Me^{\oplus} + CH_2-CH_2\ (\text{epoxide}) \rightleftharpoons \left[Y^{\ominus}-Me^{\oplus}\ \text{coordinated to}\ \delta^{\oplus}CH_2-CH_2\delta^{\ominus}\ (\text{epoxide})\right] \longrightarrow Y-CH_2CH_2-O^{\ominus}\cdots Me^{\oplus}$$

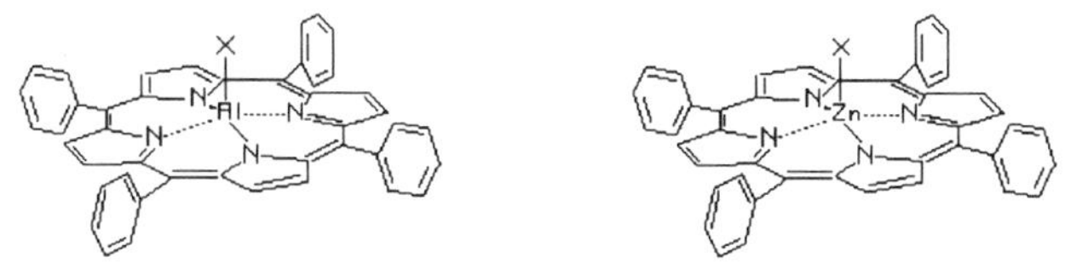

Figure 4.1. Catalysts. X = methyl, methoxy, or other groups.

where Me means metal,

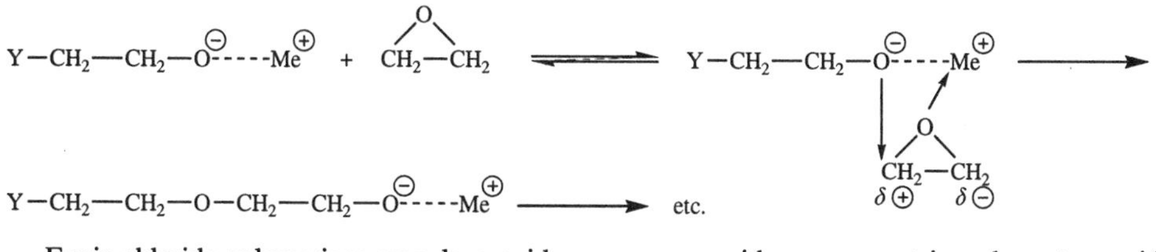

$$Y{-}CH_2{-}CH_2{-}\overset{\ominus}{O}{\cdots}\overset{\oplus}{Me} + \underset{CH_2-CH_2}{\overset{O}{\triangle}} \rightleftharpoons Y{-}CH_2{-}CH_2{-}\overset{\ominus}{O}{\cdots}\overset{\oplus}{Me} \longrightarrow$$

$$Y{-}CH_2{-}CH_2{-}O{-}CH_2{-}CH_2{-}\overset{\ominus}{O}{\cdots}\overset{\oplus}{Me} \longrightarrow \text{etc.}$$

Ferric chloride polymerizes propylene oxide, a monomer with an asymmetric carbon atom, with retention of asymmetry in the backbone.[2] The products of polymerization contain either optically active polymers or racemic mixtures, depending upon the monomers used. When only a pure optical isomer monomer is used the products are crystalline polymers composed of the same optically active units:

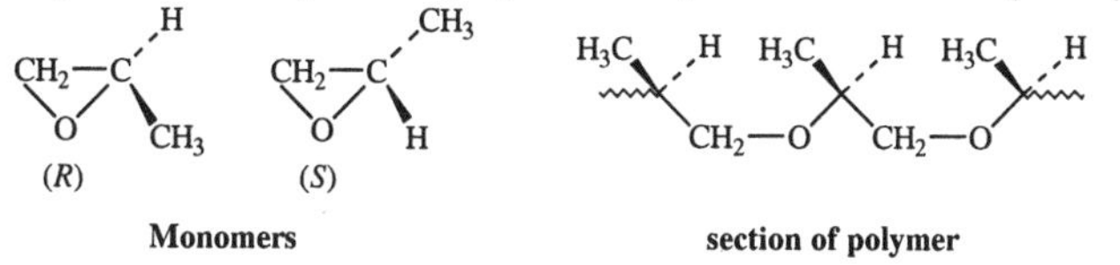

(*R*) (*S*)

Monomers **section of polymer**

The polymers are fairly high in molecular weight, approximately 100 times greater than the products from KOH initiations. Propylene oxide initially reacts with ferric chloride to form an oligomer, a chloropolyalkoxide. The material contains approximately four or five propylene oxide repeat units. This forms two different halogen sites. It can be illustrated as follows:

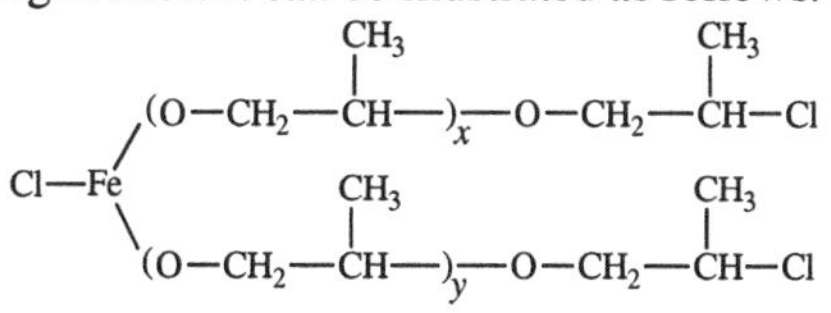

$$Cl{-}Fe\begin{cases}(O{-}CH_2{-}CH(CH_3){-})_x{-}O{-}CH_2{-}CH(CH_3){-}Cl\\(O{-}CH_2{-}CH(CH_3){-})_y{-}O{-}CH_2{-}CH(CH_3){-}Cl\end{cases}$$

The above compound may be the catalyst, or one closely related to it, for forming stereoregular polymers. Water appears to play a role, because the proportion of crystallinity increases with addition of water. When water is added in a molar ratio of 1.8:1.0 of water to iron, the proportion of crystalline to amorphous fraction increases from 0.13 to 0.86. Price and Osgen[16] suggested that the polymerization proceeds in a step-growth mechanism as follows:

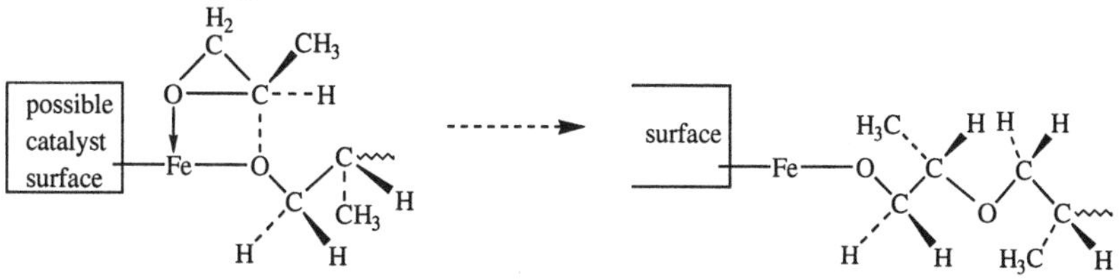

possible catalyst surface ------→ surface

The solid surface of the catalyst causes the transition state to be more compressed. Steric repulsions between the incoming monomer and the ultimate unit are minimized if the incoming monomer molecule is forced to be *trans* to the methyl group of the previous unit. Such a conformational approach also results in minimum repulsion between the incoming monomer and the bulky growing polymer chain.[17–19] Also, ferric alkoxides are associated with nonpolar solvents. A dimer may have the following structure:

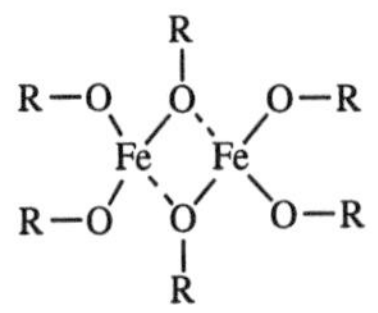

By comparison, intramolecular chelation can be expected to reduce the degree of association of the catalyst. Addition of water results in increased association after hydrolysis of the ferric alkoxide. This may explain the effect of promoting stereoregularity by addition of water.[19] The ferric alkoxide catalyst can also be made highly stereospecific by partial hydrolysis and still remain soluble in ether, the polymerization medium.[20] This led to a suggestion[21] that the catalyst may contain active Fe–O–Fe bonds. Such bonds would be formed from condensation of partially hydrolyzed alkoxide derivative. The monomer insertion between the iron–oxygen bonds can be illustrated as follows:

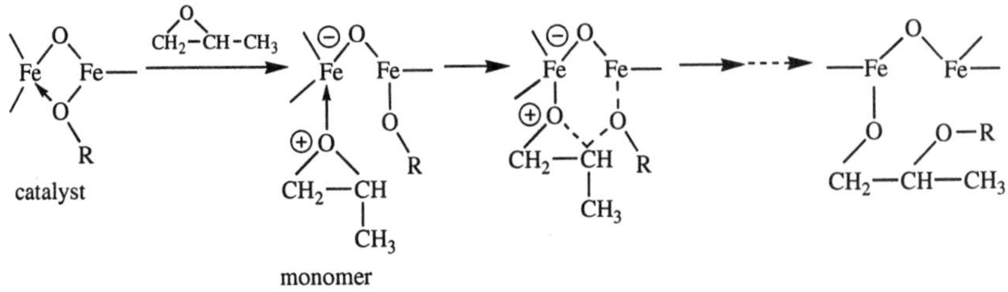

The forces of interaction between the iron atoms and the various oxygen atoms as shown above assure a *cis* opening of the epoxide ring. The mechanism of the reaction of the ferric alkoxide is S$_N$2-type. There is therefore increased restriction on the conformation of the monomer unit as it approaches the reaction center.[21]

Many other coordinated anionic catalysts that are *metal alkoxides* or *metal alkyls* are also much more reactive in the presence of water or alcohols. The function of these coreactants is to modify the catalyst itself. For instance, diethylzinc combined with water in a ratio of 1:1 yields very reactive species. The exact nature of the catalyst is still not fully established, however, the reaction product is pictured as follows[22,23]:

$$C_2H_5\text{–}Zn\text{–}O\text{–}Zn\text{–}C_2H_5$$

Several reaction mechanisms were proposed. One suggested pathway for propylene oxide polymerization pictures an initial coordination of the monomer with a cationically active center[24]:

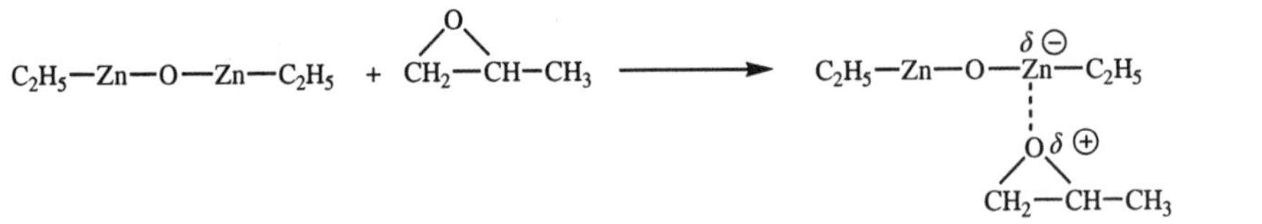

The propagation is preceded by an intramolecular rearrangement:

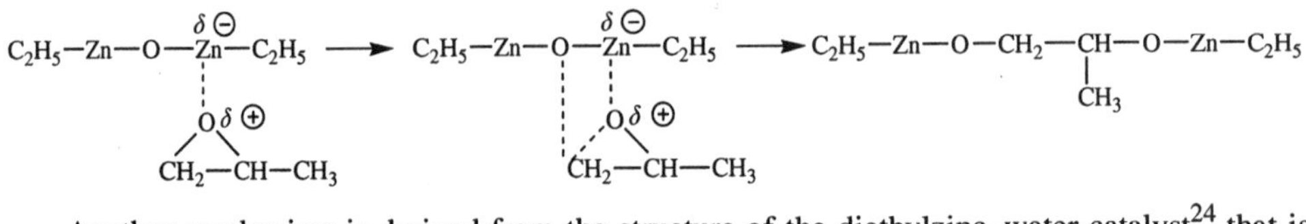

Another mechanism is derived from the structure of the diethylzinc–water catalyst[24] that is visualized as a dimer:

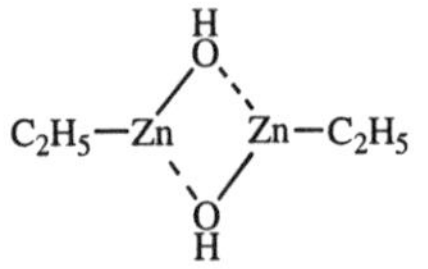

A similar structure is pictured for diethylzinc–alcohol. The asymmetric induction takes place during coordination of the monomer to the catalyst site. This is a result of indirect regulation that results from interactions between the monomer and the penultimate unit.[24]

In yet another mechanism the initial coordination and subsequent propagation steps are pictured as follows[25]:

R—O—Zn(μ-OR)₂Zn—O—R + CH_2—CH—CH_3 (O) ⟶ R—O—Zn(μ-OR)₂Zn—O—R (coordinated H, H, CH_3 C—C—H epoxide) ⟶ R—O—Zn(μ-OR)₂—O—R (with H_3C, H—C, H, H, C, O, R)

While the detailed structures of most catalyst sites are still unknown, it was established that stereoselectivity does not come from the chirality of the growing chain end. Rather, it is built into the catalyst site itself.[26,27] Normal preparations of the catalysts give equal numbers of (*R*) and (*S*) chiral catalyst sites. These coordinate selectively with (*R*) and (*S*) monomers, respectively, in the process of catalytic-site control.[22]

4.3.4. Steric Control in Polymerizations of Oxiranes

Cationic polymerizations of oxiranes are much less isospecific and regiospecific than are anionic polymerizations. In anionic and coordinated anionic polymerizations, only chiral epoxides, like propylene oxide, yield stereoregular polymers. Both pure enantiomers yield isotactic polymers when the reaction proceeds in a regiospecific manner with the bond cleavage taking place at the primary carbon.

In all polymerizations of oxiranes by cationic, anionic, and coordinated anionic mechanisms, the ring opening is generally accompanied by an inversion of the configuration at the carbon where the cleavage takes place. A linear transition state mechanism involving dissociated nucleophilic species has been proposed.[14] Yet, there are some known instances of ring-opening reactions of epoxies that are stereochemically retentive. For instance, ring opening of 2,3-epoxybutane with $AlCl_3$ results in formation of 3-chloro-2-butanol, where the *cis* and *trans* epoxides are converted to the *erythro* and *threo* chlorohydrins. Inoue and co-workers found,[18] however, that polymerizations of *cis*- and *trans*-2,3-epoxybutanes take place with inversion of configuration when aluminum 5,10,15,20-tetraphenylporphine and zinc 5,10,15,20-tetraphenyl-21-methylporphine catalysts are used. To explain the inversion, Inoue and co-workers proposed a linear transition state mechanism that involves a simultaneous participation of two porphyrin molecules.[18] One porphyrin molecule accommodates a coordinative activation of the epoxide and the other one serves as a nucleophile to attack the coordinated epoxide from the back side.

Potassium hydroxide or *alkoxide* polymerizes racemic propylene oxide with better than 95% regioselectivity of cleavage at the bond between oxygen and the carbon substituted by two hydrogens. The product, however, is atactic. Both (*R*) and (*S*) propylene oxides react at the same rate. This shows that the initiator is unable to distinguish between the two enantiomers of propylene oxide. When *t*-butyl ethylene oxide is polymerized by KOH, it yields a crystalline product. This product is different in its melting point, X-ray diffraction pattern, and solution-NMR spectra from the typical isotactic polymers. It contains alternating isotactic and syndiotactic sequences.[30] It was suggested[33] that this may be a result of the configuration of the incoming monomer being opposite to that of the penultimate unit. Chelation of the paired cation ($K^{\oplus}$) with the last and the next to last oxygen is visualized. Geometry of such a chelate is dictated by the requirement that the penultimate *t*-butyl group be in an equatorial conformation. This makes it reasonable to postulate that the necessary preference for the incoming monomer is to be opposite to that of the penultimate unit[30]:

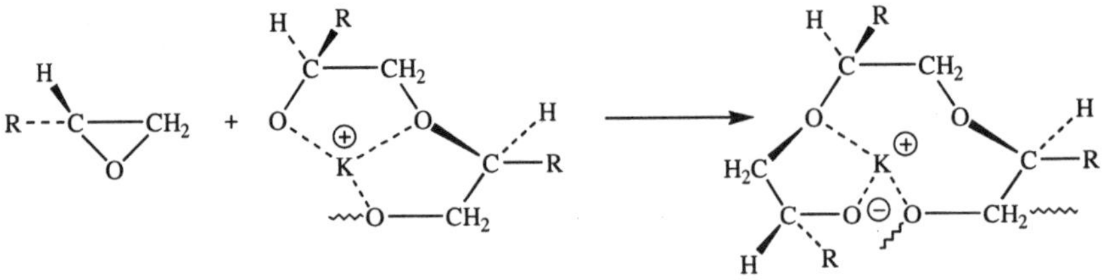

When phenyl glycidyl ethers are polymerized under the same conditions, the steric arrangement is all isotactic, rather than isotactic-syndiotactic.[30] Price explained that on the basis of the oxygen in C_6H_5–O–CH_2 seeking to coordinate to potassium ions in the transition state.[30] In the case of *t*-butylethylene oxide, on the other hand, the tertiary butyl group tends to be as far as possible away from the potassium ion.[33] This is supported by the observation that *p*-methoxy and *p*-methyl groups on phenyl glycidyl ether increase the crystalline portion of the polymer, while the *p*-chloro substituent decreases it.[30]

Most stereoselective coordination catalysts polymerize propylene oxide to yield polymers that contain high ratios of isotactic to syndiotactic sequences. Large portions of amorphous materials, however, are also present in these same products. These amorphous portions contain head-to-head units that are imperfections in the structures.[28,29] For every head-to-head placement, one (*R*) monomer is converted to an (*S*) unit in the polymer.[22] This shows that at the coordination sites abnormal ring openings occur at the secondary carbon with an inversion of the configuration and result in head-to-head placements.[22,30] Also, *erythro* and *threo* isomers units are present. The isotactic portion consists almost exclusively of the *erythro* isomer while the amorphous fraction contains 40–45% *erythro* and 55–60% *threo*.[30]

All the above information is indirect evidence that a typical catalyst, such as $(C_2H_5)_2Zn–H_2O$, contains isotactic and amorphous sites. The isotactic sites are very selective and coordinate either with (*R*) or with (*S*) monomers. The amorphous sites, on the other hand, coordinate equally well with both (*R*) and (*S*) monomers. In addition, there is little preference for attack on either the primary or the secondary carbons during the ring-opening reactions.[22]

According to a Tsuruta mechanism,[35] the first step in propylene oxide polymerization, with catalysts like zinc alcoholates, is the coordination of the ether oxygen onto a zinc atom. The second step is a nucleophilic attack at the oxirane ring by the alkoxy ion. Almost all the bond cleavage takes place at the CH_2–O bond. This results in retention of the steric configuration of the carbon atom at the C–H group. The next oxirane molecule repeats the process, coordinates with the same zinc atom, and then undergoes the ring-opening reaction to form a dimer. Repetition of this process many times yields a high molecular weight polymer[35]:

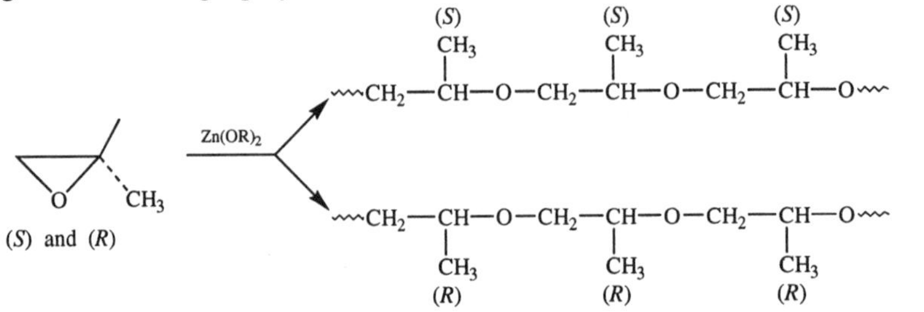

The catalyst can also be $ZnR_2–CH_3–OH$.

Special catalyst complexes, like $[Zn(OCH_3)_2 \cdot (C_2H_5OCH_3)_6]$, form through careful control of reaction conditions by adding 16 moles of methyl alcohol to 14 moles of diethylzinc in heptane under an argon atmosphere. X-ray analysis shows two different structures[35]: one of them is a centrosymmetric complex of two enantiomorphic distorted cubes that share a corner Zn atom. The two would be equivalent if they were not distorted. Another structure, also centrosymmetric, consists of two enantiomorphic distorted structures that resemble "chairs without legs," where the surfaces share a common seat. Both types of complexes are active initiators for polymerization of propylene

d cube, l cube catalyst complex + CH_2-CH-CH_3 (oxirane) → d cube, l cube catalyst complex → l transition state → l product

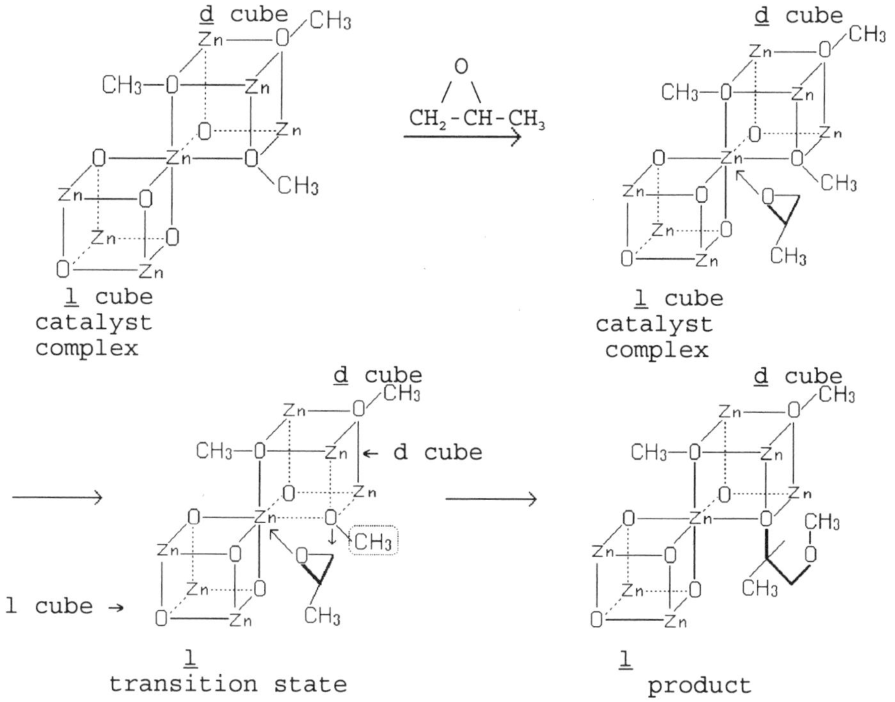

Figure 4.2. Tsuruta mechanism.

oxide. Each has two enantiomorphic sites for polymerization. Based on that knowledge, NMR spectra, and GPC curves, Tsuruta suggested the following mechanism of a monomer coordinating with the catalyst[35] (see Fig. 4.2). The bonds at the central zinc atom are loosened and a coordination takes place with methyl-oxirane molecule at the central atom. Cleavage at the O–CH_2 bond of the oxirane takes place by a concerted mechanism. If the bond loosening takes place at the *d* cube and the nucleophilic attack takes place at one of the methoxy groups on that cube, then chirality around the central zinc will favor the *L* monomer over the *D* monomer. This is the origin of the *l** catalyst site. If the bond loosening takes place in the *l* cube, the catalyst site will have *d** chirality. Because the probability of bond loosening in the *d* cube is exactly the same as in *l* cube, an equal number of *l** and *d** sites should be expected to form. These two cubes become a source of *d** and *l** chiral nature.[35]

4.4. Polymerization of Oxetanes

Oxetanes (or oxacyclobutanes) are preferably polymerized in a solvent to maintain stirring and temperature control. It is necessary to carefully purify both the solvent and the monomer, because impurities interfere with attainment of high molecular weights.

4.4.1. The Initiation Reaction

Theoretically, any Lewis acid can catalyze oxetane polymerizations. However, these acids differ considerably in their effectiveness. Boron trifluoride and its etherates are the most widely reported catalysts. Moisture must be excluded as it tends to be detrimental to the reaction.[34]

Chlorinated hydrocarbon solvents, like methylene chloride, chloroform and carbon tetrachloride, are common choices. The reaction is usually conducted at low temperatures and indications are that the lower the reaction temperature, the higher the molecular weight of the product.

It was reported that when oxetane polymerizations are carried out with boron trifluoride catalyst in methylene chloride at temperatures between 0 °C to –27.8 °C, a cocatalyst is not required.[31] The product, however, is a mixture of a linear polymer and a small amount of a cyclic tetramer. This is in agreement with an earlier observation that the polymerizations of oxetane are complicated by formations of small amounts of cyclic tetramers.[32] Other catalysts, protonic acids, capable of generating oxonium ions, will polymerize oxetane. These acids are sulfuric, trifluoracetic, and fluorosulfuric. The initiation reaction can be illustrated as follows:

$$H^{\oplus}A^{\ominus} + O\square \longrightarrow H{-}\overset{\oplus}{O}\square \;\; A^{\ominus}$$

The adduct reacts with another cyclic ether:

$$H{-}\overset{\oplus}{O}\square\;A^{\ominus} + O\square \longrightarrow HO{-}CH_2{-}CH_2{-}CH_2{-}\overset{\oplus}{O}\square\;A^{\ominus}$$

When Lewis acid complexes with active hydrogen compounds initiate the polymerizations, the complexes acts as protonic acids. On the other hand, etherates initiate by forming oxonium ions and may involve alkyl exchange reactions with the monomer:

$$(C_2H_5)_3{-}O^{\oplus}(BF_4)^{\ominus} + O\square \longrightarrow C_2H_5{-}\overset{\oplus}{O}\square\;BF_4^{\ominus} + (C_2H_5)_2O$$

4.4.2. The Propagation Reaction

The propagation takes place via tertiary oxonium ions[36,37]:

$$\sim\!\!O{-}CH_2{-}CH_2{-}CH_2{-}\overset{\oplus}{O}\square\;A^{\ominus} + :O\square \longrightarrow$$

$$\sim\!\!O{-}CH_2{-}CH_2{-}CH_2{-}O{-}CH_2{-}CH_2{-}CH_2{-}\overset{\oplus}{O}\square\;A^{\ominus}$$

A cyclic oligomer forms in some instances in addition to the polymer.[38–40] For instance, in polymerizations with BF_3 in methylene chloride at low temperatures a cyclic tetramer forms, probably by a backbiting process[41]:

$$\sim\!\!O{-}({-}CH_2{-})_3{-}[{-}O{-}(CH_2)_3{-}]_2{-}\overset{\oplus}{O}\square\;A^{\ominus} \longrightarrow \sim\!\!\overset{\oplus}{O}\langle (CH_2)_3{-}O{-}(CH_2)_3 \rangle O\langle (CH_2)_3{-}O{-}(CH_2)_3\rangle\;A^{\ominus}$$

$$\sim\!\!\overset{\oplus}{O}\langle (CH_2)_3{-}O{-}(CH_2)_3 \rangle O\langle (CH_2)_3{-}O{-}(CH_2)_3\rangle\;A^{\ominus} + O\square \longrightarrow O\langle (CH_2)_3{-}O{-}(CH_2)_3 \rangle O\langle (CH_2)_3{-}O{-}(CH_2)_3\rangle + \sim\!\!\overset{\oplus}{O}\square\;A^{\ominus}$$

The oxonium exchange reactions may occur with the polymer ether linkages as well as with cyclic tetramers that form, as shown above. The concentrations of the oxonium ions of the ether group on the polymer and on the cyclic tetramers, however, are very small.[41] Polymerizations with PF_5, on the other hand, or with $(C_2H_5)_3OPF_6$ either in bulk or in methylene chloride solutions, yield no significant amounts of cyclic oligomers.[42]

The activation energy of polymerizations of oxetane monomers is higher that that of tetrahydrofuran (see next section). This indicates that the orientation of the cyclic oxonium ion and the monomer is looser in the S_N2 transition state[41]:

$$\sim\!O\cdots C(H)(H)\cdots\cdots O^{\delta\oplus}(\text{oxetane ring})\cdots C\cdots H$$

In principle, stereospecificity should be possible in substituted polyoxycyclobutanes, such as 2-methyl, 3-methyl, and others. The 2-methyl derivative, however, yields amorphous polymers. This is due to the monomer's unsymmetrical structure.[32] NMR studies of the microstructure of polymers from 3,3-dimethyloxetane[43] and 2-methyloxetane[42] led to no conclusions about the manner of ring opening. The predominant head-to-tail structures may result from attacks at either the methylene of the methine carbons next to the oxonium ions of the propagating species.

Oxetane compounds also polymerize with the aid of aluminum trialkyl–water acetylacetone catalysts.[64,65] The reactions can take place at 65 °C in heptane and yield very high molecular weight polymers. These polymerizations, however, are ten times slower that similar ones carried out with propylene oxide, using the same catalyst. The reaction conditions and the high molecular weights of the products led to assumptions that coordinated mechanisms of polymerizations take place.[65]

4.5. Polymerization of Tetrahydrofurans

Lewis acids, carbon cations, salts of oxonium ions, and strong protonic acids initiate polymerizations of tetrahydrofuran. The reactions can be conducted in solution or without a solvent. It was originally polymerized[44–46] with a trialkyloxonium salt, $R_3O{\cdot}BF_4^{\ominus}$.

4.5.1. The Initiation Reaction

The initiations result from coordination of the cation catalysts with the oxygen of the monomers to form oxonium ions.[45,46] This weakens the oxygen–carbon bonds and leads to ring openings after reactions with second molecules of the monomer. New oxonium ions are generated in the process:

$$R_3O^{\oplus}\,BF_4^{\ominus} + O(CH_2)_4 \rightleftharpoons R{-}\overset{\oplus}{O}(CH_2)_4 \cdots BF_4^{\ominus} + R_2O$$

$$R{-}\overset{\oplus}{O}(CH_2)_4 \cdots BF_4^{\ominus} + O(CH_2)_4 \rightleftharpoons R{-}O{-}CH_2{-}CH_2{-}CH_2{-}CH_2{-}\overset{\oplus}{O}(CH_2)_4 \cdots BF_4^{\ominus}$$

Some active oxonium salts are[44–46]: $[(C_2H_5)_3O]^{\oplus}\,BF_4^{\ominus}$, $[(C_2H_5)_3O]^{\oplus}\,SbCl_6^{\ominus}$, $[(C_2H_5)_3O]^{\oplus}\,FeCl_4^{\ominus}$, and $[(C_2H_5)_3O]^{\oplus}\,AlCl_4^{\ominus}$.

Examples of carbon cations that can initiate polymerizations of tetrahydrofuran and some other cyclic ethers are:

$$HClO_4 + CH_3{-}C(=O){-}O{-}C(=O){-}CH_3 \longrightarrow CH_3{-}\overset{\oplus}{C}(=O)\cdots\cdots(ClO_4)^{\ominus} + CH_3COOH$$

$$CH_3{-}C(=O){-}Cl + AgClO_4 \longrightarrow CH_3{-}\overset{\oplus}{C}(=O)\cdots\cdots(ClO_4)^{\ominus} + AgCl$$

$$CH_3{-}C(=O){-}Cl + SnCl_4 \longrightarrow CH_3{-}\overset{\oplus}{C}(=O)\cdots\cdots(SnCl_5)^{\ominus}$$

$$C_6H_5{-}CH_2{-}Cl + FeCl_3 \longrightarrow C_6H_5{-}\overset{\oplus}{C}H_2\cdots\cdots FeCl_4^{\ominus}$$

$$RO{-}SO_3{-}R \;+\; BF_3 \longrightarrow \overset{\oplus}{R}\cdots\cdots(RO{-}SO_3BF_3)^{\ominus}$$

$$(C_6H_5)_3{-}C{-}Cl \;+\; SbCl_5 \longrightarrow (C_6H_5)_3{-}\overset{\oplus}{C}\cdots\cdots \overset{\ominus}{SbCl_6}$$

The initiation mechanisms, however, by many carbon cations, such as, for instance, by triphenylmethyl cations, are not straightforward. Initially, hydride ions are abstracted from the monomers to form triphenylmethanes.[52–54] Simultaneously, acids are released from the counterions. The acids become stabilized by complexing with monomers. After that, the complexes react slowly with additional monomers to form the propagating oxonium ions. This makes the acids the real initiators:

$$(C_6H_5)_3{-}\overset{\oplus}{C}\cdots\cdots\overset{\ominus}{X} \;+\; \text{O(CH}_2)_4 \longrightarrow \text{2,3-dihydrofuran} \;+\; (C_6H_5)_3C{-}H \;+\; HX$$

$$HX \;+\; 2\ \text{O(CH}_2)_4 \longrightarrow (CH_2)_4\overset{\oplus}{O}{-}H\cdots\cdots O(CH_2)_4 \quad (\overset{\ominus}{X}\ \text{paired with } \overset{\oplus}{O})$$

$$(CH_2)_4\overset{\oplus}{O}{-}H\cdots\cdots O(CH_2)_4 \;(\overset{\ominus}{X}) \longrightarrow HO{-}CH_2{-}CH_2{-}CH_2{-}CH_2{-}\overset{\oplus}{O}(CH_2)_4 \;(\overset{\ominus}{X})$$

Other initiators for tetrahydrofuran polymerizations also include Lewis acids in combinations with "promoters." These are complexes of Lewis acids, like BF_3, $SnCl_4$, or $C_2H_5AlCl_2$ with epirane compounds like epichlorohydrin.[41] The small-ring compounds are more reactive toward many Lewis acids, or protonic acids, than tetrahydrofuran and act as promoters of the initiation reactions. The initiations in the presence of small quantities of oxirane compounds can be illustrated as follows:

$$[\text{Cat.}]\leftarrow O\!\!<\!(CH_2{-}CH(CH_2Cl)) \;+\; \text{O(CH}_2)_4 \longrightarrow [\text{Cat.}]^{\ominus}{-}O{-}CH_2\underset{\substack{|\\ CH_2Cl}}{CH}{-}\overset{\oplus}{O}(CH_2)_4$$

or

$$H{-}\overset{\oplus}{O}\!\!<\!(CH_2{-}CH(CH_2Cl))\cdots[\text{Cat}{-}X]^{\ominus} \;+\; \text{O(CH}_2)_4 \longrightarrow HO{-}CH_2\underset{\substack{|\\ CH_2Cl}}{CH}{-}\overset{\oplus}{O}(CH_2)_4\cdots[\text{Cat}{-}X]^{\ominus}$$

Strong Bronsted acids form when diaryliodonium salts containing anions, like $BF_4^{\ominus}$, $AsF_6^{\ominus}$, $PF_6^{\ominus}$, and $SbF_6^{\ominus}$, are reduced with compounds like ascorbic acid in the presence of copper salts. Such acids also initiate polymerizations of tetrahydrofuran, cyclohexene oxide, and *s*-trioxane.[194]

4.5.2. The Propagation Reaction

The propagation process is a succession of nucleophilic attacks by free electrons on the oxygens of the monomers upon the α-carbons of the heteroatoms of the ultimate polymerizing species[1]:

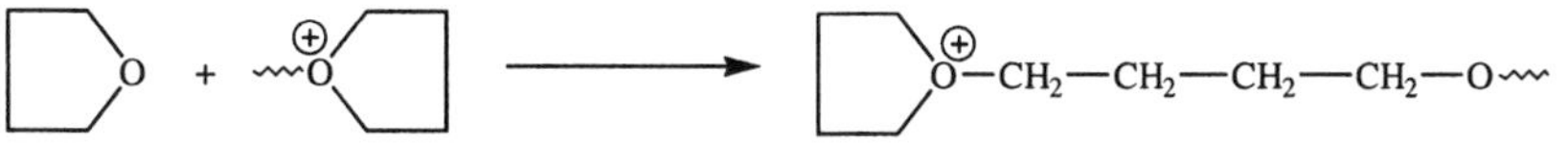

$$\text{THF} + \sim\sim O^{\oplus}(\text{ring}) \longrightarrow \text{THF}^{\oplus}\text{—}CH_2\text{—}CH_2\text{—}CH_2\text{—}CH_2\text{—}O\sim\sim$$

The polymers produced are linear. This is common to polymerization of many heterocyclics. These propagation reactions proceed by stepwise additions of monomers by the S_N2 mechanism to the growing ends of the propagating chains. NMR spectra of the growing species show only a presence of oxonium ions[47,48]:

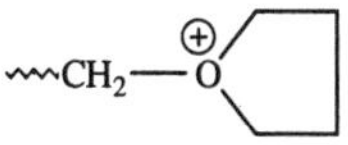

$$\sim\sim CH_2\text{—}O^{\oplus}(\text{ring})$$

The oxonium ions could, in principle, be in equilibrium with minute quantities of carbon cations, $\sim CH_2^{\oplus}$, that are more active. All evidence to date, however, shows that in tetrahydrofuran polymerizations the presence of carbon cations is negligible in the propagation process.[49] Also, the rate constant for propagation of free macroions with the counterions is equal, within experimental error, to the rate constant for macroion–counterion pairs. This does not appear to depend upon the structure of the anion studied.[49] Such information, however, was obtained on large anions. With smaller anions, differences in the rates of propagation of macrocations and those of macroion–counterion pairs has not been ruled out.

An S_N2 attack requires that the reaction occur at the oxygen–carbon bond. In such an attack, steric requirements are less restricted than they are in an anionic polymerization. In addition, positive and negative charges in the macroion pairs that contain the oxonium ions are dispersed and the anions are large. This means that the electrostatic interactions are less important in cationic polymerizations of this type than they are in anionic ones.

When the polymerization of tetrahydrofuran is carried out with the aid of CF_3SO_3H, both covalent and ionic species are present. They can be detected during propagation with NMR spectroscopy.[49] Both species exist in a mobile equilibrium. Solvent polarity apparently influences the position of such equilibria. In nitromethane, 95% of the growing chains are macroions. In carbon tetrachloride, 95% of them are macroesters. In methylene chloride both species are present in the reaction mixture, approximately in equal amounts.[50,51,56,60,61] The propagation rate of macroions, however, is 10^2 times faster than that of the macroesters. As a result, chain growth even in carbon tetrachloride is still by way of the ions. The macroesters can therefore be considered as dormant species,[51] or, as some suggest, even cases of temporary termination.[59] The much higher reactivity of the macroions is attributed to the contribution of the partially released strain in ionic species.[49] Macroions and macroesters can be illustrated as follows:

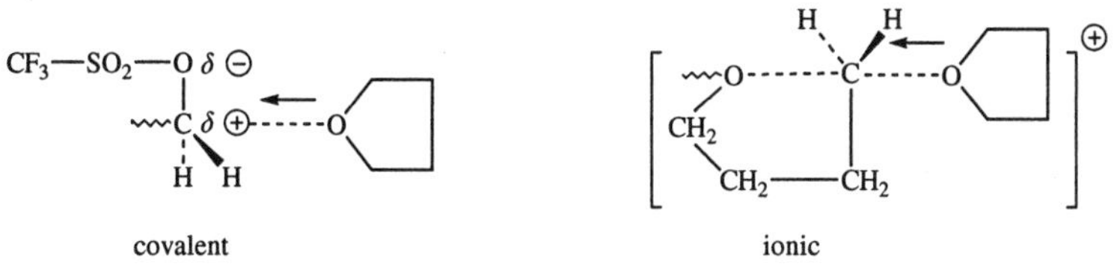

$$CF_3\text{—}SO_2\text{—}O\,\delta\ominus \cdots \sim\sim C(H)(H)\,\delta\oplus \cdots O(\text{ring}) \qquad \left[\sim\sim O \cdots C(H)(H) \cdots O(\text{ring});\ CH_2\text{—}CH_2\text{—}CH_2\right]^{\oplus}$$

covalent — ionic

4.5.3. The Termination Reaction

Terminations in tetrahydrofuran polymerizations can depend upon the choice of the counterion, particularly if the reaction is conducted at room temperature.[56] In many reactions the chain continues to grow without any considerable chain termination or transfer.[57,58] This produced the term "living" polytetrahydrofuran. Thus, in polymerizations of tetrahydrofuran[55] with $PF_6^{\ominus}$ or $SbF_6^{\ominus}$ counterions, the molecular weights of the products can be calculated directly from the ratios of the initiators to the monomers. The molecular weight distributions of the polymers from such polymerization reactions with $PF_6^{\ominus}$ and $SbF_6^{\ominus}$, however, start out as narrow, but then broaden. This is believed[56] to be due to transfer reactions with ether oxygen. It is supported by evidence that with $SbF_6^{\ominus}$ initiation, both termination and transfer reactions take place.[55] In addition, polymerizations of tetrahydrofuran, like those of the epoxides, can be accompanied by formations of some macrocyclic oligomers. This is often the case[62,63] when strong acids are used as initiators. The proposed mechanism involves backbiting and chain coupling, and results in linear polymers with hydroxy groups and oxonium ions at opposite chain ends as well as some macrocycles.

The absence of linear oligomers is due to rapid reactions of the hydroxy and oxonium ion end groups. This mechanism is quite general[63,64] for ring-opening polymerizations of cyclic ethers initiated with strong protonic acids. Substituted tetrahydrofurans generally resist polymerizations.

4.6. Polymerization of Oxepanes

Oxepanes are polymerized by various cationic initiators like $(C_2H_5)_3C^{\ominus}BF_4^{\ominus}$, $(C_2H_5)_3C^{\oplus}$-$SbCl_6^{\ominus}$, BF_3-epichlorohydrin, and $SbCl_6$-epichlorohydrin.[41] The reactions take place in chlorinated solvents, like methylene chloride. The rates of these reactions, however, are quite slow.[41] In addition, these polymerizations are reversible. The rates of propagation of the three cyclic ethers, oxetane, tetrahydrofuran, and oxepane, at 0 °C fall in the following order[41]:

oxetane > tetrahydrofuran > oxepane

At that temperature oxetane is about 35 times as reactive as tetrahydrofuran, which in turn is about 270 times as reactive as oxepane. This cannot be explained on the basis of ring strain, nor can it be explained from considerations of basic strength. Saegusa suggested[41] that the differences in the propagation rates are governed by nucleophilic reactivities of the monomers. They are also affected by the reactivities of the cyclic oxonium ions of the propagating species and also by the steric hindrances in the transition states of propagation. Higher activation energy of oxepane is explained by increased stability of the seven-membered oxonium ion. The oxepane molecule has a puckered structure, and the strain that comes from the trivalent oxygen is relieved by small deformations of the angles of the other bonds.[41]

4.7. Ring-Opening Polymerizations of Cyclic Acetals

The cationic polymerizations of cyclic acetals are different from the polymerizations of the rest of the cyclic ethers.[66] The differences arise from greater nucleophilicity of the cyclic ethers as compared to that of the acetals. In addition, cyclic ether monomers, epirane, tetrahydrofuran, and oxepane, are stronger bases than their corresponding polymers. The opposite is true of the acetals. As a result, in acetal polymerizations, active species like those of 1,3-dioxolane may exist in equilibrium with macroalkoxy carbon cations and tertiary oxonium ions.[67] By comparison, the active propagating species in polymerizations of cyclic ethers, like tetrahydrofuran, are only tertiary oxonium ions. The properties of the equilibrium of the active species in acetal polymerizations depend very much upon polymerization conditions and upon the structures of the individual monomers.

4.7.1. Polymerization of Trioxane

Trioxane is unique among the cyclic acetals because it is used commercially to form polyoxymethylene, a polymer that is very much like the one obtained by cationic polymerization of formaldehyde. Some questions still exist about the exact mechanism of initiation in trioxane polymerizations. It is uncertain, for instance, whether a cocatalyst is required with strong Lewis acids like BF_3 or $TiCl_4$.

The cationic polymerization of trioxane can be initiated by protonic acids, complexes of organic acids with inorganic salts, and compounds that form cations.[80] These initiators differ from each other in activity and in the influence on terminations and on side reactions. Trioxane can also be polymerized by high-energy radiation.[80] In addition, polymerizations of trioxane can be carried out in the solid phase, in the melt, in the gas phase, in suspension, and in solution. Some of these procedures lead to different products, however, because variations in polymerization conditions can cause different side reactions.

Polymerizations in the melt above 62 °C are very rapid. They come within a few minutes to completion at 70 °C when catalyzed by 10^{-4} mole of boron trifluoride. This procedure, however, is only useful for preparation of small quantities of the polymer, because the exothermic heat of the reaction is difficult to control.

Typical cationic polymerizations of trioxane are characterized by an induction period. During that period only oligomers and monomeric formaldehyde form. This formaldehyde apparently results from splitting the carbon cations that form in the primary steps of polymerization. The reaction starts after a temperature-dependent equilibrium concentration of formaldehyde is reached.[80]

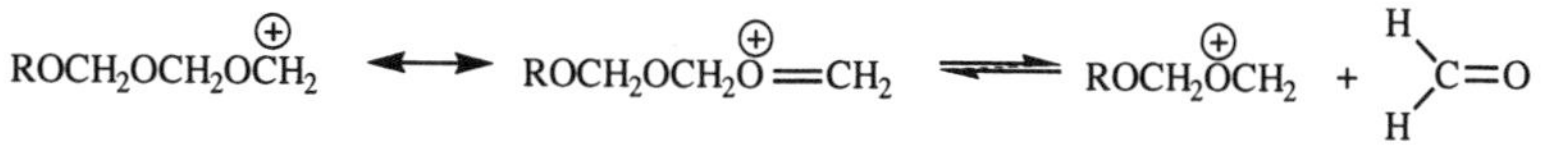

$$ROCH_2OCH_2O\overset{\oplus}{C}H_2 \longleftrightarrow ROCH_2OCH_2\overset{\oplus}{O}{=}CH_2 \rightleftharpoons ROCH_2\overset{\oplus}{O}CH_2 + H_2C{=}O$$

Several reaction mechanisms were proposed. One is based on a concept that Lewis acids, like BF_3, coordinate directly with oxygen of the acetal. This results in ring opening that is induced to form a resonance-stabilized zwitterion[68]:

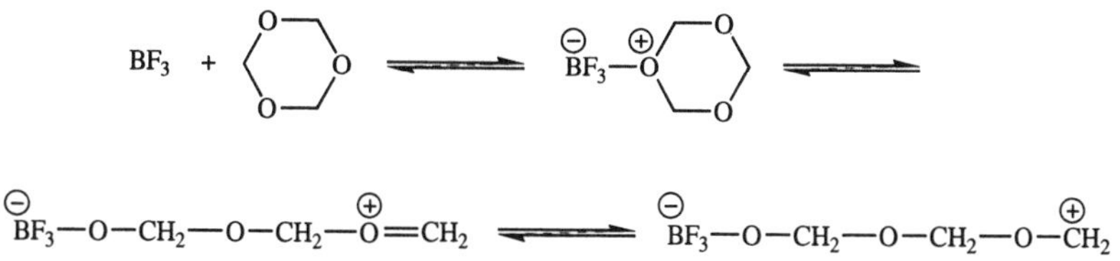

$$BF_3 + \text{trioxane} \rightleftharpoons \overset{\ominus}{B}F_3{-}\overset{\oplus}{O}(\text{trioxane ring}) \rightleftharpoons$$

$$\overset{\ominus}{B}F_3{-}O{-}CH_2{-}O{-}CH_2{-}\overset{\oplus}{O}{=}CH_2 \rightleftharpoons \overset{\ominus}{B}F_3{-}O{-}CH_2{-}O{-}CH_2{-}O{-}\overset{\oplus}{C}H_2$$

Resonance stabilizations of the adjacent oxonium ions lead to formations of carbon cations that are believed to be the propagating species.[68] Propagations consist of repetitions of the sequences of addition of the carbon cations to the monomer molecules and are followed by ring opening. The above mechanism has to be questioned, however, because rigorously dried trioxane solutions in cyclohexane fail to polymerize with $BF_3{\cdot}O(C_4H_9)_2$ catalyst.[82] The same is true of molten trioxane.[83] It appears, therefore, that BF_3–trioxane complexes do not form as suggested and do not result in initiations of the polymerizations. Additions of small quantities of water, however, do result in initiations of the polymerizations.[83]

Another mechanism is based on a concept that two molecules of BF_3 are involved in the initiation process.[69] This also appears improbable since, without water, BF_3 fails to initiate the reaction. The following scheme of initiation, based on water as the catalyst, was therefore developed[83]:

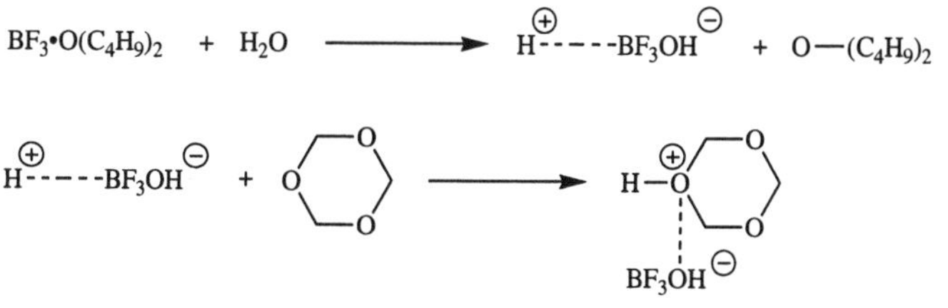

$$BF_3{\cdot}O(C_4H_9)_2 + H_2O \longrightarrow \overset{\oplus}{H}{-}{-}{-}{-}{-}BF_3OH^{\ominus} + O{-}(C_4H_9)_2$$

$$\overset{\oplus}{H}{-}{-}{-}{-}{-}BF_3OH^{\ominus} + \text{trioxane} \longrightarrow H{-}\overset{\oplus}{O}(\text{trioxane ring})\cdots BF_3OH^{\ominus}$$

Chain growth in this reaction is accompanied by formations of tetroxane and 1,3-dioxolane through backbiting.[68,70]

Complex molibdenyl acetylacetonates also act as catalysts in trioxane polymerizations. Here, a mechanism based on formation of a coordinated intermediate is visualized[81]:

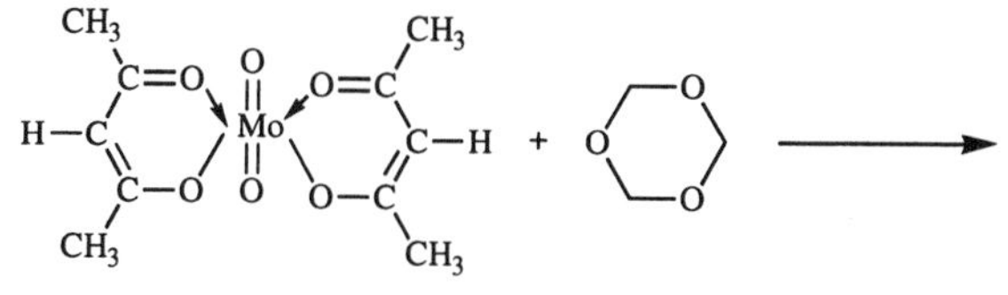

$$MoO_2(\text{acac})_2 \; [\text{bis}(CH_3{-}C{=}O \cdots H{-}C \cdots C{-}O{-}CH_3) \text{ chelate}] + \text{trioxane} \longrightarrow$$

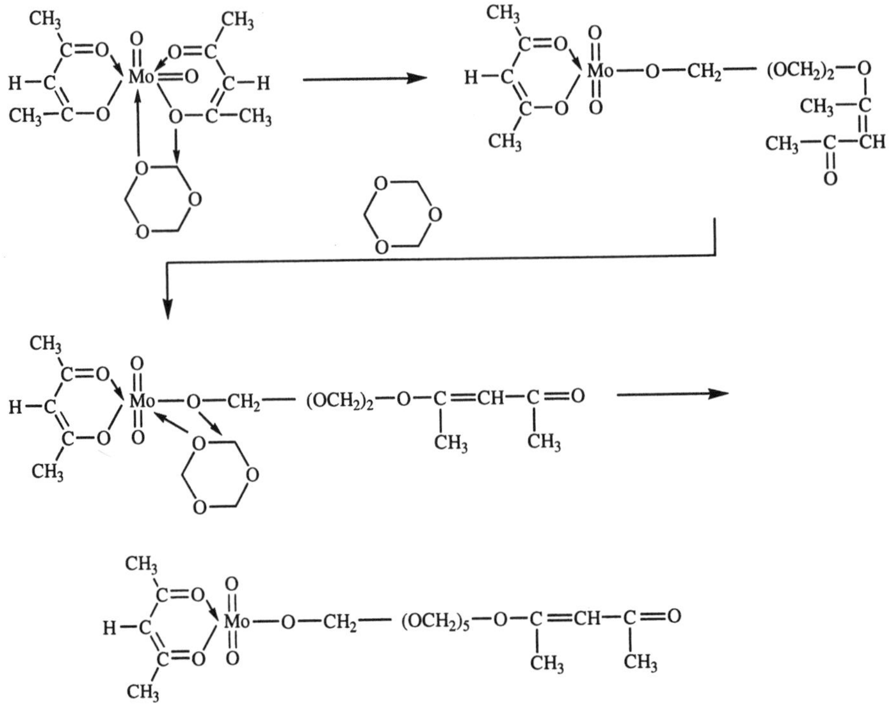

The termination mechanism and cocatalyst requirement have not yet been fully explained.

Some transfer to water takes place during the reaction. As a result, the polymer contains at least one terminal hydroxyl group.[71] Besides water, methyl alcohol and low molecular weight ethers also act as transfer agents[72]:

$$\sim\sim O\!-\!(-CH_2\!-\!O-)_n\!-\!CH_2^{\oplus} + R'\!-\!O\!-\!R \rightarrow \sim\sim O\!-\!(-CH_2\!-\!O-)_n\!-\!CH_2\!-\!OR' + R^{\oplus}$$

The new cation can initiate chain growth:

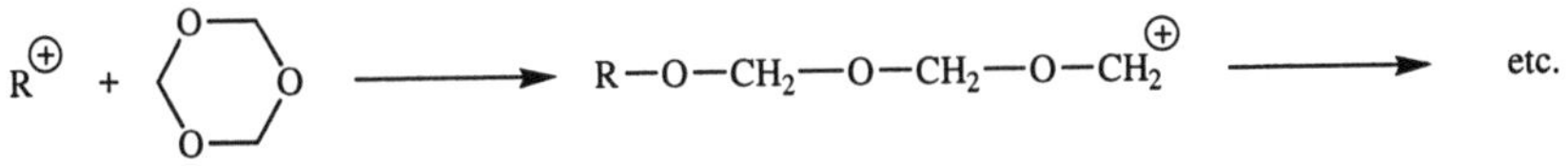

$$R^{\oplus} + \text{(trioxane)} \longrightarrow R\!-\!O\!-\!CH_2\!-\!O\!-\!CH_2\!-\!O\!-\!CH_2^{\oplus} \longrightarrow \text{etc.}$$

4.7.2. Polymerization of Dioxolane

Polymerizations of this cyclic monomer yield polymers that consists of strictly alternating oxymethylene and oxyethylene units.[73,74] The polymerization reaction can be induced by acidic catalysts, like sulfuric acid, boron trifluoride, *p*-toluenesulfonic acid, and phosphorous pentafluoride[1]:

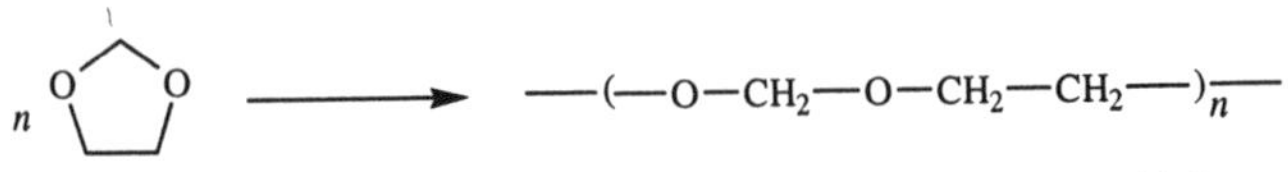

$$n\ \text{(dioxolane)} \longrightarrow -(-O\!-\!CH_2\!-\!O\!-\!CH_2\!-\!CH_2-)_n-$$

The polymers of 10,000 molecular weight are tough solids that can be cold drawn. The following mechanism was proposed for polymerizations that are initiated by catalysts prepared by combining acetic anhydride with perchloric acid[75]:

$$CH_3\!-\!\overset{O}{\overset{\|}{C}}\!-\!O\!-\!\overset{O}{\overset{\|}{C}}\!-\!CH_3 + HClO_4 \longrightarrow CH_3\!-\!\overset{O}{\overset{/\!/}{C}}{\oplus}\cdots\cdots ClO_4^{\ominus} + CH_3COOH$$

initiation:

$$CH_3\!-\!\overset{O}{\overset{/\!/}{C}}{\oplus} + \text{(dioxolane)} \longrightarrow CH_3\!-\!\overset{O}{\overset{\|}{C}}\!-\!\overset{\oplus}{O}\text{(dioxolane ring)}$$

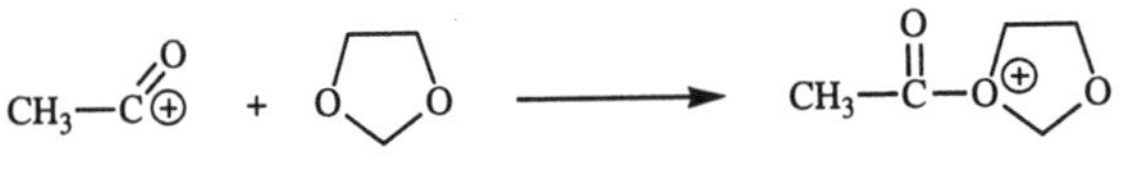

propagation:

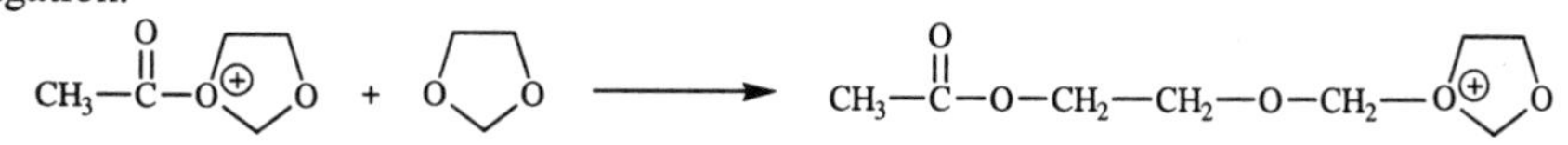

$$CH_3-\overset{\overset{O}{\|}}{C}-\overset{\oplus}{O}\langle\text{ring}\rangle O + \text{dioxolane} \longrightarrow CH_3-\overset{\overset{O}{\|}}{C}-O-CH_2-CH_2-O-CH_2-\overset{\oplus}{O}\langle\text{ring}\rangle O$$

termination:

$$CH_3-\overset{\overset{O}{\|}}{C}-(-O-CH_2-CH_2-O-CH_2-)_n-\overset{\oplus}{O}\langle\text{ring}\rangle O + CH_3-\overset{\overset{O}{\|}}{C}-O^{\ominus} \longrightarrow$$

$$CH_3-\overset{\overset{O}{\|}}{C}-(-O-CH_2-CH_2-O-CH_2-)_{n+1}-O-\overset{\overset{O}{\|}}{C}-CH_3$$

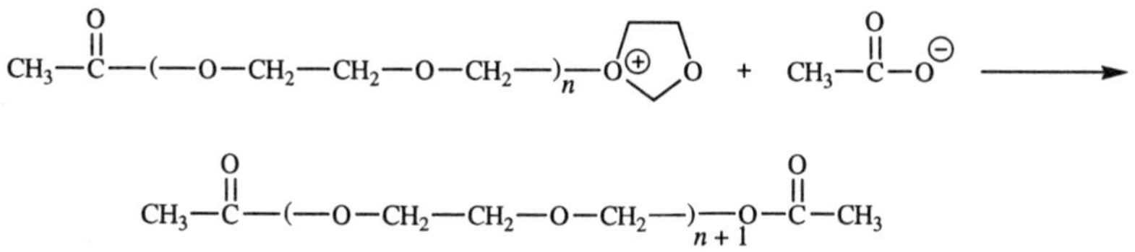

Acetate groups are present at both ends of the polymer molecules as shown above.[75] This was confirmed by analytical evidence. The initiation of dioxolane polymerization by boron trifluoride-etherate is pictured differently[84,85]:

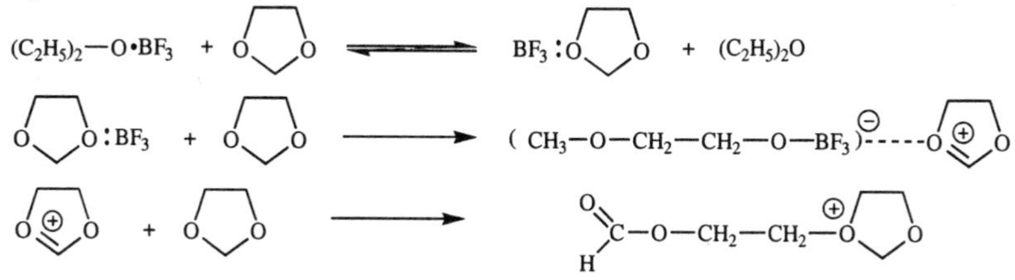

$$(C_2H_5)_2-O\cdot BF_3 + \text{dioxolane} \rightleftharpoons BF_3:\text{dioxolane} + (C_2H_5)_2O$$

$$\text{dioxolane}:BF_3 + \text{dioxolane} \longrightarrow (CH_3-O-CH_2-CH_2-O-BF_3)^{\ominus}\text{-----}\overset{\oplus}{O}\langle\text{ring}\rangle O$$

$$\text{(dioxolanium cation)} + \text{dioxolane} \longrightarrow \overset{O}{\underset{H}{>}}C-O-CH_2-CH_2-\overset{\oplus}{O}\langle\text{ring}\rangle O$$

In addition, chain cleavage can occur through BF_3 complexation with oxygen in the chain[85]:

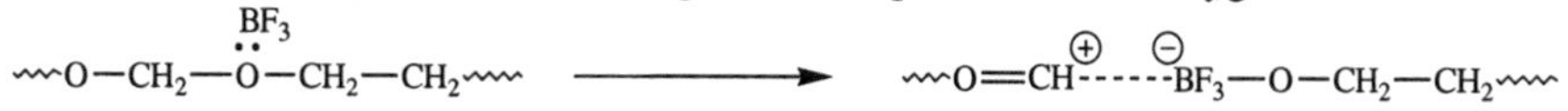

$$\sim O-CH_2-\overset{BF_3}{\ddot{O}}-CH_2-CH_2\sim \longrightarrow \sim O=\overset{\oplus}{C}H\text{------}\overset{\ominus}{B}F_3-O-CH_2-CH_2\sim$$

There is some disagreement about the nature of the end groups and there is some speculation that the polymers might possess large cyclic structures. Nevertheless, polymerizations initiated with benzoylium hexafluoroantimonate ($C_6H_5CO^{\oplus}\ SbF_6^{\ominus}$) and conducted at –15 °C in nitromethane or methylene chloride result mostly in linear polymers. The terminal end-groups come from terminating agents that are added deliberately[67]:

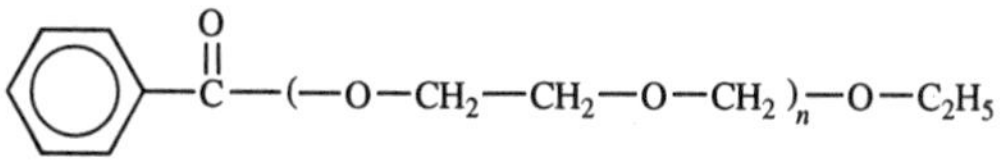

$$C_6H_5-\overset{\overset{O}{\|}}{C}-(-O-CH_2-CH_2-O-CH_2)_n-O-C_2H_5$$

Also, the polymerizations proceed without any appreciable amounts of transfer reactions affecting the DP.[67]

4.7.3. Polymerization of Dioxepane

Polymerization of six-membered cyclic formals has apparently not been reported.[1] Polymerization of 1,3-dioxepane can be initiated by camphor sulfonic acid[76,77]:

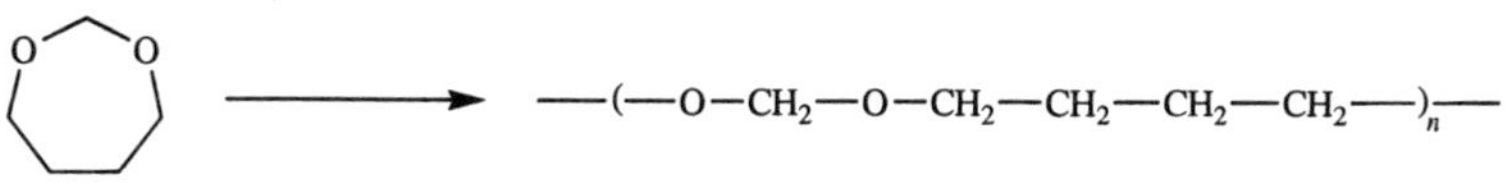

$$\text{1,3-dioxepane} \longrightarrow -(-O-CH_2-O-CH_2-CH_2-CH_2-CH_2-)_n-$$

1,3,5-trioxepane, a product of condensation of trioxane with ethylene oxide, can be polymerized by cationic mechanism both in solution and in bulk[78]:

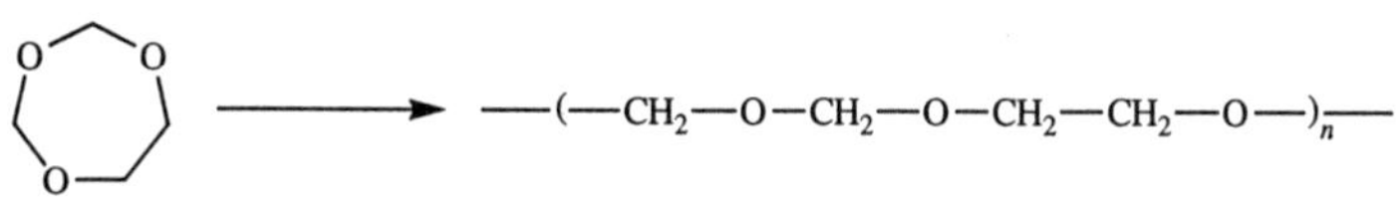

$$\text{1,3,5-trioxepane} \longrightarrow -(-CH_2-O-CH_2-O-CH_2-CH_2-O-)_n-$$

Polymerizations carried out with boron trifluoride catalyst in dichloroethane solvent result in several reactions that occur simultaneously. A polymer and a copolymer with a different cyclic monomer form side by side[78]:

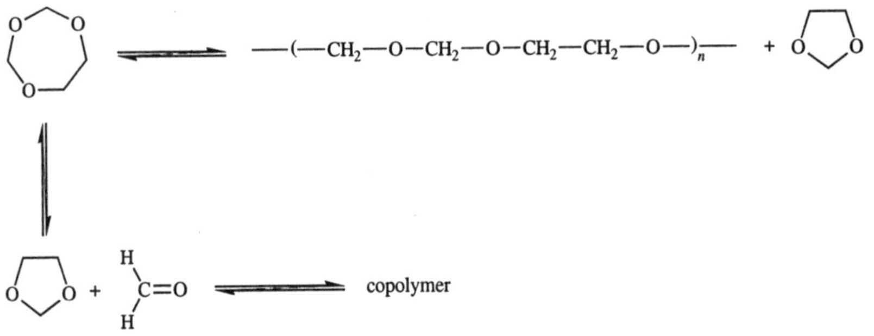

$$\text{(trioxocane)} \rightleftharpoons -(-CH_2-O-CH_2-O-CH_2-CH_2-O-)_n- + \text{(dioxolane)}$$

$$\text{(dioxolane)} + \begin{matrix}H\\ \backslash\\ C{=}O\\ /\\ H\end{matrix} \rightleftharpoons \text{copolymer}$$

At low temperatures, the amount of dioxolane that forms in the above reaction decreases considerably and can become zero.

Ring-opening polymerizations of trioxocane leads to the following polymer[78,79]:

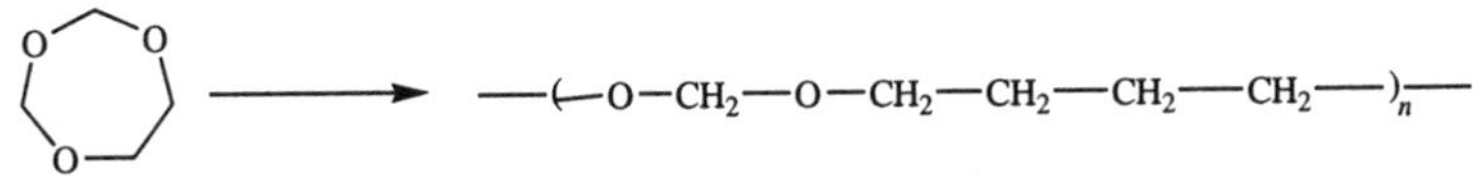

$$\text{(trioxocane)} \longrightarrow -(-O-CH_2-O-CH_2-CH_2-CH_2-CH_2-)_n-$$

So far, the nature of the end groups has not been established. Nor has it been shown that a macrocyclic structure does not form.

1,3,6,9-tetraoxacycloundecane (triethylene glycol formal) can be polymerized by several cationic initiators in solution or in bulk at varying temperatures from –20 °C to +150 °C[78]:

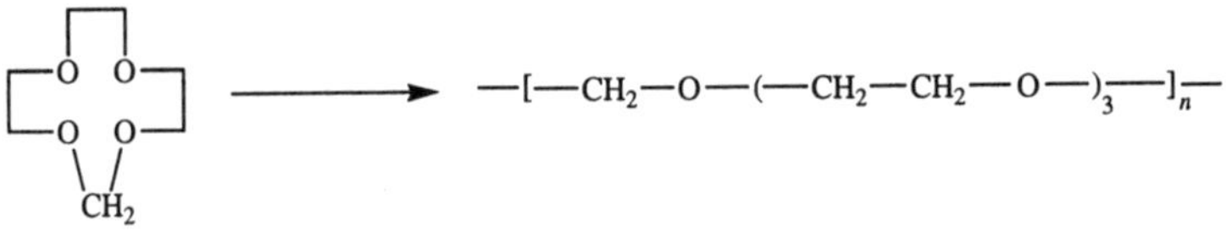

$$\text{(tetraoxacycloundecane)} \longrightarrow -[-CH_2-O-(-CH_2-CH_2-O-)_3-]_n-$$

4.8. Polymerization of Lactones

Lactones polymerize by three different mechanisms, namely, cationic, anionic, and coordinated mechanisms. Often, the mechanism by which a specific lactone polymerizes depends upon the size of the ring.

4.8.1. Cationic Polymerization of Lactones

Various reaction schemes were proposed to explain the cationic mechanism. They tend to resemble the schemes suggested for polymerizations of cyclic ethers.[87,88] The initiation step involves an equilibrium that is followed by a ring-opening reaction:

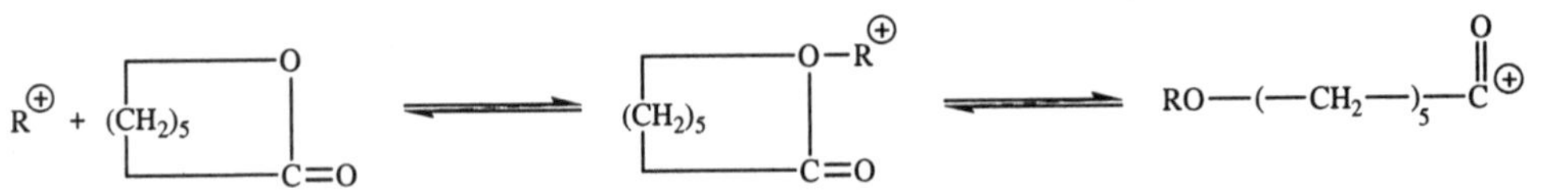

$$R^{\oplus} + (CH_2)_5\!\begin{bmatrix}O\\ C{=}O\end{bmatrix} \rightleftharpoons (CH_2)_5\!\begin{bmatrix}O^{\oplus}-R\\ C{=}O\end{bmatrix} \rightleftharpoons RO-(-CH_2-)_5-\overset{O}{\overset{\|}{C}}{}^{\oplus}$$

The propagation consists of many repetitions of the above step:

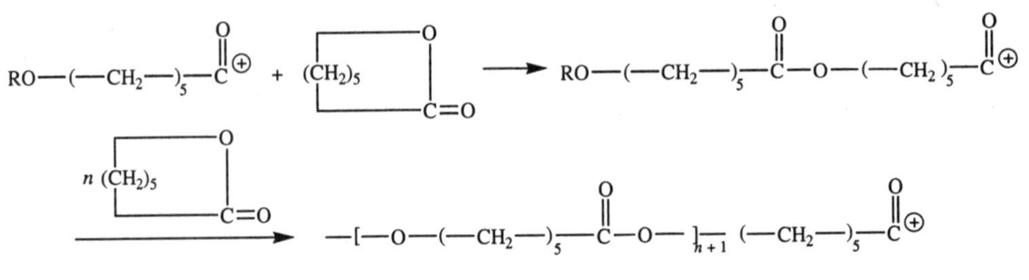

$$RO-(-CH_2-)_5-\overset{O}{\overset{\|}{C}}{}^{\oplus} + (CH_2)_5\!\begin{bmatrix}O\\ C{=}O\end{bmatrix} \longrightarrow RO-(-CH_2-)_5-\overset{O}{\overset{\|}{C}}-O-(-CH_2-)_5-\overset{O}{\overset{\|}{C}}{}^{\oplus}$$

$$\xrightarrow{n\,(CH_2)_5\begin{bmatrix}O\\ C{=}O\end{bmatrix}} -[-O-(-CH_2-)_5-\overset{O}{\overset{\|}{C}}-O-]_{n+1}(-CH_2-)_5-\overset{O}{\overset{\|}{C}}{}^{\oplus}$$

The polymerization of propiolactone in methylene chloride with an antimony pentachloride-dietherate catalyst was investigated.[86] The results show that the concentration of the active centers is dependent upon catalyst concentration and upon the initial concentration of the monomer. They also support the concept that opening of the lactone rings includes initial formation of oxonium ions[86]:

$$\sim\!\!O-(-CH_2-)_2-\overset{O}{\overset{\|}{C}}{}^{\oplus} + \underset{\square}{O-C{=}O} \xrightarrow{\text{slow}} \sim\!\!O-(-CH_2-)_2-\overset{O}{\overset{\|}{C}}-\underset{\oplus}{O}-C{=}O$$

$$\xrightarrow{\text{fast}} \sim\!\!O-(-CH_2-)_2-\overset{O}{\overset{\|}{C}}-O-(-CH_2-)_2-\overset{O}{\overset{\|}{C}}{}^{\oplus}$$

Because the carbonyl oxygen is the most basic of the oxygens in the lactone molecule, a reverse reaction is possible[86]:

$$\sim\!\!\overset{O}{\overset{\|}{C}}{}^{\oplus} + O{=}C-O \rightleftharpoons \sim\!\!\overset{O}{\overset{\|}{C}}-\overset{\oplus}{O}{=}C-O$$

Conductivity measurements during polymerizations of β-propiolactone with antimony pentafluoride-etherate or p-toluenesulfonic acid show[192] that ion triplets form during the reaction. These are:

$$\sim\!\!\overset{O}{\overset{\|}{C}}{}^{\oplus},\ A^{\ominus},\ {}^{\oplus}\overset{O}{\overset{\|}{C}}\!\!\sim$$

The triplets appear to be active centers throughout the course of the polymerizations. In addition, most of the growing chain-ends exist as ion pairs, depending upon the concentration of the monomer.[192]

4.8.2. Anionic Polymerization of Lactones

In anionic polymerizations the initiations result from attacks by the bases upon the carbonyl groups[89]:

$$\overset{\ominus}{R:} + \underbrace{(CH_2)_5\,O\,C{=}O}_{\text{ring}} \longrightarrow R-\overset{O}{\overset{\|}{C}}-(-CH_2-)_5-O^{\ominus}$$

The propagations take place by a similar process:

$$R-\overset{O}{\overset{\|}{C}}-(-CH_2-)_5-O^{\ominus} + \underbrace{(CH_2)_5\,O\,C{=}O}_{\text{ring}} \longrightarrow R-\overset{O}{\overset{\|}{C}}-(-CH_2-)_5-O-\overset{O}{\overset{\|}{C}}-(-CH_2-)_5-O^{\ominus}$$

These steps repeat themselves until the chains are built up. Anionic polymerizations can yield optically active polymers. This was observed in formations of poly(α-methyl,α-ethyl-β-propiolactone)[193] that contains asymmetric carbon atoms.

4.8.3. Polymerization of Lactones by a Coordination Mechanism

The mechanism of coordination polymerization was pictured by Young, Matzner, and Pilato[90] as being an intermediate between the above two modes of polymerization (a cationic and anionic one):

Initiation

$$R-M_T + \underbrace{(CH_2)_5\,O\,C{=}O}_{\text{ring}} \rightleftharpoons \left[\underbrace{(CH_2)_5\,O\cdots M_T\,;\,C{=}O\cdots R}_{\text{ring}}\right] \rightleftharpoons R-\overset{O}{\overset{\|}{C}}-(-CH_2)_5-OM_T$$

where M_T means metal.

$$R-\overset{O}{\overset{\|}{C}}-(-CH_2-)_5-OM_T + (CH_2)_5\langle O, C{=}O\rangle \rightleftharpoons [R-\overset{O}{\overset{\|}{C}}-(-CH_2-)_5-O\cdots M_T \cdots O-C{=}O(CH_2)_5]$$

The above mechanism, however, is incorrect when caprolactone is polymerized with tin compounds.[91] Yet, it appears to be correct for polymerizations of propiolactones with an ethylzinc monoxide catalyst.[95]

The bimetallic oxoalkoxides are useful catalysts for the polymerizations of ε-caprolactone. The general course of the reaction is quite similar to one for oxiranes. A typical anionic–coordinated mechanism is indicated from kinetic and structural data.[163] The molecular weight increases with conversion and the reaction exhibits a "living" character, because there is a linear relationship between DP and conversion. When the monomer is all used up, addition of fresh monomer to the reaction mixture results in increases in DP. By avoiding side reactions it is possible to achieve high molecular weights (up to 200,000) with narrow molecular weight distribution ($M_w/M_n \geq 1.05$).[163] The reaction proceeds through insertion of the lactone units in the Al–OR bonds. The acyl–oxygen bond cleaves and the chain binds through the oxygen to the catalyst by forming an alkoxide link rather than a carboxylate one:

$$>Al-OR + n\ (CH_2)_5\langle O, C{=}O\rangle \longrightarrow\ >Al-[-O-(CH_2)_5-\overset{O}{\overset{\|}{C}}-]_n-OR$$

There are potentially four active sites per trinuclear catalytic molecule. The number of actual sites, however, depends upon the aggregation of the oxoalkoxides. Two different types of OR groups exist, depending upon the bridging in the aggregates. Only one is active in the polymerization. This results in a catalytic star-shaped entity. The fact that the dissociated catalysts generate four growing chains per each $Al_2(CH_2)_5CO_2(OR)_4$ molecule[163] tends to confirm this.

4.8.4. Special Catalysts for Polymerizations of Lactones

Some lactones can polymerize in the presence of compounds like alcohols, amines, and carboxylic acids without additional catalysts. The reactions, however, are slow and yield only low molecular weight polymers.[91] Exceptions are polymerizations of pivalolactone in the presence of cyclic amines that yield high molecular weight polyesters at high conversion.[92] The initiating steps result from formations of adducts, amine-pivalate betaines:

$$\text{N(bicyclic amine)} + \text{pivalolactone} \xrightarrow{\text{slow}} \overset{\oplus}{N}-CH_2-\underset{CH_3}{\overset{CH_3}{C}}-COO^{\ominus}$$

$$\overset{\oplus}{N}-CH_2-\underset{CH_3}{\overset{CH_3}{C}}-COO^{\ominus} + \text{pivalolactone} \xrightarrow{\text{fast}}$$

$$\overset{\oplus}{N}-[-CH_2-\underset{CH_3}{\overset{CH_3}{C}}-\overset{O}{\overset{/\!/}{C}}-O-]-CH_2-\underset{CH_3}{\overset{CH_3}{C}}-COO^{\ominus} \xrightarrow{\text{slow}}$$

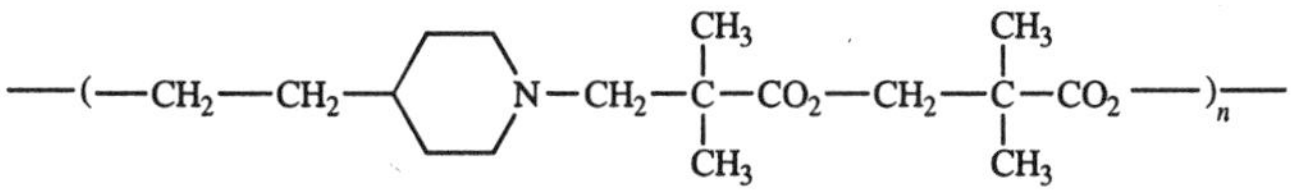

$$—(—CH_2—CH_2—\bigcirc N—CH_2—\underset{CH_3}{\overset{CH_3}{C}}—CO_2—CH_2—\underset{CH_3}{\overset{CH_3}{C}}—CO_2—)_n—$$

The above reaction appears to be restricted to highly strained lactones and may not work with larger lactones.[91] For instance, when polymerization of δ-valerolactone is initiated with ethanolamine at temperatures up to 200 °C, there is initially a rapid reaction between the amine group and the monomer:

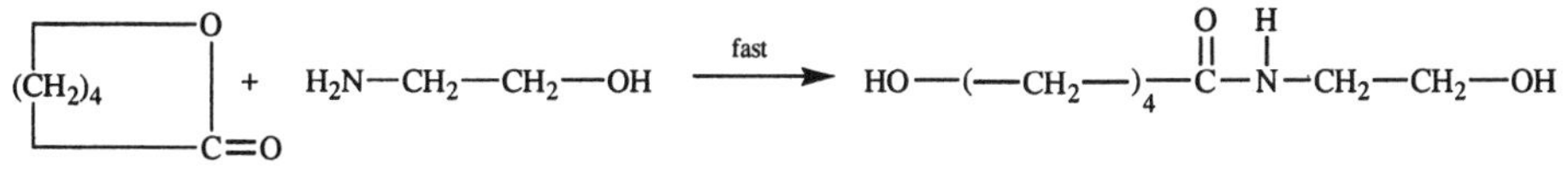

$$\underset{\text{ring: }(CH_2)_4—O—C{=}O}{} + H_2N—CH_2—CH_2—OH \xrightarrow{\text{fast}} HO—(—CH_2—)_4—\overset{O}{\overset{\|}{C}}—\overset{H}{\overset{|}{N}}—CH_2—CH_2—OH$$

The subsequent reactions, however, are slow:

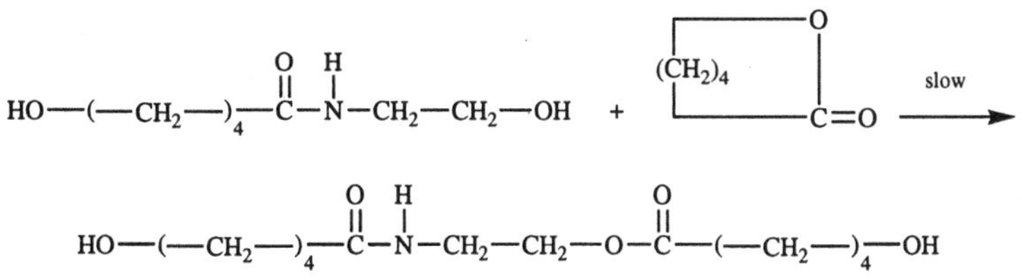

$$HO—(—CH_2—)_4—\overset{O}{\overset{\|}{C}}—\overset{H}{\overset{|}{N}}—CH_2—CH_2—OH + \underset{\text{ring: }(CH_2)_4—O—C{=}O}{} \xrightarrow{\text{slow}}$$

$$HO—(—CH_2—)_4—\overset{O}{\overset{\|}{C}}—\overset{H}{\overset{|}{N}}—CH_2—CH_2—O—\overset{O}{\overset{\|}{C}}—(—CH_2—)_4—OH$$

It was suggested that initiators, like dibutylzinc, that lack active hydrogens should be placed into a special category.[91] They can initiate polymerizations of some lactones. One of them is ε-caprolactone. Polymers form that are inversely proportional in molecular weight to the catalyst concentrations.[93] The same is true of stannic tetraacylate. High molecular weight poly(ε-caprolactone), as high as 100,000, forms. Addition of compounds that may serve as a source of active hydrogens is not necessary.[91] This group of initiators also includes dimethylcadmium, methylmagnesium bromide, and a few others that are effective in polymerizations of δ-valerolactone, ε-caprolactone, and their alkyl-substituted derivatives. The polymers that form are high in molecular weight, some as high as 250,000.[94]

Another group consists of zinc and lead salts, stannous esters, phosphines, and alkyl titanates. This group does require additions of compounds with active hydrogens. Such additives can be polyols, polyamines, or carboxylic acid compounds.[91] Molecular-weight control is difficult with the catalysts belonging to the first group. This second group, on the other hand, not only allows control over the molecular weights, but also over the nature of the end groups.[91]

4.9. Polymerization of Lactams

Polymerizations of lactams produce important commercial polymers. The polymerization reactions therefore received considerable attention. Lactam molecules polymerize by three different mechanisms: cationic, anionic, and a hydrolytic one (by water or water-releasing substances).

The lactam ring is strongly resonance-stabilized and the carbonyl activity is low. Nevertheless, the ring-opening polymerizations start with small amounts of initiators through *trans*-acylation reactions. Fairly high temperatures, however, are needed, often above 200 °C. In all such reactions, one molecule acts as the acylating agent or as an electrophile, while the other one acts as a nucleophile and undergoes the acylation.

Generally, the initiators activate the inactive amide groups, causing them to react with other lactams through successive transamidations that result in formations of polyamides. Both acids and bases catalyze the transamidation reactions. The additions of electrophiles affect increases in the electrophilicity of the carbonyl carbon of the acylating lactam. The nucleophiles, on the other hand, increase the nucleophilic character of the lactam substrate (if they are bases).

All initiators can be divided into two groups. To the first belong strong bases capable of forming lactam anions by removing the amide proton. This starts the anionic polymerization reaction. To the second group belong active hydrogen compounds capable of protonating the amide bond and thereby affecting cationic polymerization.[96]

Side reactions are common in lactam polymerizations. Their nature and extent depend upon the concentration and character of the initiators, the temperatures of the reactions, and the structures of the lactams. When cationic polymerizations of lactams are initiated by strong acids, strongly basic amidine groups can be produced. These groups bind the strong acids, inactivate the growth centers, and decrease the rate of polymerization. Use of strong bases to initiate polymerizations of lactams possessing at least one α-hydrogen also result in side reactions. Compounds form that decrease the basicity of lactams and polyamides and slow the polymerizations. Also, side reactions give rise to irregular structures, namely, branching.

The ring-opening polymerization reactions depend upon thermodynamic and kinetic factors, and on the total molecular strain energies of the particular ring structures. Six-membered δ-valerolactam is the most stable ring structure and most difficult to polymerize. Also, the presence of substituents increases the stability of the rings and decreases the ability to polymerize.

4.9.1. Cationic Polymerization of Lactams

The catalysts for cationic polymerization can be strong anhydrous acids, Lewis acids,[98] salts of primary and secondary amines, carboxylic acids, and salts of amines with carboxylic acids that split off water at elevated temperatures.[96] The initiators react by coordinating with and forming rapid pre-equilibrium lactam cations. These cations are the reactive species in the polymerizations. Initiations of this type are also possible with weakly acidic compound, but such compounds are not able to transfer protons to the lactam. They are capable, however, of forming hydrogen bonds with the lactams. The high reactivity of the lactam cations may be attributed to the decreased electron density at the carbonyl carbon atoms. This makes them more subject to nucleophilic attacks.[96]

Protonations of the amides occur at the oxygens,[97] but small fractions of N-protonated amides are also presumed to exist in tautomeric equilibrium. To simplify the illustrations, all lactams will be shown in this section as

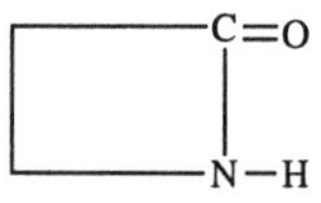

C=O
N–H

So, while the above structure commonly represents propiolactam, in this section it can mean any lactam, like a caprolactam, valerolactam, etc. Thus, the reaction can be shown as follows:

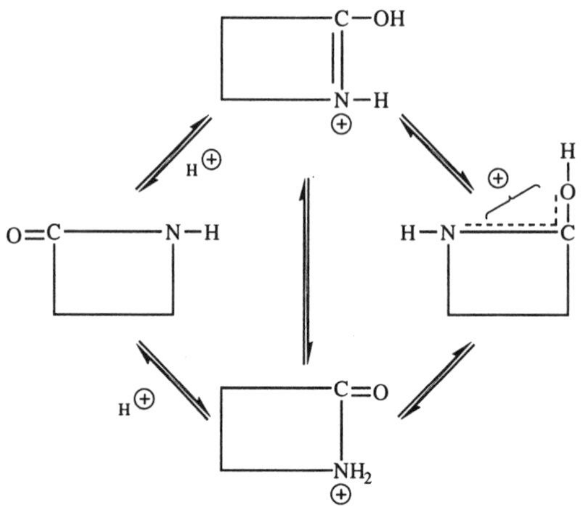

C–OH
N–H ⊕
H⊕
H
O
O=C N–H
⊕
H–N C
H⊕
C=O
NH_2 ⊕

In a reaction mixture where the initiators are strong acids, the strongest nucleophiles are the monomers. Acylations of the monomers with the amidinium cations result in formations of aminoacyllactams[98]:

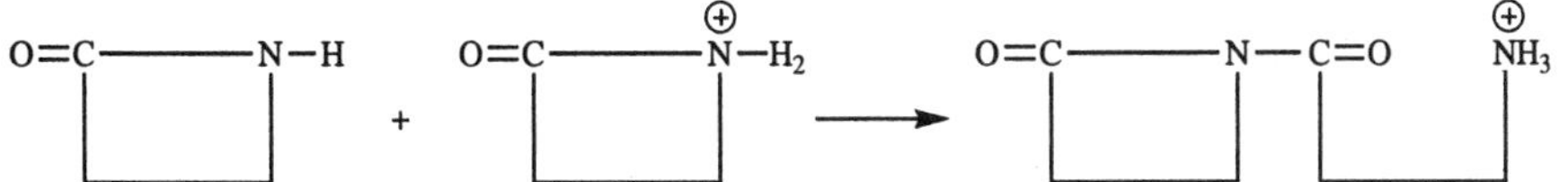

Acylation of these amine groups by molecules of other protonated lactams results in the monomers becoming incorporated into the polymers.[99] The growth centers are preserved and a molecule of lactam is protonated. This occurs in two steps[99]:

1. O=C——N—C=O $\overset{\oplus}{NH_3}$ + O=C——N—H ⇌

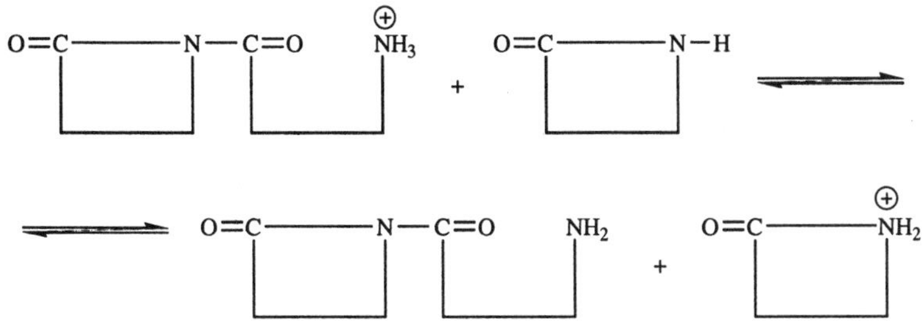

⇌ O=C——N—C=O NH_2 + O=C——$\overset{\oplus}{NH_2}$

2. ∿∿NH_2 + O=C——$\overset{\oplus}{NH_2}$ ⇌ ∿∿N(H)—C=O $\overset{\oplus}{NH_3}$

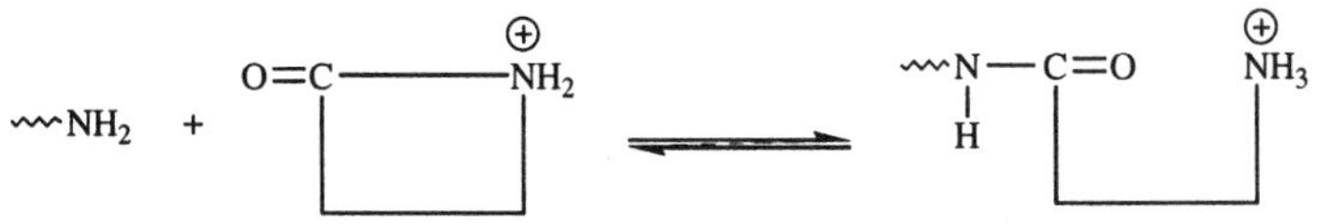

These reactions attain equilibrium quickly. Aminolyses of acyllactams, that are the reverse of the initiation reactions, proceed rapidly.[99–101] Aminolyses of aminoacyllactams actually contribute to the propagation process[102,103]:

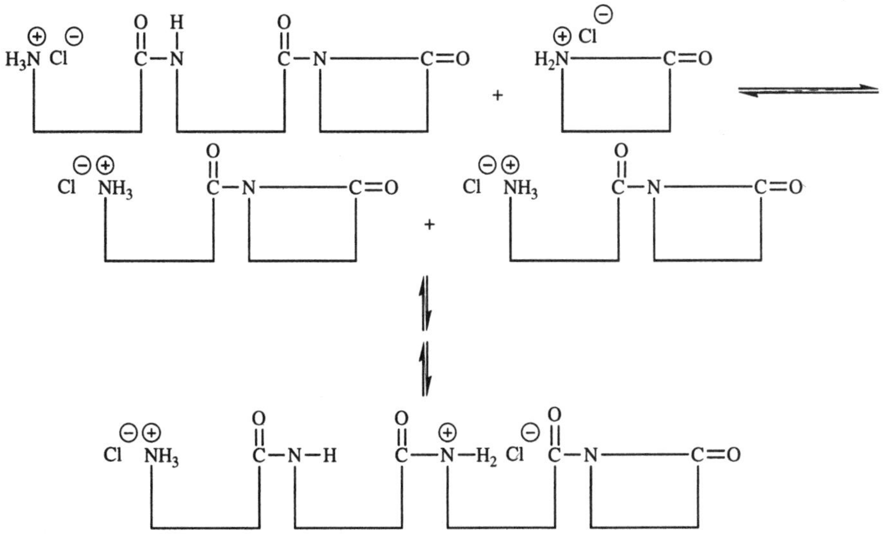

$\overset{\oplus}{H_3N}\ \overset{\ominus}{Cl}$ C(=O)—N(H) C(=O)—N——C=O + $\overset{\oplus}{H_2N}$——C=O $\overset{\ominus}{Cl}$ ⇌

$\overset{\ominus}{Cl}\ \overset{\oplus}{NH_3}$ C(=O)—N——C=O + $\overset{\ominus}{Cl}\ \overset{\oplus}{NH_3}$ C(=O)—N——C=O

⇅

$\overset{\ominus}{Cl}\ \overset{\oplus}{NH_3}$ C(=O)—N—H C(=O)—$\overset{\oplus}{N}$—H_2 $\overset{\ominus}{Cl}$ C(=O)—N——C=O

The above reaction results in the destruction of the equilibrium and a regeneration of the strongly acidic amide salt. Total lactam consumption results from repetitions of the above sequences and formations of new aminoacyllactam molecules.[98–103] Initiations of polymerizations with acid salts of primary and secondary amines result in chain growths that proceed predominantly through additions of protonated lactams to the amines[98]:

∿∿NH_2 + C(=O)——$\overset{\oplus}{NH_2}$ ⇌ ∿∿N(H)—C(=O) $\overset{\oplus}{NH_3}$

The rate at which the initiating amines are incorporated is proportional to the basicity. As the conversion progresses, the concentration of protonated lactams in the reaction mixture decreases while that of the protonated polymer amide groups increases. The latter takes part in the initiation reactions with lactam molecules and in exchange reactions with polymer molecules[98]:

In each of the initiation steps the strongest nucleophile present reacts with the lactam cation. When strong anhydrous Bronsted acids initiate the polymerizations, the free lactams are acylated first with the formation of aminoacyllactams. When the polymerizations are initiated by amine salts, the initial steps are conversions of the amines to the amino acid amides. On the other hand, hydrolytic polymerizations start formations of unsubstituted amino acids[104]:

When weak carboxylic acids or acids of medium strength initiate lactam polymerizations at anhydrous conditions, there is an induction period.[105] In addition, the rates of these reactions are proportional to the pK_a of the acids.[105] It appears that different reaction mechanisms are involved, depending upon the acid strengths.[98] The nucleophiles are present in equilibrium:

The acylation of the carboxylate anions is assumed to lead to formations of mixed anhydrides of the acids with amino acids[106] and subsequent rearrangements:

When strong acids, however, initiate the polymerizations, the strongest nucleophiles present are the lactam amide groups that undergo acylations. As a result, the acids are not incorporated into the polymers.

The propagation steps in cationic polymerizations of lactams occur by transamidation reactions between lactam rings and the ammonium groups formed during the steps of initiation. It is believed that during the reaction proton transfers take place first from the amine salts to the lactams or to the acyllactams to form cations. These in turn acylate the free amines that form with the regeneration of ammonium groups:

$$\sim NH_3^{\oplus} + \sim C(=O){-}N{-}C{=}O \rightleftharpoons \sim NH_2 + \sim C(=O){-}N^{\oplus}H{-}C{=}O$$

$$\sim NH_3^{\oplus} + \sim C(=O){-}N^{\oplus}H{-}C{=}O \rightleftharpoons \sim N(H){-}C(=O) \ \ {}^{\oplus}NH_2{-}C(=O)\sim$$

$$\rightleftharpoons \sim NH{-}C(=O)\sim + H_2{-}N^{\oplus}{-}C{=}O$$

$$\sim NH_3^{\oplus} + H_2{-}N^{\oplus}{-}C{=}O \rightleftharpoons \sim N(H){-}C(=O) \ \ NH_3^{\oplus}$$

The propagation step is very rapid when aminolysis takes place at the carbonyl groups of the activated acid derivative (like acyllactam or an acid chloride). It is slower, however, if it involves an amide group of the monomer.[96] As is typical of many carbonyl reactions, acylations are followed by eliminations[108]:

$$\sim NH_2 + O{=}C{-}NH_2^{\oplus} \rightleftharpoons \left[\sim NH{-}C(OH){-}NH_2^{\oplus}\right] \underset{+H_2O}{\overset{-H_2O}{\rightleftharpoons}}$$

$$\sim N(H){-}C{=}N^{\oplus}{-}H \longleftrightarrow \sim N^{\oplus}(H){=}C{-}N{-}H$$

The above water-elimination reaction results in formations of amidines. Acylamidinium ions can also result from dehydration of the tetrahedral intermediates during the reactions of amino groups with acyllactams. Such groups could also be present within the polymer molecules. The water that is released in these reactions hydrolyzes the acyllactams, acylamidine salts, and lactam salts to yield carboxylic acids.[96]

In the cationic polymerization of lactams the ammonium and amidinium groups form N-terminal chain ends. The C-terminal chain ends are in the form of carboxylic acid groups or alkylamide residues. This is important, because the nature of the end groups and their reactivity determine the following step in the polymerizations. This means that the different types of cationic polymerizations of lactams are the results of the different end groups that form during the initiation steps. Formation of amidines increases with increasing acidity and concentration of the initiator and with an increase in the temperature:

$$R{-}N^{\oplus}H{-}C{=}O + Cl^{\ominus} \rightleftharpoons R{-}NH{-}(CH_2)_x{-}C(Cl){=}O$$

When strong acids or amine salts initiate the polymerizations, almost all amine salt groups become converted to amidine salts shortly after the start of the initiation reaction.[108] Formation of amidinium salts leads to a decrease in the reaction rate because they initiate polymerizations of lactams less effectively than do ammonium salts.[108,109] Lewis acids act in a similar manner, unless a co-reactant is present, like water. In that case, the Lewis acids are transformed into protonic acids and the polymerizations proceed as if they were initiated by protonic acids.[96]

N-substituted lactams can generally not be polymerized. Some exceptions, however, are known when cationic mechanisms are employed[104] and when strong carboxylic or inorganic acids are used as initiators. In such cases the anions of the initiating acids, like $Cl^{\ominus}$, react with the lactam cations to yield amino acid chlorides[96]:

$$R{-}\overset{H}{\underset{|}{N^{\oplus}}}\frown C{=}O + Cl^{\ominus} \rightleftharpoons R{-}N(H^{\oplus}\cdots Cl^{\ominus})\frown C{=}O \rightleftharpoons R{-}N(\cdot HCl)\frown C(=O){-}Cl$$

Only the more strained four-, eight-, and nine-membered N-substituted lactams have so far been shown to be capable of polymerizations.[98] The 2,2-dimethylquinuclidone is highly strained and undergoes polymerizations at room temperature.[107] The propagation reaction of substituted lactams can be illustrated as follows[104]:

$$H{-}N{-}R\frown C(Cl){=}O + H{-}N{-}R\frown C(Cl){=}O \rightleftharpoons H{-}N{-}R\frown C(=O){-}N{-}R\frown C(Cl){=}O + HCl$$

$$HCl\cdot \overset{H}{N^{\oplus}}{-}R\frown C(Cl){=}O + HCl\cdot \overset{H}{N^{\oplus}}{-}R\frown C(Cl){=}O \rightleftharpoons HCl\cdot HN{-}R\frown C(=O){-}N{-}R\frown C(Cl){=}O + 2\ HCl$$

4.9.2. Anionic Polymerization of Lactams

The anionic polymerizations of lactams are initiated by strong bases[141] capable of forming lactam anions:

$$O{=}C\frown NH + B{:}^{\ominus} \rightleftharpoons O{=}C\frown N{:}^{\ominus} + BH$$

Such bases can be alkali metals, metal hydrides, organometallic compounds, and metal amides. The initiation step of ring-opening amidation can be shown as follows:

$$2\ O{=}C\frown NH + 2\ Na \longrightarrow 2\ O{=}C\frown N{:}^{\ominus} + 2\ Na^{\oplus} + H_2$$

$$O{=}C\frown N{:}^{\ominus} + O{=}C\frown NH \rightleftharpoons O{=}C\frown N{-}C{=}O\frown NH^{\ominus}$$

The primary amine anions abstract protons very rapidly from other molecules of lactams to form amino-acyllactams[110]:

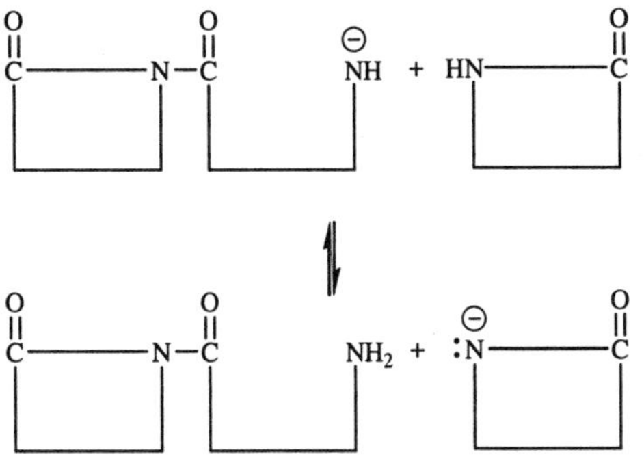

In these reactions, the propagation centers are the cyclic amide linkages of the N-acylated terminal lactam rings. Acylation of the amide nitrogens have the effect of increasing the electron deficiencies of these groups. This in turn increases the reactivities of the cyclic amide carbonyls toward attacks by the nucleophilic lactam anions[98]:

Very rapid proton exchange follows. This results in equilibrium between the lactam and the polymeric amide anions.[111]

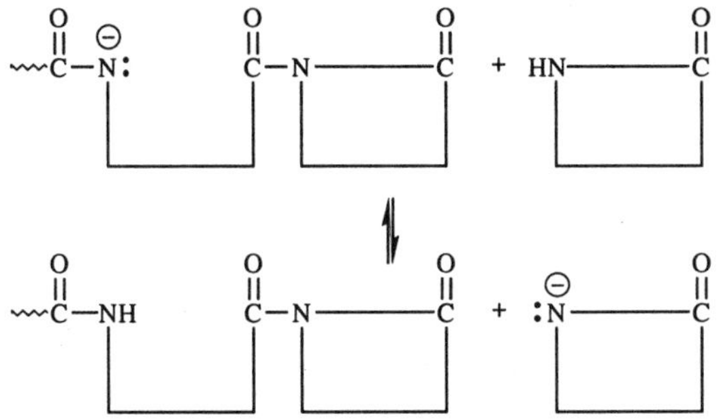

The polymer amide anions can undergo acylation by acyllactam groups with accompanying ring opening or with formation of lactam anions. In the first instance, it is an alternate path of propagation with formation of imide groups:

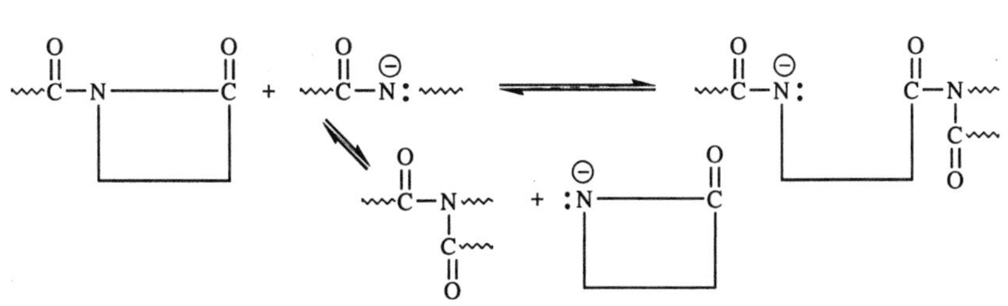

$$\sim C(=O)-N(H)\sim + \sim C(=O)-\overset{\ominus}{N}:\sim \rightleftharpoons \sim C(=O)-N(-C(=O)\sim)\sim + H-\overset{\ominus}{N}:\sim$$

$$H-N\text{—(ring)—}C(=O) + \sim C(=O)-\overset{\ominus}{N}:\sim \rightleftharpoons H-\overset{\ominus}{N}:\text{—(chain)—}C(=O)-N(-C(=O)\sim)\sim$$

The acylation reactions shown above are much faster than the initiation reactions.[111,112] As a result, there are induction periods in anionic polymerizations of lactams.[98] In addition, steep increases in molecular weights take place at the beginning of the polymerizations. Bimolecular aminolyses may contribute to that, though their contributions to the total conversions are negligible.[98]

The overall rates of polymerizations depend on the concentrations of acyllactams and diacylamine groups as well as on the lactam anions. The latter result form dissociations of the lactam salts, depending upon the nature of the metal:

$$\overset{\oplus}{Me}\ \overset{\ominus}{:N}\text{—(ring)—}C{=}O \longleftrightarrow \overset{\oplus}{Me} + \overset{\ominus}{:N}\text{—(ring)—}C{=}O$$

where Me means metal.

The alkali metals can be rated in the following order with respect to rates of initiations and propagations[98,113]:

$$Li < Na < K < Ca$$

Additions of *activators* or cocatalysts, such as acyl halides, anhydrides, or isocyanates, can result in elimination of the induction period. These additives insure formations of stabilized adducts:

$$O{=}C\text{—(ring)—}\overset{\ominus}{N}: + R-C(=O)Cl \longrightarrow O{=}C\text{—(ring)—}N-C(=O)-R + Cl^{\ominus}$$

$$O{=}C\text{—(ring)—}N-C(=O)-R + \overset{\ominus}{:N}\text{—(ring)—}C{=}O \longrightarrow O{=}C\text{—(ring)—}N-C(=O)\text{—(chain)—}\overset{\ominus}{:N}-C(=O)-R$$

$$O{=}C\text{—(ring)—}N-H + R-N{=}C{=}O \longrightarrow O{=}C\text{—(ring)—}N-C(=O)-N(H)-R$$

The structures of the activators can determine the rates of addition to the first lactam anion.[98] If, for instance, the acyl group is large, as in pivaloylcaprolactam, the decrease in the rate can be merely due to steric hindrance.[98,111] On the other hand, substituents like the benzoyl group increase the rates of additions to the first lactam anions.[98,111] In addition, the structures of the activators can also affect the course of the polymerization. This is because they become incorporated at the end of the polymeric molecules and may influence the basicity during the polymerization reactions.

Polymerizations in the presence of acylating agents are often called *activated* polymerization. If the acylating agents are absent from the reaction mixture, the reactions may be called *nonactivated*. Sometimes, the terms *assisted* and *nonassisted* are used instead.

Several reaction mechanisms were offered to explain the mechanism of anionic ring-opening polymerizations of lactams. One mechanism is based on nucleophilic attacks by the lactam anions at the cyclic carbonyl groups of N-acylated lactams. This leads to formations of intermediate symmetrical mesomeric anions that rearrange with openings of the rings[114,115]:

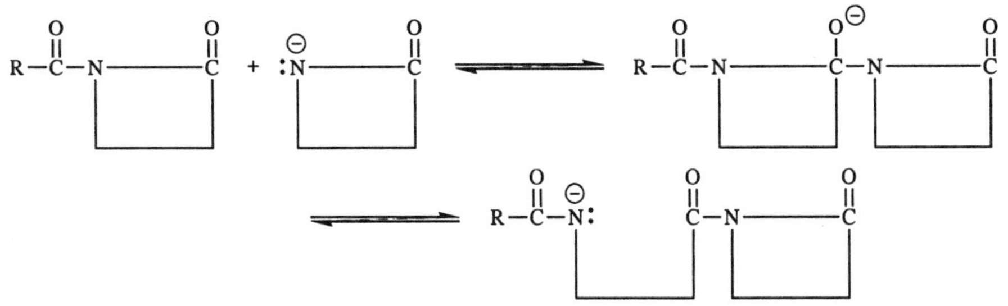

Champetier and Sekiguchi concluded that the intermediate anions are neutralized first by protons from the lactams or from the polymer amide groups. The neutral molecules subsequently rearrange with the openings of the penultimate units[116,117]:

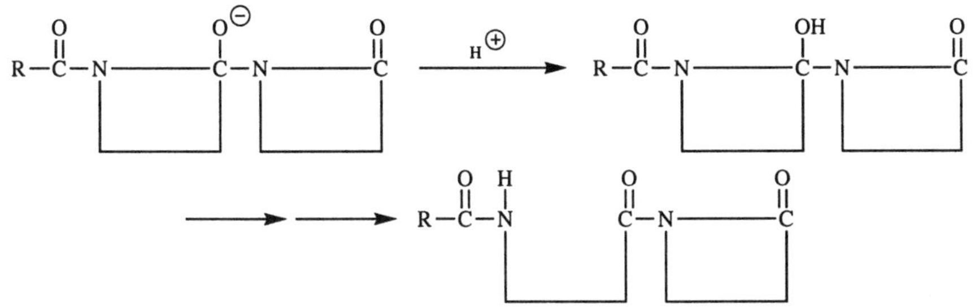

They also felt that the acylating strength of the acyllactams is enhanced by coordination of the cations with the imide carbonyl groups.[116,117] This is based on an assumption that the incorporations of the lactam units proceed through additions of lactam anions. Protonations and subsequent rearrangements follow.[116,117] This type of chain growth is termed *lactomolytic* propagation[136]:

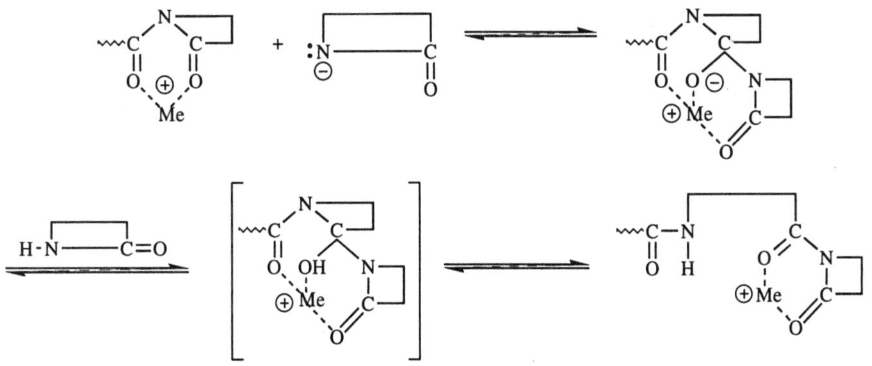

The mechanism implies that the alkaline cation is fixed to the imide group and that a nucleophilic attack (that is the rate-determining step) by the lactam anion takes place on the endocyclic carbonyl group of the imide to give a "carbinolate" anion. Proton exchange takes place between this intermediate and a lactam monomer. Intramolecular rearrangement results in ring opening of the unit that is now in the penultimate position.

In anionic activated polymerization of ε-caprolactam, chain growth involves both free anions and ion pairs.[190] Quantum-chemical calculations suggest that, in the alkali metal lactamate molecule, the negative charge is delocalized between the oxygen and the nitrogen heteroatoms. This led to a suggestion by Frunze *et al.*[190] that the acts of initiation are formations of activated intermediate chelate-type complexes between the activators and the catalyst molecules[190]:

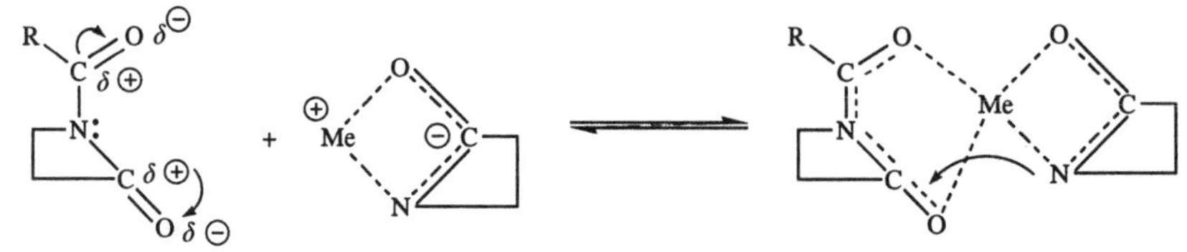

complex

where Me is a metal like Li, K, Cs, etc.

The carbinol fragment of the resulting complex, shown above, undergoes an intramolecular rearrangement. It leads to opening of the heterocyclic ring and to growth of the polymer chain by one unit:

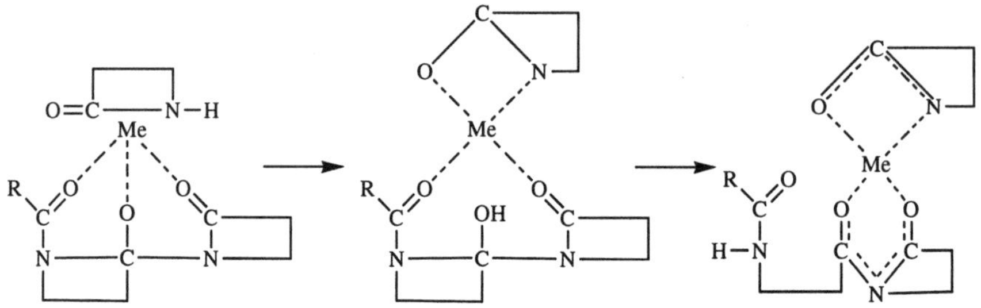

The Frunze *et al.* mechanism[190] has much in common with the "alkali lactamolytic" mechanism of Champetier and Sekiguchi,[116] except for the formation of the above-shown complex. Frunze *et al.* also believe that probably a single mechanism exists for the anionic polymerization of lactams that they describe as *ion-coordinative*.[190] The contributions of various mechanisms via ion pairs or via free ions depend upon the nature of the alkali metal counterion and upon their capacity to coordinate with electron-donating compounds (activator and monomer). The growth of ion pairs may mainly be expected from a lithium counterion, while growth by free anions may be expected from potassium or cesium.

The products of anionic lactam polymerizations can vary, depending upon reaction conditions such as the temperatures, and upon the structures of the lactams themselves. Thus, five-, six-, and seven-membered lactams polymerize at different temperatures and the products differ in molecular weights. For instance, α-pyrrolidone polymerizes readily at 30 °C to a polymer of molecular weight 15,000, while ε-caprolactam requires 178 °C to form polymers of that size or larger.[136] A third lactam, α-piperidone, is hard to polymerize to a high molecular weight polymer in good yields.[136] The regular molecules that form in activated anionic polymerizations, as already shown, are[137]

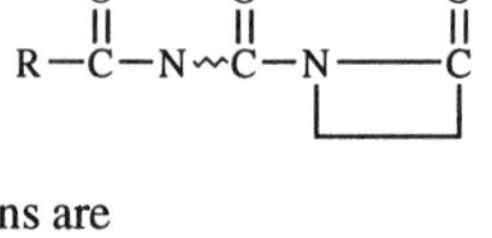

while those activated by lactam anions are

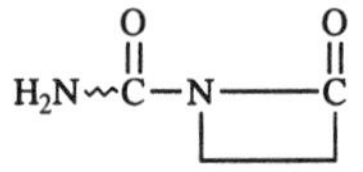

At higher polymerization temperatures, however, side reactions occur. Among them are Claisen-type condensations. They lead to two types of N-acylated β-keto imide structures and take place readily above 200 °C. Formation of these imides decreases the concentration of lactam anions:

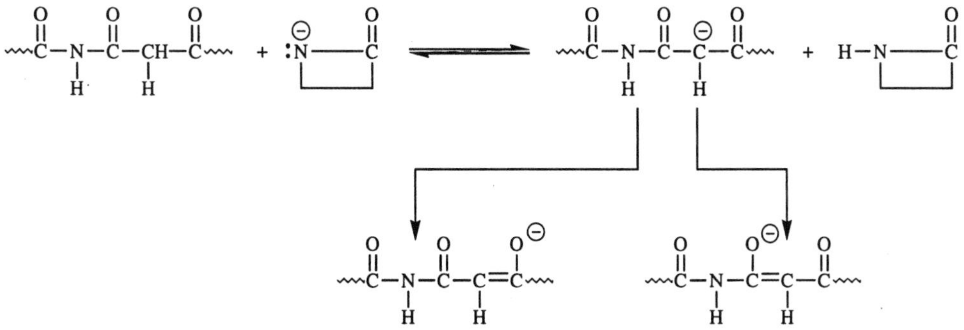

Cyclic keto imides as well as linear ones can yield active species through acylation of lactam anions. This results in formations of growth centers and keto amides:

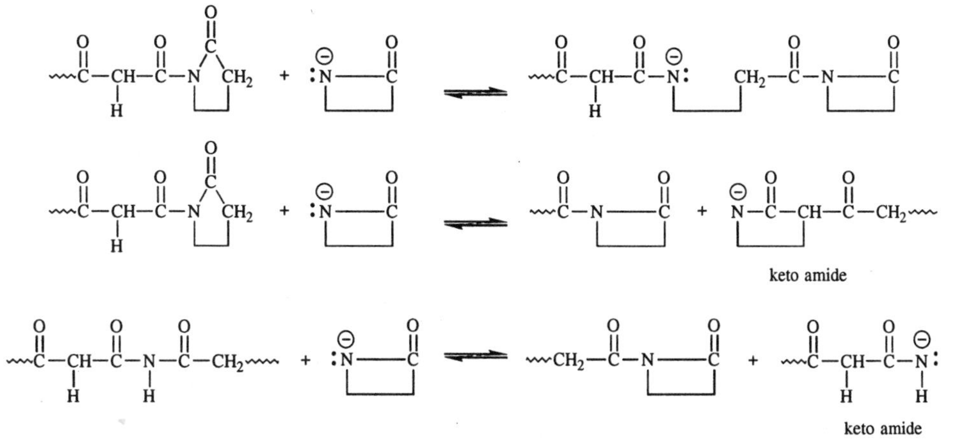

keto amide

keto amide

The acidity of keto amides with α-hydrogen atoms is much greater than that of the monomers or of polymer amide groups. Any formation of such structures therefore decreases the concentration of lactam anions.

Side reactions give rise to a variety of irregular structures that may be present either in the backbones, or at the ends of the polymer molecules, or both. Formation of branches in anionic polymerizations occurs in polymerizations of ε-caprolactam.[138,139] This lactam and higher ones polymerize at temperatures greater than 120 °C. Above 120 °C, the β-keto-amide units and possibly the n-acyl-keto-amide structures are preserved. They may, however, be potential sites for chain splitting later during polymer processing that takes place at much higher temperatures.[191]

A new group of catalysts, metal dialkoxyaluminum hydrides, for anionic polymerizations of lactams, were reported recently.[215] A different anionic mechanism of polymerization apparently takes place. When ε-caprolactam is treated with sodium dialkoxyaluminum hydride, a sodium salt of 2 (dialkoxyaluminoxy)-1-azacycloheptane forms:

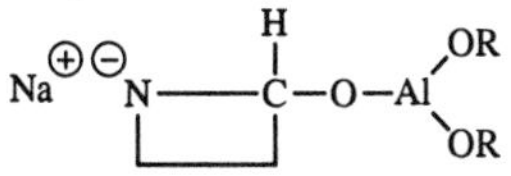

Such a compound differs in nucleophilicity from activated monomers. These salts are products of deprotonation of lactam monomers at the amides followed by reduction of the carbonyl functions. It is postulated that during lactam polymerizations, after each monomer addition, the active species form again in two steps.[215] In the first one proton exchanges take place:

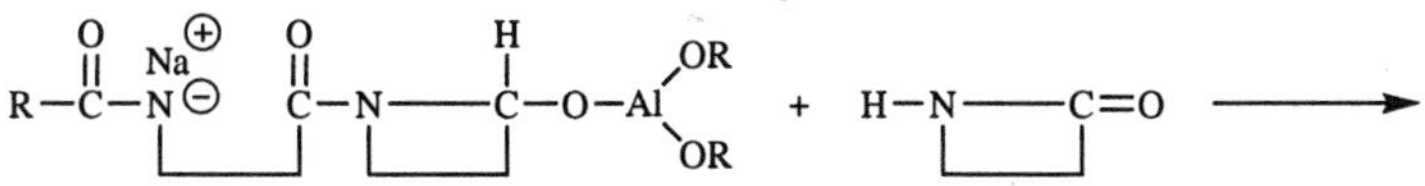

$$R{-}\overset{O}{\overset{\|}{C}}{-}\overset{H}{\overset{|}{N}}\ \ \overset{O}{\overset{\|}{C}}{-}N{-}\overset{H}{\overset{|}{C}}{-}O{-}Al(OR)_2 \quad + \quad Na^{\oplus}\,{}^{\ominus}N{-}C{=}O$$

in the second step hydrogen and dialkoxyaluminum group are transferred:

$$R{-}\overset{O}{\overset{\|}{C}}{-}\overset{H}{\overset{|}{N}}\ \ \overset{O}{\overset{\|}{C}}{-}N{-}\overset{H}{\overset{|}{C}}{-}O{\cdots}Al(OR)_2 \;\cdots\; Na^{\oplus}\,{}^{\ominus}N{-}C{=}O \longrightarrow$$

$$R{-}\overset{O}{\overset{\|}{C}}{-}\overset{H}{\overset{|}{N}}\ \ \overset{O}{\overset{\|}{C}}{-}N{-}C{=}O \quad + \quad Na^{\oplus}\,{}^{\ominus}N{-}\overset{O-Al(OR)_2}{\overset{|}{C}}{-}H$$

The third step is propagation. It consists of addition of a unit to the chain end and takes place upon reaction of the terminal acyl lactam with a sodium salt of 2(dialkoxyaluminoxy)-1-azacycloheptane shown above.[215]

4.9.3. Hydrolytic Polymerization of Lactams

This polymerization can be looked upon as a special case of cationic polymerization. It is particularly true when weak acids are added to the reaction mixture, as is often the case in industrial practice. In hydrolytic polymerizations of lactams initiated by water, the hydrolysis–condensation equilibria determine the concentrations of the amine and the carboxylic acid groups. Both functional groups participate in the propagation reactions:

$$H{-}N{-}C{=}O \quad + \quad H_2O \rightleftharpoons H_2N \quad \overset{OH}{C{=}O}$$

$$\sim\overset{O}{\overset{\|}{C}}{-}\overset{H}{\overset{|}{N}}\sim \quad + \quad H_2O \rightleftharpoons \sim\overset{OH}{\overset{|}{C}}{=}O \quad + \quad H_2N\sim$$

$$\sim\overset{OH}{\overset{|}{C}}{=}O \quad + \quad H_2N\sim\overset{OH}{\overset{|}{C}}{=}O \rightleftharpoons H_2O \quad + \quad \sim\overset{O}{\overset{\|}{C}}{-}\underset{H}{\underset{|}{N}}\sim\overset{OH}{\overset{|}{C}}{=}O$$

The concentration of these groups also determines the molecular weights of the final products.[118–128] This type of equilibria also occurs in polymerizations initiated by amino acids or by salts of carboxylic acids formed with primary and secondary amines. In the hydrolytic polymerizations of caprolactam the above reactions involve only a few percent of the total lactam molecules present.[129,130] The predominant propagation reaction is a step-growth addition of lactam molecules to the end groups. It is acid-catalyzed[129,131]:

$$\sim NH_2 \quad + \quad O{=}C{-}N{-}H \rightleftharpoons \sim\underset{H}{\underset{|}{N}}{-}\overset{O}{\overset{\|}{C}}\quad NH_2$$

The exact mechanism of this addition is uncertain. It was postulated that the addition steps are through reactions of neutral lactam molecules with ammonium cations.[132–134] Others felt, however, that the lactam molecules add to the undissociated salts.[135]

Hydrolytic polymerizations are the smoothest of all three types of polymerization reactions, because the growing species are less activated than in either cationic or anionic polymerizations. Many commercial processes utilize it in ε-caprolactam polymerizations. Formation of irregular structures, however, and even crosslinked material was detected. In addition, at elevated temperatures deamination and decarboxylation of polycaprolactam can take place.[140] Such reactions can result in formations of ketones and secondary amine groups. The ketones, in turn, can react with amines and form Schiff bases. This leads to branching and crosslinking.

In industrial preparations, most of the water used to initiate the polymerizations is removed after conversions reach 80–90% in order to attain high molecular weights. The final products contain about 8% of caprolactam and about 2% of a cyclic oligomer.[140] These are removed by vacuum or hot-water extraction. The material is then dried under vacuum at 100–200 °C to reduce moisture to about 0.1%.

4.10. Polymerization of *N*-Carboxy-α-Amino Acid Anhydrides

The polymerizations of these anhydrides (or substituted oxazolidine-2,5-diones) can be carried out with basic catalysts to yield polyamides:

These polymerization reactions are important to biochemists because the products are poly(α-amino acid)s and resemble the building blocks of naturally occurring polyamides.

When the polymerization is initiated with strong bases, the initiating step is hydrogen abstraction from the anhydride by the base. This results in formation of *activated* species:

The reaction then proceeds by the *activated mechanism*. The initiation reaction was pictured by Ballard and Bamford[142] as follows:

Each propagation step consists of an addition of one unit of the anhydride and an accompanying loss of carbon dioxide:

```
                                    H   O
                                    |  //
    H           O                   N—C              H          O
    |           ||                  |   \            |          ||
R—C—N————C—C—NH2         +          |    O  ——→  R—C—N—[—C—C—NH—]2—H   +  CO2
  |    |        / \                 |   /            |    |       / \
O=C   C=O      H   R            R—C—C           O=C   C=O     H   R
   \ /                              |  \\              \ /
    O                               H   O               O
```

When the reaction is initiated by primary amines, the first step is a nucleophilic attack by the amine on the C_5 of the anhydride.[142–144] The carbon dioxide that is released comes from the C_2 carbonyl group. The propagation proceeds by addition of terminal amine groups to the C_5 carbonyl groups of the monomers[142–144]:

```
   H   O
   |  //
   N—C                    O  R                             O  R
   |   \                  ||  |                            ||  |
   |    O  +  H2NR'  ——→  R'NH—C—C—NH—COOH  ——→  R'—NH—C—C—NH2  +  CO2
   |   /                        |                               |
R—C—C                           H                               H
   |  \\
   H   O
```

The polymerization rate depends upon the concentration of the amine and the monomer. The degree of polymerization is often, but not always, equal to the ratio of the monomer to the amine.[147] It means that the reaction may be similar to, but not identical to, a living-type polymerization. In addition, the molecular weight distribution curve may be broadened or bimodal. This may be due to some chemical termination reactions. These can be intramolecular reactions of the terminal amine group with some functional group in the side chain and lead to formation of hydantoic acid end groups.[147] It may also be due to physical termination from precipitation of the product.

Dialkylzinc initiated polymerizations apparently take place by a different mechanism. The first step is pictured by Makino, Inoue, and Tsuruta as a hydrogen abstraction by dialkylzinc from NH.[145] This is similar to the reaction with a base shown earlier. The second stage of initiation, however, is a reaction between two molecules of the activated carboxyanhydrides, and formation of zinc carbamate[145]:

```
   ⊖   O
   :N—C                    -2R'H      [   H              H    ]       [   H                     H   ]
    |   \                             [   |              |    ]       [   |                     |   ]
2   |    O  +  R'2Zn       ——→        [ R—C—N—Zn—N—C—R  ]  ⇌  [ R—C—N             N—C—R ]
    |   /                             [   |   |    |   |  ]       [   |  ||             ||  |  ]
 R—C—C                                [ O=C  C=O O=C  C=O ]       [ O=C  C—O—Zn—O—C  C=O ]
    |  \\                             [   \ /      \ /  ]       [   \ /             \ /  ]
    H   O                             [    O        O   ]       [    O               O   ]
```

The propagation is a carbonyl addition of the zinc carbamate to the activated *N*-carboxyanhydride to form a mixed anhydride. The mixed anhydride then changes into an amide group with elimination of carbon dioxide[145]:

```
     H                    H                   H   O
     |                    |                   |   ||      ,Zn-activated N-carboxyanhydride
  R—C—N               N—C—R               R—C—N—C—CH—N
2    |  ||               ||  |      ——→        |   |    |   \
  O=C  C—O—Zn—O—C  C=O                O=C   C=O  R    C=O
     \ /                 \ /                  \ /           |
      O                   O                    O            O—Zn-activated N-carboxyanhydride
```

where the activated *N*-carboxyanhydride portion is[145]

```
           R                       R
          /                       /
      N—C—H                 —N—C—H
      ||  |         ⇌           |   |
  —O—C   C=O                O=C   C=O
      \ /                       \ /
       O                         O
```

The complete reaction can be illustrated as follows:

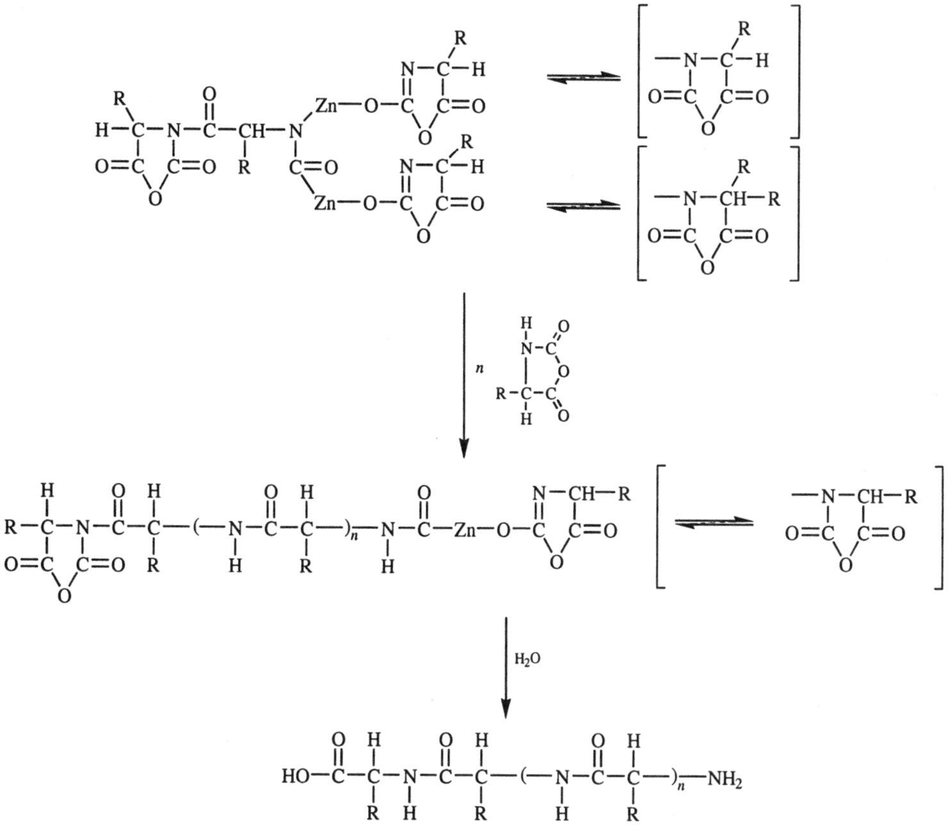

H₂O

Organotin compounds are also active as catalysts in the polymerizations of *N*-carboxy anhydrides.[146] The mechanism of the reaction was postulated by Freireich, Gertner, and Zilkha[146] to consist of addition of the organotin compound to the anhydride and formation of organotin carbamate. It subsequently decarboxylates and leaves an active –N–Sn– group that adds to another molecule of *N*-carboxyanhydride. This process is repeated in every step of the propagation[146]:

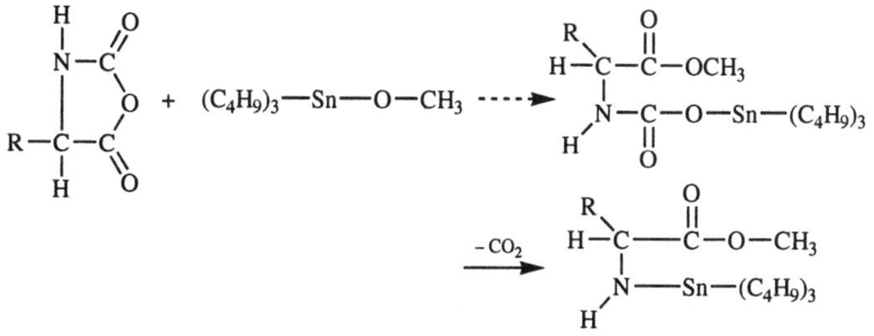

Propagation:

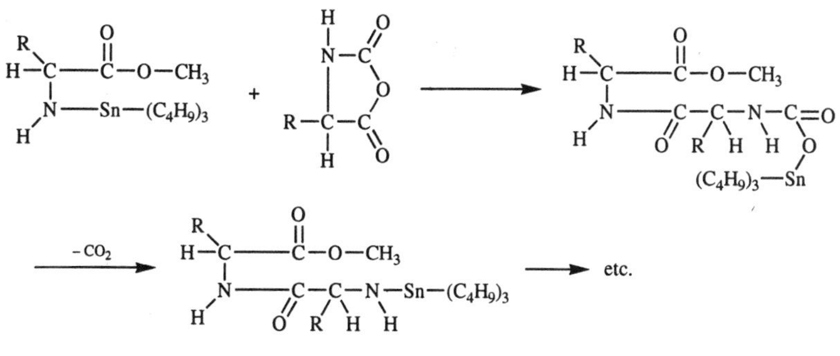

etc.

When *N*-carboxyanhydride polymerizations are initiated by secondary amines with small substituents, the amines act as nucleophiles, similarly to primary amines.[147] Secondary amines with bulky substituents, however, produce only *N*-carboxyanhydride anions. The same is true of tertiary amines. These anions in turn initiate polymerizations that proceed by the "active monomer mechanism."

4.11. Metathesis Polymerization of Alicyclics

Ring-opening polymerizations of alicyclics by Ziegler–Natta-type catalysts resulted from general studies of olefin metathesis.[148–150] These interesting reactions can be accomplished with the aid of many catalysts. The best results, however, are obtained with complex catalysts based on tungsten or molybdenum halides. One such very good catalyst forms when tungsten hexachloride is combined in the correct proportions with ethylaluminum dichloride and ethanol.

Several reaction mechanisms were proposed to explain the course of olefin metathesis. Most of the evidence supports a carbene mechanism involving metal complexes.[151–154] A typical metathesis reaction of olefins can be illustrated as follows:

$$WCl_6 + C_2H_5OH + C_2H_5AlCl_2 + 2\ R'\text{–}CH{=}CHR'' \longrightarrow$$

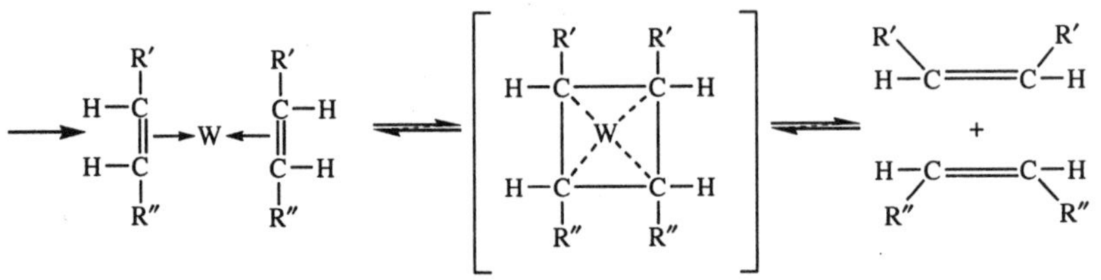

When this reaction is applied to cyclopentene, a high molecular weight polymer forms[155]:

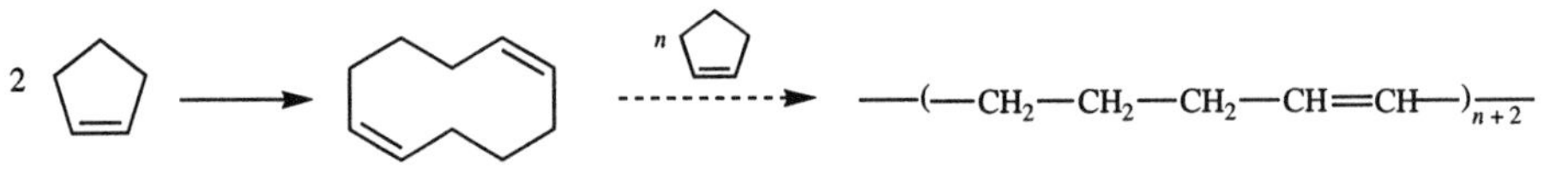

$$\text{—(—}CH_2\text{—}CH_2\text{—}CH_2\text{—}CH{=}CH\text{—)}_{n+2}\text{—}$$

Tungsten hexachloride can apparently also act as a catalyst without the aluminum alkyl. In that case it is believed to be activated by oxygen.[157] The propagation reaction based on the tungsten carbene mechanism can be shown as follows[152]:

Initiation:

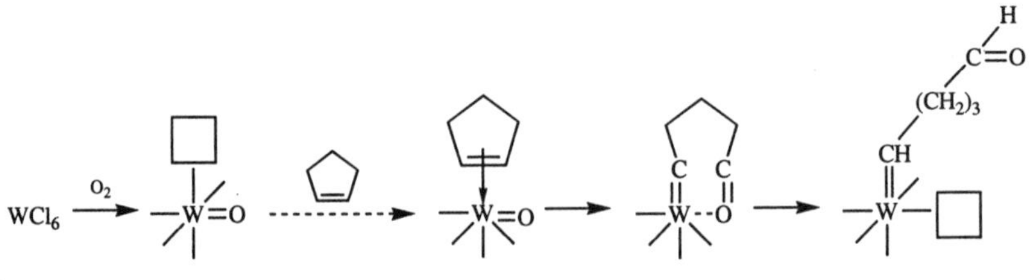

Propagation:

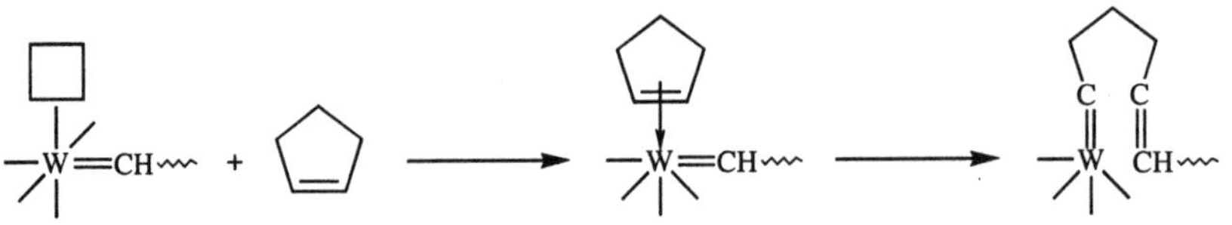

It is significant that metal carbenes can act as catalysts for this reaction. Thus, a carbene $(C_6H_5)_2$–$C{=}W(CO)_5$ will polymerize 1-methylcyclobutene to yield a polymer that is very similar in structure to *cis*-polyisoprene[155]:

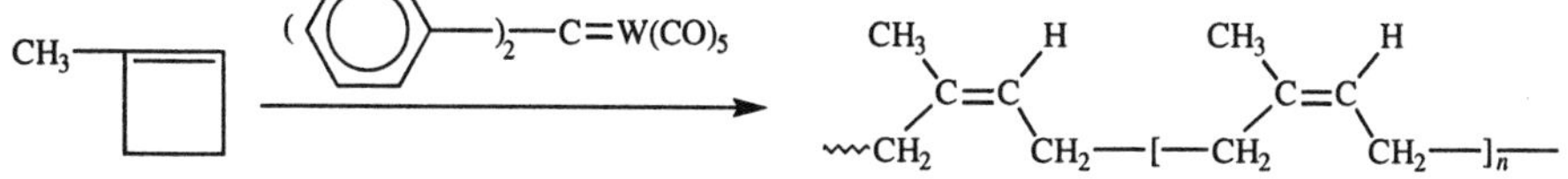

This carbene also yields high molecular weight linear polymers from bicyclo[4.2.0]octa-7-ene monomer[156]:

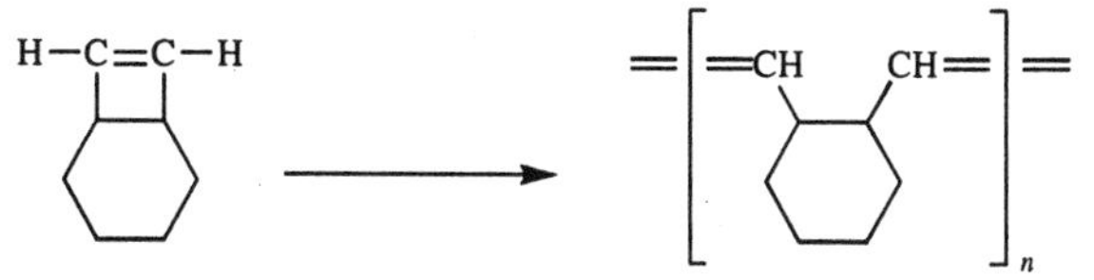

The same product can also be obtained with WCl_6 / $Sn(CH_3)_4$ catalyst. The molecular weight of the product, however, is lower.[156]

Some cycloolefins can undergo either a regular cationic polymerization or a metathesis one, depending upon the catalyst. One of them is norbornene and its derivatives. For instance, 5-methylene-2-norbornene polymerizes by a cationic mechanism with a 1:1 combination of tungsten hexachloride with tetraalkyltin. A 1:4 combination of a tungsten halide with either $C_2H_5AlCl_2$, or $MoCl_5$, or $TiCl_4$, or other acidic catalysts[158] yields the same product. The polymer that forms has the repeat units:

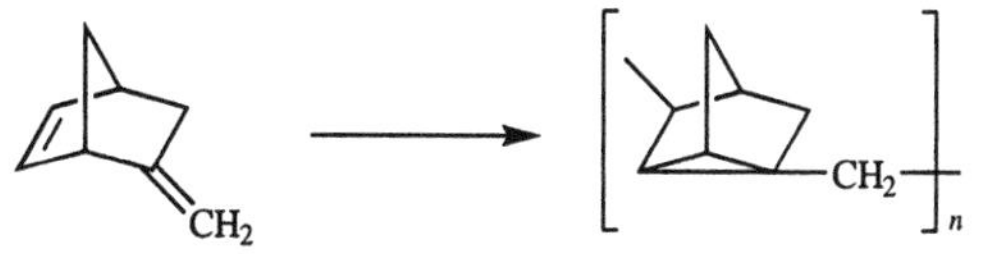

On the other hand, metathesis-type polymerization of norbornene takes place with WCL_6-$[(C_2H_5)_3Al]_{1.5}$ or WCl_6-$(CH_3)_4Sn$ to yield[158]

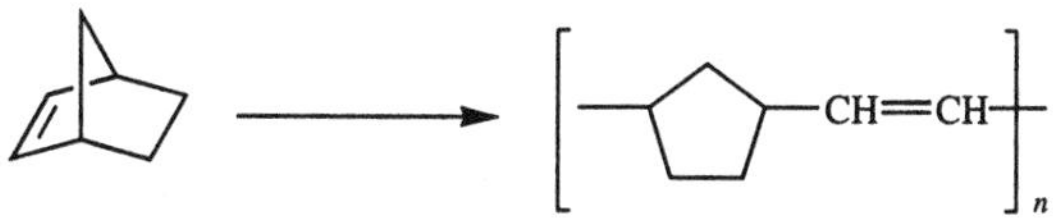

The product, poly[1,3-cyclopentylenevinylene], is a commercial synthetic specialty rubber, with a trade name of Norsorex. Reports in the literature show that there may be more than one mechanism of termination.[159,151] One of them may be a formation of cyclopropane rings. This is a typical reaction of carbenes.[151] Another one may be a reduction of the transition metal and formation of free radicals[151]:

$$\sim CH{=}WCl_x \rightarrow \sim CHCl\cdot + WCl_{x-1}$$

Still another way may be by hydrogen migration in the carbene complex[151]:

$$\sim CH_2{-}CH{=}WCl_x \rightarrow \sim CH{=}CH_2 + WCl_x$$

The chemistry of metathesis polymerization has been applied to preparation of unsaturated polycarbonates.[200] This is a case of an acyclic diene metathesis. It takes place when Lewis-acid free catalysts are employed.[201] An example of one such catalyst is $Mo[CHC(CH_3)_2Ph](N\text{-}2,6\text{-}C_6H_3\text{-}i\text{-}Pr_2)[OCCH_3(CF_3)_2]_2$. One interesting point about this process is that unconjugated dienes are polymerized to high molecular weight linear polymers without formation of any cyclic structures.

4.12. Polymerization of Cyclic Amines

The cyclic amines or imines (aziridines) polymerize only with acidic catalysts.[160–162] This reaction can be illustrated as follows:

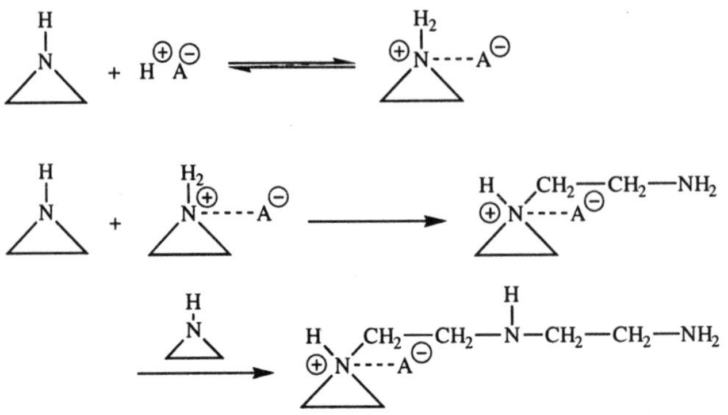

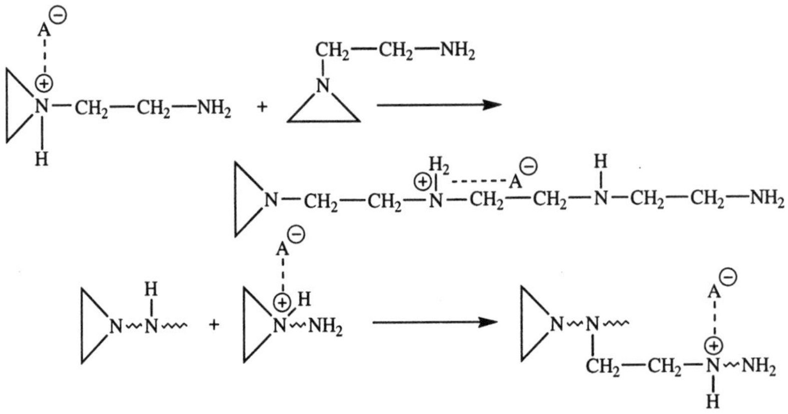

The high degree of strain in the three-membered rings causes very rapid polymerizations. A variety of cationic species acts as efficient catalysts for such reactions. The propagating species are iminium ions and the propagation steps result from nucleophilic attacks by the monomers on the ions, as shown above. Branches form due to reactions of secondary amine groups with the iminium centers. They can also result from attacks by the imine end groups of inactive polymer chains on the iminium centers of the propagating species. As the reaction progresses, it slows down, because the protons become equilibrated with various amines.[161] The polymer is also extensively cyclized due to intramolecular nucleophilic attacks of primary and secondary amines on the iminium group. The product contains cyclic oligomers and polymer molecules with large-size rings.

The termination mechanism is still not fully explained. It is believed that it may take place by proton abstractions from the iminium ions by the counterions, or by any nitrogen in the polymer chains, or by the nitrogens of the monomer units. It was also suggested[161,162] that backbiting and ring expansion terminate the reactions. Such ring expansions result in formations of relatively unreactive piperazine end groups:

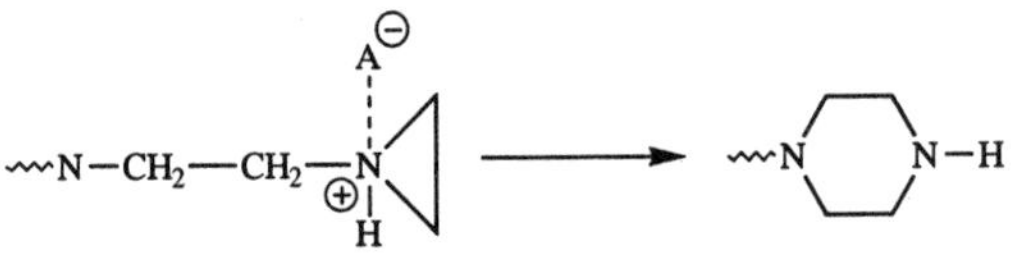

Substitution on the ethylene imine ring hinders polymerization.[161] The 2,3- and 1,2-substituted aziridines fail to polymerize. Only low molecular weight linear and cyclic oligomers form from 1- and 2-substituted ethylene imines.

Ring-opening polymerizations of cyclic sulfides can be carried out by anionic, cationic, and coordinated mechanisms.[164–166] These polymerizations are easier to carry than those of the oxygen analogs, because the sulfur–carbon bond is more polarizable. On the other hand, due to the larger size of the sulfur atoms the rings are less strained than in the oxygen compounds. As a result, the sulfur analog of tetrahydrofuran fails to polymerize. In cationic polymerizations the propagating species are sulfonium ions[166,167] and, in anionic ones, the sulfide anions. Goethals and Drijvers proposed the following cationic mechanism for the polymerization of dimethyl thiethane[166]:

1. The initiation mechanism with triethyl fluoroborate consists of alkylation of the monomer molecule and formations of cyclic sulfonium ions. The reaction occurs instantaneously and quantitatively.

$$(C_2H_5)_3O^{\oplus}\ BF_4^{\ominus} + \text{(3,3-dimethylthietane)} \longrightarrow C_2H_5{-}S^{\oplus}\text{(ring)}\ \cdots BF_4^{\ominus} + (C_2H_5)_2O$$

2. The propagation reaction probably involves nucleophilic attacks at the α-carbon atom of the cyclic sulfonium ions by the sulfur atoms from other monomer molecules:

$$\sim\!CH_2{-}S^{\oplus}\text{(ring, } CH_3, CH_3)\ \cdots BF_4^{\ominus} + S\text{(ring, } CH_3, CH_3) \longrightarrow \sim\!CH_2{-}S{-}CH_2{-}C(CH_3)_2{-}CH_2{-}S^{\oplus}\text{(ring, } CH_3, CH_3)\ \cdots BF_4^{\ominus}$$

The existence of the sulfonium ions among the propagating species was confirmed with NMR studies.[176]

3. Termination is presumed to occur through formations of unreactive sulfonium ions.

Two mechanisms of formation of sulfonium ions are possible: (1) by approaches to the catalyst's electron-accepting sites, (2) by abstraction of hydrides by methyl cations[167]:

$$CH_3^{\oplus} + R'\text{–(thiirane)} \longrightarrow [R'\text{–(thiirene)}]^{\oplus} + CH_4\uparrow$$

$$[R'\text{–(thiirene)}]^{\oplus} + S\text{(thiirane)–}R' \longrightarrow R'\text{–(thiirane)–}S^{\oplus}\text{(thiirane)–}R'$$

There are indications of a "living" chain-growth mechanism in boron trifluoride–diethyl ether initiated polymerizations of propylene sulfide[168] at conversions of 5–20%. In these early stages of polymerization the molecular weight corresponds to that calculated for typical "living" polymers. This is believed to take place through formations of stable sulfonium ions:

$$R_2O{:}BF_3 + S\text{(thiirane)–}CH_3 \rightleftharpoons BF_3\,S^{\oplus}\text{(thiirane)–}CH_3 \xrightleftharpoons{\text{monomer}} BF_3{:}S{-}CH_2{-}CH(CH_3){-}S^{\oplus}\text{(thiirane)–}CH_3$$

Higher conversions in thiirane polymerizations, however, proceed with a chain scission transfer mechanism under the influence of $BF_3{\cdot}(C_2H_5)_2O$.[168] This is indicated by a change in the molecular weight distribution, a bimodal character. When the reaction is complete, there is a marked decrease in the average molecular weight of the polymer. When thietane polymerizes with triethyl-oxonium tetrafluoroborate initiation in methylene chloride, the reaction terminates after only limited conversion.[169] This results from reactions between the reactive chain ends (cyclic sulfonium salts) and the sulfur atoms on the polymer backbone. In propylene sulfide polymerization, however, terminations are mainly due to formations of 12-membered-ring sulfonium salts from intramolecular reactions.[169]

When the polymerizations of cyclic sulfides are carried out with anionic initiators, many side reactions can occur. On the other hand, common anionic initiators, like KOH, yield optically active polymers from optically active propylene sulfide.[170] An example of a side reaction is formation[168] of propylene and sodium sulfide in sodium naphthalene initiated polymerizations. Such reactions are very rapid even at –78 °C. A similar reaction was shown to take place with ethyllithium[171]:

$$\underset{\text{(S-bridged)}}{CH_2\text{—}CH\text{—}CH_3} + C_2H_5Li \longrightarrow C_2H_5\text{—}S\text{—}Li + CH_3\text{—}CH{=}CH_2$$

Other side reactions with butyllithium-initiated polymerizations are cleavages of the polysulfides[168]:

$$\sim CH_2\text{—}CH_2\text{—}S(\leftarrow LiC_4H_9)\text{—}CH_2\text{—}CH_2\text{—}S\sim \longrightarrow \begin{cases} C_4H_{10} + \sim S\text{—}CH{=}CH_2 + Li\text{—}S\text{—}CH_2\text{—}CH_2\sim \\ C_4H_{10} + \sim CH{=}CH_2 + Li\text{—}S\text{—}CH_2\text{—}CH_2\text{—}S\sim \end{cases}$$

High molecular weight polymers can be prepared from ethylene sulfide with a diethylzinc/water catalyst.[172] The polymers form in two steps. Initially, insoluble crystalline polymers form at room temperature with a high catalyst-to-monomer ratio. These product polymers, that contain all of the catalyst, act as seeds for further polymerizations. Though the final polymers are insoluble, the molecular weights are estimated to be high. At a conversion of 20% the molecular weights are believed to be about 900,000.[172] When diethylzinc is prereacted with optically active alcohols, optically active poly(propylene sulfide)s form.[173–175] Cadmium salts are also very effective catalysts for polymerization of thiiranes. The polymers of substituted thiiranes have high stereoregularity.

4.14. Copolymerization of Cyclic Monomers

Many copolymers have been prepared from cyclic monomers. These can form through ring-opening copolymerizations of monomers with similar functional groups as well as with different ones. Some cyclic monomers can also copolymerize with some linear monomers. Only a few copolymers of cyclic monomers, however, are currently used industrially.

The composition of the copolymers depends upon the reaction conditions, the counterions, the solvents, and the reaction temperatures. The initiator system can be very important when cyclic monomers with different functional groups are copolymerized. Also, if different propagating centers are involved in the propagation process, copolymerizations can be very difficult to achieve.

Prominent among copolymers of cyclic ethers are interpolymers of oxiranes with tetrahydrofuran. Thus, ethylene oxide copolymerizes with tetrahydrofuran with the aid of a boron trifluoride–ethylene glycol catalytic system.[177] The resultant copolyether diol contains virtually no unsaturation.

Another example is a copolymer of allylglycidyl ether with tetrahydrofuran formed with antimony pentachloride catalyst[178]:

$$CH_2{-}CH(O)\,[{-}O{-}CH_2{-}CH{=}CH_2] + \text{tetrahydrofuran} \longrightarrow {-}({-}CH_2{-}CH(O{-}CH_2{-}CH{=}CH_2){-}O{-})_x{-}\;{-}({-}CH_2{-}CH_2{-}CH_2{-}CH_2{-}O{-})_y{-}$$

In addition to the above, liquid copolymers form from 1,3-dioxolane with ethylene oxide, when boron trifluoride is used as the catalyst.[1] Also, a rubbery copolymer forms from tetrahydrofuran and 3,3-diethoxycyclobutane with phosphorus pentafluoride catalyst.[179] A 3,3-bis(chloromethyl) oxacyclobutane copolymerizes with tetrahydrofuran with boron fluoride or with ferric chloride catalysis. The product is also a rubbery material.[1]

Various copolymers were reported from trioxane with dioxolane or with glycidyl ethers.[180] For instance, a copolymer of trioxane and dioxolane forms with $SnCl_4$, BF_3, or $HClO_4$ catalysts. The products from each reaction differ in molecular weights and in molecular weight distributions. Copolymerizations of trioxane with phenylglycidyl ether yield random copolymers.[180]

Different lactones can be made to interpolymerize.[181] The same is true of different lactams.[182,183,189] The products are copolyesters and copolyamides, respectively.

More interesting are copolymers from cyclic monomers of different chemical types. For instance, cyclic phosphite will copolymerize with lactone at 150 °C or above in the presence of basic catalysts[184]:

$$\text{propiolactone} + C_6H_5{-}O{-}P(O{-}CH_2CH_2{-}O) \longrightarrow {-}[{-}P(=O)(OC_6H_5){-}O{-}CH_2{-}CH_2{-}C(=O){-}O{-}CH_2{-}CH_2{-}]_n{-}$$

Aziridine copolymerizes with succinimide to form a crystalline polyamide that melts at 300 °C.[185]

$$\text{aziridine} + \text{succinimide} \longrightarrow \text{aziridinyl}{-}N{-}[{-}C(=O){-}CH_2{-}CH_2{-}C(=O){-}NH{-}CH_2CH_2{-}NH{-}]_n{-}$$

When in place of succinimide a cyclic carbonate is used, a high molecular weight polyurethane forms[186]:

$$\text{aziridine} + O{=}C(O{-}CH_2CH_2{-}O) \xrightarrow{\Delta} \text{aziridinyl}{-}N{-}[{-}(CH_2)_2{-}NH{-}C(=O){-}O{-}(CH_2)_2{-}O{-}]_n{-}H$$

Terpolymers form from epoxides, anhydrides, and tetrahydrofuran or oxetane with a trialkylaluminum catalyst[187]:

$$\text{ethylene oxide} + \text{tetrahydrofuran} + \text{phthalic anhydride} \longrightarrow$$

$$—[—CH_2—CH_2—O—CH_2—CH_2—CH_2—CH_2—O—\overset{O}{\overset{\|}{C}}—C_6H_4—\overset{O}{\overset{\|}{C}}—O—]_n—$$

Copolymerizations of caprolactone with caprolactam in various ratios take place with lithium tetraalkylaluminate as the catalyst.[188] The products are mainly random copolymers with some block homopolymers.

When lactones copolymerize with cyclic ethers, such as β-propiolactone with tetrahydrofuran, in the early steps of the reaction the cyclic ethers polymerize almost exclusively.[195] This is due to the greater basicity of the ethers. When the concentration of the cyclic ethers depletes to the equilibrium value, their consumption decreases markedly. Polymerizations of the lactams commence. The products are block copolymer.[195]

4.15. Spontaneous Alternating Zwitterion Copolymerizations

This type of copolymerization results from spontaneous interactions of nucleophilic and electrophilic monomers (M_N and M_E, respectively) without any additions of catalysts. Zwitterions form in the process that subsequently leads to formation of polymers.[202–214] The mechanism is a step-growth polymerization. It can be illustrated as follows:

$$M_N + M_E \rightarrow {}^{\oplus}M_N - M_E{}^{\oplus}$$

$${}^{\oplus}M_N - M_E{}^{\ominus} + {}^{\oplus}M_N - M_E{}^{\ominus} \rightarrow {}^{\oplus}M_N - M_E - M_N - M_E{}^{\ominus}$$

$${}^{\oplus}M_N\text{–}M_E\text{–}M_N\text{–}M_E{}^{\ominus} + {}^{\oplus}M_N\text{–}M_E{}^{\ominus} \rightarrow {}^{\oplus}M_N\text{–}(\text{–}M_E\text{–}M_N\text{–})_2\text{–}M_E{}^{\ominus} \rightarrow \text{etc.}$$

Repeated additions of the charged species and the resulting zwitterionic products lead to high polymers:

$${}^{\oplus}M_N\text{–}[\text{–}M_EM_N\text{–}]_n\text{–}M_E{}^{\ominus}$$

The initial zwitterion that forms upon combination of a nucleophilic with an electrophilic monomer is called a *genetic zwitterion.*[202] Intramolecular reactions can produce "macrocycles":

$${}^{\oplus}M_N—(—M_E—M_N—)_n—M_E{}^{\ominus} \longrightarrow \boxed{(—M_N—M_E—)_{n+1}}$$

The contribution of the cyclization reaction, however, is apparently small.[202] A reaction can also take place between a free monomer and any zwitterion at one of the ionic sites:

$${}^{\oplus}M_N—(—M_E—M_N—)_n—M_E{}^{\ominus} \xrightarrow{M_N} {}^{\oplus}M_N—M_N—(—M_E—M_N—)_n—M_E{}^{\ominus}$$

$${}^{\oplus}M_N—(—M_E—M_N—)_n—M_E{}^{\ominus} \xrightarrow{M_E} {}^{\oplus}M_N—(—M_E—M_N—)_n—M_E—M_E{}^{\ominus}$$

Such reactions disturb the alternating arrangements of the units—M_N–M_E—in the products. The reactivity of the monomers determines whether homopropagations occur as well. Alternating propagation depends upon dipole–dipole interactions between M_N and M_E monomers in preference to ion–dipole reactions between ion centers of zwitterions and monomers in homopropagations.[202]

An example of an alternating copolymerization via zwitterion intermediates is a copolymerization of 2-oxazoline with β-propiolactone. It takes place in a solution in a polar solvent like dimethylformamide at room temperature over a period of a day to yield quantitative conversions[203]:

$$\text{2-oxazoline} + \beta\text{-propiolactone} \longrightarrow -(-CH_2-CH_2-N(CHO)-CH_2-CH_2-C(=O)-O-)_n-$$

A zwitterion that forms first is the key intermediate for the polymerization. The onium ring from 2-oxazoline is opened by a nucleophilic attack of the carboxylate anion at carbon 5[202]:

$$2\ (\text{oxazolinium})^{\oplus}N-CH_2-CH_2-C(=O)-O^{\ominus} \longrightarrow (\text{oxazolinium})^{\oplus}N-CH_2-CH_2-C(=O)-O-CH_2-CH_2-N(CHO)-CH_2-CH_2-C(=O)-O^{\ominus}$$

In this reaction the number of copolymer molecules increases at first, then reaches a maximum, and finally decreases as the conversion becomes high.[202–214] When the concentration of both monomers is high, then the formation of "genetic" zwitterions is favored. As the concentration of macrozwitterions becomes high and the monomer concentration decreases, the macrozwitterions react preferentially with each other. When stoichiometry is not observed and β-propiolactone molecules predominate in the reaction mixture, the carboxylate end groups can react in various ways. They can react not only with the cyclic onium sites of the zwitterions, but also with free β-propiolactones, and incorporate more than 50% of the propiolactone units.[202]

Another example of such copolymerization is that of 2-oxazoline with acrylic acid. The reaction can be carried out by combining the two in equimolar quantities and then heating the reaction mixture to 60 °C in the presence of a free-radical inhibitor. Such an inhibitor can be p-methoxy phenol. The reaction mixture becomes viscous as an alternating copolymer forms[206]:

$$\text{2-oxazoline} + CH_2{=}CH-COOH \longrightarrow -(-CH_2-CH_2-N(CHO)-CH_2-CH_2-C(=O)-O-)_n-$$

This copolymer is identical to the one obtained from reacting 2-oxazolone with β-propiolactone. The acrylic acid is converted into the same repeat unit as the one that forms from ring opening of β-propiolactone shown in the previous example. The suggested reaction mechanism involves a nucleophilic attack by oxazolone on acrylic acid and is followed by proton migration[202]:

$$\text{2-oxazoline} + CH_2{=}CH-C(=O)OH \longrightarrow \left[(\text{oxazolinium})^{\oplus}N-CH_2-\overset{\ominus}{C}H-COOH\right] \longrightarrow (\text{oxazolinium})^{\oplus}N-CH_2-CH_2-COO^{\ominus}$$

A similar proton migration takes place in copolymerizations of acrylamide with cyclic imino ethers. The proton migration is part of the propagation process.[207] Other examples are copolymerizations of a nucleophilic monomer, 2-phenyl-1,2,3-dioxaphospholane with electrophilic monomers.[212,213] Here, too, the electrophilic monomers can be either acrylic acid or propiolactone. Identical products are obtained from both reactions[211]:

$$\text{2-phenyl-1,3,2-dioxaphospholane} + CH_2{=}CH-COOH \longrightarrow \left[(\text{dioxaphospholane})(C_6H_5)P-CH_2-\overset{\ominus}{C}H-COOH\right]$$

$$\downarrow$$

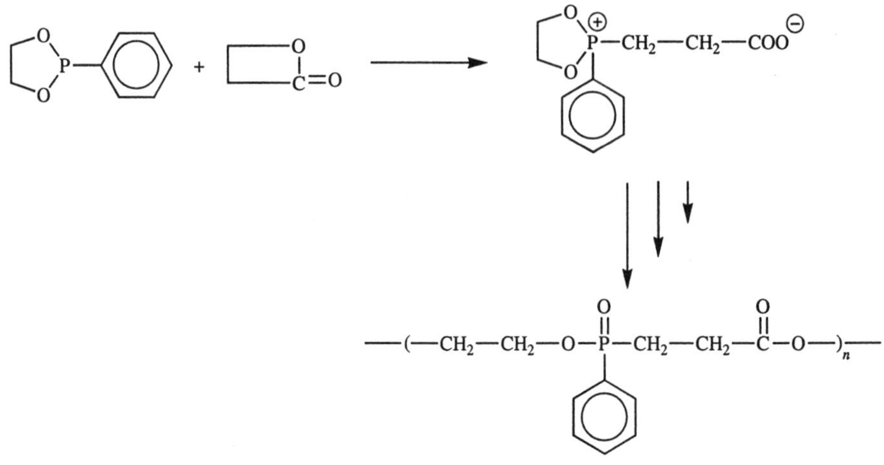

The opening of the phosphonium ring requires higher temperatures (above 120 °C) and follows the pattern of the Arbusov reaction.[211,214] Examples of some other monomers that can also act as nucleophiles in the above reaction are *p*-formyl benzoic acid,[202] acrylamide,[211] and ethylene sulfonamide. All three react in the same manner[211]:

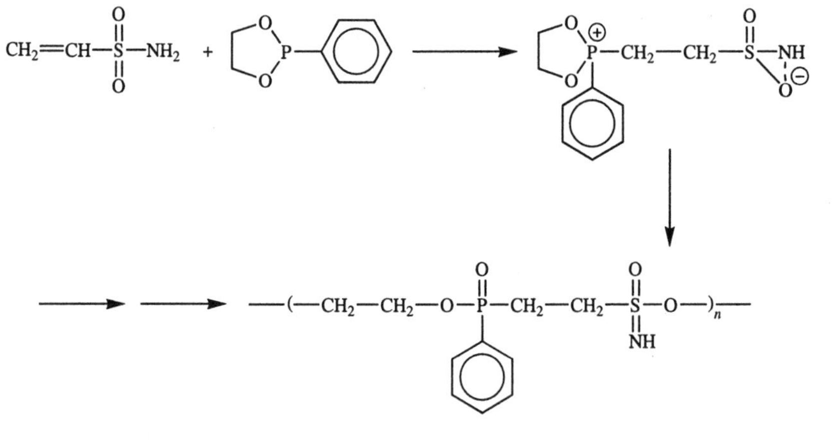

It is reasonable to expect that some compounds can act at one time as M_N monomers and at other times as M_E, depending upon the comonomer. This is the case with salicylyl phenyl phosphonite.[211] In the presence of benzoquinone it behaves as an M_N monomer and produces a 1:1 alternating copolymer at room temperature[211]:

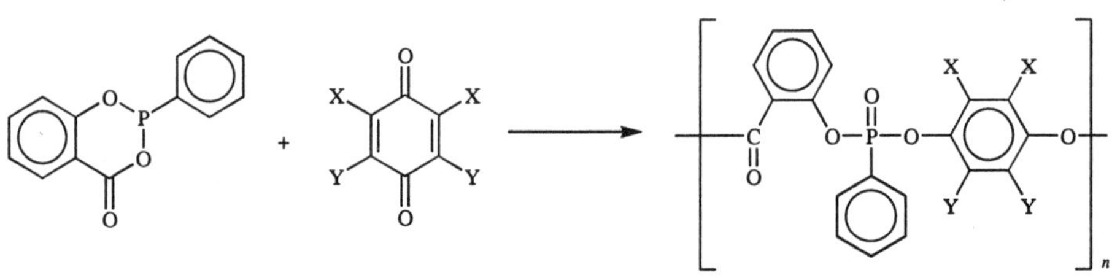

where X=Y=H; X=Y=Cl; X=Y=CH_3; X=Cl and Y=CN.

The above reaction is called a *redox copolymerization* reaction.[211] The trivalent phosphorus in the monomer is oxidized to the pentavalent state in the precess of polymerization and the quinone structure is reduced to hydroquinone. The phosphonium-phenolate zwitterion is the key intermediate:

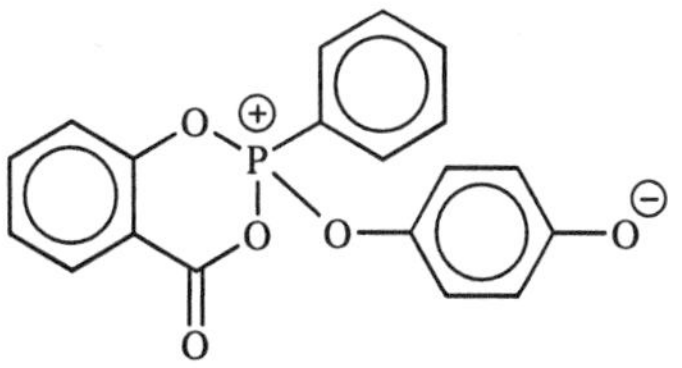

Nucleophilic attack of the phenoxide anion opens the phosphonium ring due to enhanced electrophilic reactivity of the mixed anhydride and acid structures.[211] Salicylyl phenylphosphonite, however, in combination with 2-methyl-2-oxazoline behaves as a M_E monomer.[211]

Terpolymerizations by this mechanism of sequence-ordered 1:1:1 components can also take place. The following is an example[211]:

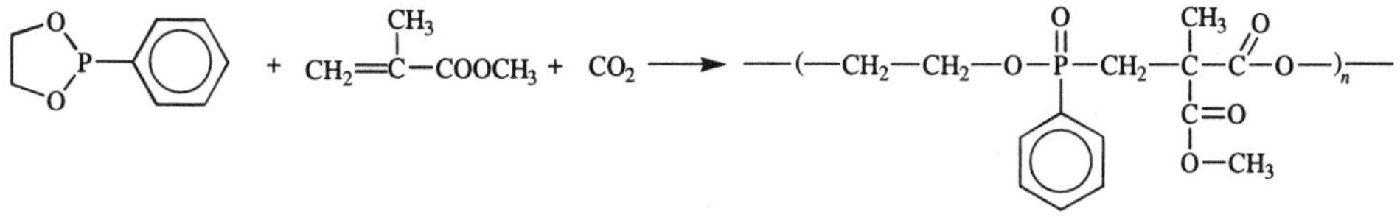

$$+ \; CH_2{=}C(CH_3){-}COOCH_3 + CO_2 \longrightarrow {-}({-}CH_2{-}CH_2{-}O{-}P(=O)(C_6H_5){-}CH_2{-}C(CH_3)(C(=O){-}O{-}CH_3){-}C(=O){-}O{-})_n{-}$$

In addition 2:1 binary copolymerizations were also observed. The following is an example of a binary copolymerization[214]:

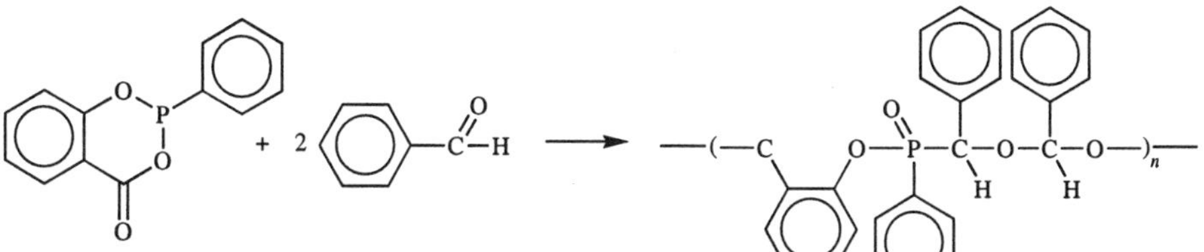

$$+ \; 2\, C_6H_5{-}C(=O){-}H \longrightarrow {-}({-}C(=O){-}C_6H_4{-}O{-}P(=O)(C_6H_5){-}C(H)(C_6H_5){-}O{-}C(H)(C_6H_5){-}O{-})_n{-}$$

4.16. Ring-Opening Polymerizations by a Free-Radical Mechanism

There are some reports in the literature of ring-opening polymerizations by free-radical mechanism. One is a polymerization of substituted vinyl cyclopropanes.[196] The substituents are radical stabilizing structures that help free-radical ring-opening polymerizations of the cyclopropane rings. This can be illustrated as follows:

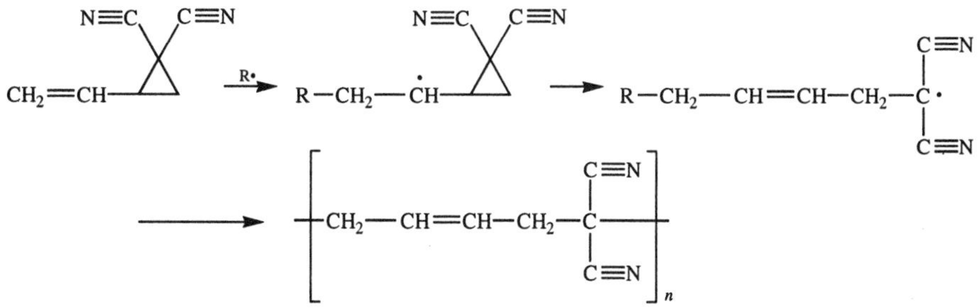

$$CH_2{=}CH{-}\triangle(C{\equiv}N)_2 \xrightarrow{R\cdot} R{-}CH_2{-}\dot{C}H{-}\triangle(C{\equiv}N)_2 \longrightarrow R{-}CH_2{-}CH{=}CH{-}CH_2{-}\dot{C}(C{\equiv}N)_2$$

$$\longrightarrow \left[{-}CH_2{-}CH{=}CH{-}CH_2{-}C(C{\equiv}N)_2{-}\right]_n$$

A high molecular weight polymer forms. In place of nitrile groups, ester groups can be utilized as well. The polymerizations of vinyl cylopropanes proceed by cationic and coordination mechanisms exclusively through the double bonds. Free-radical polymerizations of these substituted vinyl cyclopropanes, however, take place only through ring-opening polymerizations of the propane rings.

In a similar manner, ring-opening polymerizations of five-membered acetals are helped by free-radical stabilizing substituents.[197] Complete ring-opening polymerizations take place with phenyl-substituted compounds:

$$CH_2{=}C\text{(2-phenyl-1,3-dioxolane)} \xrightarrow{R\cdot} R{-}CH_2{-}\dot{C}\text{(2-phenyl-1,3-dioxolane)} \longrightarrow$$

$$R{-}CH_2{-}\overset{O}{\overset{\|}{C}}{-}O{-}CH_2{-}\dot{C}H(C_6H_5) \longrightarrow {-}({-}CH_2{-}\overset{O}{\overset{\|}{C}}{-}O{-}CH_2{-}CH(C_6H_5){-})_n{-}$$

Some other heterocyclic monomers, like acetals, also polymerize by a free-radical mechanism.[198] Particularly interesting is an almost quantitative ring-opening polymerization of a seven-membered acetal, 2-methylene-1,3-dioxepane[199]:

$$\text{2-methylene-1,3-dioxepane} \xrightarrow{R\cdot} \text{2-(R-CH}_2\text{)-1,3-dioxepan-2-yl radical} \longrightarrow R{-}CH_2{-}\overset{O}{\overset{\|}{C}}{-}O{-}CH_2{-}CH_2{-}CH_2{-}CH_2\cdot$$

$$\longrightarrow ({-}CH_2{-}\overset{O}{\overset{\|}{C}}{-}O{-}CH_2{-}CH_2{-}CH_2{-})_n{-}$$

The product is an almost pure poly(ε-caprolatone).

Review Questions

Section 4.1

1. Are the mechanisms of ring-opening polymerizations of cyclic monomers chain-growth or step-growth reactions? Explain.

Section 4.2

1. Write the rate expression for propagation in ring-opening polymerizations where there is an equilibrium between propagation and depropagation.
2. Write the kinetic expression for the total concentration of monomer segments that are incorporated into the polymer.

Section 4.3

1. Oxiranes can be polymerized by three different mechanisms. What are they? Explain.
2. Write the chemical reactions for the mechanism of polymerization of ethylene oxide with the aid of stannic chloride. Does a high molecular weight polymer form? If not, explain why.
3. Write the chemical reactions for the mechanism of polymerization of propylene oxide with boron trifluoride–water.
4. Describe the mechanism and write the chemical reactions of ring-opening polymerizations of oxiranes with potassium hydroxide. In polymerization of propylene oxide with KOH, what type of tacticity polymer forms? Explain.
5. Describe the mechanism and write the chemical equations for coordinated anionic polymerizations of propylene oxide by ferric chloride and by diethylzinc–water. Show the reaction mechanism.
6. Discuss the general characteristics of steric control in the polymerizations of oxiranes.

7. Explain the mechanism postulated by Tsuruta of steric control in polymerizations of oxiranes with the aid of organozinc compounds, giving the structure of the catalyst and the mode of monomer insertion and the mode or ring opening.

Section 4.4

1. Describe the initiation process in polymerizations of oxetanes, including initiators and reaction mechanism.
2. Describe the mechanism of propagation in polymerizations of oxetanes.

Section 4.5

1. Discuss, including chemical equations, the initiation reactions in tetrahydrofuran polymerization, including the mechanism and various initiators.
2. Discuss the propagation reaction in polymerization of tetrahydrofuran.
3. When are both ionic and covalent species present during the polymerization of tetrahydrofuran? Explain conditions that cause formation of both species and draw structures of both.
4. Describe the termination reaction in tetrahydrofuran polymerization, including living polymerization.

Section 4.6

1. How do the rates of oxepane polymerization compare to those of oxetane and tetrahydrofuran? What affects these rates?

Section 4.7

1. How and why do the cationic polymerizations of cyclic acetals differ from those of other cyclic ethers?
2. What initiators are effective in polymerizations of trioxane? Discuss polymerizations with different initiators.
3. Describe typical polymerization conditions of trioxane.
4. Explain the proposed reaction mechanisms for polymerization of trioxane, including the coordinated mechanism in polymerizations with molybdenum acetylacetonates. Illustrate all with chemical structures.
5. Discuss the polymerization of dioxalane, showing the mechanism of initiation, propagation, and terminations with different initiators.
6. How does a polymer and a copolymer form side by side in boron trifluoride initiated polymerizations of dioxepane?
7. What type of structures are obtained from ring-opening polymerizations of trioxocane? Show and explain.

Section 4.8

1. Describe cationic polymerization of lactones, showing the initiation and propagation processes.
2. Repeat the previous question for anionic polymerization.
3. Describe the coordination polymerization of lactones.

Section 4.9

1. What are the three mechanisms of polymerization of lactams?
2. Describe the catalysts that are useful in cationic polymerizations of lactams and the mechanism of polymerization.
3. Show how amidine salts form in cationic polymerizations of lactams and explain how that influences the reaction.
4. Discuss the anionic polymerization of lactams and compare that with the cationic one.
5. What is meant by lactomolytic propagation? Explain.

6. Describe the proposed mechanism for polymerizations of lactams with dialkoxyaluminum hydrides.
7. Describe hydrolytic polymerization of lactams.
8. Compare cationic, anionic, and hydrolytic polymerizations of lactams by writing out all three modes of polymerization side by side and discuss and show the side reactions that take place in each of them.

Section 4.10

1. Discuss the polymerization of N-carboxy-α-amino acid anhydrides.

Section 4.11

1. What is metathesis polymerization? Explain the mechanism and show the reaction on a disubstituted olefin.
2. Describe metathesis polymerization of methyl cyclobutene, showing the mechanisms of initiation and propagation.

Section 4.12

1. Describe the polymerization of aziridines, showing the initiation and propagation processes.

Section 4.13

1. Explain the three mechanisms by which cyclic sulfides can be polymerized. Describe each.
2. Describe the initiation and propagation reactions in cationic polymerizations of cyclic sulfides.
3. Describe the termination reaction in cationic polymerizations of cyclic sulfides.
4. What type of side reactions can occur in anionic polymerizations of cyclic sulfides?

Section 4.14

1. Discuss copolymerizations of cyclic monomers giving several examples.

Section 4.15

1. How does a spontaneous zwitterion copolymerization occur? Explain.
2. What is meant by a genetic zwitterion?
3. Give several examples of zwitterion copolymerization.

Section 4.16

1. Explain ring-opening polymerizations by a free-radical mechanism, giving two examples.

References

1. J. Furukawa and T. Saegusa, *Polymerization of Aldehydes and Oxides*, Interscience, Wiley, New York, 1963; S. Penczek and P. Kubisa, "Cationic Ring Opening Polymerization: Ethers," Chapt. 48 in *Comprehensive Polymer Science*, Vol. 3 (G.C. Eastmond, A. Ledwith, S. Russo, and P. Sigwalt, eds.), Pergamon Press, London, 1989.
2. R.D. Colclough, G. Gee, W.C.E. Higginson, J.B. Jackson, and M. Litt, *J. Polym. Sci.*, **34**, 171 (1959).
3. A.M. Eastham, *Fortschr. Hochpolym. - Forsch.*, **2**, 18 (1960).
4. D.J. Worsford and A. M Eastham, *J. Am. Chem. Soc.*, **79**, 897 (1957).
5. G.T. Merall, G.A. Latremouille, and A.M. Eastham, *J. Am. Chem. Soc.*, **82**, 120 (1960).
6. S. Searles and M. Tamres, *J. Am. Chem. Soc.*, **73**, 3704 (1951).
7. E. J. Vandenberg, *J. Am. Chem. Soc.*, **83**, 3538 (1961); *J. Polym. Sci*, **B,2**, 1085 (1964); *J. Polym. Sci.*, **A-1,7**, 525 (1969).
8. C.C. Price and R. Spector, *J. Am. Chem. Soc.*, **88**, 4171 (1966).
9. T.L. Cairns and R.M. Joyce, Jr., U.S. Patent #2,445,912 (1948).
10. E.J. Vandenberg, *J. Polym. Sci.*, **47**, 489 (1960).
11. E. Roithner, *Monatsch. Chem.*, **15**, 679 (1894); *J. Chem. Soc.*, **68** (1), 319 (1895).

12. S. Perry and H. Hibbert, *Can. J. Res.*, **8**, 102 (1953).
13. S. Perry and H. Hibbert, *J. Am. Chem. Soc.*, **62**, 2599 (1940).
14. R.E. Parker and N.S. Isaacs, *Chem. Rev.*, **59**, 737 (1959).
15. H.E. Pruitt and J. B. Baggett, U.S.Patent #2,706,1181 (1955).
16. C.C. Price and M. Osgan, *J. Am. Chem. Soc.*, **78**, 4787 (1956); *J. Polym. Sci.*, **34**, 153 (1959); C.C. Price and D.D. Carmelite, *J. Am. Chem. Soc.*, **88**, 4039 (1966).
17. R. Nomura, H. Hisada, A. Ninagawa, and H. Matsuda, *Makromol. Chem., Rapid Commun.*, **1**, 135 (1980); *ibid.*, *Makromol. Chem., Rapid Commun.*, **1**, 705 (1980).
18. Y. Watanabe, T. Yasuda, T. Aida, and S. Inoue, *Macromolecules*, **25**, 1396 (1992).
19. R. D. Colclough, G. Gee, and A.H. Jagger, *J. Polym. Sci.*, **48**, 273 (1960).
20. G. Gee, *Trans. J. Plastics Inst.*, **28**, 89 (1960).
21. C.E.H. Bawn and A. Ledwith, *Quarterly Reviews*, **16** (4), 361 (1962).
22. C.C. Price, Chapt. 1 in *Polyethers* (E.J. Vandenberg, ed.), Am. Chem. Soc. Symposium Series #6, 1975.
23. R.D. Colclough and K. Wilkinson, *J. Polym. Sci.*, **C4**, 311 (1964).
24. J. Furukawa, *Polymer*, **3**, 487 (1962).
25. T. Tsuruta, S. Inoue, M. Ishimori, and N. Yoshida, *J. Polym.* Sci., **C4**, 267 (1964).
26. T. Tsuruta, S. Inoue, N. Yoshida, and Y. Yokota, *Makromol. Chem.*, **81**, 191 (1965); T. Tsuruta, *J. Polym. Sci.*, **D**, 180 (1972).
27. A. Kassamaly, M. Sepulchre, and N. Spassky, *Polym. Bull.*, **19**, 119, (1988); A. Le Borgne, N. Spassky, C.L. Jun, and A Momtaz, *Makromol. Chem.*, **189**, 637 (1988).
28. C.C. Price and R. Spector, *J. Am. Chem. Soc.*, **87**, 2069 (1965).
29. C.C. Price and A.L. Tumolo, *J. Polym. Sci.*, **A-1,5**, 407 (1967).
30. C.C. Price, M.K. Akkapediti, B.T. deBona, and B.C. Furie, *J. Am. Chem. Soc.*, **94**, 3964 (1972).
31. T. Saegusa, Y. Hashimoto, and S. Matsumoto, *Macromolecules*, **4**, 1 (1971).
32. J. B. Rose, *J. Chem. Soc.*, 542 (1956); S. Penczek and P. Kubisa, "Cationic Ring-Opening Polymerizations," Chapters 48 and 49 in *Comprehensive Polymer Science*, Vol. 3 (G.C. Eastmond, A. Ledwith, S. Russo, and P. Sigwalt, eds.), Pergamon Press, London, 1989.
33. C.C. Price, *Accounts of Chemical Research*, **7**, 294 (1974).
34. P. Dreyfuss and M.P. Dreyfuss, "Oxetane Polymers," pp. 653–670 in *Encyclopedia of Polymer Science and Engineering*, Vol. 10, 2nd ed. (H.F. Mark, N.M. Bikales, C.G. Overberger, and G. Menges, eds.), Wiley-Interscience, New York, 1987.
35. T. Tsuruta, *J. Polym. Sci., Polym. Symp.*, **67**, 73 (1980); T. Tsuruta and Y. Kawakami, "Anionic Ring-Opening Polymerization: Stereospecificity for Epoxides, Episulfides, and Lactones," Chapt. 33 in *Comprehensive Polymer Science,* Vol. 3, (G.C. Eastmond, A. Ledwith, S. Russo, and P. Sigwalt, eds.), Pergamon Press, London, 1989.
36. T. Saegusa, H. Fujii, S. Kabayashi, H. Ando, and R. Kawase, *Macromolecules*, **6**, 26 (1973).
37. P.E. Black and D.J. Worsford, *Can. J. Chem.*, **54**, 3326 (1976).
38. P. Dreyfuss and M.P. Dreyfuss, *Polym. J.*, **8**, 81 (1976).
39. E.J. Goethals, *Adv. Polym. Sci.*, **23**, 101 (1977).
40. E.J. Goethals, *Makromol. Chem.*, **179**, 1681 (1978).
41. T. Saegusa, Chapt. 1 in *Polymerization of Heterocyclics* (O. Vogl and J. Furukawa, eds.), Dekker, New York, 1973.
42. J. Kops, S. Hvilsted, and H. Spauggaerd, *Macromolecules*, **13**, 1058 (1980).
43. M. Bucquoye and E. Goethals, *Eur. Polym. J.*, **14**, 323 (1978).
44. H. Meerwein and E. Kroning, *J. Prakt. Chem.*, **2**, **147**, 257 (1937); H. Meerwein, D. Delfs, and H. Morschel, *Angew. Chem.*, **72**, 927 (1960).
45. S. Inoue and T. Aida, "Cyclic Ethers," Chapt. 4 in *Ring Opening Polymerizations*, Vol. 1 (K.J. Ivin and T. Saegusa, eds.), Elsevier, London, 1984.
46. K. Hamann, *Angew. Chem.*, **63**, 231 (1951).
47. T. Saegusa, *Makromol. Chem.*, **175**, 1199 (1974).
48. K. Matyjaszewski and St. Penczek, *J. Polym. Sci., Polym. Chem. Ed.*, **12**, 1905 (1974).
49. St. Penszek, *Makromol. Chem., Suppl.*, **3**, 17 (1979).
50. A.M. Boyle, K. Matyjaszewski, and St. Penczek, *Macromolecules*, **10**, 269 (1977).
51. K. Matyjaszewski, T. Diem, and St. Penczek, *Makromol. Chem.*, **180**, 1917 (1979).
52. C.E.H. Bawn, C. Fitzsimmons, and A. Ledwith, *Proc. Chem. Soc.*, 391 (1964).
53. A. Ledwith, *Adv. Chem. Series*, **91**, 317 (1969).
54. I. Kuntz, *J. Polym. Sci.*, **A-1,5**, 193 (1967).
55. P. Dreyfuss, *J. Macroml. Sci.-Chem.*, **A7** (7), 1361 (1973).
56. P. Dreyfuss and M. P. Dreyfuss, *Adv. Chem. Series*, **91**, 335 (1969).
57. T. Saegusa and S. Matsumoto, *J. Polym. Sci.*, **A-1,6**, 459 (1968).

58. T. Saegusa and S. Matsumoto, *Macromolecules*, **1**, 442 (1968).
59. E.J. Goethals, *J. Polym. Sci., Polym. Symp.*, **56**, 271 (1976).
60. K. Matyjaszewski, A. Boyle, and S. Panczek, *J. Polym. Sci., Polym. Lett.*, **14**, 1 (1976).
61. S. Kobayashi, H. Danda, and T. Saegusa, *Macromolecules*, **7**, 415 (1974).
62. I.M. Robinson and G. Pruckmayr, *Macromolecules*, **12**, 1043 (1979).
63. G. Pruckmayr and T.K. Wu, *Macromolecules*, **11**, 265 (1978).
64. E.J. Vandenberg, *J. Polym. Sci.*, **A-1,7**, 525 (1969).
65. E.J. Vandenberg, and A.E. Robinson, Chapt. 7 in *Polyethers* (E.J. Vandenberg, ed.), Am. Chem. Soc. Symposium Series #6, 1975.
66. C. Penczek, *Makromol. Chem.*, **175**, 1217 (1974).
67. S. Penczek and P. Kubisa, Chapt. 5 in *Ring Opening Polymerizations* (T. Saegusa and E. Goethals, eds.), Am. Chem. Soc. Symposium Series #59, 1977.
68. M.B. Price and F.B. McAndrew, *J. Macromol. Sci.-Chem.*, **A1** (2), 231 (1967).
69. T.J. Dolce and J.A. Grates, "Acetal Resins," Chapt. 2 in *Encyclopedia of Polymer Science and Engineering*, Vol. 1, 2nd ed. (H.F. Mark, N.M. Bikales, C.G. Overberger, and G. Menges, eds.), Wiley-Interscience, New York, 1985.
70. T. Miki, T. Higashimura, S. Okamura, *J. Polym. Sci.*, **A-1,5**, 95 (1967).
71. V. Jaack and W. Kern, *Makromol. Chem.*, **62**, 1 (1963).
72. K. Weissermel, E. Fischer, and K. Gutweiler, *Kunststoffe*, **54**, 410 (1964).
73. P.H. Plesch and P. H. Westermann, *J. Polym. Sci.*, **C16**, 3837 (1968).
74. M. Okada, Y. Yamashita, and Y. Ischii, *Makromol. Chem.*, **80**, 196 (1964).
75. M. Ikeda, *J. Chem. Soc. Japan, Ind. Chem. Sect.*, **65**, 691 (1962). (From Ref. 1, p. 245.)
76. J.W. Hill and W. H. Carothers, *J. Am. Chem. Soc.*, **57**, 925 (1935).
77. A.A. Strepikheev and A. V. Volokhima, *Dokl. Akad. Nauk SSSR*, **99**, 407 (1954).
78. R.C. Schulz, K. Albrecht, C. Rentsch, and Q. V. Tran Thi, Chapt. 6 in *Ring Opening Polymerization* (T. Saegusa and E. Goethals, eds.), Am. Chem. Soc. Symposium Series #59, 1977.
79. D. Weichert. *J. Polym. Sci.*, **C16**, 2701 (1967).
80. K. Wessermel, E. Fischer, K. Gutweiler, H.D. Hermann, and H. Chedron, *Angew. Chem., Int. Ed.*, **6** (6), 526 (1967).
81. C.D. Kennedy, W.R. Sorenson, and G.G. McClaffin, *Am. Chem. Soc., Polym. Prepr.*, **7**, 667 (1966).
82. M. Iguchi, *Br. Polym. J.*, **5**, 195 (1973).
83. G. L. Collins, R. K. Greene, F. M. Bernardinelli, and W. V. Garruto, *J. Polym. Sci., Polym. Letters*, **17**, 667 (1979).
84. Y. Yamashita, M. Okada, and T. Suyama. *Makromol. Chem.*, **111**, 277 (1968).
85. B.A. Rosenberg, B.A. Kamarov, T.I. Ponomareva, and N.S. Enikolopyan, *J. Polym. Sci., Polym. Chem. Ed.*, **11**, 1 (1973).
86. E.B. Ludvig, D.K. Khomyakov, and G.S. Sanina, *J. Polym. Sci., Symp.*, **42**, 289 (1973).
87. K. Satome and I. Hodiza, *Makromol. Chem.*, **82**, 41 (1965).
88. Y.N. Sazanov, *Usp. Khim.*, **37**, 1084 (1968).
89. G.L. Brode and J.V. Koleske, *J. Macromol. Sci.*, **A6** (6), 1109 (1972).
90. R.H. Young, M. Matzner, and L.A. Pilato, Chapt. 11 in *Ring-Opening Polymerizations* (T. Saegusa and E. Goethals, eds.), Am. Chem. Soc. Symposium Series #59, 1977.
91. G.L. Brode and J.V. Koleske, Chapt. 7 in *Polymerization of Heterocyclics*, D. Vogl and J. Furukawa, eds., Dekker, New York, 1973.
92. D.R. Wilson and R.G. Bearman, *J. Polym. Sci.*, **A-1,18**, 2161 (1970).
93. R.D. Lundberg, J.V. Koleske, and K.B. Wischmann, *J. Polym. Sci.*, **A-1,7**, 2915 (1969).
94. E.F. Cox and F. Hostettler, U.S. Patent #3,021,310 (1969).
95. J. G. Noltes, F. Verbeek, H. G. J. Overmars, and J. Boersma, *J. Organometal. Chem.*, **24**, 257 (1970).
96. M. Rothe and G. Bertalan, Chapt. 9 in *Ring-Opening Polymerizations* (T. Saegusa and E. Goethals, eds.), Am. Chem. Soc. Symposium Series #59, 1977; G. Bertalan, I. Rusznak, P. Anna, M. Boros-Ivcz, and G. Marosi, *Polym. Bull.*, **19**, 539 (1989); G. Bertalan, T.T. Nagy, P. Valko, A. Boros, M. Boros-Ivcz, an P. Anna, *Polym. Bull.*, **19**, 547 (1988).
97. R.B. Homer and C.D. Johnson, p. 188 in *The Chemistry of Amides* (J. Zabicky, ed.), Wiley-Interscience, New York, 1970.
98. J. Sebenda, Chapt. 8 in *Polymerization of Heterocyclics* (O. Vogl and J Furukawa, eds.), Dekker, New York, 1973; J. Sebenda, "Anionic Ring-Opening Polymerizations of Lactams," Chapt. 35 in *Comprehensive Polymer Science,* Vol. 3 (G.C. Eastmond, A. Ledwith, S. Russo, and P. Sigwaldt, eds.), Pergamon Press, London, 1989.
99. M. Rothe , H. Boenisch, and D. Essig, *Makromol. Chem.*, **91**, 24 (1966).
100. S. Dubravszky and F. Geleji, *Makromol. Chem.*, **110**, 246 (1967).
101. S. Dubravszky and F. Geleji, *Makromol. Chem.*, **143**, 259 (1971).
102. S Dubravszky and F. Geleji, *Makromol. Chem.*, **105**, 261 (1967).

103. S. Dubravszky and F. Geleji, *Makromol. Chem.*, **113**, 270 (1968).
104. M. Rothe, G. Reinisch, W. Jaeger, and I. Schopov, *Makromol. Chem.*, **54**, 183 (1962).
105. G.M. Burnett, A.J. MacArthur, and J.N. Hay, *Eur. Polym. J.*, **3**, 321 (1967).
106. K. G. Wyness, *Makromol. Chem.*, **38**, 189 (1960).
107. M. Pracejus, *Chem. Ber.*, **92**, 988 (1959).
108. G. Bertalan and M. Rothe, *Makromol. Chem.*, **172**, 249 (1973).
109. M. Rothe and J. Mazanek, *Makromol. Chem.*, **145**, 197 (1971).
110. H. K. Hall, Jr., *J. Am. Chem. Soc.*, **80**, 6404 (1958).
111. S. Barzakay, M. Levy, and D. Vofsi, *J. Polym. Sci.*, **A-1,4**, 2211 (1966).
112. J. Sebenda, *J. Polym. Sci.*, **C-23**, 169 (1968).
113. E. Sittler and J. Sebenda, *J. Polym. Sci.*, **C-16**, 67 (1967).
114. O. Wichterle, *Makromol. Chem.*, **35**, 174 (1960).
115. O. Wichterle, J. Sebenda, and J. Kralicek, *Fortschr. Hochpolym. Forsch.*, **2**, 578 (1961).
116. G. Champetier and H. Sekiguchi, *J. Polym. Sci.*, **48**, 309 (1960).
117. H. Sekiguchi, *J. Polym. Sci.*, **48**, 309 (1960).
118. O. Fukumoto, *J. Polym. Sci.*, **22**, 263 (1956).
119. D. Heikens and P.H. Hermans, *Makromol. Chem.*, **28**, 246 (1958).
120. P.H. Hermans, D. Heikens, and P.F. van Velden, *J. Polym. Sci.*, **16**, 451 (1955).
121. H.K. Reimschuessel, *J. Polym. Sci.*, **41**, 457 (1959).
122. F. Wiloth, *Makromol. Chem.*, **14**, 156 (1954).
123. F. Wiloth, *Makromol. Chem.*, **15**, 106 (1955).
124. T.G. Majury, *J. Polym. Sci.*, **24**, 488 (1957).
125. G.B. Gechele and A. Matiussi, *Eur. Polym. J.*, **3**, 573 (1967).
126. C. Giori and B.T. Hayes, *J. Polym. Sci.*, **A-1,8**, 335 (1970).
127. C. Giori and B.T. Hayes, *J. Polym. Sci.*, **A-1,8**, 351 (1970).
128. A.V. Tobolsky and A. Eisenberg, *J. Am. Chem. Soc.*, **81**, 2302 (1959).
129. P.H. Hermans, D. Heikens, and P.F. van Velden, *J. Polym. Sci.*, **30**, 81 (1958).
130. V.V. Korshak, R.V. Kydryatsev, V.A. Sergeev, and L.B. Icikson, *Izv. Akad. Nauk S.S.R. Otd. Khim. Nauk*, 1468 (1962).
131. D. Heikens, P.H. Hermans, and G.M. van der Want, *J. Polym. Sci.*, **44**, 437 (1960).
132. H. Yumoto and N. Ogata, *Makromol. Chem.*, **25**, 71 (1957).
133. M. Rothe, H. Boenisch, and D. Essig, *Makromol. Chem.*, **91**, 24 (1966).
134. J. N. Hay, *J. Polym. Chem., Polym. Lett.*, **5**, 577 (1965).
135. O.B. Salamatina, D.K. Bonetskaya, S. M. Skuratov, and N.S. Enikolopyan, *Vysokomol. Soyed.*, **A-11**, 158, (1969).
136. H. Sekiguchi and B. Cautin, *J. Polym. Sci., Polym. Chem. Ed.*, **11**, 1601 (1973).
137. J. Sebenda and V. Kouril, *Eur. Polym. J.*, **7**, 1637 (1971).
138. G. Stea, and G.B. Gechele, *Eur. Polym. J.*, **1**, 213 (1965).
139. C.V. Goebel, P. Cefelin, J. Stehlicek, and J. Sebenda, *J. Polym. Sci.*, **A-1,10**, 1411 (1972).
140. H.K. Reimschuessel and G.J. Dege, *J. Polym. Sci.*, **A-1,8**, 3265 (1970); G. DiSilvestro, P. Sozzani, S. Bruckner, L. Malpezzi, and C. Guaita, *Makromol. Chem.*, **188**, 2745 (1987).
141. T. Makino, S. Inoue, and T. Tsuruta, *Makromol. Chem.*, **131**, 147 (1970).
142. D.G.H. Ballard and C.H. Bamford, *J. Chem. Soc.*, 381 (1956).
143. M. Szwarc, *Fortschr. Hochpolym. Forsch.*, **4**, 1 (1965).
144. Y. Iwakura and K. Uno, *J. Polym. Sci.*, **A-1,6**, 2165 (1968).
145. T. Makino, S. Inoue, and T. Tsuruta, *Makromol. Chem.*, **131**, 147 (1970).
146. S. Freireich, D. Gertner, and A. Zilkha, *Eur. Polym. J.*, **10**, 439 (1974).
147. H.R. Kricheldorf, *Makromol. Chem.*, **178**, 1959 (1977); H.R. Kricheldorf, "Anionic Ring-Opening Polymerizations of N-Carboxyanhydrides," Chap. 36 in *Comprehensive Polymer Science,* Vol. 3 (G.C. Eastmond, A. Ledwith, S. Russo, and P. Sigwalt, eds.), Pergamon Press, London, 1989.
148. N. Calderon, E.A. Ofstead, J.P. Ward, W.A. Judy, and K.W. Scott, *J. Am. Chem. Soc.*, **90**, 4133 (1968).
149. R. Alamo, J. Guzman, and J.G. Fatou, *Makromol. Chem.*, **182**, 725 (1981).
150. K.W. Scott, N. Calderon, E.A. Ofstead, W.A. Judy, and J.P. Ward, *Adv. Chem. Series*, **91**, 399 (1969).
151. B.A. Dolgoplosk, *J. Polym. Sci., Symp.*, **67**, 99 (1980).
152. J.L. Herrisson and Y. Charwin, *Makromol. Chem.*, **141**, 161 (1970).
153. M.T. Macella, M.A. Busch, and E.L. Mueteris. *J. Am. Chem. Soc.*, **98**, 1283 (1976).
154. T.J. Katz and J.M. McGinnis, *J. Am. Chem. Soc.*, **97**, 1952 (1975).
155. J.M. McGinnis, T.J. Katz, and S. Hurwitz, *J. Am. Chem. Soc.*, **98**, 606 (1976).
156. C.T. Thu, T. Bastelberger, and H. Hocker, *Makromol. Chem., Rapid Commun.*, **2**, 7 (1981).
157. A.J. Amass and T.A. McGourtey, *Eur. Polym. J.*, **16**, 235 (1980); A.J. Amass, M. Lotfipur, J.A. Zurimendi, B.J. Tighe, and C. Thompson, *Makromol. Chem.*, **188**, 2121 (1987).

158. K.J. Ivin. L.D. Theodore, B.S.R. Reddy, and J.J. Rooney, *Makromol. Chem., Rapid Commun.*, **1**, 467 (1980); S. Streck, *Chemtech*, **19**, 489 (1989).
159. A.J. Amass and J. A. 2uramendi, *Eur. Polym. J.*, **17**, 1 (1981).
160. R. A. Patsiga, *J. Macromol. Sci., Rev. Macromol. Chem.*, **C1** (2), 223 (1967).
161. G.D. Jones, D.C. Mac Williams, and N.A. Baxtor, *J. Org. Chem.*, **30**, 1994 (1965).
162. G.D. Jones, Chapt. 14 in *The Chemistry of Cationic Polymerization* (P.H. Plesch, ed.), Pergamon Press, Oxford, 1963; E.J. Goethals, "Cyclic Amines," Chap. 10 in *Ring-Opening Polymerizations,* Vol. 2 (K.J. Ivin and T. Saegusa, eds.), Elsevier, London, 1984.
163. Ph. Teyssie, J.P. Bioul, A. Hamitou, J. Heuschen, L. Hocks, R. Jerome, and T. Ouhadi, Chapt. 12 in *Ring Opening Polymerizations* (T. Saegusa and E. Goethals, eds.), Am. Chem. Soc. Symposium # 59, A.C.S., Washington, 1977.
164. P. Dreyfuss, *J. Macromol. Sci.-Chem.*, **7**, 1361 (1973).
165. M. Morton, R.F. Kammereck, and L.J. Fetters, *Macromolecules*, **4**, 11 (1971); *Brit. Polym. J.*, **3**, 120 (1971).
166. E.J. Goethals and W. Drijvers, *Makromol. Chem.*, **136**, 73 (1970); T. Aida, K. Kawaguchi, and S. Inoue, *Macromolecules*, **23**, 3887 (1990).
167. J. K. Stille and J. E. Empren, in *The Chemistry of Sulfides* (A. V. Tobolsky, ed.), Wiley-Interscience, New York, 1968.
168. F. Lautenschlaeger, in *Polymerization of Heterocyclics* (O. Vogel and J. Furukawa, eds.), Dekker, New York, 1973; N. Spassky, Chapt. 14 in *Ring-Opening Polymerizations*, T. Saegusa and E. Goethals, eds., Am. Chem. Soc. Symposium # 59, A.C.S., Washington, 1977.
169. E. J. Goethals, W. Drijvers, D. van Ooteghem, and A.M. Boyle, *J. Macromol. Sci.-Chem.*, **7**, 1375 (1973).
170. T. Tsunetsugu, J. Furukawa, and T. Fueno, *J. Polym. Sci.*, **A-1,9**, 3541 (1971).
171. M. Morton and R. F. Kammereck, *J. Am. Chem. Soc.*, **92**, 3217 (1970).
172. R.H. Gobran and R. Larson, *J. Polym. Sci.*, **C-31**, 77 (1970).
173. J. Furukawa, N. Kawabata, and A. Kato, *J. Polym. Sci., Polym. Lett.*, **5**, 1073 (1967).
174. S. Inoue, T. Tsuruta, and J. Furukawa, *Makromol. Chem.*, **53**, 215 (1962).
175. N. Spassky and P. Sigwalt, *Eur. Polym. J.*, **7**, 7 (1971).
176. E. J. Goethals and W. Drijvers, *Makromol. Chem.*, **165**, 329 (1973).
177. W. J. Murbach and A. Adicoff, *Ind. Eng. Chem.*, **52**, 772 (1960).
178. Japan Patent #10046 (1960) to du Pont Co. (from Ref. 1).
179. T. Saegusa, H. Imai, and J. Furukawa, *Makromol. Chem.*, **56**, 55 (1962).
180. H. Cherdron, in *Polymerizaton of Heterocyclics* (O. Vogl and J. Furukawa, eds.), Dekker, New York, 1973.
181. K. Tada, Y. Numata, T. Saegusa, and J. Furukawa, *Makromol. Chem.*, **77**, 220 (1964).
182. S.M. Glickman and E.S. Miller, U.S. Patent #3,016,367 (1962).
183. R.M. Hedrick, E.H. Motters, and T.M. Butler, U.S. Patent #3,120,503 (1964).
184. W. Fish, W. Hoffman, and J. Koskikallio, *J. Chem. Ind.*, 756 (1956).
185. T. Ragiua, S. Narisawa, K. Manobe, and M. Kobata, *J. Polym. Sci.*, **A-1,4**, 2081 (1966).
186. E. K. Drecksel, U.S. Patent #424,457 (1954).
187. H. L. Hsieh, *J. Macromol. Sci.-Chem.*, **7**, 1525 (1973).
188. M. L. Hsieh, Chapt. 10, in *Ring Opening Polymerization* (T. Saegusa and E. Goethals, eds.), A.C.S. Symposium Series #59, 1977.
189. H.R. Kricheldorf and W. E. Hull, *J. Polym. Sci., Polym. Chem. Ed.*, **16**, 2253 (1978).
190. T.M. Frunze, V.A. Kotelnikov, T.V. Volkova, and V.V. Kurashov, *Eur. Polym. J.*, **17**, 1079, (1981).
191. J. Roda, Z. Vortubcova, J. Klalicek, J. Stehlicek, and S. Pokorny, *Makromol. Chem.*, **182**, 2117 (1981).
192. E. B. Ludvig, B. G. Belenkaya, and A. K. Khomyakov, *Eur. Polym. J.*, **17**, 1097 (1981).
193. A. Leborgne, D. Grenier, R.E. Prud'homme, and N. Spassky, *Eur. Polym. J.*, **17**, 1103 (1981).
194. J.V. Crivello and J.H.W. Lam, *J. Polym. Chem., Polym. Chem. Ed.*, **19**, 539 (1981).
195. A. K. Khomyakov, E. B. Ludvig, and N. N. Shapetko, *Eur. Polym. J.*, **17**, 1089 (1981).
196. I. Cho and K.D. Ahn, *J. Polym. Sci., Polym. Chem. Ed.*, **17**, 3169 (1979).
197. I. Cho and M.S. Gong, *J. Polym. Sci., Polym. Lett.*, **20**, 61 (1942).
198. T. Endo and W.J. Balley, *J. Polym. Sci., Polym. Lett.*, **14**, 25 (1940).
199. W.J. Balley, Z. Ni, and S.R. Wu, *Macromolecules*, **1**, 711 (1982).
200. J.G. Nel, K.B. Wagner, and J.M. Boncella, *Macromolecules*, **24**, 2649 (1991); M. Lindmartk and K.B. Wagner, *Macromolecules*, **20**, 2949 (1987).
201. K.B. Wagner and J.T. Patton, *Macromolecules*, **26**, 249 (1993).
202. T. Saeguse, *Angew. Chem., Intern. Ed.*, **16**, 826 (1977).
203. T. Saegusa, H. Ikeda, and F. Fujii, *Macromolecules*, **5**, 354 (1972).
204. T. Saegusa, S. Kobayashi, and Y. Kimura, *Macromolecules*, **7**, 1 (1974).
205. T. Saegusa, H. Ikeda, S. Hirayanagi, Y. Kimura, and S. Kobayaski, *Macromolecules*, **8**, 259 (1975).
206. T. Saegusa, S. Kobayashi, and Y. Kimura, *Macromolecules*, **7**, 139 (1974).

207. T. Saegusa, S. Kobayashi, and Y. Kimura, *Macromolecules*, **8**, 374 (1975).
208. T. Saegusa, Y. Kimura, and S. Kobayashi, *Macromolecules*, **10**, 239 (1977).
209. T. Saegusa, S. Kobayashi, and J. Furukawa, *Macromolecules*, **9**, 728 (1976).
210. T. Saegusa, Y. Kimura, S. Sawada, and S. Kobayaski, *Macromolecules*, **7**, 956 (1974).
211. T. Saegusa, *Makromol. Chem., Suppl.*, **3**, 157 (1979).
212. T. Saegusa, Y. Kimura, N. Ishikawa, and S. Kobayashi, *Macromolecules*, **9**, 724 (1976).
213. T. Saeguse, S. Kobayashi, and J. Furukawa, *Macromolecules*, **10**, 73 (1977).
214. T. Saeguse, *Makromol. Chem., Suppl.*, **4**, 73 (1981).
215. N. Mougin, C.A. Veith, R.E. Cohen, and Y. Gnanou, *Macromolecules*, **25**, 2004, 1992.
216. T. Aida, Y. Mackawa, S. Asano, and S. Inoue, *Macromolecules,* **21**, 1195 (1988); S. Inoue and T. Aida, *Chemtech,* **24** (5), 28 (1994).
217. M. Hitota and H. Fukuda, *Makcromol. Chem.*, **188**, 2259 (1987).

5

Common Chain-Growth Polymers

5.1. Polyethylene and Related Polymers

Polyethylene is produced commercially in very large quantities in many parts of the world. The monomer can be synthesized from various sources. Today, however, most ethylene comes from petroleum by high-temperature cracking of ethane or gasoline fractions. Other potential sources can probably be found, depending upon the availability of raw materials.

Two main types of polyethylene are manufactured commercially. These are low-density (0.92–0.93 g/cm^3) and high-density (0.94–0.97 g/cm^3) polymers. The low-density material is branched while the high-density material is mostly linear and much more crystalline. The most important applications for the low-density polyethylene are in films, sheets, paper, wire and cable coatings, and in injection molding. The high-density material finds use in blow molded objects and in injection molding.

5.1.1. Preparation of Polyethylene by a Free-Radical Mechanism

Up to the late 1960s, most low-density polyethylene was produced commercially by high-pressure free-radical polymerization. Much of this has now been replaced by preparation of copolymers of ethylene with α-olefins by coordinated anionic mechanism. These preparations are discussed further in this chapter. High-pressure polymerization of ethylene, however, is still practiced in some places and is therefore discussed here. The reaction requires a minimum pressure of 500 atmospheres[1] to proceed. The branched products contain long and short branches as well as vinylidine groups. With an increase in pressure and temperature of polymerization there is a decrease in the degree of branching and in the amount of vinylidine groups.[2,3]

Free-radical commercial polymerization is conducted at 1000–3000 atmospheres pressure and 80–300 °C. The reaction has two peculiar characteristics: (1) a high exotherm, and (2) a critical dependence on the monomer concentration. In addition, at these high pressures oxygen acts as an initiator. At 2000 atmospheres pressure and 165 °C temperature, however, the maximum safe level of oxygen is 0.075% of ethylene gas in the reaction mixture. Any amount of oxygen beyond that level can cause explosive decompositions. Yet, the oxygen concentration in the monomer is directly proportional to the percent conversion of monomer to polymer, though inversely proportional to the polymer's molecular weight. This limits many industrial practices to conducting the reactions below 2000 atmospheres and below 200 °C. These reactions are therefore carried out between 1000–2000 atmospheres pressure. Small quantities of oxygen, limited to 0.2% of ethylene, are accurately metered in.[4,5] The conversion per each pass in continuous reactors is usually low, about 15–20%.

There is an induction period that varies inversely with the oxygen concentration to the power of 0.23. During this period oxygen is consumed autocatalytically. This is not accompanied by any significant decrease in pressure. A high concentration of ethylene is necessary for a fast rate of chain growth, relative to the rate of termination. Also, high temperatures are required for practical rates of initiation.

If oxygen is completely excluded and the pressure is raised between 3500–7750 atmospheres, while using relatively low temperatures of 50–80 °C, linear polyethylene forms.[6] The reactions take about 20 hours. Various solvents can be used, like benzene, isooctane, methyl or ethyl alcohols. Higher ethyl alcohol concentrations and low concentrations of the initiator result in higher molecular weights. The products range from 2000 for wax-like polymers to 4,000,000 for nearly intractable materials. Favorite free-radical initiators for these reactions are benzoyl peroxide, azobisisobutyronitrile, di-*t*-butylperoxydicarbonate, di-*t*-butyl peroxide, and dodecanoyl peroxide. The above conditions differ, however, from typical commercial ones, because such high pressures and long reaction times are impractical.

The actual commercial conditions vary, depending upon location and individual technology of each company. Often, tubular and multiple-tray autoclaves are used.[7] Good reactor design must permit dissipation of the heat of polymerization (800–1000 cal/g), with good control over other parameters of the reaction. Tubular reactors are judged as having an advantage over stirred autoclaves in offering greater surface-to-volume ratios and better control over residence time.[7] On the other hand, the stirred autoclaves offer a more uniform temperature distribution throughout the reactor.

The tubular reactors have been described as consisting of stainless steel tubes between one-half and one inch in internal diameters and about two inches in external diameters. The residence time in these tubes is from three to five minutes and they can be equipped with pistons for pressure regulation. Pressure might also be controlled by flow pulses to the reactor.[8] For the oxygen-initiated reactions, the optimum conditions are[7] 0.03–0.1% oxygen at 190–210 °C and 1500 atmospheres pressure. At this pressure the density of ethylene is 0.46 g/cm^3. This compares favorably with the critical density of ethylene, namely, 0.22 g/cm^3. Once the polymerization is initiated, the liquid monomer acts as a solvent for the polymer. Impurities, such as acetylene or hydrogen, cause chain transferring and must be carefully removed. In some processes hindered phenols are added in small quantities (between ten and one-thousand parts per million). This has the effect of reducing long-chain branching and yields film grade resins with better clarity, lower haze, and a reduced amount of microgels. Also, diluents are used in some practices. Their main purpose is to act as heat-exchanging media, but they can also help to remove the polymer from the reactor. Such diluents

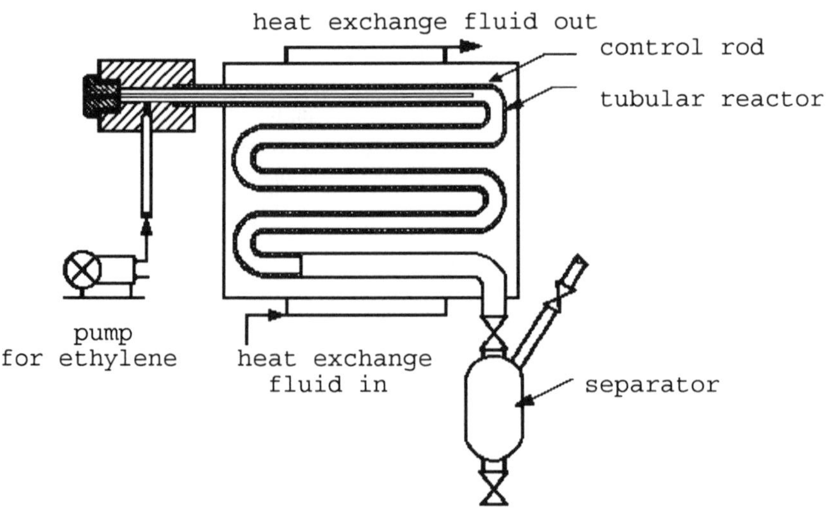

FIGURE 5.1. Tubular reactor as described in a Monsanto Co. U.S. Patent # 2,856,494.

are water, benzene, ethyl or methyl alcohols. Sometimes, chain-transferring agents like carbon tetrachloride, ketones, aldehydes, or cyclohexane might also be added to control molecular weight. A tubular reactor is illustrated in Fig. 5.1. The finished product (polymer–monomer mixture) is conveyed to a separator, where almost all of the unreacted ethylene is removed under high pressure (3500–5000 psi) and recycled. The polymer is extruded and pelletized. Ethylene conversion per pass is a limiting factor on the economics.

Polyethylene prepared in this way may have as many as 20 to 30 short branches per 10,000 carbon atoms in the chain[9] and one or two long-chain branches per molecule, due to "backbiting"[10]:

$$\sim CH_2-\overset{H}{\overset{|}{C}}(-H\;\; CH_2\bullet)(-H_2C-CH_2-CH_2-) \longrightarrow \sim CH_2-\overset{H}{\overset{|}{C}}\bullet\;\; (CH_3)(-H_2C-CH_2-CH_2-) \xrightarrow{CH_2=CH_2} \sim CH_2-\underset{\underset{CH_3}{|}}{\underset{(CH_2)_3}{\underset{|}{CH}}}-CH_2-CH_2\sim$$

The reaction results in predominantly ethyl and butyl branches. The ratio of ethyl to butyl groups is roughly 2:1.[11,12] Chain transferring to the tertiary hydrogens at the location of the short branches causes elimination reactions and formation of vinylidine groups.[13,14] This mechanism also accounts for formation of low molecular weight species[18]:

$$\sim CH_2-CH_2-\underset{\underset{CH_3}{|}}{\underset{(CH_2)_3}{\underset{|}{\overset{H}{\overset{|}{C}}}}}-CH_2-CH_2\sim \;+\; R\bullet \longrightarrow \sim CH_2-CH_2-\underset{\underset{CH_3}{|}}{\underset{(CH_2)_3}{\underset{|}{\overset{\bullet}{C}}}}-CH_2-CH_2\sim \;+\; RH$$

$$\longrightarrow \sim CH_2-CH_2-\underset{CH_2}{\underset{\|}{C}}-CH_2-CH_2\sim \;+\; C_3H_7\bullet$$

Commercial grades of low-density polyethylene vary widely in the number of short and long branches, average molecular weights, and molecular weight distributions; M_W/M_N is between 20 and 50 for commercial low-density materials. The short branches control the degree of crystallinity, stiffness, and polymer density. They also influences the flow properties of the molten material.

5.1.2. Commercial High-Density Polyethylene, Properties and Manufacture

High-density polyethylene (0.94–0.97 g/cm^3) is produced commercially with two types of catalysts:

1. Ziegler–Natta-type catalysts.
2. Transition metal oxides on various supports.

The two catalytic systems are used at different conditions. Both types have undergone evolution from earlier development. The original practices are summarized in Table 5.1.

The Ziegler process yields polyethylene as low as 0.94 g/cm^3 in density, but process modifications can result in products with a density of 0.965 g/cm^3. The transition metal oxide catalysts on support, on the other hand, yield products in the density range of 0.960–0.970 g/cm^3.

The original development by Ziegler led to what appears to be an almost endless number of patents for various Ziegler–Natta-type catalysts and processes. As described in Chapter 3, such catalysts have been vastly improved. Progress was made toward enhanced efficiency and selectivity. The amount of polymer produced per gram of the transition metal has been increased manyfold. In addition, new catalysts, based on zirconium compounds complexed with aluminoxane oligomers (sometimes called Kominsky catalysts[305]), were developed. They yield very high quantities of polyethylene per gram of the catalyst. For instance, a catalyst bis(cyclopentadienyl)zirconium

Table 5.1. Typical Conditions for the Preparation of High-Density Polyethylene as Described in the Original Patents

	Ziegler–Natta process	Chromium oxide on support	Molybdenum oxide on support
Approximate temperature	75 °C	140 °C	234 °C
Approximate pressure	60 psi	420 psi	1000 psi
Usual state of the polymer in reaction mixture	Suspension	Suspension	Suspension

dichloride combined with methylaluminoxane is claimed to yield 5000 kg of linear polyethylene per gram of zirconium per hour.[14] Polymers of the same density are still produced.

An important factor in catalyst activity is the degree of oligomerization of the aluminoxane moiety. The catalytic effect is enhanced by increase in the number of alternating aluminum and oxygen atoms. These catalysts have long storage life and offer such high activity that they need not be removed from the product, because the amount present is negligible.[14,15] This makes the workup of the product simple.

The continuous solution processes are usually carried out between 120–160 °C at 400–500 lb/in^2 pressure. The diluents may be cyclohexane or isooctane. In one-zone reactors the solid catalyst is evenly dispersed throughout the reactor. In the two-zone reactors (specially constructed) the polymerizations are conducted with stirring in the lower zone where the catalysts are present in concentrations of 0.2–0.6% of the diluent. Purified ethylene is fed into the bottom portions of the reactors. The polymers that form are carried with small portions of the catalyst to the top and removed. To compensate for the loss, additional catalysts are added intermittently to the upper "quiescent" zones.

In suspension or slurry polymerizations, various suspending agents, like diesel oil, lower petroleum fractions, heptane, toluene, mineral oil, chlorobenzene, or others, are used. The polymerization temperatures are kept between 50–75 °C at only slightly elevated pressures, like 25 lb/in^2. If these are batch reactions, they last between one and four hours.

Polymerizations catalyzed by transition metal oxides on support were described variously as employing solid/liquid suspensions, fixed beds, and solid/gas-phase operations. It appears, however, that the industrial practices are mainly confined to use of solid/liquid suspension processes. The polymerization is carried out at the surface of the catalyst suspended in a hydrocarbon diluent.

In continuous slurry processes the temperatures are kept between 90–100 °C and pressures between 400–450 lb/in^2. The catalyst concentrations range between 0.004–0.03% and typical diluents are *n*-pentane and *n*-hexane. Individual catalyst particles become imbedded in polymer granules as the reaction proceeds. The granules are removed as a slurry containing 20–40% solids.

There are variations in the individual processes. In some procedures the temperature is maintained high enough to keep the polymer in solution. In others, it is kept deliberately low to maintain the polymer in a slurry. The products are separated from the monomer that is recycled. They are cooled, precipitated (if in solution), and collected by filtration or centrifugation.

Various reactors were developed to handle different slurry polymerization processes. One model, a countercurrent-flow reactor for production of polyethylene, is shown in Fig. 5.2. The slurry is maintained in suspension by ethylene gas. The gas rises to the top and maintains agitation, while the polymer particles settle to the bottom where they are collected.

Several companies adopted loop reactors. These are arranged so that the flowing reactants and diluents continuously pass the entrance to a receiving zone. The heavier particles gravitate from the flowing into the receiving zone while the lighter diluents and reactants are recycled. To accommodate that, the settling area must be large enough for the heavy polymer particles to be collected and separated.

In addition to suspension, a gas-phase process was developed. No diluent is used in the polymerization step. Highly purified ethylene gas is combined continuously with a dry-powdery catalyst and then fed into a vertical fluidized-bed reactor. The reaction is carried out at 270 psi and 85–100 °C. The circulating ethylene gas fluidizes the bed of growing granular polymer and serves

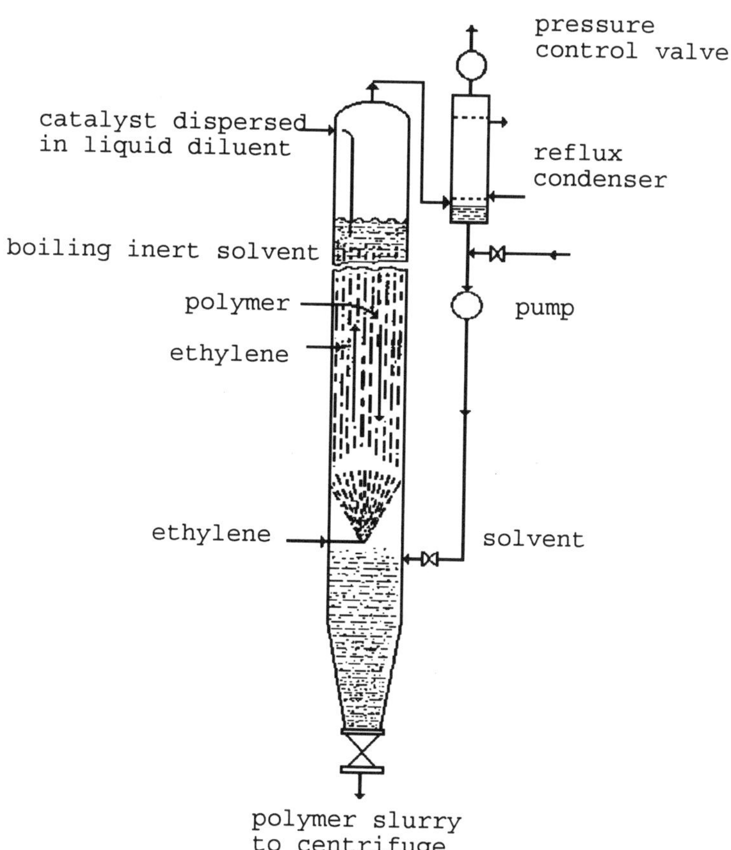

FIGURE 5.2. Countercurrent-flow reactor for slurry polymerization of ethylene with Ziegler catalysts (as illustrated in Koppers Co. Inc., British Patent # 826,563).

to remove the heat.[15] Formed polymer particles are removed intermittently from the lower sections of the vertical reactor. The product contains 5% monomer that is recovered and recycled. Control of polymer density is achieved by copolymerization with α-olefins. Molecular weights and molecular weight distributions are controlled by catalyst modifications, by varying operating conditions, and/or use of chain transferring agents,[15] such as hydrogen.[16]

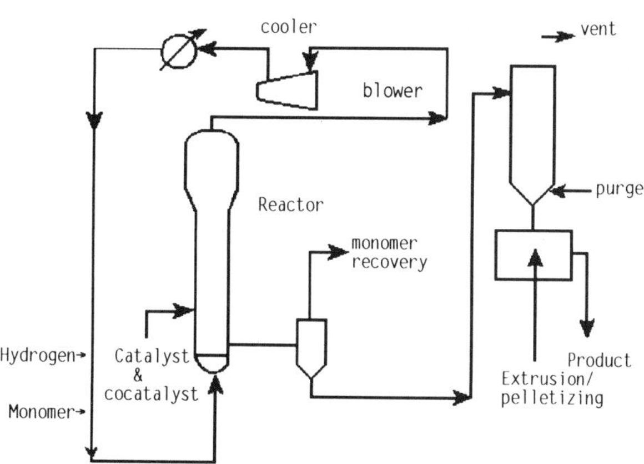

FIGURE 5.3. Gas-phase polyolefin process. From Burdett, by permission of the American Chemical Society.

Table 5.2. Properties of Commercial Polyethylenes[a]

Properties	Free-radical polymerization	Ziegler–Natta process	Metal oxides on support
Density	0.92–0.93 g/cm^2	0.94 112	0.95–0.96 g/cm^2
Melting point	108–110.7 °C	129–131 °C	136 °C
% Amorphous	43.1	25.8	25.8
Structure	20–30 ethyl and butyl branches per 1000 carbons a few long branches	Mainly linear 7 ethyl branches per 1000 carbons	Almost linear
Double bonds	0.6–2 per 1000 carbon atoms	0.1–1 per 1000 carbon atoms	Up to 3 per 1000 carbon atoms
Types of double bond	15% Terminal vinyl 68% vinylidine 17% internal *trans* olefins	43% terminal vinyl 32% vinylidine 25% internal *trans* olefins	94% terminal vinyl 1% vinylidine 5% internal *trans* olefins

[a]From various sources in the literature.

The reactors for the fluidized gas-phase process are simple in design. There are no mechanical agitators and they rely upon blowers to keep the bed fluidized and well mixed. Catalysts and cocatalysts are fed directly to the reactor. Figure 5.3 illustrates the process.[304]

The weight average molecular weights of most commercial low- and high-density polyethylenes range between 50,000–300,000. Very low molecular weight polyethylene waxes and very high molecular weight materials are also available. The molecular weight distributions for high-density polyethylene vary between 4–15. The product generally has fewer than three branches per thousand carbon atoms.[9]

Table 5.2 summarizes the properties of various polyethylenes.

Some Ziegler–Natta catalysts can be adopted to produce polyethylene homopolymers that are branched and possess low density. One example is a catalyst that is prepared by *in situ* reactions of bis(1,5-cyclooctadiene)nickel or bis(η^3-allyl)nickel with amino-bis-(imino) phosphorane.[17] These catalysts, however, are not used for commercial production of low-density polyethylene.

5.1.3. Materials Similar to Polyethylene

Materials that are quite similar to polyethylene can be obtained from other starting materials. The most prominent is formation of polymethylene and similar high molecular weight paraffin hydrocarbons from diazoalkanes. The reaction was originally carried out by Pechmann in 1899[19] when small quantities of a white flocculent powder formed in an ether solution of diazomethane. Bamberger and Tschirner[20] showed that this white powder is polymethylene $-(-CH_2-)_n-$ that melts at 128 °C. The synthesis has been improved since by the introduction of various catalysts. The reaction can yield highly crystalline polymers that melt at 136.5 °C[21] with the molecular weight in millions.[22,23] Among the catalysts, boron compounds are very efficient.[22,23] Bawn *et al.*[24] postulated the mechanism of catalytic action. It consists of initial coordination of a monomer with the initiator, BF_3. This is followed by a loss of nitrogen and a shift of a fluorine atom from boron to carbon. The successive additions of molecules of diazoalkane follow a similar path with a shift of the chain fragment to the electron-deficient carbon:

$$:\overset{\ominus}{C}H_2-\overset{\oplus}{N}\equiv N \longleftrightarrow CH_2=\overset{\oplus}{N}=\underset{\ominus}{\ddot{N}}: \xrightarrow{BF_3} F_3B \leftarrow CH_2-\overset{\oplus}{N}\equiv\overset{\ominus}{N}: \longrightarrow$$

$$F_3\overset{\ominus}{B}-\overset{\oplus}{C}H_2 + N_2\uparrow \longrightarrow F_2B-CH_2F \xrightarrow{CH_2N_2} \begin{array}{l} F_2\overset{\ominus}{B}-CH_2F \\ \quad | \\ \quad CH_2-\overset{\oplus}{N}\equiv N \end{array} \longrightarrow$$

$$\begin{array}{l} F_2\overset{\ominus}{B}-CH_2F \\ \quad | \\ \oplus CH_2 \end{array} + N_2\uparrow \longrightarrow F_2B-CH_2-CH_2F \xrightarrow{n\ CH_2N_2} F_2B-CH_2-(-CH_2-)_n-CH_2F$$

The resulting macromolecules are still reactive toward additional diazoalkanes. The above step-growth polymerization reactions can also yield block copolymers:

$$F_2B\text{—}[\text{—}CH_2\text{—}]_n\text{—}[\text{—}\underset{\displaystyle R}{\underset{|}{CH}}\text{—}]_m\text{—}F$$

Formation of polymethylene by this reaction is not practical for commercial utilization.

Colloidal gold and fine copper powder also catalyze diazoalkane polymerizations. The reaction appears to proceed by formation of alkylidene or carbene species that are bound to the surfaces of metals.[25–27] The initiations are completed by additions of diazoalkanes to the bound carbenes followed by liberations of nitrogen. Termination may take place by chain transfer, perhaps to a monomer, or to the solvent.[25–27]

Many different diazoalkanes lend themselves to these polymerization reactions. Polypentylidine, polyhexylidine, polyheptilidine, and polyoctylidine form with a gold complex catalyst, $AuCl_3$–pyridine.[28]

An entirely different route to preparation of macroparaffins is through a high-pressure reaction between hydrogen and carbon monoxide. Transition metals, like finely divided ruthenium, catalyze this reaction. At pressures of about 200 atmospheres and temperatures below 140 °C, polymethylene of molecular weight as high as 100,000 forms[29]:

$$H_2 + CO \xrightarrow{Ru} \text{—}[\text{—}CH_2\text{—}]_n\text{—}$$

5.2. Polypropylene

Propylene monomer, like ethylene, is obtained from petroleum sources. Free-radical polymerizations of propylene and other α-olefins are completely controlled by chain transferring.[30] It is therefore polymerized by anionic coordination polymerization. At present, mainly isotactic polypropylene is being used in large commercial quantities. There is some utilization of atactic polypropylene as well. Syndiotactic polypropylene, on the other hand, still remains mainly a laboratory curiosity.

The polypropylene that was originally described by Natta contained less than 50% isotactic fractions. The remainder was atactic material. Some stereoblocks composed of isotactic and atactic polypropylenes were also formed. This type of product forms when α-olefins are polymerized in inert hydrocarbons with catalysts prepared by reducing high-valency metal compounds, like $TiCl_4$, with organometallic compounds like $Al(C_2H_5)_3$.

Later, heterogeneous, highly crystalline catalysts based on transition metals (valence 3 or less), like $TiCl_2$, $TiCl_3$, $ZrCl_3$, and VCl_3, were developed that yielded stereospecific polypropylene. The metal halides were combined with selected metal alkyls. Only those alkyls were picked that would not destroy the crystalline lattice of the transition metal salts in the process of the reaction. The resultant catalysts yielded crystalline polypropylenes with high fractions of the isotactic material. The products, however, also contained some low molecular weight fractions, some amorphous and stereoblock materials. They still required costly purification and separations to obtain relatively pure isotactic polypropylene. The atactic polymer is a wax-like substance that lacks toughness. Also, the presence of amorphous materials, or very low molecular weight compounds, causes tackiness and impedes processing. Table 5.3 lists some of the catalysts and the amounts of crystallinity in polymers that were reported by Natta.[55]

To avoid costly purification of isotactic polypropylene, three-component catalyst systems were developed. Some of the original ones appear to have been reported by Natta himself, who found that addition of Lewis bases enhances the quantity of the crystalline material. Table 5.4 shows the effects of addition of Lewis bases on the amount of crystallinity, reported by Natta.[56]

Many other three-component systems have been developed since.[31–36] Also, development of more active catalysts[35,37] eliminates a need to remove them from the finished product.[15] The first improvement in catalyst productivity came from treating $TiCl_3$ [formed from $TiCl_4$ and $Al(C_2H_5)Cl_2$] with aliphatic ethers resulting in yields of 520 grams of polymer for each gram of Ti.[41] Further

Table 5.3. Polypropylenes Prepared by Natta[55]

Transition metal halide	Metal alkyl	Percent crystallinity
$TiCl_3$ (β)	$Al(C_2H_5)_3$	40–50
$TiCl_3$ (α, γ, or δ)	$Al(C_2H_5)_3$	96–98
$TiCl_3$ (α, γ, or δ)	$Al(C_2H_5)_2Cl$	96–98
$TiCl_3$ (α, γ, or δ)	$Be(C_2H_5)_2$	94–96
$TiCl_3$ (α, γ, or δ)	$Mg(C_2H_5)_2$	78–85
$TiCl_3$ (α, γ, or δ)	$Zn(C_2H_5)_2$	30–40
VCl_3	$Al(C_2H_5)_3$	73
$TiCl_2$	$Al(C_2H_5)_3$	55

improvement was achieved by supporting TiCl on $MgCl_2$ or by producing a supported catalyst by reacting $TiCl_4$ with $Mg(OC_2H_5)$ or with other magnesium compounds. This raised the productivity to over 3000 grams of polymer for every gram of Ti.[41] The products, however, contained low percentages of the isotactic isomer (20–40%). Addition of a Lewis base like *N,N,N′,N′*-tetramethylethylenediamine in solid component and ethyl benzoate in solution raised the isotactic content to 93% with a productivity of 2500 grams of polymer per gram of Ti.[41] Claims are made today for much greater catalyst activity. It was reported, for instance, that catalyst efficiencies of 40 kg of polymer per one gram of Ti can be achieved. Such yields require proper choice of catalysts and control over polymerization conditions. The isotactic fractions in the products are reported to range from 95%–97%.[42–44]

In a catalyst system $TiCl_3/MgCl_2/C_6H_5COOC_2H_5/Al(C_2H_5)_3$, the high activity was initially attributed to higher propagation rates rather than to an increase in the concentration of the active sites.[61] The higher activity of these catalysts, however, was shown instead to be due to higher numbers of active centers and only slightly higher values of K_P.[62] Subsequent trends in modifications of supported Ziegler–Natta catalysts consisted of using sterically hindered amines.[313–315] For instance, 2,2,6,6-tetramethylpiperidine might be used together with different trialkylaluminum compounds as modifier–cocatalyst systems for the supported catalysts:

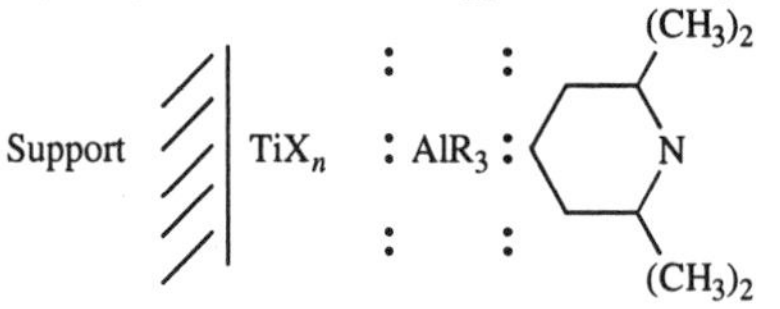

Support | TiX_n : AlR_3 : N [piperidine ring with $(CH_3)_2$ at 2 and 6 positions]

where X represents a halogen.

Other analogous amines, like 1,2,4-trimethylpiperazine and 2,3,4,5-tetraethylpiperidine, are also used in preparations of titanium halide catalysts supported on $MgCl_2$. The amine remains as a built-in modifier in the catalyst system.[314]

Recent research efforts concentrated on soluble catalytic systems, like di-η^5-cyclopentadienyldiphenyltitanium and tetrabenzylzirconium complexed with methylaluminoxane, $(CH_3)_2Al–[–O–Al(CH_3)–]_n–Al(CH_3)_2$. Such catalysts, however, yield products that contain only about 85% isotactic polypropylene,[302–308] and only if the reactions are conducted at low temperatures, –45 °C or lower.

Table 5.4. Effect of Addition of Lewis Bases on the Amount of Crystalline Fraction in Polypropylene[a,56]

Transition metal halide	Aluminum alkyl	Lewis base	Percent of crystalline material
$TiCl_3$	$2Al(C_2H_5)Br_2$	Pyridine	> 98.5
$TiCl_3$	$2Al(C_2H_5)Cl_2$	$N(C_2H_5)_3$	95
$TiCl_3$	$2Al(C_2H_5)Cl_2$	$NH(C_2H_5)_2$	93
$TiCl_3$	$2Al(C_2H_5)Br_2$	$N^{\oplus}(C_4H_9)_4I^{\ominus}$	> 99
$TiCl_3$	$2Al(C_2H_5)Cl_2$	$N^{\oplus}(C_4H_9)_4Br^{\ominus}$	96

[a]From Natta, Pasquon, Zambelli, and Gatti, by permission of Wiley and Sons, Inc.

A major breakthrough occurred when rigid chiral metallocene initiators were developed, like 1,1-ethylene-di-η^5-indenylzirconium dichloride, complexed with methylaluminoxane. In place of zirconium, titanium and hafnium analogs can also be used. These catalysts are highly isospecific[308–310] when used at low temperatures. The compounds are illustrated in Chapter 3. Typical catalysts consist of aluminum to transition metal ratios of 103 or 104 to 1. Many of them yield 98–99% isotactic fractions of the polymer. In addition, these are very active catalysts, yielding large quantities of polymer per gram of zirconium.

5.2.1. Manufacturing Techniques

The earliest commercial methods used slurry polymerizations with liquid hydrocarbon diluents, like hexane or heptane. These diluents carried the propylene and the catalyst. Small amounts of hydrogen were fed into the reaction mixtures to control molecular weights. The catalyst system consisted of a deep purple or violet-colored $TiCl_3$ reacted with diethyl aluminum chloride. The $TiCl_3$ was often prepared by reduction of $TiCl_4$ with an aluminum powder. These reactions were carried out in stirred autoclaves at temperatures below 90 °C and at pressures sufficient to maintain a liquid phase. The concentration of propylene in the reaction mixtures ranged between 10–20%. The products formed in discrete particles and were removed at 20–40% concentrations of solids. Unreacted monomer was withdrawn from the product mixtures and reused. The catalysts were deactivated and dissolved out of the products with alcohol containing some HCl, or removed by steam extraction. This was followed by extraction of the amorphous fractions with hot liquid hydrocarbons.

Later, bulk polymerization processes were developed where either liquid propylene was used as the only diluent in a loop reactor or was permitted to boil out to remove the heat of reaction. The latter was conducted in stirred vessels with vapor space at the top.

More recently, gas-phase polymerizations of propylene were introduced. The technology is similar to the gas-phase technology in ethylene polymerizations[15] described in Section 5.1.

5.2.2. Syndiotactic Polypropylene

Isotactic polypropylene received most attention because it is commercially more desirable. Nevertheless, syndiotactic polypropylene, though less crystalline, has greater clarity, elasticity, and impact resistance. It melts, however, at a lower temperature. This isomer was originally prepared with both heterogeneous, titanium-based catalysts and soluble, vanadium-based ones. The heterogeneous catalysts gave very low yields of the syndiotactic fractions. In fact, original samples contained only a few percent of the desired material, almost an impurity. The yield of syndiotactic polypropylene increased with a decrease in polymerization temperature, but still remained low.[38]

Highly syndiotactic polypropylene was prepared by Natta and co-workers[39] with homogeneous catalysts formed from VCl_4 or from vanadium triacetylacetonate, aluminum dialkyl halide, and anisole at –48 to –78 °C. No isotactic fractions formed. This led to the development of many effective soluble catalysts. The catalyst components and the conditions for their preparation are quite important in maintaining control over syndiotactic placement. For the most effective soluble catalyst the ratio of AlR_2X to the vanadium compound must be maintained between 3 and 10.[39] The organic portion of the organoaluminum compound can be either methyl, ethyl, isobutyl, neopentyl, phenyl, or methylstyryl.[39,40] In addition to VCl_4 and to vanadium triacetylacetonate,[39] various other vanadates can be used, like $[VO(OR)_xCl_{3-x}]$, where $x = 1, 2,$ or 3.[38] The exact nature of the vanadium compound, however, is very important for the resultant steric arrangement of the product. For instance, VCl_4 combined with $Al(C_2H_5)_2F$ forms a heterogeneous catalyst that yields the isotactic isomer.[38] Vanadium triacetylacetonate, on the other hand, upon reacting with $Al(C_2H_5)_2F$ forms a soluble catalyst that yields the syndiotactic isomer.[39] Additions of certain electron donors increases the amount of syndiotactic placement. These are: anisole, furan, diethyl ether, cycloheptanone, ethyl acetate, and thiophene.[40] The optimum results are obtained when an anisole-to-vanadium ratio is 1:1. Also, the highest amount of syndiotactic polymer is obtained when the soluble catalysts are

Table 5.5. Comparison of Isotatic and Syndiotactic Polypropylenes[a]

Isomer	Crystal structure	Density at 25 °C	Melting point (°C)	Typical M_w	Typical M_n
Isotactic	Monoclinic	0.92–0.43 g/cm	171–186	220–700 K	38–160 K
	Triclinic	0.943 g/cm			
	Hexagonal				
Syndiotactic	Orthorhombic	0.89–0.91 g/cm	138		

[a]From Ref. 47.

prepared and used at low temperatures. Even at low temperatures, however, like –78 °C, the amount of syndiotacticity that can be obtained with a specific catalyst decreases with time.[38,39,46] This indicates a deterioration of the syndiotactic placing sites. On the other hand, polymerization of propylene with soluble vanadium triacetylacetonate–$Al(C_2C_5)_2Cl$ system was reported to be a "living"-type polymerization.[145] The product has a narrow molecular weight distribution (M_w/M_n = 1.05–1.20). A kinetic study indicates an absence of chain transferring and termination at temperatures below –65 °C.

The most recent catalysts for syndiotactic polypropylene are complexes, like *i*-propyl(cyclopentadienyl-1-fluorenyl)hafnium dichloride with methyl aluminoxane.[49] Another, similar catalyst is *i*-propyl(η^5-cyclopentadienyl-η^3-fluorenyl)zirconium dichloride with methyl aluminoxane. These catalysts yield polymers that are high in syndiotactic material (the zirconium-based compound yields 86% of racemic pentads).[48,49] Commercial production of syndiotactic polypropylene is in the planning stage.[311] What catalytic system will be used, however, is not known at this time. Some of the properties of the two isomers, isotatic and syndiotactic polypropylenes, are compared in Table 5.5.

The molecular weights of syndiotactic polypropylenes can apparently vary from a number average molecular weight of 25,000 to 60,000, depending upon reaction conditions.[45] Also, in isotactic polypropylene there is less than one double bond per thousand carbon atoms.[47] A typical value of M_w/M_n is 5–12.

5.3 Polyisobutylene

The original commercial methods for preparing high molecular weight poyisobutylene by cationic polymerization in good yields were reported in 1940. The reaction was carried out at –40 to –80 °C in a diluent with BF_3 catalysis.[50] This developed into current commercial practices of polymerizing isobutylene at –80 to –100 °C, using liquid ethylene or methyl chloride as a diluent.[51,54] Even at these low temperatures the reaction is quite violent. Methods were therefore developed to dissipate the heat. In one of them, called the "flash polymerization process," the catalyst (a Lewis acid, like BF_3 or $AlCl_3$, for instance) is added in solution to the cooled isobutylene solution. The polymerization takes place very rapidly and is complete in a few seconds with the heat of the reaction being removed by vaporization of the diluent. Such reactions, however, are very difficult to carry out in conventional batch reactors. Two types of procedures were therefore adopted.[52] The first is built around a moving stainless steel belt contained inside a gas-tight reactor housing. Isobutylene and liquid ethylene from one source and a Lewis acid in ethylene solution (0.1–0.3 percent based on monomer) from another source are fed continuously onto the moving belt, where they are mixed and moved. The movement of the belt is adjusted at such a speed that the polymerization is complete before the polymer arrives at the end of its travel, where it is removed with a scraper and further processed.

In the second process, the polymerization is carried out in multiple kneaders or mixers. These are arranged in a series of descending steps. Here, the reaction mixture is carried from one kneader to another with the temperature being raised at each station and completed at the last one.

All commercially important polyisobutylenes are linear, head-to-tail polymers, with tertiary butyl groups at one end of the chains and vinylidine groups at the other:

$$CH_3-\underset{CH_3}{\overset{CH_3}{\underset{|}{\overset{|}{C}}}}-(-CH_2-\underset{CH_3}{\overset{CH_3}{\underset{|}{\overset{|}{C}}}}-)_n-CH_2-\overset{CH_3}{\overset{|}{C}}=CH_2$$

The differences lie in molecular weights. They range from 2000 to 20,000 for viscous liquids, to between 100,000 and 400,000 for high molecular weight elastomers that resemble unmilled crepe rubber. The polymers degrade readily from thermal abuse. They can be stabilized effectively, however, by adding small quantities (0.1–1.0%) of such stabilizers as aromatic amines, phenols, or sulfur compounds. Polyisobutylenes are soluble in many hydrocarbons and are resistant to attacks by many chemicals.

Coordinated anionic polymerizations with Ziegler–Natta catalysts yield similar polymers that range from viscous liquids to rubbery solids. At 0 °C, a catalyst with a 1:16 Ti to Al molar ratio yields a polymer with a molecular weight of 5000–6000.[53] The molecular weight, however, is dependent upon the reaction time. This contrasts with polymerizations of ethylene, propylene, and 1-butene by such catalysts, where the molecular weights of the products are independent of the reaction time. In addition, there are some questions about the exact molecular structures of the products.[53]

High molecular weight polyisobutylene has fair tensile strength, but suffers from the disadvantage of considerable cold flow. A copolymer of butylene with some isoprene for crosslinking is therefore used as a commercial elastomer and called "butyl rubber." The isoprene is present in the copolymer in only minor proportions (1.4–4.5%). The uncrosslinked material is very similar to polyisobutylene. Copolymers of isobutylene with other dienes are also called butyl rubbers. They can also be terpolymers, where the third component may be cyclopentadiene for improved ozone resistance.

The molecular weights of the copolymers vary inversely with the quantities of isoprene incorporated, the polymerization temperatures and amount of impurities present during polymerization. Impurities like *n*-butene or water act as chain-transferring agents.[54] To maintain uniform molecular weights, the conversions are usually kept from exceeding 60%.

5.4. Higher Poly(α-olefin)s

Many α-olefins were polymerized by the Ziegler–Natta catalysts to yield high polymers and many such polymers were found to be stereospecific and crystalline. Polymerizations of α-olefins of general structure $CH_2=CH-(-CH_2-)_x-R$, where x is 0 to 3 and R denotes CH_3, $CH-(CH_3)_2$, $C(CH_3)_3$, or C_6H_5, can be catalyzed by vanadium trichloride/triethyl aluminum.[57] The conversions are fairly high, though higher crystallinity can be obtained with titanium-based catalysts.[58] Addition of Lewis bases, such as $(C_4H_9)_2O$, $(C_4H_9)_3N$, or $(C_4H_9)_3P$, to the catalyst system further increases crystallinity.[59]

5.4.1. Properties of Poly(α-olefin)s

Many poly(α-olefin)s reported in the literature are not used commercially for various reasons. Table 5.6 lists some of the olefins polymerized by the Ziegler–Natta catalysts.[47,60]

5.4.2. Poly(butene-1)

Isotactic poly(butene-1) is produced commercially with three-component Ziegler–Natta-type catalysts. It is manufactured by a continuous process with simultaneous additions to the reaction vessel of the monomer solution, a suspension of $TiCl_2-AlCl_3$, and a solution of diethyl aluminum chloride.[63] The effluent containing the suspension of the product is continually removed from the

Table 5.6. Properties of Poly(α-olefins)[a]

Monomer	State	Melting point (°C)
$CH_2{=}CH_2$	Crystalline	136
$CH_2{=}CH{-}CH_3$	Crystalline	165–168
$CH_2{=}CH{-}C_2H_5$	Crystalline	124–130
$CH_2{=}CH{-}(CH_2)_2CH_3$	Crystalline	75
$CH_2{=}CH{-}(CH_2)_3CH_3$	Rubber, amorphous	—
$CH_2{=}CH{-}(CH_2)_9CH_3$	Some crystallinity	45
$CH_2{=}CH{-}(CH_2)_{15}CH_3$	Crystallinity in pendant groups	70; 100
$CH_2C{=}CH{-}CH{-}(CH_3)_2$	Crystalline, hard	240–285
$CH_2{=}CH{-}CH_2{-}CH{-}(CH_3)_2$	Crystalline, hard	200–240
$CH_2{=}CH{-}CH_2{-}C(CH_3)_3$	Crystalline, hard	300–350
$CH_2{=}CH{-}CH_2{-}C(CH_3)_2{-}CH{=}CH_2$	Crystalline, hard	350
$CH_2{=}CH{-}CH_2{-}CH(CH_3){-}C_2H_5$	Crystalline, hard	160
$CH_2{=}CH{-}(CH_2)_2{-}CH(CH_3)_2$	Rubber, amorphous	—
$CH_2{=}CH{-}(CH_2)_2{-}C_6H_5$	Crystalline, slightly rubbery	158
$CH_2{=}CH{-}CH(C_6H_5){-}CH_3$	Crystalline, intractable	360
$CH_2{=}CH{-}(CH_2)_3{-}C_6H_5$	Rubber, amorphous	—

[a]From Refs. 47 and 60.

reactor. Molecular weight control is achieved through regulating the reaction temperature. The effluent contains approximately 5 to 8% of atactic polybutene that is dissolved in the liquid carrier. The suspended isotactic fractions (92–98%) are isolated after catalyst decomposition and removal. The product has a density of 0.92 g/cm^3 and melts at 124–130 °C.

Isotactic polybutene crystallizes into three different forms. When it cools from the melt, it initially crystallizes into a metastable crystalline one. After several days, however, it transforms into a different form. Noticeable changes in melting point, density, flexural modulus, yield, and hardness accompany this transformation. The third crystalline form is due to crystallization from solution. The polymer exhibits good impact and tear resistance. It is also resistant to environmental stress-cracking.

5.4.3. Poly(4-methyl pentene-1)

Another commercially produced polyolefin is isotactic poly(4-methyl pentene-1). The polymer carries a trade name of TPX. This material is known for high transparency, good electrical properties, and heat resistance. Poly(4-methyl pentene-1) has a density of 0.83 g/cm^3. This polyolefin exhibits poor load-bearing properties and is susceptible to UV degradation. It is also a poor barrier to moisture and gases and scratches readily. This limits its use in many applications.

Poly(4-methyl pentene) is produced by the same process and equipment as polypropylene. A post-finishing deashing step, however, is required. In addition, aseptic conditions are maintained during manufacture to prevent contamination that may affect clarity.

A number of similar polyolefins with pendant side groups are known. These include poly(3-methyl butene-1), poly(4,4-dimethyl pentene-1) and poly(vinyl cyclohexane). Due to their increased cohesive energy, ability to pack into tight structures, and the effect of increasing stiffness of the pendant groups, some of these polymers have a high melting point. This can be seen from Table 5.9. Many of these polymers, however, tend to undergo complex morphological changes on standing. This can result in fissures and planes of weakness in the structure.

5.5. Copolymers of Ethylene and Propylene

Many monomers have been copolymerized with ethylene by a variety of polymerization methods. When ethylene is copolymerized with other olefins, the resultant hydrocarbon polymers have reduced regularity and lower density, lower softening point, and lower brittle point.

Copolymers of ethylene and propylene are a commercially important family of materials. They vary from elastomers that can contain 80% ethylene and 20% propylene to polypropylene that is modified with small amounts of ethylene to improve impact resistance.

5.5.1. Ethylene–Propylene Elastomers

The commercial *ethylene–propylene rubbers* typically range in propylene content from 30 to 60%, depending upon intended use. Such copolymers are prepared with Ziegler–Natta-type catalysts. Soluble catalysts and true solution processes are preferred. The common catalyst systems are based on VCl_4, $VOCl_3$, $V(Acac)_3$, $VO(OR)_3$, $VOCl(OR)_2$, $VOCl_2(OR)$, etc. with various organoaluminum derivatives. The products are predominantly amorphous. Polymerization reactions are usually carried out at 40 °C in solvents like chlorobenzene or pentane. The resultant random copolymers are recovered by alcohol precipitation. Because these elastomers are almost completely saturated, crosslinking is difficult. A third monomer, a diene, is therefore included in the preparation of these rubbers that carry the tradenames EPTR or EPDM. Inclusion of a third monomer presents some problems in copolymerization reactions. For instance, it is important to maintain constant feed mixtures of monomers to obtain constant compositions. Yet, two of the three monomers are gaseous and the third one is a liquid. Natta[64] developed a technique that depends upon maintaining violent agitation of the solvent while gaseous monomers were bubbled through the liquid phase. This was referred to as a "semi-flow technique." The process allows the compositions of gaseous and liquid phases to be in equilibrium with each other and to be more or less constant.[65] Other techniques have meanwhile evolved. All are designed to maintain constant polymerization mixtures.

The vanadium-based catalyst systems deteriorate with time and decrease in the number of catalytic centers as the polymerizations progress. The rate of decay is affected by conditions used for catalyst preparation, compositions of the catalysts, temperature, solvents, and Lewis bases. It is also affected by the type and concentration of the third monomer.[66–68] Additions of chlorinated compounds to the deactivated catalysts, however, help restore activity.[69,70] Catalyst decay can also be overcome by continually feeding catalyst components into the polymerization medium.[71]

While the third monomer can be a common diene, like isoprene, more often it is a bridged ring structure with at least one double bond in the ring. In typical terpolymer rubbers with 60 to 40 ratios of ethylene to propylene, the diene components usually comprise about 3% of the total. Some specialty rubbers, however, may contain 10% of the diene or even more. Reaction conditions are always chosen to obtain 1,2 placement of the diene. Dienes in common use are ethylidine norbornene, methylene norbornene, 1,4-hexadiene, dicyclopentadiene, and cycloocadiene:

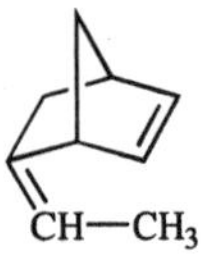

$CH_2{=}CH{-}CH_2{-}CH{=}CH{-}CH_3$

1, 4-hexadiene

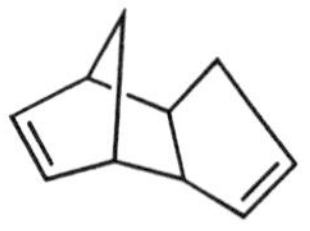

CH—CH₃ → $CH{-}CH_3$

ethylidine norbornene

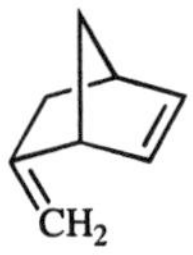

CH_2

methylene norbornene

dicylcopentadiene

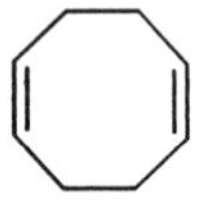

cyclooctadiene

In addition to the above, the patent literature describes many other dienes. An idealized picture of a segment of an uncrosslinked gum stock might be shown as having the following structure:

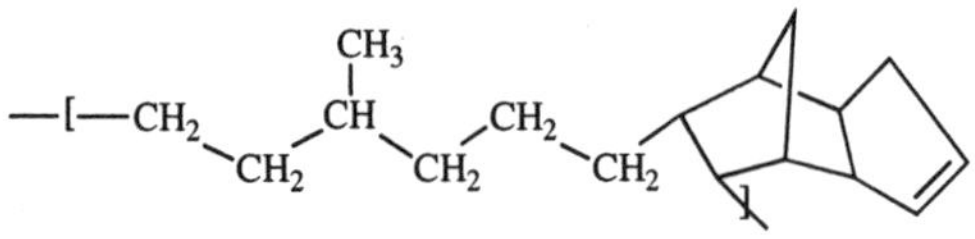

$-[-CH_2-CH_2-CH(CH_3)-CH_2-CH_2-CH_2-$

5.5.2. Copolymers of Ethylene with α-Olefins

Many copolymers of ethylene with α-olefins are prepared commercially. Thus ethylene is copolymerized with butene-1, where a comonomer is included to lower the regularity and the density of the polymer. Many copolymers are prepared with transition metal oxide catalysts on support. The comonomer is usually present in approximately 5% quantities. This is sufficient to lower the crystallinity and to markedly improve the impact strength and resistance to environmental stress cracking. Copolymers of ethylene with hexene-1, where the hexene-1 content is less than 5%, are also produced for the same reason.

In most cases, the monomers that homopolymerize by Ziegler–Natta coordination catalysts also copolymerize by them.[72] In addition, some monomers that do not homopolymerize may still copolymerize to form alternating copolymers. Because the lifetime of a growing polymer molecule is relatively long (it can be as long as several minutes), block copolymerization is possible through changes in the monomer feeds. Also, the nature of the transition metal compound influences the reactivity ratios of the monomers in copolymerizations. On the other hand, the nature of the organometallic compound has no such effect.[73] It also appears that changes in the reaction temperature between 0 and 75 °C have no effect on the r values. Copolymers can be formed using either soluble or heterogeneous Ziegler–Natta. One problem encountered with the heterogeneous catalysts is the tendency by the formed polymers to coat the active sites. This forces the monomers to diffuse to the sites and may cause starvation of the more active monomer if both diffuse at equal rates.

Many different block copolymers of olefins, like ethylene with propylene and ethylene with butene-1, are manufactured. Use of the anionic coordination catalysts enables variations in the molecular structures of the products. It is possible to vary the length and stereoregularity of the blocks. This is accomplished by feeding alternately different monomers into the reactor. When it is necessary that the blocks consist of pure homopolymers, then after each addition the reaction is allowed to subside. If any residual monomer remains, it is removed.[74] This requires a long lifetime for the growing chains and an insignificant amount of termination. The stability of the anion depends upon the catalyst system. One technique for catalyst preparation is to preform $TiCl_3$ by reducing $TiCl_4$ with diethylaluminum chloride, followed by careful washing of the product of reduction to remove the byproduct $Al(C_2H_5)Cl$. Other reports describe using the α-form of $TiCl_3$ or heat treating it to form the β- or γ-forms that yield more stereospecific products.

The transition metal oxide catalysts on support, such as the CrO_3/silica-alumina (Phillips) and MoO_3/Al_2O_3 (Standard Oil), are used to copolymerize minor quantities of α-olefins with ethylene. Such copolymerizations introduce short pendant groups into polymer backbones.

5.5.3. Copolymers of Propylene with Dienes

Although presently lacking industrial importance, alternating copolymers can be made from propylene and butadiene,[75] also from propylene and isoprene.[76] Copolymers of propylene and butadiene form with vanadium- or titanium-based catalysts combined with aluminum alkyls. The catalysts have to be prepared at very low temperature (–70 °C). Also, it was found that a presence of halogen atoms in the catalyst is essential.[75] Carbonyl compounds, such as ketones, esters, and others, are very effective additives. A reaction mechanism based on alternating coordination of propylene and butadiene with the transition metal was proposed by Furukawa.[75]

5.5.4. Copolymers of Ethylene with Vinyl Acetate

Various copolymers of ethylene with vinyl acetate are prepared by a free-radical mechanism in emulsion polymerizations. Both reactivity ratios are close to 1.[77] The degree of branching in these copolymers is strongly temperature-dependent.[78] These materials find wide use in such areas as paper coatings and adhesives. In addition, some are hydrolyzed to form copolymers of ethylene with vinyl alcohol. Such resins, available commercially in various ratios of polyethylene to poly(vinyl alcohol), can range from 30% poly(vinyl alcohol) to as high as 70%.

Vinyl acetate residues in ethylene-vinyl acetate copolymers reduce the regularity of polyethylene. This reduces crystallinity in the polymer. Materials containing 45% vinyl acetate are elastomers and can be crosslinked with peroxide.

5.5.5. Ionomers

Another group of commercial copolymers of ethylene are those formed with acrylic and methacrylic acids, where ethylene is the major component. The copolymerizations are carried out under high pressures. These materials range in comonomer content from 3–20%. Typical values are 10%. A large proportion of the carboxylic acid groups (40–50%) are prereacted with metal ions like sodium or zinc. The copolymer salts are called ionomers with a tradename like Syrlin. The materials tend to behave similarly to crosslinked polymers at ambient temperature by being stiff and tough. Yet they can be processed at elevated temperatures, because aggregation of the ionic segments from different polymeric molecules is destroyed. The material becomes mobile, but after cooling the aggregates reform. Ionomers exhibit good low-temperature flexibility. They are tough, abrasion-resistant resins that adhere well to metal surfaces.

5.6. Homopolymers of Conjugated Dienes

Many different polymers of conjugated dienes are prepared commercially by a variety of processes, depending upon the need. They are formed by free-radical, ionic, and coordinated anionic polymerizations. In addition, various molecular weight homopolymers and copolymers, ranging from a few thousand for liquid polymers to high molecular weight ones for synthetic rubbers, are on the market.

5.6.1. Polybutadiene

1,3-Butadiene, the simplest of the conjugated dienes, is produced commercially by thermal cracking of petroleum fractions and catalytic dehydrogenation of butane and butene. Polymerization of butadiene can potentially lead to three poly(1,2-butadiene)s, atactic, isotactic, and syndiotactic, and two *cis* and *trans* forms of poly(1,4-butadiene). This is discussed in Chapters 2 and 3.

Free-radical polymerizations of 1,3-butadiene usually result in polymers with 78–82% of 1,4-type placement and 18–22% of 1,2-adducts. The ratio of 1,4- to 1,2-adducts is independent of the temperature of polymerization. Moreover, this ratio is obtained in polymerizations that are carried out in bulk and in emulsion. The ratio of *trans*-1,4 to *cis*-1,4 tends to decrease, however, as the temperature of the reaction decreases. Polybutadiene polymers formed by free-radical mechanism are branched, because the residual unsaturations in the polymeric chains are subject to free-radical attacks:

```
                                        ~~CH2
                                          |
                                          CH •          ------>  etc.
          ~~CH2                           |
            |                    ~~CH2—CH
            CH               ↗            |
            ||                            CH2
~~CH2 • +   CH                            |
            |                             CH2
            CH2              ↘            |
            |                     ~~CH—CH=CH2
            CH2
            |
       ~~CH—CH=CH2
                           ~~CH2—CH=CH—CH2—CH2—CH~~
                                                 |
                                          ~~CH2—CH
                                                 |
                                                 CH2 •
```

Should branching become excessive, infinite networks can form. The products become crosslinked, insoluble, and infusible. Such materials are called *popcorn* polymers. This phenomenon is more common in bulk polymerizations. The crosslinked polymers form nodules that occupy much more volume than the monomers from which they formed and often clog up the polymerization equipment, sometimes even rupturing it.

High molecular weight homopolymers of 1,3-butadiene formed by a free-radical mechanism lack the type of elastomeric properties that are needed from commercial rubbers. Copolymers of butadiene, however, with styrene or acrylonitrile are more useful and are prepared on a large scale. This is discussed in another section.

5.6.1.1. Liquid Polybutadiene

Low molecular weight liquid homopolymers of 1,3-butadiene and also some liquid copolymers find industrial uses in many applications. These materials can range in molecular weights from 500 to 5000 depending upon the mode of polymerization. Liquid polybutadienes formed by cationic polymerizations are high in *trans*-1,4 content. Such materials find applications in industrial coatings. They are usually prepared with Lewis acids in chlorinated solvents. When the reactions are catalyzed by $AlCl_3$ at –78 °C, two types of polymers form.[80] One is soluble and the other is insoluble, depending upon the extent of conversion. $AlCl_3$, $AlBr_3$, and BF_3–Et_2O produce polymers with the same ratios of *trans*-1,4- to 1,2-adducts. These range from 4 to 5. Polymerizations carried out in ethylene chloride[80] catalyzed by $TiCl_4$ yield products with lower ratios of *trans*-1,4- to 1,2-adducts. The ratios of the two placements are affected by the solvents. They are also affected by additions of complexing agents, such as nitroethane and nitrobenzene.[80] The changes, however, are small.

Hydroxy-terminated liquid polybutadienes are prepared for reactions with diisocyanates to form elastomeric polyurethanes (see Chapter 6). Such materials can be prepared by anionic polymerizations as "living" polymers and then quenched at the appropriate molecular weight. These polybutadienes can also be formed by a free-radical mechanism. The microstructures of the two products differ, however, and this may affect the properties of the finished products. To form hydroxy terminated polymers by a free-radical mechanism, the polymerization reactions may be initiated by hydroxy radicals from hydrogen peroxide.

Liquid polybutadienes that are high in 1,2 placement are also available commercially. These range from reactive polymers containing approximately 70% of vinyl groups to very reactive ones containing more than 90% of 1,2 units. The materials are formed by anionic polymerization with either sodium naphthalene, or with sodium dispersions, or with organolithium initiators in polar solvents. Carboxyl group terminated liquid polybutadienes are predominantly used as modifiers for epoxy resins (Chapter 6). They are formed by anionic mechanisms in solution with organolithium catalysts like diphenylethanedilithium, butanedilithium, isoprenelithium, or lithium methylnaphthalene complexes. Cyclohexane is the choice solvent. The reactions are quenched with carbon dioxide to introduce the terminal carboxyl groups.

5.6.1.2. High Molecular Weight Polybutadiene

High molecular weight polybutadiene homopolymers are prepared commercially with anionic catalysts and with coordinated anionic ones. Polybutadiene formed with sodium dispersions was prepared industrially in the former USSR, and perhaps might still be produced in that area today. This sodium-catalyzed polybutadiene contains 65% of 1,2-adducts.[83] Most of the preparations by others, however, utilize either alkyllithium or Ziegler–Natta-type catalysts prepared with titanium tetraiodide or preferably containing cobalt.

Because high molecular weight polybutadiene can be prepared by different catalytic systems, the choice of catalyst is usually governed by the desired microstructure of the product. Alfin catalysts yield very high molecular weight polymers with large amounts of *trans*-1,4 structures.[53] Both the molecular weight and microstructure can be affected significantly, however, by variations in the Alfin catalyst components. These can be the alkyl groups of the organometallic compounds or alkoxide portions.

When butadiene is polymerized with lithium metal or with alkyllithium catalysts, inert solvents like hexane or heptane must be used to obtain high *cis*-1,4 placement (see Chapter 3). Based on ^{13}C NMR spectra, 1,4-polybutadiene formed with *n*-butyllithium consists of blocks of *cis*-1,4 units and *trans*-1,4 units that are separated by isolated vinyl structures[79]:

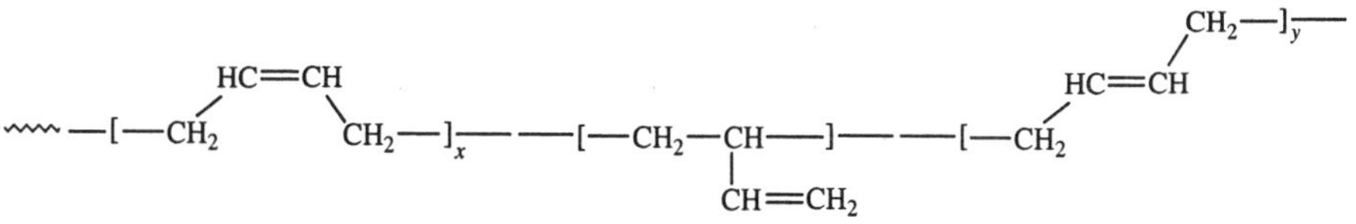

$$\sim\!\!\sim -[-CH_2-CH{=}CH-CH_2-]_x - [-CH_2-\underset{\substack{|\\ CH=CH_2}}{CH}-] - [-CH_2-CH{=}CH-CH_2-]_y-$$

The quantity of such units in the above polybutadienes is approximately 48–58% *trans*-1,4, 33–45% *cis*-1,4, and 7–10% 1,2 units.[81,82] There is little effect of the reaction temperatures upon this composition. As described in Chapter 3, however, addition of Lewis bases has a profound effect. Reactions in tetrahydrofuran solvent result in 1,2 placement that can be as high as 87%.

The microstructures of polybutadienes prepared with Ziegler–Natta catalysts vary with catalyst composition. It is possible to form polymers that are high either in 1,2 placement or in 1,4 units. The catalysts and the type of placement are summarized in Table 5.7.

Butadiene can be polymerized with chromium oxide catalyst on support to form solid homopolymers. The products, however, tend to coat the catalyst within a few hours after the start of the reaction and interfere with further polymerization. Polybutadiene can also be prepared in the presence of molybdena catalyst promoted by calcium hydride. The product contains 80% of 1,4 units and 20% of 1,2 units. Of the 1,4 units, 62.5% are *cis* and 37.5% are *trans*.[53]

Cobalt oxide on silica-alumina in the presence of alkyl aluminum also yields high *cis*-1,4 structure polymers. An all-1,2 polybutadiene can be prepared with *n*-butyllithium modified with bis-piperidino ethane. The atactic polymer can be formed in hexane at −5 to +20 °C temperature.[126]

The 100% 1,2 placement was postulated to proceed according to the following scheme[126]: First a complex forms between the base and butyllithium:

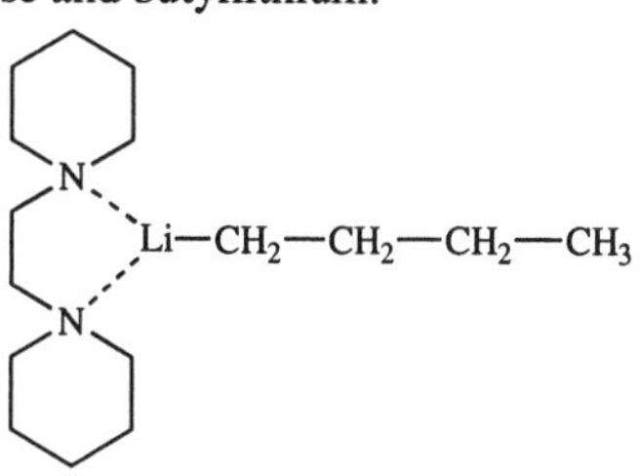

$$\text{(piperidino-N)}\cdots Li-CH_2-CH_2-CH_2-CH_3 \cdots \text{(N-piperidino)}$$

Table 5.7. Microstructures of Polybutadienes Prepared with Ziegler–Natta Catalysts[a]

Catalyst	Microstructure (%)		
	cis-1,4	*trans*-1,4	1,2
$TiI_4/Al(C_2H_5)_3$	95	2	3
$TiBr_4/Al(C_4H_9)_3$	88	3	9
β-$TiCl_3/Al(C_2H_5)_3$	80	12	8
$Ti(OC_4H_9)_4/Al(C_2H_5)_3$	—	—	99–100
$Ti(OC_6H_5)_4/R_3Al$	90–100	—	—
(π-Cyclooctadiene)$_2$Ni, CF_3CO_2H	100	—	—
Bis(π-Crotyl *NiCl*)	92	—	
Bis(π-Crotyl *NiI*)	—	94	—
$CoCl_2/Al(C_2H_5)_2Cl$	96–97	2.5	1–1.5
$CoCl_2/Al(C_2H_5)_3$	94	3	3
$CoCl_2/Al(C_2H_5)_3$/Pyridine	90–97	—	—
Co-stearate/AlR_2Cl	98	1	1
VCl_3/AlR_3	—	99	1
$VOCl_3/Al(C_2H_5)_3$	—	97–99	2–3
$VCl_4/AlCl_3$	—	95	—
$Cr(C_6H_5CN)_6/Al(C_2H_5)_3$	100	—	—
$MoO_2(OR)_3/Al(C_2H_5)_3$	—	—	75

[a]From various sources in the literature.

The above complex, upon formation, reacts with butadiene to form a new complex:

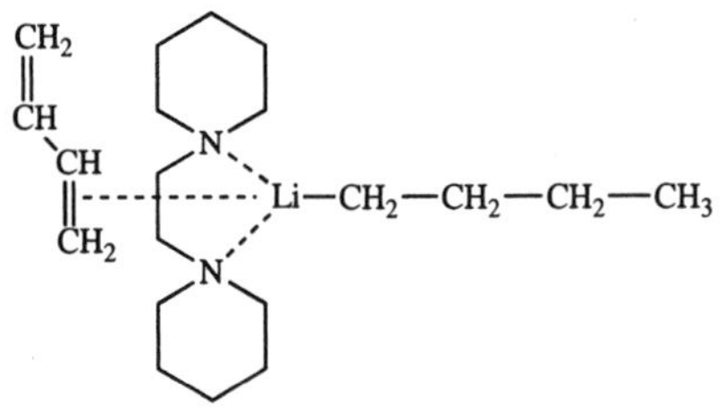

This is followed by insertion of butadiene into the carbon–lithium bond:

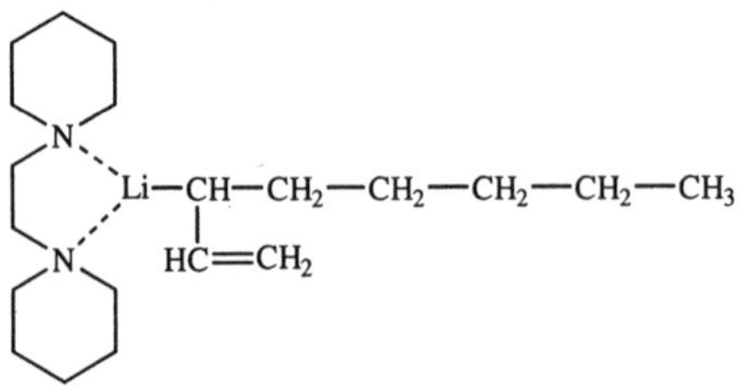

The process is repeated many times in the propagation reaction with the complex being the active species for insertion:

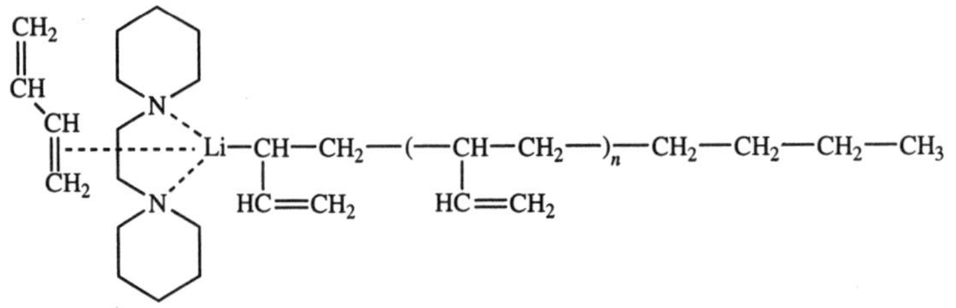

5.6.2. Polyisoprene

Polyisoprenes occur in nature. They are also prepared synthetically. Most commercial processes try to duplicate the naturally occurring material.

5.6.2.1. Natural Polyisoprenes

Rubber hydrocarbon is the principle component of raw rubber. The subject is discussed in greater detail in Chapter 7. Natural rubber is 97% *cis*-1,4-polyisoprene. It is obtained by tapping the bark of rubber trees (*Hevea brasiliensis*) and collecting the exudate, a latex consisting of about 32–35% rubber. A similar material can also be found in the sap of many other plants and shrubs. The structure of natural rubber has been investigated over 100 years, but it was only after 1920, however, that the chemical structure was elucidated. It was shown to be a linear polymer consisting of head-to-tail links of isoprene units, 98% bonded 1,4.

5.6.2.2. Synthetic Polyisoprenes

In following natural rubber, the synthetic efforts are devoted to obtaining very high *cis*-1,4-polyisoprene and to form a synthetic "natural" rubber. Two types of polymerizations yield products that approach this. One is through the use of Ziegler–Natta-type catalysts and the other through anionic polymerization with alkyllithium compounds in hydrocarbon solvents. One commercial process, for instance, uses reaction products of $TiCl_4$ with triisobutylaluminum at an Al/Ti ratio of 0.9–1.1 as the catalyst. Diphenyl ether or other Lewis bases are sometimes added as catalyst modifiers.[84–86,90] The process results in an approximately 95% *cis*-1,4-polyisoprene product. Typically, such reactions are carried out on a continuous basis, usually in hexane, and take 2–4 hours. Polymerizations are often done in two reaction lines, each consisting of four kettles arranged in series. The heat of the reaction is partially absorbed by precooling the feed streams. The remaining heat is absorbed on cooled surfaces. When the stream exits, the conversion is about 80%. Addition of a shortstop solution stabilizes the product.

Alkyllithium-initiated polymerizations of isoprene yield polymers with 92–93% *cis*-1,4 content. One industrial process uses butyllithium in a continuous reaction in two lines each consisting of four reaction kettles. The heat of the reaction is removed by vaporization of the solvent and the monomer. The catalyst solution is added to the solvent stream just before it is mixed intensively with the isoprene monomer stream and fed to the first reactor. After the stream leaves each reactor, small quantities of methanol are injected between stages into the reaction mixture. This limits the molecular weight by stopping the reaction. Fresh butyllithium catalyst is added again at the next stage in the next reactor to initiate new polymer growth.[87–89]

As described in Chapters 2 and 3, the monomer can be inserted into the polyisoprene chain potentially in nine different ways. These are the three tactic forms of the 1,2-adducts, two 1,4-adducts, *cis* and *trans*, and three tactic forms of 3,4-adducts. In addition, there is some possibility of head-to-head and tail-to-tail insertion, though the common addition is head-to-tail. Table 5.8 presents the various microstructures that can be obtained in polymerizations of isoprene with different catalysts.

Cationic polymerizations of isoprene proceed more readily than those of butadiene, though both yield low molecular weight liquid polymers. $AlCl_3$ and stannic chloride can also be used in chlorinated solvents at temperatures below 0 °C. Without chlorinated solvents, however, polymerizations of isoprene require temperatures above 0 °C.

At high conversions, cationic polymerizations of isoprene result in formations of some crosslinked material.[91] The soluble portions of the polymers are high in *trans*-1,4 structures. Alfin catalysts yield polymers that are higher in *trans*-1,4 structures than free-radical emulsion polymerizations.[92]

Chromium oxide catalysts on support polymerize isoprene like butadiene to solid polymers. Here too, however, during the polymerization process polymer particles cover the catalyst completely within a few hours from the start of the reaction and retard or stop further polymer formation. The polymerization conditions are the same as those used for butadiene. The reactions can be carried out over fixed bed catalysts containing 3% chromium oxide on SiO_2–Al_2O_3. Conditions are 88 °C

Table 5.8. Polymerization Products of Isoprene

Mode of polymerization	Solvent	Approximate % *cis*-1,4	% *trans*-1,4	% 1,2	% 3,4
Free radical	Emulsion in water	22	65	6	7
Cationic	—	37	51	4	9
	Chloroform (30 °C)	—	90	4	6
Anionic					
Lithium	Pentane	94	0		6
Ethyllithium	Pentane	94	0		6
Butyllithium	Pentane	93	0		7
Sodium	Pentane	0	43	6	51
Ethylsodium	Pentane	6	42	7	45
Butylsodium	Pentane	4	35	7	54
Potassium	Pentane	0	52	8	40
Ethylpotassium	Pentane	24	39	6	31
Butylpotassium	Pentane	20	41	6	34
Rubidium	Pentane	5	47	8	39
Cesium	Pentane	4	51	8	37
Ethyllithium	Ethyl ether	6	29	5	60
Ethylsodium	Ethyl ether	0	14	10	76
Lithium	Ethyl ether	4	27	5–7	63–65
Alfin	Pentane	27	52	5	16
Coordination catalysts					
α-$TiCl_3/AlR_3$			91		
$VCl_3/Al(C_2H_5)_3$		—	99	—	—
$TiCl_4/Al(C_2H_5)_3$			95–96		
$TiCl_4/AlR_3$ + amine		100	—	—	—
$CoCl_2/AlR_3$ + pyridine		96			
V(acetylacetonate)$_3/AlR_3$		90			
$Ti(OR)_4/Al(C_2H_5)_3$		95			

[a]From Refs. 95–107.

and 42 kg/cm^2 pressure with the charge containing 20% isoprene and 80% isobutane.[93] The mixed molibdena–alumina catalyst with calcium hydride also yields polyisoprene.

Lithium metal dispersions form polymers of isoprene that are high in *cis*-1,4 contents as shown in Table 5.8. These polymers form in hydrocarbon solvents. This is done industrially and the products are called Coral rubbers. They contain only a small percent of 3,4 structures and no *trans*-1,4 or 1,2 units. The materials strongly resemble Hevea rubber.

Use of Ziegler–Natta catalysts, as seen from Table 5.8, can yield an almost all-*cis*-1,4-polyisoprene or an almost all-*trans*-1,4-polyisoprene. The microstructure depends upon the ratio of titanium to aluminum. Ratios of Ti:Al between 0.5:1 and 1.5:1 yield the *cis* isomer. A 1:1 ratio is the optimum. At the same time, ratios of Ti:Al between 1.5:1 to 3:1 yield the *trans* structures.[53] The titanium-to-aluminum ratios affect the yields of the polymers as well as the microstructures. There also is some influence on the molecular weight of the product.[94] Variations in catalyst compositions, however, do not affect the relative amounts of 1,4 to 3,4 or to 1,2 placements. Only *cis* and *trans* arrangements are affected. In addition, the molecular weights of the polymers and the microstructures are relatively insensitive to the catalyst concentrations. The temperatures of the reactions, however, do affect the rates, the molecular weights, and the microstructures.

5.7. Methyl Rubber, Poly(2,3-dimethylbutadiene)

Early attempts at preparations of synthetic rubbers resulted in developments of elastomers from 2,3-dimethylbutadiene. The material, called "methyl rubber," was claimed to yield better elastomeric properties than polybutadiene. Methyl rubber was produced in Germany during World War I where the monomer was prepared from acetone. The polymerizations were carried out by free-radical mechanism and anionically, using sodium metal dispersions for initiation. Later, it was demonstrated that 2,3-dimethyl polybutadiene can be polymerized to very high *cis*-1,4 polymer with Ziegler–Natta catalysts.[108,109]

5.8. Chloroprene Rubber, Poly(2-chloro-1,3-butadiene)

2-Chloro-1,3-butadiene (chloroprene) was originally synthesized in 1930. The material can polymerize spontaneously to an elastomer that has good resistance to oil and weathering. Commercial production of chloroprene rubber started in 1932. Since then, many types of polymers and copolymers were developed with the trivial generic name of *neoprene*.

The monomer can be prepared from acetylene:

$$CH{\equiv}CH \xrightarrow[CCl_4]{NH_4Cl} CH_2{=}CH{-}C{\equiv}CH \xrightarrow{HCl} CH_2{=}CH{-}\underset{}{\overset{Cl}{\overset{|}{C}}}{=}CH_2$$

It can also be formed from butadiene.

Chloroprene is polymerized commercially by free-radical emulsion polymerization. The reaction is carried out at 40 °C to a 90% conversion. A typical recipe for such an emulsion polymerization is as follows[110]:

Material	Parts
Water	150
Chloroprene	100
Rosin	4 (stabilizer)
NaOH	0.8 (stabilizer)
$K_2S_2O_8$	
Sulfur	
Methylene-bis-(Na-Naphthalenesulfonic acid)	0.7

When the polymerization reaches 90% conversion, the reaction mixture is cooled to 20 °C and tetraethylthiuram disulfide is added. This is done to prevent the pendant unsaturation in the polychloroprene backbones from crosslinking or forming branches. An unmodified polymer is difficult to process even at a 70% conversion. To overcome this, a sulfur–tetraethylthiuram modification is carried out.

When the product is treated with the thiuram, an exchange reaction takes place to yield a stable, thiuram-modified polymer of reduced molecular weight. It is believed that the reaction takes place through cleavage of the sulfur links, formed during polymerization in the presence of sulfur, and formation of free radicals[111]:

$$\sim\!(CH_2{-}\overset{Cl}{\overset{|}{C}}{=}CH{-}CH_2{-})_n{-}S{-}S{-}S{-}S{-}({-}CH_2{-}\overset{Cl}{\overset{|}{C}}{=}CH{-}CH_2{-})_m\!\sim \longrightarrow$$

$$\longrightarrow \sim\!({-}CH_2{-}\overset{Cl}{\overset{|}{C}}{=}CH{-}CH_2{-})_n{-}S{-}S\cdot \; + \; \cdot S{-}S{-}({-}CH_2{-}\overset{Cl}{\overset{|}{C}}{=}CH{-}CH_2{-})_m\!\sim$$

$$\Big\downarrow \; + \begin{matrix} S{-}\overset{S}{\overset{\|}{C}}{-}N(C_2H_5)_2 \\ | \\ S{-}\underset{S}{\underset{\|}{C}}{-}N(C_2H_5)_2 \end{matrix}$$

$$\sim\!({-}CH_2{-}\overset{Cl}{\overset{|}{C}}{=}CH{-}CH_2{-})_n{-}S{-}S{-}\overset{S}{\overset{\|}{C}}{-}N(C_2H_5)_2 \; + \; (C_2H_5)_2N{-}\overset{S}{\overset{\|}{C}}{-}S{-}S\cdot$$

After the reaction with tetraethylthiuram disulfide is completed, the latex is acidified with acetic acid, short of coagulation. The rubber is then recovered at a low temperature (about –15 °C) in the form of sheets, by deposition of the latex on cooled rotating drums.[111,112]

Table 5.9. Structures of Polychloroprenes Formed by Free-Radical Polymerization

	% 1,4				
Polymerization temperature (°C)	*cis*	*trans*	% 1,2	% 3,4	Ref.
–40	5	94	0.9	0.3	112
–20	6	91.5	0.7	0.5	113
–10	7	—	—	—	112
10	9	84	1.1	1.0	112
40	10	86, 81	1.6	1.0	112
40	13	88.9	0.9	0.3	113
100	13	71	2.4	2.4	112

[a]From Refs. 110–115.

The polymer, formed by this technique, consists of about 85% *trans*-1,4 units, 10% *cis*-1,4 units, 1.5% 1,2 units, and 1.0% 3,4 units. The polymer is essentially linear with a molecular weight equal to approximately 100,000. The sulfur-modified polychloroprenes are sold under the tradename Neoprene-G. An unmodified version prepared with mercaptan chain-transferring agents (Neoprene W) is a polymer with a molecular weight of about 200,000.[111,113]

Table 5.9 lists the structures of polychloroprenes that form by free-radical polymerization at different temperatures. Chloroprene polymerizes by cationic polymerization with the aid of Lewis acids in chlorinated solvents. When aluminum chloride is used in a mixture of ethyl chloride–methylene chloride solvent mixture at –80 °C, the polymer has 50% 1,4 units.[114,115] If it is polymerized with boron trifluoride, the product consists of 50–70% 1,4-adducts. A very high *trans*-1,4-poly(2-chloro-1,3-butadiene) forms by X-ray radiation polymerization of large crystals of chloroprene at –130 to –180 °C. It is 97.8% *trans*-1,4.[113] Presumably, the mechanism of polymerization is free-radical.

5.9. Miscellaneous Polymers from Dienes

There are many reports in the literature of preparations of polymers from various other substituted dienes. Most have no commercial significance. Some are, however, interesting materials. An example is a polymer of 2-*t*-butyl-1,3-butadiene formed with $TiCl_4$ and either alkylaluminum or aluminum hydride catalysts.[116] The polymer is crystalline and melts at 106 °C. It can be dissolved in common solvents. Based on X-ray data, the monomer placement is high *cis*-1,4.

Poly(carboxybutadiene)s also form with coordination catalysts[117–119]:

R, H, H, COOR′ → [R, H, H, COOR′]$_n$

where R = CH_3; R′ = CH_3, C_2H_5, C_4H_9, or C_6H_5.

X-ray crystallography[117–119] showed that the placement is *trans*-isotactic. Based on the mode of packing of the chains in the crystalline regions and from the encumbrance of the side groups in relationship to the main chain, an *erythro* configuration can be assigned.[118] The polymers are therefore *trans-erythro*-isotactic.

Polymerization of 1,3-pentadiene can potentially result in five different insertions of the monomers. These are 1,4-*cis*, 1,4-*trans*, 1,2-*cis*, 1,2-*trans*, and 3,4. In addition, there are potentially 3-*cis*-1,4 and 3-*trans*-1,4 structures (isotactic, syndiotactic, and atactic). Formations of *trans*-1,4-isotactic, *cis*-1,4-isotactic, and *cis*-1,4-syndiotactic polymers are possible with Ziegler–Natta catalysts.[120–124] Amorphous polymers also form that are predominantly *cis*-1,4 or *trans*-1,4, but lack tactic order.[123] Stereospecificity in poly(1,3-pentadiene) is strongly dependent upon the solvent used during the polymerization. Thus, *cis*-1,4-syndiotactic polymers form in aromatic solvents and *trans*-1,2 in aliphatic ones. The preparations require cobalt halide/aluminum alkyl dichloride (or dialkyl chloride) catalysts in combinations with Lewis bases. To form a *trans*-1,4 structure a catalyst containing an aluminum-to-titanium ratio close to 5 must be used.[125]

5.10. Cyclopolymerization of Conjugated Dienes

Conjugated dienes like isoprene, butadiene, and chloroprene cyclopolymerize with catalysts consisting of aluminum alkyls, like ethylaluminum dichloride and titanium tetrachloride.[127] The ladder polymers that form contain fused cyclic structures. The products prepared in hexane are generally insoluble powders, while those prepared in aromatic solvents are soluble even when the molecular weights are high.[128] A high ratio of the transition metal halide to that of the aluminum alkyl must be used. Such a ratio might conceivably mean that the mechanism of polymerization is cationic. Also, conventional cationic initiators can be used to yield similar products. The cyclization occurs during propagation. Unsaturation in the products can vary from none to as high as 80%, depending upon the initiator used.[128]

Different mechanisms were offered to explain the cyclization of 1,3-dienes.[128,129] The cyclization might conceivably occur by a sequential process:

or, perhaps, from attacks by the propagating carbon cation on *trans*-1,4 double bonds:

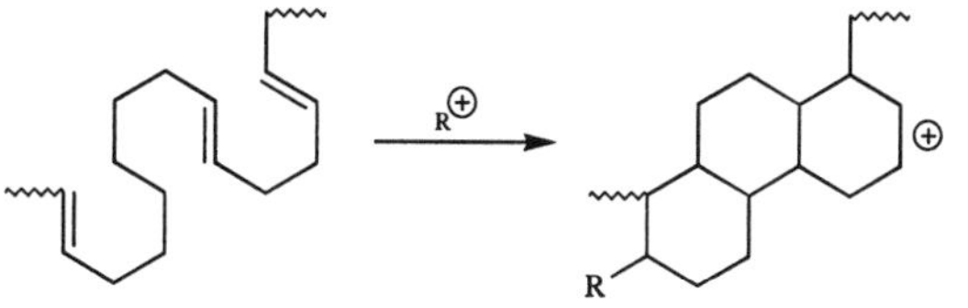

where $R^{\oplus}$ can represent either a propagating carbon cation or an initiating species. The extensive cyclization may be a result of a sequential process.[128,129]

Cyclopolymerizations typically result in low conversions and dormant reaction mixtures. When additional monomer is added, the dormant mixtures reinitiate polymerizations that again proceed to some limited conversions. If the original dormant mixtures are allowed to stand for a long time, the unreacted monomers are slowly consumed.[128]

Polymerization of 2,3-dimethylbutadiene-1,3 with Ziegler–Natta catalysts consisting of $Al(i\text{-}C_4H_9)_3$–$TiCl_4$ yields *cis*-1,4-polydimethylbutadiene, as described earlier. This, however, takes place when the aluminum alkyl is in excess. If, on the other hand, the ratio of Al to Ti is 1 or less, cyclic polymer forms instead. The product has reduced unsaturation and some *trans*-1,4 units in the chain.[131] A complex catalyst, consisting of $Al(i\text{-}C_4H_9)_3$–$CoCl_2$, yields polymers that are predominantly *cis*-1,4 with about 20% of 1,2 units. On the other hand, acid catalysts, like $Al(C_2H_5)Cl_2$, yield cyclic polymers.[130,131] A polymer formed with the aid of X-ray radiation at low temperatures also contains cyclic units and some *trans*-1,4.[132] Butadiene and isoprene also form this type of polymer under the same conditions.[132]

5.11. Copolymers of Dienes

Several different elastomers, copolymers of butadiene, are produced commercially. The major ones are copolymers of butadiene with styrene and butadiene with acrylonitrile. Some terpolymers, where the third component is an unsaturated carboxylic acid, are also manufactured. Block copolymers of isoprene with styrene and butadiene with styrene are important commercial elastomers.

Table 5.10. Typical Recipes for Preparation of Butadiene–Styrene Rubbers by Emulsion Polymerization

	"Hot" process		"Cold" process	
Material	parts	purpose	parts	purpose
Butadiene	75	Comonomer	72	Comonomer
Styrene	25	Comonomer	28	Comonomer
Water	180	Carrier	180	Carrier
Fatty acid soap	5.0	Emulsifier	4.5	Emulsifier
n-Dodecyl mercaptan	0.5	Chain transferring agent	—	—
t-Dodecyl mercaptan	—	—	0.2	Chain transferring agent
Potassium persulfate	0.3	Initiator	—	—
Auxiliary surfactant	—	—	0.3	Stabilizer
Potassium chloride	—	—	0.3	Stabilizer
p-Menthane hydroperoxide	—	—	0.06	Initiator system
Ferrous sulfate	—	—	0.01	
Ethylenediamine tetraacetic acid sodium salt	—	—	0.05	
Sodium formaldehyde sulfoxylate			0.05	

[a]From Ref. 110 and patent literature.

5.11.1. GR-S Rubber[133]

Copolymerization of butadiene with styrene by free-radical mechanism has been explored very thoroughly. The original efforts started during World War I in Germany. Subsequent work during the thirties was followed by a particularly strong impetus in the U.S. during World War II. This led to the development of GR-S rubber in the U.S. and Buna-S rubber in Germany. After World War II, further refinements were introduced into the preparatory procedures and "cold" rubber was developed. Industrially, the copolymer is prepared by emulsion copolymerization of butadiene and styrene at low temperatures in a continuous process. A typical product is a random distribution copolymer, with the butadiene content ranging from 70–75%. The diene monomer placement is roughly 18% *cis*-1,4, 65% *trans*-1,4, and 17% 1,2; M_n of these copolymers is about 100,000.

A "redox" initiator is used in the cold process, but not in the "hot" one. Also, the "hot" process is carried out at about 50 °C for 12 hours to approximately 72% conversion. The "cold" process is also carried for 12 hours, but at about 5 °C to a 60% conversion. The two recipes for preparation of GR-S rubbers are shown in Table 5.10 for comparison of the "hot" and "cold" processes.

In both polymerizations, the unreacted monomer has to be removed. In the "hot" one the reaction is often quenched by addition of hydroquinone, and in the "cold" one by addition of *N,N*-diethyldithiocarbamate. After the monomers are steam-stripped in both processes, an antioxidant like *N*-phenyl-2-naphthylamine is added. The latex is usually coagulated by addition of a sodium chloride–sulfuric acid solution.

The "cold" process yields polymers with less branching than the "hot" one, slightly higher *trans*-to-*cis* ratios, and narrower molecular weight distributions.

During the middle sixties a series of butadiene–styrene and isoprene–styrene *block copolymer elastomers* were developed. These materials possess typical rubber-like properties at ambient temperatures, but act like thermoplastic resins at elevated temperatures. The copolymers vary from diblock structures of styrene and butadiene:

$$—[—CH_2—\underset{\displaystyle C_6H_5}{CH}—]_n——[—CH_2—CH{=}CH—CH_2—]_m—$$

FIGURE 5.4. Polystyrene and polybutadiene domains.

to triblock ones, like styrene–butadiene–styrene:

$$—[—CH_2—\underset{\displaystyle C_6H_5}{CH}—]_n—\,—[—CH_2—CH{=}CH—CH_2—]_m—\,—[—CH_2—\underset{\displaystyle C_6H_5}{CH}—]_n—$$

A typical triblock copolymer may consist of about 150 styrene units at each end of the macromolecule, and some 1000 butadiene units in the center. The special physical properties of these block copolymers are due to inherent incompatibility of polystyrene with polybutadiene or polyisoprene blocks. Within the bulk material there are separations and aggregations of the domains. The polystyrene domains are dispersed in continuous matrixes of the polydienes that are the major components. At ambient temperature, below the T_g of the polystyrene, these domains are rigid and immobilize the ends of the polydiene segments. In effect, they serve both as filler particles or as crosslinks. Above T_g of polystyrene, however, the domains are easily disrupted and the material can be processed as a thermoplastic polymer. The separation into domains is illustrated in Fig. 5.4.

These thermoplastic elastomers are prepared by anionic solution polymerization with organometallic catalysts. A typical example of such preparation is polymerization of a 75/25 mixture of butadiene/styrene in the presence of *sec*-butyllithium in a hydrocarbon–ether solvent blend. At these reaction conditions butadiene blocks form first, and when all the butadiene is consumed styrene blocks form. In other preparations, monomers are added sequentially, taking advantage of the "living" nature of these anionic polymerizations.

These block copolymers have very narrow molecular weight distributions. Also, the sizes of the blocks are restricted to narrow ranges to maintain optimum elastomeric properties.

5.11.2. GR-N Rubber

Butadiene–acrylonitrile rubbers are another group of useful synthetic elastomers. These copolymers were originally developed in Germany where they were found superior in oil resistance to the butadiene–styrene rubbers. Commercially, these materials are produced by free-radical emulsion polymerization very similarly to the butadiene–styrene copolymers. Similarly, "hot" and "cold" processes are employed. "Low," "medium," and "high" grades of solvent-resistant copolymers are formed, depending upon the amount of acrylonitrile in the copolymer that can range from 25–40%. The butadiene placement in these copolymers is approximately 77.5% *trans*-1,4, 12.5% *cis*-1,4, and 10% 1,2 units. Also, the polymers formed by the "cold" process are less branched and have a narrower molecular weight distribution than those formed by the "hot" process.

An interesting alternating copolymer of butadiene and acrylonitrile was developed in Japan.[184] The copolymer is formed with coordination catalysts consisting of AlR_3, $AlCl_3$, and $VOCl_3$ in a suspension polymerization process. The product is more than 94% alternate and is reported to have very good mechanical properties and good oil resistance.

5.12. Polystyrene and Polystyrene-Like Polymers

Styrene is produced in the U.S. from benzene and ethylene by a Friedel-Craft reaction that is followed by dehydrogenation over alumina at 600 °C. Polystyrene was first prepared in 1839, though the material was confused for an oxidation product of the styrene monomer.[134] Today, polystyrene is produced in very large quantities and much is known about this material.

5.12.1. Preparation of Polystyrene

Styrene is one of those monomers that lends itself to polymerization by free-radical, cationic, anionic, and coordination mechanisms. This is due to several reasons. One is resonance stabilization of the reactive polystyryl species in the transition state that lowers the activation energy of the propagation reaction. Another is the low polarity of the monomer. This facilitates attack by free radicals, differently charged ions, and metal complexes. In addition, no side reactions that occur in ionic polymerizations of monomers with functional groups are possible. Styrene polymerizes in the dark by a free-radical mechanism more slowly than it does in the presence of light.[135] Also, styrene formed in the dark is reported to have a greater amount of syndiotactic placement.[90] The amount of branching in the polymer prepared by a free-radical mechanism increases with temperature.[136] This also depends upon the initiator used.[215]

The following information has evolved about the free-radical polymerization of styrene:

1. Styrene polymerizes thermally.[137–141] This is discussed in Chapter 2.
2. Oxygen retards polymerizations of styrene. At higher temperatures, however, the rate is accelerated due to peroxide formation.[142]
3. The rate of styrene polymerizations in bulk is initially, at low conversions, first order with respect to monomer concentrations. In solution, however, it is second order with respect to monomer.[143]

Commercial polystyrene is manufactured only by free-radical polymerization. Isotactic polystyrene formed with Ziegler–Natta catalysts was introduced commercially in the sixties, but failed to gain acceptance.

Industrially, styrene polymerizations are carried out in bulk, in emulsion, in solution, and in suspension. The clear plastic is generally prepared by mass polymerization. Because polystyrene is soluble in the monomer, mass polymerization, when carried out to completion, results in a tremendous increase in melt viscosity. To avoid this, when styrene is polymerized in bulk in an agitated kettle, the reaction is only carried out to 30–40% conversion. After that, the viscous syrup is transferred to another type of reactor for the completion of the reaction. According to one early German patent, polymerization is completed in a plate and frame filter press.[143] Water circulating through the press removes the heat of the reaction and the solid polymer is formed inside the frames. This process is still used in some places.[144]

Another approach is to use adiabatic towers. Styrene is first partially polymerized in two agitated reaction kettles at 80–100 °C. The syrup solution of the polymer in the monomer is then fed continually into the towers from the top. The temperatures in the towers are gradually increased from 100–110 °C at the top to 180–200 °C at the bottom. By the time the material reaches the bottom, in about three hours, the polymerization is 92–98% complete.[145] The unreacted monomer is removed and recycled. A modification of the process is to remove the monomer vapor at the top of the tower for reuse. An adiabatic tower for mass polymerization of styrene is illustrated in Fig. 5.5.

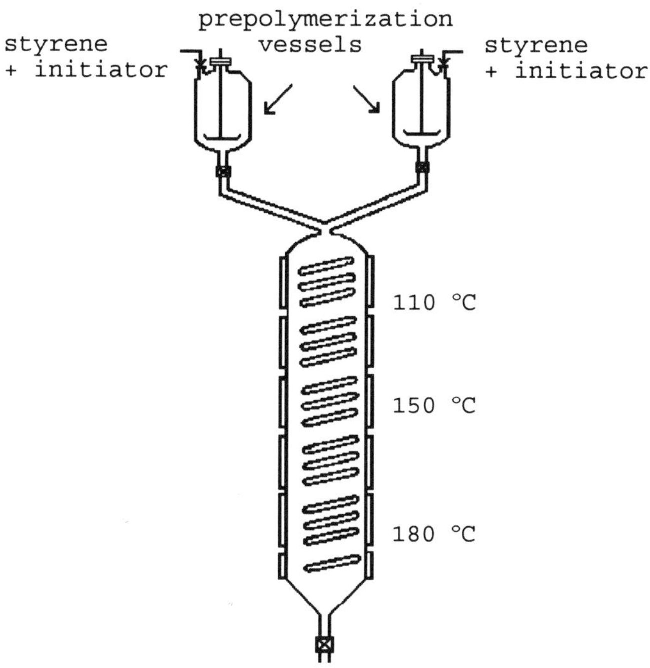

FIGURE 5.5. Adiabatic tower for mass polymerization of styrene.

An improvement in the above procedure is the use of agitated towers.[146] To avoid channeling inside the towers and for better heat transfer, three towers are arranged in series. They are equipped with slow agitators and with grids of pipes for cooling and heating.[144] Polymeric melt is heated from 95–225 °C to reduce viscosity and help heat transfer. A solvent like ethyl benzene may be added. A vacuum devolatalizer removes both monomer and solvent from the product.

5.12.2. Polystyrene Prepared by Ionic Chain-Growth Polymerization

Although styrene polymerized by ionic mechanism is not utilized commercially, much research was devoted to both cationic and anionic polymerizations. An investigation of cationic polymerization of styrene with an $Al(C_2H_5)_2Cl/RCl$ (R = alkyl or aryl) catalyst/cocatalyst system was reported by Kennedy.[147,148] The efficiency (polymerization initiation) is determined by the relative stability and/or concentration of the initiating carbocations that are provided by the cocatalyst RCl. *N*-butyl, isopropyl, and *sec*-butyl chlorides exhibit low cocatalytic efficiencies because of a low tendency for ion formation. Triphenylmethyl chloride is also a poor cocatalyst, because the triphenylmethyl ion that forms is more stable than the propagating styryl ion. Initiation of styrene polymerizations by carbocations is now well established.[149]

Anionic polymerization of styrene with amyl sodium yields an isotactic polymer.[150] Polymerizations catalyzed by triphenylmethylpotassium also yield stereospecific polystyrene.[151] The same is true of organolithium compounds.[152,153]

In butyllithium-initiated polymerizations of styrene in benzene, termination was claimed to occur by association between the propagating anions and the lithium counterions with another butyllithium molecule.[154] This was contradicted by claims that terminations result from association of two propagating chains.[155] Alfin catalysts polymerize styrene to yield stereospecific products.[156]

Coordination catalysts based on aluminum alkyls and titanium halides yield isotactic polystyrene.[157–160] The polymer matches isotactic polystyrene formed with amylsodium. It is composed of head-to-tail sequences with the main chain fold being helical. There are three monomer units per each helical fold.[157,158] The catalyst composition, however, has a strong bearing on the microstructure of the resultant polymer.[162] Syndiotactic polystyrene has so far not been synthesized.[161]

There is an interest in forming polystyrenes with vary narrow molecular-weight distributions, because of practical applications, and from purely academic interests. Several preparations of virtually monodisperse polystyrenes of $M_w/M_n = 1.06$ by anionic polymerizations were developed. The materials are available commercially[163–168] in small quantities, as standards for GPC.

5.12.3. Polymers from Substituted Styrene

Many derivatives of styrene can be readily synthesized. Some are commercially available. One of them is α-methyl styrene. It is formed from propylene and benzene by a process that is very similar to styrene preparation:

$$C_6H_6 + CH_3{-}CH{=}CH_2 \xrightarrow{AlCl_3} CH_3{-}CH(CH_3){-}C_6H_5 \xrightarrow{Al_2O_3} CH_2{=}C(CH_3){-}C_6H_5$$

Due to the allylic nature of α-methyl styrene it cannot be polymerized by free-radical mechanism. It polymerizes readily, however, by ionic mechanism. Resins based on copolymers of α-methyl styrene are available commercially. Other styrene derivatives that can be obtained commercially are:

1. Alkyl or aryl substituted styrenes

   $CH_2{=}CH{-}C_6H_3(R_1)(R_2)$

   where R_1 and R_2 are alkyl or aryl groups.
2. Halogen derivatives

   $CH_2{=}CH{-}C_6H_3(X_1)(X_2)$

   where X_1 and X_2 = F, Cl, Br, or I.
3. Polar substituted styrenes

   $CH_2{=}CH{-}C_6H_4{-}R$

   where R = CN, CHO, COOH, $OCOCH_3$, OH, OCH_3, NO_2, NH_2, and SO_3H

Vinyl toluene polymerizes readily by a free-radical mechanism at 100 °C. The absolute rate at that temperature is greater than for styrene. The activation energy for vinyl toluene polymerization is 17–19 kcal per mole, while that for styrene is 21 kcal per mole. This monomer can also be polymerized by ionic and coordination mechanisms. Earlier attempts at polymerization of α-methyl styrene with Ziegler–Natta catalysts were not successful.[169,170] Later, however, it was shown that polymerization does take place with $TiCl_4/Al(C_2H_5)_3$ at −78 °C. The activity of the catalyst and the DP depend on the ratio of aluminum to titanium, the nature of the solvent, and on the aging of the catalyst.[171] The optimum ratio of the aluminum alkyl to titanium chloride is 1.0 to 1.2. Mixing and

Table 5.11. Transition Temperatures of Poly(phenyl alkenes)

Polymer	T_g (°C)	Ref.	T_m (°C)	Ref.
$-[-CH_2-CH(C_6H_5)-]_n-$	80	178	240	176
$-[-CH_2-CH(C_6H_4\text{-}4\text{-}CH_3)-]_n-$	73, 81	178	Amorphous	176
$-[-CH_2-CH(C_6H_4\text{-}3\text{-}CH_3)-]_n-$	77	178	219	176
$-[-CH_2-CH(C_6H_3\text{-}2,4\text{-}(CH_3)_2)-]_n-$	92	178	310	176
$-[-CH_2-CH(C_6H_3\text{-}2,6\text{-}(CH_3)_2)-]_n-$	123	178	330	176
$-[-CH_2-CH(C_6H_3\text{-}3,4\text{-}(CH_3)_2)-]_n-$	91	178	240	176
$-[-CH_2-CH(C_6H_3\text{-}2,5\text{-}(CH_3)_2)-]_n-$	84	177	290	176
$-[-CH_2-CH(CH_2C_6H_5)-]_n-$	60	177	207–208	176
$-[-CH_2-CH(CH_2C_6H_4\text{-}2\text{-}CH_3)-]_n-$	80	177	2	177

(continued)

Table 5.11. (continued)

Polymer	T_g (°C)	Ref.	T_m (°C)	Ref.
$-[-CH_2-CH(CH_2C_6H_4CH_3\text{-}m)-]_n-$	35–40	177	180	177
$-[-CH_2-CH(CH_2C_6H_4CH_3\text{-}p)-]_n-$	60–65	177	240	177
$-[-CH_2-CH((CH_2)_2C_6H_5)-]_n-$	10	177	162–168	177

[a]From Refs. 176–178.

aging of the catalyst must be done below room temperature and the valence of titanium must be maintained between 3 and 4.[171] This led Sakurada to suggest that the reaction actually proceeds via a cationic rather than a coordinated anionic mechanism.[171]

Various reports in the literature describe cationic polymerizations of α-methyl styrene with Lewis acids.[172–174] The products are mostly low molecular weight polymers, some containing unsaturation with pendant phenylindane groups. A high molecular weight polymer can be prepared from α-methyl styrene by cationic polymerization at –90 to –130 °C with $AlCl_3$ in ethyl chloride or in carbon disulfide.[175] The product has a narrow molecular weight distribution. Some transition temperatures, T_g and T_m, of polystyrene-like materials, including isotactic polystyrene, are presented in Table 5.11.

5.13. Copolymers of Styrene

Many copolymers of styrene are manufactured on a large commercial scale. Because styrene copolymerizes readily with many other monomers, it is possible to obtain a wide distribution of properties. Random copolymers form quite readily by a free-radical mechanism.[179,180] Some can also be formed by ionic mechanism. In addition, graft and block copolymers of styrene are also among commercially important materials.

Most comonomers differ from styrene in polarity and reactivity. A desired copolymer composition can be achieved, however, through utilization of copolymerization parameters based on kinetic data and on quantum-chemical considerations. This is done industrially in preparations of styrene-acrylonitrile, styrene-methyl methacrylate, and styrene-maleic anhydride copolymers of different compositions.

5.13.1. High-Impact Polystyrene

For many applications the homopolymer of styrene is too brittle. To overcome that, many different approaches were originally tried. These included use of high molecular weight polymers, use of plasticizers, fillers (glass fiber, wood flour, etc.), deliberate orientation of the polymeric chains, copolymerization, and addition of rubbery substances. The effect of plasticizers is too severe for practical use and the use of high molecular weight polymers exhibits only marginal improvement. Use of fillers, though beneficial, is mostly confined to the United States. Orientation is limited to sheets and filaments, and copolymerization usually lowers the softening point too much.

Addition of rubbery materials, however, does improve the impact resistance of polystyrene. This is therefore done extensively. The most common rubbers used for this purpose are butadiene–styrene copolymers. Some butadiene homopolymers are also used, but to a lesser extent. The high-impact polystyrene is presently prepared by dissolving the rubber in a styrene monomer and then polymerizing the styrene. This polymerization is either done in bulk or in suspension. The product contains styrene–butadiene rubber, styrene homopolymer, and a considerable portion of styrene–graft copolymer that forms when polystyrene radicals attack the rubber molecules. The product has very enhanced impact resistance.

Past practice, however, has consisted in simply blending a mixture of polystyrene and rubber on a two-roll mill, or in a high shear internal mixer, or passing it through an extruder. The impact strength of the product was only moderately better than that of the unmodified polymer. Another procedure was to blend polystyrene emulsion latex with a styrene–butadiene rubber emulsion latex and then to coagulate the two together. The product is also only marginally better in impact strength than styrene homopolymer. This practice, however, may still be in existence in some places.

In high-impact polystyrene, the rubber exists in discrete droplets, less than 50 μm in diameter. In effect, the polymerization serves to form an oil-in-oil emulsion,[181] where the polystyrene is the continuous phase and the rubber is in a dispersed phase. The graft copolymer that forms serves to "emulsify" this heterogeneous polymer solution.[182]

Commercial high-impact polystyrene usually contains 5–20% styrene–butadiene rubber. The particle size ranges from 1–10 μm. High-impact polystyrene may have as much as seven times the impact strength of polystyrene, but it has only half its tensile strength, lower hardness, and a lower softening point.

5.13.2. ABS Resins

Styrene–acrylonitrile copolymers are produced commercially for use as structural plastics. The typical acrylonitrile content in such resins is between 20–30%. These materials have better solvent and oil resistance than polystyrene and a higher softening point. In addition, they exhibit better resistance to cracking and crazing and an enhanced impact strength. Although the acrylonitrile copolymers have enhanced properties over polystyrene, they are still inadequate for many applications. Acrylonitrile–butadiene–styrene polymers, known as ABS resins, were therefore developed.

Although ABS resins can potentially be produced in a variety of ways, there are only two main processes. In one of them acrylonitrile–styrene copolymer is blended with a butadiene–acrylonitrile rubber. In the other, interpolymers are formed of polybutadiene with styrene and acrylonitrile.

In the first one, the two materials are blended on a rubber mill or in an internal mixer. Blending of the two materials can also be achieved by combining emulsion latexes of the two materials together and then coagulating the mixture. Peroxide must be added to the blends in order to achieve some crosslinking of the elastomer to attain optimum properties. A wide range of blends are made by this technique with various properties.[183] Most common commercial blends of ABS resins may contain 70 parts of styrene–acrylonitrile copolymer (70/30) and 40 parts of butadiene–nitrile rubber (65/35).

In the second process, styrene and acrylonitrile are copolymerized in the presence of polybutadiene latex. The product is a mixture of butadiene homopolymer and a graft copolymer.

5.13.3. Copolymers of Styrene with Maleic Anhydride

Many styrene–maleic anhydride copolymers are produced commercially for special uses. These are formed by free-radical copolymerization and many commercial grades are partially esterified. Molecular weights of such polymers may range from 1500–50,000, depending upon the source. The melting points of these copolymers can vary from 110–220 °C, depending upon molecular weight, degree of hydrolysis and esterification, and also the ratio of styrene to maleic anhydride.

Only a small percent of styrene copolymers reported in the literature achieved industrial importance. Some of the interesting copolymers of styrene that were reported but not utilized commercially are copolymers with various unsaturated nitriles. This includes vinylidine cyanide, fumaronitrile, malononitrile, methacrylonitrile, acrylonitrile, and cinnamonitrile.[185] Often, copolymerization of styrene with nitriles yields copolymers with higher heat distortion temperature, higher tensiles, better craze resistance, and higher percent elongation.

Styrene was also copolymerized with many acrylic and methacrylic esters. Products with better weatherability often form. Copolymerization with some acrylates lowers the value of T_g.[186]

5.14. Polymers of Acrylic and Methacrylic Esters

There are many synthetic procedures for preparing acrylic acid and its esters. One way, used for a long time, is to make acrylic esters from ethylene oxide:

$$\underset{\text{(ethylene oxide)}}{CH_2{-}CH_2(O)} \xrightarrow[\text{55–60 °C, amine catalyst}]{HCN} \overset{OH\quad C{\equiv}N}{CH_2{-}CH_2} \xrightarrow{-H_2O} CH_2{=}CH{-}C{\equiv}N \xrightarrow[-NH_3]{H_2O} CH_2{=}CH{-}COOH$$

$$CH_2{=}CH{-}C{\equiv}N \xrightarrow[H_2SO_4]{ROH} CH_2{=}CH{-}C(=O){-}OR$$

Another route is through oxidation of propylene over cobalt molybdenum catalyst at 400–500 °C:

$$CH_2{=}CH{-}CH_3 \longrightarrow CH_2{=}CHCHO \longrightarrow CH_2{=}CHCOOH \xrightarrow{ROH} CH_2{=}CHCOOR$$

Many industrial preparations start with acetylene, carbon monoxide, and an alcohol or water:

$$HC{\equiv}CH + CO + ROH \xrightarrow{Ni(CO)_4} CH_2{=}CHCOOR$$

$$HC{\equiv}CH + CO + HOH \xrightarrow{Ni} CH_2{=}CHCOOH$$

One route to α-methyl acrylic acid (or methacrylic acid) and its esters is by a cyanohydrin reaction of acetone:

$$CH_3{-}C(=O){-}CH_3 + HCN \longrightarrow CH_3{-}C(C{\equiv}N)(OH){-}CH_3 \xrightarrow{H_2SO_4} CH_2{=}C(C{\equiv}N){-}CH_3 \xrightarrow[H_2SO_4]{ROH} CH_2{=}C(CH_3){-}COOR$$

$$CH_2{=}C(C{\equiv}N){-}CH_3 \xrightarrow{HOH} CH_2{=}C(CH_3){-}COOH$$

Another route to methacrylic acid and its esters is via oxidation of isobutene:

$$CH_3{-}C(CH_3){=}CH_2 \xrightarrow{O_2} CH_3{-}C(CH_3)(OH){-}COOH \xrightarrow{-H_2O} CH_2{=}C(CH_3){-}COOH$$

5.14.1. Polymerizations of Acrylic and Methacrylic Esters

Free-radical bulk polymerizations of acrylate esters exhibit rapid-rate accelerations at low conversions. This often results in formation of some very high molecular weight polymer and some crosslinked material. The crosslinking is a result of chain transferring by abstractions of labile tertiary hydrogens from already formed "dead" polymeric chains.[187] Eventually, termination by combination of the branched radicals leads to crosslinked structures. Addition of chain-transferring agents, like mercaptans (that reduce the length of the primary chains), helps prevent gel formation. There are no labile tertiary hydrogens in methacrylic esters. The growing methacrylate radicals are still capable of abstracting hydrogens from the α-methyl groups. Such abstractions, however, require more energy and are not an important problem in polymerizations of methacrylic esters.[188] Nevertheless, occasional formation of crosslinked poly(alkyl methacrylate)s does occur. This is due to chain transferring to the alcohol moiety.[189,193]

The termination reaction in free-radical polymerizations of the esters of acrylic and methacrylic acids takes place by recombination and by disproportionation.[189,190] Methyl methacrylate polymerizations, however, terminate at 25 °C predominantly by disproportionation.[188]

Oxygen inhibits free-radical polymerization of α-methyl methacrylate.[191] The reaction with oxygen results in formation of low molecular weight polymeric peroxides that subsequently decompose to formaldehyde and methyl pyruvate[192]:

$$CH_2{=}C(CH_3){-}C({=}O){-}O{-}CH_3 \; + \; O_2 \longrightarrow {-}[{-}CH_2{-}C(CH_3)(C({=}O){-}O{-}CH_3){-}O{-}O{-}]_n{-} \longrightarrow CHO \; + \; CH_3{-}C({=}O){-}C({=}O){-}OCH_3$$

Oxygen is less effective in inhibiting polymerizations of acrylic esters. It reacts 400 times faster with the methacrylic radicals than with the acrylic ones. Nevertheless, even small quantities of oxygen affect polymerization rates of acrylic esters.[216] This includes photopolymerizations of gaseous ethyl acrylate that are affected by oxygen and by moisture.[217]

Acrylic and methacrylic esters polymerize by free-radical mechanism to atactic polymers. The sizes of the alcohol portions of the esters determine the T_g values of the resultant polymers. They also determine the solubility of the resultant polymers in hydrocarbon solvents and in oils.

Solvents influence the rate of free-radical homopolymerization of acrylic acid and its copolymerization with other monomers. Hydrogen-bonding solvents slow down the reaction rates.[219] Due to the electron-withdrawing nature of the ester groups, acrylic and methacrylic ester polymerize by anionic but not by cationic mechanisms. Lithium alkyls are very effective initiators of α-methyl methacrylate polymerization yielding stereospecific polymers.[194] Isotactic poly(methyl methacrylate) forms in hydrocarbon solvents.[195] Block copolymers of isotactic and syndiotactic poly(methyl methacrylate) form in solvents of medium polarity. Syndiotactic polymers form in polar solvents, like ethylene glycol dimethyl ether, or pyridine. This solvent influence is related to Lewis basicity[196] in the following order:

tetrahydrofuran > tetrahydropyran > dioxane > diethyl ether

Furthermore, polymerizations in solvating media, like ethylene glycol dimethyl ether, tetrahydrofuran, or pyridine using biphenylsodium or biphenyllithium, yield virtually monodisperse syndiotactic poly(methyl methacrylate).[220]

The nature of the counterion in anionic polymerizations of methyl methacrylate in liquid ammonia with alkali metal amide or alkali earth metal amide catalysts is an important variable.[197] Lithium and calcium amides yield high molecular weight polymers, though the reactions tend to be slow. Sodium amide, on the other hand, yields rapid polymerizations but low molecular weight polymers. Polymers formed with sodium amide, however, have a narrower molecular weight distribution than those obtained with lithium and calcium amides. Calcium amide also yields high molecular weight polymers from ethyl acrylate and methyl methacrylate monomers in aromatic and aliphatic solvents at temperatures from –8 to 110 °C. When, however, tetrahydrofuran or acetonitrile are used as solvents, much lower molecular weight products form.[198]

Products from anionic polymerizations of methyl methacrylate catalyzed by Grignard reagents (RMgX) vary with the nature of the R and X groups, the reaction temperature, and the nature of the solvent.[196,198,199] Secondary alkyl Grignard reagents give the highest yields and the fastest rates of the reactions. Isotacticity of the products increases with the temperature. When anion radicals from alkali metal ketyls of benzophenone initiate polymerizations of methyl methacrylate, amorphous polymers form at temperatures from –78 to +65 °C.[200]

Sodium dispersions in hexane yield syndiotactic poly(methyl methacrylate).[201] A 60–65% conversion is obtained over a 24-hour period at a reaction temperature of 20–25 °C. Lithium dispersions,[202] butyllithium,[203] and Grignard reagents[204,205] yield crystalline isotactic poly(*t*-butyl acrylate). The reactions take place in bulk and in hydrocarbon solvents. Isotactic poly(isopropyl acrylate) forms with Grignard reagents.[206,207]

Coordinated anionic polymerizations of methyl methacrylate with diethyliron–bipyridyl complex in nonpolar solvents like benzene or toluene yield stereoblock polymers. In polar solvents, however, like dimethylformamide or acetonitrile, the products are rich in isotactic placement.[208]

There are many reports in the literature on polymerizations of acrylic and methacrylic esters with Ziegler–Natta catalysts.[209–212] The molecular weights of the products, the microstructures, and the rates of the polymerizations depend upon the metal alkyl and the transition metal salt used. The ratios of the catalyst components to each other are also important.[211,218]

5.14.2. Acrylic Elastomers

Polymers of lower *n*-alkyl acrylates are used in plastics to only a limited extent. Ethyl and butyl acrylates are, however, major components of *acrylic elastomers*. The polymers are usually formed by free-radical emulsion polymerization. Because acrylate esters are sensitive to hydrolysis under basic conditions, the polymerizations are usually conducted at neutral or acidic pH. The acrylic rubbers, like other elastomers, must be crosslinked or *vulcanized* to obtain optimum properties. Crosslinking can be accomplished by reactions with peroxides through abstractions of tertiary hydrogens by free radicals:

$$\begin{array}{c} O{=}C{-}O{-}C_2H_5 \\ | \\ \sim\!CH_2{-}CH\!\sim \end{array} \quad + \quad R\bullet \longrightarrow \begin{array}{c} O{=}C{-}O{-}C_2H_5 \\ | \\ \sim\!CH_2{-}\dot{C}\!\sim \end{array} \quad + \quad RH$$

$$2\ \begin{array}{c} O{=}C{-}O{-}C_2H_5 \\ | \\ \sim\!CH_2{-}\dot{C}\!\sim \end{array} \longrightarrow \begin{array}{c} O{=}C{-}O{-}C_2H_5 \\ | \\ \sim\!CH_2{-}C\!\sim \\ | \\ \sim\!CH_2{-}C\!\sim \\ | \\ O{=}C{-}O{-}C_2H_5 \end{array}$$

Another way to crosslink acrylic elastomers is through a Claisen condensation:

$$\begin{array}{c} O{=}C{-}O{-}C_2H_5 \\ | \\ \sim\!CH_2{-}C\!\sim \\ | \\ H \\ + \\ C_2H_5{-}O{-}C{=}O \\ | \\ \sim\!CH_2{-}C\!\sim \\ | \\ H \end{array} \xrightarrow{\text{base}} \begin{array}{c} O{=}C{-}O{-}C_2H_5 \\ | \\ \sim\!CH_2{-}C\!\sim \\ | \\ C{=}O \\ | \\ \sim\!CH_2{-}C\!\sim \\ | \\ H \end{array}$$

The above illustrated crosslinking reactions of homopolymers, however, form elastomers with poor aging properties. Commercial acrylic rubbers are therefore copolymers of ethyl or butyl acrylate with small quantities of comonomers that carry special functional groups for crosslinking. Such comonomers are 2-chloroethylvinyl ether or vinyl chloroacetate, used in small quantities (about 5%). These copolymers crosslink through reactions with polyamines.

Table 5.12. Typical Components of Thermoset Acrylic Resins

Monomers that contribute rigidity	Flexibilizing monomers	Monomers used for crosslinking
Methyl methacrylate	Ethyl acrylate	Acrylic acid
Ethyl methacrylate	Isopropyl acrylate	Methacrylic acid
Styrene	Butyl acrylate	Hydroxyethyl acrylate
Vinyl toluene	*i*-Octyl acrylate	Hydroxypropyl acrylate
Acrylonitrile	Decyl acrylate	Glycidyl acrylate
Methacrylonitrile	Lauryl methacrylate	Glycidyl methacrylate
		acrylamide
		aminoethyl acrylate

5.14.3. Thermoplastic and Thermoset Acrylic Resins

Among polymers of methacrylic esters, poly(methyl methacrylate) is the most important one industrially. Most of it is prepared by free-radical polymerizations of the monomer and a great deal of these polymerizations are carried out in bulk. Typical methods of preparation of clear sheets and rods consist of initial partial polymerizations in reaction kettles at about 90 °C with peroxide initiators. This is done by heating and stirring for about ten minutes to form syrups. The products are cooled to room temperature and various additives may be added. The syrups are solutions of about 20% polymer dissolved in the monomer. They are poured into casting cells where the polymerizations are completed. The final polymers are high in molecular weight, about 1,000,000.

Poly(methyl methacrylate) intended for surface coatings is prepared by solution polymerization. The molecular weights of the polymers are about 90,000 and the reaction products that are 40–60% solutions are often used directly in coatings.

A certain amount of poly(methyl methacrylate) is also prepared by suspension polymerization. The molecular weights of these polymers are about 60,000 and they are used in injection molding and extrusion.

Thermosetting acrylic resins are used widely in surface coatings. Both acrylic and methacrylic esters are utilized and the term is applied to both of them. Often, such resins are terpolymers or even tetrapolymers, where each monomer is chosen for a special function.[214] One is selected for rigidity, surface hardness, and scratch resistance; another for ability to flexibilize the film, and the third for crosslinking it. In addition, not all comonomers are necessarily acrylic or methacrylic esters or acids. For instance, among the monomers that may be chosen for rigidity may be methyl methacrylate. On the other hand, it may be styrene instead, or vinyl toluene, etc. The same is true of the other components. Table 5.12 illustrates some common components that can be found in thermoset acrylic resins.

The choice of crosslinking reaction may depend upon the desired application. It may also simply depend upon price, or a particular company that manufactures the resin, or simply to overcome patent restrictions. Some common crosslinking reactions will be illustrated in the remainder of this section. If the functional groups are carboxylic acids in the copolymer or terpolymer, crosslinking can be accomplished by adding a diepoxide.

$$\sim\!CH_2\!-\!\underset{\substack{|\\COOH}}{CH}\!\sim \;+\; \overset{\;O}{\overbrace{CH_2\!-\!CH}}\!-\!R\!-\!\overset{O}{\overbrace{CH\!-\!CH_2}} \;+\; \sim\!CH_2\!-\!\underset{\substack{|\\COOH}}{CH}\!\sim \longrightarrow$$

$$\begin{array}{c}\sim CH_2-CH\sim \qquad\qquad OH \qquad OH \qquad\qquad \sim CH-CH_2\sim \\ O=\overset{|}{C}-O-CH_2-\overset{|}{C}H-R-\overset{|}{C}H-CH_2-O-\overset{|}{C}=O\end{array}$$

Other reactions can also crosslink resins with pendant carboxylic acid groups. For instance, one can add a melamine formaldehyde condensate:

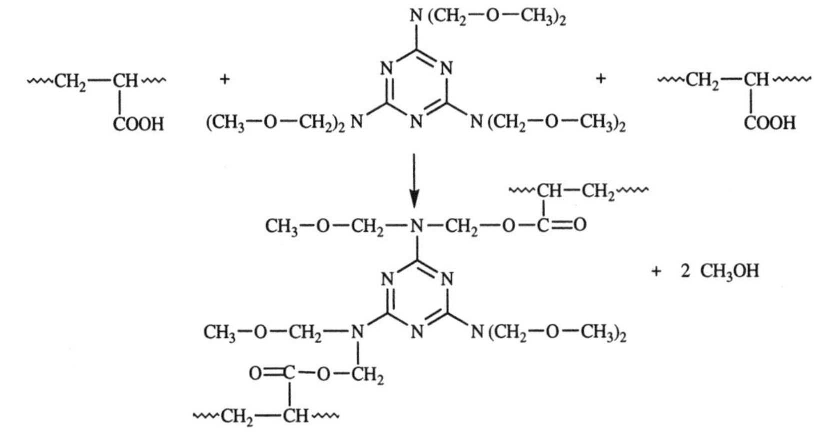

$$\sim CH_2-\underset{COOH}{CH}\sim \;+\; (CH_3-O-CH_2)_2N-C_3N_3[N(CH_2-O-CH_3)_2]_2 \;+\; \sim CH_2-\underset{COOH}{CH}\sim$$

$$\downarrow$$

$$CH_3-O-CH_2-N(-CH_2-O-\underset{\sim CH-CH_2\sim}{C}=O)-C_3N_3[N(CH_2-O-CH_3)_2]-N(CH_2-O-\underset{\sim CH_2-CH\sim}{C}=O)-CH_2-O-CH_3 \;+\; 2\;CH_3OH$$

A diisocyanate, a phenolic, or a melamine-formaldehyde resin can be used as well. Resins with pendant hydroxy groups can also be crosslinked by these materials. A diisocyanate is effective in forming urethane linkages:

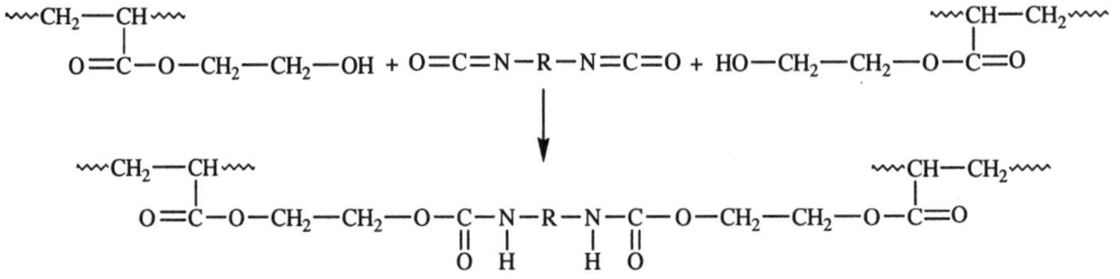

$$\sim CH_2-\underset{O=C-O-CH_2-CH_2-OH}{CH}\sim \;+\; O=C=N-R-N=C=O \;+\; HO-CH_2-CH_2-O-\underset{\sim CH-CH_2\sim}{C}=O$$

$$\downarrow$$

$$\sim CH_2-CH\sim\;\;(O=C-O-CH_2-CH_2-O-\underset{O}{\overset{\|}{C}}-\underset{H}{N}-R-\underset{H}{N}-\underset{O}{\overset{\|}{C}}-O-CH_2-CH_2-O-C=O)\;\;\sim CH-CH_2\sim$$

When the pendant groups are epoxides, like glycidyl esters, crosslinking can be carried out with dianhydrides or with compounds containing two or more carboxylic acid groups.[213] Aminoplast resins (urea-formaldehyde or melamine-formaldehyde and similar ones) are also very effective.[221]

Pendant amide groups from terpolymers containing acrylamide can be reacted with formaldehyde to form methylol groups for crosslinking[254]:

$$\sim CH_2-\underset{\underset{NH_2}{C=O}}{CH}\sim \;+\; H_2C=O \longrightarrow \sim CH_2-\underset{\underset{H-N-CH_2OH}{C=O}}{CH}\sim$$

The product of the above reaction can be thermoset like any urea-formaldehyde resin (see Chapters 6 and 8). Many crosslinking routes are described in the patent literature, because there are many different functional groups available.

5.15. Acrylonitrile and Methacrylonitrile Polymers

Polymers from acrylonitrile are used in synthetic fibers, in elastomers, and in plastic materials. The monomer can be formed by dehydration of ethylene cyanohydrin:

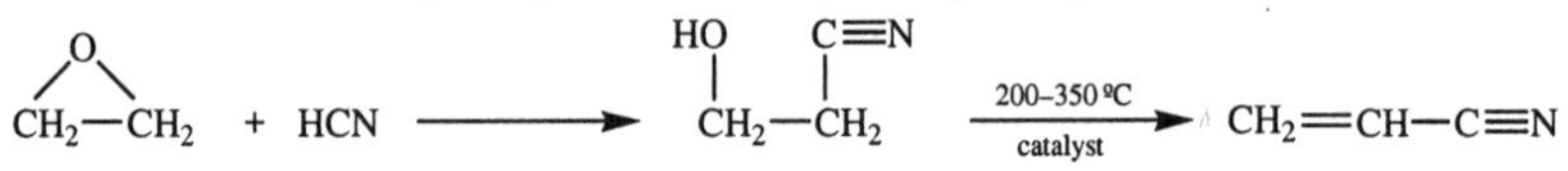

$$\underset{CH_2-CH_2}{\overset{O}{\wedge}} \;+\; HCN \longrightarrow \underset{CH_2-CH_2}{HO\quad C{\equiv}N} \xrightarrow[\text{catalyst}]{200\text{–}350\,^\circ C} CH_2{=}CH{-}C{\equiv}N$$

Other commercial processes exist, like condensation of acetylene with hydrogen cyanide, or ammoxidation of propylene[222]:

$$CH_3{-}CH{=}CH_2 + NH_3 + 3/2\;O_2 \xrightarrow{\text{catalyst}} CH_2{=}CH{-}C{\equiv}N$$

Acrylonitrile polymerizes readily by a free-radical mechanism. Oxygen acts as a strong inhibitor. When the polymerization is carried out in bulk, the reaction is autocatalytic.[223,224] In solvents, like dimethylformamide, however, the rate is proportional to the square root of the monomer concentration.[223] The homopolymer is insoluble in the monomer and in many solvents.

Acrylonitrile polymerizes also by anionic mechanism. There are many reports in the literature of polymerizations initiated by various bases. These are alkali metal alkoxides,[225] butyllithium,[226,227] metal ketyls,[228,229] solutions of alkali metals in ethers,[230,231] sodiummalonic esters,[232] and others. The propagation reaction is quite sensitive to termination by proton donors. This requires the use of aprotic solvents. The products, however, are often insoluble in such solvents. In addition, there is a tendency for the polymer to be yellow. This is due to some propagation taking place by 1,4 and 3,4 insertion in addition to the 1,2 placement[233,234]:

$$CH_2{=}CH{-}C{\equiv}N$$

$$\xrightarrow{1,2}\ {-}({-}CH_2{-}\underset{\underset{C\equiv N}{|}}{CH}{-})_n{-} \qquad \xrightarrow{1,4}\ {-}({-}CH_2{-}CH{=}C{=}N{-})_n{-} \qquad \xrightarrow{3,4}\ {-}({-}\underset{\underset{CH=CH_2}{|}}{C}{=}N{-})_n{-}$$

Another disadvantage of anionic polymerization of acrylonitrile is formation of cyanoethylate as a side reaction. This, however, can be overcome by running the reaction at low temperatures. An example is polymerizations initiated by KCN at –50 °C in dimethylformamide,[235] or by butyllithium in toluene at –78 °C.[236] Both polymerizations yield white, high molecular weight products that are free from cyanoethylation.

It was suggested that the terminations in anionic polymerizations of acrylonitrile proceed by proton transfer from the monomer. This, however, depends upon catalyst concentrations.[226,227] At low concentrations the terminations can apparently occur by a cyclization reaction[227] instead:

$$\sim CH_2{-}CH(C{\equiv}N){-}CH_2{-}C(H)(C{\equiv}N){-}CH_2{-}\overset{\ominus}{C}H(C{\equiv}N) \dashrightarrow \text{cyclic product: } \sim CH_2{-}CH,\ {}^{\ominus}N{=}C,\ CH_2,\ H{-}C(C{\equiv}N),\ CH_2,\ C{-}H(C{\equiv}N)$$

Industrially, polyacrylonitrile homopolymers and copolymers are prepared mainly by free-radical mechanism. The reactions are often conducted at low temperatures, in aqueous systems, either in emulsions or in suspensions, using redox initiation. Colorless, high molecular weight materials form. Bulk polymerizations are difficult to control on a large scale.

Over half the polymer that is prepared industrially is for use in textiles. Most of these are copolymers containing about 10% of a comonomer. The comonomers can be methyl methacrylate, vinyl acetate, or 2-vinylpyridine. The purpose of comonomers is to make the fibers more dyeable. Polymerizations in solution offer an advantage of direct fiber spinning.

Polyacrylonitrile copolymers are also used in barrier resins for packaging. One such resin contains at least 70% acrylonitrile and often methyl acrylate as the comonomer. The material has poor impact resistance and in one industrial process the copolymer is prepared in the presence of about 10% butadiene–acrylonitrile rubber by emulsion polymerization. The product contains some graft copolymer and some polymer blend. In another process the impact resistance of the copolymer is improved by biaxial orientation. The package, however, may have a tendency to shrink at elevated temperature, because the copolymer does not crystallize.

It is possible to form clear transparent polyacrylonitrile plastic shapes by a special bulk polymerization technique.[237,238] The reaction is initiated with *p*-toluenesulfinic acid–hydrogen peroxide. Initially, heterogeneous polymerizations take place. They are followed by spontaneous transformations, at high conversion, to homogeneous, transparent polyacrylonitrile plastics.[239] A major condition for forming transparent solid polymer is a continuous supply of monomer to fill the gaps formed by volume contraction during the polymerization process.[240]

Methacrylonitrile, $CH_2{=}C(CH_3)CN$, can also be prepared by several routes. Some commercial processes are based on acetone cyanohydrin intermediate and others on dehydrogenation (or oxydehydrogenation) of isobutyronitrile. It is also prepared from isobutylene by ammoxidation:

$$CH_2{=}C(CH_3){-}CH_3 + NH_3 + 3/2\ O_2 \longrightarrow CH_2{=}C(CH_3){-}C{\equiv}N$$

Just like acrylonitrile, methacrylonitrile does not polymerize thermally but polymerizes readily in the presence of free-radical initiators. Unlike polyacrylonitrile, polymethacrylonitrile is soluble in some ketone solvents. Bulk polymerizations of methacrylonitrile have the disadvantage of a long reaction time. The rate, however, accelerates with temperature. The polymer is soluble in the monomer at ambient conditions.[241]

Emulsion polymerization of methacrylonitrile is a convenient way to form high molecular weight polymers. With proper choices of emulsifiers, the rates may be increased by increasing the numbers of particles in the latexes. At a constant rate of initiation, the degree of polymerization of methacrylonitrile increases rather than decreases as the rate of polymerization rises.[242]

Methacrylonitrile polymerizes readily in inert solvents. The polymer, depending on the initiator and on reaction conditions, is either amorphous or crystalline. Polymerizations take place over a broad range of temperatures from ambient to –5 °C, when initiated by Grignard reagents, triphenylmethylsodium, or sodium in liquid ammonia.[243] The properties of these polymers are essentially the same as those of the polymers formed by a free-radical mechanism.

The homopolymer, prepared by polymerization in liquid ammonia with sodium initiator at –77 °C, is insoluble in acetone, but it is soluble in dimethylformamide.[244] When it is formed with lithium in liquid ammonia, at –75 °C, the molecular weight of the product increases with monomer concentration and decreases with initiator concentration. If, however, potassium initiates the reaction rather than lithium, the molecular weight is independent of the monomer concentration.[245,246] Polymethacrylonitrile prepared with *n*-butyllithium in toluene or in dioxane is crystalline and insoluble in solvents like acetone.[247] When polymerized in petroleum ether with *n*-butyllithium, methacrylonitrile forms a living polymer.[248] Highly crystalline polymethacrylonitrile can also be prepared with beryllium and magnesium alkyls in toluene over a wide range of temperatures.

5.16. Polyacrylamide, Poly(acrylic acid), and Poly(methacrylic acid)

Commercially, acrylamide is formed from acrylonitrile by reaction with water. Similarly, the preferred commercial route to methacrylamide is through methacrylonitrile. Acrylamide polymerizes by a free-radical mechanism.[249] Water is the common solvent for acrylamide and methacrylamide polymerizations, because the polymers precipitate out from organic solvents.

Crystalline polyacrylamide forms with metal alkyls in hydrocarbon solvents by an anionic mechanism.[250] The product is insoluble in water and in dimethylformamide.

Both acrylic and methacrylic acids can be converted to anhydrides and acid chlorides. The acids polymerize in aqueous systems by a free-radical mechanism. Polymerizations of these monomers in nonpolar solvents like benzene result in precipitations of the products.

Polymerizations of anhydrides proceed by inter–intramolecular propagations[251]:

$$CH_2{=}C(R){-}C({=}O){-}O{-}C({=}O){-}C(R){=}CH_2 \xrightarrow{R\bullet} {-}[{-}CH_2{-}\overset{R}{C}{-}CH_2{-}\overset{R}{C}{-}]_n{-} \text{ (ring: } C({=}O){-}O{-}C({=}O))$$

where R = H, CH_3. The above cyclopolymerizations produce soluble polymers rather than gels.

The *acid chlorides* of both acrylic and methacrylic acids polymerize by a free-radical mechanism in dry aromatic and aliphatic solvents. Molecular weights of the products, however, are low, usually under 10,000.[252,253] Polyacrylic and polymethacrylic acids are used industrially as thickeners in cosmetics, as flocculating agents, and when copolymerized with divinyl benzene in ion-exchange resins.

The volume of commercial fluorine containing polymers is not large when compared with other polymers, such as poly(vinyl chloride). Fluoropolymers, however, are required in many important applications. The main monomers are tetrafluoroethylene, trifluorochloroethylene, vinyl fluoride, vinylidine fluoride, and hexafluoropropylene.

5.17.1. Polytetrafluoroethylene

This monomer can be prepared from chloroform[254]:

$$CHCl_3 + 2\ HF \xrightarrow[\text{pressure up to 30 atm}]{50\text{–}180\ ^\circ C} CHClF_2 + 2\ HCl$$

$$2\ CHClF_2 \xrightarrow{700\ ^\circ C} CF_2{=}CF_2 + 2\ HCl$$

Tetrafluoroethylene boils at –76.3 °C. It is not the only product from the above pyrolytic reaction of difluorochloromethane. Other fluorine byproducts form as well and the monomer must be isolated. The monomer polymerizes in water at moderate pressures by a free-radical mechanism. Various initiators appear effective.[255] Redox initiation is preferred. The polymerization reaction is strongly exothermic and water helps dissipate the high heat of the reaction. A runaway, uncontrolled polymerization can lead to explosive decomposition of the monomer to carbon and carbon tetrafluoride[256]:

$$CF_2{=}CF_2 \rightarrow C + CF_4$$

Polytetrafluoroethylene is linear and highly crystalline.[257] Absence of terminal $CF_2{=}CF$– groups shows that few, if any, polymerization terminations occur by disproportionation but probably all take place by combination.[258] The molecular weights of commercially available polymers range from 39,000–9,000,000. Polytetrafluoroethylene is inert to many chemical attacks and is only swollen by fluorocarbon oils at temperatures above 300 °C. The T_m value of this polymer is 327 °C and T_g is below –100 °C.

The physical properties of polytetrafluoroethylene depend upon crystallinity and on the molecular weight of the polymer. Two crystalline forms are known. In both cases the chains assume helical arrangements to fit into the crystalloids. One such arrangement has fifteen CF_2 groups per turn and the other has thirteen.

Polytetrafluoroethylene does not flow even above its melting point. This is attributed to restricted rotation around the C–C bonds and to high molecular weights. The stiffness of the solid polymer is also attributed to restricted rotation. The polymer exhibits high thermal stability and retains its physical properties over a wide range of temperatures. The loss of strength occurs at about the crystalline melting point. It is possible to use the material for long periods at 300 °C without any significant loss of its strength.

5.17.2. Polychlorotrifluoroethylene

The monomer can be prepared by dechlorination of trichlorotrifluoroethane with zinc dust and ethanol.

$$CCl_2F{-}CClF_2 \xrightarrow[C_2H_5OH]{Zn} CClF{=}CF_2 + ZnCl_2$$

It is a toxic gas that boils at –26.8 °C. Polymerization of chlorotrifluoroethylene is usually carried out commercially by free-radical suspension polymerization. Reaction temperatures are kept between 0–40 °C to obtain a high molecular weight product. A redox initiation based on reactions of persulfate, bisulfite, and ferrous ions is often used. Commercial polymers range in molecular weights from 50,000–500,000.

Polychlorotrifluoroethylene exhibits greater strength, hardness, and creep resistance than does polytetrafluoroethylene. Due to the presence of chlorine atoms in the chains, however, packing cannot be as tight as in polytetrafluoroethylene, and it melts at a lower temperature. The melting point is 214 °C. The degree of crystallinity varies from 30–85%, depending upon the thermal history of the polymer. Polytrifluorochloroethylene is soluble in certain chlorofluoro compounds above 100 °C. It flows above its melting point. The chemical resistance of this material is good, but inferior to polytetrafluoroethylene.

5.17.3. Poly(vinylidine fluoride)

The monomer can be prepared by dehydrochlorination of 1,1,1-chlorodifluoroethane:

$$F_2C(Cl)-CH_3 \longrightarrow CF_2{=}CH_2 + HCl$$

or by dechlorination of 1,2-dichloro-1,1-difluoroethane[259]:

$$CF_2Cl\text{-}CH_2Cl \rightarrow CF_2{=}CH_2$$

Vinylidine fluoride boils at –84 °C. The monomer is polymerized in aqueous systems under pressure. Details of the process, however, are kept as trade secrets. Two different molecular weight materials are available commercially, 300,000 and 600,0000. Poly(vinylidine fluoride) is crystalline and melts at 171 °C. The material exhibits fair resistance to solvents and chemicals, but is inferior to polytetrafuroethylene and to polytrifluorochloroethylene.

5.17.4. Poly(vinyl fluoride)

Vinyl fluoride monomer can be prepared by addition of HF to acetylene. The monomer is a gas at room temperature and boils at –72.2 °C. Commercially, vinyl fluoride is polymerized in aqueous medium using either redox initiation or one from thermal decomposition of peroxides. Pressures of up to 1000 atmospheres may be used. Radicals generated at temperatures between 50–100 °C yield very high molecular weight polymers.

Poly(vinyl fluoride) is moderately crystalline. The crystal melting point, T_m, is approximately 200 °C. The high molecular weight polymers dissolve in dimethylformamide and in tetramethyl urea at temperatures above 100 °C. The polymer is very resistant to hydrolytic attack. It does, however, lose HF at elevated temperatures.

5.17.5. Copolymers of Fluoroolefins

Many different copolymers of fluoroolefins are possible and have been reported in the literature. Commercial use of fluoroolefin copolymers, however, is restricted mainly to elastomers. Such materials offer superior solvent resistance and good thermal stability.

The elastomers that are most important industrially are vinylidine fluoride–chlorotrifluoroethylene copolymers[260] and vinylidine fluoride–hexafluoropropylene copolymers.[261] These copolymers are amorphous due to irregularities in their structures and can range in properties from resinous to elastomeric, depending upon composition.[262] Those that contain 50–70 mole% of vinylidine fluoride are elastomers. The values of T_g range from 0 °C to –15 °C, also depending upon vinylidine fluoride content.[260] They may be crosslinked with various peroxides, polyamines,[263] or ionizing radiation. The crosslinking reactions by peroxides take place through hydrogen abstraction by primary radicals:

$$\sim\!CH_2-CF_2-CH_2-CF_2\!\sim + R\bullet \longrightarrow \sim\!CH_2-CF_2-\dot{C}H-CF_2\!\sim + RH$$

$$2\ \sim\!CH_2-CF_2-\dot{C}H-CF_2\!\sim \longrightarrow \begin{array}{c}\sim\!CH_2-CF_2-CH-CF_2\!\sim\\ |\\ \sim\!CH_2-CF_2-CH-CF_2\!\sim\end{array}$$

Copolymers of vinylidine fluoride with hexafluoropropylene are prepared in aqueous dispersions using persulfate initiators. Hexafluoropropylene does not homopolymerize but it does copolymerize. This means that its content in the copolymer cannot exceed 50%. Preferred compositions appear to contain about 80% of vinylidine fluoride. The crosslinking reactions with diamines are not completely understood. It is believed that the reaction takes place in two steps.[264,265] In the first one, a dehydrofluorination occurs:

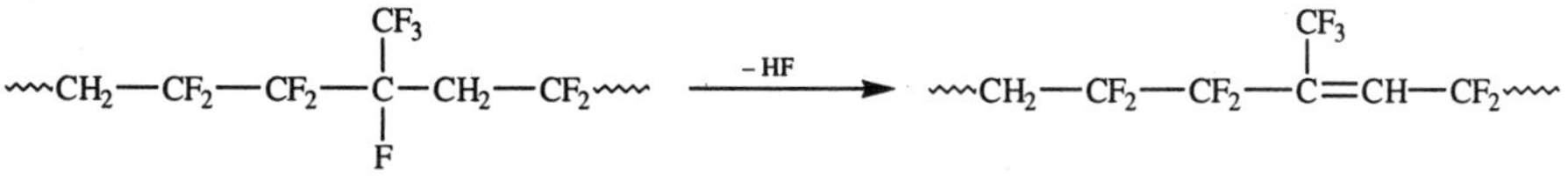

$$\sim CH_2-CF_2-CF_2-\underset{F}{\overset{CF_3}{C}}-CH_2-CF_2\sim \xrightarrow{-HF} \sim CH_2-CF_2-CF_2-\overset{CF_3}{C}=CH-CF_2\sim$$

The above elimination is catalyzed by basic materials. These may be in the form of MgO, which is often included in the reaction medium. In the second step the amine groups add across the double bonds:

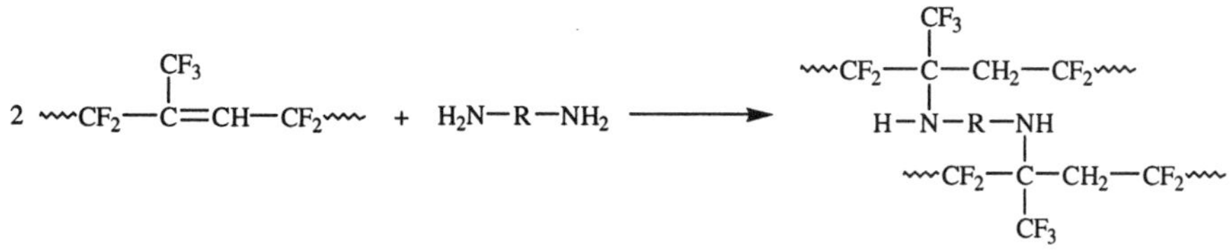

$$2 \sim CF_2-\overset{CF_3}{C}=CH-CF_2\sim + H_2N-R-NH_2 \longrightarrow \begin{array}{c} \sim CF_2-\overset{CF_3}{C}-CH_2-CF_2\sim \\ H-N-R-NH \\ \sim CF_2-\underset{CF_3}{C}-CH_2-CF_2\sim \end{array}$$

Free diamines, used for crosslinking, are too reactive and can cause premature gelation. It is therefore common practice to add these diamine compounds in the form of carbamates, like ethylenediamine carbamate or hexamethylenediamine carbamate. The above fluoroelastomers exhibit good resistance to chemicals and maintain useful properties from –50 °C to +300 °C.

Copolymers of tetrafluoroethylene with hexafluoropropylene are truly thermoplastic polyperfluoroolefins that can be fabricated by common techniques. Such copolymers soften at about 285 °C and have a continuous-use temperature of –260 °C to +205 °C. Their properties are similar to, though somewhat inferior to, polytetrafluoroethylene.

5.17.6. Miscellaneous Fluorine Containing Chain-Growth Polymers

One of the miscellaneous fluoroolefin polymers is a copolymer of trifluoronitrosomethane and tetrafluoroethylene,[266] an elastomer:

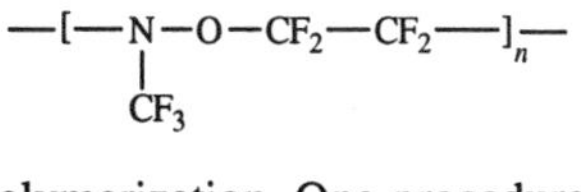

$$-[-\underset{CF_3}{N}-O-CF_2-CF_2-]_n-$$

It can be formed by suspension polymerization. One procedure is to carry out the reaction in an aqueous solution of lithium bromide at –25 °C with magnesium carbonate as the suspending agent. No initiator is added and the reaction takes about 20 hours. Because the reaction is inhibited by hydroquinone and accelerated by ultraviolet light, it is believed to take place by a free-radical mechanism. Whether it is chain-growth polymerization, however, is not certain. A 1:1 copolymer is always formed regardless of the composition of the monomer feed, and the copolymerization takes place only at low temperatures. At elevated temperatures, however, cyclic oxazetidines form instead:

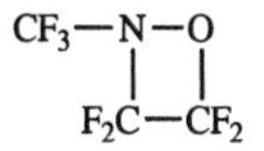

$$\begin{array}{ccc} CF_3-N & - & O \\ | & & | \\ F_2C & - & CF_2 \end{array}$$

Two polyfluoroacrylates are manufactured on a small commercial scale for some special uses in jet engines. These are poly(1,1-dihyroperfluorobutyl acrylate):

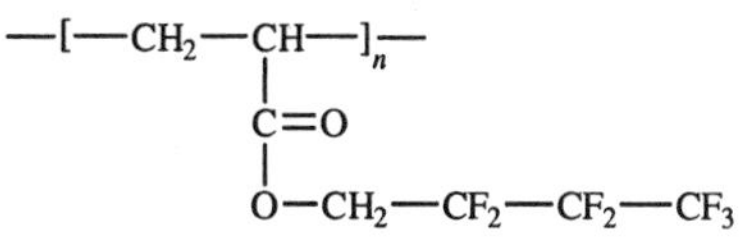

$$-[-CH_2-\underset{\underset{O-CH_2-CF_2-CF_2-CF_3}{|}}{\underset{C=O}{\underset{|}{CH}}}-]_n-$$

and poly(3-perfluoromethoxy-1,1-dihydroperfluoropropyl acrylate):

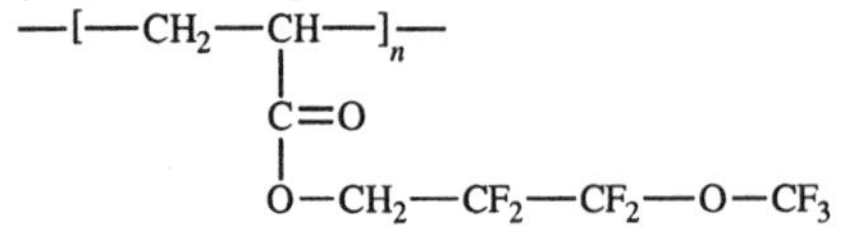

$$—[—CH_2—CH(—C(=O)—O—CH_2—CF_2—CF_2—O—CF_3)—]_n—$$

The polymers are prepared by emulsion polymerization with a persulfate initiator.

Although many other fluorine-containing polymers have been described in the literature, it is not possible to describe all of them here. They are not utilized commercially on a large scale. A few, however, will be mentioned as examples. One of them is polyfluoroprene[267]:

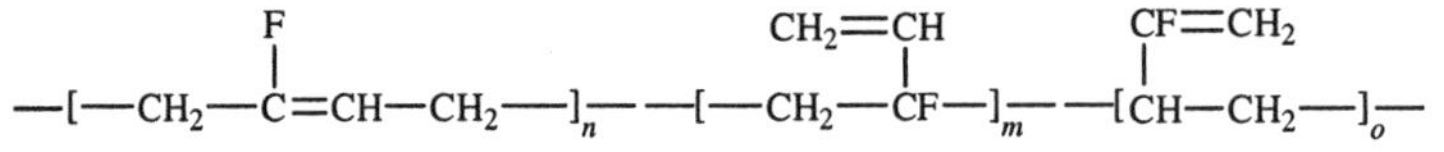

$$—[—CH_2—C(F)=CH—CH_2—]_n—[—CH_2—CF(CH_2=CH)—]_m—[CH(CF=CH_2)—CH_2—]_o—$$

The polymer is formed by a free-radical mechanism, in an emulsion polymerization using redox initiation. All three possible placements of the monomer occur.[267]

Polyfluorostyrenes are described in many publications. A β-fluorostyrene can be formed by a cationic mechanism.[268] The material softens at 240–260 °C. An α,β,β'-trifluorostyrene can be polymerized by a free-radical mechanism to yield an amorphous polymer that softens at 240 °C.[269] Ring-substituted styrenes apparently polymerize similarly to styrene. Isotactic poly(*o*-fluorostyrene) melts at 265 °C. It forms by polymerization with Ziegler–Natta catalysts.[270] The *meta* analog, however, polymerized under the same conditions, yields an amorphous material.[270]

5.17.7. Poly(vinyl chloride)

Poly(vinyl chloride) is used in industry on a very large scale in many applications, such as rigid plastics, plastisols, and surface coatings. The monomer, vinyl chloride, can be prepared from acetylene:

$$CH{\equiv}CH + HCl \rightarrow CH_2{=}CHCl$$

The reaction is exothermic and requires cooling to maintain the temperature between 100–108 °C. The monomer can also be prepared from ethylene:

$$CH_2{=}CH_2 + Cl_2 \xrightarrow{30\text{–}50\,°C} CH_2Cl—CH_2Cl \xrightarrow[\text{kaolin}]{500\,°C} CH_2{=}CHCl + HCl$$

The reaction of dehydrochlorination is carried out at elevated pressure of about 3 atmospheres.

Free-radical polymerization of vinyl chloride was studied extensively. For reactions that are carried out in bulk the following observations were made[271]:

1. The polymer is insoluble in the monomer and precipitates out during the polymerization.
2. The polymerization rate accelerates from the start of the reaction. Vinyl chloride is a relatively unreactive monomer. The main sites of initiation occur in the continuous monomer phase.
3. The molecular weight of the product does not depend upon conversion nor does it depend upon the concentration of the initiator.
4. The molecular weight of the polymer increases as the temperature of the polymerization decreases. The maximum for this relationship, however, is at 30 °C.

There is autoacceleration in the bulk polymerization rate of vinyl chloride.[272] It was suggested by Schindler and Breitenbach[273] that the acceleration is due to trapped radicals present in the precipitated polymer swollen by monomer molecules. This influences the rate of the termination that decreases progressively with the extent of the reaction, while the propagation rate remains constant. The autocatalytic effect in vinyl chloride bulk polymerizations, however, depends on the type of initiator used.[274] Thus, when 2,2′-azobisisobutyronitrile initiates the polymerization, the autocatalytic effect can be observed up to 80% of conversion. Yet, when benzoyl peroxide initiates the reaction it only occurs up to 20–30% of conversion.

When vinyl chloride is polymerized in solution there is no autoacceleration. Also, a major feature of vinyl chloride free-radical polymerization is chain transferring to monomer.[275] This is supported by experimental evidence.[272,276] In addition, the growing radical chains can terminate by chain transferring to "dead" polymer molecules. The propagations then proceed from the polymer backbone.[272] Such new growth radicals, however, are probably short-lived as they are destroyed by transfer to monomer.[277]

The ^{13}C NMR spectroscopy of poly(vinyl chloride), which was reduced with tributyltin hydride, showed that the original polymer contained a number of short four-carbon branches.[278] This, however, may not be typical of all poly(vinyl chloride) polymers formed by free-radical polymerization. It conflicts with other evidence from ^{13}C NMR spectroscopy that chloromethyl groups are the principal short-chain branches in poly(vinyl chloride).[312,313] The pendant chloromethyl groups were found to occur with a frequency of two to three per thousand carbons. The formation of these branches, as seen by Bovey and co-workers, depends upon head-to-head additions of monomers during the polymer formation. Such additions are followed by 1,2-chlorine shifts with subsequent propagations.[312,313] Evidence from still other studies also shows that some head-to-head placement occurs in the growth reaction.[279] It was suggested that this may be not only an essential step in formation of branches, but also one leading to formation of unsaturation at the chain ends[279,280]:

$$\sim\!CH_2\!-\!\underset{Cl}{CH}\!-\!\underset{Cl}{CH}\!-\!CH_2\cdot \longrightarrow \sim\!CH_2\!-\!\underset{Cl}{CH}\!-\!\underset{CH_2Cl}{CH}\cdot \xrightarrow{CH_2=CHCl} \sim\!CH_2\!-\!\underset{Cl}{CH}\!-\!\underset{CH_2Cl}{CH}\!-\!CH_2\!-\!\underset{Cl}{CH}\cdot$$

$$\sim\!CH_2\!-\!\underset{Cl}{CH}\!-\!\underset{Cl}{CH}\!-\!CH_2\cdot \longrightarrow \sim\!CH_2\!-\!\underset{Cl}{CH}\!-\!CH\!=\!CH_2 + Cl\cdot$$

$$\sim\!CH_2\!-\!\underset{Cl}{CH}\!-\!\underset{Cl}{CH}\!-\!CH_2\cdot \longrightarrow \sim\!CH_2\!-\!\underset{Cl}{CH}\!-\!\underset{Cl}{C}\!=\!CH_2 + H\cdot$$

Poly(vinyl chloride) prepared with boron alkyl catalysts at low temperatures possesses higher amounts of syndiotactic placement and is essentially free from branches.[293–295]

Many attempts were made to polymerize vinyl chloride by ionic mechanisms using different organometallic compounds, some in combinations with metal salts.[271,283–286] Attempts were also made to polymerize vinyl chloride with Ziegler–Natta catalysts complexed with Lewis bases. To date, however, it has not been established unequivocally that vinyl chloride does polymerize by ionic mechanism. Use of the above catalysts did yield polymers with higher crystallinity. These reactions, however, were carried out at low temperatures where a greater amount of syndiotactic placement occurs by the free-radical mechanism.[281] Vinyl chloride was also polymerized by $AlCl(C_2H_5)O_2H_5$ + $VO(C_3H_7O_2)$ without Lewis bases.[282] Here too, however, the evidence indicates a free-radical mechanism.

On the other hand, butyllithium–aluminum alkyl initiated polymerizations of vinyl chloride are unaffected by free-radical inhibitors.[287] Also, the molecular weights of the resultant polymers are unaffected by additions of CCl_4 that acts as a chain-transferring agent in free-radical polymerizations. This suggests an ionic mechanism of chain growth. Furthermore, the reactivity ratios in copolymerization reactions by this catalytic system differ from those in typical free-radical polymerizations.[287] An anionic mechanism was also postulated for polymerization of vinyl chloride with *t*-butylmagnesium in tetrahydrofuran.[288]

Commercially, by far the biggest amount of poly(vinyl chloride) homopolymer is produced by suspension polymerization and, to a lesser extent, by emulsion and bulk polymerization. Very little polymer is formed by solution polymerization.

One process for bulk polymerization of vinyl chloride was developed in France, where the initiator and monomer are heated at 60 °C for approximately 12 hours inside a rotating drum containing stainless steel balls. Typical initiators for this reaction are benzoyl peroxide or azobisisobutyronitrile. The speed of rotation of the drum controls the particle size of the final product. The process is also carried out in a two-reactor arrangement. In the first one, approximately 10% of the monomer is converted. The material is then transferred to the second reactor, where the polymerization is continued until it reaches 75–80% conversion. Special ribbon blenders are present in the second reactor. Control of the operation in the second reactor is quite critical.[289]

Industrial suspension polymerizations of vinyl chloride are often carried out in large batch reactors or stirred jacketed autoclaves. Continuous reactors, however, have been introduced in several manufacturing facilities.[313] Typical recipes call for 100 parts of vinyl chloride for 180 parts of water, a suspending agent, like maleic acid–vinyl acetate copolymer, a chain-transferring agent, and a monomer soluble initiator. The reaction may be carried out at 100 lb/in^2 pressure and 50 °C for approximately 15 hours. As the monomer is consumed, the pressure drops. The reaction is stopped at an internal pressure of about 10 lb/in^2 and the remaining monomer (about 10%) is drawn off and recycled. The product is discharged.

Emulsion polymerizations of vinyl chloride are usually conducted with redox initiation. Such reactions are rapid and can be carried out at 20 °C in one to two hours with a high degree of conversion. Commercial poly(vinyl chloride)s range in molecular weights from 40,000–80,000. The polymers are mostly amorphous with small amounts (about 5%) of crystallinity. The crystalline areas are syndiotactic.[290,291]

Poly(vinyl chloride) is soluble at room temperature in oxygen-containing solvents, such as ketones, esters, ethers, and others. It is also soluble in chlorinated solvents. The polymer, however, is not soluble in aliphatic and aromatic hydrocarbons. It is unaffected by acid and alkali solutions, but has poor heat and light stability. Poly(vinyl chloride) degrades at temperatures of 70 °C or higher, or when exposed to sunlight, unless it is stabilized. Heating changes the material from colorless to yellow, orange, brown, and finally black. Many compounds tend to stabilized poly(vinyl chloride). The more important ones include lead compounds, like dibasic lead phthalate and lead carbonate. Also effective are metal salts, like barium, calcium, and zinc octoates, stearates, and laurates. Organotin compounds, like dibutyl tin maleate or laurate, also belong to that list. Epoxidized drying oils are effective heat stabilizers, particularly in coatings based on poly(vinyl chloride). Some coating materials may also include aminoplast resins, like benzoguanamine–formaldehyde condensate.

The process of degradation is complex. It involves loss of hydrochloric acid. The reactions are free-radical in nature, though some ionic reactions appear to take place as well. The process of dehydrochlorination results in formations of long sequences of conjugated double bonds. It is commonly believed that formation of conjugated polyenes, which are chromophores, is responsible for the darkening of poly(vinyl chloride). In addition, the polymer degrades faster in open air than it does in an inert atmosphere. This shows that oxidation contributes to the degradation process. All effective stabilizers are hydrochloric acid scavengers. This feature alone, however, can probably not account for the stabilization process. There must be some interaction between the stabilizers and the polymers. Such interaction might vary, depending upon a particular stabilizer.

5.17.7.1. Copolymers of Vinyl Chloride

A very common copolymer of vinyl chloride is with vinyl acetate. Copolymerization with vinyl acetate improves stability and molding characteristics. The copolymers are also used as fibers and as coatings. Copolymers intended for use in moldings are usually prepared by suspension polymerization. Those intended for coating purposes are prepared by solution, emulsion, and suspension polymerizations. The copolymers used in molding typically contain about 10% poly(vinyl acetate). Copolymers that are prepared for coating purposes can contain from 10–17% poly(vinyl acetate). For coatings, a third comonomer may be included in some resins. This third component may, for instance, be maleic anhydride, in small quantities, like 1%, to improve adhesion to surfaces.

Copolymers of vinyl chloride with vinylidine chloride are similar in properties to copolymers with vinyl acetate. They contain from 5–12% of poly(vinylidine chloride) and are intended for use in stabilized calendaring.

Copolymers containing 60% vinyl chloride and 40% acrylonitrile are used in fibers. The fibers are spun from acetone solution. They are nonflammable and have good chemical resistance.

5.17.8. Poly(vinylidine chloride)

Vinylidine chloride homopolymers form readily by free-radical polymerization, but lack sufficient thermal stability for commercial use. Copolymers, however, with small amounts of comonomers find many applications.

The monomer, vinylidine chloride, can be prepared by dehydrochlorination of 1,1,2-trichloroethylene:

$$CH_2Cl{-}CHCl_2 \xrightarrow[400\,°C]{-HCl} CH_2{=}CHCl_2$$

It is a colorless liquid that boils at 32 °C. Also, it is rather hard to handle as it polymerizes on standing. This takes place upon exposure to air, water, or light. Storage under an inert atmosphere does not completely prevent polymer formation.

Poly(vinylidine chloride) can be formed in bulk, solution, suspension, and emulsion polymerization processes. The products are highly crystalline with regular structures and a melting point of 220 °C. The structure can be illustrated as follows:

$$\sim CH_2{-}CCl_2{-}CH_2{-}CCl_2{-}CH_2{-}CCl_2{-}CH_2\sim$$

This regularity in structure is probably due to little chain transferring to the polymer backbone during polymerization. Such regularity of structure allows close packing of the chains and, as a result, there are no effective solvents for the polymer at room temperature.

Copolymerization of vinylidine chloride with vinyl chloride reduces the regularity of the structure. It increases flexibility and allows processing the polymer at reasonable temperatures. Due to extensive crystallization, however, that is still present, 85:15 copolymers of vinylidine chloride with vinyl chloride melt at 170 °C. The copolymerization reactions proceed at slower rates than do homopolymerizations of either one of the monomers alone. Higher initiator levels and temperatures are therefore used. The molecular weights of the products range from 20,000–50,000. These materials are good barriers for gases and moisture. This makes them very useful in films for food packaging. Such films are formed by extrusion and biaxial orientation. The main application, however, is in filaments. These are prepared by extrusion and drawing. The tensile strength of the unoriented material is 10,000 lb/in^2 and the oriented one, 30,000 lb/in^2.

Vinylidine chloride is also copolymerized with acrylonitrile. This copolymer is used mainly as a barrier coating for paper, polyethylene, and cellophane. It has the advantage of being heat-sealable.

5.18. Poly(vinyl acetate)

Vinyl acetate monomer can be prepared by reacting acetylene with acetic acid:

$$HC{\equiv}CH + CH_3COOH \longrightarrow CH_2{=}CH{-}O{-}\overset{O}{\overset{\|}{C}}{-}CH_3$$

The reaction can be carried out in a liquid or in a vapor phase. A liquid-phase reaction requires a 75–80 °C temperature and a mercuric sulfate catalyst. The acetylene gas is bubbled through glacial acetic acid and acetic anhydride. Vapor-phase reactions are carried out at 210–250 °C. Typical catalysts are cadmium acetate or zinc acetate. There are other routes to vinyl acetate as well, based on ethylene.

Commercially, poly(vinyl acetate) is formed in bulk, solution, emulsion, and suspension polymerizations by a free-radical mechanism. In such polymerizations, chain transferring to the polymer may be as high as 30%. The transfer can be to a polymer backbone through abstraction of a tertiary hydrogen:

$$\sim R\cdot + \sim CH_2-\underset{\underset{\underset{O}{\|}}{O-C-CH_3}}{\overset{H}{C}}\sim \longrightarrow \sim R-H + \sim CH_2-\underset{\underset{\underset{O}{\|}}{O-C-CH_3}}{\overset{\cdot}{C}}\sim$$

It can also take place to the methyl proton of the acetate group:

$$\sim R\cdot + \sim CH_2-\underset{\underset{\underset{O}{\|}}{O-C-CH_3}}{\overset{H}{C}}\sim \longrightarrow \sim R-H + \sim CH_2-\underset{\underset{\underset{O}{\|}}{O-C-CH_2\cdot}}{\overset{H}{C}}\sim$$

The polymer has a head-to-tail structure and is highly branched. It is quite brittle and exhibits cold flow. This makes it useless as a structural plastic. It is, however, quite useful as a coating material and as an adhesive for wood. The polymer is soluble in a wide range of solvents and swells and softens upon prolonged immersion in water. At higher temperatures or at extended exposures to temperatures above 70 °C, the material loses acetic acid.

A number of copolymers are known where vinyl acetate is the major component. In coatings, vinyl acetate is often used in copolymers with alkyl acrylates (line 2-ethylhexyl acrylate) or with esters of maleic or fumaric acids. Such copolymers typically contain 50–20% by weight of the comonomer and are usually formed by emulsion polymerization in batch processes. They are used extensively as vehicles for emulsion paints.

5.19. Poly(vinyl alcohol) and Poly(vinyl Acetal)s

Vinyl alcohol monomer does not exist because its keto tautomer is much more stable. Poly(vinyl alcohol) can be prepared from either poly(vinyl ester)s or from poly(vinyl ether)s. Commercially, however, it is prepared exclusively from poly(vinyl acetate). The preferred procedure is through a transesterification reaction using methyl or ethyl alcohols. Alkaline catalysts yield rapid alcoholyses. A typical reaction employs about 1% sodium methoxide and can be carried to completion in one hour at 60 °C. The product is contaminated with sodium acetate that must be removed. The reaction of transesterification can be illustrated as follows:

$$\sim CH_2-\underset{\underset{\underset{O}{\|}}{O-C-CH_3}}{CH}\sim \xrightarrow{CH_3O^{\ominus}} \sim CH_2-\underset{\underset{{}^{\ominus}O\ \ O-CH_3}{O-C-CH_3}}{CH}\sim \longrightarrow \sim CH_2-\underset{O^{\ominus}}{CH}\sim + CH_3-COOCH_3$$

The branches of poly(vinyl acetate) that form during polymerization as a result of chain transferring to the acetate groups cleave during transesterification. As a result, poly(vinyl alcohol) is lower in molecular weight than its parent material.

Poly(vinyl alcohol) is very high in head-to-tail structures, based on NMR data.[292] It shows the presence of only a small amount of adjacent hydroxyl groups. The polymer prepared from amorphous poly(vinyl acetate) is crystalline, because the relatively small size of the hydroxyl groups permits the chains to line-up into crystalline domains. Synthesis of isotactic poly(vinyl alcohol) was reported from isotactic poly(vinyl ethers) like poly(benzyl vinyl ether),[296,297] poly(*t*-butyl vinyl ether),[298] poly(trimethylsilyl vinyl ether),[299] and some divinyl compounds.[300–302]

Poly(vinyl alcohol) is water soluble. The hydroxyl groups attached to the polymer backbone, however, exert a significant effect on the solubility. When the ester groups of poly(vinyl acetate) are cleaved to a hydroxyl content of 87–89%, the polymer is soluble in cold water. Further cleavage of the ester groups results in a reduction of the solubility and the products require heating of the water to 85 °C in order to dissolve. This is due to strong hydrogen bonding that also causes unplasticized poly(vinylalcohol) to decompose below its flow temperature. On the other hand, due to hydrogen bonding the polymer is very tough.

Poly(vinyl acetal)s are prepared by reacting poly(vinyl alcohol) with aldehydes. Reactions of poly(vinyl alcohol) with ketones yield ketals. These are not used commercially.

Not all hydroxyl groups participate in formations of acetals and some become isolated. A typical poly(vinyl acetal) contains acetal groups, residual hydroxyl groups, and residual acetate groups from incomplete transesterification of the parent polymer.

Poly(vinyl acetal)s can be formed directly from poly(vinyl acetate) and this is actually done commercially in preparations of poly(vinyl formal). A typical reaction is carried out in the presence of acetic acid, formalin, and sulfuric acid catalyst at 70 °C:

$$\sim CH_2-\underset{\underset{\underset{CH_3}{|}}{O-C=O}}{\underset{|}{CH}}-CH_2-\underset{\underset{\underset{CH_3}{|}}{O-C=O}}{\underset{|}{CH}}\sim \quad + \quad HCHO \xrightarrow[CH_3COOH\ +\ H_2O]{70\ °C} \sim CH_2-\underset{\underset{|}{O}}{CH}-CH_2-\underset{\underset{|}{O}}{CH}\sim \quad (\text{O and O joined by } CH_2)$$

Poly(vinyl butyral), on the other hand, is prepared from poly(vinyl alcohol) and butyraldehyde. Sulfuric acid is used as the catalyst. Commercially, only poly(vinyl formal) and poly(vinyl butyral) are utilized on a large scale in coating materials.

Review Questions

Section 5.1

1. What are the two types of polyethylene that are currently manufactured commercially?
2. Describe the chemical structure of low-density polyethylene produced by a free-radical mechanism and show by chemical equations how all the groups that are present form.
3. Describe conditions and the procedure for commercial preparation of polyethylene by a free-radical mechanism, the role of oxygen, and the problems associated with oxygen.
4. Describe a tubular reactor for preparation of polyethylene.
5. What are the industrial conditions for preparations of high-density polyethylene? Describe the continuous solution process, the slurry process, and the gas-phase process.
6. Show with chemical reactions how polymethylene forms from diazomethane.

Section 5.2

1. Discuss high-activity catalysts for the manufacturing of isotactic polypropylene, heterogenous and homogenous.
2. What are the current techniques for polypropylene manufacture?
3. How can syndiotactic polypropylene be prepared and what are its properties?

Section 5.3

1. Describe the two industrial processes for manufacturing polybutylene.

Section 5.4

1. Draw the chemical structure of isotactic poly(butene-1). How is it prepared and used?
2. What is TPX, how is it prepared, and what are its properties?

Section 5.5

1. Discuss copolymers of ethylene and propylene. How are they prepared? What catalysts are used in the preparations? How are ethylene–propylene rubbers crosslinked?
2. What are the copolymers of ethylene with higher α-olefins and why are they prepared and how?
3. Discuss the copolymers of ethylene with vinyl acetate. How are they prepared and used?
4. What are ionomers? Describe each type. How are they used?

Section 5.6

1. Discuss polybutadiene homopolymers. How are they prepared? What are their uses?
2. What are popcorn polymers? What causes their formation?
3. Discuss liquid polybutadienes. How are they prepared and used?
4. How are high molecular weight polybutadienes prepared and used?
5. Discuss polyisoprenes. What is natural rubber? Where does it come from? What are synthetic polyisoprenes? How are they prepared?

Section 5.7

1. What is methyl rubber?

Section 5.8

1. What is chloroprene rubber? How is it made and used?

Section 5.9

1. What are poly(carboxybutadiene)s?

Section 5.10

1. Discuss cyclopolymerization of conjugated dienes.

Section 5.11

1. What is SBR rubber? Explain and describe its preparation and properties.
2. What are block copolymer elastomers? How are they prepared and what gives them their unique properties?
3. What is GR-N rubber? Explain and describe its preparation and properties.

Section 5.12

1. How is polystyrene prepared commercially? Describe.
2. What polymers of substituted styrenes are available commercially? How are they prepared?

Section 5.13

1. What is high-impact polystyrene and how is it prepared?
2. Discuss ABS resins. How are they prepared?

Section 5.14

1. Discuss the chemistry of free-radical polymerization of acrylic and methacrylic esters.
2. What are acrylic elastomers and how are they vulcanized?
3. How is poly(methyl methacrylate) prepared commercially, such as Plexiglass in the form of sheets and rods? Is poly(methyl methacrylate) prepared in any other way? How? For what applications?
4. Describe the thermosetting acrylic resins used in industrial coatings. How are they prepared? How are they crosslinked?

Section 5.15

1. Discuss industrial polymers and copolymers of acrylonitrile and methacrylonitrile. How are they prepared and used?

Section 5.16

1. Describe preparation and uses of polyacrylamide, polyacrylic acid, and polymethacrylic acid.

Section 5.17

1. How is polytetrafluoroethylene prepared? What are its properties and uses?
2. Discuss the chemistry of polychlorotrifluoroethylene, poly(vinylidine fluoride), and poly(vinyl fluoride).
3. What common copolymers of fluoroolefins are used commercially?
4. Discuss the chemistry of poly(vinyl chloride) and poly(vinylidine chloride).
5. Discuss the important commercial copolymers of vinyl chloride. What are their main uses?
6. Discuss the chemistry of poly(vinylidine chloride).

Section 5.18

1. Discuss the preparation, properties, and uses of poly(vinyl acetate).

Section 5.19

1. How is poly(vinyl alcohol) prepared, used, and converted to poly(vinyl acetal)s?

References

1. E.W. Fawcett, O.R. Gibson, M.W. Perrin, J.G. Paton, and E.G. Williams, Brit. Pat. # 571,590 (1937).
2. R.O. Symcox and P. Ehrlich, *J. Am. Chem. Soc.*, **84**, 531 (1962).
3. J.C. Woodbrey and P. Ehrlich, *J. Am. Chem. Soc.*, **85**, 1580 (1963).
4. R.O. Gibson, "The Discovery of Polythene," *Roy. Inst. Chem., London Lectures*, **1**, 1 (1964).
5. E.W. Fawcett *et al.*, Brit. Pat. # 471,590 (1937).
6. R.A. Hines, W.M.D, Bryant, A.W. Larchar, and D.C. Pease, *Ind. Eng. Chem.*, **49**(7), 1071 (1957).
7. A.L.J. Raum, Chapt. 4 in *Vinyl and Allied Polymers* (P.D. Ritchle, ed.), CRC Press, Cleveland, Ohio, 1968.
8. Monsanto Co., U.S. Pat. # 2,856,395 and 2,852,501.
9. G. Downs, Chapt. 5 in *Vinyl and Allied Polymers* (P.D. Ritchle, ed.), CRC Press, Cleveland, Ohio, 1968.
10. M.J. Roedel, *J. Am. Chem. Soc.*, **75**, 6110 (1953).
11. A.H. Willbourn, *J. Polym. Sci.*, **34**, 569 (1959).
12. D.E. Dorman, E.P. Otacka, and P.A. Bovey, *Macromolecules*, **5**, 574 (1972).
13. W.G. Oakes and R.B. Richards, *J. Chem. Soc.*, 2929 (1949).
14. K.W. Doak, "Low Density Polyethylene (High Pressure)," in *Encyclopedia of Polymer Science and Technology*, Vol. 6, 2nd Ed. (H.F. Mark, N.M. Bikales, C.G. Overberger, and G. Menges, eds.), Wiley-Interscience, New York, 1986.
15. K. Weissermel, H. Cherdron, J. Berthold, B. Diedrich, K.D. Keil, K. Rust, H. Strametz, and T. Toth, *J. Polym. Sci., Symp.* **51**, 187 (1975).
16. D.M. Rasmussen, *Chem. Eng.*, 104 (Sept. 18, 1972).
17. W. Keim, R. Appel, A. Storeck, C. Kruger, and R. Goddard, *Angew. Chem., Int. Ed. Engl.*, **20** (1), 116 (1981).
18. W.C. Von Dohlen and T.P. Wilson, *J. Polym. Sci., Polym. Chem. Ed.*, **17**, 2511 (1979).
19. H. Pechmann, *Ber.*, **31**, 2643 (1898).
20. E. Bamberger and F. Tschirner, *Ber.*, **33**, 955 (1900).
21. G.D. Buckley, L.H. Cross, and N.H. Roy, *J. Chem. Soc.*, 2714 (1950).
22. A. Renfrew and P. Morgan, *Polythene*, Wiley-Interscience, New York, 1957.
23. H. Meerwein, *Angew. Chem.*, **A60**, 78 (1948).
24. C.E.H. Bawn, A. Ledwith, and P. Matthies, *J. Polym. Sci.*, **34**, 93 (1959).
25. C.E.H. Bawn and A. Ledwith, *Chem. Ind. (London)*, 1180 (1957).
26. G.D. Buckley and N.H. Ray, *J. Chem. Soc.*, 3701 (1952).
27. A.G. Nasini, G. Saini, and T. Trossarelli, *J. Polym. Sci.*, **48**, 435 (1960); *Pure Appl. Chem.*, **4**, 255 (1962).
28. M.G, Krakonyak and S.S. Skorokhodov, *Vysokomol. Soyed.*, **A11**(4), 797 (1969).
29. H. Pechler, B. Firnhaber, D. Kroussis, and A, Dawallu, *Makromol. Chem.*, **70**, 12 (1964).
30. G.A. Mortimer and L.C. Arnold, *J. Polym. Sci.*, **A2**, 4247 (1964).
31. Avison Corp. U.S. Pat. # 3,134,642 (May 26, 1964), 2,980,664 (April 18, 1961), 3,055,878 (Sept. 25, 1962), 3,216,987 (Nov. 9, 1965), 3,303,179 (Feb. 7, 1967), 3,328,375 (June 27, 1967), 3,362,916 (Jan. 9, 1964), 3,313,791 (April 11, 1967), 3,255,167 (June 7, 1966); German Pat. # 1,236,789 (March 16, 1967), 1,234,218 (April 6, 1967).
32. Eastman Kodak Co., U.S. Pat. # 3,232,919 (Feb. 1, 1966), 3,058,696 (Oct. 10, 1962), 3,081,287 (March 12, 1963), 2,962,487 (Nov. 29, 1960), 3,143,537 (July 21, 1964), 3,004,015 (Oct. 10, 1965), 3,149,097 (Sept. 15, 1964), 3,201,379 (Aug. 17, 1965), 3,213,073 (Oct. 19, 1965), 3,230,208 (Jan. 18, 1966), 3,149,097 (Sept. 15,

1964), 3,186,977 (June 1, 1965), 3,189,590 (June 15, 1965), 3,072,629 (Jan. 8, 1963), 3,178,401 (April 13, 1965), 3,284,427 (Nov. 8, 1966); British Pat. # 921,039 (March 13, 1963), 930,633 (July 3, 1963), 1,007,030 (Oct. 13, 1965), 1,00,348 (Aug. 4, 1965), 1,000,720 (Aug 11, 1965); French Pat. # 1,315,782 (Jan. 25, 1963).

33. Montecantini-Edison Sp.A., U.S. Pat. # 3,112,300 (Nov. 26, 1963), 3,259,613 (July 5, 1966), 3,139,418 (June 30, 1964), 3,141,872 (July 21, 1964), 3,277,069 (Oct. 4, 1969), 3,252,954 (May 25, 1966); British Pat. # 1,014,944 (Jan. 10, 1964); Canadian Pat. # 649,164 (Sept. 25, 1962); German Pat. # 1,238,667 (April 13, 1967), 1,214 000 (June 6, 1961); Italian Pat. # (from *Chem. Abstr.*) 646,950 (Sept. 20, 1965).
34. Phillips Petroleum Co. U.S. Patents # 3,280,092 (Oct. 18, 1966), 3,119,798 (Jan. 28, 1964), 3,147,241 (Sept. 1, 1964), 3,182,049 (May 4, 1965), 3,210,332 (Oct. 5, 1965), 3,317,502 (May 2, 1967); British Pat. # 1,017,988 (Jan. 26, 1966), 1,034,155 (June 29, 1966), 1,119,033 (March 7, 1968); Belgian Pat. # 695,060 (June 9, 1967); German Pat. # 1,266,504 (April 18, 1968).
35. Shell Oil Co. U.S. Pat. # 3,147,238 (Sept. 1, 1964), 3,240,773 (March 15, 1966), 3,282,906 (Nov. 1,1966), 3,398,130 (Aug. 20, 1968), 3,311,603 (March 28, 1967), 3,394,118 (July 23, 1968), 3,264,277 (Aug, 2, 1966); British Pat. # 1,006,919 (Oct, 6, 1965).
36. U. Giannini, A Casata, P. Lorgi, and R. Mazzocchi, German Pat. Appl. 2,346,577 (1974).
37. H.R. Sailors and J.P. Hogam, *J. Macromol. Sci.-Chem.*, **A15**, 1377 (1981).
38. J. Boor Jr. and E.A. Youngman, *J. Polym. Sci.*, **A-1,4**, 1861 (1966).
39. G. Natta, I. Pasquon, and A. Zambelli, *J. Am. Chem. Soc.*, **84**, 1488 (1962).
40. A. Zambelli, G. Natta, and I. Pasquon, *J. Polym. Sci.*, **C,4**, 411 (1963).
41. P. Pino and R. Mulhaupt, *Angew. Chem., Int. Ed., Engl.*, **19**, 857 (1980).
42. K. Weissermel, H. Cherdron, J. Berthold, B. Diedrich, K.D. Keil, K. Rust, H. Strimetz, and T. Toth, *J. Polym. Sci., Symp.*, **51**, 187 (1975).
43. S. Sivaram, *Ind. Eng. Chem., Prod. Res. Dev.*, **16**(2), 121 (1977).
44. E.E. Vermel, V.A. Zakharov, Z.K. Bukatova, G.P. Shkurina, S.G. Echevskaya, E.M. Moroz, and S.V. Sudakova, *Vysokomol. Soyed.*, **22**(1), 22 (1980).
45. Y. Doi, S. Ueki, and T. Keii, *Macromolecules*, **12**(5), 814 (1979).
46. V.W. Buls and T.L. Higgins, *J. Polym. Sci., Polym. Chem. Ed.*, **11**, 925 (1973).
47. J. Brandrup and E.H. Immergut, *Polymer Handbook*, Wiley-Interscience, New York, 1989.
48. T. Asanuma, Y. Nishimori, M. Ito, N. Uchikawa, and T. Shiomura, *Polym. Bull.*, **25**, 567 (1991).
49. J.A. Ewen, R.L. Jones, A. Razavi, and J.D. Ferrara, *J. Am. Chem. Soc.*, **110**, 6255 (1988).
50. R.M. Thomas, W.J. Sparks, P.K. Frolich, M. Otto, and M. Mueller-Cuuradi, *J. Am. Chem. Soc.*, **62**, 276 (1940).
51. R.B. Cundall, in *The Chemistry of Cationic Polymerization* (P.H. Plesch, ed.), Macmillan Co., New York, 1963.
52. A. Gandini and H. Cheradame, "Cationic Polymerization," in *Encyclopedia of Polymer Science and Engineering*, Vol. 2 (H.F. Mark, N.M. Bikales, C.G. Overberger, and G. Menges, eds.), Wiley-Interscience, New York, 1985.
53. N.G. Gaylord and H.F. Mark, *Linear and Stereoregular Addition Polymers*, Interscience, New York, 1959.
54. J.P. Kennedy, in *Polymer Chemistry of Synthetic Elastomers*, Part 1 (J.P. Kennedy and E.G.M. Tornquist, eds.), Wiley-Interscience, New York, 1968.
55. G. Natta, P. Pino, G Mazzanti, and P. Longi, *Gazz. Chim. Ital.*, **87**, 570 (1957) (from a private translation).
56. G. Natta, I. Pasquon, A. Zambelli, and G. Gatti, *J. Polym. Sci.*, **51**, 387 (1961).
57. K.R. Dunham, J. Vanderberghe, J.W.H. Faber, and L.E. Contois, *J. Polym. Sci.*, **A-1,1**, 751 (1963).
58. G. Natta, P. Carradini, and I.W. Bassi, *Gazz. Chim. Ital.*, **89**, 784 (1959).
59. H.W. Coover Jr. and F.B. Joyner, *J. Polym. Sci.*, **A-1,3**, 2407 (1965).
60. T.W. Campbell and A.C. Haven, Jr., *J. Appl. Polym. Sci.*, **1**, 73 (1959); P. Carradini, V. Busico, and G. Guerra, "Monoalkene Polymerization: Stereochemistry," Chapt. 3 in *Comprehensive Polymer Science*, Vol. 4 (G.C. Eastmond, A. Ledwith, S. Russo, and P. Sigwalt, eds.), Pergamon Press, Oxford, 1989.
61. E. Suzuki, M. Tamura, Y. Doi, and T. Keii, *Makromol. Chem.*, **180**, 2235 (1979).
62. U. Giannini, *Makromol. Chem., Suppl.*, **5**, 216 (1981).
63. W. Passmann, *Ind. Eng. Chem.*, **62**(5), 48 (1970).
64. G. Natta, G. Mazzanti, A. Valvassori, and G. Pajaro, *Chim. Ind. (Milan)*, **39**, 733 (1957).
65. W. Marconi, S. Cesea, and G. Della Foruna, *Chim. Ind. (Milan)*, **46**, 1131 (1964).
66. S. Cesea, S. Arrighetti, and W. Marconi, *Chim. Ind. (Milan)*, **50**, 171 (1968).
67. S. Cesea, G. Bertolini, G. Santi, and P.V. Duranti, *J. Polym. Sci.*, **A-1,9**, 1575 (1971).
68. E.K. Gladding, B.S. Fischer, and J.W. Collette, *Ind. Eng. Chem., Prod. Res. Dev.*, **1**, 65 (1962).
69. E.W. Duck and W. Cooper, *23rd IUPAC Congress, Macromol. Prepr.*, **2**, 722 (1971).
70. E.K. Easterbrook, T.J. Brett, F.C. Loveless, and D.N. Mathews, *23rd IUPAC Congress, Macromol. Prepr.*, **2**, 712 (1971).
71. C. Cozewith and G. Ver Strate, *Macromolecules*, **4**, 482 (1971).
72. C.A. Lukach and H.M. Spurlin, in *Copolymerization* (G.E. Ham, ed.), Interscience, New York, 1964.
73. G. Natta, G. Mazzanti, A. Valvassori, G. Sartori, and A. Barbagallo, *J. Polym. Sci.*, **51**, 411, 429 (1961).
74. G. Bier and G. Lehmann, in *Copolymerization* (G.E. Ham, ed.), Interscience, New York, 1964.

75. J. Furukawa, *Angew. Makromol. Chem.*, **23**, 189 (1972).
76. J. Furukawa, S. Tsuruki, and J. Kiji, *J. Polym. Sci., Polym. Chem. Ed.*, **11**, 1819 (1973).
77. G. Bier, *Angew. Chem.*, **73**, 186 (1961).
78. M.J. Wisotsky, A.B. Kober, and I.A. Zlochower, *J. Appl. Polym. Sci.*, **15**, 1737 (1971).
79. V.D. Mochel, *J. Polym. Sci.*, **A-1,10**, 1009 (1972).
80. T.E. Ferrington and A.V. Tobolsky, *J. Polym. Sci.*, **31**, 25 (1958).
81. J. Inomata, *Makromol. Chem.*, **135**, 113 (1970).
82. I. Kuntz and A. Gerber, *J. Polym. Sci.*, **42**, 299 (1960).
83. F.P. Gintz, Chapt. 11 in *Vinyl and Allied Polymers* (P.D. Ritchie, ed.), Vol. I, CRC Press, Cleveland, Ohio, 1968.
84. W.M. Saltman *et al.*, *J. Am. Chem. Soc.*, **80**, 5615 (1958).
85. W.M. Saltman, *J. Polym. Sci.*, **A-1,1**, 373 (1963).
86. H.E. Adams, *et al.*, *Ind. Eng. Chem.*, **50**, 1507 (1958).
87. R.S. Stearns *et al.*, *J. Polym. Sci.*, **41**, 381 (1959).
88. A.V. Tobolsky, *J. Polym. Sci.*, **40**, 73 (1959).
89. F.W. Stavely *et al.*, *Ind. Eng. Chem.*, **48**(4), 778 (1956).
90. J.C. D'Ianni, *Kautschuk Gummi Kunststoffe*, **19**(3), 138 (1966).
91. G.A. Thomas and W.H. Carmody, *J. Am. Chem. Soc.*, **55**, 3854 (1933).
92. J.D. D'ianni, F.J. Naples, and J.E. Field, *Ind. Eng. Chem.*, **41**, 95 (1950).
93. J.E. Field and M. Feller, U.S. Pat. # 2,728,758 (Dec. 27, 1955).
94. H.E. Adams, R.S. Stearns, W.A. Smith, and J.L Binder, *Ind. Eng. Chem.*, **50**, 1507 (1958).
95. S. Minekawa *et al.*, Japanese Pat. # 45-36519 (Nov. 20, 1970).
96. S. Minekawa *et al.*, Japanese Pat. # 46-03795 (Jan. 29, 1971).
97. J. Witte *et al.*, German Pat. # 1,121,813 (Jan. 11, 1962).
98. K. Neutzel *et al.*, German Pat. # 1,116,903 (Aug. 20, 1960).
99. K. Neutzel *et al.*, German Pat. # 1,143,332 (Feb. 7, 1963).
100. J. Witte *et al.*, British Pat. # 1,069,817 (May 24, 1967).
101. German Pat. # 1,223,159 (April 1, 1965).
102. G. Friedman *et al.*, French Pat. # 1,589,920 (May 15, 1970).
103. M. Gippin, U.S. Pat. # 3,442,878 (May 6, 1969).
104. L.C. Kreider, South African Pat. # 68/00327 (June 12, 1968).
105. R.H. Mayor *et al.*, U.S. Pat. # 3,047,559 (July 31, 1962).
106. E. Schoenberg, British Pat. # 1,135,991 (Dec. 11, 1968).
107. H. Mori *et al.*, Japanese Pat. # 45-6518 (Nov. 20, 1970).
108. T.F. Yen, *J. Polym. Sci.*, **33**, 535 (1959).
109. O. Solomon and C. Amrus, *Rev. Chem. (Romania)*, **9**, 150 (1958).
110. I.M. Koltoff and W.E. Harris, *J. Polym. Sci.*, **2**, 41 (1947); C.A. Uraneck, in *Polymer Chemistry of Synthetic Elastomers*, Part 1 (J.P. Kennedy and E.G.M. Tornqvist, eds.), Interscience, New York, 1968.
111. A.L. Klebanski, *J. Polym. Sci.*, **30**, 363 (1958); *Rubber Chem. Technol.*, **32**, 588 (1959).
112. J.T. Maynard and W.E. Mochel, *J. Polym. Sci.*, **13**, 235, 251 (1954).
113. R.R. Garrett, C.D. Hargreaves, II, and D.N. Robinson, *J. Macromol. Sci.-Chem.*, **A4**(8), 1679 (1970).
114. C.E.N. Bawn, A.M. North, and J.S. Walker, *Polymer*, **5**, 419 (1964).
115. C.E.N. Bawn, D.G.T. Cooper, and A.M. North, *Polymer*, **7**(3), 113 (1966).
116. W. Marconi, A. Mazzei, S. Cucinella, and M. Cesari, *J. Polym. Sci.*, **A-1,2**, 426 (1964).
117. G. Natta, M. Farina, M. Donati, and M. Peraldo, *Chim. Ind. (Milan)*, **42**, 1363 (1960).
118. G. Natta, M. Farina, and M. Donati, *Makromol. Chem.*, **43**, 251 (1961).
119. G. Natta, P. Corradini, and P. Canis, *J. Polym. Sci.*, **A-1,3**, 11 (1965).
120. G. Natta, L. Porri, P. Corradini, G. Zanini, and F. Ciampelli, *J. Polym. Sci.*, **51**, 463 (1961).
121. G. Natta, L. Porri, G. Stoppa, G. Allegra, and F. Ciampelli, *J. Polym. Sci.*, **B,1**, 67 (1963).
122. G. Natta, L. Porri, A. Carbonaro, F. Ciampelli, and G. Allegro, *Makromol. Chem.*, **51**, 229 (1962).
123. G. Natta, L. Porri, A. Carbonaro, and F. Stoppa, *Makromol. Chem.*, **77**, 114 (1964).
124. G. Natta and L. Porri, *ACS Polym. Prepr.*, **5**(2), 1163 (1964).
125. K. Bujadoux, J. Josefonvicz, and J. Neel, *Eur. Polym. J.*, **6**, 1233 (1970).
126. A.L. Halasa, D.F. Lohr, and J.E. Hall, *J. Polym. Sci., Polym. Chem. Ed.*, **19**, 1357 (1981).
127. N.G. Gaylord, I. Kossler, M. Stolka, and J. Vodehnal, *J. Polym. Sci.*, **A-1,2**, 3969 (1964).
128. N.G. Gaylord, I. Kossler, and M. Stolka, *J. Macromol. Sci.-Chem.*, **A2**(2), 421 (1968).
129. K. Hasegawa and R. Asami, *J. Polym. Sci., Polym. Chem. Ed.*, **16**, 1449 (1978); K. Hasegawa, R. Asami, and T. Higashimura, *Macromolecules*, **10**, 585, 592 (1977).
130. I.I. Yermakova, Ye.N. Kropacheva, A.I. Kol'tsov, and B.A. Dolgoplosk, *Vysokomol. Soyed.*, **A11**(7), 1639 (1969).
131. N.G. Gaylord, M. Stolka, V. Stepan, and I. Kossler, *J. Polym. Sci.*, **C**(23), 317 (1968).

132. E.R. Moore (ed.), "Styrene Polymers" in *Encyclopedia of Polymer Science and Engineering*, Vol. 16, 2nd ed. (H.F. Mark, N.M. Bikales, C.G. Overberger, and G. Menges, eds.), Wiley-Interscience, New York, 1989.
133. P.J. Canterno and G.R. Hahle, *J. Polym. Sci.*, **6**, 20 (1962).
134. E. Simon, *Ann.*, **31**, 265 (1839).
135. H. Stobbe and G. Posnjack, *Ann.*, 259 (1909).
136. G.V. Schulz, *Z. Phys. Chem. (Leipzig)*, **B44**, 277 (1939).
137. S.R. Mayo, *J. Am. Chem. Soc.*, **75**, 6133 (1953).
138. R.R. Hiatt and P.D. Bartlett, *J. Am. Chem. Soc.*, **81**, 1149 (1959).
139. J. Kurze, D.J. Stein, P. Simak, and R. Kaiser, *Angew. Makroml. Chem.*, **12**, 25 (1970).
140. F.R. Mayo, *J. Am. Chem. Soc.*, **90**, 1289 (1968).
141. J. Wiesner and P. Mehnert, *Makromol. Chem.*, **165**, 1 (1973).
142. F.A. Bovey and I.M. Kolthoff, *Chem. Rev.*, **42**, 491 (1948).
143. O. Roehm, German Pat. # 656,469 (1938).
144. N. Platzer, *Ind. Eng. Chem.*, **62**(1), 6 (1970).
145. I.G. Farbenindustri, A.-G., German Pat. # 634,278 (1936).
146. Dow Chem, Co. U.S. Pat. # 2,727,884 (1955).
147. J.P. Kennedy, *J. Macromol. Sci.-Chem.*, **A3**, 861 (1969).
148. J.P. Kennedy, *J. Macromol. Sci.-Chem.*, **A3**, 885 (1969).
149. F.S. D'yachkovskii, G.A. Kazaryan, and N.S. Yenikolopyan, *Vysokomol. Soyed.*, **A11**, 822 (1969).
150. A.A. Morton and E. Grovenstein Jr., *J. Am. Chem. Soc.*, **74**, 5435 (1952).
151. J.L.R. WIlliams, T.M. Laasko, and W.J. Dulmage, *J. Org. Chem.*, **23**, 638, 1206 (1958).
152. D. Baum, W. Betz, and W. Kern, *Makromol. Chem.*, **42**, 89 (1960).
153. R.F. Zern, *Nature*, **187**, 410 (1960).
154. K.F. O'Drlscoll and A.V. Tobolsky, *J. Polym. Sci.*, **35**, 259 (1959).
155. R.C.P. Cubbon and D. Margerison, *Proc. Roy. Soc. (London)*, **A268**, 260 (1962).
156. A.A. Morton and L.D. Taylor, *J. Polym. Sci.*, **38**, 7 (1959).
157. G. Natta, *Makromol. Chem.*, **16**, 77 (1955).
158. G. Natta, *J. Polym. Sci.*, **16**, 143 (1955).
159. C.G. Overberger, *J. Polym. Sci.*, **35**, 381 (1959).
160. A.C. Shelyakov, *Dokl. Akad. Nauk USSR*, **122**, 1076 (1958).
161. D.Y. Yoon, P.R. Sundararajan, and P.J. Flory, *Macromolecules*, **8**, 776 (1975).
162. V.A. Kargin, V.A. Kabanov, and I.I. Mardenko, *Polym. Sci., USSR*, **1**, 41 (1960).
163. F. Wenger and S.P.S. Yen, *Makromol. Chem.*, **43**, 1 (1961).
164. D.J. Worsford and S. Bywater, *Can. J. Chem.*, **38**, 1881 (1960).
165. D.P. Wyman and T. Altares Jr., *Makromol. Chem.*, **72**, 68 (1964); T. Altares Jr., D.P. Wyman, and V.R. Allen, *J. Polym. Sci.*, **A3**, 4131 (1965).
166. F. Wenger and S.P.S. Yen, *Am. Chem. Soc., Polym. Prepr.*, **3**, 163 (1962).
167. S.P.S. Yen, *Makromol. Chem.*, **81**, 152 (1965).
168. T. Altares, Jr. and E.L. Clark, *Ind. Eng. Chem., Prod. Res. Dev.*, **9**(2), 168 (1970).
169. G. Natta, F. Danusso, and D. Sianesi, *Makromol. Chem.*, **28**, 253 (1958).
170. G. Natta, F. Danusso, and D. Sianesi, *Makromol Chem.*, **30**, 238 (1959).
171. Y. Sakurada, *J. Polym. Sci.*, **A-1,1**, 2407 (1963).
172. F.S. Dalton and R.H. Tomlinson, *J. Chem. Soc.*, 151 (1953).
173. D.O. Jordan and A.R. Mathieson, *J. Chem. Soc.*, 2354 (1952).
174. D.J. Worsford and S. Bywater, *J. Am. Chem. Soc.*, **9**, 491 (1957).
175. A. Hersberger, J.C. Reid, and R.G. Heilliman, *Ind. Eng. Chem.*, **37**, 1073 (1945).
176. G. Natta, *SPE J.*, **15**, 373 (1959).
177. J.A. Price, M.R. Lytton, and B.G. Ranby, *J. Polym. Sci.*, **51**, 541 (1961).
178. R.F. Boyer, in *Encyclopedia of Polymer Science and Technology* (H.F. Mark, N.G. Gaylord, and N.M. Bikales, eds.), Vol. 13, Wiley-Interscience, New York, 1970.
179. F.R. Mayo and C. Walling, *Chem. Rev.*, **46**, 191 (1950).
180. R.H. Boundy and R.F. Boyer, *Styrene, Its Polymers, Copolymers, and Derivatives*, Reinhold, New York, 1952.
181. G.E. Molan, *J. Polym. Sci.*, **A-1,3**, 126 (1965).
182. G.E. Molan, *J. Polym. Sci.*, **A-1,3**, 4235 (1965).
183. N.E. Davenport, L.W. Hubbard, and M.R. Pettit, *Brit. Plastics*, **32**, 549 (1959).
184. T. Takamatsu, A. Ohnishi, T. Nishikida, and J. Furukawa, *Rubber Age*, 23 (1973).
185. L.J. Young, Chapt. 8 in *Copolymerization* (G.E. Ham, ed.), Interscience, New York, 1964.
186. L.A. Wood, *J. Polym. Sci.*, **28**, 319 (1958).
187. J.G. Fox and S. Gratch, *Ann. NY Acad. Sci.*, **57**, 367 (1953).
188. R.H. Wiley and G.M. Brauer, *J. Polym. Sci.*, **3**, 455, 647, (1948); *ibid*, **4**, 351 (1949).

189. J.G. Bevington, H.W. Melville, and R.P. Taylor, *J. Polym. Sci.*, **12**, 449 (1954).
190. U.D. Standt and J. Klein, *Makromol. Chem., Rapid Commun.*, **2**, 41 (1981).
191. G.E, Schildknecht, *Vinyl and Related Polymers*, Wiley, New York, 1952.
192. G.E. Barnes, R.M. Olofson, and G.O. Jones, *J. Am. Chem. Soc.*, **72**, 210 (1950).
193. H.W. Melville, *Proc. Roy. Soc. (London)*, **A167**, 99 (1938).
194. D.L. Glusker, E. Stiles, and Yoncoskie, *J. Polym. Sci.*, **49**, 197 (1961).
195. T. Fox, *J. Am. Chem. Soc.*, **80**, 1768 (1958).
196. W.E. Goode, F.H. Owens, R.P. Fellman, W.H. Snuder, and J.E. Moore, *J. Polym. Chem.*, **46**, 317 (1960).
197. W.E. Goode, W.H. Snyder, and R.G. Fettes, *J. Polym. Sci.*, **42**, 367 (1960).
198. F.J. Welch, U.S. Pat. # 3,048,572 (Aug. 7, 1962).
199. H. Nagai, *J. Appl. Polym. Sci.*, **7**, 1697 (1963).
200. S. Smith, *J. Polym. Sci.*, **38**, 259 (1959).
201. W.G. Gall and N.G. McCrum, *J. Polym. Sci.*, **50**, 489 (1961).
202. M.R. Miller and G.E. Rauhut, *J. Am. Chem. Soc.*, **80**, 4115, (1958).
203. M.R. Miller and G.E. Rauhut, *J. Polym. Sci.*, **38**, 63 (1959).
204. W.E. Goode, F.H. Owens, and W.L. Myers, *J. Polym. Sci.*, **47**, 75 (1960).
205. B. Garrett *et al.*, *J. Am. Chem. Soc.*, **81**, 1007 (1959).
206. W.E. Goode, R.P. Fellman, and F.H. Owens, in *Macromolecular Syntheses* (C.G. Overberger, ed.), Vol. I, Wiley, New York, 1963.
207. C.F. Ryan and J.J. Gormley, in *Macromolecular Syntheses* (C.G. Overberger, ed.), Vol. I, Wiley, New York, 1963.
208. A. Yamamoto, T. Shimizu, and I. Ikeda, *Polymer J.*, **1**, 171 (1979). [Information was obtained from A. Yamamoto and T. Yamamoto, *Macromol. Rev.*, **13**, 161 (1978).]
209. H. Abe, K. Imai, and M. Matsumoto, *J. Polym. Sci.*, **B3**, 1053 (1965).
210. A. Akimoto, *J. Polym. Sci., Polym. Chem. Ed.*, **10**, 3113 (1972).
211. N. Yamazaki and T. Shu, *Kogyo Kagaku Zashi*, **74**, 2382 (1971). [From A. Yamamoto and T. Yamamoto, *Macromol. Chem.*, **13**, 161 (1978).]
212. S.S. Dixit, A.B. Deshpande, and S.L. Kapur, *J. Polym. Sci.*, **A-1,8**, 1289, (1970).
213. A. Ravve and J.T. Khamis, U.S. Pat. # 3,306,883 (Feb. 28, 1967), 3,323,946 (June 6, 1967).
214. R.H. Yokum, *Functional Monomers, Their Preparations, Polymerization and Application*, Vols. I and II, Dekker, New York, 1973.
215. A. Ravve and J.T. Khamis, *J. Macromol. Sci.,-Chem.*, **A1**(8), 1423 (1967).
216. A. Ravve and K.H. Brown, *J. Macromol. Sci.,-Chem.*, **A13**(2), 285 (1979).
217. G.D. Dixon, *J. Polym. Sci., Polym. Chem. Ed.*, **12**, 1717 (1974).
218. H. Berghmans and G. Smets, *Makromol. Chem.*, **115**, 187 (1960).
219. A. Stall, *J. Polym. Sci., Polym. Lett.*, **18**, 811 (1980).
220. A. Roig, J.E. Figueruelo, and E. Liano, *J. Polym. Sci.*, **C**(16), 4141 (1968).
221. D.P. Kelly, G.J.H. Melrose, and D.H. Solomon, *J. Appl. Polym. Sci.*, **7**, 1991 (1963).
222. J.D. Idol Jr., U.S. Pat. # 2,904,580 (1959).
223. C.H. Bamford and A.D. Jenkins, *Proc. Roy. Soc. (London), Ser. A*, **A216**, 515 (1953).
224. W.M. Thomas and J.J. Pellon, *J. Polym. Sci.*, **13**, 329 (1954).
225. A. Zilkha, B.A. Feit, and M. Frankel, *J. Chem. Soc.*, 928 (1959).
226. M. Frankel, A. Ottolenghi, M. Albeck, and A. Zilkha, *J. Chem. Soc.*, 3858 (1959).
227. A. Ottolenghi and A. Zilkha, *J. Polym. Sci.*, **A-1,1**, 647 (1963).
228. S. Inoue, T. Tsuruta, and J. Furukawa, *Makromol. Chem.*, **42**, 12 (1960).
229. A. Zilkha, P. Neta, and M. Frankel, *J. Chem. Soc.*, 3357 (1960).
230. F.S. Dainton, E.M. Wiles, and A.N. Wright, *J. Polym. Sci.*, **45**, 111 (1960).
231. J.L. Down, J. Lewis, B. Moore, and G. Wilkinson, *Proc. Chem. Soc.*, 209 (1957).
232. R.B. Cundall, D.D. Eley, and R. Worrall, *J. Polym. Sci.*, **58**, 869 (1962).
233. N.S. Wooding and W.C.S. Higginson, *J. Chem. Soc.*, 774 (1952).
234. C.S.H. Chen, N. Colthup, W. Deichert, and R.L. Webb, *J. Polym. Sci.*, **45**, 247 (1960).
235. J. Ulbricht and R. Sourisseau, *Faserforsch. Textiltech.*, **12**, 547 (1961).
236. M.L. Miller, *J. Polym. Sci.*, **56**, 203 (1962).
237. N. Shavit, A. Oplaka, and M. Levy, *J. Polym. Sci.*, **A-1,4**, 2041 (1966).
238. N. Shavit, M. Konigsbuch, and A. Oplaka, U.S. Pat. #3,345,350 (1967).
239. N. Shavit and M. Konigsbuch, *J. Polym. Sci.*, **C**(16), 43 (1967).
240. S. Amdur and N. Shavit, *J. Polym. Sci.*, **A-1,5**, 1297 (1967).
241. W. Kern and H. Fernow, *Rubber Chem. Technol.*, **18**, 267 (1945).
242. L.E. Ball and J.L. Greene, in *Encyclopedia of Polymer Science and Technology* (H.F. Mark, N.G. Gaylord, and N.M. Bikales, eds.), Vol. 15, Wiley-Interscience, New York, 1971.
243. R.G. Beaman, *J. Am. Chem. Soc.*, **70**, 3115 (1948).

244. O.H. Bullitt, Jr., U.S. Pat. # 2,608,555 (1952).
245. C.G. Overberger, E.M. Pearce, and N. Mayers, *J. Polym. Sci.*, **34**, 109 (1959).
246. C.G. Overberger, H. Yuki, and N. Urakawa, *J. Polym. Sci.*, **45**, 127 (1960).
247. W.K. Wilkinson, U.S. Pat. # 3,087,919 (1963).
248. Ben-Ami Feit, E. Heller, and A. Zilkha, *J. Polym. Sci.*, **A-1,4**, 1151 (1966).
249. W.M. Thomas and D.W. Wang, "Acrylamide Polymers," in *Encyclopedia of Polymer Science and Engineering*, Vol. 1 (H.F. Mark, N.M. Bikalis, C.G. Overberger, and G. Menges, eds.), Wiley-Interscience, New York, 1985.
250. K. Butler, P.R. Thomas, and G.J. Tyler, *J. Polym. Sci.*, **48**, 357 (1960).
251. J.F. Jones, *J. Polym. Sci.*, **33**, 15 (1958).
252. R.C. Schulz, P. Elyer, and W. Kern, *Chimia*, **13**, 235 (1959).
253. L.A.R. Hall, W.J. Belanger, W. Kirk, Jr., and Y.A. Sundatrom, *J. Appl. Polym. Sci.*, **2**, 246 (1959).
254. S.V. Gangal, "Tetrafluoroethylene Polymers," in *Encyclopedia of Polymer Science and Engineering*, Vol. 16 (H.F. Mark, N.M. Bikalis, C.G. Overberger, and G. Menges, eds.), Wiley-Interscience, New York, 1989.
255. W.E. Hanford and R.M. Joyce, *J. Am. Chem. Soc.*, **68**, 2082 (1946).
256. R. Kiyama, J. Osugi, and S. Kusuhara, *Rev. Phys. Chem., Japan*, **27**, 22 (1957).
257. R.E. Moynihan, *J. Am. Chem. Soc.*, **81**, 1045 (1959).
258. K.L. Berry and J.H. Peterson, *J. Am. Chem. Soc.*, **73**, 5195 (1951).
259. E.T. McBee, H.M. Hill, and G.B. Bachman, *Ind. Eng. Chem.*, **41**, 70 (1949).
260. L.E. Robb, F.J. Honn, and D.R. Wolf, *Rubber Age*, **82**, 286 (1957).
261. J.S. Rugg and A.C. Stevenson, *Rubber Age*, **82**, 102 (1957).
262. L. Mandelkern, G.M. Martin, and F.A. Quinn, *J. Res. Natl. Bur. Stand.*, **58**, 137 (1957).
263. C.B. Griffins and J.C. Montermoso, *Rubber Age*, **77**, 559 (1959).
264. J.F. Smith, *Rubber World*, **142**(3), 102 (1960).
265. J.F. Smith and G.T. Perkins, *J. Appl. Polym. Sci.*, **5**(16), 460 (1961).
266. K.L. Paciorek *et al., J. Polym. Sci.*, **45**, 405 (1960).
267. R.J. Orr and H.L. Williams, *Can. J. Chem.*, **33**, 1328 (1955).
268. F. Bergmann, A. Kalmus, and E. Breuer, *J. Am. Chem. Soc.*, **20**, 4540 (1958).
269. D.I. Livingston, P.M. Kamath, and R.S. Corley, *J. Polym. Sci.*, **20**, 485 (1956).
270. D. Sianeai and P. Carradini, *J. Polym. Sci.*, **43**, 531 (1959).
271. G. Talanini and E. Peggion, Chapt. 5 in *Vinyl Polymerization* (G. Ham, ed.), Dekker, New York, 1967; J. Lewis, P.E. Okieimen, and G.S. Park, *J. Macromol. Sci.-Chem.*, **A17**, 1021 (1982).
272. W.I. Bengough and R.G.W. Norrish, *Proc. Roy. Soc. (London)*, **A200**, 301 (1950).
273. A. Schindler and J.W. Breitenbach, *Ric. Sci., Suppl.*, **25**, 34 (1955).
274. E.J. Alerman and W.M. Wagner, *J. Polym. Sci.*, **9**, 541 (1951).
275. K. Nozaki, *Discuss. Faraday Soc.*, **2**, 337 (1947).
276. K.J. Mead and R.M. Fuoss, *J. Am. Chem. Soc.*, **64**, 277 (1942).
277. C.H. Bamford, W.G. Barb, A.D. Jenkins, and P.F. Onyon, *The Kinetics of Vinyl Polymerization by Free-Radical Mechanism*, Butterworth, London, 1954.
278. T. Hjertberg and E. Sorvik, *J. Polym. Sci., Polym. Lett.*, **19**, 363 (1981).
279. A. Caraculacu, E.C. Burniana, and G. Robila, *J. Polym. Sci., Polym. Chem. Ed.*, **16**, 2741 (1978).
280. J. Petiaud and D.B. Pham, *Makromol. Chem.*, **178**, 741 (1977).
281. J.W.L. Fordham, P.H. Burleigh, and C.L. Sturm, *J. Polym. Sci.*, **41**, 73 (1959).
282. U. Giannini and S. Cesca, *Chim. Ind. (Milan)*, **44**, 371 (1962).
283. V.G. Gason-Zade, V.V. Mazurek, and V.P. Sklizkova, *Vysokomol. Soedin.*, **A10**(3), 479 (1968).
284. A. Gyot and Pham-Qung-Tho, *J. Chim. Phys.*, **63**, 742 (1966).
285. A. Gyot, D.L. Trung, and R. Ribould, *C.R. Acad. Sci.* Paris, 266, C 1139 (1968).
286. V. Jisova, M. Kolinsky, and D. Lim, *J. Polym. Sci.*, **A-1,8**, 1525 (1970).
287. N. Yamazaki, K. Sasaki, and S. Kambara, *J. Polym. Sci., Polym. Lett.*, **2**, 487 (1964).
288. A. Guyot and Pham Aung Tho, *J. Polym. Sci.*, **C,4**, 299 (1964).
289. L.F. Albright, *Chem. Eng.*, 85 (July 3, 1967).
290. C.S. Fuller, *Chem. Rev.*, **26**, 162 (1940).
291. G. Natta and P. Garradini, *J. Polym. Sci.*, **20**, 262 (1956).
292. R.L. Adelman and R.C. Ferguson, *J. Polym. Sci., Polym. Chem. Ed.*, **13**, 891 (1975).
293. S. Inoue, Chapt. 5 in *Structure and Mechanism of Vinyl Polymerization* (T. Tsuruta and K.F. O'Driscoll, eds.), Dekker, New York, 1969.
294. I. Rosen, P.H. Burleigh, and J.F. Gillespie, *J. Polym. Sci.*, **54**, 31 (1961).
295. I. Rosen and W.E. Marshal, *J. Polym. Sci.*, **56**, 501 (1962).
296. S. Marahashi, H. Yuki, T. Sano, U. Yonemura, H. Tadokaro, and Y. Chanti, *J. Polym. Sci.*, **62**, 77 (1962).
297. K. Fujii, *Kobunshi Kagaku*, **19**, 12 (1962) [from *Chem. Abstr.*, **58**, 585g (1963)].
298. S. Okamura, T. Kodama, and T. Higashimura, *Makromol. Chem.*, **53**, 180 (1962).

299. S. Murahashi, S. Noazakura, and M. Sumi, *J. Polym. Sci., Polym. Lett.*, 245 (1965).
300. C.G. Matsoyan, *J. Polym. Sci.*, **52**, 189 (1961).
301. K.K. Kikukawa, S. Nozakura, and S. Murahashi, *Kobunshi Kagaku* **25**, 19 (1964) [from *Chem. Abstr.*, **69**, 1968x (1968)].
302. K. Noro, Chapt. 6 in *Polyvinyl Alcohol* (C.A. Finch, ed.), Wiley, New York, 1973.
303. J.A. Brydson, *Plastic Materials*, Van Nostrand, London, 1966.
304. I.D. Burdett, *Chemtech*, **22**, 616 (1992).
305. D. O'Sullivan, *Chem. Eng. News*, p. 29 (July 4, 1983).
306. D. Cam, E. Albizzati, and P. Cinquina, *Makromol. Chem.*, **191**, 1641 (1990).
307. J.A. Ewen, *J. Am. Chem. Soc.*, **106**, 6355 (1984).
308. B. Rieger. X. Mu, D.T. Mallin, M.D. Rausch, and J.C.W. Chien, *Macromolecules*, **23**, 3559 (1990).
309. H.N. Cheng and J.A. Ewen, *Makromol. Chem.*, **190**, 823 (1989).
310. A. Grassi, A. Zambelli, L. Resconi, E. Albizzati, and R. Mazzocchi, *Macromolecules*, **21**, 617 (1988).
311. G.G. Arzonmanidis and N.M. Karayannis, *Chemtech*, **23**, 43 (1993).
312. G. Vancso, O. Egyed, S. Pekker, and A. Janossy, *Polymer*, **23**, 14 (1982).
313. K.H. Reichert and H.U. Moritz, *J. Appl. Polym. Sci.*, **36**, 151 (1981).
314. J. Lewis, P.E. Okieimen, and G.S. Park, *J. Macromol. Sci.-Chem.*, **A17**, 1021 (1982).
315. F.A. Bovey, K.B. Abbas, F.C. Schilling, and W.H. Starnes, Jr., *Macromolecules*, **8**, 437 (1975).

6

Step-Growth Polymerization and Step-Growth Polymers

6.1. Mechanism and Kinetics of Step-Growth Polymerization

Two types of monomers can undergo step-growth polymerizations.[1,4,5] Both are polyfunctional, but one type possesses only one kind of functionality. An example is adipic acid that has two functional groups, but both are carboxylic acid groups. Another is hexamethylene diamine with two amine functional groups. To the second type belong monomers that have both functional groups needed for condensation on the same molecule. An example of such a monomer is *p*-aminobenzoic acid, where both amino and carboxylic acid groups are present in the same molecule. Chain growths proceed from reactions between two different functional groups with both types of molecules. An exception is formation of polyanhydrides, where the polymeric chains are formed from two carboxylic acid groups reacting with each other and splitting out water. There are some other exceptions as well.

6.1.1. Reactions of Functional Groups

Kinetic considerations are of paramount importance in understanding the mechanism of step-growth polymerization.[1] As stated in Chapter 1, chain-growth polymerizations take place in discrete steps. Each step is a reaction between two functional groups; for instance, in a polyesterification reaction it is a reaction between –COOH and –OH. The increase in molecular weight is slow. The first step is a condensation between two monomers to form a dimer:

```
O=C—R—C=O  +  HO—R′—OH  ———→  O=C—R—C=O      +  H2O
  |     |                       |     |
  HO    OH                      HO    O—R′—OH
```

A dimer can react next with another monomer to form a trimer:

```
O=C—R—C=O        +  O—C—R—C=O  ———→  O=C—R—C=O            +  H2O
  |     |             |     |          |     |
  HO    O—R′—OH       HO    OH         HO    O—R′—O—C=O
                                                   |
                                                   R—C=O
                                                     |
                                                     OH
```

Two dimers can combine to form a tetramer:

```
2 O=C—R—C=O        ———→  O=C—R—C=O         O=C—R—C=O        +  H2O
    |     |                |      \       /       |
    HO    O—R′—OH          HO      O—R′—O         O—R′—OH
```

These step condensations continue slowly with the molecular weights of the polymers increasing with each step. In such reactions the monomeric species disappear early from the reaction mixtures, long before any large molecular weight species develop. In most step-growth polymerizations, on a weight basis, less than 1% of monomeric species remain by the time the average chain length attains the size of ten combined monomeric units.[1,3,4,6]

One important characteristic of step-growth polymerizations is that any functional group on any one molecule is capable of reacting with any opposite functional group on any other molecule. Thus, for instance, if it is a reaction of polyesterification, any carboxylic acid group on any one molecule, regardless of size, can react with any hydroxy on another one. This is true of all other step-growth polymerizations. It means that the rates of step-growth polymerizations are the sums of the rates of all reactions between molecules of various sizes. A useful assumption that can be applied here is that the reactivities of both functional groups remain the same throughout the reaction, regardless of the size of the molecules to which they are attached. This allows treating step-growth polymerizations like reactions of small molecules. General observations would suggest slower reactivity of functional groups attached to large molecules. This, however, is usually due to lower diffusion rates of large molecules. The actual reactivity of the functional groups depends upon collision frequencies (number of collisions per unit of time) of the groups and not upon the rate of diffusion. Functional groups on the terminal ends of large molecules have greater mobility than the remaining portions of the molecules as a whole. In addition, the reactivity of one given functional group in a bifunctional molecule is not altered by the reaction of the other group (if there is no neighboring group effect). This implies that the reactivities of functional groups are not altered during the polymerization.

The kinetics of step-growth polymerization can be derived from a polyesterification reaction that follows the same course as all acid-catalyzed esterifications.[2]

1. Protonation step:

$$\sim\!\!\sim C(=O)OH + HA \underset{k_2}{\overset{k_1}{\rightleftharpoons}} \sim\!\!\sim \overset{\oplus}{C}(-OH)OH + \overset{\ominus}{A}$$

2. Reaction of the protonated carboxylic acid group with the alcohol:

$$\sim\!\!\sim \overset{\oplus}{C}(-OH)OH + \sim\!\!\sim OH \underset{k_4}{\overset{k_3}{\rightleftharpoons}} \sim\!\!\sim C(OH)_2-\overset{\oplus}{O}(H)\sim\!\!\sim \xrightarrow{k_5} \sim\!\!\sim C(=O)O\sim\!\!\sim + H_2O + \overset{\oplus}{H}$$

The above polyesterifications, like many other reactions, are equilibrium reactions. They must be conducted in a way that allows the equilibrium to shift to the right to attain high molecular weights. One way is by continual removal of the byproducts. In such situations the reactions take place at nonequilibrium conditions and there is no K_4.

6.1.2. Kinetic Considerations

The rate of polymerization can be expressed as the *rate of disappearance of one of the functional groups*. In reactions of polyesterification this can be the rate of disappearance of carboxyl groups, $-d[CO_2H]/dt$.

$$R_P = \frac{-d[CO_2H]}{dt} = k_3[\overset{\oplus}{C}(OH)_2][OH]$$

In the above equation $[CO_2H]$, $[OH]$, and $[\overset{\oplus}{C}(OH)_2]$ represent carboxyl, hydroxy, and protonated carboxyl groups, respectively. Also, it is possible to write an equilibrium expression for the protonation reaction of the acid as follows:

$$K = \frac{k_1}{k_2} = \frac{[\overset{\oplus}{C(OH)_2}][A]^{\ominus}}{[COOH][HA]}$$

This equation can be combined with the above rate expression:

$$\frac{-d[COOH]}{dt} = \frac{k_1 k_3 [COOH][OH][HA]}{k_2[A^{\ominus}]} = k_3 K[COOH][OH][HA]$$

It can also be written in a different form:

$$\frac{-d[COOH]}{dt} = \frac{K k_3 [COOH][OH][H]^{\oplus}}{K_{HA}}$$

where K_{HA} is the dissociation constant of the acid HA.

If there is no catalyst present and the dicarboxylic acid acts as its own catalyst, HA is replaced by [COOH] and the expression becomes

$$\frac{-d[COOH]}{dt} = k[COOH]^2[OH]$$

In the above expression k_1, k_2, k_3, and the concentration of the $[A]^{\ominus}$ ions have been replaced by an experimentally determined rate constant, k.

In most step-growth polymerization reactions the concentrations of the two functional groups are very close to stoichiometric. This allows writing the above rate equation as follows:

$$\frac{-d[M]}{dt} = k[M]^3$$

In this equation [M] represents the concentration of each of the reacting species. They can be hydroxy and carboxylic acid groups in a polyesterification reaction, or amino and carboxylic acid groups in polyamidation reaction, and so on.

The above equation can also be written as follows:

$$kdt = -d[M]/[M]^3$$

After integrating the latter equation we get

$$2kt = 1/[M]^2 + \text{constant}$$

The constant in the above equation equals $1/[M_0]^2$, where $[M_0]$ represents the initial concentration of the reactants (of hydroxyl or carboxyl groups in a polyesterification) at time $t = 0$.

At the start of the polymerization there are $[M_0]$ molecules present. After some progress of the reaction, there are [M] molecules left; $[M_0] - [M]$ is then the number of molecules that participated in the formation of polymeric chains. The conversion, p, can be written, according to Carothers,[6] as

$$p = \frac{[M_0] - [M]}{[M_0]}$$

or the concentration of [M] at any given time t is

$$[M] = [M_0](1 - p)$$

and the degree of polymerization is given by

$$\overline{DP} = \frac{1}{1 - p}$$

It is important to realize from the above equation that in order to obtain a $\overline{DP}$ of only 50, it is necessary to achieve 98% conversion (p must equal 0.98).

The value of $\overline{\mathrm{DP}}$ at any given time t is equal to the ratio of monomer molecules that were present at the start of the reaction divided by the number of molecules that are still present at that particular time:

$$\overline{\mathrm{DP}} = \frac{[\mathrm{M}]_0}{[\mathrm{M}]}$$

By combining the above expression with Carother's equation and solving for [M], one obtains

$$[\mathrm{M}] = [\mathrm{M}_0](1 - p)$$

For a second-order rate expression, the above equation can be written as

$$1/\{[\mathrm{M}_0](1 - p)\} - 1/[\mathrm{M}_0] = kt$$

and by replacing $1/(1 - p)$ with $\overline{\mathrm{DP}}$, one obtains

$$\overline{\mathrm{DP}} = [\mathrm{M}_0]kt + 1$$

Using the above equation it is possible to calculate from the rate constant (if it is known) and the concentration of monomers the time required to reach a desired number average molecular weight. When there is no catalyst present and the carboxylic acid assumes the role of a catalyst itself, then a third-order rate expression (shown above) must be employed:

$$-d[\mathrm{M}]/dt = k[\mathrm{M}]^3$$

By integrating the third-order rate expression one obtains

$$1/[\mathrm{M}]^2 - 1/[\mathrm{M}_0]^2 = 2kt$$

and, by substituting for [M] from the Carothers equation and then rearranging the resultant equation, one obtains

$$\frac{1}{[\mathrm{M}_0]^2(1 - p)} - \frac{1}{[\mathrm{M}_0]^2} = 2kt$$

This can also be written as

$$1/(1 - p)^2 = 2kt[\mathrm{M}_0]^2 + 1$$

or

$$\overline{\mathrm{DP}}^2 = 2kt[\mathrm{M}_0]^2 + 1$$

The above equation shows that, without a catalyst, the molecular weight increases more gradually.

It can be deduced from the above discussion that a high stoichiometric balance is essential for attaining high molecular weight. This means that any presence of a monofunctional impurity has a strong limiting effect on the molecular weight of the product. The impurity blocks one end of the chain by reacting with it. This is useful, however, when it is required to limit the DP of the product. For instance, small quantities of acetic acid are sometimes added to preparations of some polyamides to limit their molecular weights.

In polymerizations of monomers with the same functional groups on each molecule, like A——A and B——B (i.e., a diamine and a diacid), the number of functional groups present can be designated as N_A^0 for A type and N_B^0 for B type. These numbers N_A^0 and N_B^0 represent the number of functional groups present at the start of the reaction. They are twice the number of A——A and B——B molecules that are present. If the number N_B^0 is slightly larger than N_A^0, then we have a *stoichiometric imbalance* in the reaction mixture. This imbalance is designated as r:

$$r = N_A^0/N_B^0$$

(It is common to define the ratio r as less than or equal to unity, so in the above B groups are present in excess.) The total number of monomers at the start of the reaction are $(N_A^0 + N_B^0/2)$ or $N_A^0(1 + 1/r)/2$.

The extent of the reaction, p, can be defined as the portion of the functional groups A that reacted at any given time. The portion of the functional groups B that reacted at the same time can be designated by rp. The unreacted portions of A and B groups can then be designated as $1 - p$ and $1 - rp$, respectively. The total number of unreacted A groups in the reaction mixture would then be $N_A^0(1 - p)$. This reaction mixture also contains $N_B^0(1 - rp)$ unreacted B groups. The total number of chain ends on the polymer molecules is the sum of the unreacted A and B groups. Because each polymer molecule has two chain ends, the total number of chain ends is then $[N_A^0(1 - p) + N_B^0(1 - rp)]/2$.

The number average degree of polymerization is equal to the total number of A——A and B——B molecules present at the start of the reaction divided by the number of polymer molecules at the end. This can be represented as follows:

$$\overline{\text{DP}}_n = \frac{N_A^0(1 + 1/r)/2}{[N_A^0(1 - p) + N_B^0(1 - rp)]/2}$$

the expression can be reduced (since $r = N_A^0/N_B^0$) to

$$\overline{\text{DP}}_n = \frac{1 + r}{1 + r - 2rp}$$

The molecular weight of the product can be controlled by precise stoichiometry of the polymerization reaction. This can be done by simply quenching the reaction mixture at a specified time when the desired molecular weight is achieved.

Flory derived a statistical method for relating the molecular weight distribution to the degree of conversion.[1,3] In these polymerizations, each reaction step links two monomer molecules together. This means that the number of mers in the polymer backbone is always larger by one than the number of each kind of functional group, A or B. If there are x monomers in a chain, then the number of functional groups that have reacted is $x - 1$. The functional groups that are unreacted remain at the ends of the chains. If we designate p as the extent of the reaction or the degree of conversion, as above, then the probability that $x - 1$ of A or B has reacted is p^{x-1}, where p is

$$p = (N_0 - N)/N_0$$

and the probability of finding an unreacted functional group is $p - 1$. The probability of finding a polymer molecule that contains x monomer units and an unreacted functional group A or B is $p^{x-1}(1 - p)$. At a given time t, the number of molecules present in the reaction mixture is N.

The fraction that contains x units can be designated as N_x and can be defined as

$$N_x = Np^{x-1}(1 - p)$$

The Carothers equation defines $N/N_0 = 1 - p$. The above expression for N_x can therefore be written as

$$N_x = N_0(1 - p)^2 p^{x-1}$$

where N_0 is, of course, the number of monomer units that are present at the start of the reaction.

To determine the molecular weight distribution of the polymeric species that form at any given degree of conversion, it is desirable to express the weight average and number average molecular weights by terms, like p. By defining M_0 as the mass of the repeating unit, the number average molecular weight is

$$\overline{M}_n = M_0 \cdot \overline{\text{DP}} = M_0[1/(1 - p)]$$

and the weight average molecular weight is

$$\overline{M}_w = \sum w_x x M_0$$

where w_x is the weight fraction of molecules containing x monomer units. This fraction is equal to xN_x/N_0 and, based on the above equation for N_x, can be written as $w_x = x(1-p)^2p^{x-1}$. The weight average molecular weight can now be expressed as

$$\overline{M}_w = M_0(1-p)^2 \sum x^2p^{x-1}$$

It can be shown that the latter summation is $\Sigma x^2p^{x-1} = (1+p)/(1-p)^3$. Hence the weight average molecular weight is

$$\overline{M}_w = M_0(1+p)/(1-p)$$

and the molecular weight distribution is

$$\overline{M}_w/\overline{M}_n = 1 + p$$

It is interesting that this equation tells us that at high conversion, when p approaches 1, the molecular weight distribution approaches 2. There is experimental confirmation of this.

Until now, this discussion was concerned with the formation of linear polymers. The presence, however, of monomers with more than two functional groups results in the formation of branched structures. An example is a preparation of a polyester from a dicarboxylic acid and a glycol, where the reaction mixture also contains some glycerol. Chain growth in such a polymerization is not restricted to two directions and the products are much more complex. This can be illustrated further on a trifunctional molecule condensing with a difunctional one:

```
   A                                   A           A
  /                                     \         /
n A    +  n B–B  ———————→                A–B–B–A              A
  \                                     /         \          /
   A                              B–B–A            A–B–B–A
                                                              \
                                                               A
```

Further growth, of course, is possible at every unreacted functional group and can lead to gelation. The onset of gelation can be predicted from a modified form of the Carothers equation.[1] This equation includes an *average functionality* factor that averages out the functionality of all the functional groups involved. An example is a reaction mixture of difunctional monomers with some trifunctional ones added for branching or crosslinking. The average functionality, f_{ave} may be $(2+2+2+3)/4 = 2.25$. The Carothers equation, discussed above, states that

$$p = (N_0 - N)/N_0$$

where N_0 and N represent the quantities of monomer molecules present initially and at a conversion point p. The number of functional groups that have reacted at that point is $2(N_0 - N)$. In the modified equation, the number of molecules that were present initially is $N_0 f_{ave}$. The equation now becomes

$$p = \frac{2(N_0 - N)}{N_0 f_{ave}}$$

If N_0/N is replaced by $\overline{DP}$, the above expression becomes

$$p = 2/f_{ave} + \overline{DP} \cdot f_{ave}$$

It is generally accepted that *gelation* occurs when the average degree of polymerization becomes infinite. At that point the second term in the above equation becomes zero. When that occurs the conversion term becomes p_c. It is the *critical reaction conversion* point:

$$p_c = 2/f_{ave}$$

Gelation, however, is less likely to be a major concern in polymerization reactions where only small quantities of tri- or multifunctional monomers are present. In the preparation of alkyds, for instance (described further in this chapter), some glycerin, which is trifunctional, is usually present. If the amount of glycerin is small, then the product is only branched. In addition, there might be only one branch per molecule:

```
~~A——BB—BA—AB—BA—AB—┬—BA—AB—BA—AB—BA~~
                    |
                    B
                    A
                    |
                    A
                    B
                    |
                    B
```

Statistical methods were developed for the prediction of gelation.[7] These actually predict gelation at a lower level than does the Carothers equation shown above. As an example we can use a reaction of three monomers, A, B, and C. We further assume that the functionality of two monomers, f_A and f_B, is equal to two, while that of f_C is greater than two. The critical reaction conversion can then be written as

$$p_c = 1/[r + rP(f_c - 2)]^{1/2}$$

6.1.3. Ring Formation in Step-Growth Polymerization

Step-growth polymerization can be complicated by cyclization reactions that accompany formations of linear polymers. Such ring formations can occur in reactions of monomers with either the same type of functional groups or with different ones. Some illustrations of cyclization reactions follow:

```
                               R—N—H
                              /     \
2 H2N—R—CO2H  ———→       O=C         C=O   +  H2O
                              \     /
                               H—N—R

                          ┌────────C=O
H2N—(CH2)x—CO2H  ———→    (CH2)x     |
                          └────────N—H

                               R—O
                              /   \
2 HO—R—CO2H  ———→        O=C       C=O
                              \   /
                               O—R

                          ┌────────C=O
HO—(CH2)x—CO2H  ———→     (CH2)x     |
                          └────────O
```

Similarly, dicarboxylic acids can cyclize into anhydrides.

Whether ring formation is likely to take place or not, depends upon the size of the ring that can form. If cyclization results in rings with strained bond angles or repulsions due to crowding, the probability of their formation is low. So small rings, with less than five members, do not form readily. Five-membered rings, however, are essentially free from bond-angle distortion and have a greater chance to form. Greater than five-membered rings are not planar and six- and seven-membered rings can form freely, though not quite as easily as five-membered ones. Six-membered rings are more favored than seven-membered ones. Rings with eight to twelve members are relatively strain-free from bond-angle distortion, but they are thermodynamically unstable. This is because substituents (hydrogens or others) are forced into positions of repulsion due to crowding. Also, there is little ring formation with eight to twelve members.

Whether cyclization will take place or not during polymerization also depends on the kinetic feasibility to cyclize. This feasibility is a function of the probability that functional end groups on a molecule will approach each other. As the size of the monomer increases, so does the size of the potential ring. An increase in the size of the monomer, however, also means an increase in the number of different configurations that the monomer molecule can assume. Very few of these configurations are such that the two ends become adjacent.[4] With fewer chances of the end groups encountering each other there is decreased probability of ring formation. From practical considerations, ring formations are mainly a problem when five-, six-, or seven-membered rings can form. Formation of large rings with more than twelve members is seldom encountered.[5]

6.1.4. Techniques of Polymer Preparation

Many step-growth polymerizations are carried out by mass or bulk-type polymerization. This is commonly done not only for convenience, but also because it results in minimum contamination. Few step-growth reactions are highly exothermic, so thermal control is not hard to maintain. Because equilibrium considerations are very important, the reactions are usually carried out in a way that allows continuous removal of the byproduct. Occasionally, the polymerizations are carried out in dispersion in some convenient carriers. Solution polymerizations are sometimes used as a way of moderating the reactions.

Step-growth polymerizations can also be carried out with certain monomers at low temperature by a technique known as *interfacial polymerization* or *interfacial polycondensation*.[28] The reactions (applicable only to fast reactions) are conducted at the interface between two immiscible liquids. Usually, one of the liquids is water and the other an organic solvent. An example may be a Schotten–Baumann polyamidation reaction. In such an interfacial polymerization, the diamine would be in the aqueous phase and the diacid chloride in the organic phase. The strong reactivity of acid chloride groups with amines allows the reaction to be carried out at room temperature:

$$n\ H_2N{-}(CH_2)_x{-}NH_2 \ + \ \underset{Cl}{O{=}C}{-}({-}CH_2{-})_y{-}\underset{Cl}{C{=}O} \longrightarrow$$

$$-[{-}NH{-}({-}CH_2{-})_x{-}NH{-}\overset{O}{\overset{\|}{C}}{-}({-}CH_2{-})_y{-}\overset{O}{\overset{\|}{C}}{-}]_n{-} \ + \ n\ HCl$$

Addition of a base to the aqueous phase removes the hydrochloric acid that forms and catalyzes the reaction. The choice of organic solvent is important, because it appears that the reaction occurs on the organic side of the interface.[28]

There are several important differences between interfacial polymerizations and high-temperature condensations. Much higher molecular weight products form from polymerizations at the interface. This is probably due to the high speed of the reactions between the diamines that diffuse into the organic phase and the diacid chloride chain ends.[28] Exact stoichiometry is not necessary to attain high molecular weights in interfacial polycondensation. The opposite is true in high-temperature polymerizations.

Interfacial polycondensation is an interesting procedure that is often used in demonstrations in polymer chemistry courses. Polyamides are prepared rapidly, in front of the class, from diacid chlorides and diamines. The products are removed quickly as they form, by pulling them out as a string from the interface.[47] Polyesters can also be prepared from diacid chlorides and bisphenols. On the other hand, preparation of polyesters from glycols and diacid chlorides is usually unsuccessful due to low reactivity of the dialcohols. The diacid chlorides tend to undergo hydrolysis instead. Commercially, this procedure is so far confined mainly to preparations of polycarbonates (discussed further in this chapter).

6.2. Polyesters

The class of compounds called polyesters consists of all heterochain macromolecular compounds that possess repeat carboxylate ester groups in the backbones. This excludes all polymers with ester groups located as pendant groups, like acrylic and methacrylic polymers, poly(vinyl esters), and esters of cellulose, or starch. What remains, however, is still a large group of polymeric materials that can be subdivided into saturated and unsaturated polyesters.

6.2.1. Linear Saturated Polyesters

The saturated polyesters that find commercial applications are mostly linear, except for some specially prepared branched polymers used in the preparation of polyurethanes. The linear polyesters became commercially important materials early in this century and still find many uses in industry. The earliest studies reported condensations of ethylene, trimethylene, hexamethylene, and decamethylene glycols with malonic, succinic, adipic, sebacic, and orthophthalic acids.[6] Later studies showed that such condensations yield high molecular weight compounds.[44] Nevertheless, these polyesters exhibit poor hydrolytic stability and are generally low-melting. Subsequently, however, it was found that aromatic dicarboxylic acids yield polymers with high melting points, and poly(ethylene terephthalate), which melts at 265 °C, is now an important commercial material.

Physical properties of linear polyesters follow the general observation of the relationships between physical behavior and chemical structures of polymers (see Chapter 1). Aromatic diacids and\or glycols with aromatic rings in the structures yield polyesters with high melting points, while the aliphatic ones yield low melting solids or viscous liquids. In addition, hydrogen bonding, dipole interactions, polarizations, stiff interchain bonds, molecular symmetry or regularity, and the ability of polymeric chains to undergo close packing raise the melting points. Conversely, bulky side chains and flexible interchain bonds lower the melting points.

6.2.1.1. Synthetic Methods

The synthetic methods that are in general use for the preparations of linear polyesters[15] can be summarized as follows:

(1) Dibasic acids are reacted with glycols. Stoichiometric balance is maintained rigidly. When low boiling glycols are used, however, they are often added in a slight excess and the excess gradually removed by vacuum[6] or by sweeping an inert gas over or through the reaction mixture.[6,8] This procedure is useful with dicarboxylic acids that otherwise require high temperatures and strong catalysts. Running the reaction at high temperatures can cause the glycols to condense into ethers and the dicarboxylic acids to decompose.

(2) Diesters or half esters of dicarboxylic acids or amine salts of the acids are reacted with glycols[9]:

$$n\ \mathrm{R{-}O{-}\overset{O}{\overset{\|}{C}}{-}R'{-}\overset{O}{\overset{\|}{C}}{-}O{-}R} \;+\; n\ \mathrm{HO{-}R''{-}OH} \longrightarrow \mathrm{{-}[{-}O{-}\overset{O}{\overset{\|}{C}}{-}R'{-}\overset{O}{\overset{\|}{C}}{-}O{-}R''{-}]_{n}{-}} \;+\; \mathrm{ROH}$$

The above transesterification reaction is practical for use with high melting and poorly soluble dicarboxylic acids. In addition, less energy is needed to remove alcohol than water.

(3) Hydroxy acids, like *p*-hydroxyethoxybezoic acid or ω-hydroxydecanoic acid, are capable of self-condensation to form polyesters[10]:

$$n\ \mathrm{HO{-}CH_2{-}CH_2{-}C_6H_4{-}\overset{O}{\overset{\|}{C}}{-}OH} \longrightarrow \mathrm{{-}[{-}O{-}CH_2{-}CH_2{-}C_6H_4{-}\overset{O}{\overset{\|}{C}}{-}]_{n}{-}}$$

(4) Although not practical commercially, polyesters can be prepared in the laboratory by reacting aliphatic dibromides with silver salts of dibasic acids[6]:

$$\mathrm{Br{-}R{-}Br} + \mathrm{Ag{-}O{-}\overset{O}{\overset{\|}{C}}{-}R'{-}\overset{O}{\overset{\|}{C}}{-}O{-}Ag} \longrightarrow \mathrm{{-}[{-}R{-}O{-}\overset{O}{\overset{\|}{C}}{-}R'{-}\overset{O}{\overset{\|}{C}}{-}O{-}]_{n}{-}}$$

(5) Polyesters also form from reactions of dicarboxylic acid anhydrides with glycols[10]:

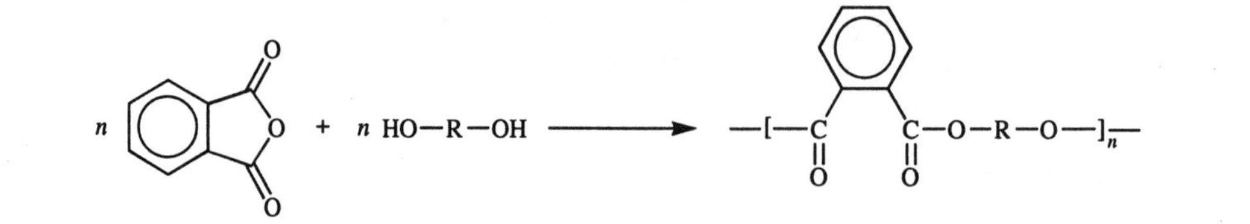

$$n\ \text{(phthalic anhydride)} + n\ HO{-}R{-}OH \longrightarrow {-}[{-}\overset{O}{\overset{\|}{C}}{-}C_6H_4{-}\overset{O}{\overset{\|}{C}}{-}O{-}R{-}O{-}]_n{-}$$

(6) Glycol carbonates undergo ester interchange reactions with dibasic acids[11]:

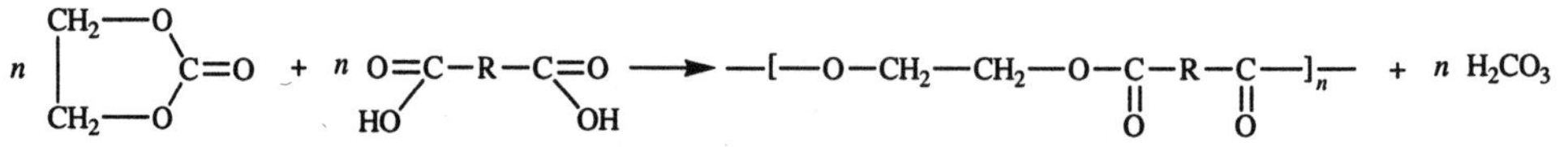

$$n\ \begin{matrix}CH_2{-}O\\ |\quad\ \ \ \ \ \ C{=}O\\ CH_2{-}O\end{matrix} + n\ \underset{HO}{O{=}C}{-}R{-}\underset{OH}{C{=}O} \longrightarrow {-}[{-}O{-}CH_2{-}CH_2{-}O{-}\overset{O}{\overset{\|}{C}}{-}R{-}\overset{O}{\overset{\|}{C}}{-}]_n{-} + n\ H_2CO_3$$

(7) Ester interchange reactions between glycol acetates[12,13] or diphenol diacetates[12] and dicarboxylic acids:

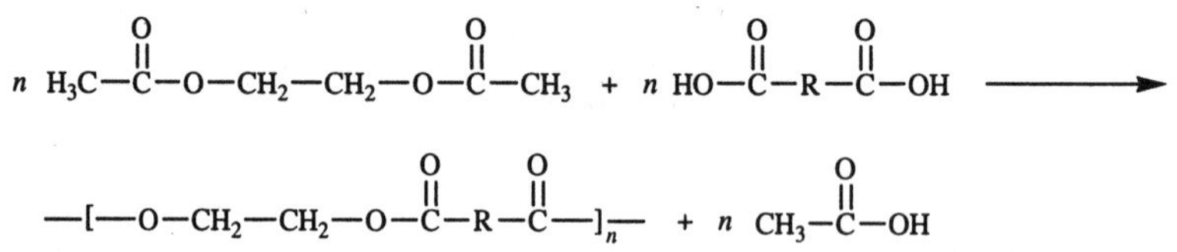

$$n\ H_3C{-}\overset{O}{\overset{\|}{C}}{-}O{-}CH_2{-}CH_2{-}O{-}\overset{O}{\overset{\|}{C}}{-}CH_3 + n\ HO{-}\overset{O}{\overset{\|}{C}}{-}R{-}\overset{O}{\overset{\|}{C}}{-}OH \longrightarrow$$

$${-}[{-}O{-}CH_2{-}CH_2{-}O{-}\overset{O}{\overset{\|}{C}}{-}R{-}\overset{O}{\overset{\|}{C}}{-}]_n{-} + n\ CH_3{-}\overset{O}{\overset{\|}{C}}{-}OH$$

(8) Acid chlorides react with diphenols to form polyesters. The reaction is quite efficient when scavengers of HCl are added to the aqueous phase. Such scavengers can be tertiary amines[15]:

$$Cl{-}\overset{O}{\overset{\|}{C}}{-}R{-}\overset{O}{\overset{\|}{C}}{-}Cl + HO{-}R'{-}OH \longrightarrow {-}[{-}\overset{O}{\overset{\|}{C}}{-}R{-}\overset{O}{\overset{\|}{C}}{-}O{-}R'{-}O{-}]_n{-}$$

When, in place of dicarboxylic acid chlorides, phosgene is used, polycarbonates form:

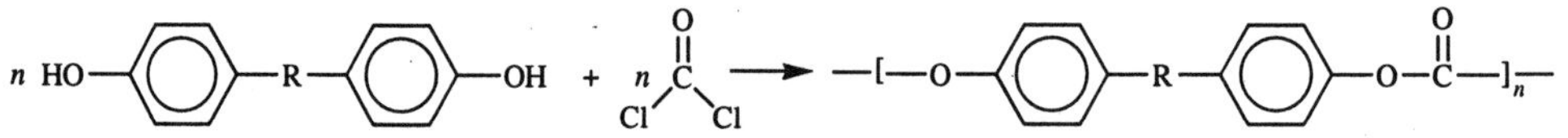

$$n\ HO{-}C_6H_4{-}R{-}C_6H_4{-}OH + n\ \underset{Cl\quad Cl}{\overset{O}{\overset{\|}{C}}} \longrightarrow {-}[{-}O{-}C_6H_4{-}R{-}C_6H_4{-}O{-}\overset{O}{\overset{\|}{C}}{-}]_n{-}$$

(9) Polyesters also form in ring-opening polymerizations of lactones. This is discussed in Chapter 4.

Many modifications of the above reactions are known. For instance, poly(propylene phthalate) can be prepared from phthalic anhydride and propylene oxide.[14] The reaction is catalyzed by tertiary amines that probably form carboxylate ion intermediates:

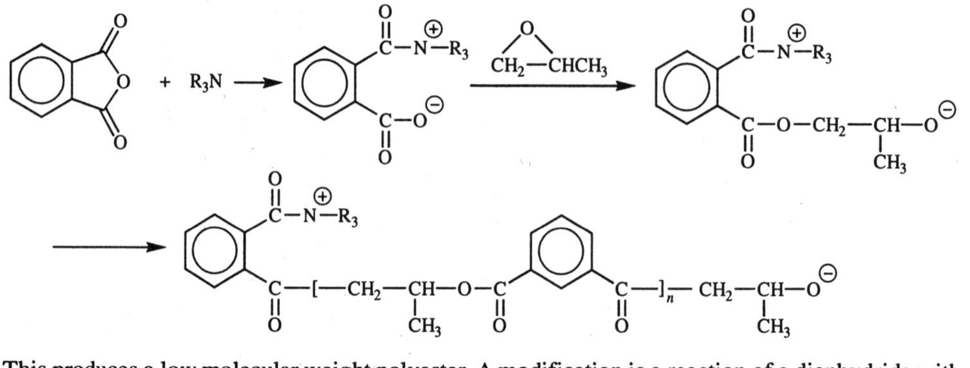

$$\text{(phthalic anhydride)} + R_3N \longrightarrow C_6H_4(\overset{O}{\overset{\|}{C}}{-}\overset{\oplus}{N}{-}R_3)(\underset{O}{\underset{\|}{C}}{-}O^{\ominus}) \xrightarrow{\overset{O}{CH_2{-}CHCH_3}} C_6H_4(\overset{O}{\overset{\|}{C}}{-}\overset{\oplus}{N}{-}R_3)(\underset{O}{\underset{\|}{C}}{-}O{-}CH_2{-}\underset{CH_3}{\underset{|}{CH}}{-}O^{\ominus})$$

$$\longrightarrow C_6H_4(\overset{O}{\overset{\|}{C}}{-}\overset{\oplus}{N}{-}R_3)(\underset{O}{\underset{\|}{C}}{-}[{-}CH_2{-}\underset{CH_3}{\underset{|}{CH}}{-}O{-}\underset{O}{\underset{\|}{C}}{-}C_6H_4{-}\underset{O}{\underset{\|}{C}}{-}]_n{-}CH_2{-}\underset{CH_3}{\underset{|}{CH}}{-}O^{\ominus})$$

This produces a low molecular weight polyester. A modification is a reaction of a dianhydride with a glycol:

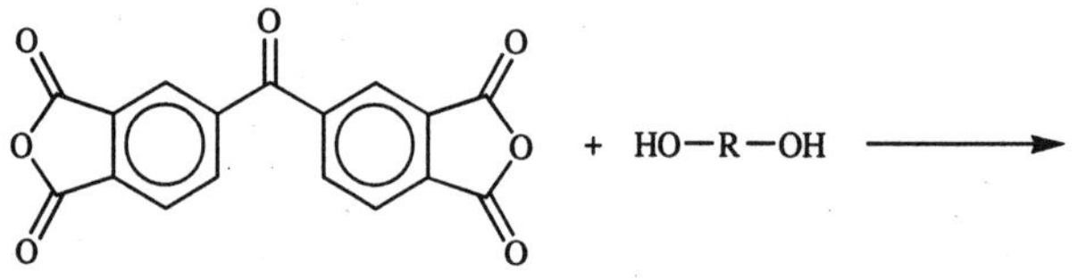

$$\text{(benzophenone tetracarboxylic dianhydride)} + HO{-}R{-}OH \longrightarrow$$

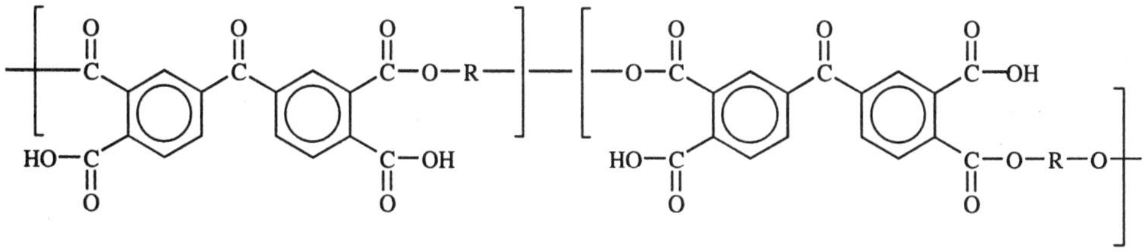

In all esterification reactions, catalysts increase the speed of condensations. Such catalysts are either acids or bases.

Beyond the above, there are many other polyester syntheses that can be found in the literature but are not in common use. For instance, polyesters form from additions of carboxylic acids to divinyl ethers[29]:

$$n\ HO{-}C(=O){-}R{-}C(=O){-}OH + n\ H_2C{=}CH{-}O{-}R'{-}O{-}CH{=}CH_2 \longrightarrow {-}[{-}R{-}C(=O){-}O{-}CH(CH_3){-}O{-}R'{-}O{-}CH(CH_3){-}O{-}C(=O){-}]_n{-}$$

Polyesters also form by the Tischenko reaction from dialdehydes.[26,27] The intramolecular hydride transfer reaction is typically catalyzed by bases:

$$O{=}CH{-}R{-}CH{=}O \xrightarrow{C_4H_9Li} {-}[{-}O{-}CH_2{-}R{-}C(=O){-}]{-}$$

Another reported procedure consists of condensations of nitiriles with glycols. The resultant poly(iminoether hydroxide)s hydrolyze to polyesters.[32]

$$n\ NC{-}R{-}CN + n\ HO{-}R'{-}OH \longrightarrow {-}[{-}R{-}C(={NH_2}^{\oplus}\ Cl^{\ominus}){-}O{-}R'{-}]_n{-} \xrightarrow{H_2O} {-}[{-}R{-}C(=O){-}O{-}R'{-}]_n{-}$$

Carbon suboxide condenses with glycols to form polyesters[26]:

$$n\ O{=}C{=}C{=}C{=}O + n\ HO{-}R{-}OH \longrightarrow {-}[{-}O{-}R{-}O{-}C(=O){-}CH_2{-}C(=O){-}]_n{-}$$

Also, an alternating free-radical addition copolymerization of cyclohexene and formic acid, perhaps via charge transfer, donor–acceptor complexes, yields polyesters[30]:

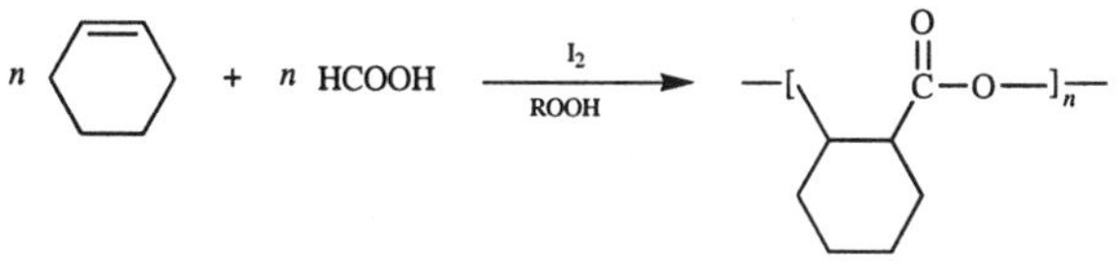

$$n\ \text{cyclohexene} + n\ HCOOH \xrightarrow[ROOH]{I_2} {-}[{-}C_6H_{10}{-}C(=O){-}O{-}]_n{-}$$

Cyclic carbonates, oxalates, and glycolates polymerize by ring-opening polymerizations to yield polyesters. An example is a conversion of a cyclic carbonate into a low molecular weight polymer (about 4000)[39,40]:

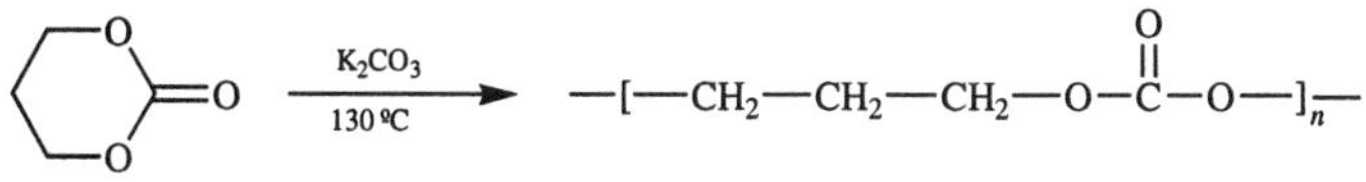

$$\text{(cyclic trimethylene carbonate)} \xrightarrow[130\,^{\circ}C]{K_2CO_3} {-}[{-}CH_2{-}CH_2{-}CH_2{-}O{-}C(=O){-}O{-}]_n{-}$$

Another example is a ring-opening polymerization of an oxalate.[39] Again, only low molecular weight polymers result:

$$\begin{array}{c} H_2C-O-C{=}O \\ | \quad\quad | \\ H_2C-O-C{=}O \end{array} \longrightarrow -[-CH_2-CH_2-O-\overset{O}{\overset{\|}{C}}-\overset{O}{\overset{\|}{C}}-O-]_n-$$

Lactides, on the other hand, when subjected to ring-opening polymerization with Lewis acids yield high molecular weight polymers[41]:

$$\begin{array}{c} O{=}C-O-CH_2 \\ | \quad\quad | \\ H_2C-O-C{=}O \end{array} \longrightarrow -[-CH_2-\overset{O}{\overset{\|}{C}}-O-]_n-$$

Also, α-hydroxycarboxylic acids can polymerize to high molecular weight polymers through formation of anhydrosulfides[42]:

$$HO-\underset{CH_3}{\overset{CH_3}{C}}-\underset{OH}{C}{=}O \xrightarrow{SOCl_2} \begin{array}{c} H_3C \quad CH_3 \\ O-C-C{=}O \\ | \quad\quad | \\ O{=}S-\!\!-\!\!-O \end{array} \longrightarrow -[-\underset{H_3C\quad CH_3}{C}-\overset{O}{\overset{\|}{C}}-O-]_n-$$

Recently, it was reported[49] that polyesterification reactions are possible at room temperature. High molecular weight polyesters form directly from carboxylic acids and phenols. These solution polymerization reactions proceed under mild conditions, near neutral pH. Equimolar mixtures of acids and alcohols condense as the reactions are being driven by additions of water across carbodiimide groups. Substituted ureas form as byproducts:

$$RN{=}C{=}NR + R'COOH + R''OH \longrightarrow R'COOR'' + \underset{H}{RN}-\overset{O}{\overset{\|}{C}}-\underset{H}{NR}$$

The reaction is useful in preparations of isoregic ordered chains with translational polar symmetry. It can also be applied in polymerizations of functional or chiral monomers.

6.2.1.2. Commercial Linear Saturated Polyesters

Many linear aliphatic polyesters are produced commercially. They are relatively low in molecular weight, less than 10,000. The main use of these materials is as plasticizers for poly(vinyl chloride) polymers and copolymers. Such polyesters are usually formed from dicarboxylic acids and glycols. Often, monocarboxylic acids or monohydroxy compounds are added toward the end of the reaction, in small quantities, to control molecular weight and to cap the reactive end groups. The condensation reactions are carried out at 200–250 °C in an inert atmosphere. To obtain a molecular weight of about 1000 these reactions are run for only several hours. For higher molecular weights, however, the glycols are added in excess and the initial products heated under vacuum (about 1 mm Hg) for several hours.

Among the high molecular weight aliphatic–aromatic polyesters, the highest commercial volume material is poly(ethylene terephthalate). Most of it is prepared from dimethylene terephthalate and ethylene glycol by a transesterification reaction:

$$CH_3-O-\overset{O}{\overset{\|}{C}}-C_6H_4-\overset{O}{\overset{\|}{C}}-O-CH_3 + HO-CH_2-CH_2-OH$$

$$\longrightarrow -[-O-CH_2-CH_2-O-\overset{O}{\overset{\|}{C}}-C_6H_4-\overset{O}{\overset{\|}{C}}-]_n- + CH_3OH\uparrow$$

Often the reaction is carried out in two steps. An excess of two moles of ethylene glycol is used and the first stage of the reaction is carried out at 150–210 °C to form bis(2-hydroxyethyl)terephthalate, a small amount of an oligomer, and methanol, which is removed:

$$CH_3-O-\overset{O}{\overset{\|}{C}}-C_6H_4-\overset{O}{\overset{\|}{C}}-O-CH_3 + 2\,HO-CH_2-CH_2-OH \longrightarrow$$

$$HO-CH_2-CH_2-O-\overset{O}{\overset{\|}{C}}-C_6H_4-\overset{O}{\overset{\|}{C}}-O-CH_2-CH_2-OH \longrightarrow$$

$$HO-CH_2-CH_2-O-[-\overset{O}{\overset{\|}{C}}-C_6H_4-\overset{O}{\overset{\|}{C}}-O-CH_2-CH_2-O-]_4-H$$

In the second stage the temperature is maintained at 270–285 °C and the reaction is carried out under vacuum at about 1 mm Hg. The bis(2-hydroxyethyl)terephthalate undergoes a transesterification reaction and the excess glycol is removed:

$$HO-CH_2-CH_2-O-[-\overset{O}{\overset{\|}{C}}-C_6H_4-\overset{O}{\overset{\|}{C}}-O-CH_2-CH_2-O-]_4-H \longrightarrow$$

$$-[-CH_2-CH_2-O-\overset{O}{\overset{\|}{C}}-C_6H_4-\overset{O}{\overset{\|}{C}}-O-]_n- + HO-CH_2-CH_2-OH$$

Various metal oxide or acetate catalysts are employed in the first stage. These are antimony, barium, cadmium, calcium, cobalt, lead, manganese, titanium, or zinc oxides or acetates. Carbonates, alcoholates, and alkanoates can also be used. Based on disclosures in the patent literature, it appears that antimony compounds, particularly the trioxide, dominate the field of catalysts for the second stage of this reaction.[16] The exact mechanism by which the antimony compounds act as catalysts for the syntheses of polyesters is still being investigated. It was shown that bis(2-hydroxyethyl)terephthalate competes successfully with oligomer end groups for Sb_2O_3 and that a complex of this compound with the metal oxide is unreactive in these polymerizations.[17] In addition, during polymerizations under vacuum there is an increase in metal surfaces of antimony oxide.[18] It was suggested that a reaction of antimony trioxide with ethylene glycol results in formation of antimony glycolates with a ligand number of 3[19,20]:

$$(OCH_2CH_2O)Sb-O-CH_2-CH_2-O-Sb(OCH_2CH_2O) \xrightarrow{HO-CH_2-CH_2-OH}$$

$$(OCH_2CH_2O)_2Sb-O-CH_2-CH_2-OH \rightleftharpoons \left[Sb\left(\begin{matrix}O\\ O\end{matrix}\right)_3\right]H^{\oplus}$$

A study of antimony glycolates as effective catalysts for preparation of poly(ethylene terephthalate) with a varying number of hydroxyethoxy ligands rated them in the following decreasing order of effectiveness[20]:

$$5 = 3 > 2 > 1 >> 0$$

This led to a suggestion by Maerov[20] that a key step in condensation may involve a chemical reaction that ties up a hydroxy chain end with the catalyst molecule. Introduction of a second hydroxyethoxy chain end is followed by the right electronic bond shift:

The role of antimony is to establish a favorable spatial configuration for the transition state.[20] An earlier study, however, resulted in a conclusion that antimony's activity is inversely proportional to the hydroxyl group concentration.[21]

Based on model reactions for the preparation of poly(ethylene terephthalate) by ester interchange, the optimum molar ratio of ethylene glycol to dimethyl terephthalate is 2.4 to 1. This ratio allows complete removal of methanol.[22] The overall polyesterification reaction is third order.[22,23] In addition, high molecular weight polymerizations of poly(ethylene terephthalate) invariably produce some cyclic oligomers as byproducts.[24,25] Eight different cyclic species were identified in one commercial polymer:

The most important side reaction, however, is formation of diethylene glycol. It becomes included in the polymer as a di(oxyethylene)oxy link. Commercial polymerizations are carried out until molecular weights of about 20,000 are reached, for use in fibers, and higher ones, for use in injection moldings or extrusions. These materials may contain up to 2–4 mol% of di(oxyethylene)oxy units.[54] The presence of such units influences the degree of whiteness of the polyesters and the melting temperature.

A process was developed[54] to reduce the presence of di(oxyethylene)oxy units in poly(ethylene terephthalate). The condensation reactions are still carried out in two steps. In the first one, or during the precondensation, the material is prepared in the melt as described previously. In second step, however, the reaction is carried out below the melting temperature. This still yields high molecular weight polymers. The products, however, are low in di(oxyethylene)oxy linkages.

There has been continued interest in developing a process for direct esterification of terephthalic acid with ethylene glycol. It does not appear, however, that this is currently practiced on a commercial scale in the U.S. In Japan, a process was commercialized where terephthalic acid is reacted with two moles of ethylene oxide to form the dihydroxy ester *in situ*, as the starting material. One mole of ethylene glycol is then removed under vacuum in the subsequent condensation process. Also, it was reported[25] that the polymer can be prepared by direct esterification at room temperature in the presence of picryl chloride. The reaction can also be performed at about 120 °C in the presence of diphenyl chloro-phosphate or toluenesulfonyl chloride.[25] This is done in solution, where pyridine is either the solvent or the cosolvent. Pyridine acts as a scavenger for HCl, that is a byproduct of the reaction, and perhaps also as an activator (by converting the acid into a reactive ester intermediate).

Another commercially important, high molecular weight polyester is poly(butylene terephthalate), also called poly(tetramethylene terephthalate). The polymer is prepared by a catalyzed ester interchange of dimethyl terephthalate and 1,4-butane diol:

$$n\ CH_3{-}O{-}\overset{O}{\overset{\|}{C}}{-}C_6H_4{-}\overset{O}{\overset{\|}{C}}{-}O{-}CH_3 + n\ HO{-}CH_2{-}CH_2{-}CH_2{-}CH_2{-}OH \longrightarrow$$

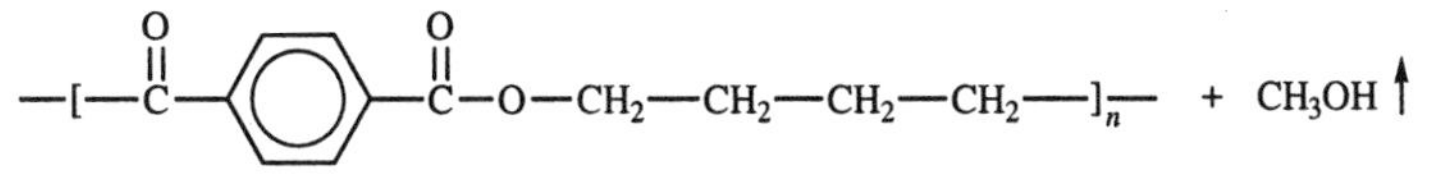

$$-[-\overset{O}{\overset{\|}{C}}-C_6H_4-\overset{O}{\overset{\|}{C}}-O-CH_2-CH_2-CH_2-CH_2-]_n- \quad + \quad CH_3OH\uparrow$$

This synthesis is also carried out in two stages. In stage one an excess of the diol is reacted with dimethyl terephthalate (about 1.3:1) to insure complete removal of methanol. Zinc acetate is favored as a catalyst for this reaction. A prepolymer mixture of bis(hydroxybutyl)terephthalate and higher oligomers forms. Stage two is conducted in vacuum at 1 mm Hg and high enough temperature (usually at least 60 °C above the melting temperature of the polyester) to remove excess diol and reach high molecular weight. Zinc oxide is favored as the catalyst for this stage.

Prolonged heating of the reaction mixture at excessive temperatures results in formation of large proportions of tetrahydrofuran. This is objectionable because it affects the properties of the product. It also results in lower molecular weight polyesters.

A polyester from terephthalic acid and 1,2-dimethylol cyclohexane is produced mainly for use in fibers. This polymer is also formed from dimethyl terephthalate and the diol by a transesterification reaction. The material has the following structure:

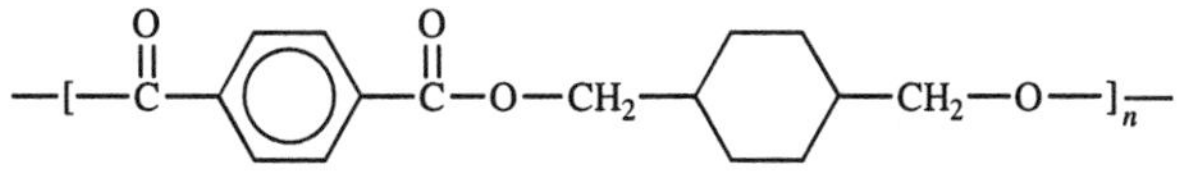

$$-[-\overset{O}{\overset{\|}{C}}-C_6H_4-\overset{O}{\overset{\|}{C}}-O-CH_2-C_6H_{10}-CH_2-O-]_n-$$

The polymer is stiffer than poly(ethylene terephthalate) and higher melting.

A polyester is being manufactured in Japan from a methyl ester of *p*-3-hydroxyethoxybenzoic acid by transesterification:

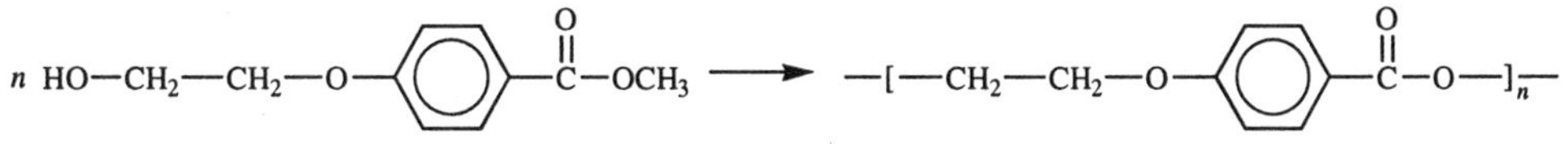

$$n\ HO-CH_2-CH_2-O-C_6H_4-\overset{O}{\overset{\|}{C}}-OCH_3 \longrightarrow -[-CH_2-CH_2-O-C_6H_4-\overset{O}{\overset{\|}{C}}-O-]_n-$$

The product is used as a fiber.

6.2.1.3. Copolyesters

Mixed dicarboxylic acids are usually used to form copolyesters. For instance, terephthalic and isophthalic acids are reacted together with 1,4-dimethylol cyclohexane to form a copolyester. The product is amorphous and transparent. Another copolyester is manufactured from terephthalic, isophthalic, and an aliphatic dicarboxylic acid like adipic with either 1,4-butanediol or 1,6-hexanediol. The aliphatic dicarboxylic acid is used in minor quantities. Many such copolyesters are used as high strength adhesives.

In addition, there are thermoplastic polyester elastomers. These are produced by equilibrium melt transesterification of dimethyl terephthalate, 1,4-butanediol, and a poly(tetramethylene oxide) glycol (molecular weight about 1000). Because equilibrium conditions exist in the melt, the products are random copolymers:

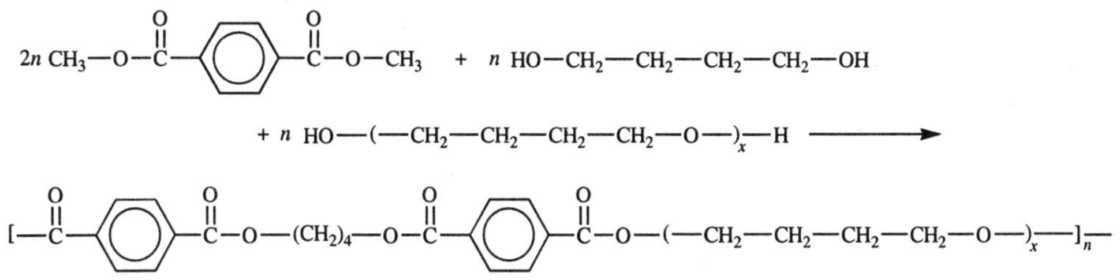

$$2n\ CH_3-O-\overset{O}{\overset{\|}{C}}-C_6H_4-\overset{O}{\overset{\|}{C}}-O-CH_3 \quad + \quad n\ HO-CH_2-CH_2-CH_2-CH_2-OH$$

$$+\ n\ HO-(-CH_2-CH_2-CH_2-CH_2-O-)_x-H \longrightarrow$$

$$[-\overset{O}{\overset{\|}{C}}-C_6H_4-\overset{O}{\overset{\|}{C}}-O-(CH_2)_4-O-\overset{O}{\overset{\|}{C}}-C_6H_4-\overset{O}{\overset{\|}{C}}-O-(-CH_2-CH_2-CH_2-CH_2-O-)_x-]_n-$$

Wholly aromatic polyesters are produced for high-temperature applications. The materials must also have good abrasion resistance. One such commercial polyester is prepared from *p*-hydroxybenzoic acid:

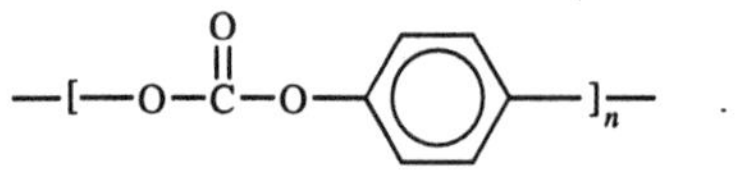

$$-[-O-\overset{O}{\overset{\|}{C}}-O-C_6H_4-]_n-$$

Another one is a copolyester prepared from *p*-hydroxybenzoic acid, *p,p'*-biphenol, and terephthalic or isophthalic acids:

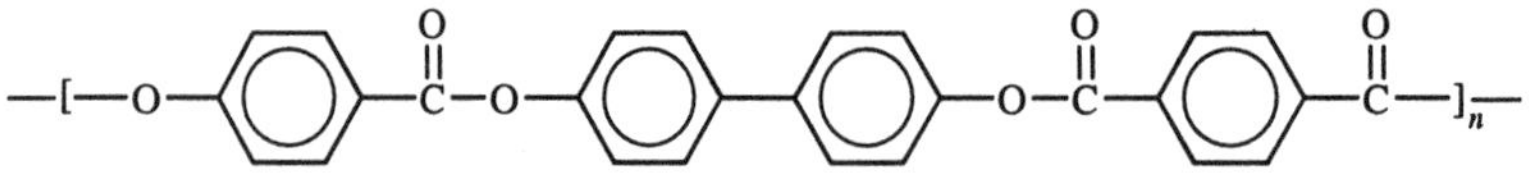

$$-[-O-C_6H_4-\overset{O}{\overset{\|}{C}}-O-C_6H_4-C_6H_4-O-\overset{O}{\overset{\|}{C}}-C_6H_4-\overset{O}{\overset{\|}{C}}-]_n-$$

The above shown polyester does not melt and decomposes at 550 °C.

The properties of various polyesters were summarized by Wilfong.[15] Table 6.1 presents T_m and T_g values of some polyesters based on information from Wilfong and other sources in the literature.

Among new methods of forming polyesters is a preparation of completely aromatic polyesters by direct condensation of hydroxyaromatic acids (like hydroxybenzoic) with the aid of triphenylphosphorus compounds or dichlorophenylphosphine.[38]

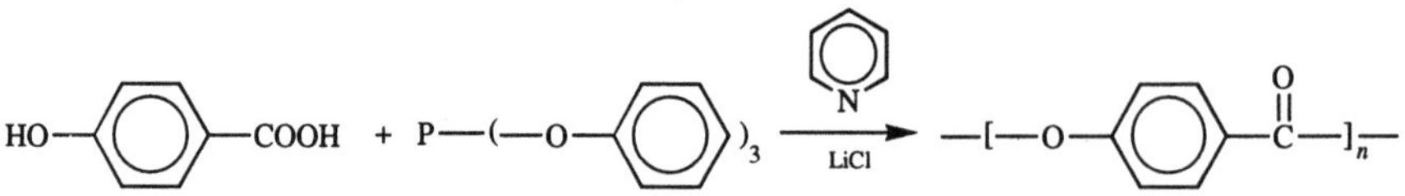

$$HO-C_6H_4-COOH + P-(-O-C_6H_5)_3 \xrightarrow[\text{LiCl}]{\text{pyridine}} -[-O-C_6H_4-\overset{O}{\overset{\|}{C}}-]_n-$$

Hexachlorotriphosphatriazine can also be used to attain direct polycondensations of hydroxybenzoic acid.[43]

6.2.2. Linear Unsaturated Polyesters

The materials in this group are linear copolyesters. One of the dicarboxylic acids is an aliphatic unsaturated diacid. The unsaturation is introduced into the polymer backbone for the purpose of subsequent crosslinking. Unsaturated polyester technology was developed for use in glass fiber laminates, thermosetting molding compositions, casting resins, and solventless lacquers.

Propylene glycol is often used as the diol. To a lesser extent, other glycols, like diethylene glycol, are also used for greater flexibility, or neopentyl glycol for a somewhat better thermal resistance. Bisphenol A (2,2 bis(4-hydroxyphenyl) propane) is used when better chemical resistance is needed. Use of mixed diols is common. Many unsaturated dicarboxylic acids can be used, but maleic (as an anhydride) or fumaric acids are the most common. Chloromaleic or chlorofumaric acids are also employed.

The saturated dicarboxylic acids act as modifiers. While aliphatic dicarboxylic acids can be used, the most common one is orthophthalic acid (added to the reaction mixture as an anhydride). The acid improves compatibility with styrene that is polymerized in the presence of the polyester to form hard, rigid, crosslinked materials. Other modifiers are used to obtain special properties. When a flexible product is needed, adipic or sebacic acids may be used instead. For better heat resistance endomethylene tetrahydrophthalic anhydride (nadic anhydride) may be utilized. Flame retardancy is achieved by using chlorinated dicarboxylic acids, like tetrachlorophthalic.

Styrene is the most common monomer used in crosslinking unsaturated polyesters. When special properties are required, other monomers like methyl methacrylate may be employed. Sometimes this is done in combination with styrene. Diallyl phthalate and triallyl cyanurate form better heat-resistant products.

An example of a typical batch preparation of a polyester is one where 1.2 moles of propylene glycol, 0.67 mole of maleic anhydride, and 0.33 mole of phthalic anhydride are combined. Propylene glycol is used in excess to compensate for loss during the reaction. The condensation at 150–200 °C lasts for 6–16 hours, with constant removal of water, the byproduct. An aromatic solvent, like toluene or xylene, is often added to the reaction mixtures to facilitate water removal by azeotropic distillation. Esterification catalysts, like toluene sulfonic acid, reduce the reaction time. In addition,

Table 6.1. Approximate Melting Points of Polyesters[a]

Dicarboxylic Acid	Glycol	T_g (°C)	T_m (°C)
HOOCCOOH	$HOCH_2CH_2OH$	—	172
HOOCCOOH	$HOCH_2CH_2CH_2OH$	—	89
$HOOCCH_2CH_2COOH$	$HOCH_2CH_2OH$	—	108
$HOOCCH_2CH_2COOH$	$HOCH_2CH_2CH_2OH$	—	52
$HOOC(CH_2)_4COOH$	$HOCH_2CH_2OH$	—	50
$HOOC(CH_2)_4COOH$	$HOCH_2CH_2CH_2OH$	—	46
$HOOC(CH_2)_8COOH$	$HOCH_2CH_2OH$	—	79–80
$HOOC(CH_2)_8COOH$	$HOCH_2CH_2CH_2OH$	—	58
HOOC—C_6H_4—OH (para)		—	> 350 decomp.
HOOC—C_6H_4—CH_2CH_2OH (para)		—	185
HOOC—C_6H_4—COOH (para)	$HOCH_2CH_2OH$	69	265–284
HOOC—$C_{10}H_6$—COOH (2,6-naphthalene)	$HOCH_2CH_2OH$	119	270
HOOC—C_6H_4—C_6H_4—COOH (para)	$HOCH_2CH_2OH$	—	355
HOOC—C_6H_4—CH_2CH_2—C_6H_4—COOH (para)	$HOCH_2CH_2OH$	—	220
HOOC—C_6H_{10}—COOH (1,4-cyclohexane)	$HOCH_2CH_2OH$	—	*trans* 120 *cis* < 30
HOOC—C_6H_4—$(CH_2)_4$—C_6H_4—COOH (para)	$HOCH_2CH_2OH$	—	170
HOOC—C_6H_4—S—$(CH_2)_2$—S—C_6H_4—COOH (para)	$HOCH_2CH_2OH$	—	200
HOOC—C_6H_4—O—$(CH_2)_4$—O—C_6H_4—COOH (para)	$HOCH_2CH_2OH$	—	252
HOOC—C_6H_4—NH—$(CH_2)_2$—NH—C_6H_4—COOH (para)	$HOCH_2CH_2OH$	—	273
HOOC—C_6H_4—COOH (meta)	$HOCH_2CH_2OH$	51	143
HOOC—C_6H_4—C_6H_4—COOH (3,3')	$HOCH_2CH_2OH$	—	156

(continued)

Table 6.1. (continued)

Dicarboxylic acid	Glycol	T_g (°C)	T_m (°C)
$HOOC{-}C_{10}H_6{-}COOH$	$HOCH_2CH_2OH$	113	260
$HOOC{-}C_6H_4{-}C_6H_4{-}COOH$	$HOCH_2CH_2OH$	—	132
$HOOC{-}C_6H_4{-}COOH$	$HOCH_2CH_2CH_2OH$	< 80	226–232
$HOOC{-}C_6H_4{-}COOH$	$HO{-}({-}CH_2{-})_6{-}OH$	—	154–161
$HOOC{-}C_6H_4{-}COOH$	$HO{-}({-}CH_2{-})_8{-}OH$	< 45	129–132
$HOOC{-}C_6H_4{-}COOH$	$HO{-}({-}CH_2{-})_{10}{-}OH$	< 25	130–138
$HOOC{-}C_6H_4{-}O{-}C_6H_4{-}COOH$	$HOCH_2CH_2OH$	—	152
$HOOC{-}C_6H_4{-}COOH$ (meta)	$HO{-}C_6H_4{-}O{-}C_6H_4{-}OH$	173	283
$HOOC{-}C_6H_4{-}COOH$ (meta)	$HO{-}C_6H_4{-}CH_2{-}C_6H_4{-}OH$	150	348
$HOOC{-}C_6H_4{-}COOH$ (meta)	$HO{-}C_6H_4{-}SO_2{-}C_6H_4{-}OH$	279	330
$HOOC{-}C_6H_4{-}COOH$ (meta)	$HO{-}C_6H_4{-}C_6H_4{-}OH$	164	315

[a]From Refs. 15, 27, and other literature sources.

the reactions are blanketed by inert gases, like nitrogen or carbon dioxide, to prevent discoloration from oxygen at high temperatures. When molecular weights of 1000–2000 are reached, the products are cooled to 90 °C and blended with vinyl monomers. Often, the blends are mixtures of equal weights of the polyesters and the monomers. The structure of the above-described unsaturated polyester can be illustrated as follows:

$$-[-\underset{\|}{\overset{}{C}}(=O)-C_6H_4-C(=O)-O-CH(CH_3)-CH_2-O-C(=O)-CH{=}CH-C(=O)-O-CH(CH_3)-CH_2-O-]_n-$$

6.2.3. Network Polyesters for Surface Coatings

The original polyesters for coatings were prepared from phthalic anhydride and glycerol and were referred to as *glyptals* or *glyptal* resins:

$$\text{phthalic anhydride} + \begin{array}{l} CH_2-OH \\ | \\ CH-OH \\ | \\ CH_2-OH \end{array} \longrightarrow \text{network: } \sim O-CH_2-CH(-O-C(=O)-C_6H_4-C(=O)-O\sim)-CH_2-O\sim$$

The products from the above polyesterifications are brittle materials. They are therefore modified with oils, either drying or nondrying. Such oil-modified resins bear the names of *alkyds*. While glycerol is widely used, other polyhydroxy compounds (polyols) are also utilized. These may be trimethylolpropane, pentaerythritol, sorbitol, or others. Phthalic anhydride is usually used in alkyd preparations. Other dicarboxylic compounds, however, may also be included for modification of properties. Common modifiers might be isophthalic, adipic, or sebacic acids, or maleic anhydride. In addition, many other acid modifiers are described in the patent literature.

The oils in alkyd resins are usually of vegetable origin. They are classified by the type and amount of residual unsaturation into drying, semidrying, and nondrying oils. The drying oils contain most residual unsaturation, while the nondrying ones contain mostly saturated fatty acids.

Alkyds are also classified by the quantity of modifying oil that they contain into *short*, *medium*, or *long* oil alkyds. Short oil alkyds contain 30–50% oil and are usually baked to obtain a hard dry surface. Medium oil (50–65%) and long oil (65–75%) alkyds will air-dry upon addition of metal dryers.

There are two main methods for preparation of alkyd resins. In the first one, called the *fatty acid process*, a free fatty acid is coesterified directly with the dibasic acid and the polyol at 200–240 °C. The reaction may be carried out without a solvent by first heating in an inert atmosphere. At the end, an inert gas may be blown into the resin from the bottom of the reaction kettle to remove water and unreacted materials. As a modification of this, a small quantity of a solvent may be used to remove water of esterification continuously by azeotropic distillation with the aid of moisture traps.

In the second method, known as the *alcoholysis process*, the drying oil is heated with the glycerol in the first stage of the reaction, at about 240 °C. This is usually done in the presence of a transesterification basic catalyst to form monoglycerides:

$$\begin{array}{l} CH_2-OOCR \\ | \\ CH-OOCR \\ | \\ CH_2-OOCR \end{array} + 2\begin{array}{l} CH_2-OH \\ | \\ CH-OH \\ | \\ CH_2-OH \end{array} \longrightarrow 3\begin{array}{l} CH_2-OH \\ | \\ CH-OH \\ | \\ CH_2-OOCR \end{array}$$

After the first stage is complete, phthalic anhydride, with or without another dibasic acid, is added and a copolyesterification is carried out in the same manner as in the first stage of the reaction.

The conditions under which the reaction is carried out and the rate at which the temperature is raised during the condensation affects the molecular weight distribution of the final product. In addition, the monoglyceride content of the original reaction mixture determines the microgel content of the alkyd and the dynamic properties of the dried film.[31] The finished alkyd resin can be illustrated as follows:

$$\sim C(=O)-C_6H_4-C(=O)-O-CH_2-CH(O-C(=O)-R)-CH_2-O-C(=O)-C_6H_4-C(=O)-O-CH_2-CH(O-C(=O)-R)-CH_2-O\sim$$

6.2.4. Polycarbonates

A special group of polyesters of carbonic acid are known as *polycarbonates*. The first polycarbonates were prepared as early as 1898 by Einhorn, by reacting phosgene with hydroquinone and with resorcinol.[33,34] These materials lack desirable properties and remain laboratory curiosities. During the fifties, however, new polymers were developed from 4,4′-dihydroxydiphenyl alkanes. These polycarbonates have high melting points and good thermal and hydrolytic stability. Nevertheless, to date only one polycarbonate has achieved significant commercial importance. It is based on 2,2′-bis(4-hydroxyphenyl)propane.

There are two main methods for preparing polycarbonates, one by direct reaction of phosgene with the diphenol and the other by an ester interchange. The direct phosgenation is a form of a Schotten–Baumann reaction that is carried out in the presence of a base:

$$n\ COCl_2 + n\ HO-C_6H_4-C(CH_3)_2-C_6H_4-OH \longrightarrow -[-O-C_6H_4-C(CH_3)_2-C_6H_4-O-C(=O)-]_n-$$

The reaction may be carried out in the presence of pyridine that acts as a catalyst and as an HCl scavenger. Often, a chlorinated solvent is used as a diluent for the pyridine. Phosgene is bubbled through a solution of the diphenol at 25–35 °C. The pyridine hydrochloride precipitates out and, after washing the pyridine solution with dilute HCl and water, the polymer is precipitated with a nonsolvent.

An interfacial polymerization procedure is also employed in direct phosgenations. A caustic solution of the diphenol is dispersed in an organic chlorinated solvent containing small quantities of a tertiary amine. Phosgene is bubbled through the reaction mixture at 25 °C. When the reaction is complete, the organic phase contains the polymer. It is separated and the product isolated as above.

The ester interchange method is carried out between the diphenol and diphenyl carbonate:

$$n\ HO-C_6H_4-C(CH_3)_2-C_6H_4-OH + n\ C_6H_5-O-C(=O)-O-C_6H_5 \longrightarrow$$

$$-[-O-C_6H_4-C(CH_3)_2-C_6H_4-O-C(=O)-]_n- + 2n\ C_6H_5-OH$$

To obtain high molecular weights by this method, almost complete removal of the phenol is required. The reaction is carried out with typical basic catalysts, like lithium hydride, zinc oxide, or antimony oxide under an inert atmosphere. The initial reaction temperature is 150 °C. It is raised over a one-hour period to 210 °C while the pressure is reduced to 20 mm Hg. The reaction mixture is then heated to about 300 °C for 5–6 hours at 1 mm Hg. Heating is stopped when the desired viscosity is reached.

A synthetic route to polycarbonates was reported that uses crown ethers. Crown ethers generally form stable complexes with metal cations and, by increasing the dissociation of ion pairs, provide highly reactive, unsolvated anions.[35] This led to direct preparations of new polycarbonates from α,ω-dibromo compounds, carbon dioxide and potassium carbonate, or salts of the diols[36]:

Table 6.2. Melting Points of Some Polycarbonates[a]

Structure of the parent diphenol	Approx. T_g (°C)	Approx. T_m(°C)
$HO-C_6H_4-C(CH_3)_2-C_6H_4-OH$	145–150	265–270
$HO-C_6H_4-CH(C(CH_3)_3)-C_6H_4-OH$	125–130	185–200
$HO-C_6H_4-C(C_3H_7)_2-C_6H_4-OH$	145–150	170–180
$HO-C_6H_4-CH_2-C_6H_4-OH$	147	223–225
$HO-C_6H_4-C(C_6H_5)_2-C_6H_4-OH$	121	210–230
$HO-C_6H_4-O-C_6H_4-OH$	—	180–190
$HO-C_6H_4-O-C_6H_4-O-C_6H_4-OH$	—	205–210
$HO-C_6H_4-O-C_6H_4-O-C_6H_4-O-C_6H_4-OH$	—	250–260
$HO-C_6H_4-CH(C_6H_{11})-C_6H_4-OH$	170	240–250

[a]From various literature sources.

$$n\ \mathrm{Br{-}CH_2{-}C_6H_4{-}CH_2{-}Br} + 2n\ \mathrm{CO_2} + \mathrm{K_2CO_3} \xrightarrow{\text{18 crown-6-ether}}$$

$$\mathrm{Br{-}CH_2{-}C_6H_4{-}CH_2{-}(-O{-}\overset{O}{\overset{\|}{C}}{-}O{-}CH_2{-}C_6H_4{-}CH_2-)_n{-}Br} + \mathrm{KBr}$$

When a potassium salt of a diol is present in the reaction mixture, mixed polyesters form[36]:

$$n\ \mathrm{Br{-}CH_2{-}C_6H_4{-}CH_2{-}Br} + n\ \mathrm{KO{-}C_6H_{10}{-}OK} \xrightarrow{CO_2}$$

$$\xrightarrow{\text{18 crown-6-ether}} \mathrm{-[-O{-}\overset{O}{\overset{\|}{C}}{-}O{-}CH_2{-}C_6H_4{-}CH_2{-}O{-}\overset{O}{\overset{\|}{C}}{-}O{-}C_6H_{10}-]_n-} + 2n\ \mathrm{KBr}$$

In addition to crown ethers, cryptates and polyglyme exhibit similar behavior.[37] Table 6.2 presents some T_g and T_m values of some polycarbonates picked from the literature.

6.2.5. Polyesters from Lactones

Polyesters that are obtainable by ring-opening polymerization of lactones (see Chapter 4) are not produced commercially on a large scale. Judging from the patent literature, however, there is a continuing interest in these materials, particularly in Japan. Because of fairly good hydrolytic stability, polypivalyllactone is at the most advanced stages of development:

$$\mathrm{-[-\overset{O}{\overset{\|}{C}}{-}O{-}CH_2{-}C(CH_3)_2-]_n-}$$

The polymer can be spun into an elastic yarn of very fine denier. It is also claimed to exhibit good mechanical properties for molding and compares favorably with commercial polyesters and nylons. Also, polycaprolactone was reported to be used in some medical applications in biodegradable surgical sutures and postoperative support pins and splints.[39] Similar uses are also found for two other polyesters, poly(lactic acid) and poly(glycolic acid).[40] The two polymers form from their cyclic dimers by cationic ring-opening polymerizations with the aid of Lewis acids:

$$\text{lactide} \longrightarrow \mathrm{-\left[-\underset{CH_3}{\underset{|}{CH}}{-}\overset{O}{\overset{\|}{C}}{-}O-\right]-}$$

$$\text{glycolide} \longrightarrow \mathrm{-\left[-CH_2{-}\overset{O}{\overset{\|}{C}}{-}O-\right]-}$$

6.3. Polyamides

The family of synthetic polymeric materials with amide linkages in their backbones is large. It includes synthetic linear aliphatic polyamides, which carry the generic name of *nylon*, aromatic polyamides, and fatty polyamides used in adhesives and coatings. In addition to the synthetic materials, there is also a large family of naturally occurring polymers of α–amino acids, called *proteins*. The latter are discussed in Chapter 7 with the rest of the naturally occurring polymers. The nylons include polyamides produced by ring-opening polymerizations of lactams and condensation products of diamines with dicarboxylic acids.

6.3.1. Nylons

The nylons are named by the number of carbon atoms in the repeat units. The materials formed by polymerizations of lactams therefore carry only one number in their names like, for instance, nylon 6 that is formed from caprolactam. By the same method of nomenclature, a nylon prepared by condensing a diamine with a dicarboxylic acid like, for example, hexamethylene diamine with adipic acid, is called nylon 6,6. It is customary for the first number to represent the number of carbons in the diamine and the second number to represent the number of carbons in the diacid.

A discussion of various individual nylons follows. Not all of them are industrially important.

At present, we only know how to prepare *Nylon 1* by anionic polymerization of isocyanates. This reaction is discussed in Chapter 3:

$$R{-}N{=}C{=}O \xrightarrow[\text{DMF}]{\text{NaCN}} {-}[{-}\underset{R}{N}{-}\overset{O}{\overset{\|}{C}}{-}]_n{-}$$

Potassium and sodium cyanide catalyze the reaction. It can be carried out between –20 and –100 °C. An example is the following preparation[47]:

$$n\ CH_3{-}O{-}C_6H_4{-}N{=}C{=}O \xrightarrow{\text{KCN}} {-}[{-}\underset{C_6H_4{-}O{-}CH_3}{N}{-}\overset{O}{\overset{\|}{C}}{-}]_n{-}$$

The resultant polymer has a molecular weight of one million. When the methyl group is replaced by butyl, the product is a tough film former, but depolymerizes in the presence of some catalysts. Many other interesting high molecular weigh polymers with various substitutions form by this reaction at low temperature.

Polymerization of *N*-carboxy-α-amino acid anhydrides result in formations of *Nylon 2*. This reaction is discussed in Chapter 4. These polymers are mainly of interest to protein chemists in model studies of naturally occurring poly(α-amino acids).

Nylon 3 can be synthesized by intramolecular hydrogen transfer polymerization of acrylamide[48] (see chapter 3):

$$CH_2{=}CH{-}\overset{O}{\overset{\|}{C}}{-}NH_2 \xrightarrow{B^{\ominus}} {-}({-}CH_2{-}CH_2{-}\overset{O}{\overset{\|}{C}}{-}\underset{H}{N}{-})_n{-}$$

The fibers from nylon 3 are reported to resemble natural silk.[50] They possess high water absorbency and good light and oxygen stability. The polymer, however, is too high melting for melt spinning, or for molding and extrusion.[46] Nylon 3 fibers can be spun, however, from special solutions containing formic acid.[52]

It is difficult to synthesize β-propiolactam. A synthetic route, however, was found for substituted propiolactams like β-butyrolactams.[50] The compounds form by nucleophilic additions of carbonyl-sulfamoyl chloride to olefins[51]:

$$CH_3{-}CH{=}CH_2 + ClO_2S{-}N{=}C{=}O \longrightarrow \begin{array}{l} CH_3{-}CH{-}CH_2 \\ \quad\ \ | \qquad\ \ | \\ ClO_2S{-}N{-}C{=}O \end{array} \longrightarrow \begin{array}{l} CH_3{-}CH{-}CH_2 \\ \quad\ \ | \qquad\ \ | \\ H{-}N{-}C{=}O \end{array}$$

The above β-butyrolactam polymerizes readily by anionic mechanism, yielding a very high molecular weight polymer. This lactam preparation reaction is quite general. It can be carried out on propene, 1-butene, 1-hexene, and styrene.[50] Generally, substituted lactams are harder to polymerize,[53] however, the four-membered lactams exhibit such a strong tendency toward ring opening that even substituted β-propiolactams polymerize well.[50] The rate of polymerization, however, does tend to decrease with the number of substituents.

Nylon 4 or polypyrrolidone is an attractive polymer for use in fibers. The original syntheses of nylon 4 from 2-pyrrolidone were carried out by alkaline catalyzed ring-opening polymerizations promoted by *N*-acylpyrrolidone.[49] The products from these reactions melt between 260–265 °C. They are unstable at these temperatures and cannot be melt spun. Fibers, however, were prepared by dry spinning from hydrocarbon suspensions.[49] Later, it was found that when the anionic ring-opening polymerizations of 2-pyrrolidone are activated by CO_2 in place of the N-acyl derivative, the resultant higher molecular weight product has much better heat resistance.[54] This "new" nylon 4, reportedly, can be melt spun.

Nylon 5 or poly(α-piperidone) can be prepared by ring-opening anionic polymerization of valerolactam.[53] The reaction requires very pure monomer to yield a high molecular weight polymer[53]:

$$\text{(CH}_2)_4\text{[C=O, N–H ring]} \longrightarrow -[-CH_2-CH_2-CH_2-CH_2-\overset{O}{\overset{\|}{C}}-\underset{H}{\underset{|}{N}}-]_n-$$

One route to valerolactam is from cyclopentadiene:

$$\text{cyclopentadiene (CH–CH, CH=CH, CH}_2) \xrightarrow{[H]} \text{cyclopentane} \xrightarrow{[O]} \text{cyclopentanone (C=O)}$$

$$\xrightarrow{H_2NOH} \text{cyclopentanone oxime (C=N–OH)} \xrightarrow[\substack{250\,°C \\ H_2SO_4}]{\text{Beckmann rearrangement}} \text{(CH}_2)_4\text{[C=O, N–H ring]}$$

Valerolactam can also be polymerized with the aid of coordination catalysts to a high molecular weight polymer using alkali metal–$Al(C_2H_5)_3$ catalysts or alkali metal alkyl–$Al(C_2H_5)_3$ catalysts.[55] The polymerizations require relatively long times.

Nylon 6 is obtained via ring-opening polymerization of caprolactam:

$$n\ \text{(CH}_2)_5\text{[C=O, N–H ring]} \longrightarrow -[-CH_2-CH_2-CH_2-CH_2-CH_2-\overset{O}{\overset{\|}{C}}-N-]_n-$$

This polymer developed over the years into an important commercial material. As a result, many preparatory routes were developed for the starting material and the polymerization reaction was studied thoroughly.

The most common starting materials for preparations of caprolactam are phenol, cyclohexane, and toluene. Some caprolactam is also made from aniline. In these synthetic processes, the key material is cyclohexanone oxime. The route based on phenol can be shown as follows:

$$\text{phenol (–OH)} \xrightarrow{[H]} \text{cyclohexanol (–OH)} \xrightarrow{[O]} \text{cyclohexanone (=O)}$$

$$\xrightarrow{H_2NOH} \text{cyclohexanone oxime (=NOH)} \xrightarrow[\substack{250\,°C \\ H_2SO_4}]{\text{Beckmann rearrangement}} \text{(CH}_2)_5\text{[C=O, N–H ring]}$$

A byproduct of the above reaction is ammonium sulfate. To avoid the necessity of disposing of ammonium sulfate, many caprolactam producers sought other routes to the oxime. One approach is to form it directly by reacting cyclohexanone with ammonia and hydrogen peroxide in the presence of tungstic acid based catalyst[56]:

$$\text{cyclohexanone}{=}O + NH_3 + H_2O_2 \xrightarrow{H_3PW_2O_{40}\cdot 8H_2O} \text{cyclohexane}{=}NOH + H_2O$$

The reaction is conducted in water and the product oxime is extracted with an organic solvent.

Another process is based on photo-nitrosyl chlorination. Here, cyclohexane is converted in one step to cyclohexanone oxime hydrochloride[57]:

$$NOCl \xrightarrow{h\nu} NO\bullet + Cl\bullet$$

$$\text{cyclohexane} + Cl\bullet \longrightarrow \text{cyclohexyl}\bullet + HCl$$

$$\text{cyclohexyl}\bullet + NO\bullet \longrightarrow \text{cyclohexane}{=}NOH\bullet HCl$$

Another process uses ketene to form cyclohexene acetate[58]:

$$\text{cyclohexanone}{=}O \xrightarrow{H_2C{=}C{=}O} \text{cyclohexenyl}{-}O{-}C({=}O)CH_3 \xrightarrow{HNO_3} \text{2-nitrocyclohexanone } (NO_2, {=}O)$$

$$\xrightarrow{H_2O} O_2N{-}CH_2{-}CH_2{-}CH_2{-}CH_2{-}CH_2{-}COOH \xrightarrow{[H]} (CH_2)_5\text{[}C{=}O / N{-}H\text{] (caprolactam)}$$

Among some more recent developments is a one-step synthesis of caprolactam from cyclohexanol[59]:

$$2\ \text{cyclohexyl}{-}OH \xrightarrow[Ca(SCN)_2,\ 90\text{–}140\,^{\circ}C]{LiCl} (CH_2)_5\text{[}C{=}O / N{-}H\text{]} + \text{cyclohexanone}{=}O$$

There are other processes for caprolactam synthesis, however, a thorough discussion of this subject is beyond the scope of this book.

The mechanism of the reaction of ring-opening polymerization of caprolactam is discussed in Chapter 4. Several important side reactions accompany this polymerization. One is formation of cyclic oligomers.[58] The cyclic oligomers, soluble in water and alcohol mixtures, range in size from cyclic dimers to cyclic nonamers.[60–62] Formation of these compounds may be governed by equilibrium.[63] The polyamide will also thermally oxidize upon prolonged exposure to heat and air. Another important side reaction is decarboxylation that occurs at high temperatures. This is a result of interaction of a carboxyl group with a molecule of caprolactam or with an amide group[58]:

$$\sim C({=}O){-}N(H){-}(CH_2)_5{-}COOH$$

$$\sim C({=}O){-}OH + (CH_2)_5\text{[}C{=}O / N{-}H\text{]} \longrightarrow \begin{cases} \sim C{=}N{-}(CH_2)_5{-}C({=}O){-}O \text{ (ring)} \\ \sim C({=}O){-}O{-}C({=}N)(CH_2)_5 \text{ (ring)} \end{cases} \longrightarrow \sim C{=}N(CH_2)_5 \text{ (ring)} + CO_2$$

Polymerizations of caprolactam should be conducted in inert atmospheres to prevent oxidative decompositions. These can result in formations of carbon monoxide, carbon dioxide, acetaldehyde, formaldehyde, and methanol. Caprolactam can even oxidize in air at temperatures between 70–100 °C,[64] according to the following scheme:

$$\text{(CH}_2)_5\text{[C=O, N-H]} \xrightarrow{[O]} (\text{CH}_2)_4\text{[H-C-O-O-H, N-H, O=C]} \xrightarrow{-\text{H}_2\text{O}} (\text{CH}_2)_4\text{[C=O, N-H, C=O]} \longrightarrow (\text{CH}_2)_4\text{[COOH, COOH]}$$

Much of nylon 6 is used in producing fibers. Polycaprolactam prepared by water-catalyzed polymerizations is best suited for this purpose. It can also be used in molding, though anionically polymerized caprolactam can be used as well.[65] The polymerizations are carried out both in batch and in continuous processes. Often, tubular flow reactors are employed.

A typical polymerization reaction is carried out as follows. Caprolactam, water (5–10% by weight of monomer), and acetic acid (about 0.1%) are fed into the reactor under nitrogen atmosphere. The reaction mixture is heated to about 250 °C for 12 hours. Internal pressure is maintained at 15 atmospheres by venting off steam. The product of polymerization is extruded as a ribbon, quenched, and chopped into chips. It consists of about 90% polymer and about 10% low molecular weight compounds and monomer. The polymer is purified by either water leaching at 85 °C or by vacuum extraction of the undesirable byproducts at 180 °C.

Castings of nylon 6 are commonly formed *in situ* in molds. Here, the preparation of the polymer by anionic mechanism is preferred. The catalyst systems consist of 0.1–1.0 mole percent of acetyl caprolactam and 0.15–0.5 mole percent sodium caprolactam. The reaction temperature is kept between 140–180 °C. An exotherm can raise it as much as 50 °C as the polymerization proceeds.

Nylon 7 and *nylon 9* are part of a process developed in Russia to form polyamides for use in fibers. The process starts with telomerization of ethylene.[66] A free-radical polymerization of ethylene is conducted in the presence of chlorine compounds that act as chain-transferring agents. The reaction is carried out at 120–200 °C temperature and 400–600 atmospheres pressure. The preferred chain-transferring agents for this reaction are CCl_4 and $COCl_2$[66]:

$$CH_2{=}CH_2 \;+\; R\bullet \longrightarrow R{-}CH_2{-}CH_2\,\bullet \xrightarrow{CCl_4} R{-}CH_2{-}CH_2Cl \;+\; CCl_3\bullet$$

$$CCl_3\bullet \;+\; x\;CH_2{=}CH_2 \longrightarrow Cl_3C{-}({-}CH_2{-}CH_2{-})_{x-1}CH_2{-}CH_2\,\bullet$$

$$\longrightarrow Cl_3C{-}({-}CH_2{-}CH_2{-})_x{-}Cl$$

The resultant chloroalkanes are then hydrolyzed with sulfuric acid:

$$Cl_3C{-}({-}CH_2{-}CH_2{-})_x{-}Cl \xrightarrow[H_2O]{H_2SO_4} O{=}C(HO){-}({-}CH_2{-}CH_2{-})_x{-}Cl$$

After hydrolyses, the products are treated with ammonia:

$$O{=}C(HO){-}({-}CH_2{-}CH_2{-})_x{-}Cl \xrightarrow{NH_3} O{=}C(O^{\ominus}){-}({-}CH_2{-}CH_2{-})_x{-}\overset{\oplus}{N}H_3$$

The amino acid is condensed to a lactam and subsequently polymerized. Table 6.3 shows the composition of the telomers in the above free-radical polymerization.[66] As seen from this table, the

Table 6.3. Composition of Telomers[a,66]

x	Fraction %
1	5
2	44
3	28
4	15
>4	8

[a]From Nesmeyanov and Freundlina, by permission of Tetrahedron Letters, Elsevier Science, Ltd.

economics of producing nylon 7 by this process is not as favorable as one may wish. An advantage, however, to producing nylon 7 is that the polymer contains little monomer and can be spun without washing or extraction, as is required with nylon 6.

An early synthesis of *nylon 8* used cyclooctatetraene, which was formed from acetylene and then converted to nylon 8 as follows:

$$4\ HC{\equiv}HC \longrightarrow \text{cyclooctatetraene} \xrightarrow{[H]} \text{cyclooctene} \xrightarrow{[O]}$$

$$\text{cyclooctene oxide} \xrightarrow{\text{weak base}} \text{cyclooctanone} \longrightarrow \text{cyclooctanone oxime (=NOH)}$$

$$\xrightarrow[H_2SO_4]{\text{Beckmann rearrangement}} \underbrace{(CH_2)_7\!\!<\!\!\begin{array}{l}C{=}O\\ |\\ N{-}H\end{array}}_{\text{ring}} \longrightarrow \text{etc.}$$

The acetylene was later replaced by butadiene for economic reasons. Butadiene is cyclodimerized, then hydrogenated to cyclooctane, and the oxime is prepared directly from cyclooctane by photonitrosation:

$$\text{cyclooctane} \xrightarrow[h\nu]{NOCl} \text{cyclooctanone oxime}{=}NOH \cdot HCl \xrightarrow[H_2SO_4]{\text{Beckmann rearrangement}} (CH_2)_7\!\!<\!\!\begin{array}{l}C{=}O\\ |\\ N{-}H\end{array}$$

The resultant capryllactam is used predominantly in forming copolymers, like nylon 6/8/12.

Nylon 9 or poly(ω-pelargonamide) is produced in Russia together with nylon 7, poly(aminoenanthic acid) as described above. In the U.S., Kohlhase, Pryde, and Cowan[67] developed a route to nylon 9 via ozonolysis of unsaturated fatty acids like those that can be obtained from soybean oil. The glycerol fatty acid esters of oleic, linoleic, and linolenic acids are transesterified with methanol to form methyl esters. The esters are then cleaved via ozonolysis to yield methylazelaldehyde and byproducts that are removed. The purified product is reacted with ammonia and then reduced over Raney nickel to yield a methyl ester of the amino acid:

$$H_2N{-}({-}CH_2{-})_8{-}\overset{\overset{\displaystyle O}{\|}}{C}{-}OCH_3$$

After hydrolysis and purification, the free amino acid is converted to high molecular weight polymers.[68]

To date, nylon 9 has not been commercialized in the U.S., though the polymer has a high melting point of 209 °C and is more flexible than nylon 6. It is also lower in water absorption.

Nylon 11 was originally synthesized in France. The monomer, ω-aminoundecanoic acid, is obtained from methyl ricinoleate that comes from castor oil. Methyl ricinoleate is first cleaved thermally to heptaldehyde and methyl undecylenate:

$$H_3C{-}({-}CH_2{-})_5{-}\overset{\overset{\displaystyle OH}{|}}{C}H{-}CH_2{-}\overset{\overset{\displaystyle H}{|}}{C}{=}\overset{\overset{\displaystyle H}{|}}{C}{-}({-}CH_2{-})_7{-}\overset{\overset{\displaystyle O}{\|}}{C}{-}OCH_3 \xrightarrow{\Delta}$$

$$C_6H_{13}\overset{\overset{\displaystyle O}{\|}}{C}{-}H + H_2C{=}CH{-}({-}CH_2{-})_8{-}\overset{\overset{\displaystyle O}{\|}}{C}{-}OCH_3$$

The ester is then hydrolyzed and converted to an amino acid:

$$CH_2{=}CH{-}({-}CH_2{-})_8{-}\overset{O}{\overset{\|}{C}}{-}OCH_3 \xrightarrow{HOH} CH_2{=}CH{-}({-}CH_2{-})_8{-}\overset{O}{\overset{\|}{C}}{-}OH \xrightarrow[\text{peroxide}]{HBr}$$

$$\xrightarrow[\text{addition}]{\text{counter Markovnikov}} Br{-}CH_2{-}({-}CH_2{-})_9{-}\overset{O}{\overset{\|}{C}}{-}OH \xrightarrow{NH_3} H_3\overset{\oplus}{N}{-}({-}CH_2{-})_{10}{-}\overset{O}{\overset{\|}{C}}{-}O^{\ominus}$$

Polycondensation is conducted in the melt under nitrogen at 215 °C for several hours. The polymer is transparent in its natural form. It has high impact resistance, low moisture absorption, and good low-temperature flexibility. It is now also manufactured in the U.S.

Nylon 12 is produced in the U.S., Japan, and Europe with the original development coming from Europe. All current manufacturing processes of this polyamide, formed by ring-opening polymerization of lauryllactam, are based on cyclododecatriene. This ring compound can be obtained by trimerization of butadiene using Ziegler–Natta-type catalysts. One patent reports using polyalkyltitanate and dialkylaluminum monochloride[69]:

$$3\ CH_2{=}CH{-}CH{=}CH_2 \longrightarrow \text{cyclododecatriene}$$

The cyclododecatriene is then converted to lauryllactam by different processes. One of them consists of hydrogenation of the cyclic triene, followed by oxidation to a cyclic ketone, conversion to an oxime, and rearrangement by the Beckmann reaction to the lactam:

$$\text{cyclododecatriene} \xrightarrow{[H]} (CH_2)_{11}\ CH_2 \xrightarrow{[O]} (CH_2)_{11}\ C{=}O \longrightarrow \text{etc.}$$

Another process utilizes photonitrosation:

$$\text{cyclododecatriene} \xrightarrow{[H]} (CH_2)_{11}\ CH_2 \xrightarrow[NOCl]{h\nu} (CH_2)_{11}\ C{=}N{-}OH \xrightarrow[H_2SO_4]{\text{Beckmann rearrangement}} (CH_2)_{11}\begin{matrix} C{=}O \\ | \\ N{-}H \end{matrix}$$

There are still other processes, but they lack industrial importance. Nylon 12, like nylon 11, exhibits low moisture absorbency, good dimensional stability, and good flexibility at low temperatures.

Preparations of other nylons were reported from time to time in the literature. For one reason or another, however, they have not developed into industrially important materials. Thus, for

Table 6.4. Approximate Melting Points of Polyamides[a]

Nylon	Repeat unit	Melting point (°C)
3	$-(-CH_2-)_2-CO-NH-$	320–330
4	$-(-CH_2-)_3-CO-NH-$	260–265
5	$-(-CH_2-)_4-CO-NH-$	260
6	$-(-CH_2-)_5-CO-NH-$	215–220
7	$-(-CH_2-)_6-CO-NH-$	225–230
8	$-(-CH_2-)_7-CO-NH-$	195
9	$-(-CH_2-)_8-CO-NH-$	197–200
10	$-(-CH_2-)_9-CO-NH-$	173
11	$-(-CH_2-)_{10}-CO-NH-$	185–187
12	$-(-CH_2-)_{11}-CO-NH-$	180
13	$-(-CH_2-)_{12}-CO-NH-$	173

[a]From various literature sources.

instance, for some time now it has been known that *nylon 13* can be prepared from erucic acid that is found in crambe and rapeseed oils. The polymer is supposed to be quite similar to nylon 11, though lower melting.

The melting points of the nylons describe above are summarized in Table 6.4.

Nylon 6,6 is a condensation product of hexamethylenediamine and adipic acid. This polyamide was originally synthesized in 1935 and first produced commercially in 1938. It is still one of the major commercial nylons produced today. Because high molecular weight is required for such polymers to possess good physical properties, it is necessary to follow exact stoichiometry of the reactants in the condensation. To achieve that, the practice is to initially form a "*nylon salt*," prior to the polymerization. To do this, equimolar quantities of adipic acid and hexamethylenediamine are combined in aqueous environment to form solutions of the salt. The end point is controlled electrochemically. An alternate procedure is to combine the diacid with the diamine in boiling methanol. A 1:1 adduct precipitates out, is filtered off, and dissolved in water.

$$n\ H_2N\text{—}(\text{—}CH_2\text{—})_6\text{—}NH_2 + n\ HOOC\text{—}(\text{—}CH_2\text{—})_4\text{—}HOOC \longrightarrow n \left[\begin{array}{c} {}^{-}OOC(CH_2)_4COO^{-} \\ {}^{+}H_3N(CH_2)_6NH_3{}^{+} \end{array} \right]$$

A 60–75% solution of the salt in water is then fed into a reaction kettle. In a typical batch process some acetic acid may also be added if it is desired to limit molecular weight (10,000–15,000). The temperature in the reaction kettle is raised to 220 °C and, due to water and steam in the reactor, internal pressure of about 20 atmospheres develops. After one to two hours, the temperature is raised to 270–280 °C. Some steam is bled off to maintain internal pressure at 20 atmospheres. The temperature is maintained and the bleeding out of the steam is continued for two hours. During that period, the internal pressure is gradually reduced to atmospheric. In some processes, vacuum is applied at this point to the reaction kettle if a high molecular weight product is desired. When the reaction is complete, the molten polymer is ejected from the kettle by applying pressure with nitrogen or carbon dioxide.

In one continuous process, the desired conditions are maintained while the reaction mixture moves through various zones of the reactor. Tubular reactors are also often employed in continuous polymerizations.

Nylon 6,10 is prepared by the same procedure as nylon 6,6 from a salt of hexamethylenediamine and sebacic acid, while *nylon 6,9* is prepared from a salt of hexamethylene diamine and azelaic acid.

The melting points of various nylons that are formed from diamines and dicarboxylic acids are presented in Table 6.5.

One commercial polyamide is prepared by condensation of a cycloaliphatic diamine with a twelve-carbon dicarboxylic acid. The diamine, bis(*p*-aminocyclohexyl)methane, is prepared from aniline:

$$2\ H_2N\text{—}C_6H_5 + H_2C{=}O \longrightarrow H_2N\text{—}C_6H_4\text{—}CH_2\text{—}C_6H_4\text{—}NH_2$$

$$\xrightarrow[\text{Ru catalyst}]{[H]} H_2N\text{—}C_6H_{10}\text{—}CH_2\text{—}C_6H_{10}\text{—}NH_2$$

The diamine is then condensed with dodecanedioic acid, which is obtained from cyclododecatriene. The structure of this polyamide is as follows:

$$\text{—}[\text{—}\underset{}{N(H)}\text{—}C_6H_{10}\text{—}CH_2\text{—}C_6H_{10}\text{—}N(H)\text{—}\overset{O}{\overset{\|}{C}}\text{—}(\text{—}CH_2\text{—})_{10}\text{—}\overset{O}{\overset{\|}{C}}\text{—}]_n\text{—}$$

The polymer has a T_m value of 280–290 °C and a T_g value of 120 °C. It exhibits lower moisture pick-up than nylon 6 and nylon 6,6, increased hardness and tensile strength, though lower impact strength. The bulk of this polymer is used in fiber production. The fibers, with a trade name of "Quiana," are claimed to exhibit high luster and a silk-like feel.

Table 6.5. Melting Points of Nylons[a]

Nylon	Repeat unit	Melting point (°C) salt	Melting point (°C) polymer
4,6	$-NH-(CH_2)_4-NH-CO-(CH_2)_4-CO-$	204	278
4,7	$-NH-(CH_2)_4-NH-CO-(CH_2)_5-CO-$	138	233
4,9	$-NH-(CH_2)_4-NH-CO-(CH_2)_7-CO-$	175	223
4,10	$-NH-(CH_2)_4-NH-CO-(CH_2)_8-CO-$		236
5,10	$-NH-(CH_2)_5-NH-CO-(CH_2)_8-CO-$	129	195
6,6	$-NH-(CH_2)_6-NH-CO-(CH_2)_4-CO-$	183	250
6,9	$-NH-(CH_2)_6-NH-CO-(CH_2)_7-CO-$		205
6,10	$-NH-(CH_2)_6-NH-CO-(CH_2)_8-CO-$	170	209
6,12	$-NH-(CH_2)_6-NH-CO-(CH_2)_{10}-CO-$		212
8,6	$-NH-(CH_2)_8-NH-CO-(CH_2)_4-CO-$	153	235
8,10	$-NH-(CH_2)_8-NH-CO-(CH_2)_8-CO-$	164	197
9,6	$-NH-(CH_2)_9-NH-CO-(CH_2)_4-CO-$	125	204–205
9,10	$-NH-(CH_2)_9-NH-CO-(CH_2)_8-CO-$	159	174–176
10,6	$-NH-(CH_2)_{10}-NH-CO-(CH_2)_4-CO-$	142	230
10,10	$-NH-(CH_2)_{10}-NH-CO-(CH_2)_8-CO-$	178	194
11,10	$-NH-(CH_2)_{11}-NH-CO-(CH_2)_8-CO-$	153	168–169
12,6	$-NH-(CH_2)_{12}-NH-CO-(CH_2)_4-CO-$	144	208–210
12,10	$-NH-(CH_2)_{12}-NH-CO-(CH_2)_8-CO-$	157	171–173
13,13	$-NH-(CH_2)_{13}-NH-CO-(CH_2)_{11}-CO-$		174

[a]From Ref. 70 and other literature sources.

A number of *copolyamides* are manufactured commercially to suit various needs. One of them is a polyamide formed by condensation of trimethylhexamethylenediamine with terephthalic acid. The diamine is a mixture of 2,4,4 and 2,2,4 isomers:

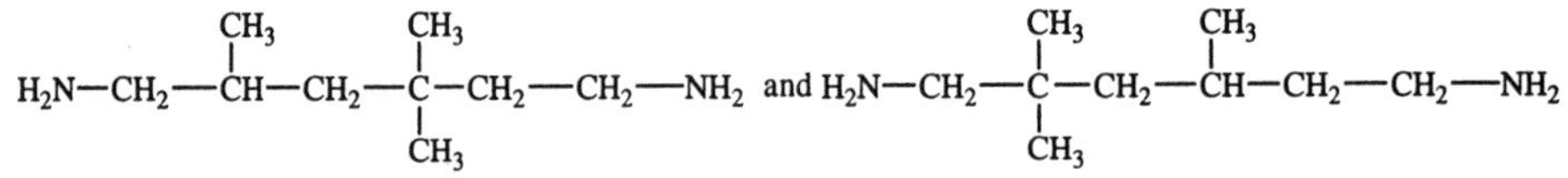

$$H_2N-CH_2-CH(CH_3)-CH_2-C(CH_3)_2-CH_2-CH_2-NH_2 \text{ and } H_2N-CH_2-C(CH_3)_2-CH_2-CH(CH_3)-CH_2-CH_2-NH_2$$

The mixture of the two isomers is synthesized from isophorone according to the following scheme:

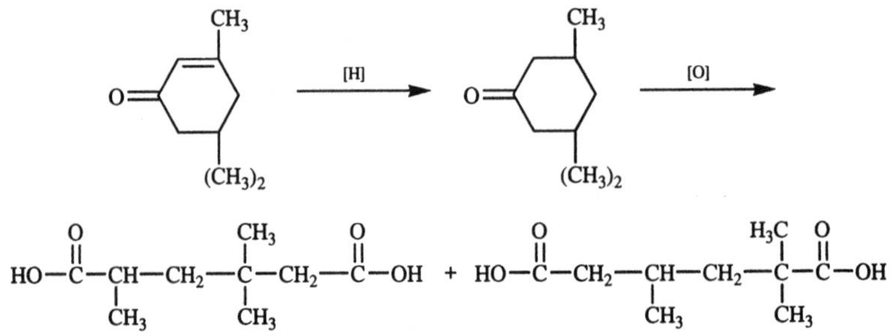

$$\text{O=}\underset{(CH_3)_2}{\overset{CH_3}{\text{(cyclohexenone ring)}}} \xrightarrow{[H]} \text{O=}\underset{(CH_3)_2}{\overset{CH_3}{\text{(cyclohexanone ring)}}} \xrightarrow{[O]}$$

$$HO-C(=O)-CH(CH_3)-CH_2-C(CH_3)_2-CH_2-C(=O)-OH + HO-C(=O)-CH_2-CH(CH_3)-CH_2-C(CH_3)_2-C(=O)-OH$$

The mixture of the isomers of trimethyladipic acids is treated with ammonia, converted to amides, dehydrated to nitriles, and reduced to amines:

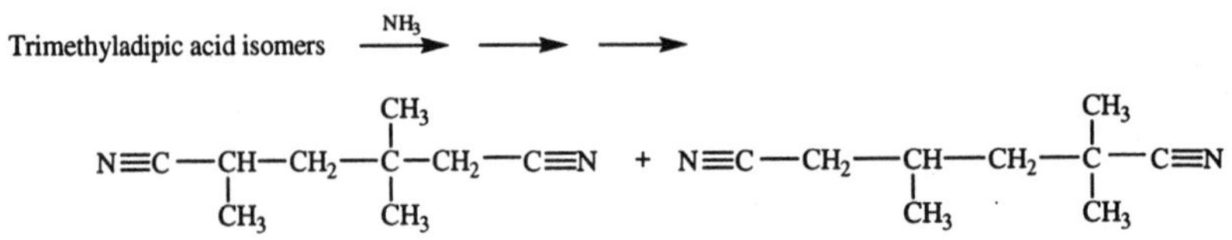

$$\text{Trimethyladipic acid isomers} \xrightarrow{NH_3} \longrightarrow \longrightarrow$$

$$N\equiv C-CH(CH_3)-CH_2-C(CH_3)_2-CH_2-C\equiv N + N\equiv C-CH_2-CH(CH_3)-CH_2-C(CH_3)_2-C\equiv N$$

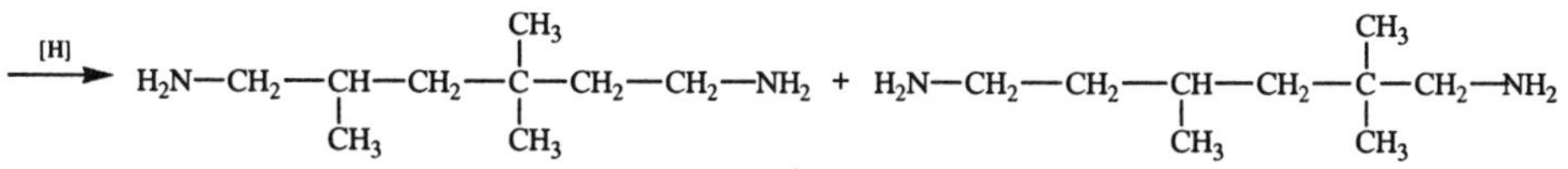

$$\xrightarrow{[H]} H_2N{-}CH_2{-}\underset{CH_3}{CH}{-}CH_2{-}\underset{CH_3}{\overset{CH_3}{C}}{-}CH_2{-}CH_2{-}NH_2 + H_2N{-}CH_2{-}CH_2{-}\underset{CH_3}{CH}{-}CH_2{-}\underset{CH_3}{\overset{CH_3}{C}}{-}CH_2{-}NH_2$$

This polyamide is prepared somewhat differently. Salts of the diamine isomers with terephthalic acid are only partially polycondensed and the reaction is completed during extrusion,[71] because the melt viscosity of the polymer is very high. The product is amorphous and exhibits greater light trasmittancy. It melts at 200 °C and is sold under the trade name of Trogamid T.

Many commercial nylon copolymers are also formed by melt mixing different nylons. Amide interchange reactions occur at melt conditions. At first block copolymers form, but prolonged heating and stirring results in formation of random copolymers. Nylon copolymers are also prepared directly from mixed monomers.

Nylon polymers generally exhibit high impact strength, toughness, good flexibility, and abrasion resistance. The principal structural differences between many nylons is in the length of the aliphatic segments between the amide linkages. As a result, the differences in properties depend mainly upon the amount of hydrogen bonding that is possible between the functional groups and the amount of crystallinity. Also, due to high cohesive energy, nylons are soluble in only a few solvents. The melt viscosity of these materials, however, is generally low.

In any one series of melting points of polyamides, polymers that contain even numbers of methylene groups between amide linkages fall on a higher curve than those that contain odd ones.[72] This is due to the crystalline arrangement of the polymeric chains.[72,73] A zigzag planar configuration of polymers with even number of methylene linkages allows only 50% of the functional groups to form hydrogen bonds. This same configuration, however, allows polymers with odd numbers of methylene linkages to form 100% hydrogen bonding.[72,73] Polyamides, like nylon 6,6 or nylon 6,10, arrange themselves in pleated sheets during crystallization and hydrogen bonds form between the N–H group of one molecule and the C=O moieties from a neighboring one.

6.3.2. Fatty Polyamides

The fatty polyamides are produced by reacting di- and polyfunctional amines with polybasic acids that result from condensations of unsaturated vegetable-oil acids. The most commonly used amines are ethylene diamine and diethylene triamine. The dicarboxylic acids are synthesized by heating mixtures of unsaturated vegetable fatty acids. The starting materials may come from linseed, soybean, or tung oil. The fatty acids are heated for several hours at 300 °C. If a catalyst is used, the heating is done at a lower temperature. After condensation, the volatile fractions are removed by vacuum. The residues, called *dimer acids*, are then condensed with di- or polyamines. A formation of one such dimer acid from linoleic acid can be illustrated as follows:

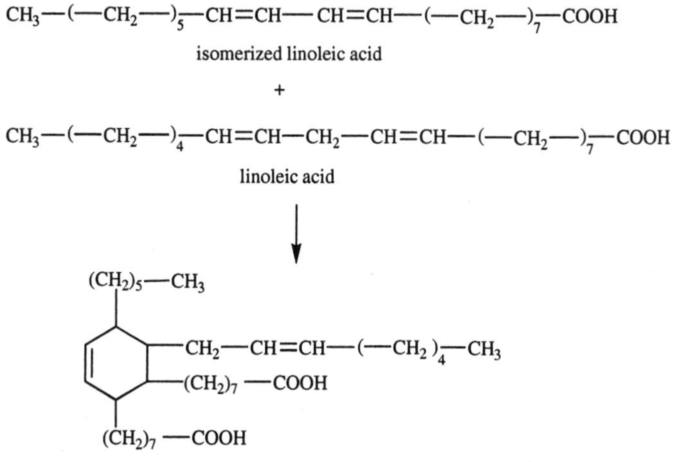

$$CH_3{-}({-}CH_2{-})_5{-}CH{=}CH{-}CH{=}CH{-}({-}CH_2{-})_7{-}COOH$$

isomerized linoleic acid

+

$$CH_3{-}({-}CH_2{-})_4{-}CH{=}CH{-}CH_2{-}CH{=}CH{-}({-}CH_2{-})_7{-}COOH$$

linoleic acid

$$\downarrow$$

$$(CH_2)_5{-}CH_3$$

$$CH_2{-}CH{=}CH{-}({-}CH_2)_4{-}CH_3$$

$$(CH_2)_7{-}COOH$$

$$(CH_2)_7{-}COOH$$

The polyamide from the above dimer acid condensed with a diamine, like ethylene diamine, can be illustrated as follows:

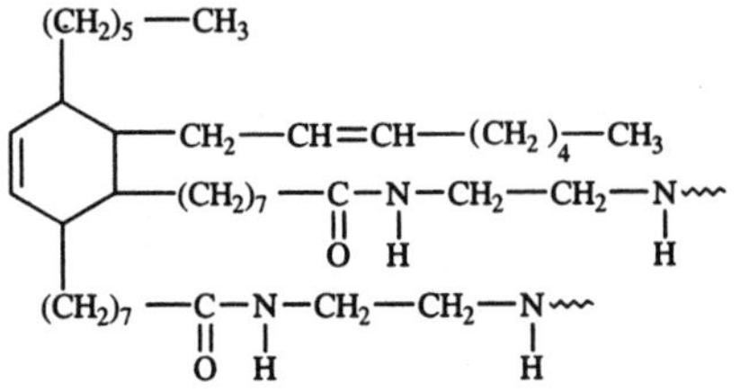

Two types of fatty polyamides are available commercially, solid and liquid. The solid polymers are mostly linear condensation products of diacids and diamines that range in molecular weights from 2000–15000. The liquid ones are highly branched, low molecular weights materials produced by condensations of the dimer acids with triamines and even higher number polyamines.

6.3.3. Special Reactions for Formation of Polyamides

There are occasional reports in the literature on the use of special reactions to form polyamides. One is a synthesis via enamines. In this case diisocyanates are condensed with cyclopentanone enamines of morpholine or piperidine[77]:

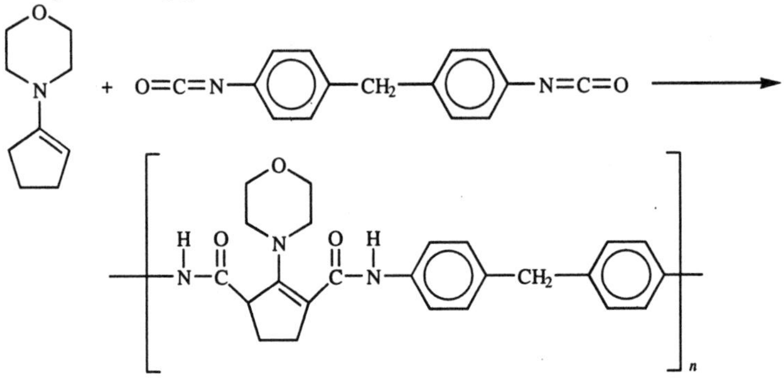

The molecular weights of the polymers decrease when the ring sizes of the ketone components increase. Excess diisocyanate yields branched and crosslinked polymers. The enamine units in the polymers can be hydrolyzed with formic acid to the corresponding ketones.[77]

Recently, a similar reaction was reported from aromatic isocyanates and imidazoles. 1,4-tetramethylene-*N,N′*-diimidazoles were reacted with aromatic polyisocyanates to form thermoset polyamides[45]:

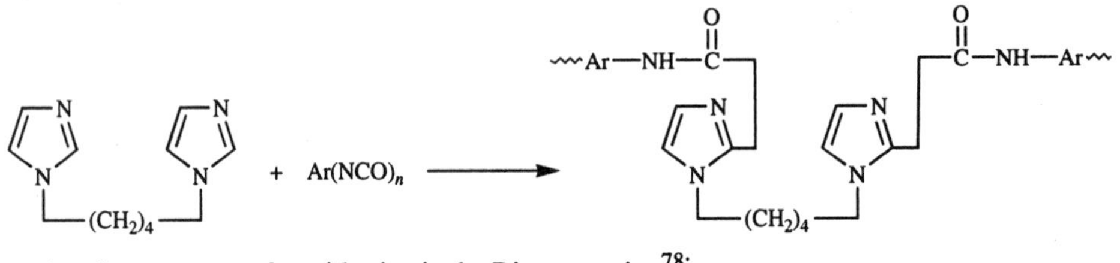

$$+ \; Ar(NCO)_n \longrightarrow$$

Another route to polyamides is via the Ritter reaction[78]:

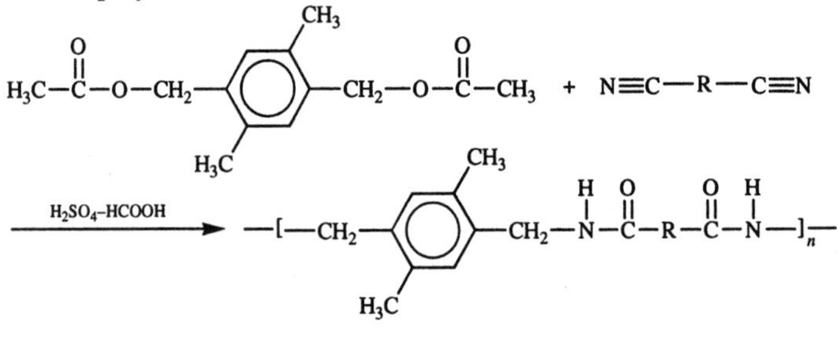

$$H_3C{-}\overset{O}{\overset{\|}{C}}{-}O{-}CH_2{-}Ar(CH_3)_2{-}CH_2{-}O{-}\overset{O}{\overset{\|}{C}}{-}CH_3 \; + \; N{\equiv}C{-}R{-}C{\equiv}N \xrightarrow{H_2SO_4-HCOOH} {-}[{-}CH_2{-}Ar(CH_3)_2{-}CH_2{-}\overset{H}{N}{-}\overset{O}{\overset{\|}{C}}{-}R{-}\overset{O}{\overset{\|}{C}}{-}\overset{H}{N}{-}]_n{-}$$

When aromatic dinitriles are used, high melting polymers with good thermal stability form.[78]

Direct polycondensation of various dicarboxylic acids with diamines is possible[81] under mild conditions by using a catalytic system of an enol phosphite in the presence of imidazole. One such enol phosphite is diethyl,1-methyl-3-oxo-1-butenyl phosphite:

$$(EtO)_2{-}P{-}O{-}\overset{\overset{\displaystyle CH_3}{|}}{C}{=}CH{-}\overset{\overset{\displaystyle O}{\|}}{C}{-}CH_3.$$

Polymers with inherent viscosities of 1–0.25 form. Among the organic bases, imidazole is most effective.[84] The reaction is applicable to both aliphatic and aromatic dicarboxylic acids and diamines.

6.3.4. Aromatic Polyamides

In this section we discuss not only wholly aromatic polyamides, but also some mixed polyamides, prepared from aromatic diacids and aliphatic diamines, or vice versa. One such material was already described in Section 6.3.2. Another one, called Nylon 6T, is formed by interfacial polymerization of terephthaloyl chloride and hexamethylenediamine:

$$-[-\overset{\overset{\displaystyle O}{\|}}{C}-C_6H_4-\overset{\overset{\displaystyle O}{\|}}{C}-\underset{\underset{\displaystyle H}{|}}{N}-CH_2-CH_2-CH_2-CH_2-CH_2-CH_2-\underset{\underset{\displaystyle H}{|}}{N}-]_n-$$

The polymer has good heat stability and the strength is unaffected by heating up to 185 °C for five hours. The polyamide melts at 370 °C. When hexamethylenediamine is replaced with tetramethylenediamine, the melting point rises to 430 °C. The condensation product from isophthalic acid and tetramethylenediamine melts at 250 °C.

Fully aromatic polyamides form from reactions of aromatic diacid chlorides and aromatic diamines.[79,80] An example is formation of poly(*m*-phenylenediamine isophthalamide):

$$-[-\overset{\overset{\displaystyle H}{|}}{N}-C_6H_4-\overset{\overset{\displaystyle H}{|}}{N}-\overset{\overset{\displaystyle O}{\|}}{C}-C_6H_4-\overset{\overset{\displaystyle O}{\|}}{C}-]_n-$$

The polymer can be prepared in dimethylacetamide from isophthaloyl chloride and *m*-phenylenediamine in the presence of an acid scavenger at room temperature:

$$Cl-\overset{\overset{\displaystyle O}{\|}}{C}-C_6H_4-\overset{\overset{\displaystyle O}{\|}}{C}-Cl \; + \; H_2N-C_6H_4-NH_2 \longrightarrow \text{polymer}$$

Because the polymer is soluble in dimethylacetamide containing 5% LiCl, fibers can be spun directly from solution. This polyamide melts at 371 °C with degradation. It is fire resistant. The fibers are sold under a trade name of Nomex.

Very similar, wholly aromatic polyamides, with very regular structures, which reportedly result in better flexibility and higher temperature resistance, were reported.[74,75] Preparation of these *ordered copolyamides* can be illustrated as follows. *N,N-m*-phenylene–bis(*m*-aminobenzamide) is formed first and then reacted with isophthaloyl chloride by interfacial condensation techniques to yield a product that melts at about 410 °C:

$$H_2N-C_6H_4-\overset{\overset{\displaystyle O}{\|}}{C}-\overset{\overset{\displaystyle H}{|}}{N}-C_6H_4-\overset{\overset{\displaystyle H}{|}}{N}-\overset{\overset{\displaystyle O}{\|}}{C}-C_6H_4-NH_2 \; + \; Cl-\overset{\overset{\displaystyle O}{\|}}{C}-C_6H_4-\overset{\overset{\displaystyle O}{\|}}{C}-Cl \longrightarrow$$

$$\left[-\overset{\overset{\displaystyle H}{|}}{N}-C_6H_4-\overset{\overset{\displaystyle O}{\|}}{C}-\overset{\overset{\displaystyle H}{|}}{N}-C_6H_4-\overset{\overset{\displaystyle H}{|}}{N}-\overset{\overset{\displaystyle O}{\|}}{C}-C_6H_4-\overset{\overset{\displaystyle H}{|}}{N}-\overset{\overset{\displaystyle O}{\|}}{C}-C_6H_4-\overset{\overset{\displaystyle O}{\|}}{C}-\right]_n$$

The above condensation, carried out with terephthaloyl chloride, yields a polymer that melts at 450 °C. Preparations of many other wholly aromatic polyamides from aromatic diacid chlorides and aromatic diamines were reported in the literature.[79,80] In addition, several polymers are manufactured from both fully and partially substituted (*para*) structures. They carry the trade name of HT4.

As described in the previous section, direct polycondensation of various nylon salts is possible under mild conditions in the presence of polyphosphates and organic bases.[84] This reaction, useful in forming aromatic polyamides, takes place also in the presence of thionyl chloride[84]:

HO—C(=O)—C₆H₄—C(=O)—OH + H₂N—C₆H₄—NH₂ —(2 SOCl)→ [—C(=O)—C₆H₄—C(=O)—N(H)—C₆H₄—]ₙ

Metal salts, like lithium chloride, significantly enhance reactions of carboxylic acids with amines promoted by triphenyl phosphite.[82] This allows direct polycondensation of dicarboxylic acids with diamines and self-condensation of *p*-aminobenzoic acid.[82] The presence of a solvent markedly enhances the reaction with the best results being obtained in *N*-methylpyrrolidone. High molecular weight polyamides form. Mixed solvents, like pyridine and *N*-methylpyrrolidone, can be used to form polyisophthalamides.[82] This combination of solvents, however, yields only low molecular weight polyterephthalamides. On the other hand, when the reaction is carried out in the presence of polymeric matrices of poly(ethylene oxide) or poly(4-vinylpyridine), high molecular weight polyterephthalamides form[82]:

n HO—C(=O)—C₆H₄—C(=O)—OH + n H₂N—C₆H₄—NH₂ —[(C₆H₅—O)₃P | N-methyl-pyrrolidone | LiCl; poly(4-vinylpyridine)]→

—[—C(=O)—C₆H₄—C(=O)—N(H)—C₆H₄—N(H)—]ₙ— + HO—P—(—O—C₆H₅)₂ + C₆H₅—

Recently, the Heck reaction was extended to carbonylation of aromatic dibromides with aromatic diamines in the presence of carbon monoxide.[112] High molecular weight aromatic polyamides form with the help of palladium catalysis:

H₂N—C₆H₄—O—C₆H₄—NH₂ + Br—C₆H₄—Br + 2 CO —(Pd cat; - HBr)→

—[—NH—C₆H₄—O—C₆H₄—NHCO—C₆H₄—CO—]ₙ—

The polymerization reaction takes place in a homogeneous dimethylacetamide solution, with catalytic amounts of $PdCl_2(PPh_3)_2$ and an HBr scavenger. The carbonylation polycondensation proceeds rapidly at 115 °C and is almost complete in 1.5 hours.[112] This reaction was also used to prepare many aromatic–aliphatic polyamides from corresponding aliphatic diamines with aromatic dibromides.

Palladium is a relatively high-priced catalyst and it would be preferable if a lower-priced nickel catalyst could be used instead. All attempts, however, to form polymers by nickel-catalyzed carbonylation polycondensations of aromatic diamines with aromatic dibromides failed to yield high molecular weight materials.[112]

Trimethylsylil-substituted amines undergo a variety of reactions with electrophiles.[113] This reaction was extended recently to preparations of high molecular weight aromatic polyamides by low temperature solution polycondensation. N-trimethylsilylated aromatic diamines were condensed with aromatic diacid chlorides[114] at –10 °C in an amide solvent:

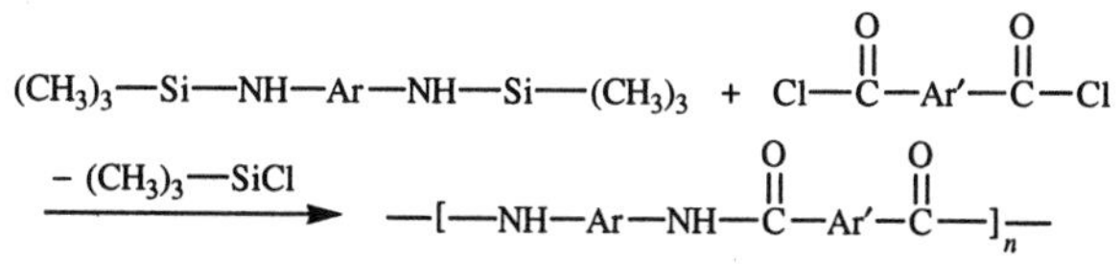

$$(CH_3)_3\text{—Si—NH—Ar—NH—Si—}(CH_3)_3 + \text{Cl—}\overset{O}{\overset{\|}{C}}\text{—Ar}'\text{—}\overset{O}{\overset{\|}{C}}\text{—Cl} \xrightarrow{-\,(CH_3)_3\text{—SiCl}} \text{—[—NH—Ar—NH—}\overset{O}{\overset{\|}{C}}\text{—Ar}'\text{—}\overset{O}{\overset{\|}{C}}\text{—]}_n\text{—}$$

Preparations of poly-*p*-phenyleneterephthalamide by polycondensations with N-silylated diamine proceed more rapidly than with the parent diamine.[113] In addition, the products have higher molecular weights than a similar commercial material made from the parent diamine and sold under the trade name of Kevlar.

Silylated diamines can also condense with diphenyl esters of aromatic dicarboxylic acids[114]:

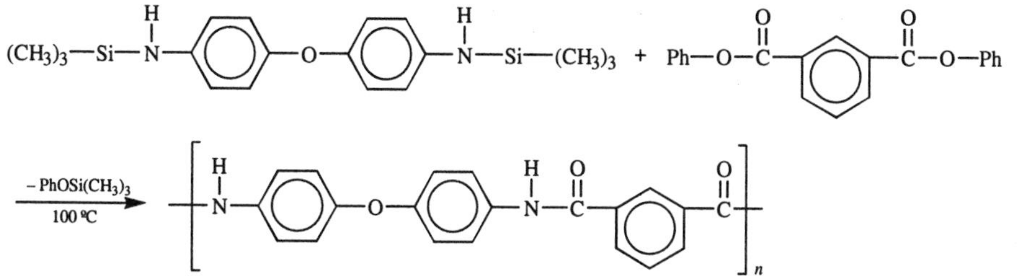

$$(CH_3)_3\text{—Si—NH—}C_6H_4\text{—O—}C_6H_4\text{—NH—Si—}(CH_3)_3 + \text{Ph—O—}\overset{O}{\overset{\|}{C}}\text{—}C_6H_4\text{—}\overset{O}{\overset{\|}{C}}\text{—O—Ph}$$

$$\xrightarrow[100\,^{\circ}C]{-\,PhOSi(CH_3)_3} \left[\text{—NH—}C_6H_4\text{—O—}C_6H_4\text{—NH—}\overset{O}{\overset{\|}{C}}\text{—}C_6H_4\text{—}\overset{O}{\overset{\|}{C}}\text{—}\right]_n$$

The product, the aromatic polyamide shown above, was reported to be of sufficiently high molecular weight to be a useful material.[113]

An example of a specialty aromatic polyamide is a fluorinated polyamide. It was prepared in an attempt to form a polymer with superior heat stability and resistance to hydrolytic attacks[76]:

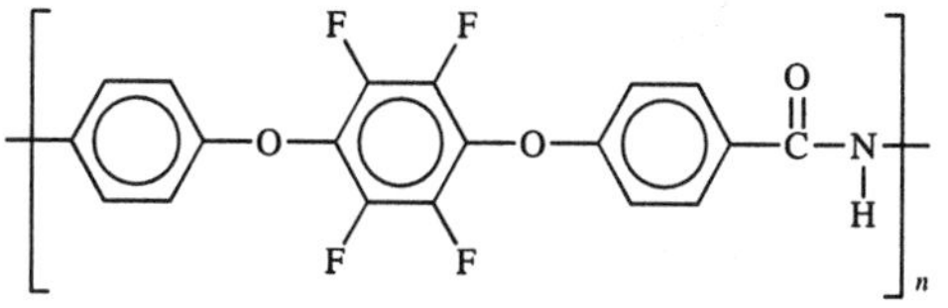

$$\left[\text{—}C_6H_4\text{—O—}C_6F_4\text{—O—}C_6H_4\text{—}\overset{O}{\overset{\|}{C}}\text{—NH—}\right]_n$$

6.4. Aromatic Polyamide-Imides and Aromatic Polyester-Imides

The aromatic polyamide-imides are related to the aromatic polyamides described in the previous section. Aliphatic materials of this type were reported originally in 1947. They were formed by reacting tricarboxylic acids with diamines[83]:

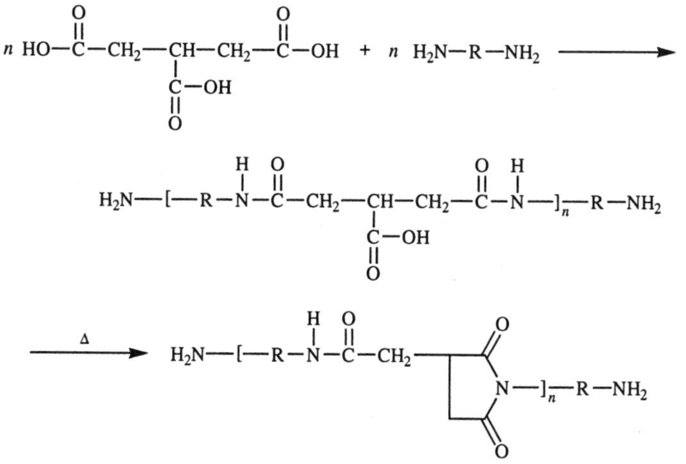

$$n\ \text{HO—}\overset{O}{\overset{\|}{C}}\text{—CH}_2\text{—CH(COOH)—CH}_2\text{—}\overset{O}{\overset{\|}{C}}\text{—OH} + n\ \text{H}_2\text{N—R—NH}_2 \longrightarrow$$

$$\text{H}_2\text{N—[—R—NH—}\overset{O}{\overset{\|}{C}}\text{—CH}_2\text{—CH(COOH)—CH}_2\text{—}\overset{O}{\overset{\|}{C}}\text{—NH—]}_n\text{—R—NH}_2$$

$$\xrightarrow{\Delta} \text{H}_2\text{N—[—R—NH—}\overset{O}{\overset{\|}{C}}\text{—CH}_2\text{—(succinimide ring)N—]}_n\text{—R—NH}_2$$

The aliphatic polyamide-imides prepared to date do not possess desirable properties. When aromatic diacids are employed, however, the products exhibit good heat stability and toughness. This led to the development of a number of useful materials.

Three general methods are employed to form aromatic polyamide-imides.[88] The first one consists of an initial reaction of a mole of a diacid chloride with two moles of a diamine. The product is then reacted with a dianhydride and after that condensed to an imide:

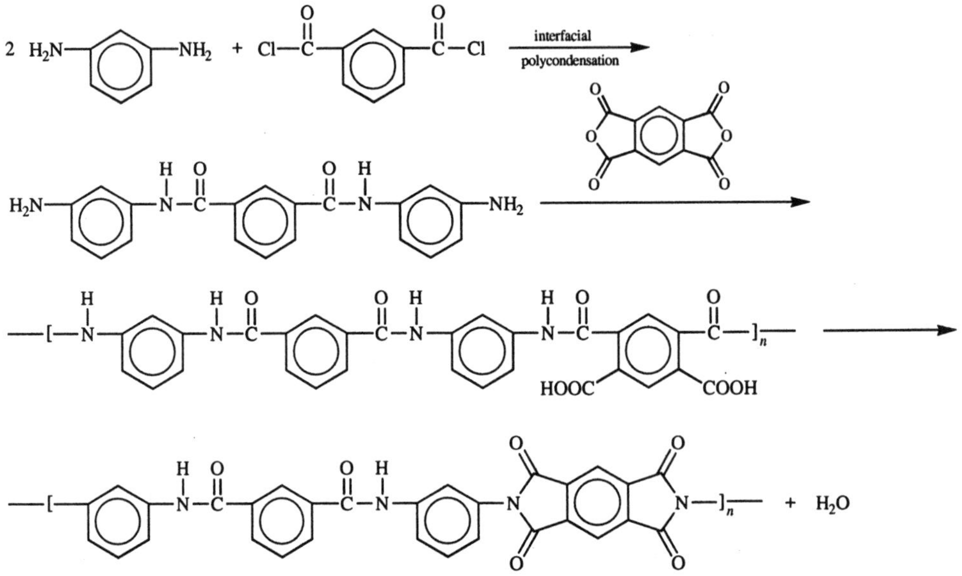

interfacial polycondensation

In the second method a dianhydride is prereacted with an excess of a diamine. The product is then reacted with a diacid chloride by interfacial polymerization technique.

In the third method an anhydride, such as trimellitic anhydride, is condensed with a diamine to form a preliminary condensate. An acetylated diamine can be used in this initial condensation[85]:

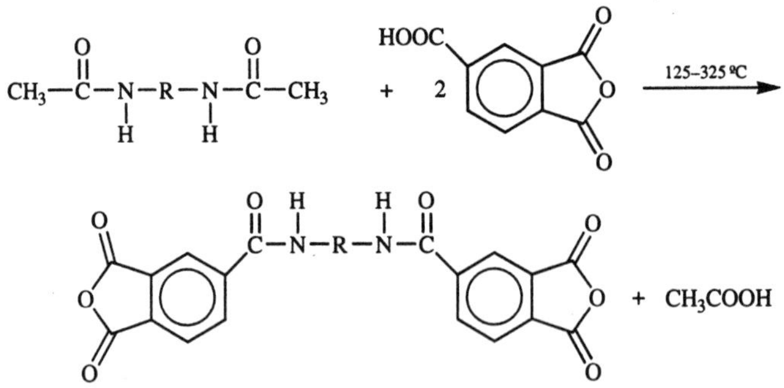

125–325 °C

This is followed by a reaction with the same or a different diamine:

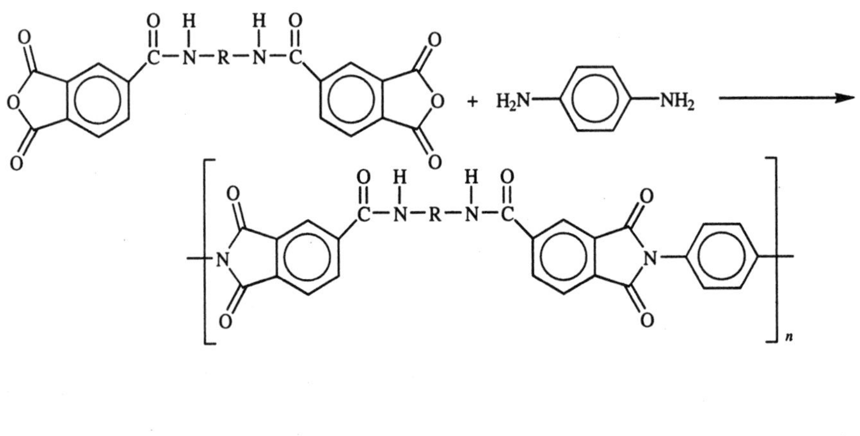

The same techniques are applied to preparations of polyester-imides. A diester can first be formed from trimellitic anhydride:

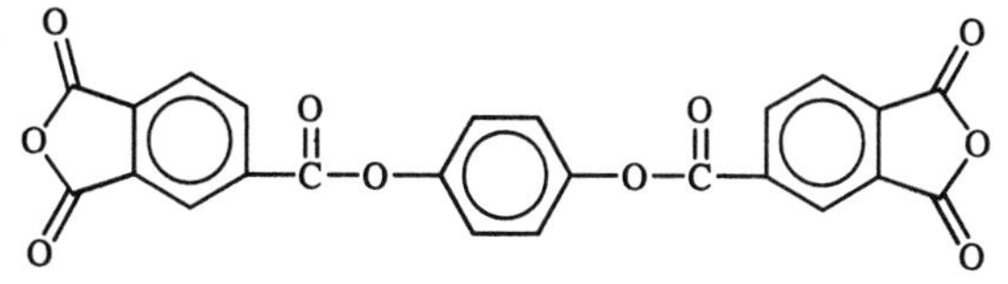

The product is then condensed with a diamine to form a polyester-imide:

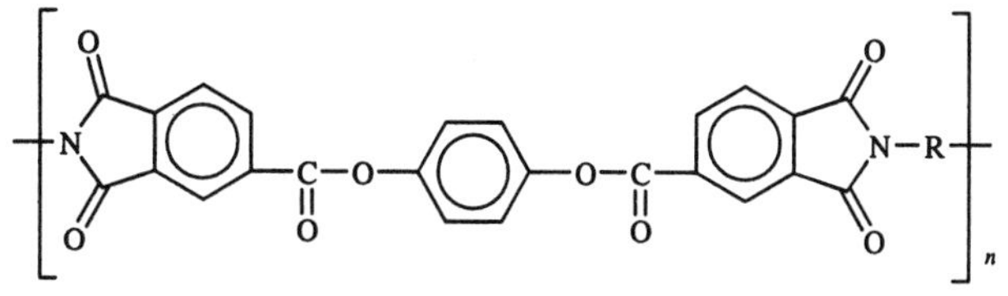

Several polyamide-imides are available commercially. Two are based on trimellitic anhydride and methylenedianiline. They are, however, prepared by two different processes. In the first, the polymer is formed from the anhydride and a diamine:

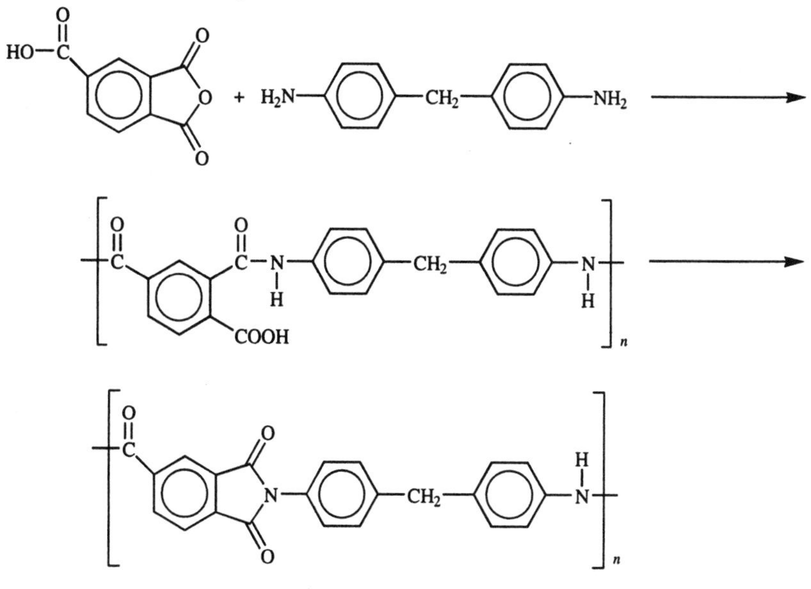

In the second, trimellitic anhydride is reacted with a diisocyanate:

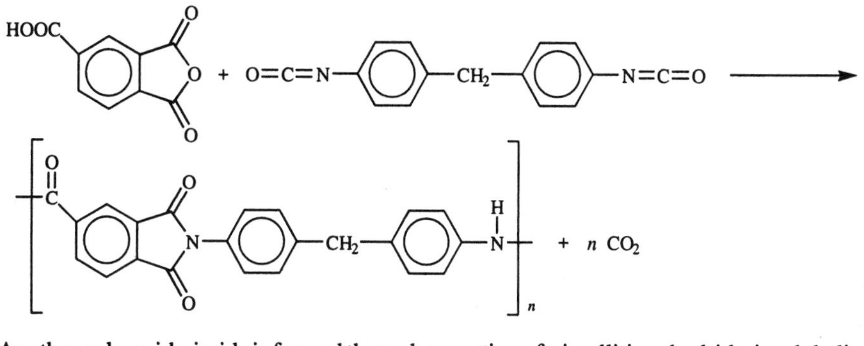

Another polyamide-imide is formed through a reaction of trimellitic anhydride, isophthalic acid, and diisocyanate. It has the following structure:

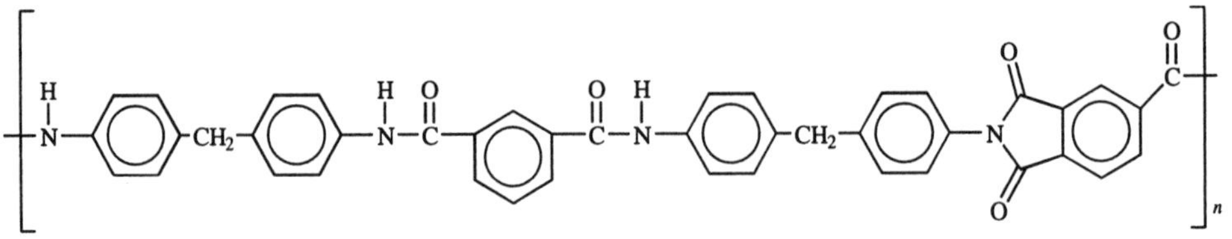

Most polyamide-imides are not as heat resistant as are the polyimides discussed in the next section. They are, however, easier to process. The polyester-imides might be considered as "upgraded" polyesters, though properties vary, depending upon chemical structure.[86]

6.5. Polyimides

It is interesting that formation of a linear aromatic polyimide was observed as early as 1908 when a polyimide was formed by heating 4-aminophthalic anhydride:

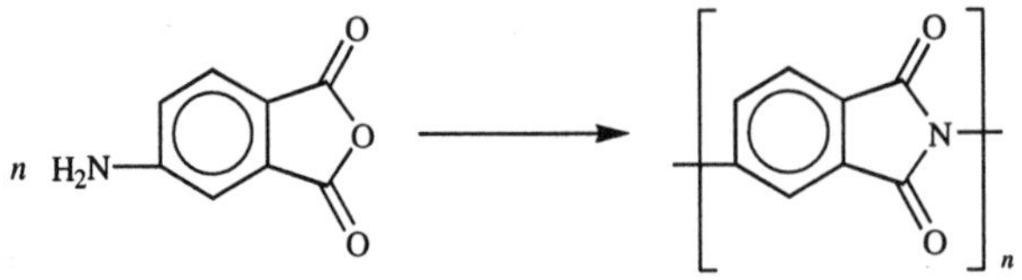

Formation of polymers, however, was at the time considered undesirable, so the material was not pursued.[87] It was learned since that an imide link is more thermally stable than an amide one and that polyimides can be very useful materials. Many polyimides have since been developed.[88] Aromatic structures in the polymeric backbone raise the melting temperatures and yield stiffer and tougher materials. Most sought-after polyimides are therefore products from aromatic tetraacids (or dianhydrides) and aromatic diamines.

Many commercial preparations of aromatic polyimides include a preliminary step of forming polyamic acids first[88]:

$$n\ \text{(dianhydride of R)} + H_2N{-}R'{-}NH_2 \longrightarrow -[-CO-R(COOH)_2-CO-NH-R'-NH-]_n-$$

This is followed by imidation, often after the polymer has been applied to a substrate as a coating or was cast as a film:

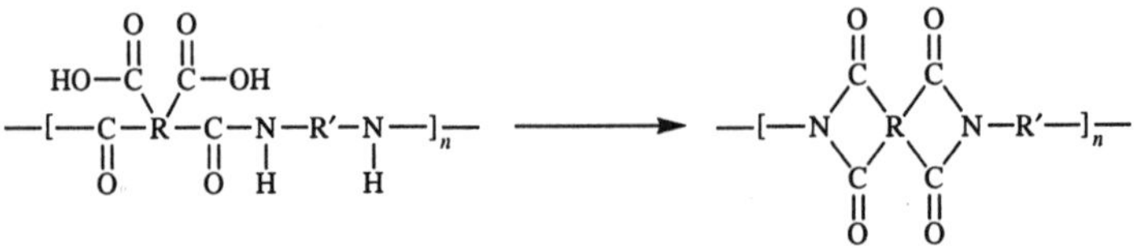

$$-[-CO-R(COOH)_2-CO-NH-R'-NH-]_n- \longrightarrow -[-N(CO)_2R(CO)_2N-R'-]_n-$$

The polyamic acids are usually prepared in solution. Suitable solvents are *N,N*-dimethylformamide, dimethyl sulfoxide, and *n*-methyl-2-pyrrolidone. The reactions require anhydrous conditions at relatively low temperatures, like 50 °C (or lower). Some, however, need high temperatures, as high as 175 °C.[88] The two reagents are combined in solution. The order of addition and reagent purity can influence the molecular weight of the products that may range from 13,000–55,000.[88] Some imidation accompanies the first step. It is desirable that during polyamic acid formation the degree of imidation not exceed 50%.

The step of conversion of polyamic acids to polyimides can take place at about 300 °C in thin films. With cyclizing agents, however, it can take place at much lower temperatures.[88]

Commercially, the most commonly used aromatic dianhydrides are pyromellitic dianhydride and benzophenone tetracarboxylic dianhydride. The common amines in industrial practices are *meta*-phenylenediamine, methylenedianiline, and oxy-dianiline. A typical polyimide preparation can be shown as follows:

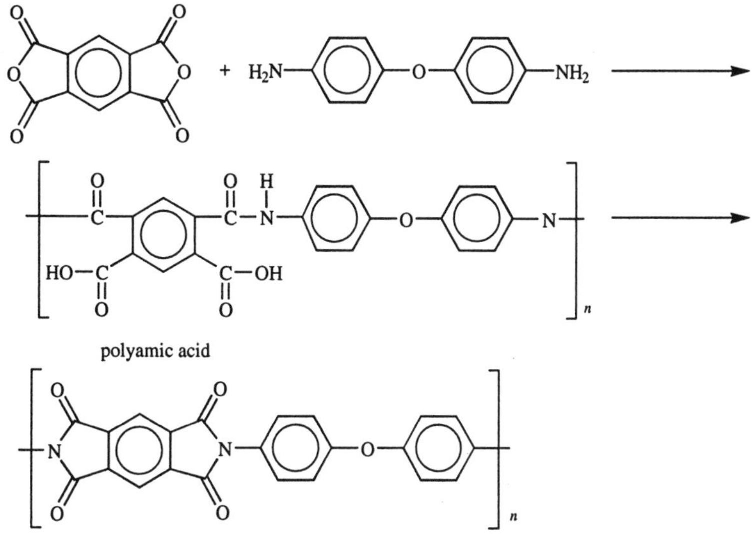

polyamic acid

The above polymer melts above 600 °C and is heat-stable up to 500 °C in an inert atmosphere. It is sold under the trade name of Kapton.

Another route is through reactions of diisocyanates with dianhydrides[89,90]:

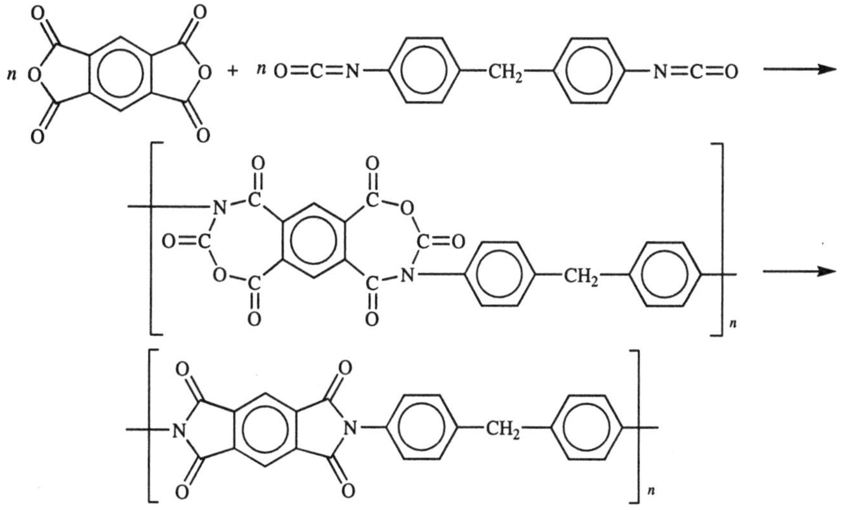

The reaction is kept between 120–200 °C.[91a] In place of isocyanates it is also possible to use aldimines, ketimines, or even *N,N'*-bis(trimethylsilyl) derivatives[91]:

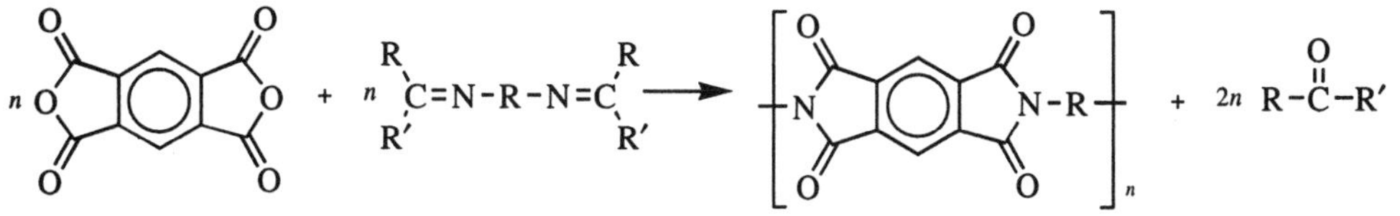

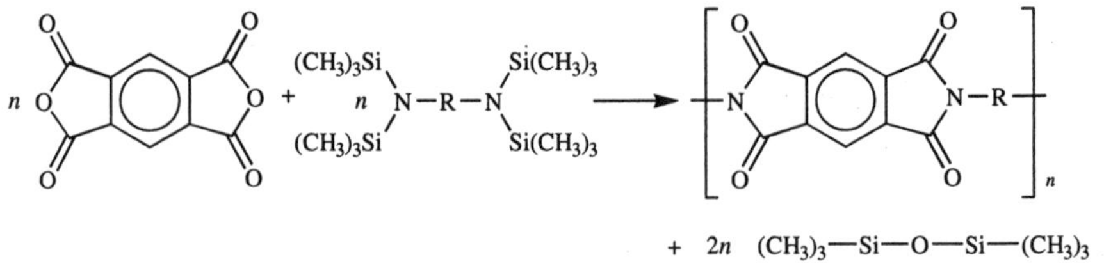

+ 2n $(CH_3)_3—Si—O—Si—(CH_3)_3$

When silylated diamines are used, trimethylsilyl esters of polyamic acids are formed first and then desilylated with methanol. Due to increased solubility of silylated aromatic amines, the initial condensations and formations of polyamic acid trimethylsilyl esters can be conducted in various solvents, yielding high molecular weight polymers. The highest molecular weights are obtained in dimethylacetamide at 50 °C. Other solvents like tetrahydrofuran and chloroform can be used as well, though they appear to yield slightly lower molecular weight products.[91]

The films of the silylated precursors of polyamic acid convert directly by heat treatment to yellow, transparent, and tough films of aromatic polyimides with the elimination of trimethylsilanol.[91]

Other preparations of polyimides include the use of di-half esters of tetracarboxylic acids[92]:

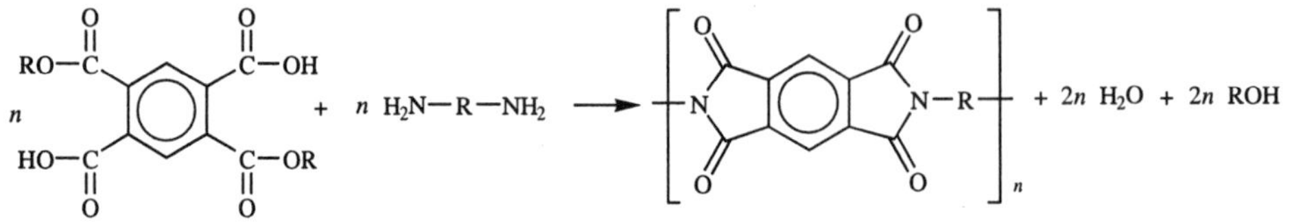

Polyimides can also be prepared by reactions of diimides with dihalides[93]:

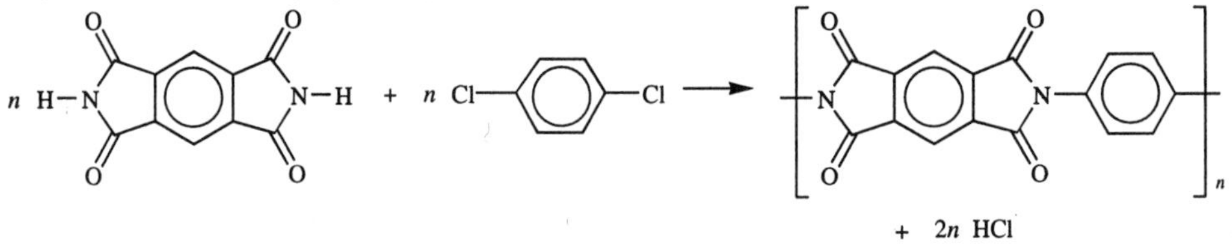

+ 2n HCl

Reactions of bis-maleimides with compounds containing active hydrogens can also lead to formations of polyimides[94]:

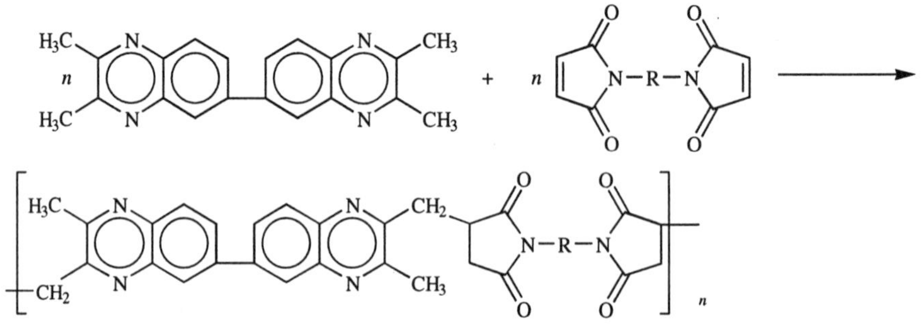

Also, unsaturated diimides can be reacted with sulfur halides to form polyimides[95]:

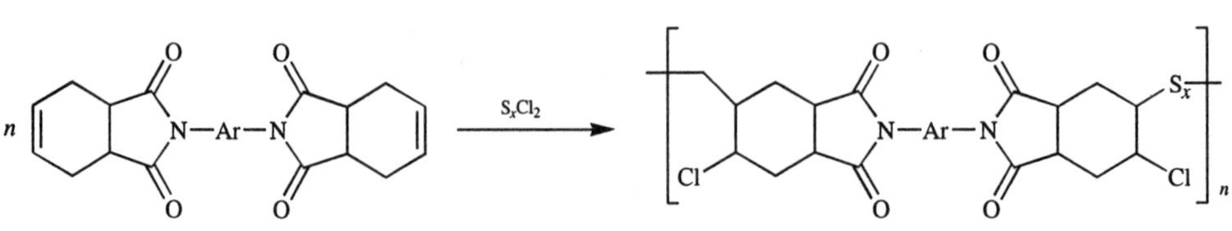

where $x = 1$ or 2

Diepoxides can add to pyromellitimide in the presence of a base to form polyimides with pendant hydroxyl groups[96]:

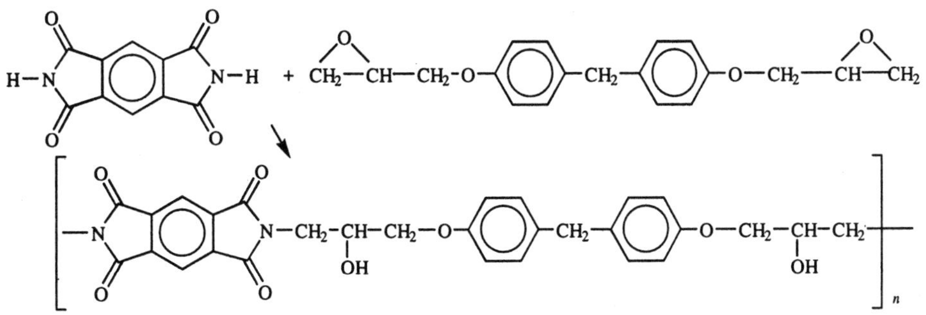

Tertiary amines and quaternary ammonium halides catalyze this reaction. Acetylation of the pendant hydroxyl groups of the product yields polymers that are soluble in solvents like dioxane and dimethylformamide.

Pyromellitimide can also add to double bonds to form polyimides[97]:

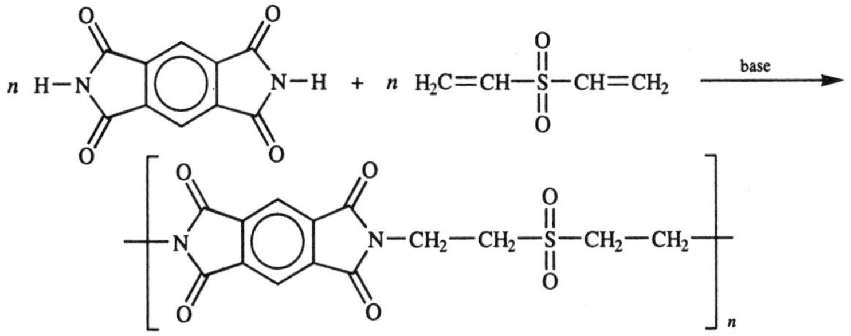

Polyimides also form by photoadditions of aliphatic or aromatic bismaleimides to benzene.[98] The reactions involve 2+2 cycloadditions that yield homoannular diene intermediates. Diels-Alder additions follow and result in formations of the polymers:

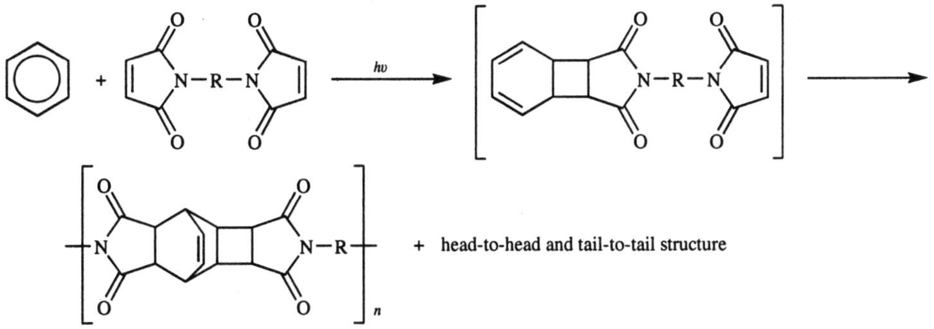

+ head-to-head and tail-to-tail structure

Diels-Alder reactions yield other polyimides, such as the following[99]:

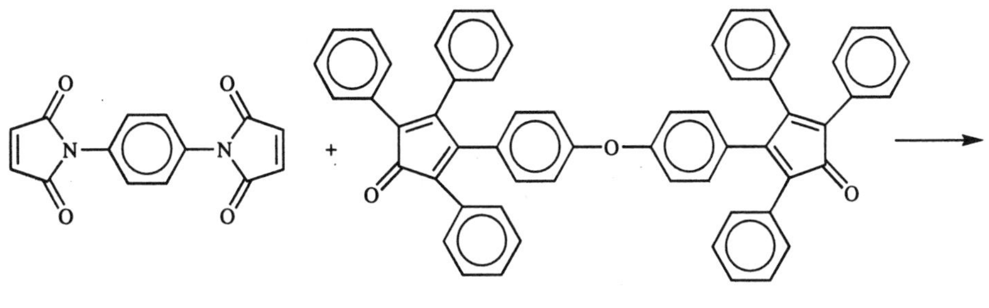

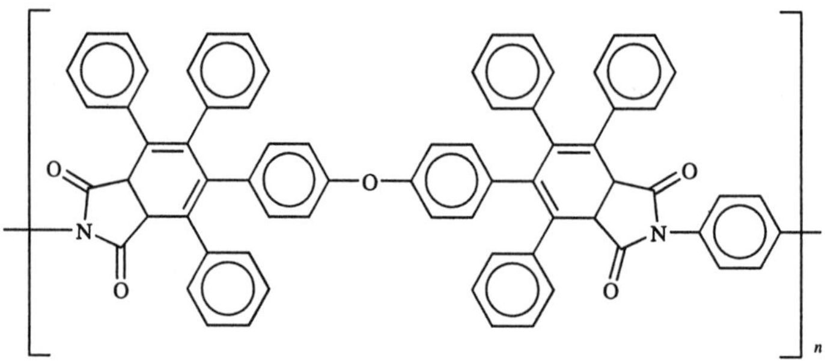

Also, *N,N*-bis(ethoxycarbonyl) pyromellitimide condenses with diamines to yield polyimides[100]:

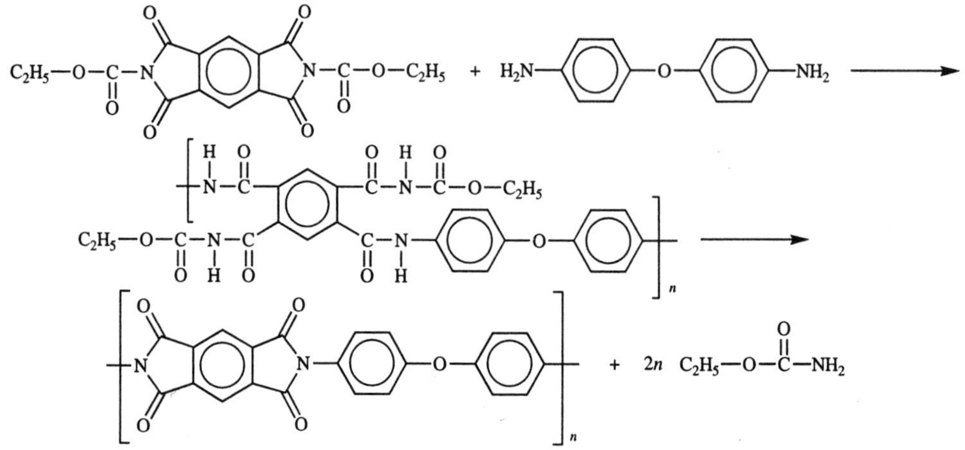

The first step in the above preparation takes place in solution. After casting a film, the second step takes place at 240 °C under vacuum. Interesting polyimides also form from reactions of 2,2′,6,6′-biphenyltetracarboxylic acids anhydride[101] with aromatic diamines, like 4,4′-diaminodiphenyl ether:

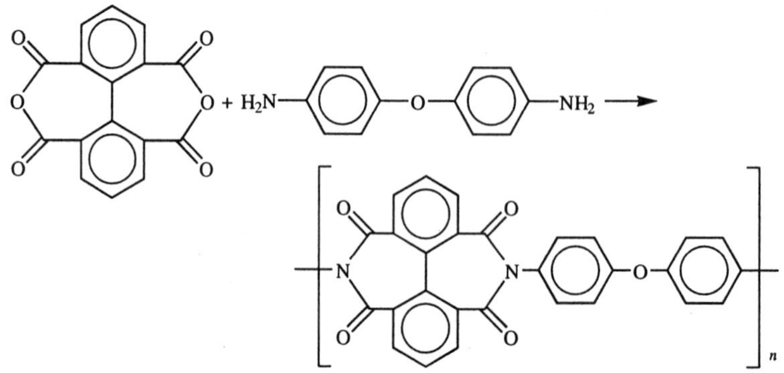

As in the previous cases, the polyamic acid forms first in solution at approximately 40 °C. It is converted to the polyimide by heating in acetic anhydride at reflux for 18 hours.

Heterocyclic dianhydrides, like pyrazinetetracarboxylic dianhydride, also react with diamines to form polyimides.[102] Such polymers are harder to form, however.

Two types of thermoset polyimides are currently prepared commercially. They are based on low molecular weight bisimides such as bismaleimides or bis-5-norbornene-2,3-dicarboximides. Due to unsaturations, the materials crosslink by a free-radical mechanism into tight networks. Michael-type additions of primary and secondary amines to the bismaleimides are often used to chain-extend them before crosslinking. This reduces the crosslinking density and the brittleness.[115] The materials are designated by the term PMR, for polymerizable monomeric reactants.

6.6. Polyethers

Polyethers that form by chain-growth polymerizations of carbonyl compounds and by ring-opening polymerizations of cyclic ethers and acetals are discussed in Chapter 5. In this section, poly(phenylene oxide)s and phenoxy resins are discussed.

6.6.1. Poly(phenylene oxide)s

These polymers are known for their good thermal stability and good mechanical properties. Commercially, these aromatic polyethers are prepared by oxidative coupling of phenols.[102] To obtain linear polymers and achieve high molecular weights, 2 and 6 positions of the phenol must be protected by substituents. This causes the aromatic rings to couple in 1,4 positions. When 2,6-dimethylphenol is used, the reaction takes place at room temperature. Oxygen is bubbled through a solution of the phenol in the presence of an amine-cuprous salt catalyst:

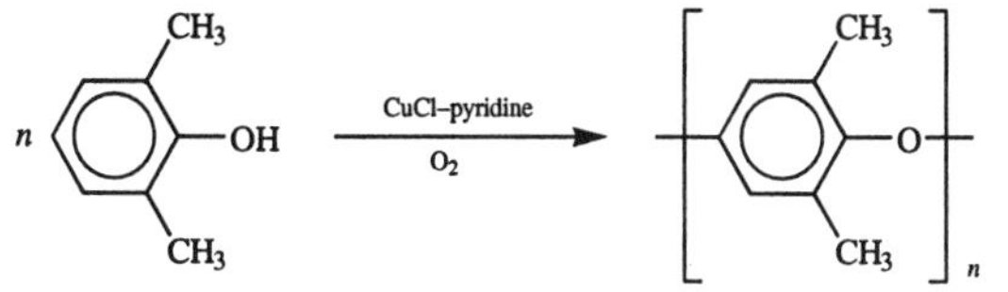

Phenols with halogen substituents require higher temperatures. Large substituents can lead to carbon-to-carbon coupling instead of carbon-to-oxygen:

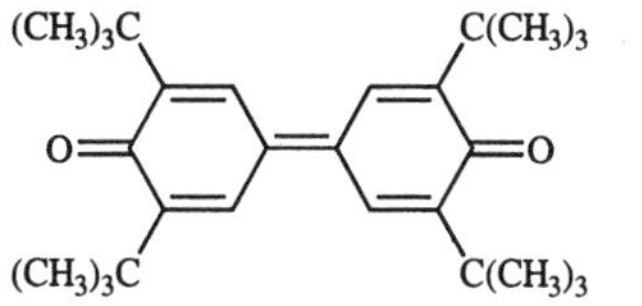

The active catalyst is believed to be a basic cupric salt that forms through oxidation of cuprous chloride followed by complexation with two molecules of the amine[102]:

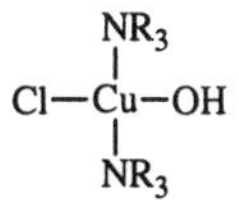

This is a step-growth polymerization involving phenoxy radicals. The polymer formation can be illustrated as follows:

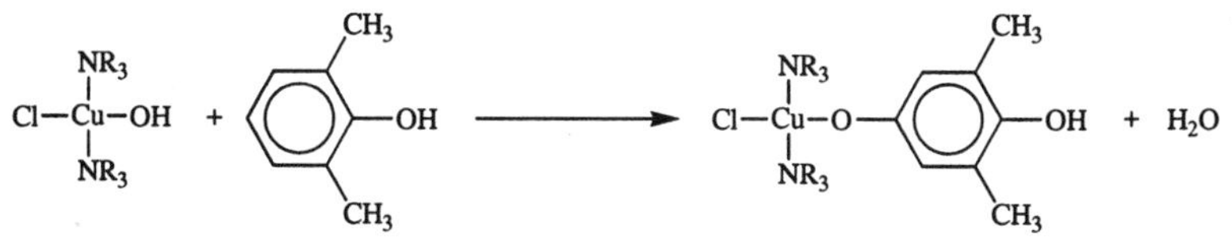

Dissociation leads to aryloxy radicals or to two new radicals that couple. Quinone ketals are formed initially. They dissociate to yield the original aryloxy radicals and then couple[102]:

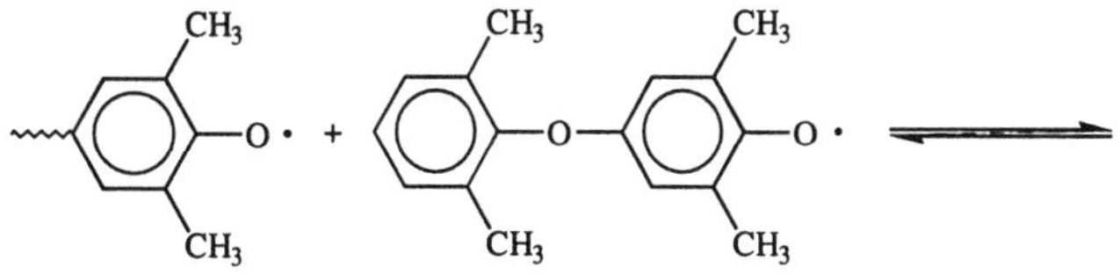

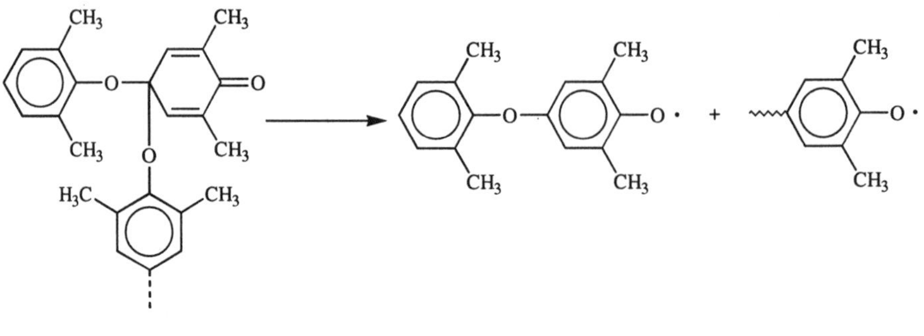

Formation of aryloxy radicals as intermediates was established with ESR spectroscopy studies that showed the presence of both monomeric and polymeric radicals in the reaction mixture.[103] Coupling occurs by two paths: one of them through rearrangements and the other through redistribution. In the redistribution process, two aryloxy radicals couple to yield an unstable quinone ketal as shown above.[102] This ketal decomposes rapidly either back into the original aryloxy radicals or into two different aryloxy radicals as follows:

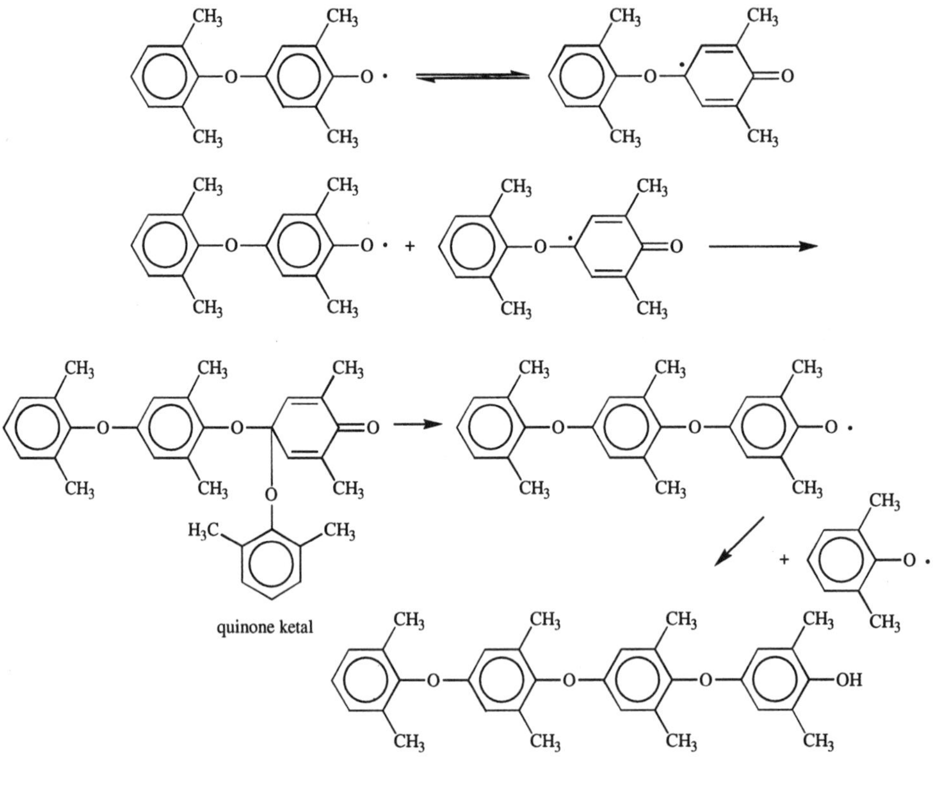

quinone ketal

The redistribution process leads to production of polymers from low molecular weight radicals. It appears that this process is unlikely to take place with high molecular weight radicals because there are too many steps involved in the production of monomer radicals. Quinone ketals are the intermediates in the rearrangement. The carbonyl oxygen of a ketal is within bonding distance of the *para* position of the next succeeding benzene ring.[104] The rearrangement can therefore give rise to a new ketal in which the second ring carries the carbonyl oxygen. The carbonyl oxygen finally ends up on a terminal unit[104] and is reduced to OH:

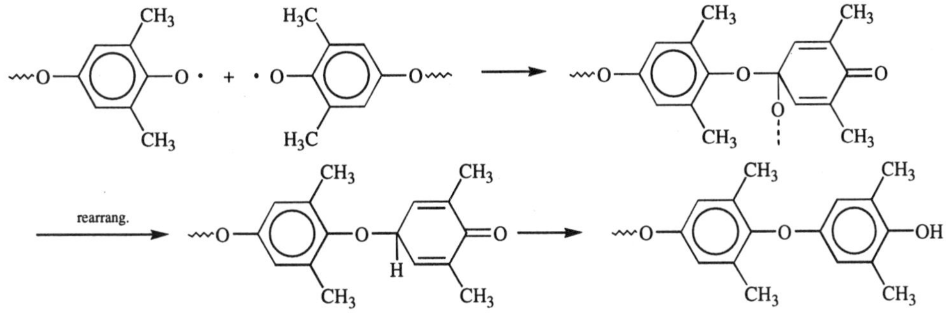

The quinone rearranges to a phenol through enolization. The product is identical to one that is obtained by direct head-to-tail coupling of two aryloxy radicals.[102]

It is important to realize that both processes, redistribution and rearrangement reactions, can occur within the same polymer molecule. At any point during the rearrangement there may be dissociation into aryloxy radicals. Also, redistribution does not have to occur by transfer of only a single unit. Rearrangement, followed by dissociation, allows any number of monomer units to be transferred in an essentially single step.

2,6-Diphenylphenol and 2,6-dimethylphenol can copolymerize by oxidative coupling.[104] If the diphenyl derivative is polymerized first and subsequently the dimethyl derivative is added to the reaction mixture, block copolymers form. If, however, the order is reversed or both phenols are polymerized together, a random copolymer results.[104]

Poly(phenylene oxide)s can also be formed by oxidative displacement of bromides from 4-bromo-2,6-dimethylphenol.[102,105] Compounds, like potassium ferricyanide, lead oxide, or silver oxide, catalyze this reaction:

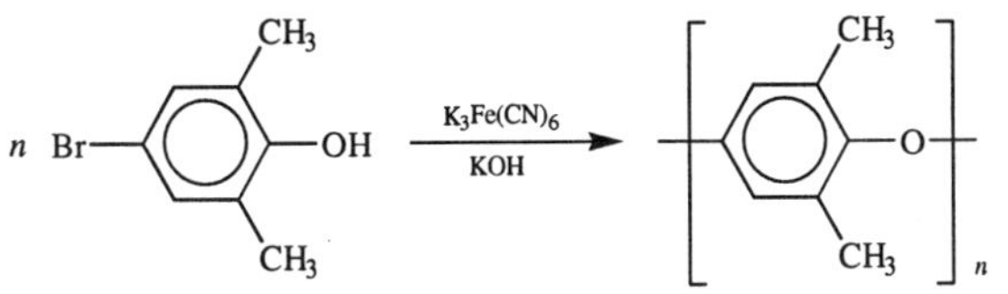

$$n\ \text{Br}-C_6H_2(CH_3)_2-OH \xrightarrow[\text{KOH}]{K_3Fe(CN)_6} \left[C_6H_2(CH_3)_2-O \right]_n$$

Poly(phenylene oxide)s also form through photodecomposition of benzene-1,4-diazooxides[102]:

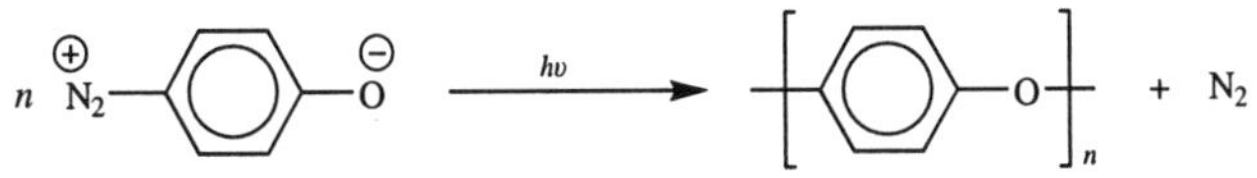

$$n\ \overset{\oplus}{N_2}-C_6H_4-\overset{\ominus}{O} \xrightarrow{h\nu} \left[C_6H_4-O \right]_n + N_2$$

Oxidative coupling is the only process used commercially. Although poly(phenylene oxide) is an important commercial material, there was initially a processing problem when the material was introduced. Currently, a large portion of the polymer is sold as a blend with polystyrene (probably high-impact) to make it more attractive economically and easier to process. The ratios of poly(2,6-dimethylphenylene oxide) to polystyrene range from approximately 1:1 to 1:2 and the material is sold under the trade name of Noryl. A fire-retardant grade, containing about 5% of an additive, believed to be triphenyl phosphate, is also on the market.

6.6.2. Phenoxy Polymers

These materials are part of the technology of epoxy resins that are discussed in a separate section, further on in this chapter. The polymers bridge a gap between thermosetting resins and thermoplastic polymers and are used in both forms commercially.[116] An idealized picture of phenoxy polymers can be shown as follows:

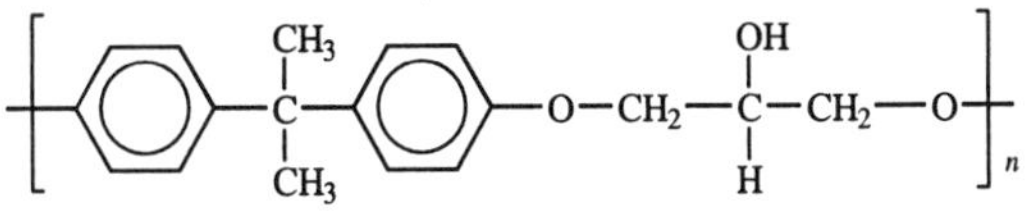

$$\left[C_6H_4-C(CH_3)_2-C_6H_4-O-CH_2-CH(OH)-CH_2-O \right]_n$$

The polymer forms through caustic-catalyzed condensations of diphenols with epichlorohydrin. Any diphenol or combination of diphenols can undergo this reaction. In commercial practice, however, mainly 4,4-isopropylidinediphenol, commonly called Bisphenol A, is used.

Theoretically, the phenoxy resins should form in equimolar reactions of epichlorohydrin with the diphenol. There are, however, a number of side reactions that accompany the condensation. To get around them and to obtain high molecular weight polymers, the syntheses are carried out in two steps. In the first, an excess of epichlorohydrin is used to form diepoxide:

$$HO-C_6H_4-C(CH_3)_2-C_6H_4-OH + Cl-CH_2-\overset{O}{CH-CH_2} \xrightarrow{OH^{\ominus}}$$

$$\overset{O}{CH_2-CH}-CH_2-O-C_6H_4-C(CH_3)_2-C_6H_4-O-CH_2-\overset{O}{CH-CH_2}$$

In the second step, equimolar quantities of the diepoxide are reacted with the diphenol:

$$\overset{O}{CH_2-CH}-CH_2-O-C_6H_4-C(CH_3)_2-C_6H_4-O-CH_2-\overset{O}{CH-CH_2} + HO-C_6H_4-C(CH_3)_2-C_6H_4-OH$$

$$\longrightarrow \left[-C_6H_4-C(CH_3)_2-C_6H_4-O-CH_2-CH(OH)-CH_2-O-\right]_n$$

Both reactions are conducted in solution, where methyl ethyl ketone is the choice solvent. The commercial resins range in molecular weights from 15,000–200,000.

6.7. Polyacetals and Polyketals

Polyacetals and polyketals are polyethers that form (1) through condensations of glycols with carbonyl compounds, (2) by exchange reactions of acetals or ketals, and (3) by additions of diols to dialkenes[109]:

$$n\ R-\overset{O}{\overset{\|}{C}}-R + n\ HO-R'-OH \longrightarrow -[-CR_2-O-R'-O-]_n-$$

An acid-catalyzed exchange reaction of glycols and acetals yields polyacetals as follows:

$$n\ R\text{–}O\text{–}CH_2\text{–}O\text{–}R + n\ HO\text{–}R'\text{–}OH \rightarrow \text{–}[\text{–}O\text{–}CH_2\text{–}O\text{–}R'\text{–}]_n\text{–} + 2\ ROH$$

These reactions often lead to cyclic acetals that interfere with the formation of high molecular weight products. Useful polyacetals, however, can be formed from pentaerythritol and acetals of dialdehydes:

$$n\ HO-CH_2-C(CH_2-OH)_2-CH_2-OH + n\ (R'O)_2CH-R-CH(OR')_2 \longrightarrow \left[-CH\overset{O-CH_2}{\underset{O-CH_2}{<}}C\overset{CH_2-O}{\underset{CH_2-O}{>}}CH-R-\right]_n$$

Acetals generally exhibit poor resistance to hydrolytic attack. Some, however, are much more resistant than others, depending upon the glycol. The following formal was reported to exhibit good hydrolytic stability under both acidic and basic conditions[109]:

$$\left[-O-C_4(CH_3)_4-O-CH_2-\right]_n$$

Pentaerythritol yields spiropolymers by this reaction. The products offer superior thermal resistance[117]:

$$C(CH_2OH)_4 + O{=}C_6H_8{=}O \longrightarrow \text{spiropolymer}$$

The addition of diols to dialkenes can be illustrated on addition of a glycol to a divinyl ether:

$$CH_2{=}CH{-}O{-}R{-}O{-}CH{=}CH_2 + HO{-}R'{-}OH \xrightarrow{H^{\oplus}} \left[\underset{|}{\overset{CH_3}{}}CH{-}O{-}R{-}O{-}\overset{CH_3}{CH}{-}O{-}R'{-}O \right]_n$$

Commercially, large volume acetals are only produced by polymerizations of formaldehyde and by ring-opening polymerizations of trioxane. These reactions are discussed in Chapters 3 and 4. Two such materials are manufactured in this country. One is a homopolymer of formaldehyde, polyoxymethylene. It is sold under the trade name of Delrin. The material is end-capped to prevent depolymerization by acetylating the terminal hydroxyl groups. The other one, a copolymer of formaldehyde with small quantities of a comonomer, is sold under the trade name of Celcon. Copolymerization accomplishes the same objective as end-capping. It also makes the product more resistant to attacks by bases. Polyoxymethylene is highly crystalline. This is due to easy packing of the simple, polar chains. The crystallinity is estimated to be 60–77%. Polyoxymethylenes is a strong material with good resistance to creep, fatigue, and abrasion.

6.8. Poly(*p*-xylylene)s

The original polymerization of *p*-xylylene was carried out by vacuum pyrolysis of *p*-xylene at 900–950 °C. The intermediate, *p*-xylylene, polymerizes spontaneously upon condensation on cooler surfaces[106]:

$$n\ H_3C{-}C_6H_4{-}CH_3 \xrightarrow[\text{vacuum}]{\Delta} n\ CH_2{=}C_6H_4{=}CH_2 \longrightarrow \left[CH_2{-}C_6H_4{-}CH_2 \right]_n$$

The process was improved by using di-*p*-xylylene as an intermediate.[106] This dimer converts to a polymer under milder conditions quantitatively. Both methylene bridges cleave to form *p*-xylylene, which is a reactive intermediate:

$$\text{di-}p\text{-xylylene} \xrightarrow[\text{vacuum } (< 1\text{mm Hg})]{600\,^{\circ}C} 2\,(CH_2{=}C_6H_4{=}CH_2) \longrightarrow$$

$$\cdot CH_2{-}C_6H_4{-}CH_2{-}\left[{-}CH_2{-}C_6H_4{-}CH_2{-} \right]_n{-}CH_2{-}C_6H_4{-}CH_2\cdot$$

The molecular weights of these polymers were estimated to be as high as 500,000. The total process is sometimes called *transport polymerization*. Poly(*p*-xylylene) films are produced commercially. The value of T_m for this polymer, which is crystalline, is 400 °C and it carries the trade name of Parylene. Films of poly(*p*-xylylene) have only fair thermal stability and are brittle, but exhibit good chemical resistance and are very good electrical insulators. Pyrolysis of xylene in steam at 950 °C yields the dimer intermediate. The yield is reported to be 15%.[88]

It is possible to form substituted poly(*p*-xylylene)s by starting with substituted structures. Among the compounds reported were chlorinated and brominated compounds as well as some containing alkyl, cyano, acetyl, and carboxymethyl derivatives. When the di-*p*-xylylenes are unsymmetrically substituted, two homogeneous polymers form during pyrolysis, because the two condense with spontaneous polymerization at two different temperatures.[88]

Poly(*p*-xylylene)s can also be prepared by other reactions. Among them is the condensation of trimethyl(*p*-methylbenzyl) ammonium halide in the presence of a base[47]:

$$n\ H_3C-C_6H_4-CH_2-N^{\oplus}(CH_3)_3\ Br^{\ominus} \xrightarrow{NaOH} \left[-CH_2-C_6H_4-CH_2-\right]_n$$

A route to a halogen-substituted polymer is through a reaction of dihalo-*p*-xylylenes with caustic. It results in 1,6 elimination of HCl and formation of a chlorine-substituted poly(*p*-xylylene)[47]:

$$n\ ClCH_2-C_6H_4-CH_2Cl \xrightarrow{NaOH} H_2C{=}C_6H_4{=}CHCl \longrightarrow \left[-CH_2-C_6H_4-CH(Cl)-\right]_n$$

With excess caustic, however, all chlorine is removed and presumably an all conjugated polymer forms[47]:

$$\left[-CH_2-C_6H_4-CH(Cl)-\right]_n \longrightarrow \left[-C_6H_4-CH{=}CH-\right]_n$$

p-Bis(trichloromethyl) benzene can be pyrolyzed over copper gauze at 300–600 °C temperatures to form $\alpha,\alpha,\alpha',\alpha'$-tetrachloro-*p*-xylylene. This monomer polymerizes at temperatures just below 140 °C[107]:

$$Cl_3C-C_6H_4-CCl_3 \xrightarrow{Cu} Cl_2C{=}C_6H_4{=}CCl_2 + CuCl$$

$$Cl_2C{=}C_6H_4{=}CCl_2 \longrightarrow \left[-C_6H_4-CCl_2-CCl_2-\right]_n$$

p-Xylylene polymers also form electrolytically[88] in dioxane–water solution with mercury or lead cathodes and carbon rod anodes. During the reaction a quinonedimethide intermediate probably forms.[88] Many substituted polymers also form by this reaction.[88] The following is an example of one of them:

$$n\ BrF_2C-C_6H_4-CF_2Br \longrightarrow \left[-CF_2-C_6H_4-CF_2-\right]_n$$

Another interesting synthesis is through an equilibration reaction of suitably substituted styrenes to form *p*-xylylenes that spontaneously polymerize to linear polymers.[108] A suitable substituent can be a *p*-cyanophenylmethyl group. Such compounds undergo complete ionization and rearrange in the presence of a strong base:

$$H-C(C{\equiv}N)(C_6H_5)-C_6H_4-CH{=}CH_2 \xrightarrow{-H^{\oplus}} (-)C(C{\equiv}N)(C_6H_5)-C_6H_4-CH{=}CH_2 \xrightarrow{+H^{\oplus}} (C_6H_5)(N{\equiv}C)C{=}C_6H_4{=}C(CH_3)-H$$

The *p*-xylylene intermediate then polymerizes:

$$(C_6H_5)(N{\equiv}C)C{=}C_6H_4{=}C(CH_3)-H \longrightarrow \left[-C(C{\equiv}N)(C_6H_5)-C_6H_4-CH(CH_3)-\right]_n$$

There are, however, competing side reactions, such as dimerization[108]:

$$(-)\underset{\text{C}_6\text{H}_5}{\overset{\text{C}\equiv\text{N}}{\text{C}}}-\text{C}_6\text{H}_4-\text{CH}=\text{CH}_2 + \text{H}-\underset{\text{C}_6\text{H}_5}{\overset{\text{C}\equiv\text{N}}{\text{C}}}-\text{C}_6\text{H}_4-\text{CH}=\text{CH}_2 \xrightarrow{+\text{H}}$$

$$\text{H}-\underset{\text{C}_6\text{H}_5}{\overset{\text{C}\equiv\text{N}}{\text{C}}}-\text{C}_6\text{H}_4-\text{CH}_2-\text{CH}_2-\underset{\text{C}_6\text{H}_5}{\overset{\text{C}\equiv\text{N}}{\text{C}}}-\text{C}_6\text{H}_4-\text{CH}=\text{CH}_2$$

6.9. Sulfur-Containing Polymers

Industrially important sulfur-containing polymers are polysulfones and polysulfides. The materials differ considerably in properties and in use.

6.9.1. Polysulfones

These materials are an important group of engineering plastics. Aliphatic polysulfones were first synthesized at the end of the last century.[109] That synthesis was based on reactions of SO_2 with olefins:

$$\text{CH}_2=\text{CHR} + \text{SO}_2 \longrightarrow -[-\text{CH}_2-\underset{\text{R}}{\text{CH}}-\text{SO}_2-]-$$

Aliphatic sulfones, however, lack good thermal stability and are not commercially important. Aromatic sulfones, on the other hand, have many desirable physical properties. They are clear, rigid, tough materials, with a high value of T_g. Several aromatic sulfones are prepared commercially.

The original preparation of aromatic polysulfones was described in 1958.[110] This was followed by investigations of many different structures of polysulfones. One current commercial material is a condensation product of 2,2′-bis(hydroxyphenyl) propane with 4,4′-bis(chlorophenyl) sulfone. It forms by a Williamson synthesis, because the reactivity of the halogens is enhanced by the sulfone groups[47]:

$$n\ \text{NaO}-\text{C}_6\text{H}_4-\underset{\text{CH}_3}{\overset{\text{CH}_3}{\text{C}}}-\text{C}_6\text{H}_4-\text{ONa} + n\ \text{Cl}-\text{C}_6\text{H}_4-\underset{\text{O}}{\overset{\text{O}}{\text{S}}}-\text{C}_6\text{H}_4-\text{Cl} \longrightarrow$$

$$\left[-\text{O}-\text{C}_6\text{H}_4-\underset{\text{CH}_3}{\overset{\text{CH}_3}{\text{C}}}-\text{C}_6\text{H}_4-\text{O}-\text{C}_6\text{H}_4-\underset{\text{O}}{\overset{\text{O}}{\text{S}}}-\text{C}_6\text{H}_4-\right]_n + 2n\ \text{NaCl}$$

The condensation takes place at 160 °C in an inert atmosphere and in some suitable solvents, like chlorobenzene. Commercially, polymers are available in molecular weight ranges from 20,000–40,000. Much higher molecular weight materials, however, form readily. This polymer is tough and high melting.

Aromatic polysulfones also form by the Friedel-Craft reaction,[111] such as

$$n\ \text{C}_6\text{H}_5-\text{O}-\text{C}_6\text{H}_4-\underset{\text{O}}{\overset{\text{O}}{\text{S}}}-\text{Cl} \xrightarrow[\text{acid}]{\text{Lewis}} \left[-\text{C}_6\text{H}_4-\text{O}-\text{C}_6\text{H}_4-\underset{\text{O}}{\overset{\text{O}}{\text{S}}}-\right]_n + \text{HCl}$$

or

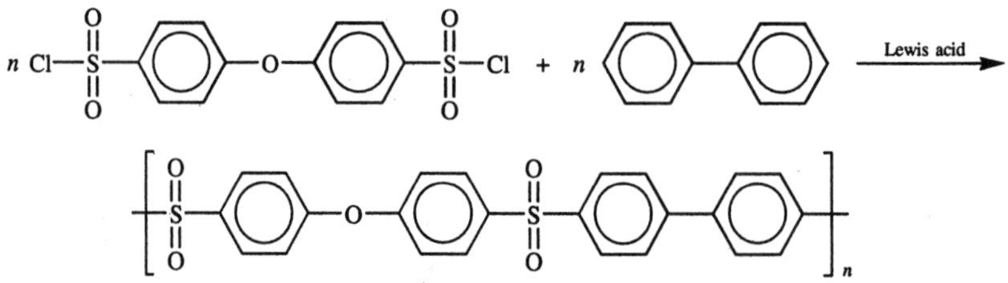

The reaction takes place at 80–250 °C in a solution or in the melt. Lewis acid catalysts are used in concentrations of 0.1–1.0 mole percent. Crosslinking through polysubstitution does not appear to be a problem. Some chain branching, however, does occur because *ortho* substitution is possible:

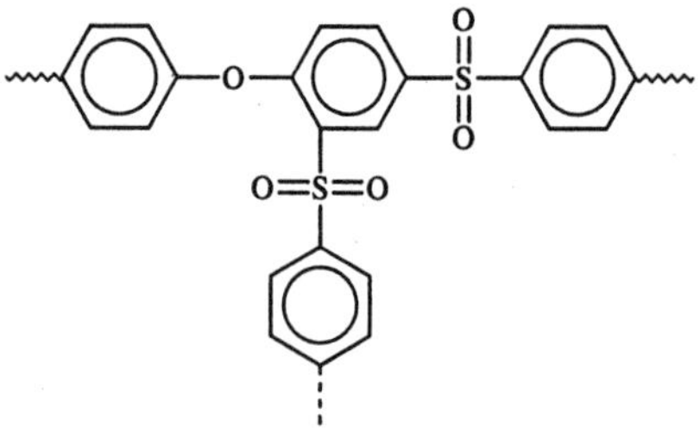

It is estimated, however, that the *ortho* substitution amounts to not more than 5–10%. Removal of the catalyst after the reaction is tedious. Molecular weights of commercial polymers range between 30,000–40,000. Table 6.6 presents the T_g values of some aromatic polysulfones.

Table 6.6. T_g values of Some Aromatic Polysulfones[118–121]

Polymer repeat unit	T_g (°C)
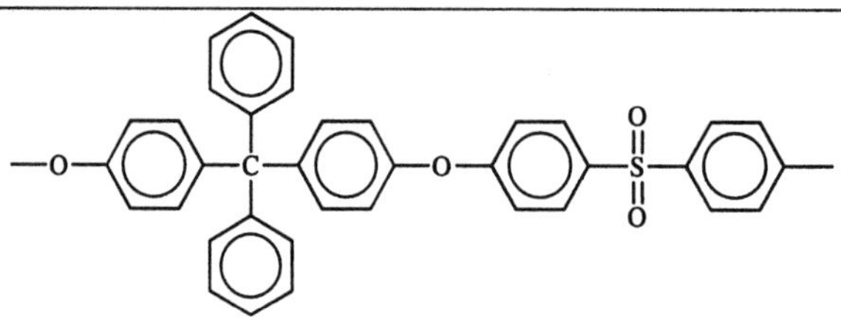	220
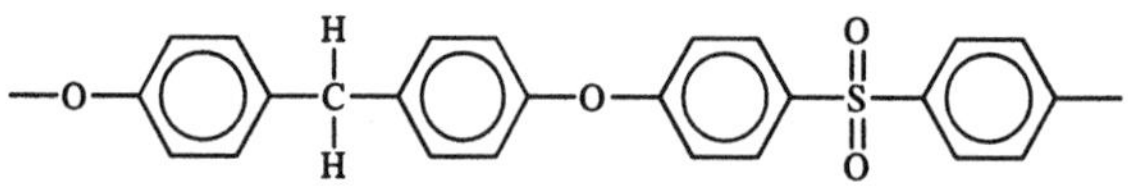	180
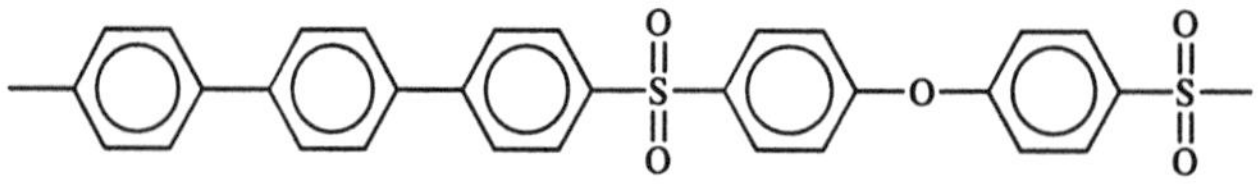	290
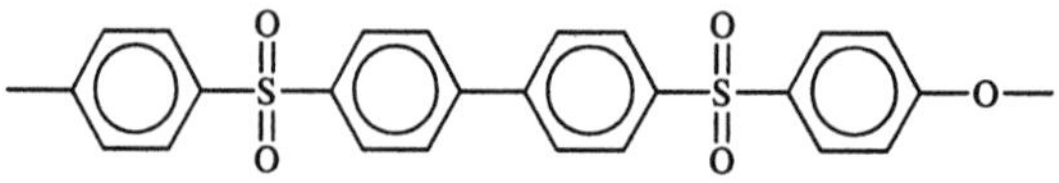	280
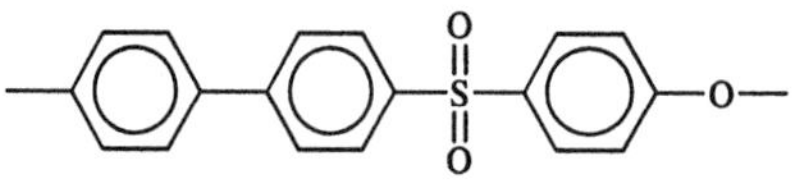	265
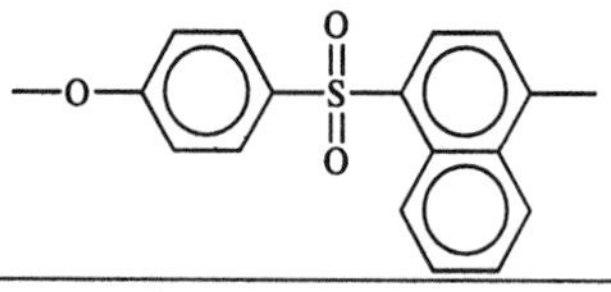	230

6.9.2. Polythioethers and Polymercaptals

These polymers form from reactions of dithiols with aldehydes or ketones:

$$n\ \text{HS—R'—SH} + n\ \text{R(H)(R'')C=O} \longrightarrow \text{—[—R'—S—C(R(H))(R')—S—]}_n\text{—}$$

The reaction is used commercially to prepare an aromatic polythioether from *p*-bis(mecapto-methyl)benzene:

$$n\ \text{HS—CH}_2\text{—C}_6\text{H}_4\text{—CH}_2\text{—SH} + n\ \text{H}_2\text{C=O} \longrightarrow \text{—[—CH}_2\text{—C}_6\text{H}_4\text{—CH}_2\text{—S—CH}_2\text{—S—]}_n\text{—}$$

The polymer melts at 150 °C and can be spun into fibers.[109]

Another material, poly(phenylene sulfide), can be prepared by several routes. This polymer forms from sodium *p*-bromothiophenol at 250–305 °C.[109] It has good thermal resistance and melts at 287 °C.

$$n\ \text{Br—C}_6\text{H}_4\text{—SNa} \longrightarrow \text{[—C}_6\text{H}_4\text{—S—]} + \text{NaBr}$$

It also forms by a reaction of dichlorobenzene with sodium sulfide.[110a] To date, the exact mechanism of this polymerization, which is carried out commercially in *N*-methylpyrrolidone solution, has not been fully established. Recent evidence indicates that in this solvent an ionic, step-growth S_NAr mechanism predominates[110b]:

$$\text{Cl—C}_6\text{H}_4\text{—Cl} + \text{Na}_2\text{S} \xrightarrow{\text{solvent}} \left[\text{Cl, S—Na Meisenheimer complex, Na}^{\oplus}\right] \longrightarrow \text{NaS—C}_6\text{H}_4\text{—Cl}$$

$$\xrightarrow{\text{Cl—C}_6\text{H}_4\text{—Cl}} \text{Cl—C}_6\text{H}_4\text{—S—C}_6\text{H}_4\text{—Cl} \xrightarrow{\text{Na}_2\text{S}} \text{[—C}_6\text{H}_4\text{—S—]}_n$$

The polymer possesses a broad, high melting point and a glass transition temperature of about 85 °C.

An important characteristic of poly(phenylene sulfide) is its ability to undergo changes upon heating.[110a] This change is complex and not completely understood. It appears to involve varying degrees of oxidation, crosslinking, and chain scission. When heated from 315 °C to 415 °C the polymer melts, thickens, gels, and eventually solidifies to a dark infusible solid. This curing phenomenon makes the polymer useful in many applications that range from coatings (powder or slurry) to molding (by injection, compression, or sintering). Poly(phenylene sulfide) also becomes highly conductive electrically when a dopant is added. Conducting polymers are discussed in Chapter 8.

Elastomers, based on *poly(alkylene sulfide)s*, are still another group of sulfur-containing polymers. They can be represented by a general structural formula of

$$\text{–[–R–S}_x\text{–]}_n\text{–} \qquad \text{where } x = 2\text{–}4$$

The most widely used methods of preparation are based on reactions of sodium polysulfides with alkyl dichlorides[110]:

$$n\ \text{Cl–R–Cl} + \text{Na}_2\text{S}_x \rightarrow \text{–[–R–S}_x\text{–]}_n\text{–} + 2n\ \text{NaCl}$$

These reactions are usually carried out in dispersions. An aqueous sodium polysulfide containing a surfactant, like alkyl aryl sulfonate, sodium hydroxide, and magnesium chloride, is heated to 80 °C.

The magnesium chloride forms magnesium hydroxide and acts as a nucleating agent. Bis(2-chloroethyl) formal is then added over two hours with stirring and external cooling of the exothermic reaction to about 90 °C. After addition, this temperature and stirring are usually maintained for an additional two hours to complete the process. The product contains a distribution of polysulfide groups.

The polysulfide anions can interchange continuously:

$$\sim S{-}S{-}R{-}S{-}R{-}S{-}S{-}R\sim + {}^{\ominus}S{-}S^{\ominus} \rightleftharpoons \sim S{-}S{-}R{-}S{-}S{-}S{-}S^{\ominus} + {}^{\ominus}S{-}R{-}S{-}S{-}R\sim$$

This allows building up the molecular weights of the polymers by additions of excess polysulfide. Also, low molecular weight fractions can be washed out. By such manipulations, molecular weights of 500,000 are readily achieved. It is interesting that in this particular step-growth polymerization, in order to obtain high molecular weights, strict stoichiometry is not only not required, but one of the components is deliberately added in excess.

Several different grades of poly(alkyl sulfide)s are available commercially. One form, hydroxy terminated, is formed by coagulating the formed polymer from the aqueous dispersion with sulfuric acid. The terminal halogens hydrolyze in the process to hydroxyl groups. For easier processing, these elastomers are usually reacted with disulfide like benzothiazyl disulfide. This reduces the molecular weight through chain cleavage:

$$\sim R{-}S{-}S{-}R\sim + (\text{benzothiazolyl}){-}C{-}S{-}S{-}C{-}(\text{benzothiazolyl}) \longrightarrow 2\ \sim R{-}S{-}S{-}C(\text{benzothiazolyl})$$

The elastomers are then chain-extended again with metal oxides, like zinc oxide, that couple the terminal hydroxy groups. The same thing can also be done by using diisocyanates.

Another group consists of thiol-terminated, low molecular weight polymers. They form from heating alkyl sulfides in aqueous dispersion of sodium bisulfite and sodium sulfite for about one hour at 80 °C. This results in mercaptide and thiothiol terminal groups:

$$\sim S{-}S{-}R{-}S{-}S{-}R{-}S{-}S{-}R\sim + SH^{\ominus} \rightleftharpoons \sim S{-}S{-}R{-}S{-}SH + {}^{\ominus}S{-}R{-}S{-}S{-}R\sim$$

The sulfite ion prevents the reversal of the equilibrium by splitting of the sulfur from the thiothiol:

$$\sim S{-}S{-}R{-}S{-}SH + SO_3^{\ominus\ominus} \rightarrow \sim S{-}S{-}R{-}SH + S_2O_3^{\ominus\ominus}$$

These polymers crosslink by oxidative coupling of the mercaptide groups:

$$\sim SH + [O] + HS\sim \rightarrow \sim S{-}S\sim + H_2O$$

and by reactions with metal peroxides like lead peroxide. They also react with epoxy resins.

Poly(alkylene sulfide)s are exceptionally oil-resistant elastomers. They also exhibit good resistance to solvents and to weathering. On the other hand, these elastomers lack the strength of synthetic rubbers and possess an unpleasant odor.

Aliphatic polysulfides also form from reactions of dithiols with alkyl dihalides. High molecular weight polymers, however, are hard to form[109]:

$$n\,NaS{-}R{-}SNa + n Br{-}R'{-}Br \rightarrow -[-S{-}R{-}S{-}R'-]_n- + 2n\,NaBr$$

Similar polymers form through addition of dithiols to diolefins.[109] The reaction can take place either by a free-radical mechanism or by an ionic one. A free-radical mechanism requires the presence of peroxides to achieve counter-Markownikoff additions across the double bonds:

$$n\,HS{-}({-}CH_2{-})_6{-}SH + n\,CH_2{=}CH{-}({-}CH_2{-})_2{-}CH{=}CH_2 \xrightarrow{\text{peroxide}} -[-S{-}({-}CH_2{-})_6{-}S{-}({-}CH_2{-})_6{-}]_n-$$

This is a step-growth polymerization process. The additions to the double bonds involve a series of hydrogen exchange reactions[108]:

$$HS{-}R{-}SH + I \rightarrow HS{-}R{-}S^{\bullet} + HI$$

$$CH_2{=}CH{-}(CH_2)_2{-}CH{=}CH_2 + {\bullet}S{-}R{-}SH \rightarrow CH_2{=}CH{-}(CH_2)_2{-}{}^{\bullet}CH{-}CH_2{-}S{-}R{-}SH \xrightarrow{HS-R-SH}$$

$$CH_2{=}CH{-}({-}CH_2{-})_2{-}CH_2{-}CH_2{-}S{-}R{-}SH + {\bullet}S{-}R{-}SH \rightarrow \text{etc.}$$

When additions take place by ionic mechanisms, they are catalyzed by either acids or bases. These additions follow the Markownikoff rule:

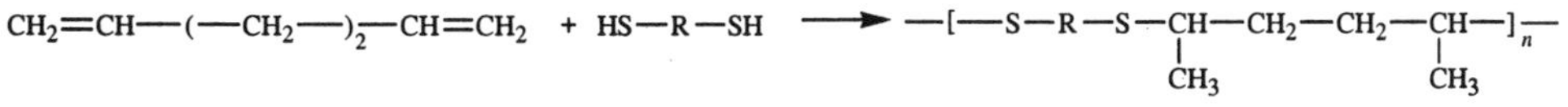

$$CH_2{=}CH{-}({-}CH_2{-})_2{-}CH{=}CH_2 + HS{-}R{-}SH \longrightarrow {-}[{-}S{-}R{-}S{-}\underset{CH_3}{\underset{|}{CH}}{-}CH_2{-}CH_2{-}\underset{CH_3}{\underset{|}{CH}}{-}]_n{-}$$

6.10. Polyurethanes

The polyurethanes[126,132] are sometimes also called "isocyanate polymers." They are characterized by the urethane linkage. Other functional groups, however, may be present in the polymer as well. The urethane linkages can be produced in polymers by several different routes. Among these, the most common are through reactions of the isocyanate groups with compounds bearing hydroxy groups. Such compounds may be glycols, dihydroxy-terminated polyethers or polyesters, and others. Difunctional reactants will produce linear polyurethanes:

$$HO{-}R'{-}OH + O{=}C{=}N{-}R{-}N{=}C{=}O \longrightarrow {-}[{-}R'{-}O{-}\overset{O}{\overset{\|}{C}}{-}\underset{H}{\underset{|}{N}}{-}R{-}\underset{H}{\underset{|}{N}}{-}\overset{O}{\overset{\|}{C}}{-}O{-}]_n{-}$$

6.10.1. Preparations of Polyfunctional Isocyanates

Polyfunctional isocyanates can be formed in many ways.[122] Commercially, the most important way is through reactions of phosgene with amines or amine salts. Other reactions, however, like that of carbon monoxide with nitro compounds, are now also becoming important. Addition of isocyanic acid to olefins is also gaining prominence.

6.10.2. Commercial Polyisocyanates

Two types of diisocyanates are employed in polymer preparations: aromatic and aliphatic. The most commonly used aromatic diisocyanates are toluene diisocyanate and 4,4′-diphenylmethane diisocyanate. Commercial toluene diisocyanate often comes as a mixture of 2,4 and 2,6 isomers in ratios of 80/20 or 65/35. When the reaction takes place at room temperature, the 4 position is 8 to 10 times more reactive than the 2 position. At elevated temperatures, however, this difference in reactivity decreases, and at 100 °C the reactivity of the isocyanate groups in both positions is approximately equal.

Phosgenation of aniline-formaldehyde condensates yields isocyanates with a functionality that averages 2.6–2.8. The structure of the product can be shown as follows:

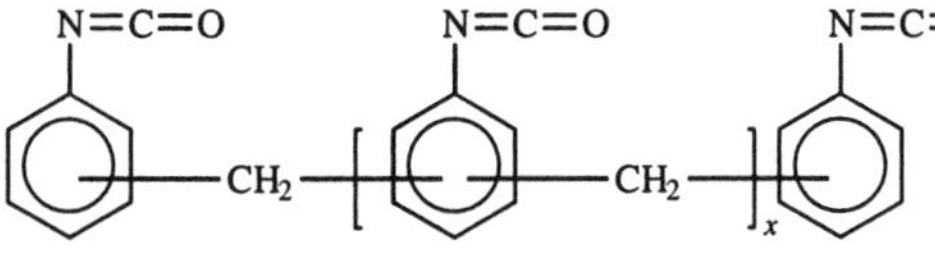

$$OCN{-}C_6H_4{-}CH_2{-}[{-}C_6H_3(NCO){-}CH_2{-}]_x{-}C_6H_4{-}NCO \quad \text{where } x = 2 \text{ or } 3$$

Among other aromatic diisocyanates in commercial use are *p*-phenylene diisocyanate, *m*-phenylene diisocyanate, 1-chloro-2,4-phenylene diisocyanate, 3,3′-dimethyl-4,4′-bisphenylene diisocyanate, 4,4′-bis(2-methylisocyanophenyl)methane, and 4,4'-bis(2-methoxyisocyanophenyl)methane.

The common aliphatic diisocyanates are hexamethylene diisocyanate, hydrogenated (H_{12}) 4,4′-dipenylmethane diisocyanate, isophorone diisocyanate, 2,2,4-trimethylhexamethylene diisocyanate, and 2,4,4-trimethylhexamethylene diisocyanate. Other aliphatic diisocyanates in various stages of commercialization are lysine diisocyanate, methylcyclohexyl diisocyanate, isopropylidine bis-(4-cyclohexyl isocyanate), and tetramethylene diisocyanate. Many additional polyfunctional isocyanates are described in the literature.

6.10.3. Reactions of the Isocyanates

These reactions can be divided into two categories. They are additions to compounds with active hydrogens and self-condensations. The reactions are well described in organic chemistry textbooks, so there is little reason to describe them here. It is noteworthy, though, that the uncatalyzed reactions of isocyanates with various active hydrogen compounds are probably broadly similar. Among them, the most investigated reactions are those of alcohols with isocyanates[123]:

$$\mathrm{R{-}N{=}C{=}O} + \mathrm{R'{-}OH} \rightleftharpoons \left[\begin{array}{c}\mathrm{R{-}N{=}C{-}\ddot{O}{:}^{\ominus}} \\ \uparrow\oplus \\ \mathrm{H{-}O{-}R'}\end{array}\right] \rightleftharpoons \left[\begin{array}{c}\mathrm{R{-}\overset{\ominus}{\ddot{N}}{-}C{=}O} \\ \uparrow\oplus \\ \mathrm{H{-}O{-}}\end{array}\right]$$

$$\xrightarrow{\text{ROH}} \left[\begin{array}{c}\mathrm{H\cdot\cdot O{-}R'} \\ \vdots\quad\vdots\ \ominus \\ \mathrm{R{-}N{-}C{-}O{:}} \\ \uparrow\oplus \\ \mathrm{H{-}O{-}R'}\end{array}\right] \longrightarrow \mathrm{R{-}\underset{|}{\overset{H}{N}}{-}\overset{O-R'}{\underset{|}{C}}{=}O} + \mathrm{R'{-}OH}$$

Electron-withdrawing substituents increase the positive charge on the isocyanate carbon and move the negative charge further away from the site for the reaction:

$$\mathrm{O_2N{-}C_6H_4{-}N{=}C{=}O} \rightleftharpoons \mathrm{(O)(O^{\ominus})N{=}C_6H_4{=}N{-}\overset{\oplus}{C}{=}O}$$

This makes an attack by an electron donor easier and yields a faster reaction. Electron-donating groups therefore have an opposite effect.

6.10.4. The Effect of Catalysts

Catalysts exert strong influence on the rates of reactions of isocyanates with active hydrogens compounds. Those most widely used are tertiary amines and metal salts, particularly tin compounds. The mechanism of catalysis by tertiary amines is believed[124–126] to proceed according to the following scheme:

$$\mathrm{R_3N} + \mathrm{R'{-}N{=}C{=}O} \rightleftharpoons \mathrm{R'{-}N{=}\underset{\oplus\,|\,N-R_3}{C}{-}O^{\ominus}} \rightleftharpoons \mathrm{R'{-}\overset{\ominus}{N}{-}\underset{\oplus\,|\,N-R_3}{C}{=}O} \xrightarrow{\text{HOR}''}$$

$$\left[\begin{array}{c}\mathrm{H\cdot\cdot O{-}R''} \\ \vdots\quad\vdots\ \ominus \\ \mathrm{R'{-}N{-}C{-}O} \\ \oplus| \\ \mathrm{N{-}R_3}\end{array}\right] \longrightarrow \mathrm{R{-}\underset{|}{\overset{H}{N}}{-}\overset{O-R'}{\underset{|}{C}}{=}O} + \mathrm{R_3N}$$

The catalytic activity of the tertiary amines generally parallels their base strength, except when steric hindrance is pronounced. Tin compounds exert much stronger catalytic effects on the reactions than do tertiary amines. This is illustrated in Table 6.7. The mechanism of catalysis by metal salts is believed to operate as follows[128]:

Table 6.7. Relative Effects of Catalysts on Reactivity of Phenylisocyanate[126,127,132]

Catalyst	Relative rates of reactions with *n*-butyl alcohol
None	1.0
N-Methyl morpholine	4.0
Triethyl amine	8.6
Triethylene diamine	120.0
Tributyltin acetate	30,000.0
Dibutyltin diacetate	60,000.0

$$\mathrm{R{-}N{=}C{=}O + MX_2 \longrightarrow \left[R{-}N{=}C{=}\overset{\oplus}{O}{-}\overset{\ominus}{M}X_2 \rightleftharpoons R{-}N{=}\overset{\oplus}{C}{-}O{-}\overset{\ominus}{M}X_2 \right]}$$

$$\mathrm{\left[R{-}N{=}\overset{\oplus}{C}{-}O{-}\overset{\ominus}{M}X_2 \right] + R'{-}OH \longrightarrow \left[R{-}N{=}\overset{\oplus}{C}{-}O{-}\overset{\ominus\ominus}{M}X_2{-}\overset{\oplus}{O}(R'){-}H \right] \longrightarrow}$$

$$\mathrm{\left[R{-}\overset{\oplus}{N}(H){=}C(OR'){-}O{-}\overset{\ominus}{M}X_2 \rightleftharpoons R{-}N(H){-}C(OR'){=}\overset{\oplus}{O}{-}\overset{\ominus}{M}X_2 \right] \longrightarrow R{-}N(H){-}C(OR'){=}O + MX_2}$$

6.10.5. Polyurethane Fibers

Originally, the main interest in polyurethanes was in preparation of fibers. They tend to resemble aliphatic polyamides, though these fibers are harder to dye, they are wiry, and difficult to handle. The common preparatory procedure is to add continuously the diisocyanate to the glycol while letting the temperature rise slowly to 200 °C. The reaction is exothermic and excess heat must be removed. An inert nitrogen atmosphere is maintained over the reaction mixture. Table 6.8 lists T_m values of some aliphatic polyurethanes. Today there is much less interest in polyurethane fibers. One polyurethane fiber that was commercialized sometime between 1960 and 1965 is a unique elastomeric material for expandable textiles.[133] It is an alternating block copolymer of "soft" and "hard" segments. The "soft" segments form from hydroxy-terminated aliphatic polyethers or polyesters of molecular weight between 1000–4000. They are linked with "hard" segments by urethane linkages. The "hard" segments form from aromatic diisocyanates and aliphatic or aromatic diamines. Hydrazine or hydrazine derivatives are sometimes used as chain extenders in place of the diamines. A preparation can be described as follows. Macroglycols are reacted first with an excess of the diisocyanate:

$$\mathrm{HO{\sim\sim}OH + \underset{excess}{OCN{-}R'{-}NCO} \longrightarrow OCN{-}R'{-}NH{-}\overset{O}{\overset{\|}{C}}{-}O{\sim\sim}O{-}\overset{O}{\overset{\|}{C}}{-}NH{-}R'{-}NCO}$$

The reaction mixture is then chain-extended with a diamine:

$$\mathrm{OCN{-}R'{-}NH{-}\overset{O}{\overset{\|}{C}}{-}O{\sim\sim}O{-}\overset{O}{\overset{\|}{C}}{-}NH{-}R'{-}NCO + OCN{-}R'{-}NCO + H_2N{-}R''{-}NH_2 \longrightarrow}$$

$$\mathrm{{\sim}NH{-}\overset{O}{\overset{\|}{C}}{-}O{\sim\sim}O{-}\overset{O}{\overset{\|}{C}}{-}NH{-}R'{-}NH{-}\overset{O}{\overset{\|}{C}}{-}({-}NH{-}R''{-}NH{-}O{-}\overset{O}{\overset{\|}{C}}{-}NH{-}R'{-}NH{-}\overset{O}{\overset{\|}{C}}{-}NH{-}R''{-})_m{\sim}}$$

prepolymer

Table 6.8. T_m Values of Some Fiber-Forming Polyurethanes[126,132]

Diisocyanate	Glycol	T_m (°C)
$OCN(CH_2)_4NCO$	$HO(CH_2)_4OH$	190
$OCN(CH_2)_4NCO$	$HO(CH_2)_6OH$	180
$OCN(CH_2)_4NCO$	$HO(CH_2)_{10}OH$	170
$OCN(CH_2)_5NCO$	$HO(CH_2)_4OH$	159
$OCN(CH_2)_6NCO$	$HO(CH_2)_4OH$	183
$OCN(CH_2)_6NCO$	$HO(CH_2)_5OH$	159
$OCN(CH_2)_8NCO$	$HO(CH_2)_4OH$	160
$OCN(CH_2)_8NCO$	$HO(CH_2)_6OH$	153

The prepolymers are spun into fibers and subsequently crosslinked, or crosslinked while they are being spun, to obtain resilient elastomeric fibers.

6.10.6. Polyurethane Elastomers

Solid elastomers can be divided into three categories, namely cast, millable, and thermoplastic. The cast elastomers are formed by casting liquid reaction mixtures of low molecular weight prepolymers into heated molds, where they crosslink and convert to high molecular weight materials. Slightly branched polyesters are combined first with diisocyanates, like toluene diisocyanate, to form the prepolymers, and degassed in a vacuum at elevated temperature (about 70 °C). The reaction mixtures are then poured into molds and heated for several hours at about 110 °C to form solid elastomers. These elastomers are soft and resilient, but they lack good mechanical strength.

Elastomers with better mechanical strength form from linear, hydroxy-terminated polyesters or polyethers. These macroglycols are also prereacted first with the diisocyanates, similarly to the procedure used for expandable fibers. The products, however, are mixed with low molecular weight glycols or diamines and then heated in molds at 110 °C for 24 hours. Slightly less than stoichiometric amounts of glycols or diamines are used so that the polymers are terminated with isocyanate groups. These terminal isocyanate groups react in the mold with urethane hydrogens to form allophanate crosslinks. Trimerization of the isocyanate groups might also take place, though this usually requires a catalyst to form at temperatures below 130 °C:

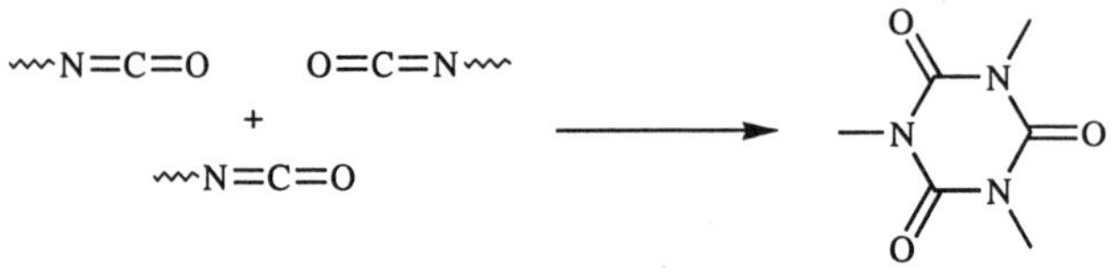

$$\sim N{=}C{=}O \quad O{=}C{=}N\sim \; + \; \sim N{=}C{=}O \longrightarrow \text{(isocyanurate ring: three N, three C=O)}$$

The concentrations of the allophanate links varies with the time of cure.[129] Also, if the crosslinking reactions are conducted in inert nitrogen atmospheres, very little scission of crosslinks takes place and a network structure forms during the cure. In open air, however, the scissions of crosslinks are extensive[129] and the products have poorer physical properties.

A drawback to the cast elastomers is limited shelf-life and a need to store them in the absence of moisture. As a result, millable elastomers were developed. These are produced by first forming hydroxy-terminated linear polyurethanes through reactions of linear aliphatic polyesters or polyethers with diisocyanates. The prepolymers are rubber-like gums that can be compounded on rubber mills with other ingredients and crosslinked. Crosslinking is accomplished by adding either more diisocyanates, or sulfur, or peroxides. Diisocyanates dimers that dissociate at about 150 °C are often used:

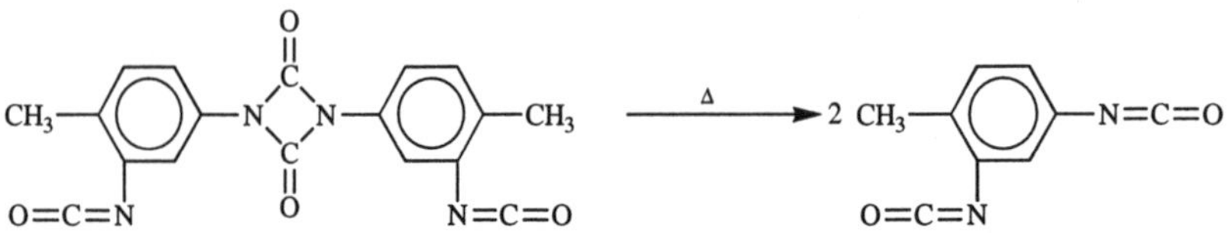

$$\text{toluene diisocyanate dimer (uretidinedione, } CH_3\text{, } N{=}C{=}O\text{)} \xrightarrow{\Delta} 2\; CH_3C_6H_3(N{=}C{=}O)_2$$

Unsaturated prepolymers crosslink with peroxides or sulfur. This unsaturation can be present in the backbone or in the pendant groups. Vulcanization or crosslinking of elastomers with sulfur or peroxides is discussed in Chapter 8.

Thermoplastic elastomers exhibit physical properties that are similar to those of cast and millable elastomers at ambient temperatures. These materials, however, are not crosslinked and flow at elevated temperatures. They are fabricated like other thermoplastic polymers, are high in molecular weight, and are hydroxy-terminated. Such polymers form from linear hydroxy-terminated polyester or polyethers that are condensed with diisocyanates and glycols. Strict stoichiometry must be maintained to achieve high molecular weights.

A structure study was carried out on a model compound of one elastomer[130] prepared with 4,4′-diphenylmethane diisocyanate and butanediol hard segments. It was shown that the chains are probably linked together in stacks through C=O· · ·H–N hydrogen bonds between the urethane groups. This bonding stabilizes the overall structure in both directions, perpendicular to the chain axis. Such an arrangement of the molecules was also proposed earlier.[131]

6.10.7. Polyurethane Foams

These foams are chemically very similar to other polyurethane materials, except that gas evolutions during the reactions take place simultaneously with chain lengthening and crosslinking. This results in formation of cellular structures. The degree of crosslinking determines to a great extent the rigidity of the foam. In addition, linear or only slightly branched polymers produce flexible foams, while more highly branched polymers form rigid ones. The foaming is caused by liberation of CO_2 from reactions between added water and isocyanate groups:

$$\sim N{=}C{=}O + H_2O \longrightarrow \sim N(H){-}C(OH){=}O \longrightarrow \sim N{-}H(H) + CO_2$$

$$2 \sim N{=}C{=}O + H_2O \longrightarrow \sim NH{-}C(=O){-}O{-}C(=O){-}NH\sim \longrightarrow \sim NH{-}C(=O){-}NH\sim + CO_2$$

In addition, in many industrial practices, additional carbon dioxide or freon gas (a major sources of environmental pollution) may be introduced into the system as it cures. For rigid foams a low boiling liquid may be added to form additional bubbles. Appropriate catalysts and foam stabilizers or surfactants are added to control foam formation, cell size, and cure. The catalysts are either tin compounds or tertiary amines. The surfactants that are necessary to control the cell size are usually based on siloxanes.

6.11. Epoxy Resins

These resins comprise a general class of crosslinkable, low molecular weight materials with epirane rings as the main functional group.[134,135] It does not include polyethers formed through ring-opening polymerizations of ethylene and propylene oxides.

6.11.1. Preparation of Commercial Epoxy Resins

The earliest developed commercial epoxy resins were diglycidyl ethers of 4,4′-isopropylidine-diphenol (Bisphenol A) formed by reacting epichlorohydrin with the diphenol. The reaction sequences involve formations of alkoxide ions, followed by nucleophilic additions to the least hindered carbons[134]:

$$HO{-}C_6H_4{-}C(CH_3)_2{-}C_6H_4{-}OH + OH^{\ominus} \rightleftharpoons HO{-}C_6H_4{-}C(CH_3)_2{-}C_6H_4{-}O^{\ominus} + HOH$$

$$HO{-}C_6H_4{-}C(CH_3)_2{-}C_6H_4{-}O^{\ominus} + H_2C\overset{O}{—}CH{-}CH_2Cl \longrightarrow$$

$$HO{-}C_6H_4{-}C(CH_3)_2{-}C_6H_4{-}O{-}CH_2{-}C(O^{\ominus})(H){-}CH_2Cl$$

This is followed by ring closures through internal displacement of chloride ions:

Table 6.9. The Effect of Molar Ratios on the Molecular Weight of the Product[134–136]

Ratio of epichlorohydrin to phenol	Molecular weight
10.0 :1	370
2.0 :1	451
1.4 :1	791
1.33:1	802
1.25:1	1133
1.2 :1	1420

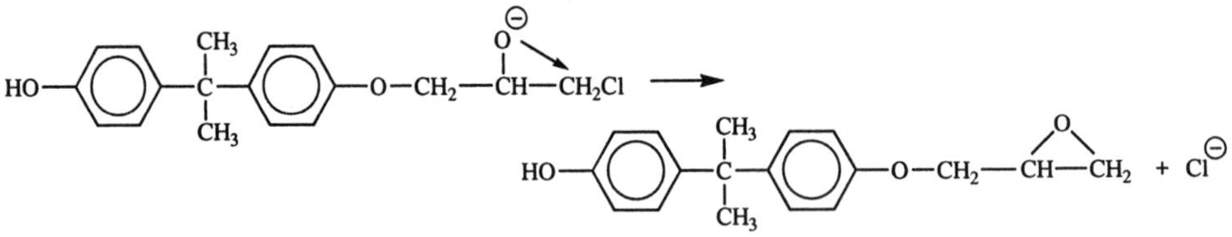

$$HO-C_6H_4-C(CH_3)_2-C_6H_4-O-CH_2-CH(O^{\ominus})-CH_2Cl \longrightarrow HO-C_6H_4-C(CH_3)_2-C_6H_4-O-CH_2-\underset{\diagdown O \diagup}{CH-CH_2} + Cl^{\ominus}$$

As explained in the discussion on phenoxy resins, a reaction of two moles of epichlorohydrin with one mole of the diphenol yields higher molecular weight byproducts and only 10% of the diether:

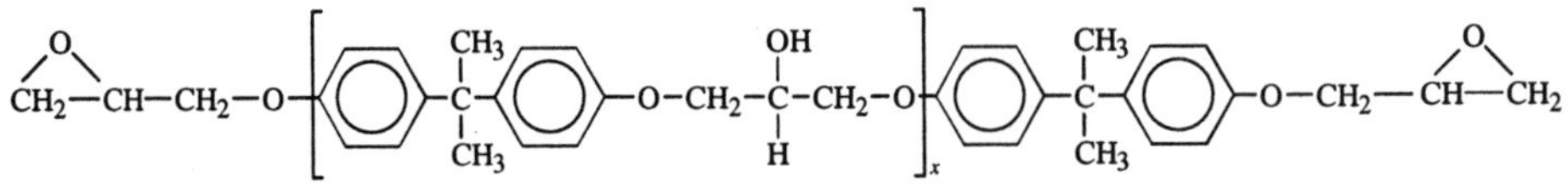

$$\underset{\diagdown O \diagup}{CH_2-CH}-CH_2-O-\left[C_6H_4-C(CH_3)_2-C_6H_4-O-CH_2-\underset{H}{\overset{OH}{C}}-CH_2-O\right]_x-C_6H_4-C(CH_3)_2-C_6H_4-O-CH_2-\underset{\diagdown O \diagup}{CH-CH_2}$$

where $x = 1–10$.

To obtain high yields of the diether, the quantities of epichlorohydrin in the reaction mixtures must be doubled or tripled.[133] This can result in yields as high as 70%. Another advantage of a large excess of epichlorohydrin in the reaction mixtures is that it serves as a solvent.[134]

A different route to the diglycidyl ethers is via Friedel-Craft reactions[135]:

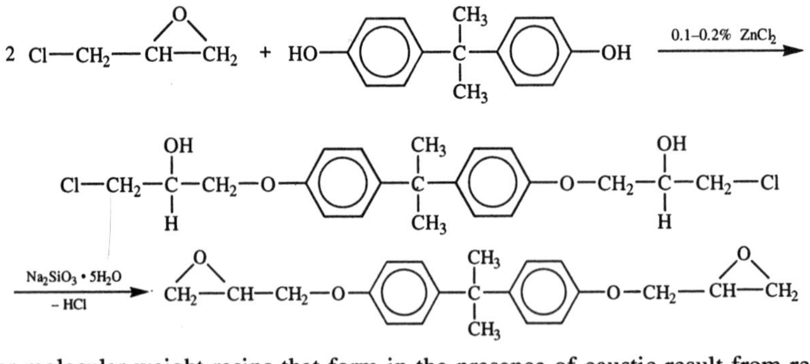

$$2\ Cl-CH_2-\underset{\diagdown O \diagup}{CH-CH_2} + HO-C_6H_4-C(CH_3)_2-C_6H_4-OH \xrightarrow{0.1\text{–}0.2\%\ ZnCl_2}$$

$$Cl-CH_2-\underset{H}{\overset{OH}{C}}-CH_2-O-C_6H_4-C(CH_3)_2-C_6H_4-O-CH_2-\underset{H}{\overset{OH}{C}}-CH_2-Cl$$

$$\xrightarrow[-\ HCl]{Na_2SiO_3 \cdot 5H_2O} \underset{\diagdown O \diagup}{CH_2-CH}-CH_2-O-C_6H_4-C(CH_3)_2-C_6H_4-O-CH_2-\underset{\diagdown O \diagup}{CH-CH_2}$$

The higher molecular weight resins that form in the presence of caustic result from reactions of glycidyl ethers with phenoxy anions:

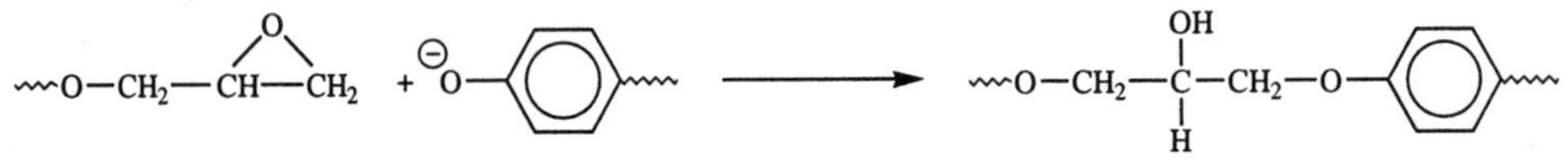

$$\sim O-CH_2-\underset{\diagdown O \diagup}{CH-CH_2} + {}^{\ominus}O-C_6H_4\sim \longrightarrow \sim O-CH_2-\underset{H}{\overset{OH}{C}}-CH_2-O-C_6H_4\sim$$

The number of repeat units in the above resin is determined by the molar ratios of the reactants. This is illustrated in Table 6.9.

6.11.2. The Crosslinking Reactions

Diglycidyl ethers of bisphenol A cannot be crosslinked through heating alone. Chemical crosslinking agents must be added. The most commonly used compounds are tertiary amines, polyfunctional amines, and acid anhydrides. Lewis acid, phenols, and compounds like dicyandiamide, however, are also used.

The reactions between tertiary amines and epoxy groups result in formations of quaternary bases:

$$R_3N\!: + \overset{O}{\widehat{CH_2-CH}}\sim\sim \longrightarrow R_3\overset{\oplus}{N}-CH_2-\underset{}{CH}(O^{\ominus})\sim\sim$$

The product reacts with hydroxy compounds to form anions:

$$R_3\overset{\oplus}{N}-CH_2-CH(O^{\ominus})\sim\sim + HO-R' \longrightarrow R_3\overset{\oplus}{N}-CH_2-CH(OH)\sim\sim + \overset{\ominus}{O}-R'$$

The anions in turn initiate polymerizations of the epoxy groups:

$$R-\overset{\ominus}{O} + \overset{O}{\widehat{CH_2-CH}}\sim\sim \longrightarrow R-O-CH_2-CH(O^{\ominus})\sim\sim$$

$$R-O-CH_2-\underset{H}{C}(O^{\ominus})\sim\sim \xrightarrow{\overset{O}{\widehat{CH_2-CH}}\sim\sim} R-O-CH_2-CH(O-CH_2-CH(O^{\ominus})\sim\sim)\sim\sim$$

$$\xrightarrow{\overset{O}{\widehat{CH_2-CH}}\sim\sim} R-O-CH_2-CH(O-CH_2-CH(O-CH_2-CH(O^{\ominus})\sim\sim)\sim\sim)\sim\sim \longrightarrow \text{etc.}$$

Because the monomer is a diepoxide, a three-dimensional lattice results. Similar three-dimensional products form from reactions with some other crosslinking materials, like boron trifluoride-etherate, boron trichloride-amine complexes, and imidazole derivatives. All these compounds initiate polymerizations of the epirane ring.

Crosslinking of the epoxy resins with primary and secondary amines is somewhat different because they react by nucleophilic addition to the epirane ring:

$$R-NH_2 + \overset{O}{\widehat{CH_2-CH}}\sim\sim \longrightarrow R-NH-CH_2-CH(OH)\sim\sim \xrightarrow{\overset{O}{\widehat{CH_2-CH}}\sim\sim} R\begin{cases}CH_2-CH(OH)\sim\sim\\ CH_2-CH(OH)\sim\sim\end{cases}$$

A considerable amount of evidence suggests that this reaction is accelerated by proton donors. Such donors may be water, phenols, or alcohols[136]:

$$R-NH_2 + \overset{O}{\widehat{CH_2-CH}}\sim\sim + HOH \longrightarrow \left[R-\underset{H}{\overset{H}{N}}\cdots\cdots\underset{H\ H}{C}(\cdots O\cdots H-OH)-CH\sim\sim\right] \longrightarrow$$

$$\left[R-\overset{\oplus}{N}H(H\cdots\overset{\ominus}{O}H)-CH_2-CH(OH)\sim\sim\right] \longrightarrow R-NH-CH_2-CH(OH)\sim\sim + HOH$$

Highly idealized pictures of reaction products from bisphenol A diglycidyl ethers with diamines can be found in the literature.[137] In actuality, the products are probably more complex. They are certainly complex when reactions involve higher molecular weight epoxy resins.

Many different aliphatic and aromatic polyamines are available for crosslinking epoxy resins. Some of these are ethylenediamine, diethylenetriamine, triethylenetetramine, tetraethylenepentamine, and many others. Among the aromatic polyamine are *p*-phenylenediamine, *m*-phenylenediamine, 4,4′-diaminodiphenylmethane, and diaminodiphenylsulfone.

A special crosslinking agent for epoxy resins is dicyanodiamide, also referred to by its trade name as cyanoguanidine. It is used for high-temperature cures and it is believed that the compound condenses with the epirane structures to form 2-aminooxazoline derivatives[137a]:

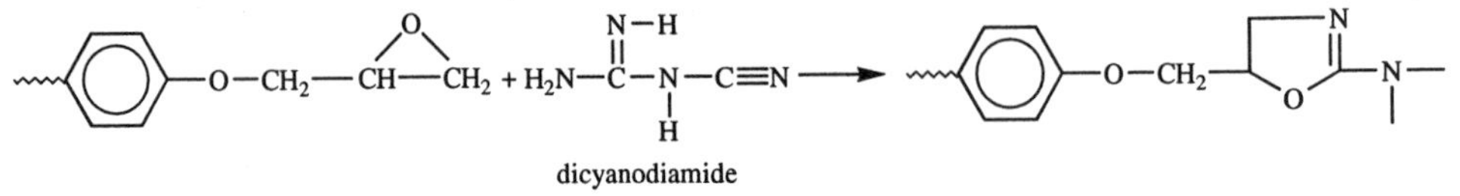

dicyanodiamide

Earlier, however, it was speculated that guanidyl urea forms.[140]

The curing mechanism of epoxy resins with imidazol was investigated. The following mechanism was proposed[137b]:

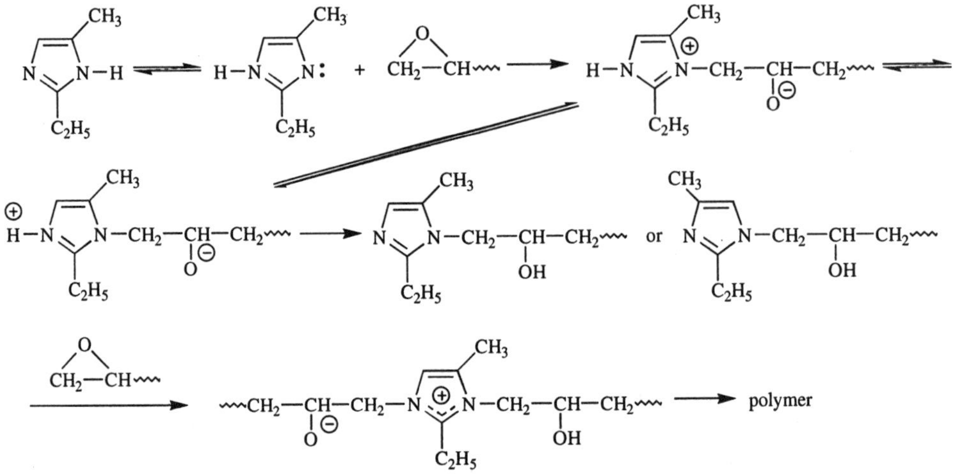

or

polymer

Many cyclic acid anhydrides are used industrially for crosslinking epoxy resins. These are mono- and polyanhydrides. Some typical, commercially used anhydrides are shown.

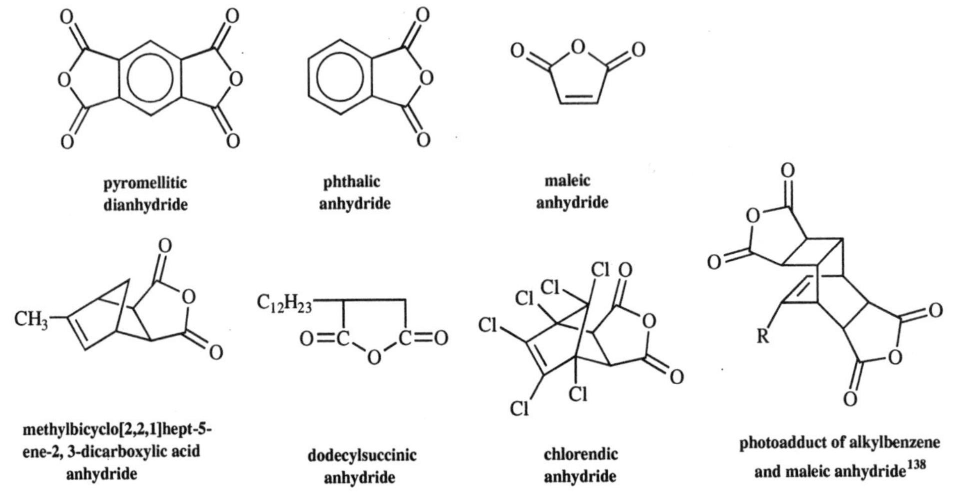

pyromellitic dianhydride

phthalic anhydride

maleic anhydride

methylbicyclo[2,2,1]hept-5-ene-2, 3-dicarboxylic acid anhydride

dodecylsuccinic anhydride

chlorendic anhydride

photoadduct of alkylbenzene and maleic anhydride[138]

The anhydrides react with traces of moisture or with pendant hydroxyl groups first. This opens the anhydride groups and frees the carboxylic acids for reactions with the epoxy rings:

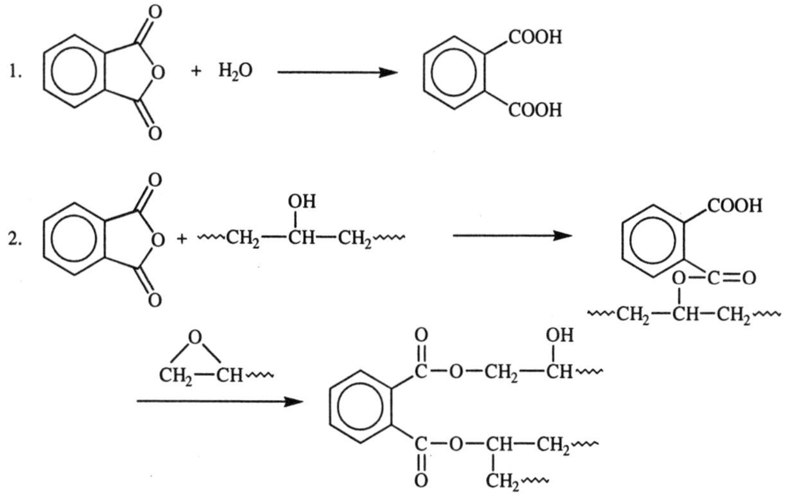

When only epoxy and hydroxy groups are present, very little reaction takes place even at elevated temperatures, as high as 200 °C. Proton donors, however, catalyze the reaction.[139] When anhydrides are present in the reaction mixtures, diesters form as a result of termolecular transition states. Kinetic data support that[136]:

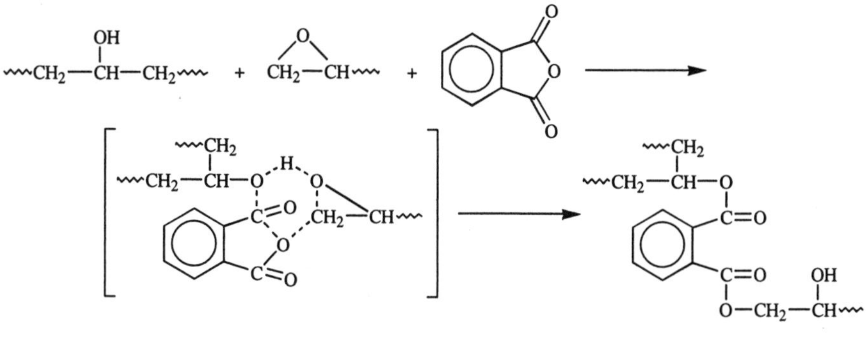

Tertiary amines are very effective in catalyzing reactions of anhydrides with epoxies:

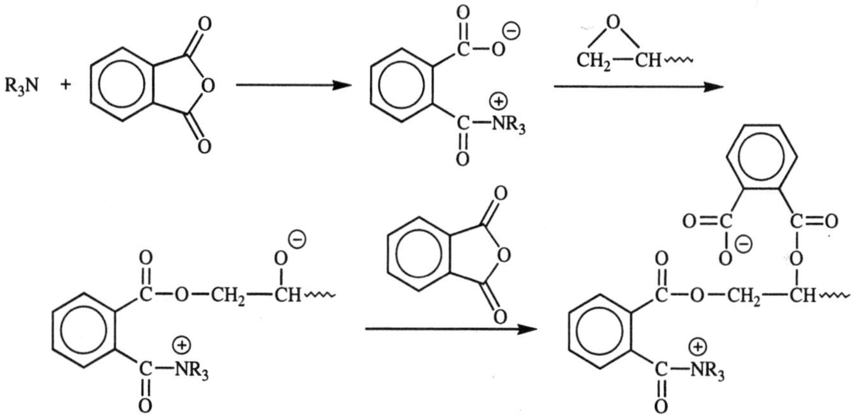

Reactions in the presence of alcohols take a somewhat different path[142,143]:

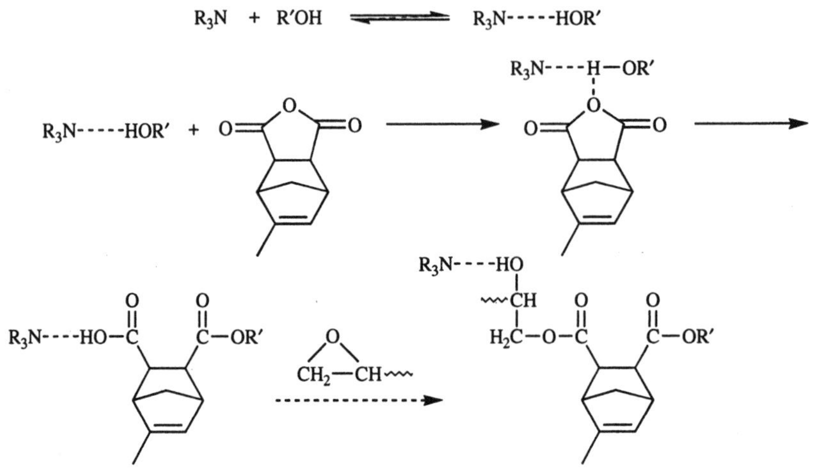

$R_3N + R'OH \rightleftharpoons R_3N\text{----}HOR'$

where R′OH represents both epoxide molecules with hydroxyl function and other alcohols in the system.

Another, similar group of epoxy resins, called *epoxy novolacs*, forms from reactions of epichlorohydrin with low molecular weight phenolic novolacs (phenolic novolac resins are discussed in the next section):

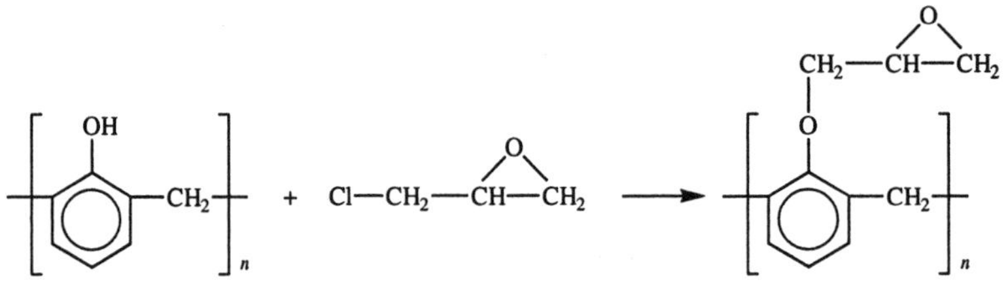

also

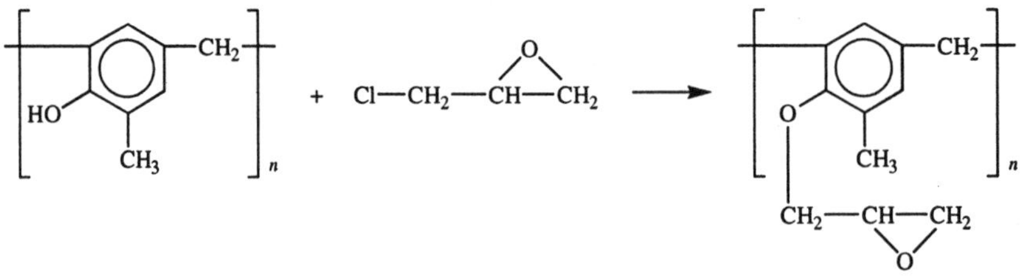

where n is typically 2.2–3.8 for liquid epoxy novolacs and 3–7 for solid resins. Epoxy resins are also prepared commercially from bisphenol F, which is a blend of *ortho* and *para* diphenolmethanes:

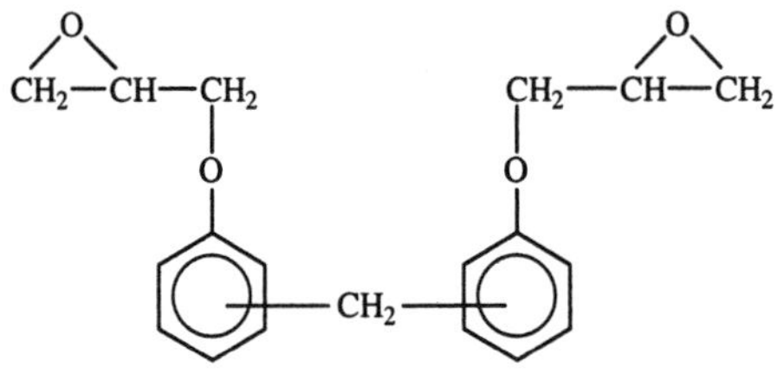

Another material is based on a condensation product of glyoxal with phenol:

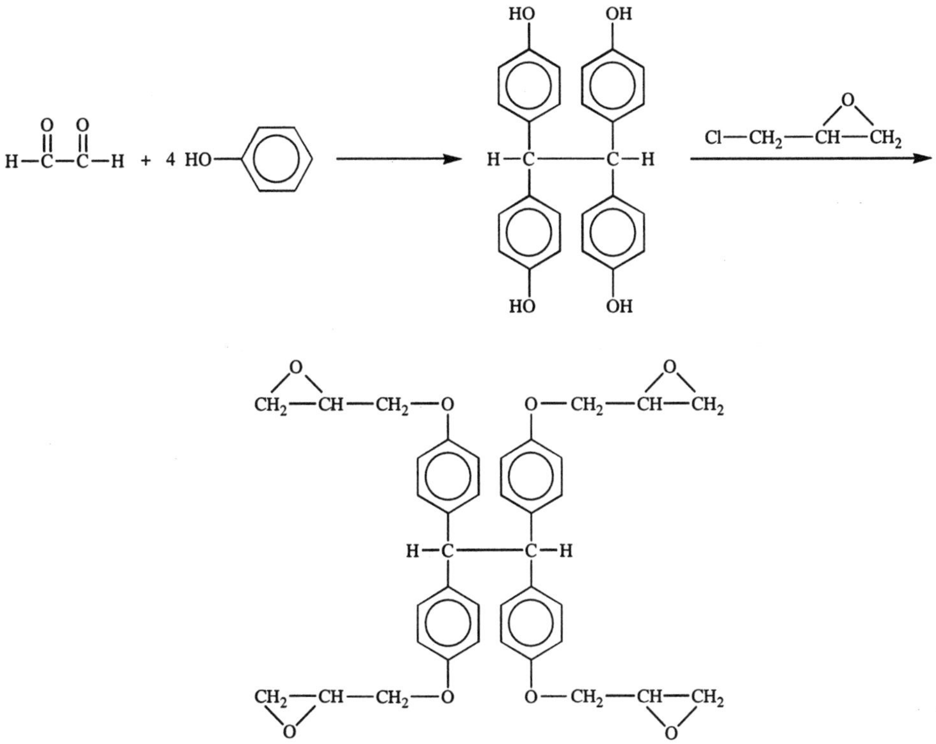

Epoxy resins formed by condensations of epichlorohydrin with resorcinol-based phenolic resins are also formed commercially:

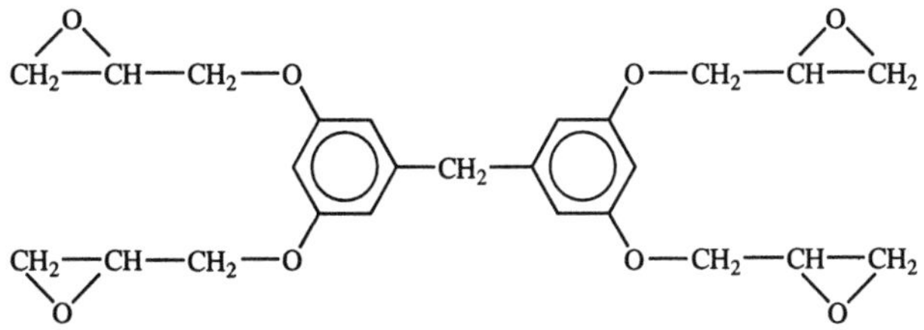

Several nitrogen-containing aromatic epoxy resins were also commercialized. These are condensation products of aromatic amines with epichlorohydrin. Some examples follow[140]:

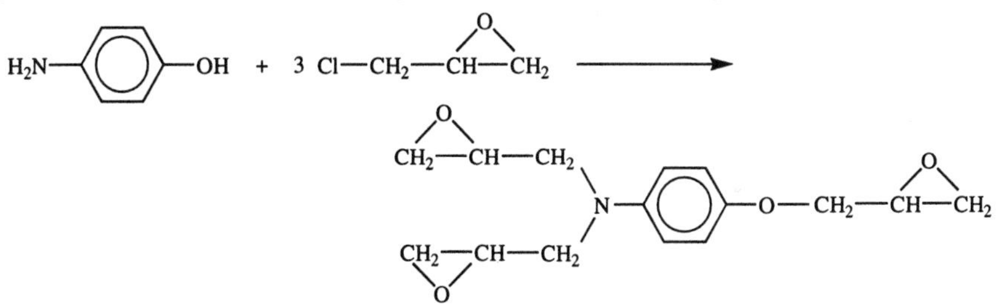

as well as

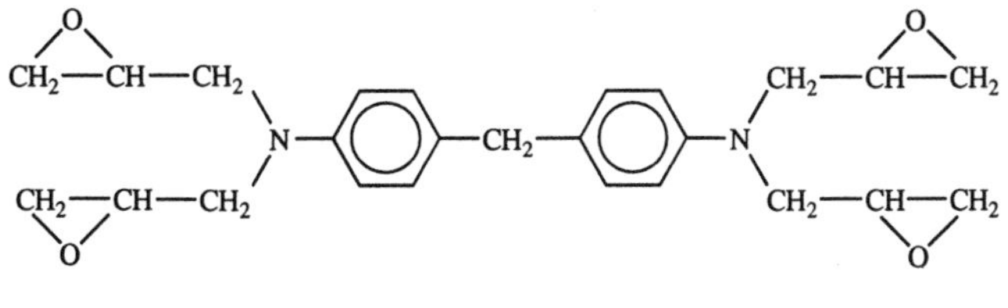

Recently,[141] the crosslinking reactions of tetrafunctional epoxy resins with aromatic primary diamines was investigated. The crosslinked polymers were characterized by UV-visible and fluorescence spectroscopies after gelation. The amount of tertiary amine fluorescence intensity of the spectra shows significant amounts of such amines in the finished products. The infrared spectra confirm the overall reaction of epoxides with amines, but also show that ether formation becomes significant only late in the cure. In addition, during the cure, especially in air, some oxidations and degradations occur.[141] This results in color formation.

6.11.3. Cycloaliphatic Epoxides

Many cycloaliphatic epoxies are products of oxidation of cyclic olefins with peracids, like peracetic. Many were commercialized over the years for use as active diluents, though later some were withdrawn. One commercial group of resins is obtained from cyclopentadiene:

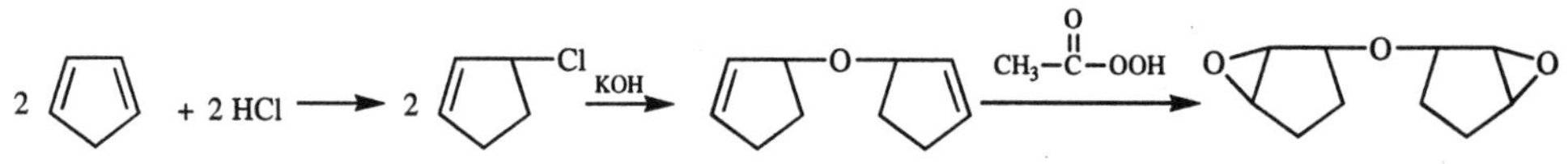

2 (cyclopentadiene) + 2 HCl ⟶ 2 (chlorocyclopentene) $\xrightarrow{KOH}$ (dicyclopentenyl ether) $\xrightarrow{CH_3-C(=O)-OOH}$ (diepoxide)

bis(2,3-epoxycyclopentyl) ether

Another group of cycloaliphatic epoxy resins are prepared via Diels-Alder additions, followed by a Tischenko reaction, and completed by epoxidation. A preparation, for instance, may start with butadiene and acrolein:

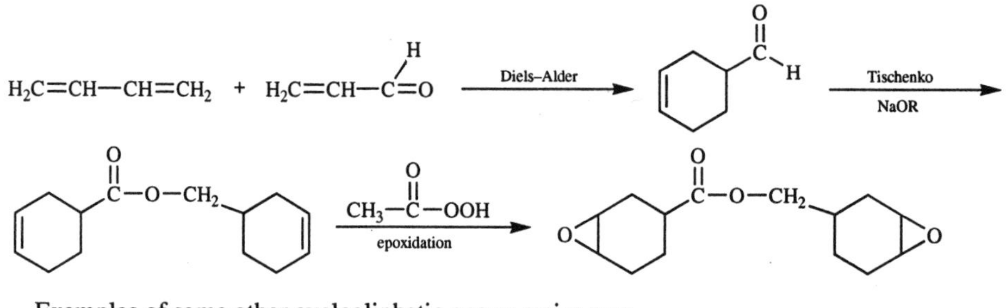

$H_2C{=}CH{-}CH{=}CH_2$ + $H_2C{=}CH{-}CH{=}O$ $\xrightarrow{\text{Diels-Alder}}$ (cyclohexene-carboxaldehyde) $\xrightarrow[\text{NaOR}]{\text{Tischenko}}$

(cyclohexenyl-C(=O)-O-CH₂-cyclohexenyl) $\xrightarrow[\text{epoxidation}]{CH_3-C(=O)-OOH}$ (diepoxide)

Examples of some other cycloaliphatic epoxy resins are:

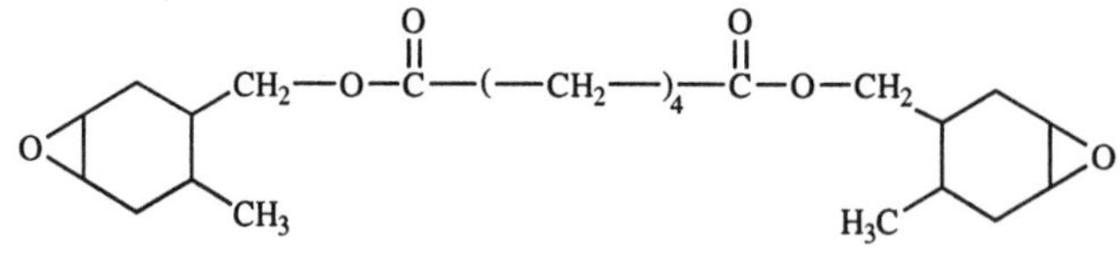

bis(3,4-epoxy-6-methyl cyclohexylmethyl)adipate

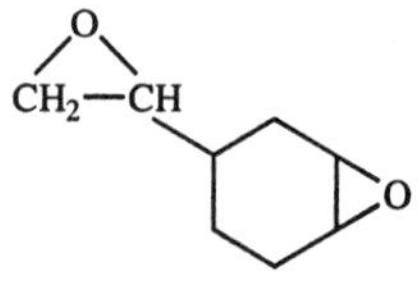

vinyl cyclohexene dioxide

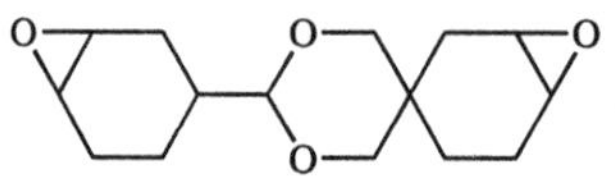

2-(3,4-epoxycyclohexyl)-5, 5-spiro(3,4-epoxy)cyclohexane-*m*-dioxane

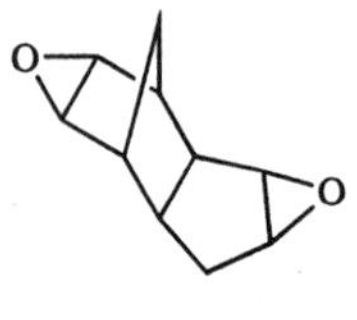

dicyclopentadiene dioxide

Acid anhydrides are more effective curing agents for cycloaliphatic epoxy resins than are the amines. In addition, the amines might also react with ester groups that are present in some of these materials and form undesirable byproducts.

6.12. Phenol-Formaldehyde Resins

The phenolic resins are condensation products of phenol and formaldehyde.[144–146,148] These materials were among the earliest commercial synthetic plastics. Two different methods[144–146] are used to prepare them. In the first, the condensations are base-catalyzed, while in the second, they are acid-catalyzed. The products formed with basic catalysts are called *resols* and with acidic catalysts, *novolacs*. Phenolic resins are used widely in coatings and laminates. The pure resins are too friable for use as structural materials by themselves. They become useful plastics, however, when filled with various fillers.

6.12.1. Resols

These thermosetting resins form in reactions of phenols with formaldehyde in water in the presence of catalytic amounts of bases. Under these conditions phenol exists as a resonance-stabilized anion:

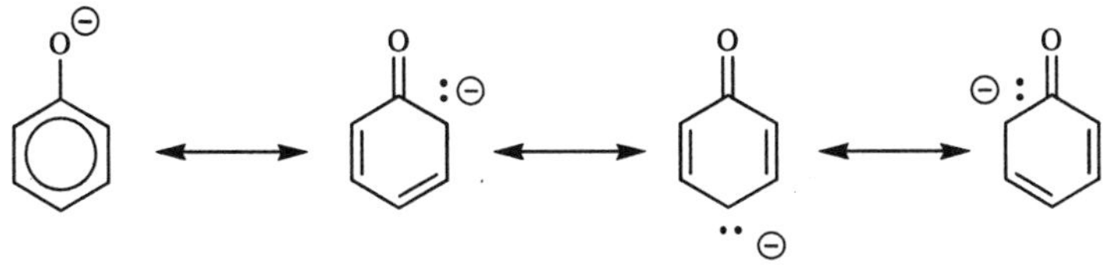

The addition of phenol anion to formaldehyde is a typical nucleophilic reaction:

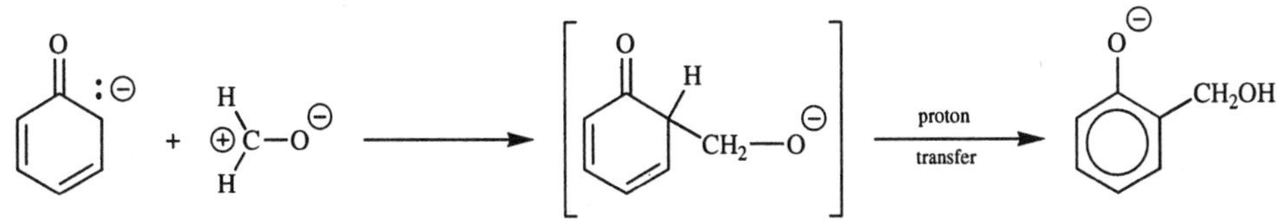

Both *ortho* and *para* methylolphenols form in the above reaction. Phenol is very reactive and monosubstituted phenols are difficult to isolate from the reaction mixture, because di- and trisubstitution occurs rapidly. No substitutions were ever shown to take place in the *meta* position. The overall reaction is as follows:

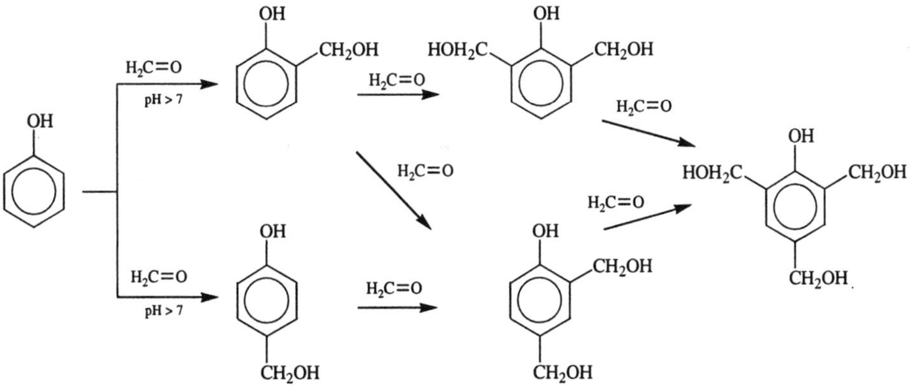

When an aqueous reaction of phenol and formaldehyde, catalyzed by sodium hydroxide, is carried out at 30 °C for 5 hours, the products are[147]:

2,4,6-trimethylolphenol	37%
2,4-dimethylolphenol	24%
2,6-dimethylolphenol	7%
p-methylolphenol	17%
o-methylolphenol	12%

The remaining 3% is unreacted phenol. As the reaction continues, methylolphenols condense with each other to form methylene bridges:

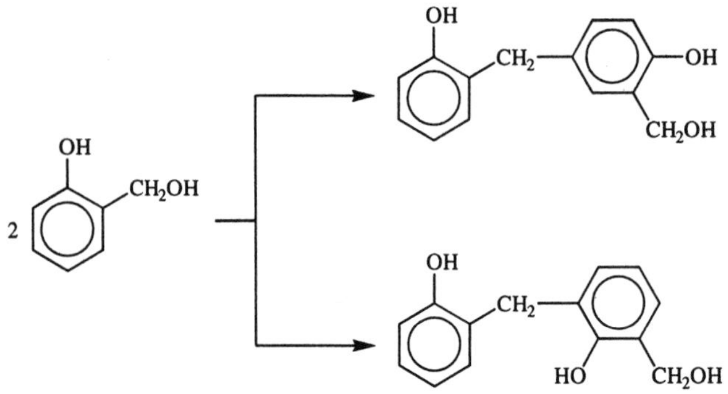

The *para*-substituted methylolphenols, of course, react in the same manner:

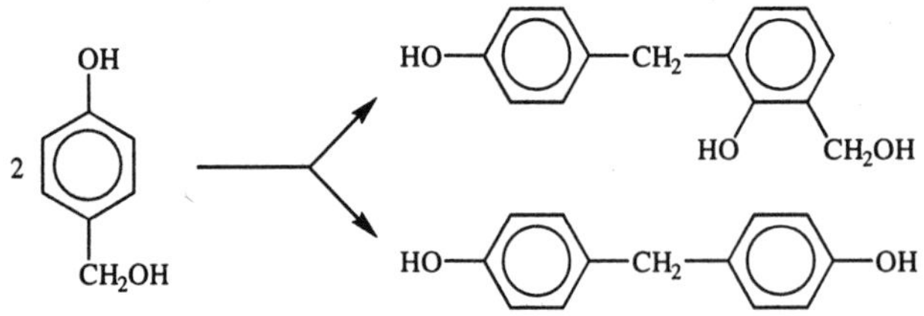

Formation of methylene bridges takes place by one of two mechanisms. One is a direct S*N*2 displacement:

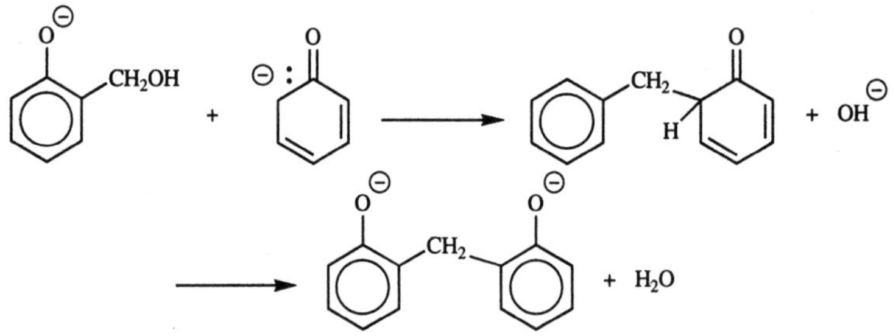

The other is addition of methylolated phenols to molecules of quinonemethides that form at typical reaction conditions, particularly when the temperatures are elevated[144–146]:

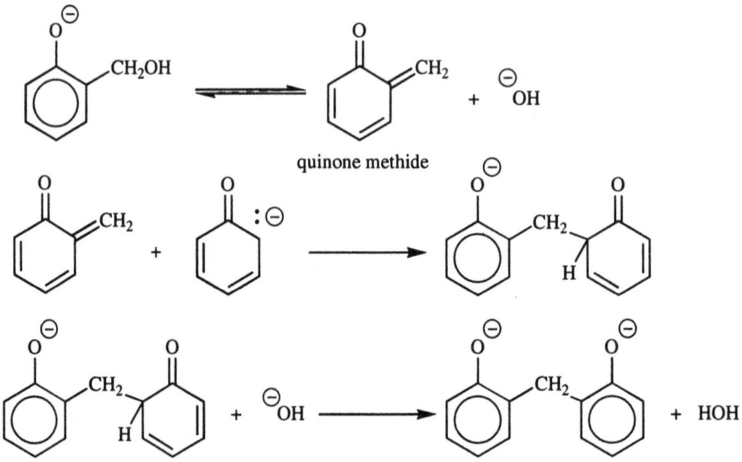

quinone methide

Two methylol-substituted phenols react with each other by the same mechanism:

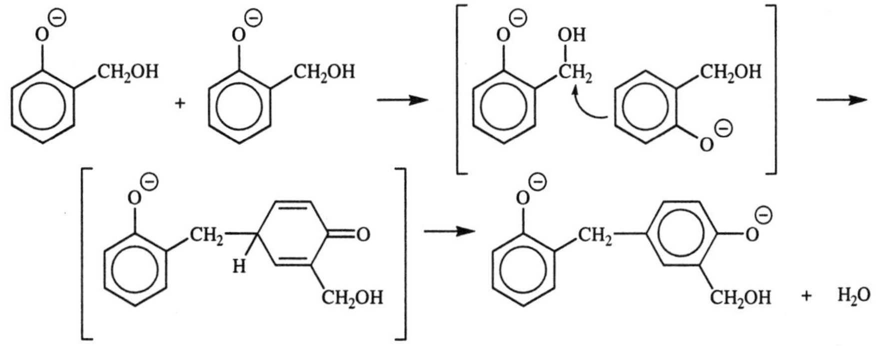

The same can be shown for *para*-substituted methylolphenols. As the reaction continues, it leads to formation of trinuclear and tetranuclear phenolic resins.

A typical liquid resole is quite low in molecular weight. It may contain no more than two or three benzene rings. Carried a little further, the condensation yields a solid resole. The pH is usually adjusted to neutral before the resoles are heated further for crosslinking. Under neutral or slightly acidic conditions, the methylol groups tend to form dibenzyl ethers:

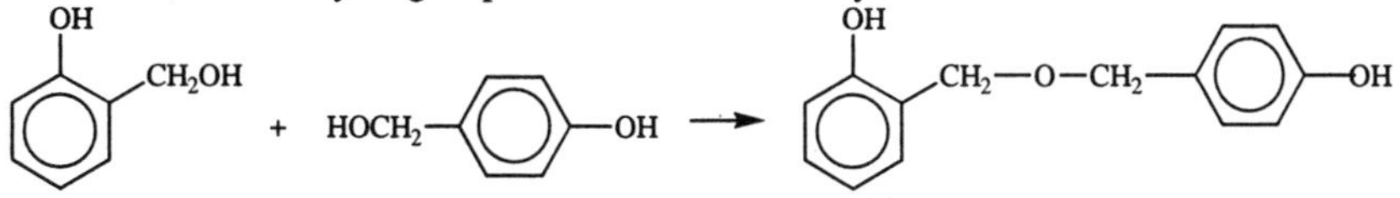

These dibenzyl ethers are unstable at higher temperatures, such as 150 °C, and decompose to yield methylene bridges and formaldehyde[147]:

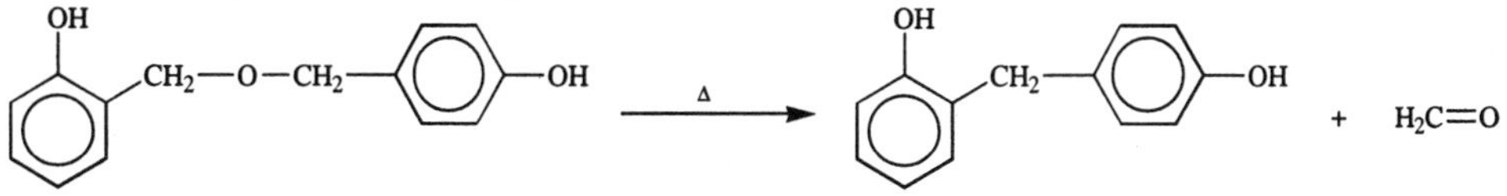

The structure of a typical resole contains both dimethylene ether and methylene bridges as well as methylol groups. Fusible and soluble resols are called A-stage resins. Further reactions cause these resins to pass through a rubbery stage, where they can still be swollen by solvents. This is called the B stage. The finally crosslinked material is called a C-stage resin. The crosslinking process involves complex and competing reactions. Each may be influenced by reaction conditions. When crosslinking of resoles take place at neutral or slightly acidic conditions, both methylene and ether linkages form. Upon heating, the ethers in turn split out formaldehyde, as shown above. The dibenzyl ethers[144–146,148] also break down at elevated temperatures and form quinonemethides:

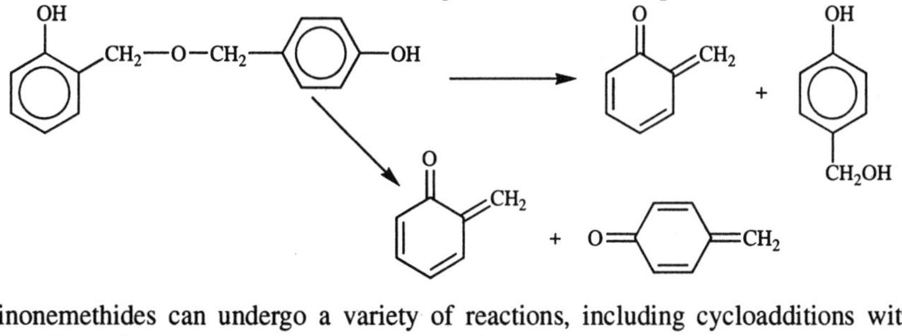

The quinonemethides can undergo a variety of reactions, including cycloadditions with other methides to form chroman groups:

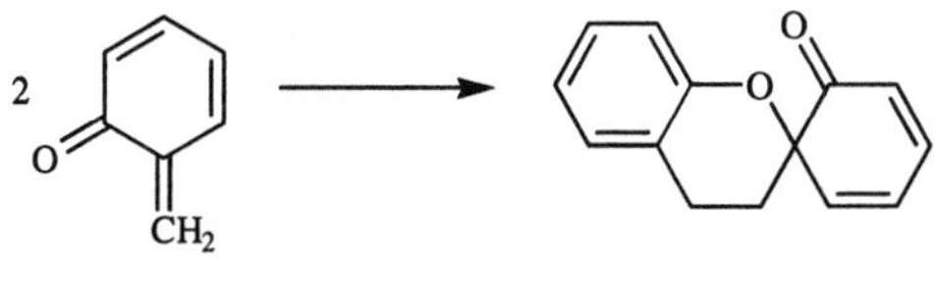

The *para*-quinone methides can couple:

$$2\ O{=}C_6H_4{=}CH_2 \longrightarrow HO{-}C_6H_4{-}CH{=}CH{-}C_6H_4{-}OH$$

6.12.2. Novolacs

The phenolic resins that form in acid-catalyzed condensations of phenols with formaldehyde are different from resols. At pH below 7, protonation of the carbonyl group of formaldehyde takes place first and is followed by electrophilic aromatic substitution at the *ortho* and *para* positions of the phenol. The initial steps of the reactions also take place in water. Here, however, a molar excess of phenol (1.25:1) must be used, because reactions on equimolar basis under acidic conditions form crosslinked resins. At a ratio of 8 moles of formaldehyde to 10 moles of phenol, novolacs of approximate molecular weight 850 form.[148] When the ratio of formaldehyde to phenol is 9:10, a molecular weight of approximately 1000 is reached. This appears to be near the limit, beyond which crosslinking results. The reaction is as follows:

$$H_2C{=}O + H^{\oplus} \longrightarrow \left[H_2C{=}\overset{\oplus}{O}H \longleftrightarrow H_2\overset{\oplus}{C}{-}OH \right]$$

$$C_6H_5OH + {}^{\oplus}CH_2OH \longrightarrow \left[\text{protonated 4-H, 4-}CH_2OH\text{ cyclohexadienone} \right] \rightleftharpoons p\text{-}HOC_6H_4CH_2OH + H^{\oplus}$$

or

$$C_6H_5OH + {}^{\oplus}CH_2OH \longrightarrow \left[\text{protonated 6-H, 6-}CH_2OH\text{ cyclohexadienone} \right] \rightleftharpoons o\text{-}HOC_6H_4CH_2OH + H^{\oplus}$$

The reaction is taking place at pH below 7, so the above *p*- and *o*-methylolphenols are transitory and are present in only small concentrations. Hydrogen ions convert them to benzylic carbocations that react rapidly with free phenol. This can be illustrated as follows:

$$HOC_6H_4CH_2OH + H^{\oplus} \rightleftharpoons HOC_6H_4CH_2\overset{\oplus}{O}H_2 \rightleftharpoons HOC_6H_4\overset{\oplus}{C}H_2 + H_2O$$

$$HO{-}C_6H_4{-}\overset{\oplus}{C}H_2 + C_6H_5OH \longrightarrow HO{-}C_6H_4{-}CH_2{-}C_6H_4{-}OH + H^{\oplus}$$

$$\searrow\ HO{-}C_6H_4{-}CH_2{-}C_6H_4(o\text{-}OH) + H^{\oplus}$$

The *ortho*-substituted mehylolphenols react in the same manner.

Further metholylation of dihydroxy diphenyl methanes takes place until all the formaldehyde is used up. Methylol groups react with each other quickly and form methylene bridges. The *para* position is more reactive than the *ortho*[144–146] at pH below 3. The opposite is true, however, at pH 5–6, where the *ortho* position is more reactive. Typical novolacs formed in these reactions are not very high in molecular weight and contain no more than 6–10 benzene rings. If divalent metal salts, like zinc acetate in acetic acid, are used to catalyze the reaction, then the *ortho* positions become considerably more reactive.[150] *Ortho*-methylene bridges predominate in the products from these reactions:

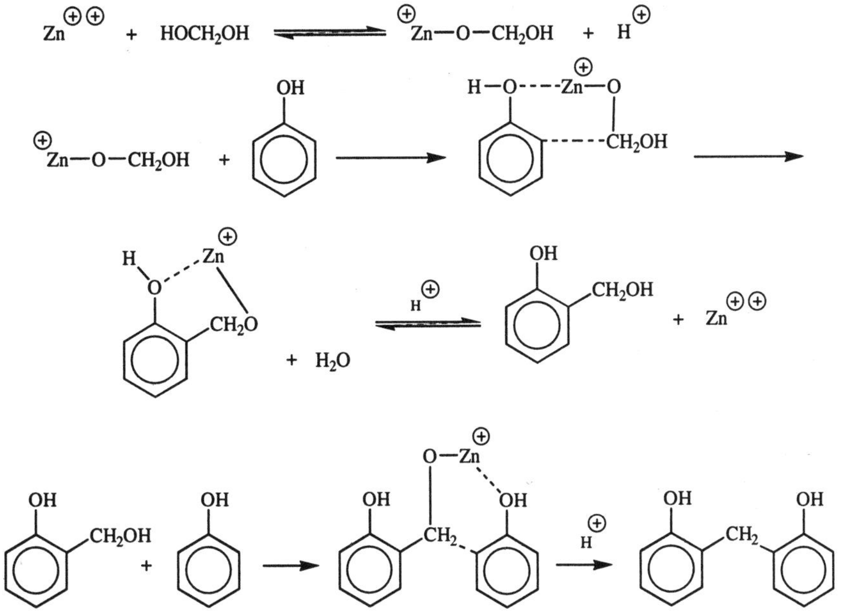

$$Zn^{\oplus\oplus} + HOCH_2OH \rightleftharpoons {}^{\oplus}Zn{-}O{-}CH_2OH + H^{\oplus}$$

It is possible to form high molecular weight novolacs by carrying out the reactions of alkyl phenyl ethers with formaldehyde in acetic acid in the presence of perchloric acid[149]:

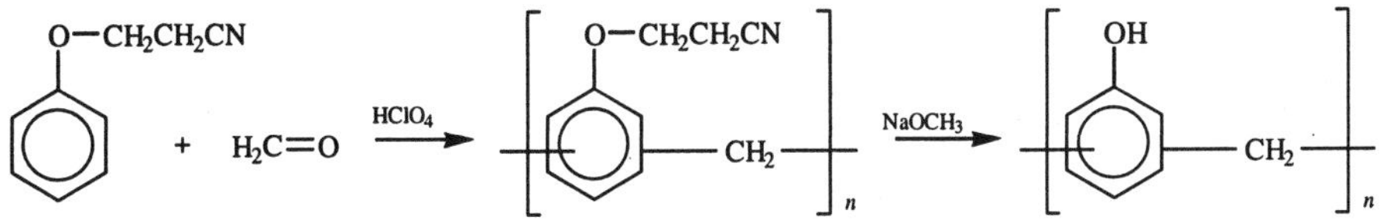

Novolacs are crosslinked by additions of more formaldehyde to the soluble, thermoplastic materials. The additional formaldehyde can be in the form of paraform, an oligomer of formaldehyde that decomposes to formaldehyde upon heating. It can also come from hexamethylenetetramine, a condensation product of formaldehyde with ammonia:

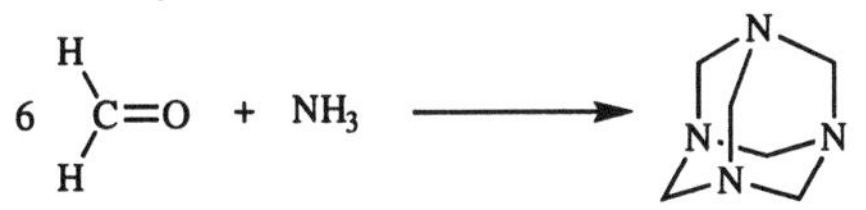

$$6\ H_2C{=}O + NH_3 \longrightarrow$$

Hexamethylenetetramine decomposes back to formaldehyde and ammonia upon heating. Some of the ammonia is picked up by the novolacs, with the result that there are some benzylamine bridges in the product:

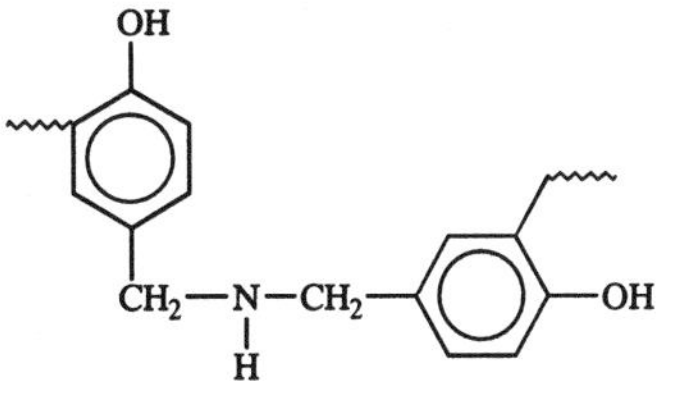

The mechanism of this reaction is discussed in the next section.

6.12.3. Ammonia-Catalyzed Phenolic Resins

These resins differ from the other resols, because there are some benzylamine bridges present (as shown above) in their structure. The reactions result in early losses of water and allow higher molecular weight buildups before the resins gel. Nitrogen-containing resols are darker in color than regular resols. The dibenzylamine bridges form as a result of reactions of the methylol groups with ammonia or amines[145]:

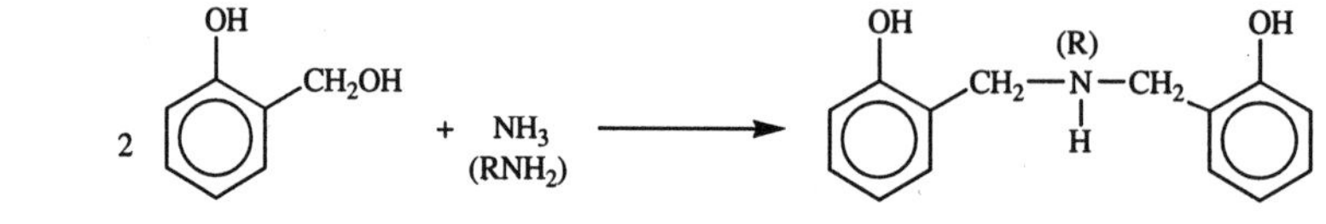

$$2\ HO\text{-}C_6H_4\text{-}CH_2OH + NH_3\ (RNH_2) \longrightarrow HO\text{-}C_6H_4\text{-}CH_2\text{-}N(H)(R)\text{-}CH_2\text{-}C_6H_4\text{-}OH$$

The overall mechanism can be shown as a special case of a Mannich reaction:

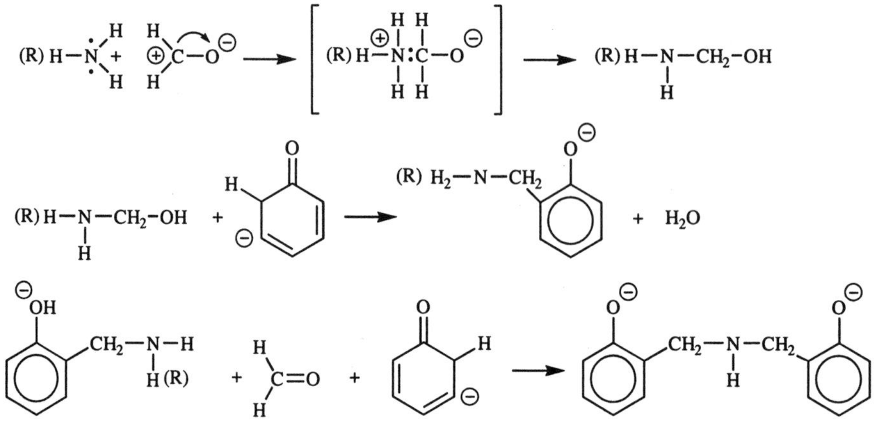

$$(R)H\text{-}NH_2 + H_2C^{\oplus}\text{-}O^{\ominus} \longrightarrow \left[(R)H\text{-}N^{\oplus}H_2\text{:}CH_2\text{-}O^{\ominus}\right] \longrightarrow (R)H\text{-}N(H)\text{-}CH_2\text{-}OH$$

$$(R)H\text{-}N(H)\text{-}CH_2\text{-}OH + \text{(phenoxide carbanion)} \longrightarrow (R)H_2\text{-}N\text{-}CH_2\text{-}C_6H_4\text{-}O^{\ominus} + H_2O$$

$$^{\ominus}O\text{-}C_6H_4\text{-}CH_2\text{-}N(H)\text{-}H(R) + H_2C{=}O + \text{(phenoxide carbanion)} \longrightarrow {}^{\ominus}O\text{-}C_6H_4\text{-}CH_2\text{-}N(H)\text{-}CH_2\text{-}C_6H_4\text{-}O^{\ominus}$$

In amine or ammonia-catalyzed reactions[145] the additions and condensations occur almost simultaneously with each other. Methylol groups are still present in the finished resins to the extent of 15–30 groups per 100 phenol residues. The structures are branched and the degree of branching depends upon the amine used.

6.12.4. Typical Commercial Preparations

The resols are usually prepared in reaction kettles, using 1.5–2.0 moles of formaldehyde per mole of the phenol. The reactions are rapid and the condensations to resoles might be accomplished in one hour. Formaldehyde is often added in the form of formalin. The quantity of the added caustic or ammonia might comprise 1% of the phenol in the reaction mixture. These reactions are carried out at water reflux for a specified time. The pH is then lowered to neutral and the water distilled off, usually at reduced pressure. The progress of the condensation is followed by measuring the melting point, the gel time (time required for the material to become thermoset at a specified temperature), solubility, or free phenol content.

Better quality novolacs and resoles are prepared in stainless steel resin kettles. For novolacs, a typical recipe might call for a mole of phenol to 0.8 mole of formaldehyde (usually added as formalin, a 37% solution in water). Acid catalysts, like oxalic, hydrochloric, or others, are added in amounts of 1–2% by weight of the phenol. Oxalic acid is favored over hydrochloric, sulfuric, or phosphoric acid due to corrosion problems. In addition, vapors of hydrochloric acid tend to react with vapors of formaldehyde and form a carcinogenic compound, 1,1′-dichlorodimethyl ether. The reactions are conducted at the reflux temperature of water for 2–4 hours. Maleic acid is sometimes used to form high-melting novolacs. In a typical preparation of novolacs, molten phenol (usually kept at 65 °C) is introduced into the reaction kettle and heated to 95 °C. The catalyst is then added. This is followed by addition of the formaldehyde solution to the kettle with stirring, at a rate that allows a gentle reflux. After addition, heating and stirring are continued until almost all the formaldehyde is used up. At that point, the resins separate from the aqueous phase. Water is distilled off and the temperature raised in the process to about 160 °C. The unreacted phenol is removed by vacuum distillation. The end of the reaction may be determined by the melting point of the product or by its melt viscosity.

Cresols are also often used in preparations of phenolic resins. These may be individual isomers or mixtures of all three. Cresilic acids, mixtures of all three isomers, rich in *m*-cresol and low in *o*-cresol, are preferred.

Xylenols (all six isomers) are now also in common use to form alkali-resistant grades of phenolic resins. High 3,5-xylenol mixtures are preferred. Also, resorcinol, which forms very reactive phenolic resins, is used in preparations of cold-setting adhesives. Higher homologs of phenol, like bisphenol A, are used to prepare special phenol-formaldehyde condensates.

6.13. Aminopolymers

Currently, the bulk of the commercial polymers that would fit into this category are urea-formaldehyde and melamine-formaldehyde resins.[151] Over the years, however, many other materials that might fit into this group were prepared but not adopted for use for various reasons.

6.13.1. Urea-Formaldehyde Resins

These thermosetting resins find applications in coatings, adhesives, laminating, and molding compositions. The materials are formed in water at a pH above 7 at the start of the reaction, because the methylol derivatives that form condense rapidly at acidic conditions. The initial step, where urea undergoes a nucleophilic addition of formaldehyde, can be shown as follows:

$$2\ H_2N{-}\overset{O}{\overset{\|}{C}}{-}NH_2 + 3\ \underset{H}{\overset{H}{>}}C{=}O \longrightarrow H_2N{-}\overset{O}{\overset{\|}{C}}{-}NH{-}CH_2{-}OH$$

In the past, it was believed by some that further condensations that take place at pH below 7 include formations of cyclic intermediates. This, however, was never demonstrated.[151] NMR spectra of urea-formaldehyde resins show[152] that condensations under acidic conditions proceed via formations of methylene linkages:

$$H_2N{-}\overset{O}{\overset{\|}{C}}{-}\underset{H}{\underset{|}{N}}{-}CH_2OH + H_2N{-}\overset{O}{\overset{\|}{C}}{-}\underset{H}{\underset{|}{N}}{-}CH_2OH \longrightarrow H_2N{-}\overset{O}{\overset{\|}{C}}{-}\underset{H}{\underset{|}{N}}{-}CH_2{-}\underset{H}{\underset{|}{N}}{-}\overset{O}{\overset{\|}{C}}{-}\underset{H}{\underset{|}{N}}{-}CH_2OH$$

Under alkaline conditions, on the other hand, dimethylene ether groups form instead[152]:

$$H_2N{-}\overset{O}{\overset{\|}{C}}{-}\underset{H}{\underset{|}{N}}{-}CH_2OH + H_2N{-}\overset{O}{\overset{\|}{C}}{-}\underset{H}{\underset{|}{N}}{-}CH_2OH \longrightarrow H_2N{-}\overset{O}{\overset{\|}{C}}{-}\underset{H}{\underset{|}{N}}{-}CH_2{-}O{-}CH_2{-}\underset{H}{\underset{|}{N}}{-}\overset{O}{\overset{\|}{C}}{-}\underset{H}{\underset{|}{N}}{-}CH_2OH$$

In addition, the more highly condensed water-soluble resins contain hemiformal groups.

Further reactions may not result in formation of polymeric materials.[153] This is especially true when the ratios of formaldehyde to urea are low. Some are of the opinion that linear oligomeric condensates form instead. These urea-formaldehyde condensates separate as colloidal dispersions that are stabilized by association with excess formaldehyde.[153] The crosslinking reaction consists of an agglomeration of colloidal particles with an accompanying release of formaldehyde. This opinion is supported by several observations: (1) When one plots the logarithm of solution viscosity against time during the polymerization, the plot exhibits a sharp break. Also, the plot differs from similar ones for phenol-formaldehyde condensation reactions that show continuous increases in viscosity. (2) Scanning electron micrographs of the fully cured resins show surface characteristics that resemble more the surfaces of coagulated and coalesced colloidal particles than those of high molecular weight polymers. (3) X-ray diffraction patterns and laser Raman spectra of the crosslinked resins show that there are crystalline areas in the material and absence of water. Similar patterns are obtained from hydrogen-bonded proteins with close chain packing. On the other hand, FT-IR studies[154] show that methylene and ether crosslinks are present in the cured resin. There are also indications of the presence of cyclic ether units. The above information also suggests that the final structure of the urea-formaldehyde resin may be a function of the feed ratio and the pH at which it was formed.

Urea-formaldehyde resins for surface coatings are commonly modified for solubility in organic solvents by reacting them with alcohols to form ether groups. Usually *n*-butyl alcohol is used. The reaction is carried out under basic conditions, before acidification:

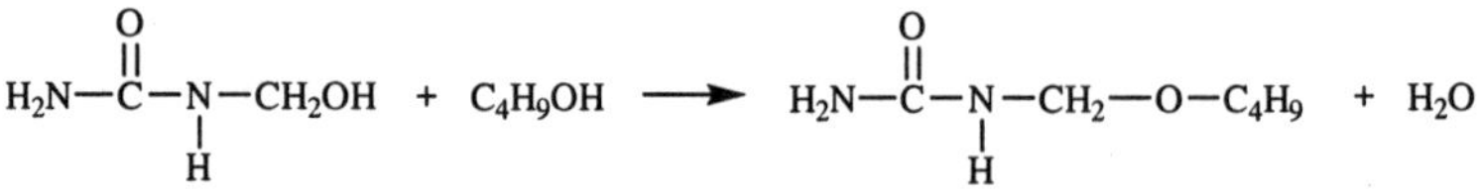

$$H_2N-\overset{O}{\overset{\|}{C}}-\underset{H}{\underset{|}{N}}-CH_2OH + C_4H_9OH \longrightarrow H_2N-\overset{O}{\overset{\|}{C}}-\underset{H}{\underset{|}{N}}-CH_2-O-C_4H_9 + H_2O$$

After etherification, the reaction mixture is acidified and the resin further reacted to acquire the desired degree of condensation. A typical butylated urea-formaldehyde resin contains 0.5–1.0 mole of butyl ether groups per mole of urea.

6.13.2. Melamine-Formaldehyde Resins

These resins are quite similar to urea-formaldehyde condensates and, probably for that reason, find similar applications. Melamine reacts with formaldehyde under slightly alkaline conditions to form mixtures of various methylolmelamines[155]:

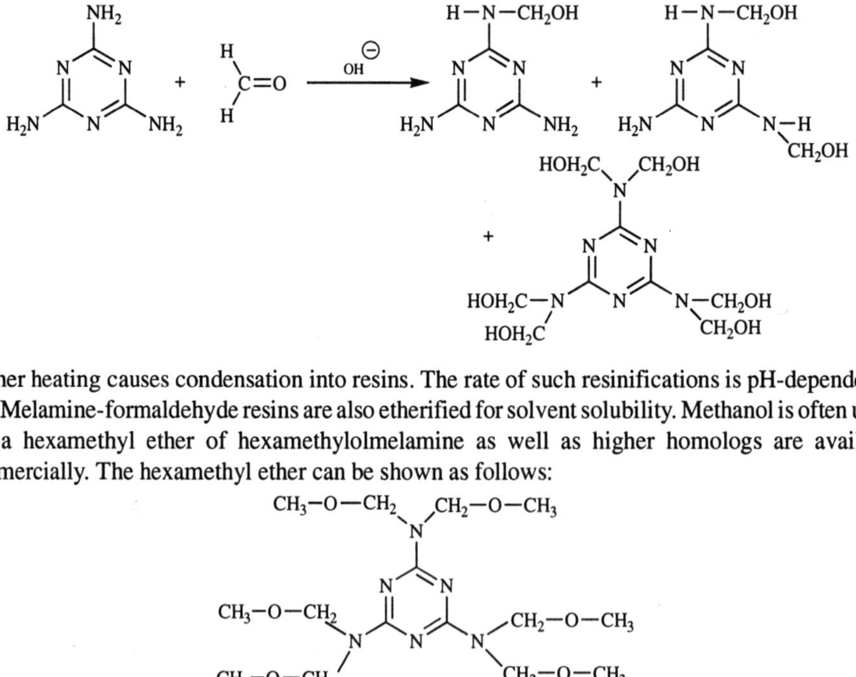

$$\text{melamine} + \underset{H}{\overset{H}{>}}C{=}O \xrightarrow{OH^{\ominus}} \text{monomethylolmelamine (H–N–CH}_2\text{OH)} + \text{dimethylolmelamine (H–N–CH}_2\text{OH, N–H/CH}_2\text{OH)} + \text{hexamethylolmelamine (three } N(CH_2OH)_2 \text{ groups)}$$

Further heating causes condensation into resins. The rate of such resinifications is pH-dependent.

Melamine-formaldehyde resins are also etherified for solvent solubility. Methanol is often used, and a hexamethyl ether of hexamethylolmelamine as well as higher homologs are available commercially. The hexamethyl ether can be shown as follows:

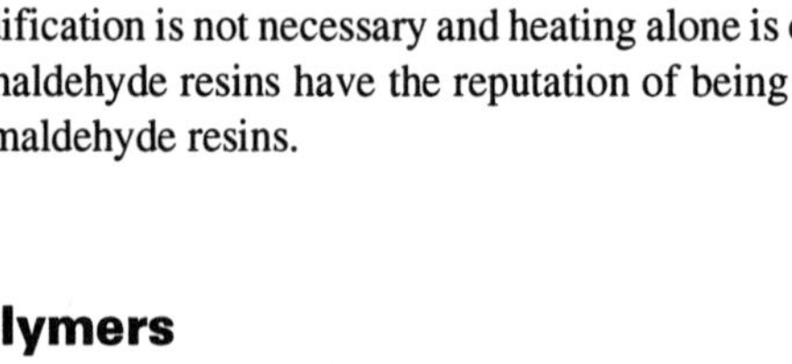

$$C_3N_3[N(CH_2-O-CH_3)_2]_3$$

The ethers cleave upon acidification and network structures form. For metholylated melamines that are not etherified, acidification is not necessary and heating alone is often adequate for network formation. Melamine-formaldehyde resins have the reputation of being harder and more moisture resistant than the urea-formaldehyde resins.

6.14. Silicone Polymers

These semi-inorganic materials are important industrially.[156–160] They exhibit good thermal stability, good electrical insulating characteristics, water repellency, and can act as release coatings for some materials. In addition, these properties are maintained over a wide range of temperatures. As a result, these polymers have many diverse uses.

6.14.1. Polysiloxanes

The silicon atom is below carbon in the periodic table with a similar electronic arrangement, which in silicon is: $1s^2\,2s^2\,2p^6\,3s^2\,3p^2$. The larger atomic radius, however, makes the silicon–silicon single bond much less energetic. Because of this, silanes (Si_nH_{2n+2}) are much less stable than alkanes. The opposite, however, is true of silicon–oxygen bonds that are more energetic (about 22 kcal/mole) than the carbon–oxygen bonds. Polysiloxanes therefore have recurring Si–O linkages in the backbones.

The starting materials can be prepared through hydrolyses of alkyl or arylsilicone halides.[156–160] Organosilicone halides, in turn, are made commercially by heating alkyl or aryl halides with silicon at 250–289 °C. This reaction is catalyzed by copper:

$$RCl + Si \xrightarrow[Cu]{250\text{–}280\,^\circ C} SiCl_4 + RSiCl_3 + R_2SiCl_2 + R_3SiCl$$

The same materials can also be formed by the Grignard reaction:

$$RMgCl + SiCl_4 \longrightarrow RSiCl_3 + MgCl_2$$

$$RMgCl + R_2SiCl_2 \longrightarrow R_3SiCl + MgCl_2$$

$$(CH_3)_3SiCl + C_6H_5\text{–}MgCl \longrightarrow CH_3\text{–}Si(CH_3)_2\text{–}C_6H_5 + MgCl_2$$

Alkyl silanes can also be prepared by additions of trichlorosilanes to ethylene or acetylene:

$$HSiCl_3 + CH_2{=}CH_2 \rightarrow CH_3\text{–}CH_2\text{–}SiCl_3$$

$$HSiCl_3 + CH{\equiv}CH \rightarrow CH_2{=}CH\text{–}SiCl_3$$

Trichlorosilane reacts with aromatic compounds in the presence of boron trichloride:

$$HSiCl_3 + 2\,C_6H_6 \xrightarrow{BCl_3} (C_6H_5)_2\text{—}SiCl_2$$

The siloxane linkages can result from hydrolysis of the halides. The products of hydrolyses, silanols, are unstable and condense:

$$Cl\text{—}SiR_2\text{—}Cl \xrightarrow{H_2O} HO\text{—}SiR_2\text{—}OH \longrightarrow \text{—[—}SiR_2\text{–O–}SiR_2\text{—O—]—} + H_2O$$

The above reaction is one possible route to siloxane polymers. As a general method, however, this approach is not very satisfactory, because ring formations accompany the reactions. Some rings that form from hydrolyses of thrichlorosilanes are structurally complex.[161] They may even possess three-dimensional structures.[162,163] High molecular weight polymers, however, form readily by ring-opening polymerizations. Such polymerizations can be applied to the simple rings that form from dihalides or complex rings from trihalides. Ring-opening polymerizations, carried out on purified (by distillation) cyclic intermediates, are catalyzed by either acids or bases,[164] leading to linear siloxane polymers:

$$n\ [(CH_3)_2SiO]_3 \xrightarrow{H_2SO_4} \text{—[—}Si(CH_3)_2\text{—O—]}_{3n}\text{—}$$

Acid-catalyzed polymerizations yield lower molecular weight polymers that are mostly oils. The molecular weights of these oils can be controlled by additions of hexamethyldisiloxane during the polymerization reactions. When catalyzed by bases, high molecular weight elastic polymers form.

$$n\,\begin{matrix}(CH_3)_2-Si-O-Si-(CH_3)_2\\ \quad\ | \qquad\ \ |\\ \quad\ O \qquad\ O\\ \quad\ | \qquad\ \ |\\ (CH_3)_2-Si-O-Si-(CH_3)_2\end{matrix} + 2\,NaOH \longrightarrow NaO-[-Si(CH_3)_2-O-]_{4n}- + H_2O$$

$$\xrightarrow{CO_2} HO-[-Si(CH_3)_2-O-]_{4n}- + Na_2CO_3$$

Another method of forming polydialkyldisiloxanes is by reacting difunctional oligomers with cyclic organisilicon compounds[164]:

$$n\,\begin{matrix}(CH_3)_2-Si-O-Si-(CH_3)_2\\ \quad\ | \qquad\ \ |\\ \quad\ O \qquad\ O\\ \quad\ | \qquad\ \ |\\ (CH_3)_2-Si-O-Si-(CH_3)_2\end{matrix} + (CH_3)_2SiCl_2 \longrightarrow Cl-[-Si(CH_3)_2-O-]_{4n}-Si(CH_3)_2-Cl$$

$$Cl-[-Si(CH_3)_2-O-]_{4n}-Si(CH_3)_2-Cl \xrightarrow{2\,H_2O} HO-[-Si(CH_3)_2-O-]_{4n}-Si(CH_3)_2-OH + HCl$$

Molecular weights of polydimethylsiloxanes can reach 700,000 or higher. Within the range of molecular weights between 4000–25,000, the materials are fluids of various viscosities. The most common commercial liquid polydimethylsiloxanes are prepared from dimethyldichloro-siloxane. Many elastomer are also based on dimethylsiloxane. Special polymers are prepared with other substituents.

6.14.2. Silicone Elastomers

The elastomers from the high molecular weight silicone polymers must be crosslinked to obtain rubber-like properties. One way to accomplish this is through hydrogen abstraction by free radicals that are generated by decomposition of added peroxides. 2,4-Dichlorobenzoyl peroxide is often used for this purpose. It is decomposed between 110–150 °C. The reaction can be shown as follows:

$$2\ \sim O-Si(CH_3)_2-O-Si(CH_3)_2-O-Si(CH_3)_2-O-Si(CH_3)_2\sim \; + \; 2\ \cdot O-C(=O)-C_6H_3Cl_2\ (2,4\text{-dichlorophenyl}) \longrightarrow$$

$$\sim O-Si(CH_3)_2-O-Si(CH_3)(\dot{C}H_2)-O-Si(CH_3)_2-O-Si(CH_3)_2\sim \; + \; \sim O-Si(CH_3)_2-O-Si(CH_3)(\dot{C}H_2)-O-Si(CH_3)_2-O-Si(CH_3)_2\sim \; + \; 2\ Cl-C_6H_3(Cl)-COOH$$

$$\downarrow$$

$$\begin{matrix}\sim O-Si(CH_3)_2-O-Si(CH_3)_2-O-Si(CH_3)(CH_2-)-O-Si(CH_3)_2\sim\\ |\\ \sim O-Si(CH_3)_2-O-Si(CH_3)(CH_2-)-O-Si(CH_3)_2-O-Si(CH_3)_2\sim\end{matrix}$$

The same can be accomplished by replacing a small quantity of the methyl groups with vinyl ones. This can be done by including a small amount of vinyl methyldichlorosilane into the monomer mix (about 0.1%). The product with a small quantity of pendant vinyl groups crosslinks readily by free-radical mechanism. When a portion of the methyl groups on the polysiloxane backbone are replaced with phenyl structures, the elastomer exhibits particularly good low-temperature properties.[159]

Room-temperature crosslinkable polysiloxane elastomers (commonly called *RTVs*) are prepared by two techniques. In the first, chlorosiloxanes with functionality larger than two are added to hydroxy-terminated prepolymers. The products are subsequently crosslinked by a second addition of polyfunctional compounds like tetraalkoxysilane in the presence of tin catalysts, like stannous octoate. Crosslinking occurs at room temperature. This reaction may vary from 10 minutes to 24 hours:

$$2\ {\sim}O{-}Si(CH_3)_2{-}OH + 2\ HO{-}Si(CH_3)_2{-}O{\sim} + (C_2H_5O)_4Si \longrightarrow Si[{-}O{-}Si(CH_3)_2{-}O{\sim}]_4$$

In the second technique, polysiloxanes terminated by hydroxy groups are either: (1) acetylated, (2) converted to ketoximes, or (3) etherified. The crosslinking is activated by reaction with atmospheric moisture:

$$(1)\quad {\sim}Si(CH_3)_2{-}O{-}C(=O){-}CH_3 + HO{-}Si(CH_3)_2{-}O{\sim} \longrightarrow {\sim}Si(CH_3)_2{-}O{-}Si(CH_3)_2{\sim} + CH_3COOH$$

$$(2)\quad R_2C{=}N{-}O{-}Si(CH_3)_2{-}O{\sim} + {\sim}Si(CH_3)_2{-}OH \longrightarrow {\sim}Si(CH_3)_2{-}O{-}Si(CH_3)_2{-}O{\sim} + HON{=}CR_2$$

$$(3)\quad {\sim}Si(CH_3)_2{-}OR' + HO{-}Si(CH_3)_2{-}O{\sim} \longrightarrow {\sim}Si(CH_3)_2{-}O{-}Si(CH_3)_2{\sim} + R'OH$$

The acetate capped RTVs exhibit good adhesion to substrates, but the released acetic acid can be corrosive.

6.14.3. Polysiloxane Coating Resins

These materials are usually prepared in two stages. In the first, low molecular weight intermediates are formed. In the second, the intermediates are reacted with other resins possessing functional groups, like alkyds, polyesters, or acrylic resins. The most common low molecular weight intermediates are:

$$CH_3{-}({-}O{-}Si(CH_3)_2{-})_x{-}O{-}CH_3 \quad \text{and} \quad \text{[cyclic siloxane: four Si(R) linked by Si–O–Si bridges, with two terminal R(HO)Si groups]}$$

where $x = 3$–6.

6.14.4. Fluorosilicone Elastomers

The elastomers are based on polysiloxanes that contain trifluoropropyl methyl siloxane units. The materials are used as sealants, elastomers, and fluids for aerospace applications. The monomers are prepared according to the following scheme:

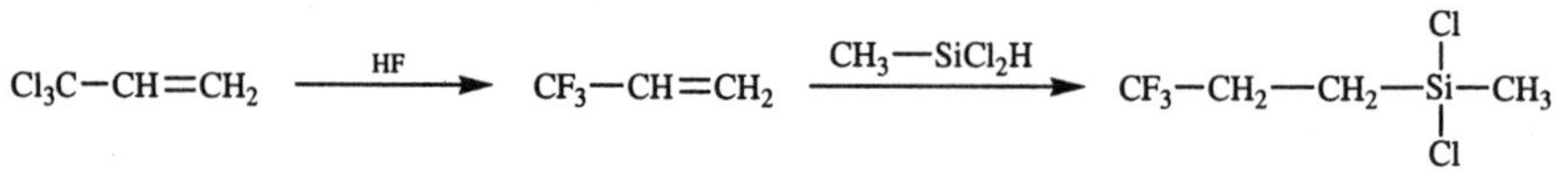

$$Cl_3C{-}CH{=}CH_2 \xrightarrow{HF} CF_3{-}CH{=}CH_2 \xrightarrow{CH_3{-}SiCl_2H} CF_3{-}CH_2{-}CH_2{-}SiCl_2{-}CH_3$$

The above dichlorosilane is converted to a cyclic trimer:

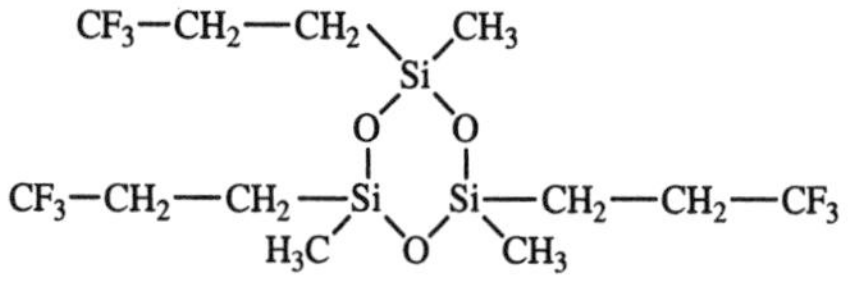

$$[(CF_3{-}CH_2{-}CH_2)(CH_3)Si{-}O]_3$$

Ring-opening polymerizations with basic catalysts convert the trimers to polymers at elevated temperatures. Some vinyl silane is usually copolymerized with the material for subsequent crosslinking. These elastomers are reported to be capable of maintaining their original strengths at temperatures as high as 205 °C for long periods of time.

6.14.5. Polyarylsiloxanes (Also see Section 6.17.4)

Many different polyarylsiloxanes were reported in the literature. Only a few of them are in general industrial use at present, though many exhibit interesting physical properties and might be used in the future. Preparation of one such material[164] starts with a reaction of aniline with dichlorosilane in the presence of an HCl scavenger:

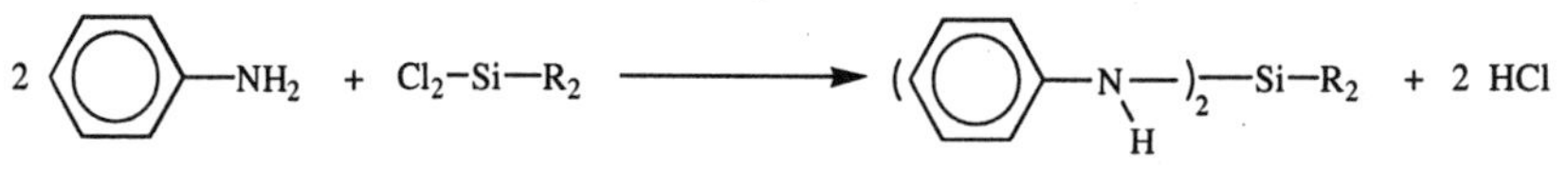

$$2\,C_6H_5{-}NH_2 + Cl_2{-}Si{-}R_2 \longrightarrow (C_6H_5{-}NH{-})_2{-}Si{-}R_2 + 2\,HCl$$

The product, dianilinosilane, is reacted with diphenols, like hydroquinone:

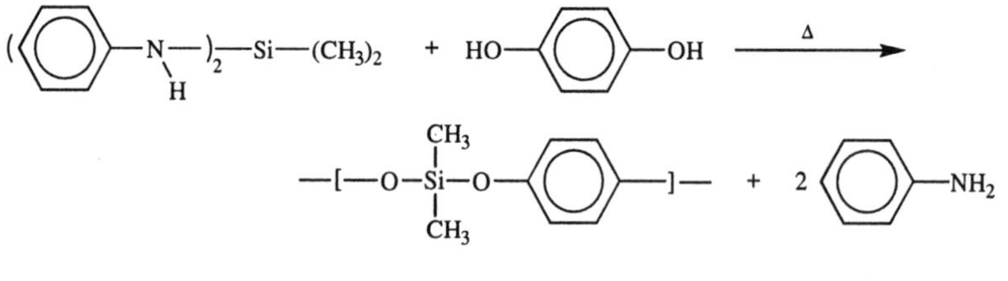

$$(C_6H_5{-}NH{-})_2{-}Si{-}(CH_3)_2 + HO{-}C_6H_4{-}OH \xrightarrow{\Delta} {-}[{-}O{-}Si(CH_3)_2{-}O{-}C_6H_4{-}]{-} + 2\,C_6H_5{-}NH_2$$

Polymers prepared by the above procedure have molecular weights up to 80,000.[164] It is also possible to start with diphenoxysiloxane and catalyze the reaction with sodium or lithium metals. Reactions of cyclic silazanes with arylene disalanols yield polymers with molecular weights as high as 900,000[164]:

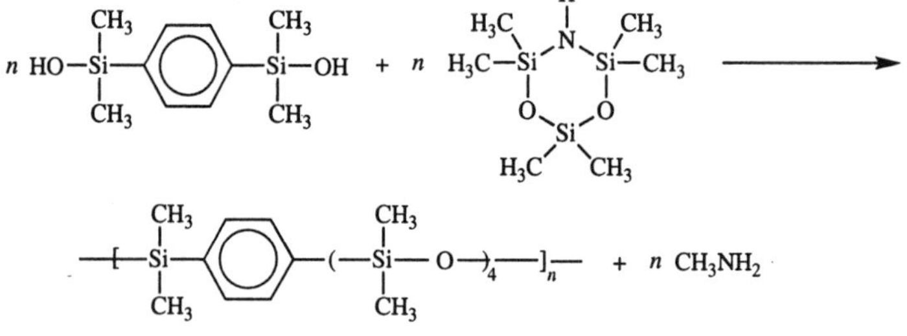

$$n\,HO{-}Si(CH_3)_2{-}C_6H_4{-}Si(CH_3)_2{-}OH + n\,[(CH_3)_2Si{-}NH{-}Si(CH_3)_2{-}O{-}Si(CH_3)_2{-}O] \longrightarrow {-}[{-}Si(CH_3)_2{-}C_6H_4{-}({-}Si(CH_3)_2{-}O{-})_4{-}]_n{-} + n\,CH_3NH_2$$

6.15. Polysilanes

Polymers with silicon–silicon single bonds in the backbone have been known for some time. It was only within the last 10–15 years, however, that high molecular weight materials were developed.[165] Behind the current interest in these materials is a realization that they have various potential applications in ceramic fibers,[166] in microlithography,[165,167] in photoconduction[168] and as nonlinear optical materials.[169]

The polymers are prepared from disubstituted dichlorosilanes by reacting them with alkali metal dispersions in a reductive coupling process. The polymerizations appear to have the characteristics of chain-growth rather than step-growth reactions[170]:

$$RR'SiCl_2 \xrightarrow{Na} \left[-Si(R)(R')- \right] + NaCl$$

The above illustrated reaction with sodium dispersions requires a greater than 80 °C temperature to proceed satisfactorily. When mixtures of different dialkylsubstituted dichlorosilanes are reacted in this manner, copolymers form.[171]

Recently, an ambient-temperature sonochemical reductive coupling process was developed.[172] The reaction is carried out in the presence of an ultrasound and results in relatively high (M_n = 50,000–100,000) molecular weight materials with narrow molecular weight distributions. In addition, it was reported[172] that polymers can also be formed by anionic ring-opening polymerization of cyclotetrasilanes to yield polymers with molecular weights of 10,000–100,000.

6.16. Phosphonitrilic Polymers

These polymers, also called *polyphosphazenes*, are useful materials when they are substituted with organic compounds.[173a] They are prepared from hexachlorocyclotriphosphazenes by ring-opening polymerizations:

$$n\ (NPCl_2)_3 \longrightarrow -[-N{=}P(Cl)(Cl)-]_n-$$

It is believed that the mechanism of polymerization involves an attack by an electron-rich nitrogen of one cyclic monomer upon another. At first, a cation forms through ionization of a phosphorus–chloride bond[173b]:

$$(NPCl_2)_3 \xrightarrow{\text{ionization}} (NPCl_2)_2(N{=}P^{\oplus}Cl)\ Cl^{\ominus}$$

This is followed by an attack on another monomer molecule:

$$(NPCl_2)_2(N{=}P^{\oplus}Cl)\ Cl^{\ominus} + (NPCl_2)_3 \longrightarrow (NPCl_2)_2(N{=}PCl)-(N{=}P-Cl_2)_2N{=}P^{\oplus}-Cl_2\ Cl^{\ominus} \longrightarrow$$

$$(NPCl_2)_2(N{=}PCl)-(N{\equiv}P-Cl_2)_{(3n+2)}-N{=}P^{\oplus}-Cl_2\ Cl^{\ominus}$$

Alkyl, aryl, alkoxy, and aminocyclotriphosphazenes fail to polymerize. This is believed to be due to an absence of easily ionizable halogen to phosphorus bonds. At the same time, materials that presumably facilitate ionization of the phosphorus to halogen bonds, such as water, alcohols, carboxylic acids, and metals, accelerate the polymerization.[173b]

The above all-inorganic polymer decomposes readily at elevated temperatures and is very sensitive to hydrolytic attack. Quantitative replacements of the halogen groups, however, are possible with alkoxy, aryloxy, alkyl, aryl, or amino groups to yield much more stable materials. The replacements are achieved by refluxing the inorganic polymer in an ether solvent for several hours with sodium alkoxide or aryloxide, a metal alkyl or aryl, or with a primary or a secondary amine. Of particular interest are substitutions with fluoroalkoxy groups, like the following:

$$-[-N=P(Cl)_2-]_n- \; + \; 2n\; C_3F_7CH_2O^{\ominus}\,Na^{\oplus} \xrightarrow{\text{THF}} -[-N=P(OCH_2C_3F_7)_2-]_n-$$

Mixed fluoroalkoxy compounds can be used to obtain a variety of properties.[226] The materials find applications as elastomers, because they exhibit good chemical resistance and good thermal stability. In addition, many retain their useful elastomeric properties at low temperature. As a result, among other applications, they are attractive for use as sealants and as fuel lines in an arctic environment.

Phosphonitrilic polymers are self-extinguishing or fire-retardant. This led to the development of flame- and heat-resistant polyimide composites that are prepared from maleimide-substituted phosphazenes.[231,232] The maleimide group is used for crosslinking. The substituted phosphazene can be illustrated as follows:

A new approach to the syntheses of polyphosphazenes was reported recently.[179] It is based on condensation of suitable Si–N–P precursors:

$$(CH_3)_3-Si-N=P(R)(R')=X \xrightarrow[-(CH_3)_3SiX]{\Delta} -[-N=P(R)(R')-]-$$

where R,R′ = alkyl, aryl; X = OCH_2CF_3, O–Ph.

In this preparation the desired substituents are introduced before the polymerization. The resultant polymers[179] are soluble in various solvents. Their molecular weight distributions vary from 1.4–3.5 and M_w from 50,000–150,000.

6.17. High-Performance Polymers

The polymers that are found in common commercial use, such as fibers, films, or structural resins, fail to withstand elevated temperatures above 250 °C for long periods of time and decompose. There is a need, however, in various technologies for materials that can tolerate temperatures over 300 °C for reasonably long periods. Many such materials might be too high priced for common commercial use. They find application, however, in specialized areas that include space, aeronautic, or military technologies, where a higher price is justified by greatly enhanced performance.

In developing tough, heat-resistant polymers the chemists pursued several goals.[191] These were:

1. To improve heat stability of the available polymers by introducing structural modifications.
2. To develop new macromolecules based on chemical structures capable of withstanding high temperatures.
3. To develop inorganic and inorganic–organic polymeric materials, because many inorganic molecules are more thermally stable than the organic ones.

Improvement of the thermal stability of existing polymers has to be based on the following considerations: (1) The primary bond energy between atoms in a polymeric chain is the greatest source of thermal stability. The strength of these bonds therefore imposes an upper limit on the vibrational energy that a molecule may withstand without bond ruptures. In cases of cyclic repeating units, as in ladder polymers, a rupture of one bond in a ring may not lead to loss of molecular weight. In such polymers two bonds would have to break within the same ring for the chain to rupture, and the probability of that is low. This means that ladder polymers should exhibit greater heat stability than single stranded chains.[183b] (2) Secondary bond forces or the cohesive energies contribute additional stability to the molecule. (3) The resonance energy of aromatic and heterocyclic structures contributes an additional amount of thermal stability and bond strength. (4) Polymers with high melting or softening temperatures are generally more heat-resistant.

6.17.1. Fluorine-Containing Aromatic Polymers

The high strength of the carbon–fluorine bonds and the shielding effect of the highly electronegative fluorine atoms improve heat stability. One such material was synthesized by heating perfluoroalkylamidines above their melting points. Ammonia evolves and triazine ring-containing polymers form[174]:

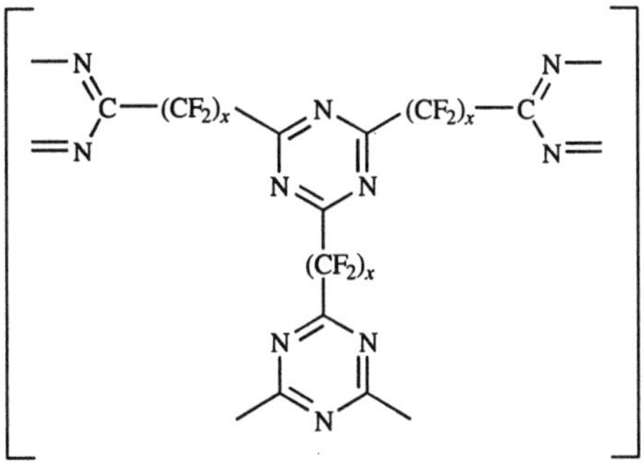

$$\left[\begin{array}{c} \text{—N=C(—N=)—}(CF_2)_x\text{—(triazine)—}(CF_2)_x\text{—C(=N—)—N=} \\ | \\ (CF_2)_x \\ | \\ \text{(triazine)} \end{array}\right]$$

The thermal stability of these polymers in vacuum is about equivalent to polytetrafluoroethylene.[174]

Among other fluorine-containing materials reported in the literature are a group of polyesters[175]:

$$1.\quad \text{—[—}\overset{O}{\overset{\|}{C}}\text{—}C_6H_4\text{—}\overset{O}{\overset{\|}{C}}\text{—O—}CH_2\text{—(—}CF_2\text{—)}_4\text{—}CH_2\text{—O—]}_n\text{—}$$

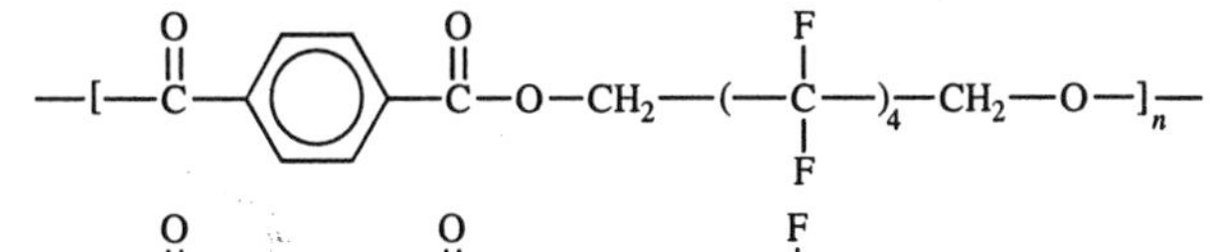

$$2.\quad \text{—[—}\overset{O}{\overset{\|}{C}}\text{—CH=CH—}\overset{O}{\overset{\|}{C}}\text{—O—}CH_2\text{—(—}CF_2\text{—)}_4\text{—}CH_2\text{—O—]}_n\text{—}$$

$$3.\quad \text{—[—}\overset{O}{\overset{\|}{C}}\text{—}(CH_2)_4\text{—}\overset{O}{\overset{\|}{C}}\text{—O—}CH_2\text{—(—}CF_2\text{—)}_4\text{—}CH_2\text{—O—]}_n\text{—}$$

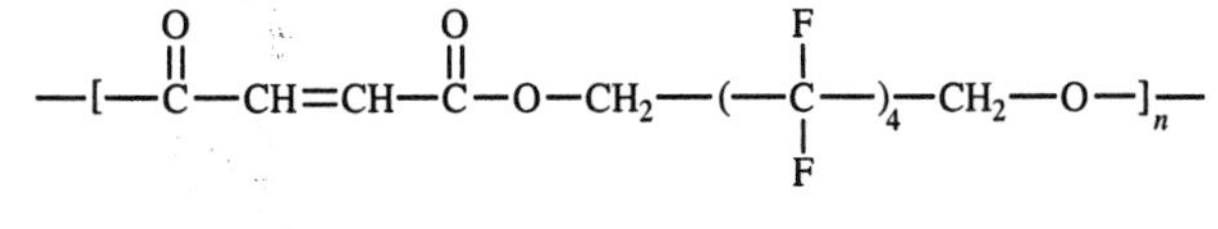

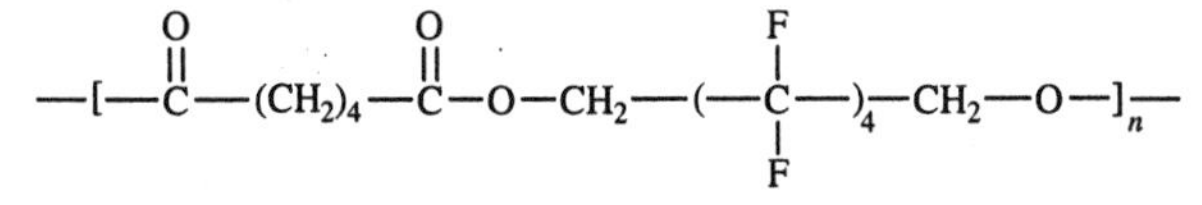

Their heat stability and other physical properties, however, do not appear to be superior to conventional materials. The Ullmann reaction can be used to prepare polyperfluorophenylene.[176] The products, however, are low in molecular weight:

$$n\ \mathrm{I{-}C_6F_4{-}I} \xrightarrow{\text{Cu powder}} -[\mathrm{C_6F_4}]_n-$$

Higher molecular weight materials, like poly(perfluorophenylene ether) form from potassium pentafluorophenoxide,[177] where M_n is about 12,500:

$$n\ \mathrm{F{-}C_6F_4{-}OK} \longrightarrow -[\mathrm{C_6F_4{-}O}]_n-$$

Also, low molecular weight fluorine-containing polymers form from perfluoroaromatic compounds through a loss of aromaticity when they react with bis(fluoroxy)difluoromethane.[180] More interesting is the formation of poly($\alpha,\alpha,\alpha',\alpha'$-tetrafluoro-*p*-xylylene) by a polymerization technique that closely resembles the preparation of poly(*p*-xylylene) by vacuum pyrolysis of a dimer[181]:

$$\text{dimer } [\mathrm{H_2C{-}C_6F_4{-}CH_2}]_2 \xrightarrow[\text{600–700 °C}]{\text{vacuum}} -[\mathrm{CH_2{-}C_6F_4{-}CH_2}]-$$

The resultant polymers maintain useful mechanical properties for up to 3000 hours in air at 250 °C.

Fluorinated epoxy resins are another group of materials that might potentially possess improved thermal stability. This, however, has not been demonstrated. It was shown,[190] though, that these materials can be key intermediates in organic coating and plastics that require special properties, like hydrophobicity, oleophobicity, light stability, and low friction.

6.17.2. Polyphenylene

This polymer is completely aromatic in character.[182] Polymerization of benzene to polyphenylene was therefore investigated quite thoroughly.[178,184,185] Benzene[186] and other aromatic structures[184,185] polymerize by what is believed to be a radical-cationic mechanism. In this type of polymerization, benzene polymerizes under mild conditions in the presence of certain Lewis acids combined with oxidizing agents[186–188]:

$$n\ \mathrm{C_6H_6} + 2n\ \mathrm{CuCl_2} \xrightarrow{\mathrm{AlCl_3}} \mathrm{H{-}[C_6H_4]_n{-}H} + 2n\ \mathrm{CuCl_2} + 2n\ \mathrm{HCl}$$

The yield of polymers reaches a maximum value (close to quantitative) at an aluminum chloride to cupric chloride molar ratio of 2:1.[180] Solvents, concentrations, and temperatures affect the molecular weights of the products.[189] Other Lewis acids that are effective in benzene polymerizations are $MoCl_5$, $FeCl_3$, and $MoOCl_4$. The products, however, possess greater degrees of structural irregularity.[183a]

Theoretical considerations indicate[183b] that during the polymerization, the benzene rings become associated in a stacked end-to-end arrangement. As a result, the radical cation becomes delocalized over the entire chain:

$$\mathrm{C_6H_6}^{\bullet\oplus} \xrightarrow{\mathrm{C_6H_6}} \bullet\bullet[\mathrm{C_6H_6}\bullet\bullet\mathrm{C_6H_6}\bullet\bullet\mathrm{C_6H_6}\bullet\bullet\mathrm{C_6H_6}\bullet\bullet]^{\bullet\oplus} \longrightarrow$$

$$(\cdot\mathrm{C_6H_5})[\mathrm{C_6H_4}]_n(\mathrm{C_6H_5}^{+}) \dashrightarrow \xrightarrow{[\mathrm{O}],\ (-\mathrm{H}^+,\ -e)} \mathrm{C_6H_5}[\mathrm{C_6H_4}]_n\mathrm{C_6H_5}$$

The σ-bond formation shown above can also be accompanied by simultaneous depropagation and loss of benzene molecules. Chain buildup stops when the radical cation on the terminal phenyl group becomes too small to promote further association.

The polymer is very stable up to 500–600 °C and oxidizes very slowly. It is, however, quite insoluble with a very high melting point that makes it difficult to process and even to determine its molecular weight. Introduction of irregularity into the polymeric structure by copolymerizing terphenyl, biphenyl, or triphenylbenzene with benzene results in the formation of soluble products. Their molecular weights, however, are low up to 3000.[192] They melt between 300–400 °C and contain phenyl branches and some fused rings. The copolymers can be crosslinked with xylylene glycol or with benzene-1,3,5-trisulfonyl chloride. Recently, soluble alkyl-substituted poly(*p*-phenylene)s were prepared[193] by a coupling process, using palladium catalysts:

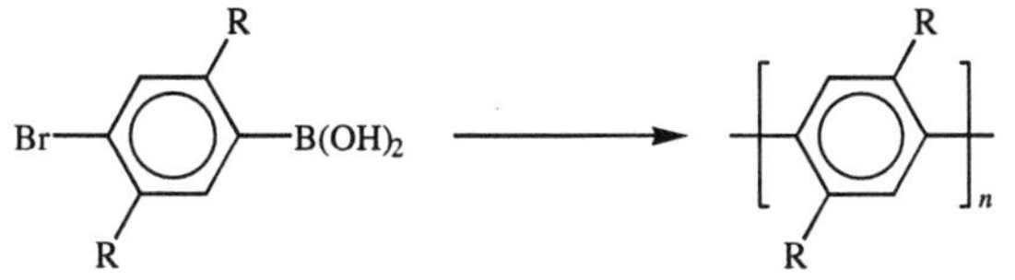

R R Br B(OH)2 → n

Syntheses also include formations of copolymers with other aromatics compounds that lack substituents. The polymers with hexyl or longer side chains are soluble in toluene.[193]

6.17.3. Diels-Alder Polymers

The Diels-Alder reaction has been used to form many polymeric materials. One such material, for instance, forms from a reaction of diethynylbenzene with cyclopentadienone.[194–196] The products, phenylated polyphenylenes, reach molecular weights M_n up to 40,000:

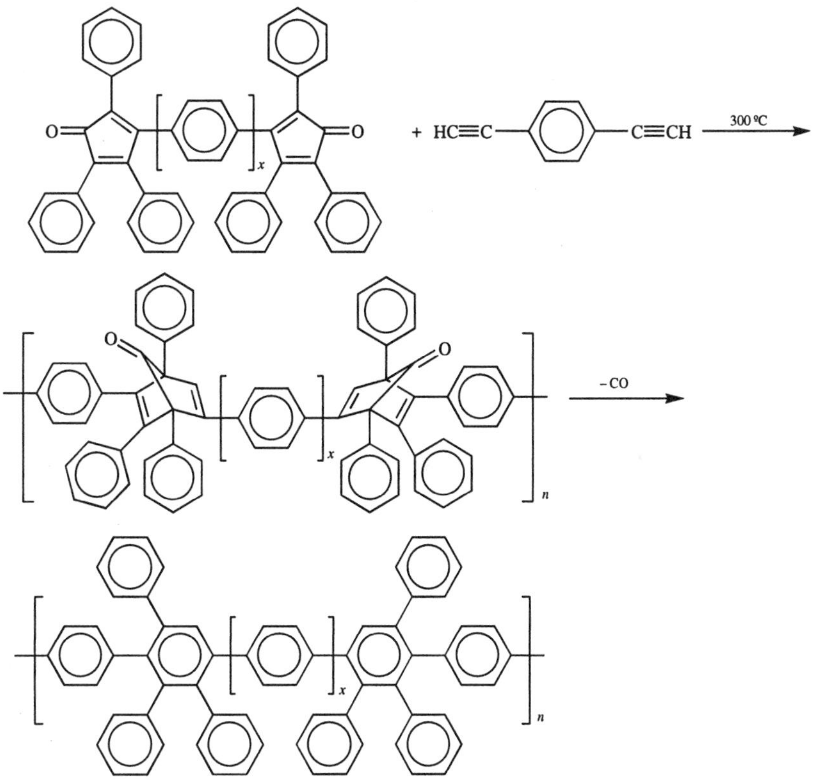

O x O + HC≡C C≡CH 300 °C →

O O x n −CO →

x n

where $x = 1, 2$. These polymers are amorphous and form clear films that are stable in air up to 550 °C.

Another example is condensation of 2-vinyl-1,3-butadiene with *p*-benzoquinone. The product is a ladder, or a double stranded polymer[197]:

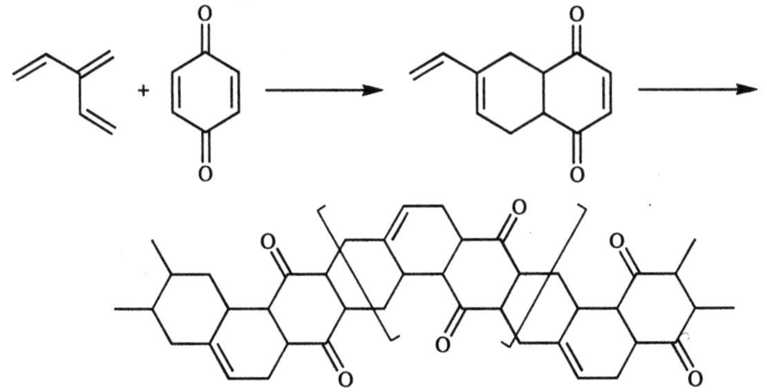

This double-stranded polymer, shown above, is soluble only in hexafluoroisopropyl alcohol[198] and is infusible. When, however, some of the rings are made flexible by condensing 2-vinyl-1,3-butadiene with a large-ring bisfumarate, the solubility improves.[198]

Some stable polymers also form in 1,3-dipole addition reactions.[190] Bis-sydnones, for instance, condense with diacetylene to form pyrazole rings in the polymer backbone. The reactions presumably proceed through Diels-Alder intermediates[199]:

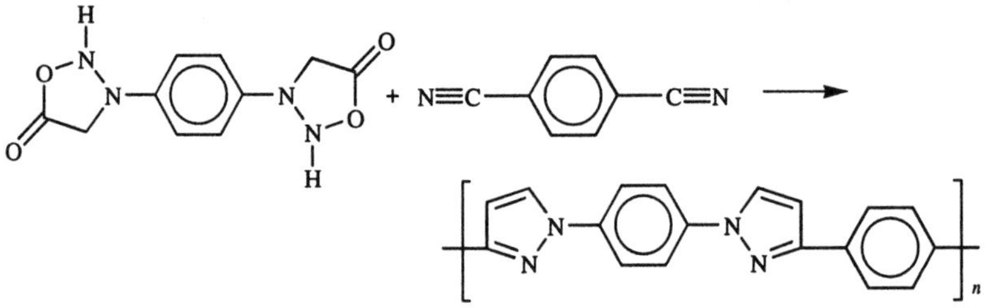

Similarly, bis-sydnones condense with quinone[199]:

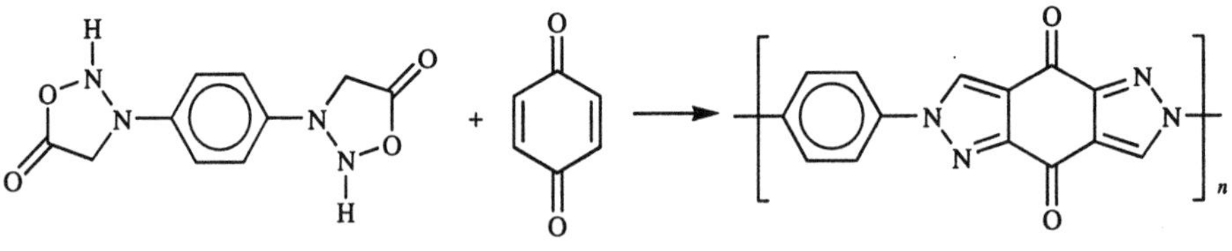

The resultant polymers are not high in molecular weights and only slightly soluble in solvents like dimethylformamide. Their powders decompose near 420 °C in air and near 500 °C in a nitrogen atmosphere.

1,3-Dipolar additions of bisnitrilimines (generated from tetrazoles with diynes or with dinitriles) result in formations of polypyrazoles and polytriazoles.[200] Some examples of these products are:

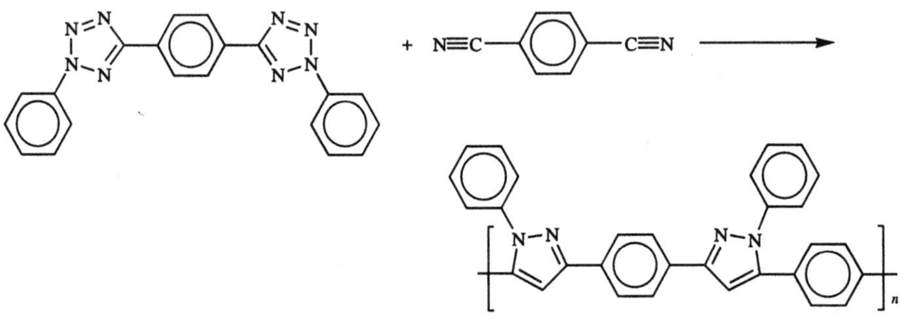

and

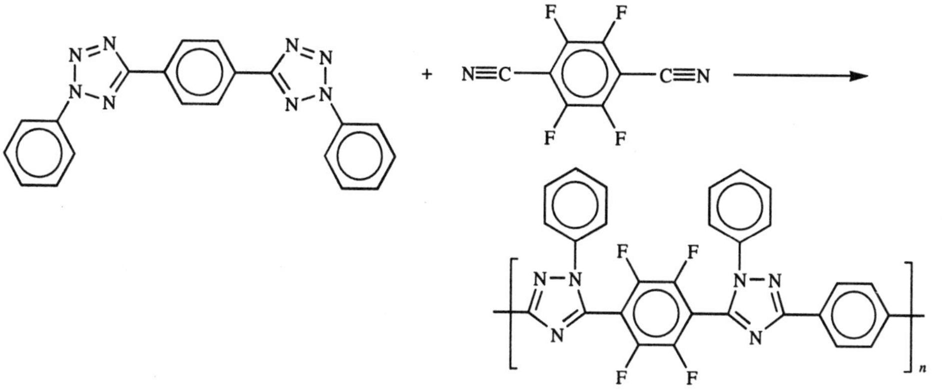

By this condensation high molecular weight polymers can form.[200] They decompose near 500 °C in air or in nitrogen atmosphere.

Cyclopentadienone derivatives condense by the Diels-Alder reaction in homopolycycloaddition. The reactions involve a series of steps consisting of initial cycloaddition, followed by loss of carbon monoxide through an expulsion of a bridge carbonyl group[201]:

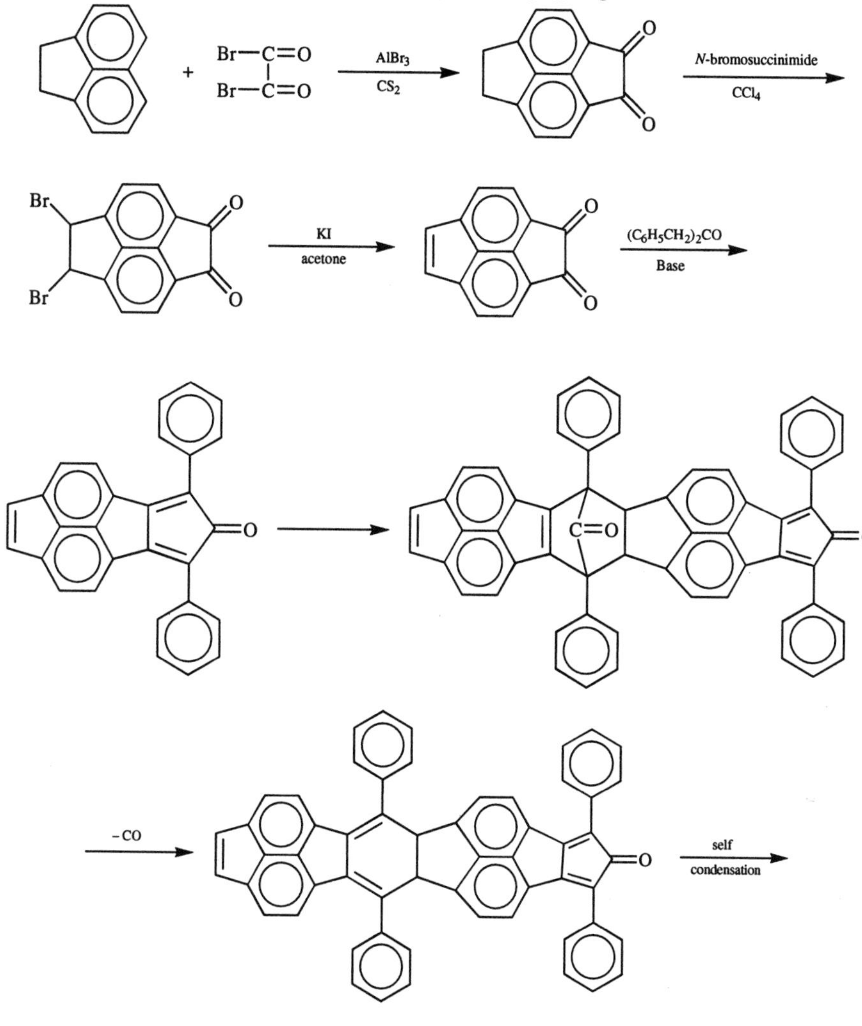

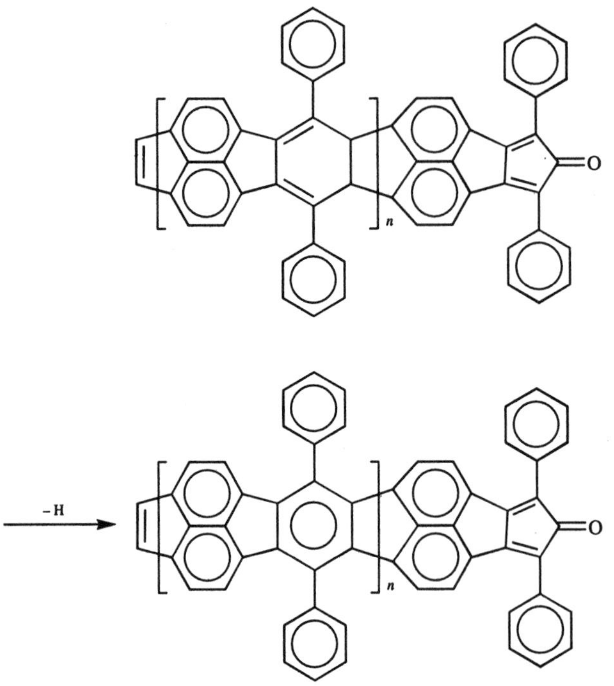

The products are ladder polymers of varying molecular weights. They lose approximately 30%[201] of their weight at 700 °C in a nitrogen atmosphere.

Another example is condensation of bisdienophiles with dienes[202]:

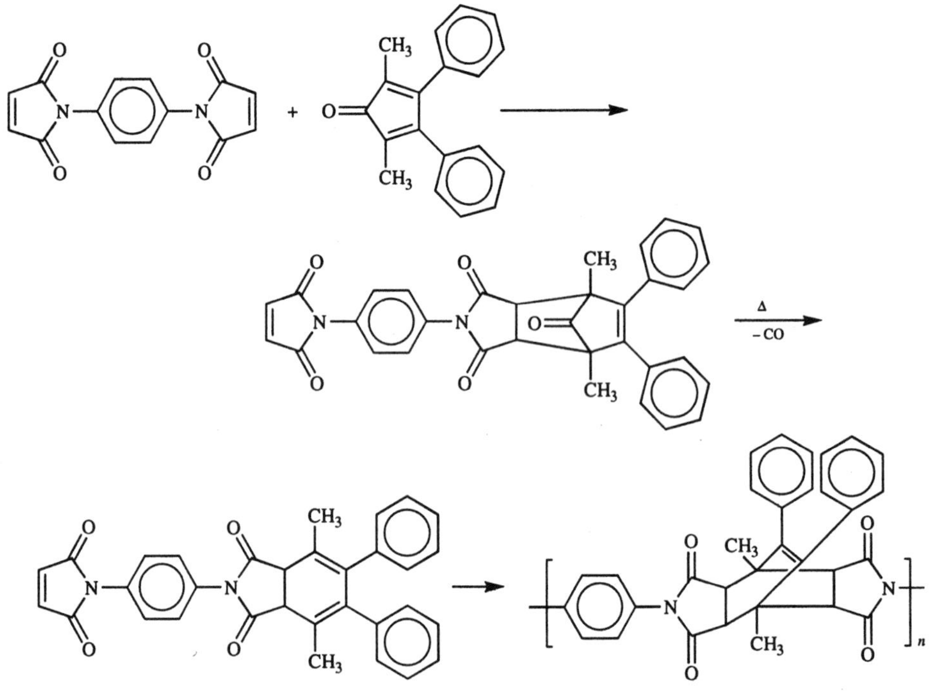

Only low molecular weight polymers form, however, together with some insoluble, possibly crosslinked material. Bismaleimides also condense with bisfulvene.[202] The products of these condensations depolymerize reversibly. Also, reactions of maleic anhydride with dienes, like bicyclopentene, bicyclohexene, dicyclopentenyl ether, and dicyclohexenyl ether, yield soluble, low molecular weight polymers[204]:

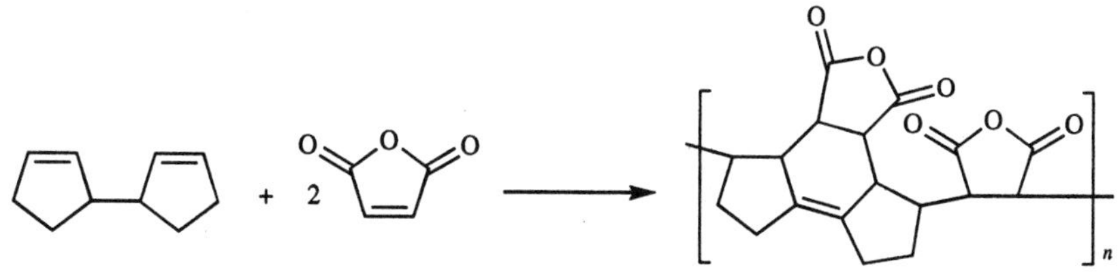

Photocycloaddition reactions also form low molecular weight polymers with heterocyclic rings in the backbone. The reactions are photoinitiated with benzophenone to obtain repeated $2\pi + 2\pi$ photocycloadditions of bismaleimides[205]:

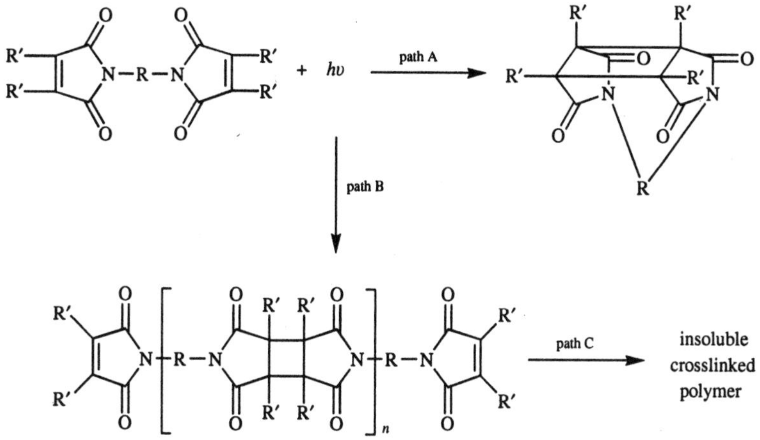

path A

path B

path C

insoluble crosslinked polymer

where R = $(CH_2)_x$. These polymeric materials form transparent, flexible films.[205]

Diels-Alder reaction can also be used to modify the chemical structures of some aromatic polyamides. This improves their heat stability, raises their T_g values, and makes them rigid at higher temperatures.[203]

6.17.4. Silicone-Containing Aromatic Polymers

These materials have the potential of being stable at high temperatures. One typical preparatory procedure consists of condensing bisphenols with suitable silicon derivatives[227]:

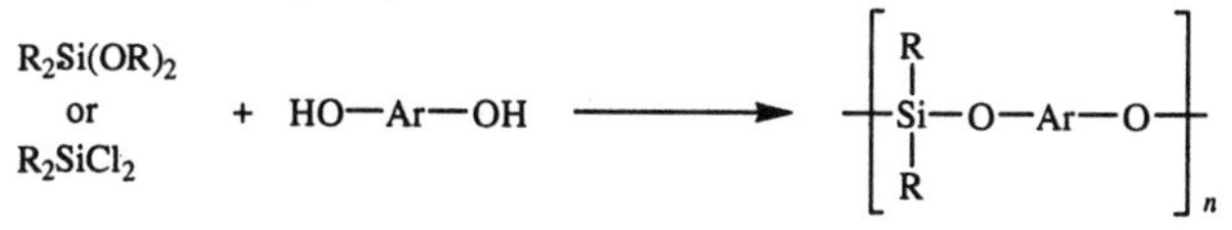

$$\begin{matrix} R_2Si(OR)_2 \\ \text{or} \\ R_2SiCl_2 \end{matrix} + HO{-}Ar{-}OH \longrightarrow \left[\begin{matrix} R \\ | \\ -Si{-}O{-}Ar{-}O- \\ | \\ R \end{matrix} \right]_n$$

Many polymers that were synthesized in this manner were indeed found to be soluble and heat-stable. One example is a material shown below:

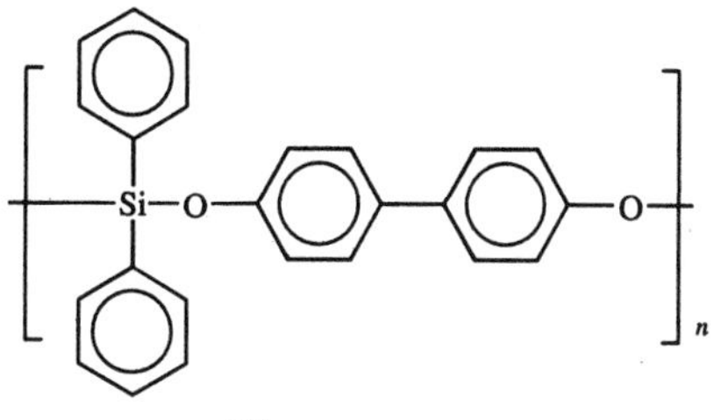

It loses only 10% of its weight at 600 °C.[191]

The starting macromolecules can also be silanols, as, for instance, in a synthesis of polycarbonates.[227] Preformed bis-silanols are used in this particular example:

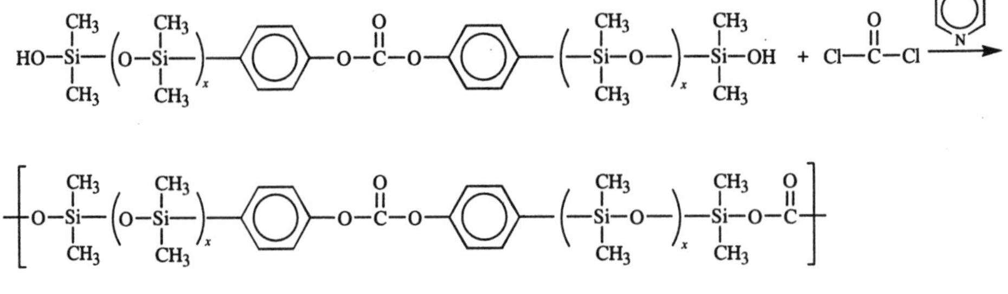

where $x = 0$ or 1.

One publication describes syntheses of copolyesters and copolyamides that contain phenoxsilin rings[228]:

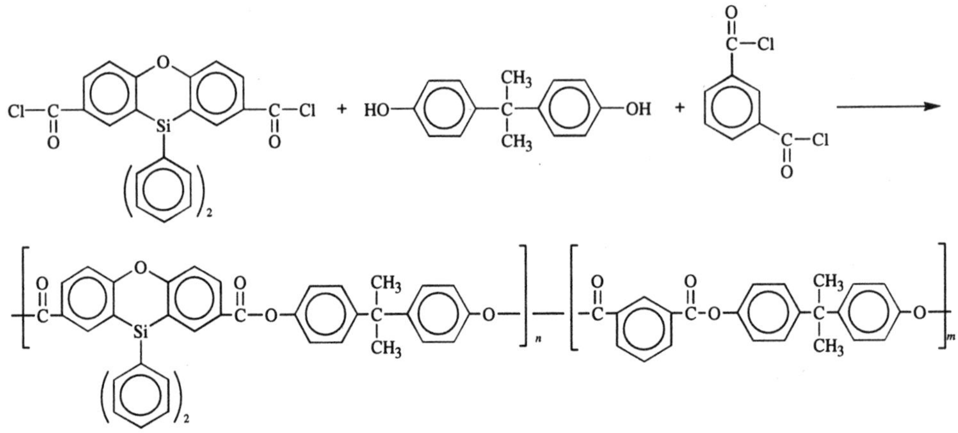

The copolymers show little degradation at temperatures up to 400 °C. In addition, the copolymers with high phenoxsilin content are soluble in several solvents.[228]

There are also reports in the literature of attempts to modify polysiloxane backbones to increase heat stability. These modifications consist of making changes in the electronic character of the Si–O bonds to prevent rearrangements at high temperatures to low molecular weight cyclic products. It is known that by making the bonds more ionic than covalent increases heat stability.[191] A metal that is more electropositive is therefore used to form metal–oxygen linkages and form polymers that are more ionic in character. Such metals are aluminum, titanium, tin, and boron. The results, however, so far are disappointing.

6.17.5. Direct Condensation Polymers

Many polymers with enhanced heat stability can be prepared simply by direct condensation. These aromatic polymers often contain a heterocyclic unit. The materials are high melting, somewhat infusible, and usually low in solubility. Many aromatic polyimides belong here. Polyimides, as a separate class of polymers, were discussed in an earlier section, because many are common commercial materials. On the other hand, the materials described in this section might be considered special and, perhaps, at this point, still too high priced for common usage.

6.17.5.1. Polybenzimidazoles, Polybenzthiazoles, and Related Polymers

Many *polybenzimidazoles* are prepared by direct condensation. They are colored polymers that mostly melt above 400 °C.[221] One such material is formed from 3,3′-diaminobenzidine and dipheyl isophthalate by heating the two together at 350–400 °C in an inert atmosphere[206]:

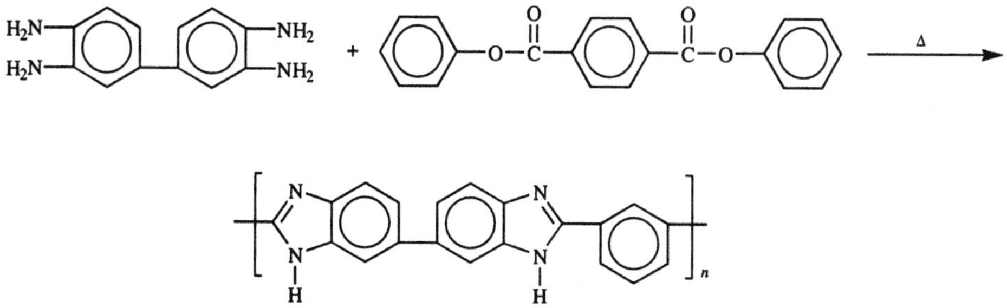

Films and fibers from this material exhibit good mechanical properties up to a temperature of 300 °C. Above that temperature, however, they degrade rapidly in air.[206]

Similar polymers are *polybenzoxazoles*[207,208]:

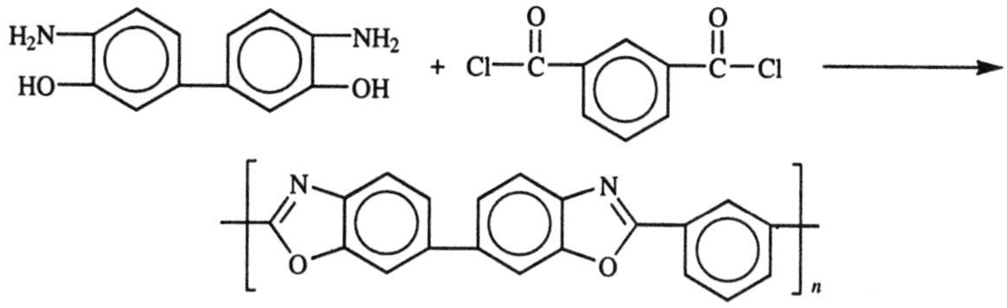

The same is true of *polybenzothiazoles*[209]:

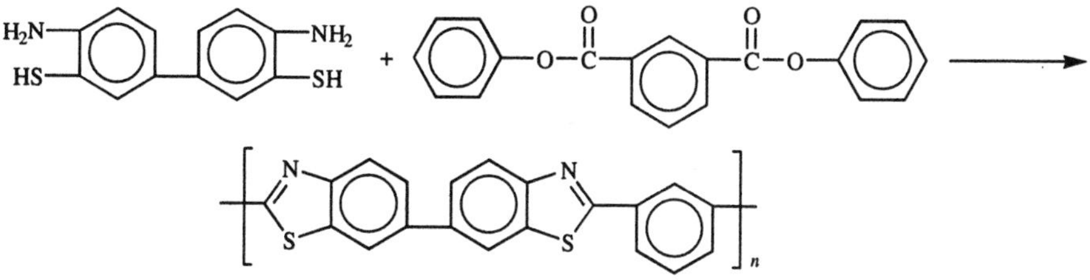

Polyoxidiazoles also belong to this general class of materials[210]:

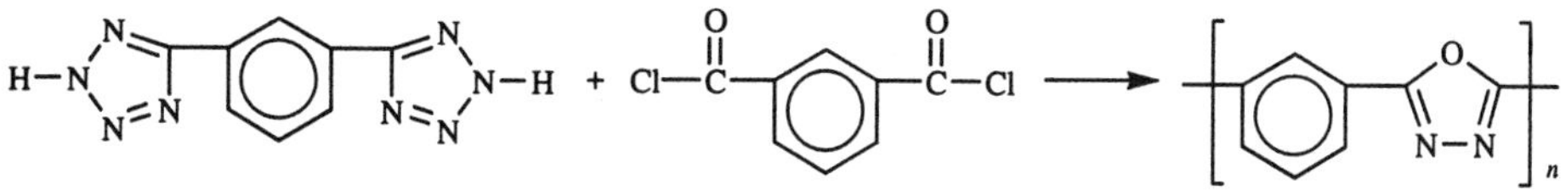

and also *polybenzotriazoles*[211]:

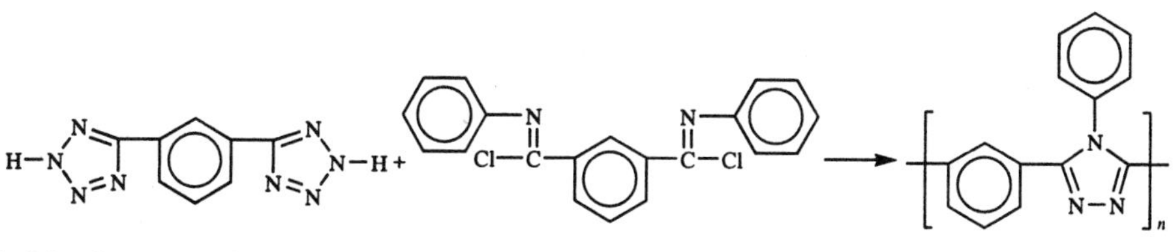

All of the above materials maintain useful properties up to 300 °C in air and can be fabricated into fibers.

Some polymeric materials are completely free from hydrogens. An example is *polysulfodiazole*,[212] a polyimide prepared from pyrazine-1,2,4,5-tetracarboxylic acid anhydride and diaminothiazine.[213] This material exhibits particularly good heat stability[213]:

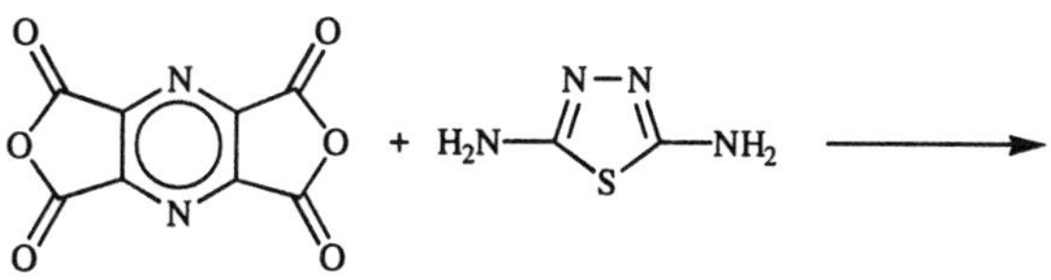

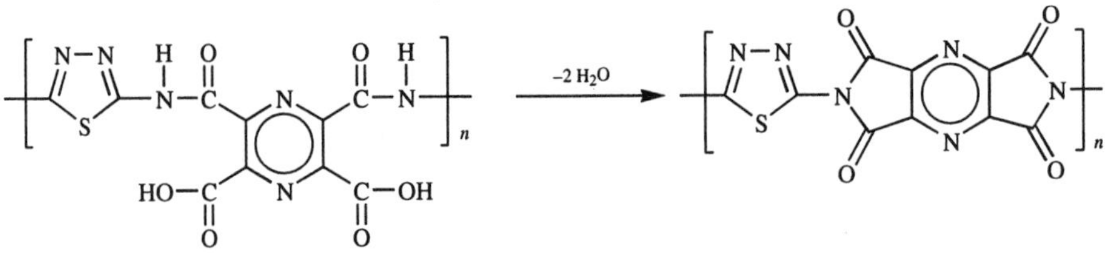

Films from this polysulfodiazole maintain their strength and stability at 592 °C. Preparation of several other, similar polyimides was reported.[219] A polyimide, however, prepared from diaminothiazole with pyromellitic dianhydride chars at 320 °C in air. The chemistry and preparations of the principal types of polyheteroarylenes were reviewed by Krongauz.[214]

Recently, a series of thermally stable, organic solvent-soluble polyimides were synthesized by reacting 3,7-diaminophenothiazinium chloride (thionine) with four different dianhydrides.[243] These polyimides can be illustrated as follows:

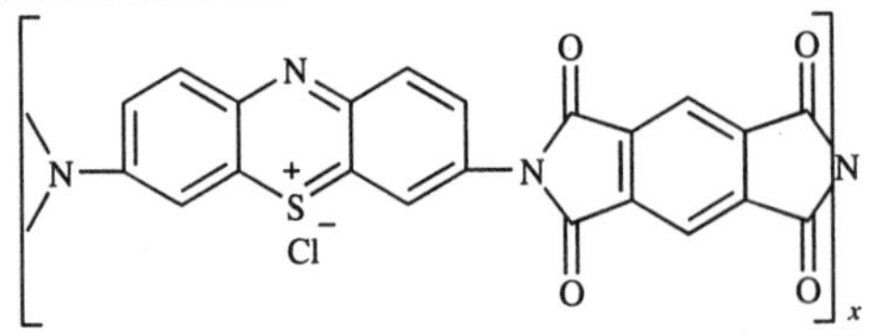

Many polymers that are described in this section can be prepared in either one or two steps. In the one-step process, polyphosphoric acid is often employed as a solvent. It is a proton donor, promotes condensation, and acts as a cyclodehydrating agent, frequently yielding polymers of high molecular weight. Shaping the resulting polymers, however, can be a problem due to infusibility and insolubility. When prepared in two steps, the polymer can be shaped into films or fibers first, before much cyclization takes place, while the material is still fusible and solvent-soluble. This is followed by further heating to complete the process. Cyclization in this process occurs in solid polymers that become increasingly rigid as the reaction progresses. As a result, the products are not as fully cyclized as in the one-step process in phosphoric acid solution.

16.17.6. Oligomers with Terminal Functional Groups

The above-described two-step processes yield polymers that evolve volatiles upon further heating because the condensations continue. To overcome this drawback, prepolymers were developed that undergo addition-type reactions at fairly moderate time–temperature schedules.[215,216] Such prepolymers are *terminated by functional groups.* The following is an example of one such material, an oligomer, polyquinoxaline-terminated by acetylene groups:

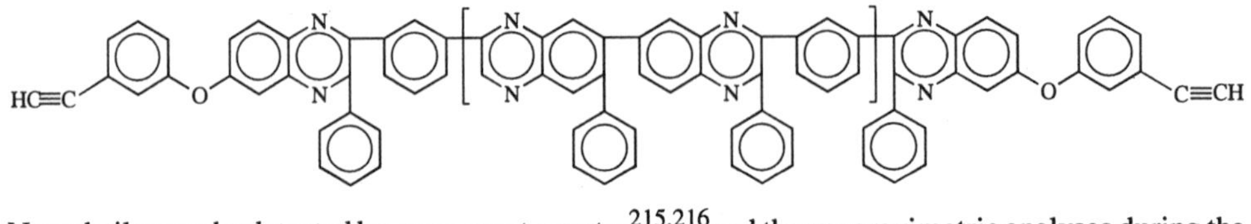

No volatiles can be detected by mass spectrometry[215,216] and thermogravimetric analyses during the crosslinking reaction. The thermooxidative stability of the resultant polymers is at least equivalent to polyphenylquioxalines not terminated by acetylene. The crosslinking reaction was shown on a model compound to be an intramolecular cyclization[215–217]:

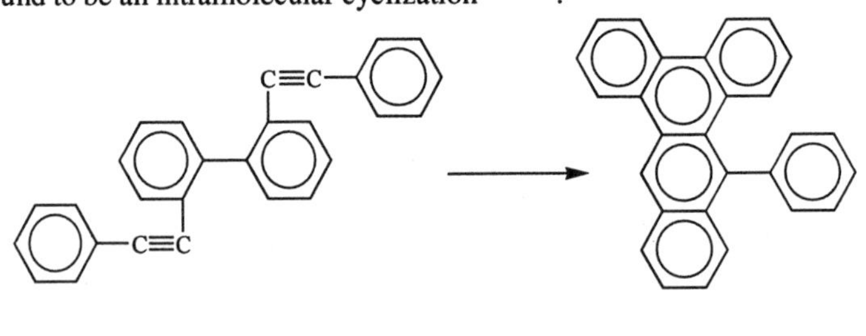

It is not necessary for the acetylenic groups to be on the terminal ends of the prepolymers. They can also be located as pendant structures.[216]

Aromatic polyamides with terminal acetylenic groups[218] were formed from 2,2′-diiodo-diphenyl-4,4′-dicarbonyl chloride reacted with aromatic diamines. The phenylethynyl groups were introduced by reacting the iodine moieties with copper phenyl acetylide. Thermal treatment converted the prepolymers to 9-phenyl dibenzanthracene based rigid-rod polymers that fail to melt below 500 °C.

High molecular weight polyquinoxaline polymers were prepared from 3,3′,4,4′-tetraamino-biphenyl that was reacted with aromatic bis(α-diketones) and/or ethynyl-substituted aromatic bis(α-diketones).[219] The polymers contain 0, 5, 10, 30, and 100% pendant groups. Also, ethynyl-substituted diketones were synthesized by the following procedure:

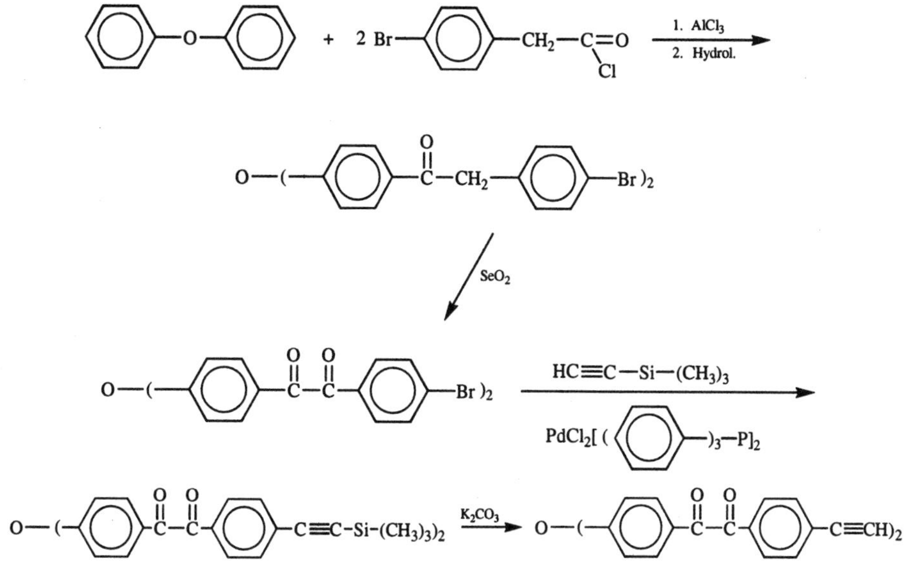

The synthesis is completed by condensation with a tetramine:

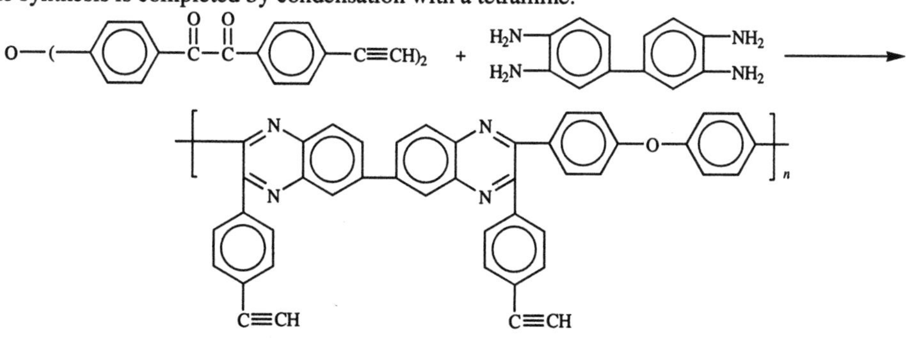

The above reactions yield polymers with high T_g values. The materials, however, exhibited lowered thermooxidative stability. The same was found to be true when the ethynyl moieties were replaced by phenylethynyl groups.[219]

Other functional groups that were investigated[229] are phenylethynyl, phenylbutadiynyl, phenyl-butenyl, biphenylene, styryl, maleimide, nadimide (5-norbornene-2,3-dicarboximide), cyanate, and *N*-cyanourea.[230] The advantage of terminally capped prepolymers is that they melt at lower temperatures and can be dissolved in different solvents. Heating of these materials converts them to thermally stable polymeric networks.

A recent paper reports the preparation of quinoline oligomers that were end-capped with 4-acetylbenzocyclobutene, 6-acetyl-8-phenyl-1,2-dihydro-[3,4]cyclobuta-[1,2-b]quinone (CBQ) and 8-acetyl-6*b*,10*b*-dihydrobenzo-[*j*]cyclobuta-[1,2-α]-acenaphthalene(BCBAN).[230] The structures of the two crosslinking groups are:

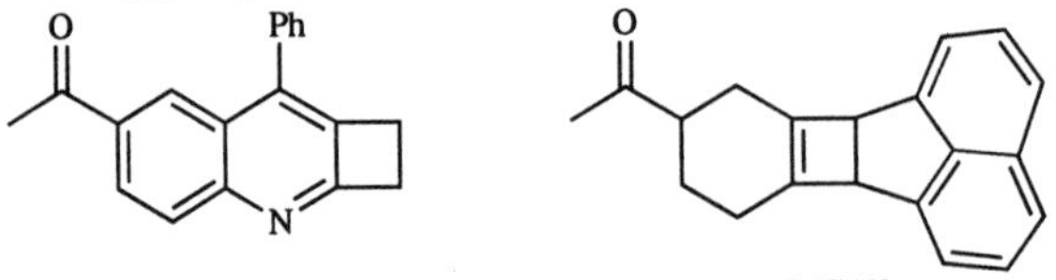

CBQ

BCBAN

It was reported[230] that the oligomer capped with BCBAN yields a cured film that exhibits good flexural moduli and superior heat stability in air at 400 °C.

6.17.7. Cardo Polymers

These are a special group[201,208,220] of polymeric materials. The name cardo comes from Latin, which means loop.[201,208] The polymers contain cyclic structures that may be perpendicular to the aromatic backbones. An example would be a cardo polybenzimidazole:

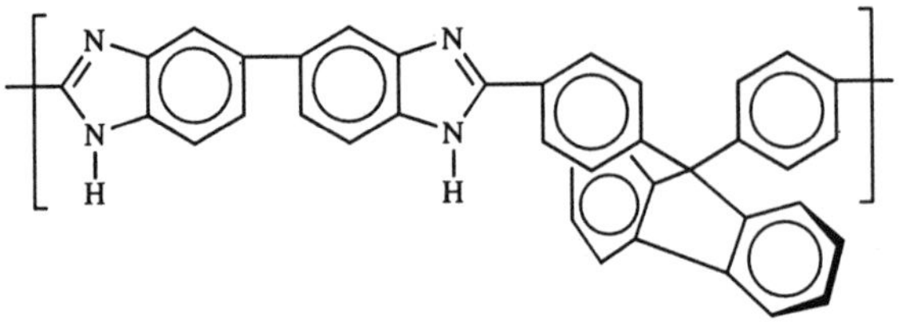

Another example is a cardo polyester[201,208]:

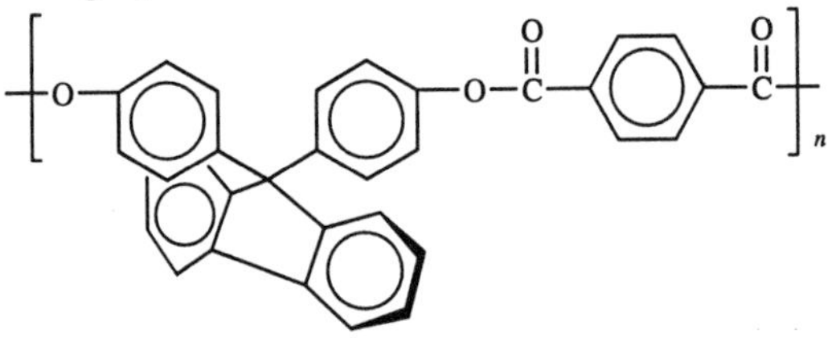

Many cardo polymers exhibit improved solubility in different solvents with little sacrifice in properties. A survey was made of the physical properties of different cardo copolyimides with varying microstructures.[220] As might be expected, copolyimides with increasing aliphatic fragment content were shown to exhibit the highest impact and flexural strength.

6.17.8. Double Stranded Polymers

Preparations of many double stranded polymers were reported. For instance, polypyrrolones were formed from 2,3,5,6-tetraaminobenzene and various dianhydrides or tetracarboxylic acid compounds.[213] Two examples follow:

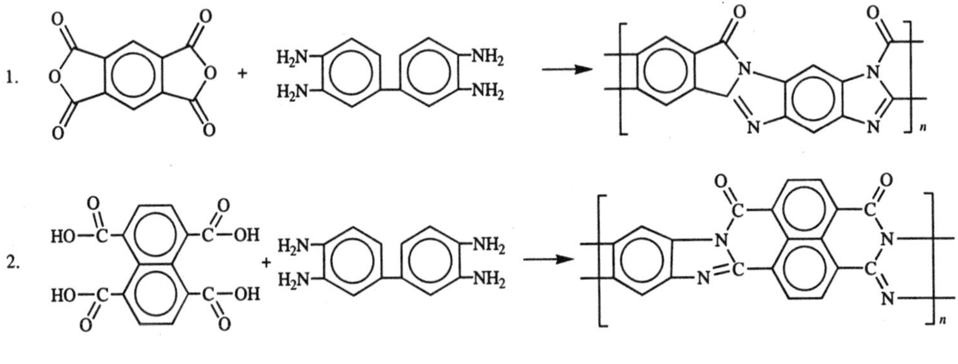

1.

2.

Different ladder polyquinoxalines were prepared as well. One example is shown below[233]:

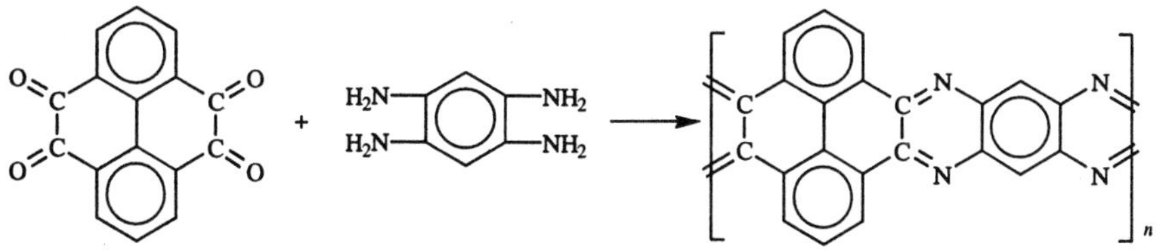

Some ladder polyquinoxalines were found to be stable in air at 460 °C, and in nitrogen up to 683 °C.

Not all attempts at formations of ladder polymers yielded completely cyclized fused ring structures. For instance, an attempt to form a polymer from tetraaminonaphthalene with naphthalene tetracarboxylic acid dianhydride failed to yield complete cyclizations.[222]

An interesting polymer containing macrocyclic rings was formed from pyromellitic tetranitrile by condensation with dianilino ether[223]:

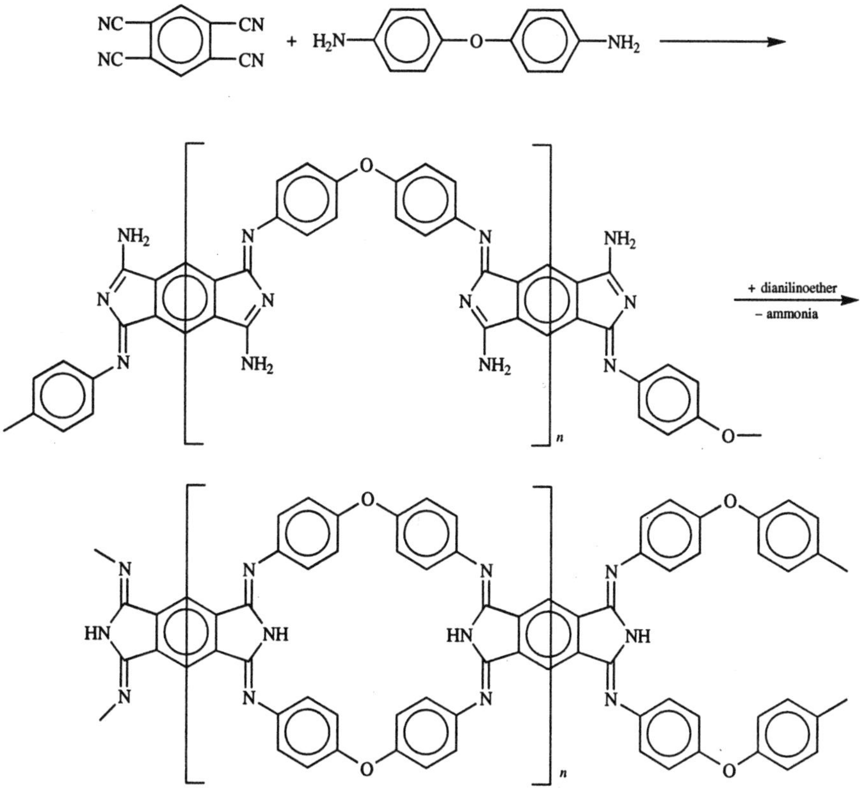

The description of all the double stranded polymers that were synthesized and reported in the literature is beyond the scope of this book. This section is concluded by a mention of a silicon-containing material prepared by an alkali-catalyzed polymerization of phenyltrichlorosilane[224]:

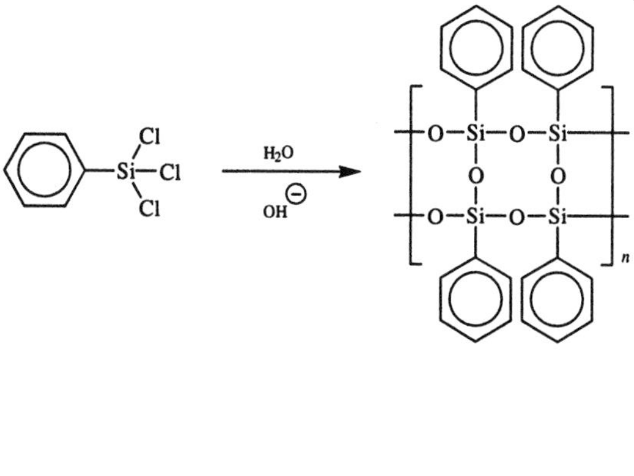

The above polymer is *cis*-syndiotactic[225] with a *cis–anti–cis* arrangement of the phenyl groups. It is stable at temperatures up to 525 °C.

6.17.9. Poly(arylene ether)s and Poly(arylene ether ketone)s

The latest trends in high-performance polymeric materials is towards the development of poly(arylene ether)s and poly(arylene ether ketone)s. They can be used as structural resins because in composite fabrications they offer an attractive combination of chemical, mechanical, and physical properties.

Commercial poly(ether ketone)s that are also poly(arylene ether)s are formed from diacid chloride by the Friedel-Craft reaction.[234]

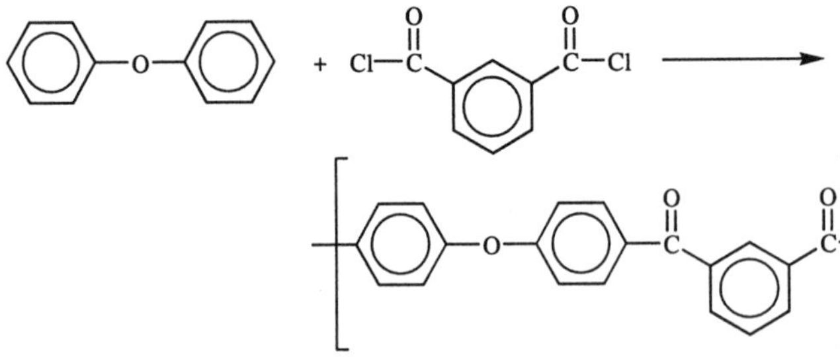

Similar polyketones form from dicarboxylic acids rather than dicarboxylic acid chlorides, when P_2O_5–methanesulfonic acid is used as a catalyst.[235] All these materials form a group of tough, high-melting resins.

Some current preparations of poly(arylene ether)s are carried out by nucleophilic displacements of activated aromatic dihalides or dinitro groups by alkali metal bisphenates. The reactions take place in polar aprotic solvents. The glass transition temperatures, tensile strengths, and tensile moduli of these materials tend to increase when heterocyclic units are incorporated into the backbones. Poly(arylene ether)s containing imide,[236] phenylquinoxaline,[237–239] imidazoles,[240] pyrazoles,[241] 1,3,4-oxadiazoles,[240] benzoxazoles,[240] and benzimidazoles[240] groups were prepared.

The preparation of such polymeric materials can be illustrated as follows[241]:

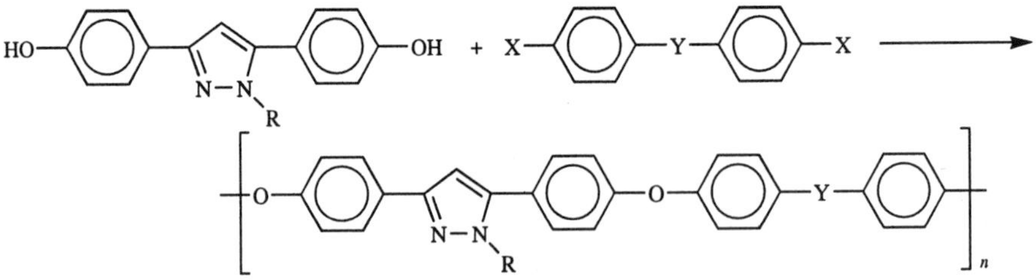

where R = H, Ph; X = Cl, F; Y = carbonyl, sulfone, or a diketone aryl group.

The above medium and high molecular weight polymers exhibit good solubility in solvents like dimethyl acetamide, and good thermal stability. The same is true of the other poly(arylene ether)s mentioned above. As a result, these high-performance thermoplastics have the potential of being useful in low-cost composite fabrications.

One recent paper[242] describes the preparation of poly(imidoaryl ether ketone)s and poly(imidoaryl ether sulfone)s that exhibit particularly good heat stability. The synthesis can be illustrated as follows:

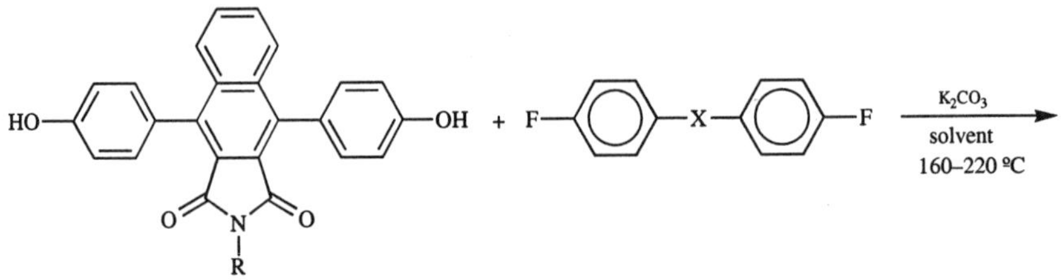

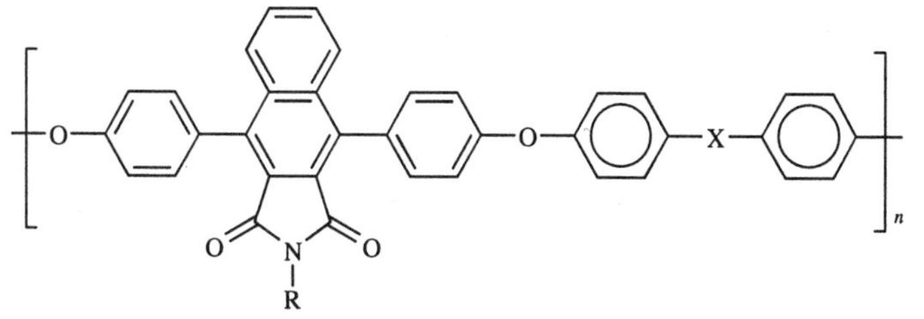

where R = CH_3, $CH_3(CH_2)_{11}$, Phenyl, and X = SO_2, CO; Isophthaloyl. Based on thermogravimetric analyses, the phenyl-substituted polymers lose only 10% of their weight in air and in nitrogen at 550 °C. In addition, the polymers have high glass transition temperatures and remain soluble in common solvents, like chloroform and methylene chloride. These are high molecular weight polymers that can be cast from solution to give tough, flexible films.[242]

Review Questions

Section 6.1

1. Describe the types of monomers that can undergo step-growth polymerizations.
2. Illustrate step-growth polymerization on formation of poly(butylene adipate), showing dimers, tetramers, etc.
3. Does the size of the molecule influence the reactivity of the functional group? Explain.
4. How would you express the rate of disappearance of one of the functional groups?
5. Assuming that the concentrations of the two reacting functional groups are stoichiometrically equal, write the rate equation.
6. Write the equation for the degree of polymerization in step-growth polymerization.
7. What is stoichiometric imbalance and how is it designated?
8. Write the expression for the molecular weight average of the product and also for the molecular weight distribution.
9. By including the average functionality factor, write the equation for the degree of conversion.
10. Discuss ring formations that can accompany step-growth polymerizations.
11. Explain how step-growth polymers are formed in the melt and by interfacial polycondensation. Why isn't the preparatory technique applicable to the preparation of aliphatic polyesters yet works well in the preparation of aliphatic polyamides?
12. In trying to form a polyester from gamma-hydroxybutyric acid, what percent conversion is required to obtain a molecular weight of 25000.

Section 6.2

1. Write chemical equations for eight common synthetic methods for preparing polyesters. Can you discuss the advantages and disadvantages of each synthesis?
2. Describe commercial preparations of poly(ethylene terephthalate), giving chemical equations, catalysts, and reaction conditions.
3. Explain, including chemical structures, how polyester elastomers are formed.
4. What are the linear unsaturated polyesters? Explain. Show by chemical reactions how they are prepared. How are they crosslinked? Explain with the help of chemical reactions.
5. Describe network polyesters. Explain how they are prepared by two different techniques.
6. What are short, medium, and long oil alkyds?
7. There are two different techniques for forming polycarbonates. Describe each.
8. What is the synthetic route to polycarbonates with the aid of crown ethers? Explain with the help of chemical equations.
9. What are the commercial uses for polycaprolactone?

Section 6.3

1. Discuss nylon nomenclature.
2. Discuss the chemistry of preparation of nylons 1, 3, 4, and 5 showing all the equations.
3. Discuss the common synthetic routes to caprolactam.
4. Describe conditions for the preparation of nylon 6.
5. Describe, with chemical equations, the preparations of nylon 7 and 9. What are the shortcomings of the process used in Russia? Explain.
6. Describe with chemical equations the synthetic route to nylons 8 and 12.
7. How is nylon 11 produced from methyl ricinoleate? Show all the steps.
8. Describe with chemical equations the commercial synthesis of Trogamid T.
9. What are fatty polyamides and how are they formed?
10. What is nylon 6T and how is it produced?
11. Discuss the chemistry of preparation of fully aromatic polyamides, showing chemical structures.
12. Explain, with an example, how the Heck reaction can be extended to form aromatic polyamides.

Section 6.4

1. Discuss the chemistry of aromatic polyamide-imides and aromatic polyester-imides. Give examples.

Section 6.5

1. What are the most common commercial procedures for preparations of aromatic polyimides? Describe with illustrations.
2. What dianhydrides are most commonly used commercially in preparations of polyimides?
3. Illustrate with chemical reactions how polyimides can be formed from aromatic diisocyanates and aromatic dianhydrides. Do the same for ketimines and *N,N'*-bis(trimethylsilyl) compounds.
4. How can polyimides be formed from reactions of diimides with dihalides? Show the chemical reactions.
5. Show the reactions for the formation of polyimides from reactions of sulfur halides with unsaturated diimides and from diepoxides with diimides.
6. Illustrate and explain the photochemical reactions of bismaleimides with benzene to form polyimides.
7. Illustrate how polyimides can form by Diels-Alder reactions.

Section 6.6

1. Discuss and illustrate with chemical equations the formation of poly(phenylene oxide) by the oxidative coupling reaction of 2,6-disubstituted phenols.
2. What is the commercial material Noryl? Explain.
3. What are phenoxy resins? Describe how they are prepared and explain how they are used.

Section 6.7

1. Discuss polyacetals and describe the polyacetals available commercially today.

Section 6.8

1. Describe how poly(*p*-xylylene) was originally prepared.
2. Discuss transport polymerization and explain how it is currently practiced commercially.
3. Describe the other routes to polyxylylenes. What are the properties of polyxylylenes?

Section 6.9

1. What are the important industrial sulfur-containing polymers?
2. Show the synthetic routes by which aromatic sulfones can be prepared.

3. Describe the preparation of poly(phenylene sulfide), its properties, and uses.
4. How is poly(alkylene sulfide) prepared and used commercially?

Section 6.10

1. Illustrate with chemical equations the routes to forming polyisocyanates.
2. What are the mechanisms of the reactions of isocyanates with hydrogen donors? How are these reactions catalyzed? Discuss the mechanism.
3. Discuss polyurethane fibers. How are they made? What are their properties?
4. Discuss polyurethane elastomers, including their properties and preparation.
5. Discuss polyurethane foams. How are they prepared?

Section 6.11

1. Discuss the chemistry of epoxy resins based on diglycidyl ethers of bisphenol A, their preparations, and crosslinking reactions with amines, dianhydrides, and dicyanodiamide.
2. What are epoxy novolacs and what are some new epoxy resins containing nitrogen?
3. Discuss the chemistry of cycloaliphatic epoxides.

Section 6.12

1. Discuss the chemistry of resoles, showing by chemical reactions how they form. What are quinone methides, and what is meant by stage A, B, and C resins?
2. Describe the chemistry of novolacs, how they are formed and crosslinked.
3. Explain what the products are in an ammonia or amine catalyzed condensation of phenol with formaldehyde.
4. Describe typical commercial preparations of resols and novolacs.

Section 6.13

1. Discuss the chemistry of urea-formaldehyde resins, their preparation and uses.
2. Repeat question 1 for melamine-formaldehyde resins.

Section 6.14

1. How do silicon compounds differ from carbon compounds?
2. How can the starting materials be prepared for the silicone resins?
3. How are high molecular weight silicone resins formed by ring-opening polymerization? What are the products from acid catalysis and basic catalysis?
4. Describe silicone elastomers. What are RTVs? Explain and show how they are crosslinked.
5. Discuss fluorosilicone elastomers.
6. Discuss polyarylsiloxanes.

Section 6.15

1. Why is there interest in polysilanes? Show by chemical reactions how are they formed.

Section 6. 16

1. What are polyphosphazines, and how are they formed and used?

Section 6.17

1. What chemical options are available to improve heat stability and toughness of polymeric materials?
2. Discuss fluorine-containing aromatic polymers.
3. Discuss the chemistry of preparation of polyphenylene.

4. Discuss Diels-Alder polymers giving at least four examples and showing all the structures of the starting materials and the products.
5. Discuss silicon-containing aromatic polymers.
6. What are direct condensation polymers? How are polybenzimidazoles, polybenzoxazoles, polybenzthiazoles, polyoxidiazoles, polybenzotriazoles, and polysulfodiazoles prepared? Illustrate with chemical equations.
7. How can polyimides be formed from thioneine?
8. Discuss the chemistry of oligomers that are terminated by functional groups to form thermoset, high heat resistant materials.
9. What are cardo polymers, what are their advantages, and how are they prepared?
10. Discuss double stranded polymers and how they are prepared.
11. Discuss the chemistry of poly(arylene ether)s and poly(arylene ketone)s.

References

1. P.J. Flory, *Principles of Polymer Chemistry*, Chapters 2,3,8, and 9, Cornell University Press, Ithaca, New York, 1953; P.E.M. Allan and C.R. Patrick, *Kinetics and Mechanism of Polymerization Reactions*, Chapter 5, Wiley-Interscience, New York, 1974.
2. I. Vansco-Szmercsanyi and E. Makay-Bodi, *Eur. Polym. J.*, **5**, 145, 155 (1969).
3. P.J. Flory, *Chem. Rev.*, **39**, 137 (1946).
4. D.H. Solomon (ed.), *Step-Growth Polymerization*, Marcel Dekker, New York, 1972.
5. G. Odian, *Principles of Polymerization*, 3rd ed., Wiley, New York, 1991.
6. W.H. Carothers, *J. Am. Chem. Soc.*, **51**, 2548 (1929); W.H. Carothers and J.A. Arvin, *J. Am. Chem. Soc.*, **51**, 2560 (1929); W.H. Carothers and J.W. Hill, *J. Am. Chem. Soc.*, **54**, 1559, 1577 (1932).
7. W.H. Stockmayer, *J. Polym. Sci.*, **9**, 69 (1952); *ibid*, **11**, 424 (1953).
8. E. Heisenberg and A.J. Watzl, Canadian Pat. # 570,148 (1959).
9. J.T. Dickson, H.P.W. Huggill, and J.C. Welch, British Pat. # 590,451 (1947).
10. J.R. Whinfield, *Nature*, **158**, 930 (1946).
11. J.G.N. Drewitt and J. Lincoln, U.S. Pat. # 2,799,667 (1957).
12. J.G. Cook, British Pat. # 590,417 (1947).
13. F. Reeder and E.R. Wallsgrove, British Pat. # 651,762 (1951).
14. R.F. Fischer, *J. Polym. Sci.*, **44**, 155 (1960).
15. R.F. Wilfong, *J. Polym. Sci.*, **54**, 385 (1961).
16. H. Ludwig (ed.), *Polyester Fibers, Chemistry and Technology*, Wiley-Interscience, New York, 1971.
17. R.W. Stevenson and H.R. Nettleton, *J. Polym. Sci.*, **A-1,6**, 889 (1968).
18. R.W. Stevenson, *J. Polym. Sci.*, **A-1,7**, 395 (1969).
19. D.M. Chay, C.C. Cumbo, M.J. Randolph, and P.C. Yates, U.S. Pat. # 3,676,477 (July 11, 1972).
20. S.B. Maerov, *J. Polym. Sci., Polym. Chem. Ed.*, **17**, 4033 (1979).
21. S.G. Hovenkamp, *J. Polym. Sci.*, **A-1,9**, 3617 (1971).
22. C. M. Fontana, *J. Polym. Sci.*, **A-1,6**, 2343 (1968); K. Ravindranath and R.A. Mashelkar, *J. Appl. Polym. Sci.*, **26**, 3179 (1981).
23. S.D. Hamann, D.H. Solomon, and J. Swift, *J. Macromol. Sci.-Chem.*, **A2**(1), 153 (1968).
24. I. Goodman and B.F. Nesbitt, *Polymer*, **1**, 384 (1960); *J. Polym. Sci.*, **48**, 423 (1960); L.H. Peebles, Jr., M.W. Huffman, and C.T. Ablett, *J. Polym. Sci.*, **A-1,7**, 479 (1969); E. Ito and S. Okamura, *Polym. Lett.*, **7**, 483 (1969).
25. H. Tanaka, Y. Iwanaga, G. Wu, K. Sanui, and N. Ogata, *Polym. J.*, **14**, 648 (1982); F. Higashi, A. Hoshio, Y. Yamada, and M. Ozawa, *J. Polym. Sci., Polym. Chem. Ed.*, **23**, 69 (1985); F. Higashi, N. Akiyama, I. Takahashi, and T. Koyama, *J. Polym. Sci. Polym. Chem. Ed.*, **22**, 1653 (1984).
26. V.V. Korshak, S.V. Rogozin, and V.I. Volkov, *Vysokomol. Soed.*, **1**, 804 (1959).
27. K. Harashi, *Macomolecules*, **3**, 5 (1970).
28. P.W. Morgan, *Condensation Polymers: By Interfacial and Solution Methods*, Wiley-Interscience, New York, 1965; P.W. Morgan, *J. Macromol. Sci.*, **A-15**, 683 (1981); F. Millich and C.E. Carraher Jr. (eds.), *Interfacial Synthesis*, M. Dekker, New York, 1975 and 1977.
29. N.G. Gaylord, *Polyethers*, Part I, Wiley-Interscience, New York, 1963.
30. C.G. Gebelein, *J. Polym. Sci.*, **A-1,10**, 1763 (1972).
31. T. Nagata, *J. Appl. Polym. Sci.*, **13**, 2601 (1969).
32. E.N. Zilberman, A.E. Kulikova, and N. M. Teplyakov, *J. Polym. Sci.*, **56**, 417 (1962).

33. W.F. Christopher and D.W. Fox, *Polycarbonates*, Reinhold, New York, 1962; H. Schnell, *Chemistry and Physics of Polycarbonates*, Wiley-Interscience, New York, 1964.
34. H. Vernaleken, in *Interfacial Syntheses* (E.F. Millich and C.E. Carreher Jr., eds.), Marcel Dekker, New York, 1975.
35. C.J. Pedersen and H.K. Frensdorff, *Angew. Chem., Int. Ed. Engl.*, **11**, 16 (1972).
36. K. Saga, Y. Toshida, S. Hosoda, and S. Ikeda, *Makromol. Chem.*, **178**, 2747 (1977); K. Saga, S. Hosoda, and S. Ikeda, *J. Polym. Sci., Polym. Lett.*, **15**, 611 (1977); *J. Polym. Sci., Polym.. Chem. Ed.*, **17**, 517 (1979).
37. R. Rokicki, W.Kuran, and J. Kielkiewicz, *J. Polym. Sci., Polym. Chem. Ed.*, **20**, 967 (1982).
38. F. Higashi, N. Kokubo, and M. Goto, *J. Polym. Sci., Polym. Chem. Ed.*, **18**, 2879 (1980).
39. H.J. Sanders, *Chem. Eng. News*, **63**(13), 30 (1985).
40. K.S. Devi and P. Vasudevan, *J. Macromol. Sci., Rev. Macromol Chem. Phys.*, **C25**, 325 (1985).
41. C.E. Lowe, U.S. Pat. # 2,688,162 (1954).
42. C.E.N. Bawn and A. Ledwith, *Chem. Ind. (London)*, 1180 (1957).
43. F. Higashi, K. Kubota, M. Sekizuka, and M. Higashi, *J. Polym. Sci., Polym. Chem. Ed.*, **19**, 2681 (1981).
44. F.W. Billmaeyer Jr. and A.D. Eckard, *Macromolecules*, **2**, 103 (1969).
45. U.H. So, *Polym. Prepr., Am. Chem. Soc.*, **32**(1), 369 (1991).
46. N. Ogata, *Bull. Chem. Soc. Japan*, **33**, 906 (1960).
47. W.R. Sorensen and T.W. Campbell, *Preparative Methods of Polymer Chemistry*, 2nd ed., Wiley-Interscience, New York, 1968; G.C. East and S. Hassell, *J. Chem. Ed.*, **60**, 69 (1983).
48. D.S. Breslow, G.A. Hulsa, and A.S. Metlack, *J. Am. Chem. Soc.*, **79**, 3760 (1957).
49. J.S. Moore and S.I. Stupp, *Macromolecules*, **23**, 65 (1990).
50. H. Bestian, *Angew. Chem., Int. Ed. Engl.*, **7**, 278 (1968).
51. R. Graf, *Liebig Ann. Chem.*, **661**, 111 (1963).
52. J. Masamoto, K. Sasaguri, C. Ohizumi, K. Yamaguchi, and H. Kabayashi, *J. Appl. Polym. Sci.*, **14**, 667 (1970).
53. H.K. Hall Jr., *J. Am. Chem. Soc.*, **80**, 6404 (1958).
54. S.G. Havenkamp and J.P. Munting, *J. Polym. Sci., Polym. Chem. Ed.*, **8**, 679 (1970); M. Drocher, *J. Appl. Polym. Sci.*, **36**, 217 (1981).
55. H. Tani and T. Konomi, *J. Polym. Sci.*, **A-1,6**, 2295 (1968).
56. S. Tsuda, *Chem. Economic and Engineering Reviews*, 39, (1970).
57. E. Muller, *Angew. Chem.*, **71**, 229 (1959).
58. H.K. Reinschuessel, *Macromol. Rev.*, **12**, 65 (1977).
59. British Pat. #1,391,135 (1975) (from Ref. 58).
60. P.H. Hermans, *Recl. Trav. Chim. Pays-Bas*, **72**, 798 (1953).
61. I. Rothe and M. Rothe, *Chem. Ber.*, **88**, 284 (1955).
62. M. Rothe, *J. Polym. Sci.*, **30**, 227 (1958).
63. M. Rothe, *Makromol. Chem.*, **35**, 183 (1960).
64. A. Reiche nad W. Schon, *Chem. Ber.*, **99**, 3238 (1967); *Kunststoffe*, **57**, 49 (1967).
65. I. Kohan, *Nylon Plastics*, Wiley, New York, 1973.
66. A.N. Nesmeyanov and R.K. Freundlina, *Tetrahedron*, **17**, 65 (1962).
67. W.L. Kohlhase, E.H. Pryde, and J.C. Cowan, *J. Am. Oil Chem. Soc.*, **47**, 183 (1970).
68. W.R. Miller, E.H. Pryde, R.A. Awl, W.L. Kohlhase, and D.J. Moore, *Ind. Eng. Chem., Prod. Res. Dev.*, **10**, 442 (1971).
69. U.S. Pat. # 3,476,820 (1968).
70. D.D. Coffman, G.J. Berchet, W.R. Peterson, and E.W. Spangel, *J. Polym. Sci.*, **2**, 306 (1947).
71. U.S. Pat.# 3,145,193; 3,150,113; 3,150,117 (1964); # 3,198,771 (1965); # 3,294,758 (1966).
72. R. Hill and E.E. Walker, *J. Polym. Sci.*, **3**, 609 (1948).
73. C.W. Bunn and E.V. Garner, *Proc. Roy. Soc.(London)*, **A189**, 39 (1947).
74. J. Preston and W.B. Black, *J. Polym. Sci., Polym. Lett.*, **3**, 845 (1965).
75. J. Preson and W.B. Black, *J. Polym. Sci., C-23*, 441 (1968).
76. B.F. Malichenko and V.V. Senkova, *Vysokomol. Soyedin Ser. B*, **14**(6), 423 (1972).
77. W.H. Daly and W. Kern, *Makromol. Chem.*, **108**, 1 (1967); J.T. Chapin, B.K. Onder, and W.J. Farrissey, *Polym. Prepr., Am. Chem. Soc.*, **21**(2), 130 (180).
78. N. Ogata, K. Sanui, and M. Harada, *J. Polym. Sci., Polym. Chem. Ed.*, **17**, 2401 (1979).
79. Y.P. Khanna, E.M. Pearce, B.D. Forman, and D.A. Bini, *J. Polym. Sci., Polym. Chem. Ed.*, **19**, 2799 (1981).
80. Y.P. Khanna, E.M. Pearce, J.S. Smith, D.T. Burkitt, H. Njuguna, D.M. Hinderlang, and B.D. Forman, *J. Polym. Sci., Polym. Chem. Ed.*, **19**, 2817 (1981).
81. M. Ueda, S. Aoyama, M. Konno, and Y. Imai, *Makromol. Chem.*, **179**, 2089 (1978).
82. F. Higashi, M. Goto, and H. Kakinoki, *J. Polym. Sci., Polym. Chem. Ed.*, **18**, 1711 (1980); N. Yamazaki, M. Matsumoto, and F. Higashi, *J. Polym. Sci., Polym. Chem. Ed.*, **13**, 1373 (1975); F. Higashi and Y. Tagushi, *J.*

Polym. Sci., Polym. Chem. Ed., **18**, 2875 (1980); F. Higashi, N. Akiyama, and S.I. Ogata, *J. Polym. Sci., Polym. Chem. Ed.*, **21**, 1025 (1983).

83. C.J. Frosch, U.S. Pat. # 2,421,024 (1947).
84. K. Yaniguchi, G. Ricicki, W. Kawanobe, S. Nakahama, and N. Yamazaki, *J. Polym. Sci., Polym. Chem. Ed.*, **20**, 118, (1982).
85. D.F. Loncrini, U.S. Pat. # 3,182,073 (1965).
86. J.L. Nieto, *Makromol. Chem.*, **183**, 557 (1982).
87. M.T. Bogert and R.R. Renshaw, *J. Am. Chem. Soc.*, **30**, 1135 (1908).
88. H. Lee, D. Stoffey, and K. Neville, *New Linear Polymers*, McGraw-Hill, New York, 1967; K.L. Mittal (ed.), *Polyimides*, Vol. 1 and 2, Plenum Press, New York, 1984; M.I. Beesonov, M.M. Koton, V.V.Kudriavtsev, and L.A. Laius, *Polyimides—Thermally Stable Polymers*, Plenum Press, New York, 1987; M.W. Ranney, *Polyimide Manufacture*, Noyes Data Corp, Park Ridge, N.J., 1971; C. Feger, M.K. Kojastech, and J.E. McGrath (eds.), *Polyimides: Chemistry, Characterization and Materials*, Elsevier, Amsterdam, 1989; D. Wilson, P. Hergenrother, and H. Stenzenberger (eds.), *Polyimides*, Chapman and Hall, London, 1990.
89. R.A. Meyers, *J. Polym. Sci.*, **A-1,7**, 2757 (1969); W.J. Farissey Jr., J.S. Rose, and P.S. Carleton, *J. Appl. Polym. Sci.*, **14**, 1093 (1970); P.S. Carleton, W.J. Farissey Jr., and J.S. Rose, *J. Appl. Polym. Sci.*, **16**, 2983 (1972).
90. W.M. Alvino and L.E. Edelman, *J. Appl. Polym. Sci.*, **19**, 2961 (1975); N.D. Ghatge and U.P. Mulik, *J. Polym. Sci., Polym. Chem. Ed.*, **18**, 1905 (1980).
91. (a) R. Merten, *Angew. Chem., Int. Ed. Engl.*, **10**, 294 (1971); (b) Y. Oishi, M. Kakimoto, and Y. Imai, *Macromolecules*, **24**, 3475 (1991).
92. V.L. Bell, *J. Polym. Sci., Polym. Lett.*, **5**, 941 (1966).
93. S. Nishizaki and A. Fukami, *Kogyo Kogaku Zasshi*, **68**, 383 (1965); *Chem. Abstr.*, **63**, 3057 (1965).
94. A.A. Berlin, T.V. Zelenetskaya, and R.M. Aseeva, *Zh. Vses. Khim. Obschest.*, **15**, 591 (1966).
95. R.A. Meyers and E.R. Wilson, *J. Polym. Sci., Polym. Lett.*, **6**, 531 (1968).
96. Y. Iwakura and F. Hayano, *J. Polym. Sci.*, **A-1,7**, 597 (1969).
97. M. Russo and L. Mortillaro, *J. Polym. Sci.*, **A-1,7**, 3337 (1969).
98. Y. Musa and M. P. Stevens, *J. Polym. Sci.*, **A-1,10**, 319 (1972).
99. F. W. Harris and J. K. Stille, *Macromolecules*, **1**, 463 (1968).
100. Y. Imai, *J. Polym. Sci., Polym. Lett.*, **8**, 555 (1970).
101. O.K. Goins and R.L. Van Deusen, *J. Polym. Sci., Polym. Lett.*, **6**, 821 (1968).
102. A.S. Hay, *Adv. Polym. Sci.*, **4**, 496 (1967); R.O. Johnson and H.S. Burlhis, *J. Polym. Sci., Polym. Symp.*, **70**, 129 (1983).
103. W.G.B. Huysmans and W.A. Waters, *J. Am. Chem. Soc.*, **89**, 1163 (1967).
104. G.D. Cooper and A. Katchman, Chapt. 43 in *Addition and Condensation Polymerization Processes*, ACS Publication, *Adv. Chem. Ser.* **91**, 1969; J.G. Bennet Jr. and G.D. Cooper, *Macromolecules*, **3**, 101 (1970).
105. C.J. Kurian and C.C. Price, *J. Polym. Sci.*, **49**, 267 (1961).
106. L.A. Errede and M. Szwarc, *Quart. Rev. (London)*, **12**, 301 (1958); L.A. Errede and S. Gregorian, *J. Polym. Sci.*, **60**, 21 (1962); L.A. Errede and N. Knoll, *J. Polym. Sci.*, **60**, 33 (1962); L.A. Errede, British Pat. #920,515 (1963).
107. H. Gilch, *Angew. Chem., Int. Ed. Engl.*, **4**, 598 (1965); *J. Polym. Sci.*, **A-1,4**, 43 (1966).
108. F. Kluiber, *J. Org. Chem.*, **6**, 2037 (1965).
109. R.B. Akin, *Acetal Resins*, Reinhold, New York, 1962; N.G. Gaylord. (ed.), *Polyethers*, Part III, Wiley-Interscience, New York, 1962.
110. (a) R.B. Seymour and G.S. Kirshenbaum (eds.), *High Performance Materials: Their Origin and Development*, Elsevier, New York, 1986; D.G. Brady, *J. Appl. Polym. Sci., Symp.*, **36**, 231, 1981; (b) D.F. Fahey and C.E. Ash, *Macromolecules*, **24**, 4242 (1991).
111. H. A. Vogel, *J. Polym. Sci.*, **A-1,8**, 2035 (1970).
112. M. Yoneyama, M. Kakimoto, and Y. Imai, *Macromolecules*, **21**, 109, (1988); M. Yoneyama, M. Kakimoto, and Y. Imai, *J. Polym. Sci., Polym. Chem. Ed.*, **27**, 1985 (1989); M. Yoneyama, T. Konishi, N. Kakimoto, and Y. Imai, *Makromol. Chem., Rapid Commun.*, **11**, 381 (1990); Y. Imai, *Polym. Prepr., Am. Chem. Soc.*, **32**(1), 331 (1991).
113. J.F. Klebe, *Adv. Org. Chem.*, **8**, 97 (1972).
114. Y. Oshi, M. Kakimoto, and Y. Imai, *Macromolecules*, **20**, 703 (1987); *Ibid.*, **21**, 547 (1988); Y. Imai, *Polym. Prepr., Am. Chem. Soc.*, **32**(1), 397 (1991).
115. H.D. Stenzenberger, M. Herzog, W. Romer, R. Scheiblich, and N.J. Reeves, *Br. Polym. J.*, **15**, 2 (1983); I.K. Varma and S. Sharma, *Eur. Polym. J.*, **20**, 1101 (1984).
116. C.A. May (ed.), *Epoxy Resins: Chemistry and Technology*, 2nd ed., Dekker, New York, 1988; B. Sedlacek and J. Kahovek (eds.) *Crosslinked Epoxies*, de Gruyler, New York, 1986; R.S. Bauer (ed.), *Epoxy Resin Chemistry*, 2 vols. *ACS Symp. Ser.*, **114** and **221**, Am. Chem. Soc., Washington, D.C., 1979 and 1983.
117. W.J. Bailey and A.A.Volpe, *J. Polym. Sci.*, **A-1,8**, 2109 (1970).
118. R.N. Johnson, A.G. Faruham, R.A. Glendenning, W.F. Hall, and C.N. Merriam, *J. Polym. Sci.*, **A-1,5**, 2375 (1967).

119. J.B. Rose, *Polymer*, **15**, 456 (1974).
120. T.E. Attwood, A.B. Newton, and J.B. Rose, *Br. Polym. J.*, **4**, 391 (1972).
121. A.B. Newton and J.B. Rose, *Polymer*, **13**, 465 (1972).
122. S. Ozaki, *Chem. Rev.*, **72**, 457 (1972).
123. K.C. Frisch and L.P. Rumao, *J. Macromol. Sci., Rev. Macromol. Chem.*, **C-5**, 103 (1970).
124. J.W. Baker and J.B. Haldsworth, *J. Chem. Soc.*, 713 (1974).
125. J.W. Baker and J. Gaunt, *J. Chem. Soc.*, 9, 19, 24, 27 (1949).
126. K.N. Edwards (ed.), *Urethane Chemistry and Applications, ACS Symp. Ser.*, **172**, Am. Chem. Soc., Washington, D.C., 1981.
127. F. Hostettler and E.F. Cox, *Ind. Eng. Chem.*,**52**, 609 (1960).
128. J.W. Britain and P.G. Gemeinhardt, *J. Appl. Polym. Sci.*, **4**, 207 (1960).
129. M. Furukawa and T. Yokoyama, *Makromol. Chem.*, **182**, 2201 (1981).
130. J. Blackwell and K.H. Gardner, *Polymer*, **20**, 13 (1979).
131. R. Bonart, *Angew. Makromol. Chem.*, **58/59**, 259 (1977).
132. B.O. Dombrow, *Polyurethanes*, Reinhold, New York. 1957.
133. E.N. Boyle, *The Development and Use of Polyurethane Products*, McGraw-Hill, New York, 1969.
134. W.G Potter, *Epoxy Resins*, Springer-Verlag, New York, 1970; C.A. May (ed.), *Epoxy Resins: Chemistry and Technology*, 2nd ed., Dekker, New York, 1988.
135. P.F. Bruins, *Epoxy Resin Technology*, Wiley-Interscience, New York, 1968; B. Sedlacek and J. Kahovek (eds.), *Crosslinked Epoxies*, de Gruyter, New York, 1986.
136. Y. Tanaka and H. Kakuichi, *J. Appl. Polym. Sci.*, **7**, 1063, 2951 (1963).
137. (a) M. Fischer, F. Lohse, and R. Schmid, *Makromol. Chem.*, **181**, 1251, (1980); (b) F. Riccardi, M.M. Joullied, W.A. Ramanchick, and A.A. Giscavage, *J. Polym. Sci., Polym. Lett.*, **21**, 127 (1982).
138. J.S. Bradshaw and M.P. Stevens, *J. Appl. Polym. Sci.*, **10**, 1809 (1966).
139. L. Shechter and J. Wynstra, *Ind. Eng. Chem.*, **48**, 86 (1956).
140. T.F. Saunders, M.F. Levy, and J.F. Serino, *J. Polym. Sci.*, **A-1,5**, 1609 (1967).
141. E. Pyun and C.S.P. Sung, *Macromolecules*, **24**, 855 (1991).
142. M. DiBenedetto, *J. Coat. Tech.*, **52**, 65 (1980).
143. M.K. Autoon and J.L. Koenig, *J. Polym. Sci., Polym. Chem. Ed.*, **19**, 549 (1981).
144. D.F. Gould, *Phenolic Resins*, Reinhold, New York, 1959.
145. N.J.L. Megson, *Phenolic Resin Chemistry*, Academic Press, New York, 1958; A. Knop and L.A. Plato, *Phenolic Resins*, Springer-Verlag, New York, 1979.
146. R.W. Martin, *The Chemistry of Phenolic Resins*, Wiley-Interscience, New York, 1956.
147. A. Zinke and St. Pucher, *Monatsh. Chem.*, **79**, 26 (1948).
148. T.S. Carswell, *Phenoplasts*, Wiley-Interscience, New York, 1947.
149. A.L. Cupples, H. Lee, and D.G. Stoffey, Chapt. 15 in *Advances in Chemistry Series*, # 92 (R.F. Gould, ed.), Am. Chem. Soc., 1970.
150. A Ninagawa and H. Matsuda, *Makromol. Chem., Rapid Commun.*, **2**, 449 (1981); A. Ninagawa, Y. Ohnishi, H. Takeuchi, and H. Matsuda, *Macromol. Chem., Rapid. Commun.*, **6**, 793 (1985).
151. C.P. Vale, *Aminoplastics*, Cleaver-Hume Press, London, 1950; C.P. Vale and W.G.K. Taylor, *Aminoplatics*, Iliffe, London, 1964.
152. S.M. Kambanis and R.C. Vasishth, *J. Appl. Polym. Sci.*, **15**, 1911 (1971).
153. A.K. Dunker, W.E. Johns, R. Rammon, and W.L. Plagemann, *J. Adhesion*, **17**, 275 (1985); A.K. Dunker, W.E. Johns, R. Rammon, B. Farmer, and S.J. Johns, *J. Adhesion*, **19**, 153 (1986).
154. S.S. Jada, *J. Appl. Polym. Sci.*, **35**, 1573 (1988).
155. K. Koeda, *J. Chem. Soc. Japan, Pure Chem. Soc.*, **75**, 571 (1954).
156. E.G. Rochow, *An Introduction to the Chemistry of Silicones*, 2nd ed.,Wiley-Interscience, New York, 1951.
157. R.R. McGregor, *Silicones and Their Uses*, McGraw-Hill, New York, 1954.
158. R.N. Meals and F.M. Lewis, *Silicones*, Reinhold, New York, 1959.
159. W. Noll, *Chemistry and Technology of Silicones*, Academic Press, New York, 1968.
160. G.G. Freeman, *Silicones, An Introduction to their Chemistry and Applications*, Iliffe, London, 1962.
161. K.A. Andrianov and Zhdanov, *Izv. Akad. Nauk SSSR, Otdel Khim. Nauk*, 1033 (1954).
162. M.M. Sprung and G. Guenther, *J. Am. Chem. Soc.*, **77**, 6045 (1955); *ibid,* **77**, 3996 (1955).
163. E.E. Bostick, Chapt. 8 in *Ring Opening Polymerizations* (K.C. Frisch and S.L. Reegen, eds.), Marcel Dekker, New York, 1969.
164. K.A. Andrianov, *Methods of Synthesis of Organometalloid Polymers*, Intern.Symp. on Inorg. Polymers, Nottingham, England, 1961.
165. R.D. Miller, *Adv. Chem. Ser.*, **224**, 413 (1990).
166. S. Yajima, J. Hayashi,and M. Omori, *Chem. Lett.*, 931 (1975).
167. R.D. Miller and J. Michi, *Chem. Rev.*, **89**, 1359 (1989).

168. R. West *J. Organomet. Chem.*, **300**, 327 (1986).
169. M.A. Abkowitz, M. Stolka, R.J. Weagley, K.M. McGrane, and F.E. Knier, *Adv. Chem. Ser.*, **224**, 467, (1990).
170. K. Matyjaszewski, *Polym. Prepr., Am. Chem. Soc.*, **31**(2), 224 (1990).
171. A.R. Wolf, I. Nozue, J. Maxka, and R. West, *J. Polym. Sci., Polym. Chem. Ed.*, **26**, 701 (1988).
172 K. Metyjaszewski and Y.L. Chen, *J. Organomet. Chem.*, **340**, 7, (1988).
173. (a) H.R. Allcock, *Chem. Rev.*, **72**, 315 (1972); (b) H.R. Allcock, *Angew. Chem., Int. Ed. Engl.*, **16**, 124 (1977); *Science*, **193**, 1214 (1976); *Polymer*, **21**, 673 (1980); T. L Evans and H.R. Allcock, *J. Macromol. Sci.-Chem.*, **A16**, 409 (1981); H.R. Allcock, *J. Polym. Sci., Polym. Symp.*, **70**, 71 (1983).
174. H.C. Brown, *J. Polym. Sci.*, **44**, 9 (1960).
175. J. M. Cox, B.A. Wright, and W.W. Wright, *J. Appl. Polym. Sci.*, **8**, 2935 (1964).
176. M. Hellman, A.J. Bilbo, and W.J. Plummer, *J. Am. Chem. Soc.*, **77**, 3650 (1955).
177. W.J. Plummer and L.A. Well, 145th ACS Meeting, 1963.
178. G.A. Edwards and G. Goldfinger, *J. Polym. Sci.*, **16**, 589 (1955).
179. R.H. Neilson, P. Wisian-Neilson, *Chem Rev.*, **88**, 541 (1988); R.H. Neilson, P. Wisian-Neilson, J.J. Meister, A.K. Roy, and G.L. Hagnauer, *Macromolecules*, **20**, 910 (1987); R.H. Neilson, D.J. Jinkerson, S. Karthikeyan, R. Samuel, and C.E. Wood, *A.C.S. Polym. Prepr., Am. Chem. Soc.*, **32**(3), 483 (1991).
180. I.J. Hotchkiss, R. Tephens and J.C. Tatlow, *J. Fluorine Chem.*, **10**, 541 (1977).
181. S.W. Chow, W.E. Loeb, and C.E. White, *J. Appl. Polym. Chem.*, **13**, 2325 (1969).
182. C.S. Marvel, *Appl. Polym. Symp.*, **22**, 47 (1973).
183. (a) J.G. Speight, P. Kovacic, and F.W. Koch, *J. Macromol Sci.,-Rev. Chem.*, **C5**, 295 (1971); (b) S.A. Milosevich, K. Saichek, L. Hunchey, W.B. England, and P. Kovacic, *J. Am. Chem. Soc.*, **105**, 1088 (1983).
184. M.B. Jones, P. Kovacic, and D. Lanska, *J. Polym. Sci., Polym. Chem. Ed.*, **19**, 89 (1981); M.B. Jones, P. Kovacic, and R.F. Howe, *J. Polym. Sci,. Polym. Chem. Ed.*, **19**, 235 (1981); C.F.Hsing, M.B. Jones, and P. Kovacic, *J. Polym. Sci., Polym. Chem. Ed.*, **19**, 973 (1981); B.S. Lamb and P. Kovacic, *J. Polym. Sci., Polym. Chem. Ed.*, **18**, 2423 (1980).
185. I. Khoury and P. Kovacic, *J. Polym. Sci., Polym. Lett.*, **19**, 395 (1981).
186. P. Kovacic and C. Wu, *J. Polym. Sci.*, **47**, 45 (1960); P. Kovacic and A. Kyriakis, *Tetrahedron Lett.*, 467 (1962); P. Kovacic and A. Hyriakis, *J. Am. Chem. Soc.*, **75**, 454 (1963).
187. P. Kovacic and R.J. Hopper, *J. Polym. Sci.*, **4**, 1445 (1966); P. Kovacic and J. Ozionick, *J. Org. Chem.*, **29**, 100 (1964).
188. P. Kovacic and L.C. Hsu, *J. Polym. Sci.*, **4**, 5 (1966).
189. G.G. Emgstrom and P. Kovacic, *J. Polym. Sci., Polym. Chem. Ed.*, **15**, 2453 (1977).
190. J.R. Griffith and R.F. Brady, Jr.,*Chemtech*, **19**(6), 370 (1989); H. S-W. Hu and J.R. Griffith, *Polym. Prepr., Am Chem. Soc.*, **32**(3), 216 (1991).
191. P.E. Cassidy, *Thermally Stable Polymers*, M. Dekker, New York, 1980; J.P. Critchy, G.J. Knight, and W.W. Wright, *Heat Resistant Polymers*, Plenum Press, New York, 1983; A.H. Frazer, *High Temperature Resistant Polymers*, Wiley-Interscience, New York, 1968.
192. N. Bilow and L.J. Miller, *J. Macromol. Sci., Chem.*, **A1**, 183 (1967); *ibid*, **A3**, 501 (1969).
193. M. Rahahn, A.D. Schluter, and G. Wegner, *Makromol. Chem.*, **191**, 1991 (1990).
194. H. Mukamal, F.W. Harris, and J.K. Stille, *J. Polym. Sci.*, **A-1,5**, 2721 (1967).
195. J.K. Stille and G.K. Noren, *J. Polym. Sci., Polym. Lett.*, **7**, 525 (1969).
196. C.L. Schilling Jr., J.A. Reed, and J.K. Stille, *Macromolecules*, **2**, 85 (1969).
197. W.J. Bailey, J. Economy, and M.E. Hermes, *J. Org. Chem.*, **27**, 3259 (1962).
198. W.J. Bailey and A.A. Volpe, *J. Polym. Sci.*, **A-1,8**, 2109 (1970).
199. J.K. Stille and M.A. Bedford, *J. Polym. Sci.*, **A-1,6**, 2331 (1968); J.K. Stille, *J. Macromol. Sci.*, **A3**, 1043 (1969).
200. J.K. Stille and L.D. Gotter, *J. Polym. Sci.*, **A-1,7**, 2493 (1969).
201. J.K. Stille, G.K. Noren, and L. Green, *J. Polym. Sci.*, **A-1,8**, 2245 (1970); J.K. Stille, R.M. Harris, and S.M. Podaki, *Macromolecules*, **14**, 486 (1981).
202. E.A. Kraiman, U.S. Pat. # 2,890,206 (1959); 2,890,207 (1959); 3,074,915 (1963).
203. V. Saukaran and C.S. Marvel, *J. Polym. Sci., Polym. Chem. Ed.*, **18**, 1835 (1980).
204. K. Meyersen and J.Y.C. Wang, *J. Polym. Sci.*, **A-1,5**, 1845 (1967).
205. F.C. De Schryver, W.J. Feast, and G. Smets, *J. Polym. Sci.*, **A-1.8**, 1939 (1970).
206. H. Vogel and C.S. Marvel, *J. Polym. Sci.*, **50**, 511 (1961).
207. C.J. Abshire and C.S. Marvel, *Makromol. Chem.*, **44–46**, 388 (1961).
208. V.V. Korshak, E.S. Krongauz, and A.L. Rusanov, *J. Polym. Sci.*, **C,16**, 2635 (1967); V.V. Korshak, S.V. Vinogradova, and Y.S. Vygodskii, *Rev. Makromol. Chem.*, **12**, 45 (1974–1975).
209. P.M. Hergenrother, W. Wrasidlo, and H.H. Levine, *J. Polym. Sci.*, **A-1,3**, 1665 (1965).
210. T. Kubota and R. Nakanishi, *J. Polym. Sci., Polym. Lett.*, **2**, 655 (1964).
211. M.R. Lilyquist and J.R. Holsten, *J. Polym. Sci.*, **C,19**, 77 (1967).
212. A.H. Frazer and W.P. Fitzgerald Jr., *J. Polym. Sci.*, **C,19**, 95 (1967).

213. S.S. Hersch, *J. Polym. Sci.*, **A-1,7**, 15 (1969).
214. E.S. Krongauz, *Uspekhi Khim.*, **42**, 857 (1973).
215. R.F. Kovar, G.F.L. Ehlers, and F.E. Arnold, *Polym. Prepr., Am Chem. Soc.*, **16**(2), 246 (1975).
216. F.L. Heldberg and F.E. Arnold, *Polym. Prepr., Am. Chem. Soc.*, **16**(1), 677 (1975).
217. C. Arnold Jr., *J. Polym. Sci., Macromol. Rev.*, **14**, 265 (1979).
218. V. Saukaran, S.C. Lin, and C.S. Marvel, *J. Polym. Sci., Polym. Chem. Ed.*, **18**, 495 (1980).
219. P.M. Hergenrother, *Macromolecules*, **14**, 891, 898 (1981).
220. S.V. Vinogradova and J.S. Visodskii, *Uspekhi Khim.*, **42**, 1225 (1973); V.V. Korshak, S.V. Alekseeva, and I.Y. Slonium, *Makromol. Chem.*, **184**, 235 (1983).
221. A.W. Chow, S.P. Butler, P.E. Penwell, D.J. Osborne, and J.R. Wolfe, *Macromolecules*, **22**, 3514 (1989).
222. F.E. Arnold and R.L. Van Deusen, *J. Polym. Sci., Polym. Lett.*, **6**, 815 (1968).
223. R.J. Gaymans, K.A. Hodd, and W.A. Holmes-Walker, *Polymer*, **12**, 400 (1971).
224. K.A. Andrianov, *Metallorganic Polymers*, Wiley-Interscience, New York, 1965.
225. J.F. Brown, *J. Polym. Sci.*, **C,1**, 83 (1963).
226. H.R. Allcock, *Chem. Rev.*, **72**, 315 (1972).
227. H. Rosenberg, T.T. Tsai, and N.K. Ngo, *J. Polym. Sci., Polym. Chem. Ed.*, **20**, 1 (1982).
228. H. Kondo, M. Sato, and M. Yokoyama, *Eur. Polym. J.*, **18**, 679 (1982).
229. F.W. Harris and H.J. Spinelli, *Reactive Oligomers*, ACS Symp. Ser. **282**, Am. Chem. Soc., Washington, D.C., 1985.
230. T.A. Upshaw, J.K. Stille, and J.P. Droske, *Macromolecules*, **24**, 2143 (1991).
231. D. Kumar, *J. Polym. Sci., Polym. Chem. Ed.*, **23**, 1661 (1985).
232. D.Kumar, G.M. Fohlen, and J. A. Parker, *J. Polym. Sci., Polym. Chem. Ed.*, **22**, 927, 1141 (1984).
233. J.K. Stille and M.E. Feeburger, *J. Polym. Sci., Polym. Lett.*, **5**, 989 (1967).
234. K. Niume, F. Toda, K. Uno, Y. Iwakara, *J. Polym. Sci., Polym. Lett.*, **15**, 283 (1977).
235. M. Ueda, T. Kano, T. Waragai, and H. Sugito, *Makromol. Chem., Rapid. Commun.*, **6**, 847 (1985).
236. D.M. White, T. Takekoshi, F.J. Williams, H.M. Relees, P.E. Donohue, H.J. Klopfer, G.R. Toncks, J.S. Manello, T.O. Mathews, and R.W. Schlueny, *J. Polym. Sci., Polym. Chem. Ed.*, **19**, 1635 (1981).
237. J.L. Hedrick and J.W. Labadie, *Macromolecules*, **21**, 1883 (1988).
238. J.W. Connell and P.M. Hergenrother, *Polym. Prepr., Am. Chem. Soc.*. **29**(1), 172 (1988).
239. F.W. Harris and J.E. Korleski, *Polym. Mater. Sci. Eng. Proc.*, **61**, 870 (1989).
240. J.G. Smith, Jr., J.W. Connell, and P.M. Hergenrother, *Polym. Prepr., Am. Chem. Soc.*, **32**(3), 193 (1991).
241. R.G. Bass and K.R. Srinivasan, *Polym. Prepr., Am. Chem. Soc.*, **32**(1), 619 (1991).
242. M. Strukelj and A.S. Hay, *Macromolecules*, **24**, 6870 (1991).
243. H.M. Gajiwala and R. Zand, *Macromolecules*, **26**, 5976 (1993).

7

Naturally Occurring Polymers

7.1. Polymers That Occur in Nature

There are many naturally occurring polymeric materials. Many are quite complex. It is possible, however, to apply an arbitrary classification and to divide them into six main categories. These are:

1. Polysaccharides. This category includes starch, cellulose, chitin, pectin, alginic acid, natural gums, and others.
2. Proteins or naturally occurring polyamides found in animal and vegetable sources.
3. Polyisoprenes or natural rubbers and similar materials that are isolated from saps of plants.
4. Polynucleotides include all the DNAs and all the RNAs found in all living organisms.
5. Lignin or polymeric materials of coniferyl alcohol and related substances.
6. Naturally occurring miscellaneous polymers, such as shellac, a resin secreted by the lac insect. This is a complex crosslinked polyester of 9,10,16-trihydroxy-exadecanoic acid (aleuritic acid). The structure also includes some unsaturated long-chain aliphatic acids together with other compounds.[1]

7.2. Polysaccharides

Fischer[2] carried out some of the original investigations of the monomeric species of many polysaccharides during the last century. He was able to demonstrate the configurational relationships within some monosaccharides.

The monomers in these naturally occurring polymers are five- or six-carbon sugars. There is considerable variety among the polysaccharides and the polymers generally tend to be polydisperse, depending upon the source.

7.2.1. Hemicelluloses

Hydrolyses of hemicelluloses yield mixtures of glucose, glucuronic acid, xylose, arabinose, galactose, galacturonic acid, mannose, and rhamnose. Some common polymers of pentoses, also known as pentosans, are xylan, galactan, araban, and others. Pentosans are found in large amounts (20–40%) in cereal straws and in brans. Large-scale industrial preparations of furfural, for instance, are based on these materials.

Xylan, one of the better known hemicelluloses, is a component of plant cell membranes. This pentosan occurs in association with cellulose. The structure of xylan was shown to be 1,4-polyxylose[3]:

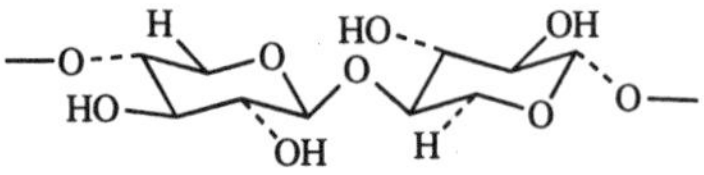

Another hemicellulose, *Galactan*, is a minor component of some coniferous and deciduous woods. Larch wood was shown to contain about 8% of this polymer.[4]

Araban, or polyarabinose, is found in plant saps. All *pectins* also belong to the family of hemicelluloses. These are gelatinizing substances that are found in many plants, particularly in fruit juices. Crude pectins contain pentosans, galactosans, and similar materials. Purified pectins yield, on hydrolysis, galacturonic acid and methanol. These high molecular weight polymers are therefore believed to consist, to a good extent, of poly(galacturonic acid), partially esterified with methyl alcohol. In addition, the polymers contain galactose and arabinose molecules. The polymer is probably linear[5–10] with a 1,4-glucosidal linkage between monomers. The relative amount of various components depends upon the source of the pectin. Citrus pectin, for instance, is rich in galacturonic acid but poor in galactose and arabinose.

Plant gums and mucilages are high molecular weight polysaccharides composed of hexoses and pentoses. They also contain some uronic acid units. Among the gums are gum arabic, gum tragacanth, and many others.

7.2.2. Starch

This is the most widely distributed substance in the vegetable kingdom and is the chief reserve carbohydrate of plants. Starch consists of single repeat units of D-glucose linked together through 1 and 4 carbons by α-linkages (*cis*).[10,11] There are two types of starch molecules, *amylose* and *amylopectin*. The first is mainly a linear polymer; its molecular weight can range from 30,000–1,000,000, though its mostly 200,000–300,000. Amylose is often pictured in a spiral form due to the conformation of the α-glucoside bonds:

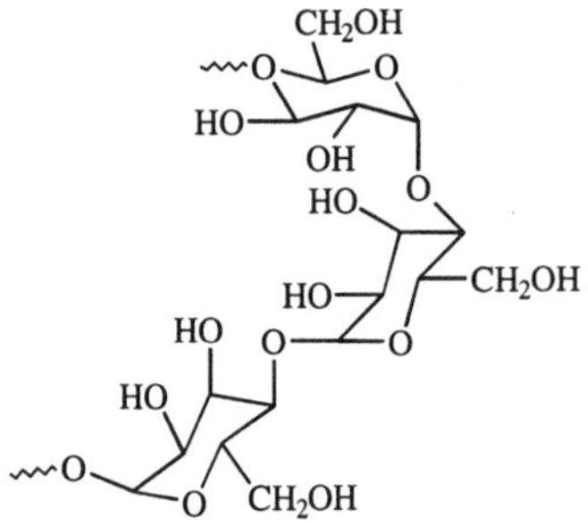

Amylopectin, on the other hand, is branched through carbon 6:

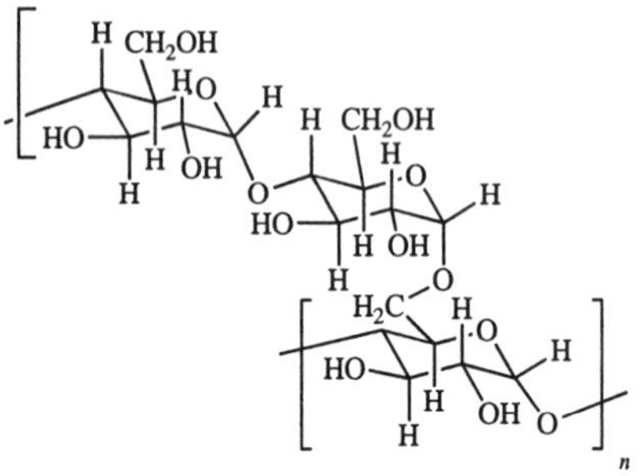

The ratios of amylopectin to amylose in many natural starches are about 3:1. The main commercial source of starch in this country is corn. Lesser amounts of industrial starch are obtained from potatoes, wheat, and tapioca (not necessarily in that order). The extraction of starch from plant material is done by grinding the plant tissues in water. The slurry is then filtered to obtain a suspension of starch granules. These granules are then collected with the aid of a centrifuge and dried.

When a water suspension of starch granules is heated to 60–80 °C, the granules swell and rupture. This results in the formation of a viscous colloidal dispersion containing some dissolved starch molecules. Cooling this dispersion results in the formation of a gel due to aggregation of the amylose molecules. It is essentially a crystallization phenomenon, known as *retrogration*. By comparison, amylopectin molecules cannot associate so readily due to branching and will not gel under these conditions.

Starches are modified chemically in various ways. Some acetate and phosphate esters are produced commercially, as well as hydroxyalkyl and tertiary aminoalkyl ethers. Both unmodified and modified starches are used principally in paper making, paper coating, paper adhesives, textile sizes, and as food thickeners. There are many reports in the literature on graft copolymers of starch. The work is often conducted is search of biodegradable materials for packaging and agricultural mulches. Most chemical modifications of starch parallel those of cellulose.

7.2.3. Cellulose

This polysaccharide is found widely in nature. It is a major constituent of plant tissues (50–70%, depending upon the wood), fibers, and leaf stalks. Chemically, cellulose is 1,4-β-poly-anhydroglucose[12] (*trans*):

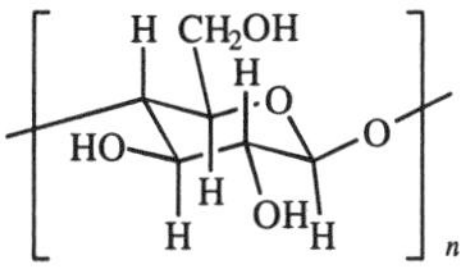

H CH_2OH H O HO O H H OH H H n

where n represents several thousand units. Hydrolysis of cellulose yields 95–96% D-glucose. This establishes its structure. Acetolysis of cellulose, however, yields cellobiose, a disaccharide, 4-*O*-β-D-glucopyranosyl-D-glucopyranose:

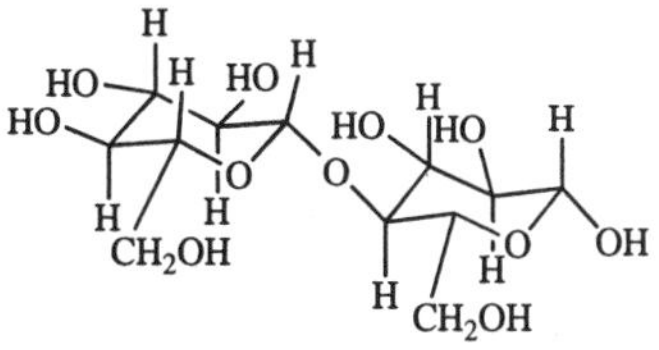

H HO H HO H HO H HO HO H O O H H CH_2OH O OH H H CH_2OH

The structure of cellulose is therefore officially based on cellobiose units. Careful molecular weight measurements by many[12] established that the DP of cellulose ranges from 2000–6700, depending upon the source. The polymer is highly crystalline and is characterized by a very high degree of intermolecular and intramolecular hydrogen bonding. This prevents it from being thermoplastic as it decomposes upon heating without melting.

7.2.3.1 Regenerated Cellulose

Cellulose is used in many forms. Often, it is modified chemically to render it soluble in organic solvents. In other modifications, it is treated in a manner that allows forming it into desired shapes, like films or fibers, followed by restoration of its chemically insoluble form. The material is then called regenerated cellulose.

Several processes have evolved for the preparation of regenerated cellulose. One, developed as far back as 1884, converts it first to a nitrate ester. The nitrated material is dissolved in a mixture of ethyl alcohol and diethyl ether and extruded into fibers. The fibers are then denitrated by treatment with ammonium hydrogen sulfide at about 40 °C. The product is called *Chardonnet silk*. It appears that this process is no longer practiced anywhere.

In another process, cellulose is dissolved in ammoniacal cupric hydroxide ($Cu(NH_3)_4(OH)_2$). The solution is then spun as a fiber into a dilute sulfuric acid solution to regenerate the cellulose. The product is called *Cuprammonium rayon*. The material may still be manufactured on a limited scale.

The third, probably major, commercial process used today forms a material that is known as *Viscose rayon*. The regenerated cellulose is prepared and sold as a fiber as well as a film, known as *cellophane*. The viscose, or more properly referred to as the *xanthate* process, consists of forming cellulose xanthate by reacting alkali cellulose with carbon disulfide:

In a typical procedure, cellulose is steeped in an approximately 20% aqueous sodium hydroxide solution at room temperature for anywhere from 20 minutes to a whole hour. It is believed that this treatment results in formation of sodium alcoholate at every hydroxymethyl group. The resultant material is pressed out to remove excess liquid, shredded, and aged for 2–3 days. The aging is known to cause some molecular weight reduction. After aging, the alkali cellulose is treated with carbon disulfide for 2–4 hours to form cellulose xanthate. The amount of xanthate groups in the product average out to one per every two glucose units. The material is dissolved in a dilute sodium hydroxide solution and again aged for 2–5 days. During the aging period some xanthate groups decompose. The solution is then spun into dilute sulfuric acid to regenerate the cellulose and form fibers:

The rayon fibers are washed, bleached, and submitted to other various treatments, like dyeing, etc., depending upon the intended use.

When cellulose xanthate is extruded through narrow slits into acid baths, cellophane films form. These films are usually plasticized by washing in baths containing some glycerin.

7.2.3.2. Derivatives of Cellulose

Many *derivatives of cellulose* have been synthesized over the years.[12–14] These include esters of both organic and inorganic acids, ethers, and various graft copolymers. Only some of them, however, achieved commercial importance.

One of the earliest commercial esters of cellulose was *cellulose nitrate*. It was originally prepared as an explosive (*guncotton*) in the middle of the last century, and later as a medical aid (*collodion*, for covering wounds). Later, films from cellulose nitrate were used in photography, called *celluloid*. Nitrocellulose was also probably the first successful commercial plastic, used to form many articles. Today it is generally displaced by other materials. Cellulose nitrate, however, is still being used in some surface finishes, though here too it is gradually being displaced.

Cellulose is nitrated by mixtures of nitric and sulfuric acids. The type of acid mixture used depends on the intended products. For the preparation of plastic grade materials, 25% nitric acid is combined with 55% sulfuric acid and 20% water. The dried cellulose is soaked for 20–60 minutes at 30–40 °C. There is little change in appearance as the structure of the cellulose is maintained. The bulk of the acid is then removed, usually by spinning in a centrifuge, and the remaining acid washed out with copious amounts of water. The product is often bleached with sodium hypochlorite and washed.

The degree of nitration is controlled by reaction conditions and particularly by the amount of water in the nitrating bath. Products with 1.9–2.0 nitrate groups per each glucose unit are used in plastics and lacquers. Some materials, however, with a nitrate content as high as 2.0–2.4 groups per each glucose have been used in some lacquers. The higher nitrate content of 2.4–2.8 groups per each glucose is in materials intended for use as explosives. The esterification reaction can be illustrated as follows:

$$\left[\text{cellulose unit: HO},\ \text{OH},\ \text{CH}_2\text{OH}\right]_n \xrightarrow{HNO_3/H_2SO_4} \left[\text{HO},\ \text{ONO}_2,\ \text{CH}_2\text{ONO}_2\right]_n$$

The molecular weights of cellulose nitrates used in plastics and lacquers is usually reduced. This is done by heating the slurry of the polymer in water at about 130–160 °C for up to 30 minutes under pressure.

Cellulose acetate was also prepared originally in the last century. Commercial development, however, started early in this century. In the 1920s, acetate rayon and acetate fibers were developed and cellulose acetate became an important molding material. At about the same time cellulose lacquers were also developed. Today, however, many of these materials have been replaced by other polymers.

The acetylation reaction of cellulose is often prepared by forming a solution in a mixture of acetic anhydride and sulfuric acid. This results in the formation of a triacetate. When a lower degree of esterification is desired, the triacetate is partially hydrolyzed. A two-step procedure is needed, because it is not possible to control the degree of esterification in the reaction with acetic anhydride and sulfuric acid. In a typical process, dry cellulose is pretreated with 300 parts acetic anhydride, one part sulfuric acid, and 400 parts methylene chloride. The reaction mixture is agitated while the temperature is maintained at 25–35 °C for 5–8 eight hours. By the end of that period all the cellulose is dissolved and the cellulose triacetate has formed in the solution.

Partial hydrolysis is accomplished by adding to the methylene chloride solution aqueous acetic acid (50%). The solution is then allowed to stand in order to reach the desired degree of hydrolysis. This usually takes about 72 hours at room temperature. Sulfuric acid, still present from acetylation, is then neutralized by addition of sodium acetate and most of the methylene chloride is distilled off. The partially hydrolyzed cellulose acetate is then precipitated by addition of water and washed thoroughly. The washing also includes a two-hour wash with very dilute sulfuric acid to remove hydrogen sulfate esters that cause polymer instability.

The process can be illustrated as follows:

$$\left[\text{HO},\ \text{OH},\ \text{CH}_2\text{OH}\right]_n \xrightarrow[H_2SO_4]{(CH_3CO)_2O} \left[\text{O–C(=O)CH}_3,\ \text{O–C(=O)CH}_3,\ \text{CH}_2\text{–O–C(=O)–CH}_3\right]_n \xrightarrow[H^{\oplus}]{H_2O} \left[\text{H–O},\ \text{O–C(=O)CH}_3,\ \text{CH}_2\text{–O–C(=O)–CH}_3\right]_n$$

Cellulose triacetate is also prepared by a heterogeneous process in the presence of benzene, a nonsolvent. The triacetate that forms in both processes is hard to mold, but it can be used in films and fibers. The diacetate is more suited for plasticization and molding.

Many other esters of cellulose were prepared at various times, including some mixed esters. Various cellulose acetate-butyrates are manufactured today and are perhaps the best known of the mixed esters. They are synthesized in the same manner as cellulose acetate. Mixed anhydrides are used in esterification reactions catalyzed by sulfuric acid. The products are then slightly hydrolyzed. The butyryl groups enhance flexibility and moisture resistance. The materials have the reputation of being tough plastics and are used in such applications as tool handles. Lower molecular weight grades are also used in surface finishes.

Several *cellulose ethers* are also prepared commercially. The original patents for preparation of cellulose ethers date from 1912. In spite of that, cellulose ethers never attained the industrial importance of cellulose esters. The ethers are prepared by reacting alkali cellulose with an alkyl halide or with an epoxide:

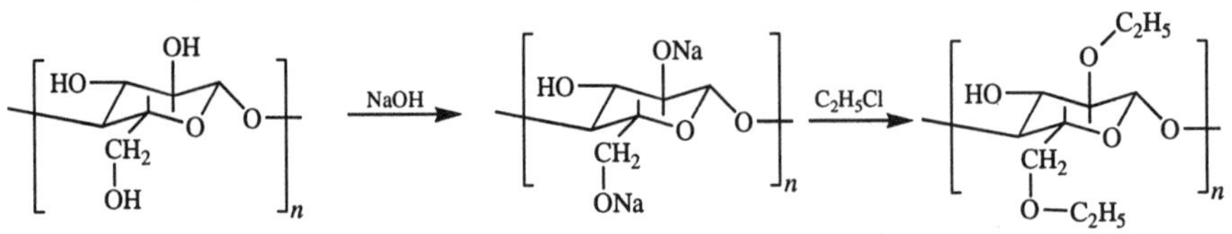

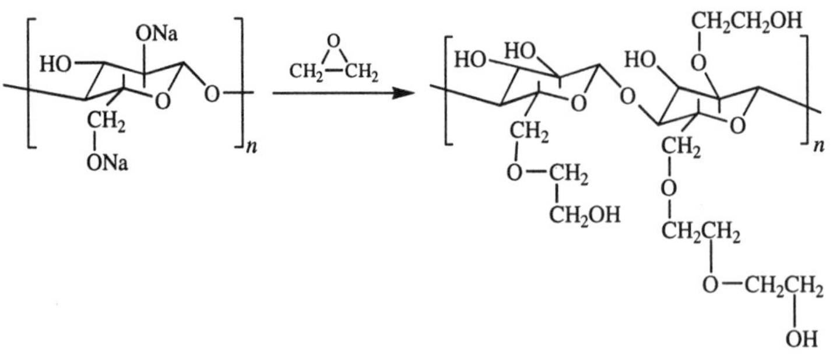

Typical commercial ethers are methyl, ethyl, hydroxyethyl, hydroxypropyl, carboxymethyl, aminoethyl, and benzyl.

Ethyl cellulose is used industrially as a plastic similarly to cellulose acetate. The water-soluble ethers, like methyl, carboxymethyl, and hydroxyethyl, are used as thickeners in foods and in paper manufacturing.

Cellulose can be reacted with acrylonitrile to form a cyanoethylether. The Michael condensation takes place with alkali cellulose:

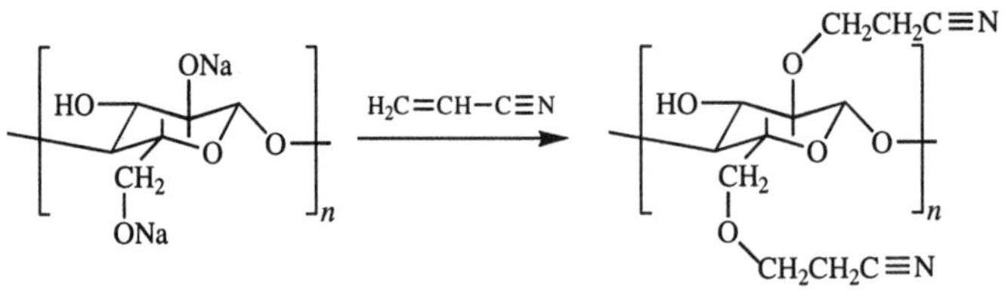

Cyanoethylated cellulose does not appear to be used commercially in any quantity.

Very stable silyl ethers form when cellulose is treated with trimethylchlorosilane or with bis(trimethylsilyl)acetamide[15]:

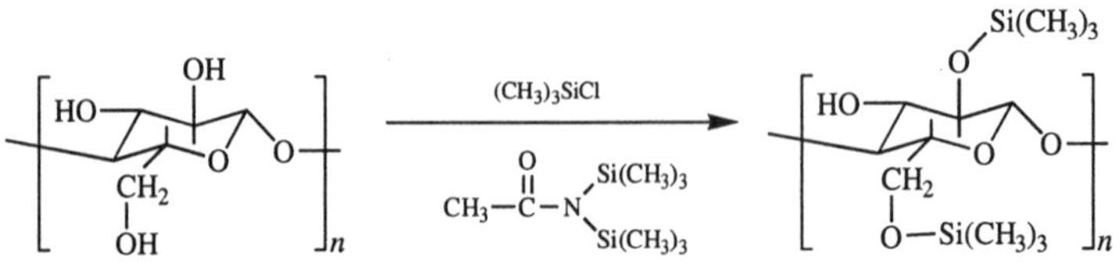

Some interesting approaches to cellulose modification are possible via formations of double bonds in the glucopyranosine unit at the 5,6 positions.[16] This is accomplished by dehydrohalogenating a previously formed 6-iodocellulose:

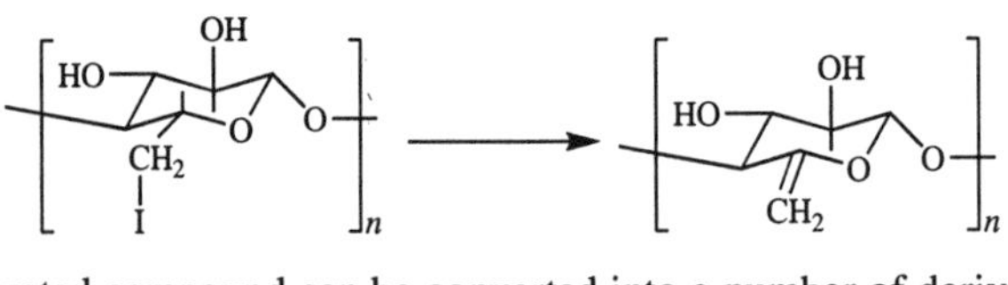

The resultant unsaturated compound can be converted into a number of derivatives. Examples of some of them are:

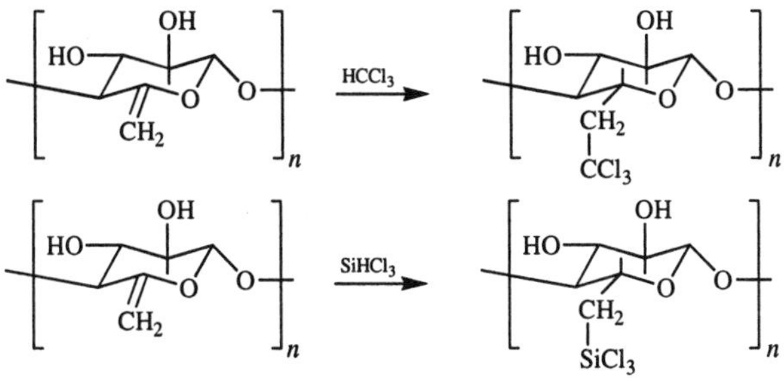

Other compounds that can be added across the double bonds are carbon tetrachloride, phosphorus trichloride, and methyl alcohol. Many graft copolymers of cellulose were reported. This is discussed in Chapter 8.

7.2.4. Miscellaneous Polysaccharides

Other polysaccharides found in nature include *alginic acid* that is isolated from certain brown seaweeds.[17] The monomers of this polymer, similarly to cellulose, are linked *trans* or β to each other through 1,4 positions:

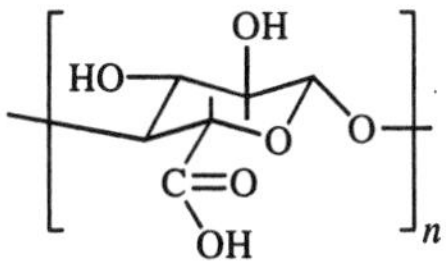

A sulfate group bearing polysaccharide is isolated from another seaweed that is red in color. This polymer is called *carrageenan.* It consists of two fractions.[17] The first has the galactose units linked through 1 and 3 or 1 and 4 carbons. A sulfate group is found at carbon 2 on some units:

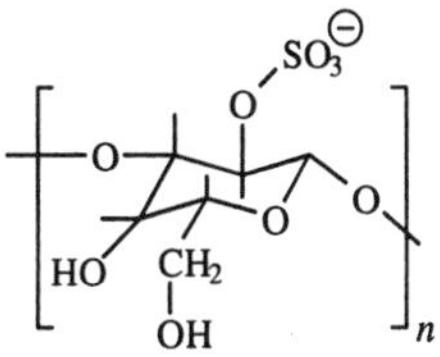

The second fraction has the galactose units linked 1 and 4 and an ether group linking carbons 3 and 6:

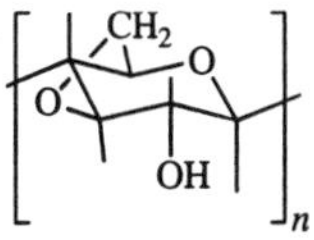

A similar polysaccharide is also obtained from seaweeds and is called *agar*. It is similar in structure, but has less sulfate groups per chain.

Crab and shrimp shell wastes are an abundant source of *chitin,* a nitrogen atom containing polysaccharide:

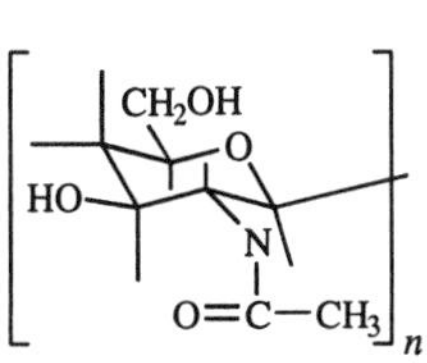

The polymer can be deacetylated to yield an amine group bearing polysaccharide.

7.3. Lignin

These polymers are also constituents of wood (about 25–30%).[18,19] It is uncertain what the molecular weights of the polymers are as the materials are quite complex in structure. The extraction processes of lignins result in considerable loss of molecular weights. The structures of lignins vary, depending upon the source. Generally, they are polymers of coniferyl alcohol:

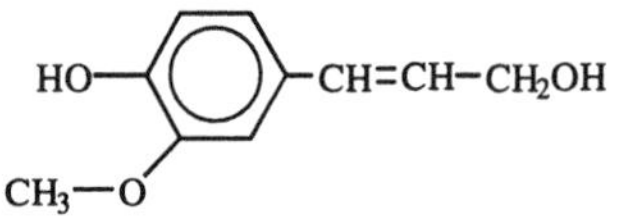

An idealized picture of lignin has been published in the literature. Figure 7.1 is an illustration of what a small segment of this complex polymer might look like, according to Freudenberg.[20]

Many attempts were made to convert lignin to a useful material for coatings and adhesives. Only very limited success, however, has been achieved. A reaction product with formaldehyde can be used as a wood adhesive. In addition, lignin obtained from wood pulping by the sulfate process (as a sulfonate) has been utilized to a limited extent as an asphalt extender and as an oil-well drilling mud additive.

7.4. Polyisoprene

Natural rubber is polyisoprene.[21] It is obtained commercially from the sap of trees called *Hevea brasilensis* and sometimes referred to as *hevea rubber.* These trees yield a latex containing approximately 35% rubber hydrocarbon and 5% nonrubber solids, like proteins, lipids, and inorganic salts. The remaining 60% of the latex is water. The hydrocarbon polymer consists of 97% *cis*-1,4 units, 1% *trans*-1,4 units, and 2% 3,4 units, in a head-to-tail structure. Molecular weights of naturally occurring polyisoprene range from 200,000–500,000. A verity of shrubs and small plants, including some weeds like dandelion and milkweed, also contain polyisoprene in their sap. The guayule shrub, which grows in Mexico and in southern United States, is a good potential source of natural rubber. Work is now proceeding in various places to cultivate this shrub for potential rubber production.

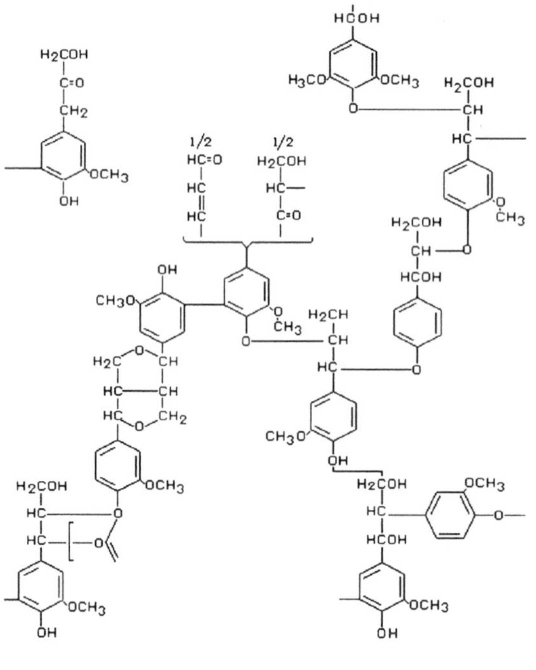

FIGURE 7.1. Structure of lignin (from Freudenberg[20], by permission of the American Association for the Advancement of Science).

An almost all-*trans*-1,4 polymer, called *gutta-percha*, is found in the exudations of various trees of the genus *Palaquium*, *Sapotaceae*, and *Habit*. The molecular weights of these polymers range from 42,000–100,000. *Balata* and *chicle*, also mainly *trans*-1,4 polyisoprenes, are found in saps of some plants in West Indies, Mexico, and South America.

Chapter 8 deals with various reactions of polymers including those of natural and synthetic rubber. That includes vulcanization of rubber. While there are very many commercial applications of the *cis* isomer, gutta-percha utilization is limited to wire coatings, impregnation of textile belting, and as a component of some varnishes. Its use is limited, because it is considerably harder than natural rubber.

7.5. Proteins

These materials are building blocks of animal tissues.[22,26,28,31] To a lesser extent they are also found in vegetable sources. Because the major constituents of animal bodies, including skins, hairs, and blood, are proteins, they are of much greater interest to the biochemists. Nevertheless, some proteins are important commercial materials. These include animal glues, silk, and wool. It is beyond the scope of this book, however, to render a thorough discussion of the proteins. For that reason, only some basic principles are presented here.

Proteins are naturally occurring polyamides, polymers of α-amino acids. The structure can be illustrated as follows:

$$\left[-\underset{\substack{| \\ H}}{N}-\underset{\substack{| \\ H}}{\overset{\substack{R \\ |}}{C}}-\overset{\substack{O \\ \|}}{C}- \right]_n$$

where R represents many different groups, so many different combinations of α-amino acids are possible and the proteins are very complex molecules. The arrangement or sequence of amino acids in proteins is referred to as their *primary structure*. The amide linkage is referred to in biochemistry as the *peptide linkage* or the *peptide group*. A dipeptide then is a compound consisting of two amino acids, a tripeptide of three, etc. *Polypeptide* refers to proteins, though the term is often reserved for lower molecular weight fractions, usually less than 10,000. Many proteins are monodisperse. This distinguishes them from many other, naturally occurring polymers, such as polysaccharides, that are polydisperse.

7.5.1. α-Amino Acids

Twenty-five known, naturally occurring amino acids were isolated from various proteins by hydrolysis. All but one of them, glycine, possess an asymmetric carbon. Table 7.1 lists the naturally occurring amino acids and gives their structures.[26,28,31] Among the above amino acids, a certain number are known as *essential amino acids*. They are not synthesized by human bodies and must be ingested for our metabolism.

All the optically active amino acids (i.e., all except glycine) have an L configuration. In addition, all amino acids exist as zwitterions.

7.5.2. Structures and Chemistry of Proteins

Proteins can be separated into two major groups, *fibrous proteins* and *globular proteins*, depending upon their shapes. The fibrous proteins are long molecules that function as structural materials in animal tissues. Hydrogen bonding holds these water-insoluble molecules together to form extended coiled chains. To this group belong: *collagen*, protein of the connecting tissues, *myosin*, protein of the muscles, *keratin*, protein found in hair, nails, horns, and feathers, and *fibroin*, protein of silk fibers.

Table 7.1 Naturally Occurring Amino Acids[26,28–31]

Name	Structure	Optical Rotation
Glycine	$H_2N–CH_2–COOH$	
Alanine	$CH_3–CH(NH_2)–COOH$	(+)
Valine	$(CH_3)_2CH–CH(NH_2)–COOH$	(+)
Leucine	$(CH_3)_2CH–CH_2–CH(NH_2)–COOH$	(–)
Isoleucine	$CH_3–CH_2–CH(CH_3)–CH(NH_2)–COOH$	(+)
Serine	$HO–CH_2–CH(NH_2)–COOH$	(–)
Threonine	$H_3C–CH(OH)–CH(NH_2)–COOH$	(–)
Cysteine	$HS–CH_2–CH(NH_2)–COOH$	(–)
Methionine	$H_3CS–CH_2–CH_2–CH(NH_2)–COOH$	(–)
Cystine	$HOOC–CH(NH_2)–CH_2–S–S–CH_2–CH(NH_2)–COOH$	(–)
Amino Acids with Aromatic Groups		
Phenylalanine	$C_6H_5–CH_2–CH(NH_2)–COOH$	(–)
Tyrosine	$HO–C_6H_4–CH_2–CH(NH_2)–COOH$	(–)
Diiodotyrosine	$HO–C_6H_2I_2–CH_2–CH(NH_2)–COOH$	(+)
Thyroxine	$HO–C_6H_2I_2–O–C_6H_2I_2–CH_2–CH(NH_2)–COOH$	(+)
Amino Acids with Heterocyclic Structures		
Proline	H, N, COOH	(–)
Hydroxyproline	H, N, COOH, HO	(–)

Table 7.1 (continued)

Name	Structure	Optical Rotation
Acidic Amino Acids		
Aspartic Acid	$HOOC-CH_2-CH(NH_2)-COOH$	(+)
Asparagine	$H_2N-C(=O)-CH_2-CH(NH_2)-COOH$	(–)
Glutamic acid	$HOOC-CH_2-CH_2-CH(NH_2)-COOH$	(+)
Glutamine	$H_2N-C(=O)-CH_2-CH_2-CH(NH_2)-COOH$	
Basic Amino Acids		
Lysine	$H_2N-(CH_2)_4-CH(NH_2)-COOH$	(+)
Hydroxylysine	$H_2N-CH_2-CH(OH)-(CH_2)_2-CH(NH_2)-COOH$	(–)
Arginine	$HN=CH(NH_2)-NH-(CH_2)_3-CH(NH_2)-COOH$	(+)
Tryptophane	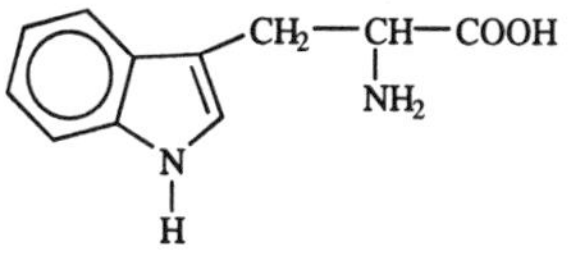 (indole ring, N–H)$-CH_2-CH(NH_2)-COOH$	(–)
Histidine	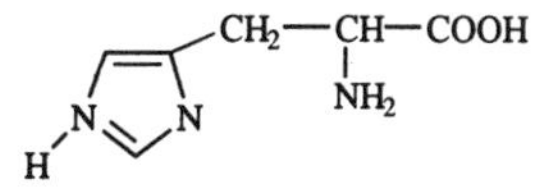 (imidazole ring, N–H)$-CH_2-CH(NH_2)-COOH$	(–)

Globular proteins are held by strong intramolecular hydrogen bonds in spherical or elliptical forms. Their intermolecular forces are weak, and they are soluble in water and in dilute salt solutions. To this group of proteins belong enzymes, many hormones, egg albumin, and hemoglobin.

Some protein also contains a nonpeptide portion that is attached chemically to the polyamide chain. The nonpeptide moieties are called *prosthetic groups*, and the proteins with such groups are called *conjugated proteins*. Examples are hemoglobin and myoglobin that consist of polypeptide portions with iron-porphyrin prosthetic groups attached. This particular prosthetic group, called *heme*, is illustrated in Figure 7.2. There are also a number of proteins that are associated with nucleic acid. They are known as *nucleoproteins*.

Numerous studies of protein structures have shown that the common conformations of the protein chains (fibrous) can be either as an α-helix, β-sheets, or random coils.[26] The steric arrangement or the conformation of the proteins are referred to as the *secondary structure*, while the composition of α-amino acids in the polypeptide chains is called the *primary structure*.[26] Based on X-ray crystallography data, Pauling et al.[23,24], deduced that an α-*helix* type configuration is formed because it accommodates hydrogen bonding of each nitrogen to a carbonyl oxygen (see Fig. 7.3.).

FIGURE 7.2. Prosthetic group heme.

It allows space for all bulky substituents in amino acids and stabilizes the structure. The α-helix is probably the most important secondary structure in proteins.[26]

The α-helix on the left shows a right-handed helix. It is interesting to note that an α-helix conformation may also occur in water solutions. This is due to van der Waal interactions,[25] because water molecules interfere with hydrogen bonding that holds the helix together, as shown in Fig. 7.3.

Not all proteins, however, form helical structures. If the substituent groups on the amino acids are small, as found in silk fibroin, then the polypeptide chains can line up side by side and form sheet-like arrangements. The chains tend to contract to accommodate hydrogen bonding and form pleated sheets. This is called a *β-arrangement*. Such an arrangement can be parallel and antiparallel. The identity period of the parallel one is 6.5 Å and that of the antiparallel, 7.0 Å.

The secondary structures of proteins do not describe completely the arrangement of these macromolecules. For instance, there may be sections that exhibit some irregularity. Alternatively, some sections may be linked chemically by sulfur–sulfur bonds of cystine groups. There may also be areas where the folding of the helix is such that it allows hydrogen bonding between distant sites. The overall, three-dimensional picture of a protein structure is referred to as the *tertiary structure*. Disruption of the tertiary structure in proteins is called *denaturation*. When the protein is composed of more than a single peptide chain, the arrangement is called a *quaternary structure*. This association results from noncovalent interactions.

There is a relationship between the primary structures, or the amino acid content of many proteins, and the secondary structures.[27] The helical contents are inversely proportional to the amount of serine, threonine, valine, cysteine, and proline in the molecule. Conversely,[28] valine,

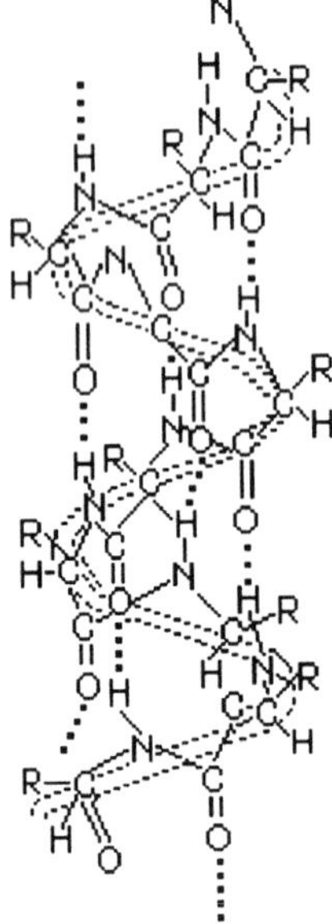

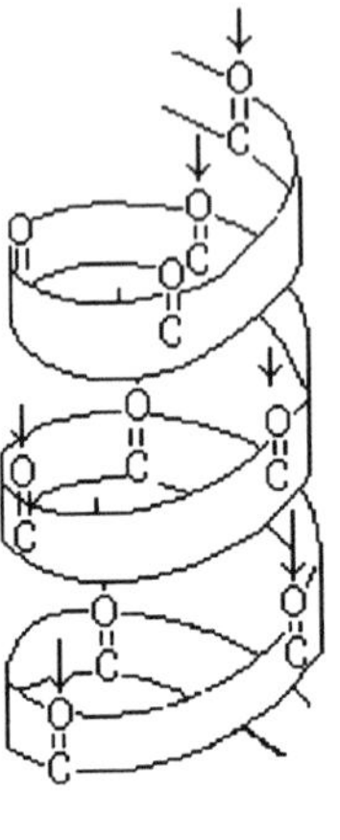

FIGURE 7.3. α-Helix structures of proteins.

isoleucine, serine, cysteine, and threonine are nonhelix-forming amino acids. Proline, due to its specific configuration, actually disrupts the helical structure when it is present in the polypeptide.[29] In addition, proteins that are composed of low ratios of polar to nonpolar amino acids have a tendency to aggregate.[30] Also, in an aqueous environment, the globular protein will tend to form shapes with nonpolar groups located inside the structure. This is due to the thermodynamic nature of the hydrophobic side chains. The polar ones, on the other hand, tend to be located outside, toward the water.[31]

To date, much more information is available on some proteins than on others. Some of the more thoroughly explored proteins will be mentioned below.

Keratins are proteins that are found in wool, hair, fur, skin, nails, horns, scales, feathers, etc. They are insoluble because the peptide chains are linked by disulfide bonds.[32,33] Many keratins contain coils of α-helixes.[34–36] Some keratines, however, were found to consist of complicated β-helical structures. This apparently has not been fully explained. Wool keratin was shown to range in molecular weight from 8,000–80,000.[37] The extensibility of α-keratins is believed to be due to the helical structures. The extent of keratin hardness (in claws, horns, and nails) is believed to be due to the amount of sulfur links.

Silk fibers, which are obtained from the secretion of the silkworm, are double filaments that are enclosed by a coating of a gum (sericin) as they are secreted.[40] The amino acid sequence of the silk protein was shown to be (glycine-serine-glycine-alanine-glycine-alanine)$_n$. The polypeptide chains are bound into antiparallel pleated β-sheet structures by hydrogen bonding.[31,39,42] The structures are also held together by van der Waal forces.[31,38]

The protein of skins, and extracellular connective tissues in animals is *collagen*. The polymer is rigid and crosslinked. Mild hydrolysis disrupts the rigid secondary valence forces and produces gelatin.[26] The fundamental unit of collagen exists as a triple helix.[41] Three left-handed helices twist together to form a right-handed threefold superhelix.[31] Collagen is composed mainly of glycine, proline, and hydroxyproline. Some other amino acids are also present in minor amounts.

A protein that is similar to collagen is *elastin*, which is present in elastic tissues, such as tendons and arteries. Hydrolyses of elastin, which has rubber-like properties, however, do not yield gelatin. Mildly hydrolyzed elastin can be fractionated into two proteins.[26]

Among the most studied globular proteins are *myoglobin* and *hemoglobin*. Myoglobin consists of a single chain of 153 amino acid residues and a prosthetic group that contains iron, called *heme*. Myoglobin polypeptides have eight helical segments that consist of right-handed α-helices which are interrupted by corners and nonhelical regions. The overall shape resembles a pocket into which the heme group just fits. The pocket is hydrophobic, because all but two side groups are nonpolar. The heme groups's two carboxylic acids protrude at the surface and are in contact with water.[43] The hemoglobin is similar to myoglobin but more complex.[44] There are four heme groups enclosed in the hemoglobin structure. Detailed conformational analysis has shown that hemoglobin is built up from 2×2 myoglobin-like subunits, α_2 and β_2.[45,46]

Casein is present in several animal and vegetable sources. Commercially, however, casein is obtained primarily from milk that contains about 3% of this protein. The polymer is isolated either by acid coagulation or with the help of enzymes obtained from animal stomachs. It is very heterogeneous. The molecular weight of a large portion of bovine casein is between 75,000–100,000. It consists of two components, α and β. Casein belongs to a group of proteins that are identified as *phosphoproteins*, because the hydroxy residues of the hydroxyamino acids are esterified with phosphoric acid.

One other group of proteins that has so far not been fully identified is *glycoproteins*. This group of proteins contains a prosthetic group that is either a carbohydrate or a derivative of a carbohydrate. Glycoproteins are found in mucous secretions.

Very special proteins are called *enzymes*. These are biological catalysts. Their primary function is to increase the rate of reactions in organisms and they are found in all living systems. Many enzymes, like *pepsin* or *trypsin*, are relatively simple proteins. Others are conjugated proteins containing prosthetic groups often known as *coenzymes*. Because of their extreme importance to biochemists, enzymes and their actions are being investigated extensively. The full structures of several enzymes have been determined. One such enzyme is lysozyme.

Lysozyme enzymes occur in many species of plants and animals and the chemical behavior may differ. The enzyme found in egg white has a peptide chain consisting of two sections, approximately equal in size. The two sections are separated by a deep cleft. This enzyme performs its function by binding the substrates within this cleft with hydrogen bonds. The substrate is then hydrolyzed with the aid of glutamine (35th amino acid) and aspertine (52nd amino acid). Egg lysozyme's primary structure contains 129 amino-acid residues. The polymer is a single polypeptide chain that is crosslinked at four places by disulfide bonds.[47]

7.5.3. Synthetic Methods for the Preparation of Polypeptides

Studies of protein structures and their functions in nature or mode of actions, as in the case of enzymes, is only part of various investigations.[51–55] Much effort is also put into syntheses of different polypeptide. Such polymers can actually be formed from a mixture of various amino acids by simply heating them together. The products, however, are complex polymeric materials with a random distribution of amino acids and do not resemble any naturally occurring materials.

Base-catalyzed ring-opening polymerization reactions of carboxyanhydrides also result in formations of polypeptides:

$$\text{H-N}\!<\!\text{ring: }R_2\text{-C, C(=O), O, C(=O)} \xrightarrow{\text{base}} \left[\!-\overset{O}{\overset{\|}{C}}-\overset{R}{\underset{R}{C}}-NH-\right] + CO_2$$

(for the mechanism of reaction, see Chapter 4).[48] Over the years, many polypeptides were synthesized by this reaction. These, however, were homopolymers of individual amino acids. Copolymerization leads only to block copolymers. The ability to form random copolymers with controlled sequences of amino acids, which would match naturally occurring proteins, appears to be beyond reach by this method.[49,50]

Duplication of naturally occurring polypeptides is needed, however, to understand the details of structures that lead to biological activities. One of the early works consisted of assembling 23 amino acids to form synthetic pig corticotropin.[43] The molecules were built stepwise.[44]

One technique used in these syntheses is to protect the terminal amino nitrogen by forming protective derivatives that can subsequently be easily cleaved. This is often done by converting them to amide groups:

$$C_6H_5-CH_2-O-\overset{O}{\overset{\|}{C}}-Cl + H_2N-\overset{R}{\underset{H}{C}}-COOH \longrightarrow C_6H_5-CH_2-O-\overset{O}{\overset{\|}{C}}-\underset{H}{N}-\overset{R}{\underset{H}{C}}-COOH$$

p-Toluenesulfonyl chloride is used in the same manner. It is also possible to form imides by reactions with phthalic anhydride:

$$\text{phthalic anhydride} + H_2N-\overset{R}{\underset{H}{C}}-COOH \longrightarrow \text{phthalimido-}N-\overset{R}{\underset{H}{C}}-COOH$$

The condensations can be carried out by a number of different techniques.[26,28] A few of them will be shown below. One is to carry out an aminoacyl halide reaction:

$$C_6H_5-CH_2-O-\overset{O}{\overset{\|}{C}}-\underset{H}{N}-\overset{R}{\underset{H}{C}}-COOH \xrightarrow[\text{or } PCl_5]{SOCl_2} C_6H_5-CH_2-O-\overset{O}{\overset{\|}{C}}-\underset{H}{N}-\overset{R}{\underset{H}{C}}-\underset{Cl}{C}=O$$

$$\xrightarrow{H_2N-CH(R')-COOH} C_6H_5-CH_2-O-C(=O)-NH-CH(R)-C(=O)-NH-CH(R')-COOH$$

The Schotten–Baumann reaction is used in many peptide syntheses. It is usually carried out in the presence of common bases to remove the halogen acid. Another reaction also utilized often is an acid azide condensation:

$$\sim NH-CH(R)-C(=O)-OH + H_2N-NH_2 \longrightarrow \sim NH-CH(R)-C(=O)-NH-NH_2$$

$$\xrightarrow{HNO_2} \sim NH-CH(R)-C(=O)-N_3 \xrightarrow{H_2N-CH(R')-COOR''} \sim NH-CH(R)-C(=O)-NH-CH(R')-C(=O)-OR''$$

A unique way of coupling carboxylic acids with an amine group is by using an N,N'-dicyclohexylcarbodiimide method. This can be illustrated as follows:

$$\sim NH-CH_2-C(=O)-NH-CH(CH_2C_6H_5)-COOH + H_2N-CH_2-C(=O)-O-C_2H_5 \xrightarrow{C_6H_{11}-N=C=N-C_6H_{11}}$$

$$\sim NH-CH_2-C(=O)-NH-CH(CH_2C_6H_5)-C(=O)-NH-CH_2-C(=O)-O-C_2H_5 + C_6H_{11}-NH-C(=O)-NH-C_6H_{11}$$

In addition to the above-mentioned, rather painstaking, techniques of polypeptide syntheses, a very elegant technique was developed by Merrifield.[53] This *solid-phase peptide synthesis* automates the reaction sequences. The method makes use of an insoluble crosslinked polymer substrate with pendant reactive groups for attachment of peptide chains. Chloromethylated polystyrene microgels are often used (see Chapter 8 for more discussions on the use of chloromethylated polystyrene for reactions of polymers). The chloromethyl moieties serve as the initiating sites for formation of the polypeptides:

$$\sim CH_2-CH(C_6H_4-CH_2Cl)\sim + {}^{\ominus}O-C(=O)-CH(R)-NH-C(=O)-O-C(CH_3)_3 \longrightarrow$$

$$\sim CH_2-CH(C_6H_4-CH_2-O-C(=O)-CH(R)-NH-C(=O)-O-C(CH_3)_3)\sim \xrightarrow{H^{\oplus}} \sim CH_2-CH(C_6H_4-CH_2-O-C(=O)-CH(R)-NH_2)\sim$$

A new amino acid with a protective group can now be added. The sequence of additions can be controlled and repeated many times to build up the desired polypeptide chain. Unwanted byproducts of the syntheses are washed away or filtered off before each new step.

This method lends itself so well to automation that automatic peptide synthesizers are now available commercially. One automatic peptide synthesizer was employed, for instance, in the synthesis of ribonuclease, an enzyme. In another instance an enzyme, ferredoxin, that consists of 55 amino acid residues, was synthesized.[26]

The most recent approach to protein syntheses is to use templates for spontaneous self-assembly of multiple copies of a derivatized peptide.[70] The resultant structure, however, is not a conventional linear-chain protein. Instead, oxime bonds are formed between aminooxyacetyl groups on the peptides and aldehyde groups on the template.The method is claimed to have many potential applications.

7.5.4. Chemical Modification of Proteins

Proteins have been utilized commercially from ancient times, either in their naturally occurring form or modified in some manner. Use of silk and wool fibers, for instance, goes back a very long time. Many animal glues, prepared from bones and hides of cattle or sheep, have also been around for a very long time. Today, such glues are being replaced rapidly by synthetic materials. Those that are still manufactured are usually formed by steaming the bones and the hides under pressure and then treating them with hot water in several cycles. This degrades the collagen and makes it soluble. The aqueous solution is concentrated by vacuum evaporation of the water. The material that gels is dried and pulverized. Milder hydrolysis yields gelatin that is used commercially in foods.

Casein, the milk protein (less readily available casein from vegetable sources is hardly ever used), is also used in adhesives. Here, too, synthetics are gradually taking over. At one time it was used to produce a fiber and a plastic that was formed by crosslinking with formaldehyde. The crosslinking reaction was carried out by immersing the proteins in a formaldehyde solution (4–5%) at 55–65 °C for long periods of time, such as days and even months, depending upon the size of the article. The crosslinking reaction involved pendant amino groups and is quite similar to the reactions of urea- and melamine-formaldehyde resins (see Chapter 6). Some condensation and formation of methylene bridges may also involve amide nitrogens. It does not appear likely that casein fibers or plastics are still being produced anywhere.

7.6. Nucleic Acids

These are protein-bound polymers that are essential in many biological processes.[56–68] They perform such functions as directing the syntheses of proteins in living cells and constitute the chemical basis of heredity.[56,57] The polymers are polyphosphate esters of sugars that contain pendant heterocyclic amines, called "bases":

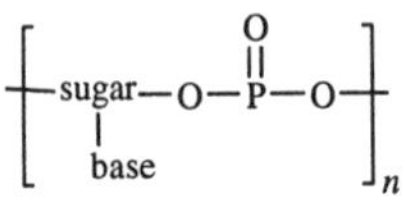

$$\left[\underset{\displaystyle \text{base}}{\text{sugar}} - O - \overset{\displaystyle O}{\overset{\|}{P}} - O \right]_n$$

There are two principle types of nucleic acid with two different sugars. One is D-*2-deoxyribose* found in *deoxyribonucleic acid* (DNA):

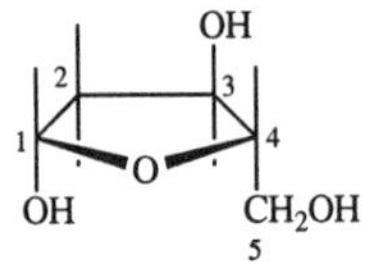

OH; 2; 3; 1; 4; O; OH; CH_2OH; 5

The other one, D-*ribose* is found in *ribonucleic acids* (RNA):

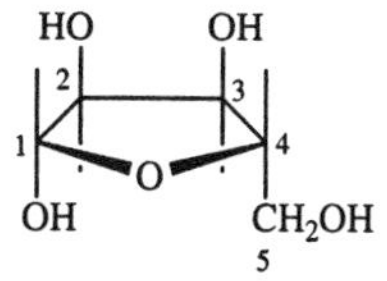

The sugars are in the furanose form. They are linked through the hydroxy groups on carbons 3 and 5 as phosphate esters. The heterocyclic amine "bases" are attached at carbon 1, replacing the hydroxy group.

A sugar molecule with a base attached to it is referred to as a *nucleoside*:

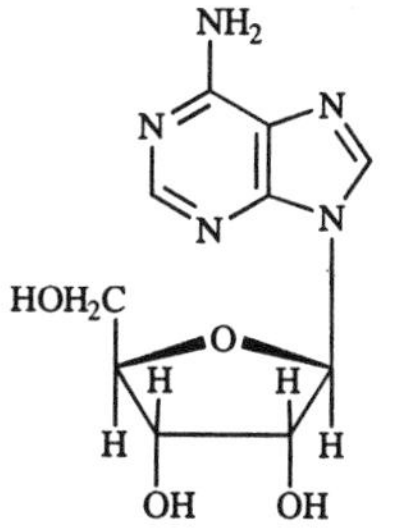

A nucleoside esterified with phosphoric acid is called a *nucleotide*:

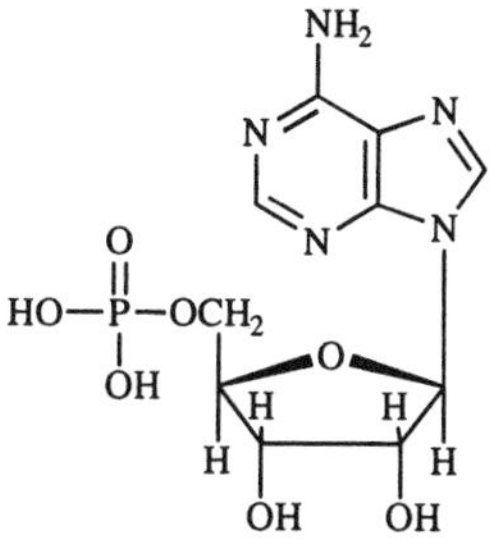

All the heterocyclic amines that occur in nucleic acids (DNA and RNA) are derivatives of either pyrimidine or purine. These are:

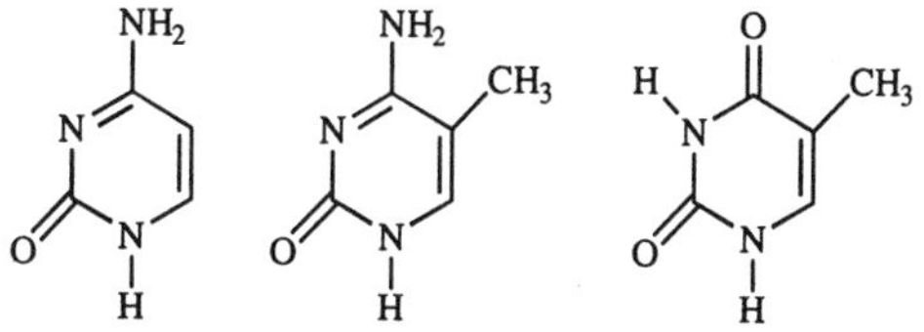

cytosine 5-methylcytosine thymine

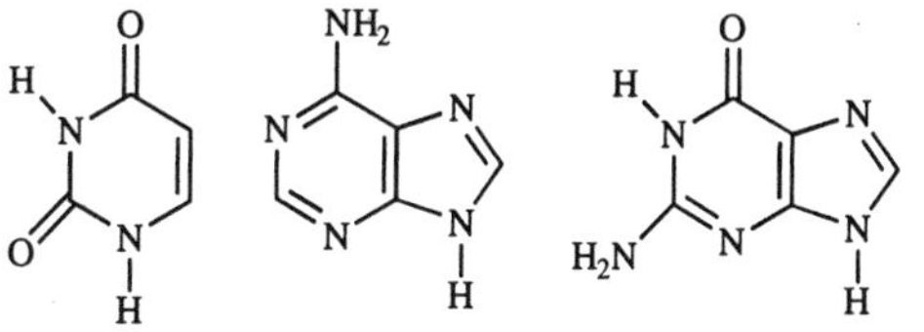

uracil adenine guanine

The naming of nucleosides depends upon the sugars. Thus, adenine attached to ribose is called adenosine. When it is attached to deoxyribose it is called deoxyadenosine.

Hydrolysis of nucleoproteins separates the acids from the proteins. Further hydrolysis yields the components of nucleic acids, namely sugars, bases, and phosphoric acid. The nucleic acids differ from each other, depending upon the source, in chain lengths, sequences, and distributions of bases. As in the proteins, the primary structure of nucleic acids is determined by partial and sequential hydrolysis.

7.6.1. DNA and RNA

Deoxynucleic acids have been isolated from all types of living cells and it was established that their main function is to carry genetic information.[57] These are very high molecular weight polymeric materials. Some were found to be as high as 100 million. Analyses of DNA structures show that the numbers of adenine bases are always the same as the number of thymine bases. Also, the numbers of guanine bases always equal the numbers of cytosines. Based on the information from various analyses and an X-ray investigation of the structure, Watson and Crick concluded that the secondary structure of DNA must be a double helix.[58] Two separate right-handed helical chains wind around each other and are held together by hydrogen bonding between base pairs. The bases that are paired off are adenine with thymine and guanine with cytosine:

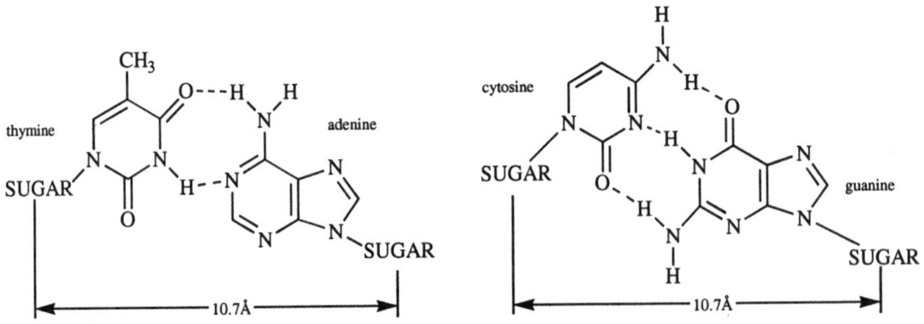

The base pairs are extended perpendicularly toward the center and the deoxyribose-phosphate ester chains are located on the outside of the helix. The two strands are antiparallel to each other. One turn of the helix corresponds to ten nucleotide pairs, 34 Å in length. The width of the helix is 20 Å. Evidence was presented that some DNAs in their native forms are cyclic[59] and may even occur as two interlocking rings. While most known DNA molecules form a right-handed helix, a left-handed helix can be prepared synthetically in the laboratory.[52] It was speculated that in some instances left-handed helixes may exist in nature and have a biological function.[53] These DNA conformations were named Z-DNAs because the backbones zigzag down the molecule.[60–68]

There is less information about the secondary structures of RNAs. It is known that the RNA molecules are lower in molecular weight than are the DNA molecules. In addition, it is known that there are three main types of RNAs in living cells. These are *ribosomal* RNA (r-RNA), *transfer* RNA (t-RNA), and *messenger* RNA (m-RNA). The molecular weight of the three forms on the average are about 1,000,000, 25,000, and 500,000, respectively. RNA molecules, with the help of hydrogen bonding, take three-dimensional cloverleaf structures.[54] The molecules's three-dimensional shape also assumes an L-shape, into which the cloverleaf is bent.

7.6.2. Synthetic Methods for the Preparation of Nucleic Acids

Over the last twenty to thirty years, methods were developed to synthesize short deoxyribonucleotide chains.[57] One synthetic procedure can be illustrated as follows:

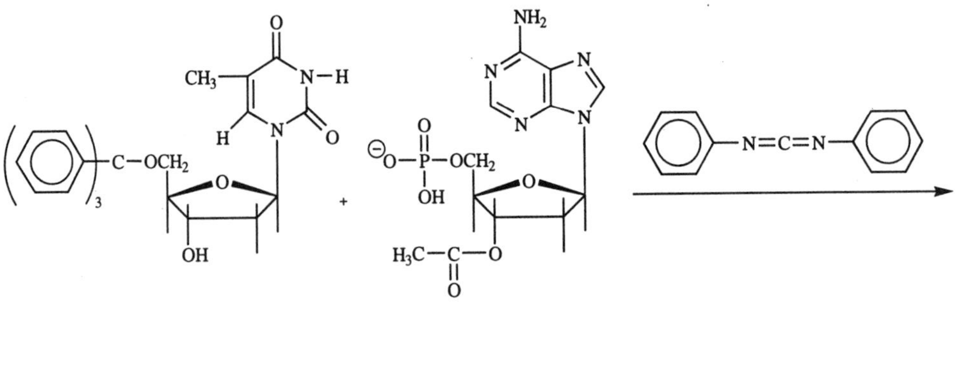

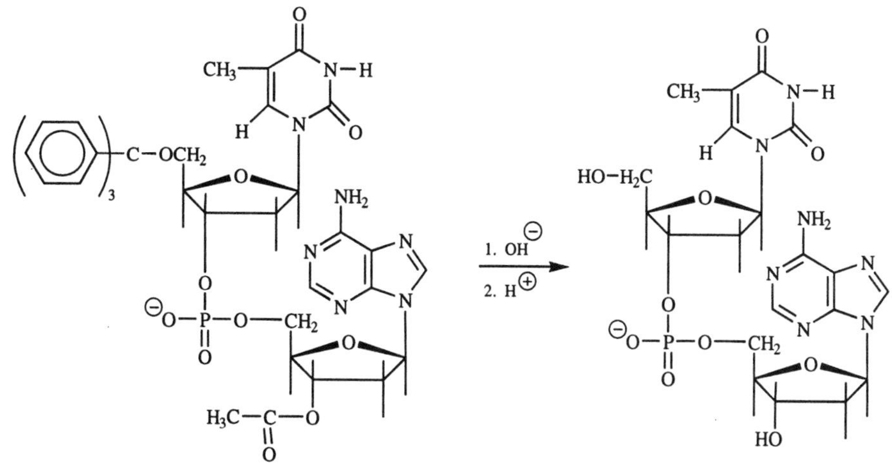

The bulky triphenylmethyl moiety functions to block the 5′ hydroxy groups and is removed when necessary. The same is true of the acetyl portion that also serves to block the 3′ hydroxy position. The product can be used for further expansion of the chain.

Another approach to the syntheses of nucleic acids is to use polymeric supports as in the syntheses of polypeptides. The preparation of protecting groups for attachment to carbon-5 of deoxyribose on the surface of crosslinked polystyrene can be illustrated as follows[57]:

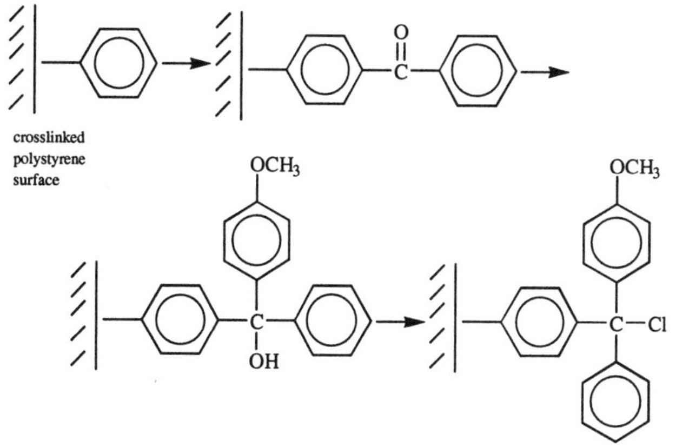

crosslinked polystyrene surface

Another protective group that has been used is dimethoxytrityl. The carbon-3 of deoxyribose has been protected with phosphoramide. Benzoyl groups are used to protect the adenine and guanine bases. Lately, in place of crosslinked polystyrene, controlled pore glass supports have become popular.

Commercial synthesizing machines are available today for polynucleotide syntheses. These are similar to the synthesizers used in polypeptide syntheses.

7.7 Polyalkanoates

Many bacteria are a potential source of naturally occurring polyesters, mainly *poly(β-hydroxyalkanoate)s*, with general structure

$$\mathrm{HO{-}\underset{\underset{R}{|}}{CH}{-}CH_2{-}\underset{\underset{O}{\|}}{C}{-}O}\left[\mathrm{\underset{\underset{R}{|}}{CH}{-}CH_2{-}\underset{\underset{O}{\|}}{C}{-}O}\right]_n\mathrm{\underset{\underset{R}{|}}{CH}{-}CH_2{-}COOH}$$

when the polymer is poly(β-hydroxybutyrate), R = CH_3.[69] The material is found in the form of hard crystalline granules in many bacterial cells. The most common one, poly(β-hydroxybutyrate), was discovered back in the 1920s and identified as a polymer of D-(–)-β-hydroxybutyric acid. In the native state this polymer may reach molecular weights of 1,000,000 or higher. It forms a compact right-handed helix with a twofold screw axis and a repeat unit of 5.96 Å. Because poly(β-hydroxybutyrate) is stereoregular, it is highly crystalline. The substitution in the β-position makes it thermally unstable. This limits its use in plastics. It was found, however, that oxygen starvation of bacterial cultures results in formation of a copolymer of β-hydroxybutyric acid with β-hydroxyvaleric acid instead of a homopolymer. Further investigations showed that as many as 11 different β-hydroxyacid constituents are present in different, naturally occurring polyalkanoates, depending upon the bacterial source and conditions of growth. Today, a family of products, marketed under the trade name Biopol, is available commercially with a range of properties, such as melting point, toughness, flexibility, and so on. The melting points range from 80–180 °C.

Recently, it was reported that the *Pseudomonas oleovorans* microorganism can be forced to produce a thermoplastic elastomer by growing it on a substrate containing sodium octanoate.[71] The product is poly(β-hydroxy octanoate). It contains crystalline regions that act as physical crosslinks.

Review Questions

Section 7.1

1. List the six main categories of naturally occurring polymers.
2. Draw the chemical structures of the repeat units of each category of naturally occurring polymers.

Section 7.2

1. Describe hemicellulose. This should include an explanation of what are xylan, galactan, araban, and plant gums.
2. In discussing starches, explain what are amylose and amylopectin. Explain and draw structures.
3. Discuss cellulose. How does cellulose differ from starch?
4. What is cellobiose? Draw the structure and give the chemical name.
5. What is regenerated cellulose? Explain what is chardonnet silk, cuprammonium rayon, and viscose rayon, and how they are prepared.
6. Discuss the chemistry of cellulose nitrates. How are they prepared and used?
7. Discuss the chemistry of cellulose acetate. How is it prepared and used? Describe mixed cellulose esters.
8. Discuss the chemistry of cellulose ethers.
9. Show the reaction of cellulose with acrylonitrile.
10. Iodocellulose can be dehydrohalogenated to form double bonds in the polymer. This can be used to form new derivatives. Give two examples.
11. Draw the structures of alginic acid and chitin.

Section 7.3

1. Discuss the chemistry of lignin, drawing the structure of coniferyl alcohol. Can you think of a useful product from lignin?

Section 7.4

1. Describe natural rubber. How is it obtained? What is the chemical structure?
2. What are gutta-percha, balata, and chicle? Explain.

Section 7.5

1. Explain what is meant by polypeptides.
2. Explain the difference between fibrous proteins and globular proteins.

3. What are nucleoproteins? What is a prosthetic group? Give an example.
4. Explain what is meant by a secondary structure of a protein and an α-helix.
5. What is meant by a tertiary structure of a protein?
6. What is an enzyme? How does it function?
7. Discuss with the aid of chemical equations the synthetic routes to polypeptides.
8. Discuss chemical modifications of proteins for commercial purposes.

Section 7.6

1. What is the basic structure of a unit in nucleic acids?
2. How do the sugars differ in DNA from RNA?
3. Draw the structures of a nucleoside and a nucleotide.
4. Discuss the syntheses of nucleic acids.

References

1. E. Hicks, *Shellac, Its Origins and Applications*, Chemical Publishing Co., New York, 1961; S. Maiti and M.D.S. Rahman, *J. Macromol. Sci., Rev. Macromol. Chem. Phys.*, **C26**, 441 (1986).
2. E. Fischer, *Ber.*, **24**, 1836,2683 (1881); *ibid.*, **29**, 1377 (1896).
3. W.N. Haworth , E.L. Hirst, and H.A. Hampton, *J. Chem. Soc.*, 1739 (1929).
4. T. Svedberg, *J. Phys. Colloid. Chem.*, **51**, 1 (1947).
5. K.H. Meyer and H. Mark, *Der Aufbau der Hochpolymeren Organischen Naturstoffe*, Akademie Verlagagesell-schaft, Leipzig, 1930.
6. S. Morell and K.P. Link, *J. Biol. Chem.*, **100**, 385 (1933).
7. S. Morell, L. Bauer, and K.P. Link, *J. Biol. Chem.*, **105**, 1 (1934).
8. P.A. Levene and L.C. Kreider, *J. Biol. Chem.*, **129**, 591 (1937).
9. P.A. Levene, G.M. Meyer, and M. Kuna, *Science*, **89**, 370 (1939).
10. R.L. Whistler and J.N. BeMiller (eds.), *Industrial Gums*, 2nd ed., Academic Press, New York, 1973.
11. R.L. Whistler and E.F. Paschall (eds.), *Starch, Chemistry and Technology*, Vols. I and II, Academic Press, New York, 1965.
12. N.M. Bikales and L. Segal (eds.), *Cellulose and Cellulose Derivatives*, Wiley-Interscience, New York, 1971; A. Hebeish and J. T. Guthrie, *The Chemistry and Technology of Cellulose Derivatives*, Springer-Verlag, New York, 1981.
13. E.M. Fetters (ed.), *Chemical Reactions of Polymers*, Wiley-Interscience, New York, 1964.
14. Z.A. Rogovin, *Vysokomol. Soedin.*, **A13** (2), 437 (1971).
15. J.F. Klebe and H.L. Finkbeiner, *J. Polym. Sci.*, **A-1,7**, 1947 (1969).
16. L.G. Nikologovskaya, L.S. Galbraikh, Y.S. Kozlova, and Z.A. Rogovin, *Vysokomol. Soedin.*, **A12**, 2762 (1970).
17. R.L. Wistler and J.N. Miller (eds.), *Industrial Gums*, 2nd ed., Academic Press, New York, 1973.
18. E.A. MacGregor and C.T. Greenwood, *Polymers in Nature*, Wiley, New York, 1980.
19. I.A. Pearl, *The Chemistry of Lignin*, Dekker, New York, 1967.
20. K. Freudenberg, *Science*, **148**, 595 (1965).
21. F.R. Eirich (ed.), *Science and Technology of Rubber*, Academic Press, New York, 1978; A.D. Roberts (ed.), *Natural Rubber Science and Technology*, Oxford University Press, New York, 1988.
22. M. Joly, *A Physico-Chemical Approach to the Denaturation of Proteins*, Academic Press, New York, 1965.
23. L. Pauling and R.B. Corey, *Proc. Natl. Acad. Sci.*, **37**, 729 (1951); *ibid.*, **39**, 253 (1953).
24. L. Pauling, R.B. Corey, and H.R. Branson, *Proc. Natl. Acad. Sci.*, **37**, 205 (1951).
25. A.M. Liquori, *Acta Cryst.*, **8**, 375 (1955).
26. S. Seifter and P.M. Gallop, in *The Proteins*, Vol. 4, 2nd ed. (H. Neurath, ed.), Academic Press, New York, 1966; A.G. Walton, *Polypeptides and Protein Structure*, Elsevier, New York, 1981.
27. D.R. Davis, *J. Mol. Biol.*, **9**, 605 (1964).
28. E.R. Blout in *Polyamino Acids, Polypeptides, and Proteins* (M.A. Stahmann, ed.), University of Wisconsin Press, Madison, 1962
29. M.F. Perutz, J.C. Kendrew, and H.C. Watson, *J. Mol. Biol.*, **13**, 669 (1965).
30. R.T. Hatch, *Nature*, **206**, 777 (1965).
31. R.E. Dickerson and I. Geis, *The Structure and Action of Proteins*, Harper and Row, New York, 1969.
32. C.H. Bamford and A. Elliott, in *Fiber Structure* (J.W.S. Hearle and R.N. Peters, eds.), Butterworth, London, 1963.

33. B. Jergenson, *Optical Rotary Dispersion of Proteins and Other Macromolecules*, Springer-Verlag, New York, 1969.
34. F.H. Crick, *Acta Cryst.*, **6**, 689 (1953).
35. R.D.B. Fraser and T.P MacRae, *Nature*, **195**, 1167 (1962).
36. C. Cohen and K.C. Holmes, *J. Mol. Biol.*, **6**, 423 (1963).
37. W.B. Ward and H. P. Lundgren, *Adv. Protein Chem.*, **9**, 243 (1954).
38. R.H. Peters, *Textile Chemistry*, Vol. I, Elsevier, New York, 1963.
39. F. Lucas, J.T.B. Shaw, and S.G. Smith, *Biochem. J.*, **66**, 468 (1957).
40. R.B.D. Fraser, T.P. MacRae, and G.E. Rogers, *Nature*, **193**, 1052 (1962).
41. A. Rich and F.H. Crick, *J. Mol. Biol.*, **3**, 483 (1961).
42. R.B.D. Fraser, T.P. MacRae, D.A.D. Parry, and E. Suzuki, *Polymer*, **10** (10), 810 (1969).
43. J.C. Kendrew, *Science*, **139**, 1259 (1963).
44. M.F. Peruz, *Science*, **140**, 863 (1964).
45. G. Buse, *Angew. Chem., Int. Ed. Engl.*, **10** (10), 663 (1971).
46. M.F. Peruz, G. Will, and A.T.C. North, *Nature*, **185**, 416 (1960).
47. P. Jolles, *Angew. Chem., Int. Ed. Engl.*, **8** (4), 227 (1969).
48. E. Katalski and M. Sela, *Adv. Protein Chem.*, **13**, 243 (1958).
49. M. Idelson and E.R. Blout, *J. Am. Chem. Soc.*, **80**, 2387 (1958).
50. Y. Shalitin and E. Katchalski, *J. Am. Chem. Soc.*, **82**, 1630 (1960).
51. K. Hofmann, *Chem. Eng. News*, p.145 (Aug. 7, 1967).
52. R. Schwyzer and P. Sieber, *Helv. Chim. Acta*, **49**, 134 (1966).
53. R.B. Merrifield, *J. Am. Chem. Soc.*, **85**, 2149 (1963).
54. *Chem. Eng. News*, p. 28 (April 22, 1968).
55. B. Gutte and R.B. Merrifield, *J. Am. Chem. Soc.*, **91**, 501 (1961).
56. J.N. Davidson, *The Biochemistry of Nucleic Acids*, 7th ed., Academic Press, New York, 1972.
57. N.K. Kochetkov and E.I. Budovskii (eds.), *Organic Chemistry of Nucleic Acids*, Plenum Press, London, Part A—1971, Part B—1972; L.B. Townsend and R.S. Tipson (eds.), *Nucleic Acid Chemistry*, Wiley-Interscience, New York, 1986.
58. J.D. Watson and F.H. Crick, *Nature*, **171**, 734, 964 (1953).
59. G. Felsenfeld and H.T. Miles, *Ann. Rev. Biochem.*, **36**, 407 (1967).
60. J.L. Fox, *Chem. Eng. News*, p. 14 (Dec. 14, 1979).
61. G. Kolata, *Science*, **214**, 1108 (1981).
62. L.B. Townsend and R.S. Tipson, ed., *Nucleic Acid Chemistry*, Wiley-Interscience, New York, (1986).
63. F.L. Suddath, G.J. Quigley, A. McPherson, J.L. Sussman, A.H.J. Wang, N.C. Seeman, and A. Rich, *Science*, **185**, 435 (1974).
64. S.E. Chang and D. Ish-Horowicz, *J. Mol. Biol.*, **84**, 375 (1974).
65. K.L. Agarwal, A. Yamazaki, P.J. Cashion, and H.G. Khorana, *Angew. Chem., Int. Ed. Engl.*, **11** (6), 451 (1972).
66. J.M. Frechet and Schurch, *J. Am. Chem. Soc.*, **93**, 492 (1971).
67. M. Lemoigne, *C.R. Acad. Sci. Paris*, **180**, 1539 (1925); *Bull. Soc. Chim. Biol.*, **8**, 1291 (1926).
68. J. Cornibert, R.H. Marchessault, H. Benoit, and G. Weil, *Macromolecules*, **3**, 741 (1970).
69. R.H. Marchessault, *Polym. Prepr., Am. Chem. Soc.*, **29** (1), 584 (1988).
70. K. Rose, *J. Am. Chem. Soc.*, **116**, 30 (1994).
71. K.G. Gagnon, R.W. Lenz, R.J. Farris, and R.C. Fuller, *Rubber Chem. Tech.*, **65**, 761 (1992).

8

Reactions of Polymers

8.1. Reactivity of Macromolecules

In consideration of various chemical reactions of macromolecules, the reactivity of their functional groups must be compared to those of small molecules. The comparisons have stimulated many investigations and led to conclusions that functional groups exhibit equal reactivity in both large and small molecules, if the conditions are identical. These conclusions are supported by theoretical evidence.[1,2] Specifically, they apply to the following situations[1]:

1. Reactions that take place in homogeneous fluid media with all reactants, intermediates, and end products fully soluble. These conditions exits from the start to the end of the reactions.
2. All elementary steps involve only individual functional groups. The other reacting species are small and mobile.
3. The steric factors in the low molecular weight compounds selected for comparison must be similar to those of the large molecules.

The above can be illustrated by a few examples. For instance, the rates of photochemical *cis–trans* isomerization of azobenzene residues on the backbones of flexible polymeric chains are analogous to those of small molecules.[3] Another example is the activation energy for *cis–trans* isomerization of azoaromatic polyuria. It is the same for low molecular weight analogs.[4] A third example is an experiment in comparing conformational transitions of some eximers in large and small molecules. A sandwich complex forms between an excited aromatic chromophore, *Ar*, and a similar chromophore in the ground state when irradiated with light of an appropriate wavelength. The conformation required by such an excimer can correspond to a prohibitive energy requirement for the unexcited molecule. All conformational transitions must take place during the lifetime of the excited state of the chromophore that is of the type[5]:

$$\mathrm{Ar^{*}{-}CH_2{-}X{-}CH_2{-}Ar} \longrightarrow \mathrm{Ar\cdots{*}\cdots Ar\ (CH_2{-}X{-}CH_2\ ring)}$$

The ratio of the fluorescence intensity of an excimer and a normal molecule is a measure of the probability that the conformational transition takes place during the excited lifetime. A polyamide with only a small proportion of the following units was used for comparison:

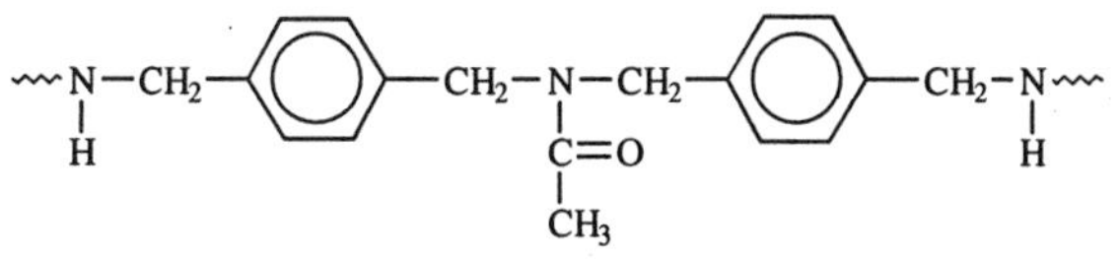

$$\sim\mathrm{NH{-}CH_2{-}C_6H_4{-}CH_2{-}N(C({=}O)CH_3){-}CH_2{-}C_6H_4{-}CH_2{-}NH}\sim$$

Emission spectra of dilute solutions of the above polyamide and its low molecular weight analog were measured over a range of temperatures. They showed that the activation energies of the conformational transitions required for excimer formations are essentially the same for both materials.[5]

In addition, all bimolecular activation controlled reactions are independent of the degree of polymerization.[6] Simple S_N2 reactions between reactive groups attached to chain ends of monodispersed macromolecules in a wide range of molecular weights are independent of the DP[7,8] in the range 20–2000.[7] This was shown on three different reactions. In the first one, the reactivities of chlorine-terminated low and high molecular weight polystyrenes towards polystyryllithium are equal in benzene and cyclohexane solvents:

$$\sim CH_2-\underset{C_6H_5}{CH^{\ominus}}\,Li^{\oplus} + \sim CH_2-\underset{C_6H_5}{CCl} \longrightarrow \sim CH_2-\underset{C_6H_5}{CH}-\underset{C_6H_5}{CH}-CH_2\sim$$

In the second one, the reactivity of primary amine-terminated polyoxyethylenes with sulfonyl chloride-ended polyoxyethylenes in chloroform are also the same:

$$\sim OCH_2-CH_2-NH_2 + \sim OCH_2-SO_2-C_{10}H_6-SO_2Cl \longrightarrow \text{Adduct}$$

In the third one, chain-length dependence of the propagation rates was measured in polymerizations of methyl methacrylate. In the range of DPs from 130–14,200 the propagation constant was shown to be independent of the chain length.[8]

On the other hand, unequal reactivity was observed:

1. In bimolecular reactions that are diffusion controlled.
2. When neighboring group participations become significant.
3. When the properties of the polymers in solution are altered by gelation.
4. When the tacticities of the polymers affect neighboring group interactions.
5. When heterogeneous conditions affect accessibility to the reactive sites.

There are special situations that can occur. For instance, electrostatic charges carried by the polymers may extend over long distances in solutions and may manifest themselves in reactions with charged reagents. Sometimes, chain flexibility or folding can cause functional groups to come together and interact, though they may be located well apart on the polymer backbone. Polymer solutions of this type are comparable to dispersions of individual droplets of concentrated solutions.

Some statements above may require additional clarification. The following elaboration therefore follows:

8.1.1. Diffusion-Controlled Reactions

Reactions that are bimolecular can be affected by the viscosity of the medium.[9] The translational motions of flexible polymeric chains are accompanied by concomitant segmental rearrangements. Whether this applies to a particular reaction, however, is hard to tell. For instance, two dynamic processes affect reactions, like termination rates, in chain-growth polymerizations. If the termination processes are controlled by translational motion, the rates of the reactions might be expected to vary with the translational diffusion coefficients of the polymers. Termination reactions, however, are not controlled by diffusions of entire molecules, but only by segmental diffusions within the coiled chains.[10] The reactive ends assume positions where they are exposed to mutual interaction and not affected by the viscosity of the medium.

8.1.2. Paired Group and Neighboring Group Effects

When *random, irreversible*, and *intramolecular* reactions occur on polymeric backbones with the functional groups adjacent to each other, they can be expected to react. There is, however, an upper limit to conversions. This upper limit is due to a statistical probability that some functional groups are bound to become isolated. The limit for conversion was calculated to be 86.5%.[11]

Theoretically, quantitative conversions should be possible with *reversible* reactions of paired functional groups on macromolecular backbones. The ability, however, of isolated reactive groups to find each other and then pair off depends either upon particularly high driving forces, or upon the time required to accomplish complete conversions.[12] For reactions initiated randomly, at different sites, the probability is high that two groups on the terminal units will eventually meet and react. Since the reactions are reversible, at least in theory, very high conversions are possible.

Neighboring group participation can usually be deduced from three types of evidence:

1. If the reactions occur more rapidly during the rate-determining step than can be expected from other considerations.
2. If the stereochemistry of the reactions suggests neighboring group involvement.
3. If molecular rearrangements occur and the groups remain bonded to reaction centers, but break away from the atoms to which they were originally attached on the substrates.

There are many examples in the literature that describe neighboring group effects in reactions of polymers. One example is hydrolysis of poly(*p*-nitrophenyl methacrylate-co-acrylic acid). The high reaction rate at a neutral pH is due to attacks by the carboxylic moieties upon the neighboring carbonyl carbons.[13–15] Decomposition rates of *t*-butyl acrylate-styrene copolymers[16] can serve as another example. Experimental data show pronounced acceleration for all samples. This is interpreted in terms of both intra- and intermolecular interactions of the esters and the carboxylic groups. It follows a suggestion of Litmanovich and Cherkezyan[16,17] that the instantaneous reactivity of any group depends on its microenvironment. That includes (for reactions of polymer in molten condition) two nearest units on the same chain (internal neighbors) and two units belonging to two different chains (external neighbors).

Another example of the neighboring group effect is the behavior of polyacrylamides in hydrolyses. There are two distinct and successive rates.[18] After conversions of up to 40–50% are reached, the reactions slow down. This is due to accumulations of negative electrostatic charges on the polymeric backbones.[18] In alkaline media, the increasing negative charges along the chains exert electrostatic repulsions toward the hydroxy ions. This results in rate decreases.

8.1.3. Effect of Molecular Size

An example is the effect of DP on the rates of alkaline hydrolyses of poly(vinyl acetate)s. Rapid increases in the rates can be seen[19] in large, but not in small, molecules, as the reactions progress. Solvents that are good for the products, like acetone–water mixtures, are used in these reactions. These are, however, poor solvents for the staring materials with high DP. Low molecular weight molecules are more soluble. This means that at the start of the reaction the large molecules are coiled up and the reactive sites not readily available. As the reactions progress, the chains unravel and the sites became more accessible with accompanying increases in the reaction rates. Because the small molecules are more soluble, the reactive sites are accessible from the start of the reactions, and the rates are constant.

There are many reports in the literature on the *effects of chain conformation*.[19–25] One example is radical bromination of poly(methyl styrene)[20] with *N*-bromosuccinimide-benzoyl peroxide or

Br_2–K_2CO_3 light. ^{13}C NMR spectroscopy shows differences in reactivities of the methyl groups in the 3 and 4 positions on the benzene rings of isotactic and atactic polystyrenes.[20]

Differences in reactivities in poly(vinyl alcohol)s between isotactic (*meso*) and syndiotactic (*dl*-diol) portions of the polymers and between *cis* and *trans* acetals[26–28] form another example. In extending this to model compounds, reactions of stereoisomers of pentane-2,4-diol and heptane-1,4,6-triol with formaldehyde take place much faster for the *meso* than for the *dl*-diol portions.[26–28] Even more important are the steric effects imposed by restricted rotations. For instance, quaternizations of chloromethylated polyether sulfones exhibit decreasing rates at high degrees of substitution. This can be attributed to restricted rotations of the polymeric chains, because this phenomenon is not observed with a more flexible chloromethylated polystyrene under identical conditions.[23,24]

8.1.4. Effects of Changes in Solubility

Changes in solubility can occur during the courses of various reactions. Such changes are observed, for instance, during the chlorination of polyethylene in aromatic and chlorinated solvents.[29] There is an increase in the solubility until 30% conversion is reached. After that, solubility decreases and reaches a minimum at 50–60% chlorine content. Following that it increases again. This, however, is not typical of many reactions of polymers in solutions. More common is that the starting material is soluble, but not the product, or the opposite is true. Higher conversions are usually expected when the polymers are solvated and the chains are fully extended. In such situations, the reagents have ready access to the reactive sites.[29] If the products are insoluble in the reaction medium and tend to precipitate as the reaction progresses, the sites become increasingly less accessible. This results in low conversions and premature terminations. If the opposite is true and the product is more soluble than the starting material, homogeneous limited reactions can be controlled. When the starting material is incompatible with the product, mutual precipitation or coiling of the chains can take place. This can result in limited reactions. Also, only minor differences in the constitutions of two polymers can result in incompatibility. For instance, among methacrylate polymers there are incompatibilities in benzene solutions that result from differences only in the amount of branching of the alkyl groups.[29]

Problems with solvent incompatibility can sometimes be overcome by using mixtures of solvents. Those that are good for the starting materials can be combined with those good for the products. With careful experimentation it may be possible to develop a mixture of solvents that will keep all components in solution.[30] In some instances, however, insolubility of the products might be an advantage. This is the case with alcoholyses of poly(vinyl acetate), where the polymer precipitates during the reactions and in doing so absorbs the catalyst with it. The phenomenon permits complete alcoholyses, particularly with the higher molecular weight species that precipitate first.

Secondary reactions, like crosslinking and gelation, result in precipitations from solution. The extent of the reactions, however, is not necessarily limited, because diffusions of low molecular weight species are still possible. Isolation of useful products, however, often becomes very difficult.

8.1.5. Effects of Crystallinity

Crystallinity can only affect reactivity when the reactions are carried out on polymers in the solid state and at heterogeneous conditions. The differences in accessibility to the reactive sites vary with the amount of crystallinity. Cellulose, for instance, is often reacted in the solid state and the degree of crystallinity is expressed in terms of reactivity to various reagents.[31] The progress of a reaction can sometimes be monitored by a loss of crystallinity. What is more significant, however, is that greater accessibility to amorphous regions results in reaction products with special properties. An example is heterogeneous and homogeneous chlorination of polyethylene. Two different products are obtained.[32] The material from heterogeneous chlorination is much less randomly substituted and remains crystalline up to a chlorine content of 55%. The products from the homogeneous reactions, on the other hand, are amorphous after 35% substitution.

8.1.6. Reactions That Favor Large Molecules

Hydrophobic interactions play important roles in many polymeric reactions. They are, for instance, significant in the hydrolyses of low molecular weight esters when catalyzed by polymeric sulfonic acid reagents, like poly(styrene sulfonic acid). In theses reactions, the hydrogen ions are located close to the macromolecules.[19] The hydrolytic cations are located in the regions of the macromolecules and not in the bulk of the solution. The rates of the reactions are high. Low molecular weight catalysts, on the other hand, like HCl, have all the hydrogen ions distributed evenly throughout the reaction medium. As a result, the rates are lower. Adsorption of the ester groups to the polymeric sites is accompanied by an increase in the apparent rate constant, as compared to reactions with HCl. Examples are hydrolyses of methyl and butyl acetates.[19] Another example is formation of eximers and exiplexes in polyesters and methacrylate polymers that always favor large molecules over small ones.[33] Proton transfer reaction of poly(vinylquinoline)[34] can serve as a third example. The emission, excitation, and absorption spectra of this polymer in a mixture of dioxane and water can be compared to that of 2-methylquinoline. The emissions coming from the protonated heterocyclic rings in the polymer occur sooner than from the low molecular weight compound.[34]

8.2. Addition Reactions of Polymers

Polymers with double bonds in the backbones or in the pendant groups can undergo numerous addition reactions. Some are discussed in this section.

8.2.1. Halogenation

Hydrochlorination of natural rubber is often accompanied by cyclization[35,36]:

$$\sim CH_2C(CH_3){=}CH{-}CH_2{-}CH_2{-}C(CH_3){=}CH{-}CH_2\sim \; + \; H^{\oplus} \longrightarrow \sim CH_2{-}\overset{\oplus}{C}(CH_3){-}CH_2{-}CH_2{-}CH_2{-}C(CH_3){=}CH{-}CH_2\sim$$

$$\longrightarrow \text{(cyclohexyl cation with } CH_3, \sim CH_2, \sim CH_2 \text{ and } \oplus CH_3\text{)} \xrightarrow{Cl^{\ominus}} \text{(cyclohexane ring with } CH_3, \sim CH_2, \sim CH_2, CH_3 \text{ and } Cl\text{)}$$

Trans-1,4- and 1,2-polybutadiene can be hydrohalogenated under mild conditions with gaseous HCl.[51] The same is true of copolymers of butadiene with piperylene and also of isotactic *trans*-1,4-piperylene. The addition of HCl to the asymmetric double bond is *trans* for polypiperylene and occurs in a stereoselective way, judging from the ^{13}C NMR[51] spectra.

Polysilanes with alkene substituents add HCl and HBr in the presence of Lewis acids.[58] The products are the corresponding chlorine and bromine containing polymers with little degradation of the polysilane backbone:

$$-[-Si(CH_3)((CH_2)_2C_6H_9)-]--[-Si(R)(CH_3)-]- \xrightarrow[\text{Lewis Acid}]{HX} -[-Si(CH_3)((CH_2)_2C_6H_{10}X)-]--[-Si(R)(CH_3)-]-$$

Chlorinations of rubber, however, are fairly complex, because several reactions occur simultaneously. These appear to be: (1) additions to the double bond; (2) substitutions; (3) cyclizations; and (4) crosslinkings. As a result, the additions of halogens to the double bonds are only a minor portion of the overall reaction scheme.[37,38] In CCl_4 the following steps are known to occur:

$$\sim CH_2C(CH_3)=CH-CH_2-CH_2C(CH_3)=CH-CH_2\sim + Cl_2 \xrightarrow{\text{substitution}} \sim CHClC(CH_3)=CH-CH_2-CHClC(CH_3)=CH-CH_2\sim + HCl$$

$$\xrightarrow{\text{cycliz.}} \text{(cyclohexene ring: } CH_3, Cl, Cl, CH_3, -CH_2-CH_2\sim) + Cl_2 \xrightarrow{\text{substitution}} \text{(cyclohexene ring: } CH_3, Cl, Cl, Cl, CH_3, -CH_2-CH_2\sim) + HCl\uparrow$$

$$\xrightarrow[Cl_2]{\text{addition}} \text{(cyclohexane ring: } CH_3, Cl, Cl, Cl, Cl, Cl, CH_3, -CH_2-CH_2\sim)$$

Halogenation reactions of unsaturated polymers follow two simultaneous paths, ionic and free radical. Ionic mechanisms give soluble products from chlorination reactions of polybutadiene.[42] The free-radical mechanisms, on the other hand, cause crosslinking, isomerization, and addition products. If the free-radical reactions are suppressed, soluble materials form. Natural rubber can be chlorinated in benzene with addition of as much as 30% by weight of chlorine without cyclization.[39,40] Also, chlorination of polyalkenamers, both *cis* and *trans*, yields soluble polymers. X-rays show that the products are partly crystalline.[43,44] The crystalline segments obtained from 1,4-*trans*-polyisoprene are diisotactic poly(*erythro*-dichlorobutamer)s while those obtained from the 1,4-*cis* isomer are diisotactic poly(*threo*-1,2-dichlorobutamer)s.[45]

Additive-type chlorination of natural rubber can also be carried out with phenyl iododichloride or with sulfuryl chloride.[39,40] Traces of peroxides must be present to initiate the reactions. This suggests a free-radical mechanism. Some cyclization accompanies this reaction as well.[40] In CCl_4, for the first 25 chlorine atoms that add per each 100 isoprene units, 23 double bonds disappear and only a small quantity of HCl forms. The subsequent 105 chlorinations, however, cause a loss of only 53 double bonds.

Rubber can be *brominated* at 30 °C. If traces of alcohol are present, the reaction appears to go on entirely by addition.[39,40] Without alcohol, substitutions take place rapidly and simultaneously with the additions to the double bonds.[41] Exomethylene groups and intramolecular cyclic structures form in the process. Slow additions of bromine to vinylidene double bonds result in formations of tribromides, $-C_5H_7Br_3$. Also, *cis* and *trans* isomers of polyisoprene[41] brominate differently. Substitution reactions take place in brominations with *N*-bromosuccinimide. They are accompanied by cyclizations.[39]

Brominations of polybutadienes with *N*-bromosuccinimide yield α-brominated polybutadienes[46,47] that may also contain butanediylidene units. The products act as typical alkyl halides and can undergo Grignard–Wurz reactions:

$$\sim\!=CH-CH_2-CH_2-CH\!=\!\sim + \text{(N-bromosuccinimide)} \longrightarrow \sim\!=CH-CH_2-CH(Br)-CH\!=\!\sim +$$

$$\xrightarrow{C_6H_5MgBr} \sim\!=CH-CH_2-CH(C_6H_5)-CH\!=\!\sim$$

The bromination reaction is accompanied by shifts of the double bonds that are coupled with the sites of substitution. Several different substituents can form. The polymers may contain pentanediylidene, hexanediylidine, and heptanediylidene units.[46,47]

By contrast, chlorination of polybutadiene in benzene is a straightforward addition reaction of the halogens to the double bonds[48,49]:

$$\sim\!CH_2{-}CH{=}CH{-}CH_2\!\sim + Cl_2 \xrightarrow{\text{benzene}} \sim\!CH_2{-}\underset{Cl}{CH}{-}\overset{\oplus}{CH}{-}CH_2\!\sim$$

$$\xrightarrow{Cl^{\ominus}} \sim\!CH_2{-}\underset{Cl}{CH}{-}\underset{Cl}{CH}{-}CH_2\!\sim$$

Very little HCl is liberated until all the double bonds are consumed. When CCl_4 is used in place of benzene, some substitutions occur during the latter stages of the reactions. If crosslinking occurs at the same time, the substitutions may not be extensive. The crosslinking reactions are believed to involve carbocationic intermediates.

Polybutadiene can be halogenated readily in tetrahydrofuran with iodine chloride or bromine.[49] The products are glassy polymers. These products dehalogenate in reactions with organolithium compounds, like *n*-butyllithium, *sec*-butyllithium, and polystyryllithium in tetrahydrofuran solution. Dehalogenation of poly(iodochlorinated butadiene) with *n*-butyllithium yields product with different *cis/trans* ratios. Also, this is accompanied by partial crosslinking. The reactions may involve[49] halogen–metal exchanges that are followed by intra- and intermolecular elimination of lithium halide. In brominations of polybutadienes both couplings and eliminations take place. Both iodo-chlorinated and brominated polybutadienes form graft copolymers when reacted with polystyryl-lithium in tetrahydrofuran.[50] Gel formation, however, accompanies the grafting reaction.

8.2.2. Hydrogenation

1,4-polybutadiene and syndiotactic 1,2-polybutadiene can be hydrogenated at 100 °C and 50 bar pressure of hydrogen with a soluble catalyst $\{[(Ph)_3{-}P]_3RhCl\}$. Complete saturation of double bonds results.[52] Butadiene acrylonitrile copolymers can also be hydrogenated quantitatively with this rhodium catalyst under mild conditions.[53] The kinetics are consistent with a mechanism where the active Rh catalyst interacts with the unsaturation at the polymer in the rate-determining step. The nitrile group, however, appears to also interact with the catalyst and inhibit the rates.[53]

Hydrogenations of carbon-to-nitrogen double bonds in polymer backbones and in the pendant groups can be carried out with lithium borohydride[86]:

$$-[-CH_2{-}\underset{CH=N{-}C_6H_5}{CH}{-}]{-}-[-CH_2{-}CH{=}CH{-}\underset{C_6H_5}{N}{-}]{-} + LiBH_4 \xrightarrow[\text{reflux}]{\text{THF}}$$

$$-[-CH_2{-}\underset{CH_2{-}NH{-}C_6H_5}{CH}{-}]{-}-[-CH_2{-}CH_2{-}CH_2{-}\underset{C_6H_5}{N}{-}]{-}$$

8.2.3. Addition of Carbenes

Polyisoprenes and polybutadienes can also be modified by reactions with carbenes. Dichloro-carbene adds to natural rubber dissolved in chloroform in a phase transfer reaction with aqueous NaOH[54] and a phase transfer reagent. Solid sodium hydroxide can be used without a phase transfer reagent. There is no evidence of *cis–trans* isomerization and the distribution of the substituents is random.[54]

Difluorocarbene, generated under mild neutral conditions, adds to 1,4-*cis*- and 1,4-*trans*-poly-butadienes to give materials containing cyclopropane groups.[55] The addition takes place randomly, to give atactic stereosequence distributions[55]:

$$-[-CH_2-CH{=}CH-CH_2-]_n- + :CF_2 \longrightarrow -[-CH_2-CH-CH-CH_2-]_n- \quad (\text{cyclopropane ring with } CF_2)$$

Fluorocarbene, formed from phenyl(fluorodichloromethyl)mercury by thermolysis *in situ*, also adds to 1,4-*cis*- and *trans*-polybutadienes. The carbene can add at various levels.[57] The addition is stereospecific and preserves the alkene geometry of the parent polybutadiene. Also, the addition is random, showing that the reactivity of the double bonds is independent of the sequence environment.[57]

Dichlorocarbene, generated *in situ* from an organomercury precursor, phenyl(bromodichloromethyl)mercury (Seyferth reagent), adds to polybutadiene in a similar manner.[56] The reactions take place under homogeneous conditions. They can be carried out on 1% solutions of the polymer in benzene, using 10–20% mole excess of the reagent.

8.2.4. Electrophilic Additions of Aldehydes

These are additions to double bonds, like the *Prins reaction*. They can be carried out on natural and synthetic rubbers.[59,60] They take place rapidly in the presence of acid catalysts. Aqueous formaldehyde,[61] or paraform in CCl_4,[62] can be used. The catalysts are inorganic acids or anhydrous Lewis acids, like boron trifluoride in acetic acid solution[63]:

$$>C{=}C< + H-\overset{+}{C}(OH)H \longrightarrow -\overset{+}{C}-C(-CH_2OH)- \xrightarrow{Cl^-} -C(Cl)-C(-CH_2OH)-$$

The reaction takes a different path in the absence of a catalyst[62]:

$$-\overset{+}{C}(CH_3)-\overset{-}{C}H- + H-\overset{+}{C}H-O^- \longrightarrow -\overset{+}{C}(CH_3)-CH(CH_2O^-)- \longrightarrow$$

$$H_2C\cdots H\cdots O-CH_2 \;(-\overset{+}{C}-CH-) \longrightarrow -C(=CH_2)-CH(CH_2OH)-$$

The products of the Prins reaction with rubbers are thermoplastic polymers that possess fair resistance to acids and bases. Free hydroxyl groups in the products are available for crosslinking with diisocyanates[64] or by other means. The Prins reaction can be carried out directly on rubber latexes.[65] It is also possible to just mill the rubber together with formaldehyde and then heat the resultant mixture in the presence of anhydrous metal chlorides[64] to get similar results.[66]

Higher aldehydes also react with natural rubber.[67] The reaction works best with purified rubber. Additions take place without a catalyst at 180 °C or in the presence of $AlCl_3$–NaCl at 120 °C. These reactions can be carried out in the solid phase by milling the rubber with an aldehyde, like glyoxal.[68] Heating in a pressure vessel above 175 °C is required to complete the reaction. Infrared spectra of the products from reactions in solution show presence of ether, carbonyl, and hydroxyl groups.[69] Two types of additions appear to take place[69]:

1. $$-[-CH_2-\overset{+}{C}(CH_3)-\overset{-}{C}H-CH_2-]- + {}^{-}O-\overset{+}{C}H-CH{=}O \longrightarrow -[-CH_2-\overset{+}{C}(CH_3)-CH(-O^--CH(CHO)-)-CH_2-]- \longrightarrow -[-CH_2-C(=CH_2)-CH(CH(OH)CHO)-CH_2-]-$$

2. $$-[-CH_2-C(CH_3){=}CH-CH_2-]- + {}^{-}O-\overset{+}{C}H-CH{=}O \longrightarrow -[-CH(CH(OH)CHO)-C(CH_3){=}CH-CH_2-]-$$

Products from reactions of rubber with glyoxal have a strong tendency to become spontaneously insoluble. This is probably due to a presence of residual aldehyde groups, because a treatment of the product with 2,4-dinitrophenylhydrazine eliminates spontaneous gelation.

Chloral adds to polyisoprene similarly. The reaction is catalyzed by Lewis acids.[70] Both $AlCl_3$ and BF_3 are efficient catalysts. Less crosslinking is encountered with aluminum chloride. Infrared spectra of the products shows the presence of hydroxyl groups, chlorine atoms, and vinylidene unsaturation.[70]

8.2.5. Polar Additions

A number of polar additions to unsaturated polymers are known. These include Michael additions, hydroboration, 1,3-dipolar additions, ene reaction, the Ritter reaction, Diels-Alder additions, and others.

8.2.5.1. Michael Condensation

Among polar additions to unsaturated polymers are reactions of amines and ammonia with unsaturated polyesters in the form of a Michael condensation. Thus, for instance, additions to poly(1,6-hexanediol maleate) and poly(1,6-hexanediol fumarate)[71] show a difference in the reactivity of the two isomers. The maleate polyester reacts with ammonia to yield a crosslinked product at room temperature, when stoichiometric quantities or an excess ammonia in alcohol is used. At the same reaction conditions, the fumarate isomer only adds a few percent of ammonia.[71] In a 1:1 mixture of chloroform and ethanol, however, approximately half of the fumarate double bonds react. Also, the maleate polyester reacts differently with piperidine or cyclohexylamine. In butyl alcohol at 60 °C the polymer initially isomerizes and precipitates. After the isomerization is complete and the temperature is raised to 80 °C, the polymer redissolves. An exothermic reaction follows and Michael-type adducts form.[71]

8.2.5.2. Hydroboration

Polymers and copolymers of butadiene or isoprene with styrene can react with diborane.[72] A suitable solvent for this reaction is tetrahydrofuran. Subsequent hydrolyses result in introductions of hydroxyl groups into the polymer backbones. The reactions with diborane are very rapid. Some side reactions, however, also occur.[72]

8.2.5.3. 1,3-Dipolar Additions

Cyclic structures form on polymer backbones through 1,3-dipolar additions to carbon-to-carbon or carbon-to-nitrogen double bonds.[73] Because many 1,3-dipoles are heteroatoms, such additions can lead to formations of five-membered heterocyclic rings. An example is addition of nitrilimine to an unsaturated polyesters[73]:

$$-[-R'-CH=CH-R''-]- \;+\; n\; C_6H_5-\overset{\oplus}{C}=N-\overset{\ominus}{N}-C_6H_5 \longrightarrow -[-R'-\underset{\text{pyrazoline ring: } C(C_6H_5)=N-N(C_6H_5)}{CH-CH}-R''-]-$$

Also, iodine isocyanate adds to polyisoprene. The product can be converted to methyliodocarbamate or to iodourea derivatives[74]:

$$-[-CH_2-\overset{CH_3}{\overset{|}{C}}=CH-CH_2-]- \xrightarrow{I-N=C=O} -[-CH_2-\overset{CH_3}{\overset{|}{C}}(N=C=O)-CH(I)-CH_2-]- \xrightarrow{CH_3OH}$$

$$-[-CH_2-\overset{CH_3}{\overset{|}{C}}(NH-C(=O)-O-CH_3)-CH(I)-CH_2-]- \xrightarrow{heat} -[-CH_2-\overset{CH_3}{\overset{|}{C}}\underset{H-N-C(=O)-O}{\;}CH-CH_2-]-$$

Iodine isocyanate additions result in approximately 40% yields. The products can undergo typical reactions of the isocyanate group,[74] as above, and also:

$$-[-CH_2-C(CH_3)(N{=}C{=}O)-CH(I)-CH_2-]- \xrightarrow{[HX]} -[-CH_2-C(CH_3)(NH_2.HX)-CH(I)-CH_2-]-$$

$$\xrightarrow{NaOH} -[-CH_2-C(CH_3)\text{—}CH-CH_2-]- \text{ (aziridine, NH bridging C and CH)}$$

as well as:

$$-[-CH_2-C(CH_3)(N{=}C{=}O)-CH(I)-CH_2-]- \xrightarrow{NH_3} -[-CH_2-C(CH_3)(NH-C(=O)-NH_2)-CH(I)-CH_2-]-$$

$$\longrightarrow -[-CH_2-C(CH_3)\text{(oxazoline ring N, O; } NH_2\text{)}-CH_2-]-$$

where HX is a halogen acid. The products exhibit enhanced heat stability.[74]

Dipolar cycloadditions take place when nitrones or nitrile oxides add to polybutadiene. Some of the products contain isoxazolidine rings[75]:

$$R-CH{=}O + C_6H_5-NH-OH \xrightarrow{H_2O} [R-CH{=}N(C_6H_5)\rightarrow O] \xrightarrow{\text{polybutadiene}} -[-CH_2-\text{(isoxazolidine: R, O, N-}C_6H_5)-CH_2-]-$$

$$C_6H_5-C(Cl){=}N-OH \xrightarrow[-\,HCl]{(C_2H_5)_3N} [C_6H_5-CN\rightarrow O] \xrightarrow{\text{polybutadiene}} -[-CH_2-\text{(isoxazoline: }C_6H_5\text{, N, O)}-CH_2-]-$$

The above modification of butadiene rubber can be carried out to the extent of 3.1 mol%. The product is higher in tensile modulus values and is greater in strength than the parent compound.[75]

A final example is a 1,3 dipolar addition to pendant azide groups.[87] The reaction takes place with phenyl vinyl sulfoxide in dimethylformamide. Forty-eight hours at 110 °C are required for the azide groups to become undetectable by infrared spectroscopy. The product precipitates out with addition of ether[87]:

$$-[-CH_2-CH(C_6H_4-CH_2-N_3)-]- + C_6H_5-S(=O)-CH{=}CH_2 \longrightarrow -[-CH_2-CH(C_6H_4-CH_2-N\text{(triazole, N=N)})-]- + C_6H_5-S(=O)-$$

8.2.5.4. The Ene Reaction

The polymers of conjugated dienes can also be modified via the ene reactions,[76] as, for instance:

$$\sim CH_2-C(CH_2-H){=}CH-CH_2\sim + X{=}Y \longrightarrow \sim CH_2-C({=}CH_2)-CH(Y-XH)-CH_2\sim$$

where X=Y can be O=N–, –N=N–, >C=S, >C=O, or >C=C<. An example of this is an addition of triazolidones[76]:

This results in formation of pendant urazole groups. The exact structure of the products, however, has not been fully established. The tensile strength of polymers improves considerably, but it is accompanied by a dramatic loss in molecular weight.[76] Nevertheless, ene reagents like C-nitroso and activated azo compounds are very efficient in adding to rubber. They add in a few minutes at temperatures between 100–140 °C. In the case of the azo compound the addition can be greater than 90%.

Substituted aryl sulfonyl azides decompose at elevated temperatures to nitrenes and add to natural rubber:

Novel types of polyamines and cationic polyelectrolytes form from polymers of conjugated dienes[77] in reactions with carbon monoxide, amines, and water at 150 °C and 1000–1500 psi pressure. The reaction can be illustrated as follows[77]:

$$>C{=}C< \; + \; CO \; + \; H{-}NRR' \xrightarrow{H_2O} >CH{-}CH_2{-}CH_2{-}NRR'$$

8.2.5.5. The Ritter Reaction

This reaction can be carried out on natural rubber and on synthetic polyisoprenes[78]:

$$-[-CH_2-C(CH_3){=}CH-CH_2-]- \xrightarrow{H^{\oplus}} -[-CH_2-C^{\oplus}(CH_3)-CH_2-CH_2-]-$$

$$\xrightarrow{RCN} -[-CH_2-C(CH_3)(NH{-}C({=}O){-}R)-CH_2-CH_2-]-$$

The carbon cation apparently reacts with any nucleophile present. When the reaction is carried out in dichloroacetic acid, chlorine atoms can be detected in the product.[78]

8.2.5.6. Diels-Alder Condensations

Crotonic acid esters of cellulose undergo addition reaction with cycloaliphatic amines, like morpholine or piperidine and with aliphatic primary amines.[79] Unsaturated polymers can also undergo Diels-Alder reactions. One example is a reaction of hexachlorocyclopentadiene with polycyclopentadiene[80]:

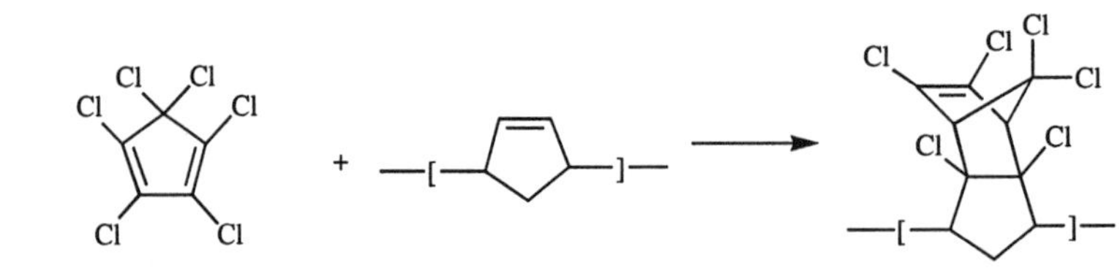

The addition takes place in an inert atmosphere at 140–150 °C. Over 90 mol% conversion is achieved in six hours.

Diels-Alder condensations of fumaric and maleic acid polyesters with various dienes[81] can serve as another example. These reactions require 20 hours at room temperature. Diels-Alder condensations can also be carried out on polymers of 1,3,5-hexatriene, 1,3,5-heptatriene, and 2,4,6-octatriene.[82] Sulfonate-substituted maleic anhydride adds to low-functionality hydrocarbon elastomers, like EPDM, presumably via an Alder-ene-type reaction[83]:

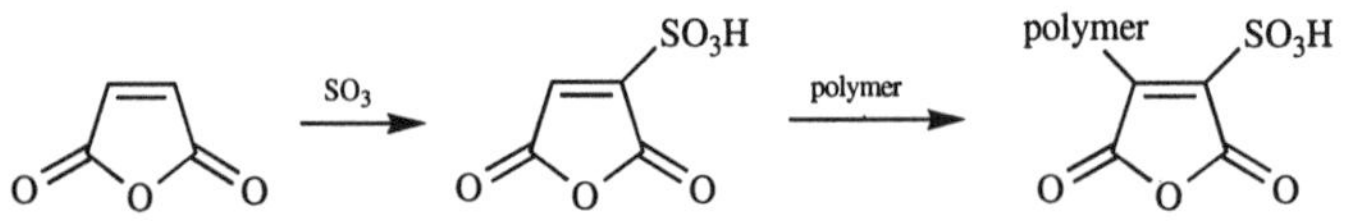

Thiols add to diene rubbers by a free-radical mechanism.[84] Thus antioxidants, like 4-(mercapto-acetamido)-diphenylamine, add –SH groups to the double bonds of *cis*-polyisoprene and polybutadiene in the presence of free-radical initiators.[84]

Thiol compounds also add photochemically to polymers containing double bonds. For instance, unsaturation can be introduced into polyepichlorohydrin by a partial elimination reaction. The product then reacts with mercaptans, aided by a photosensitizer (like benzophenone) and ultraviolet light[85]:

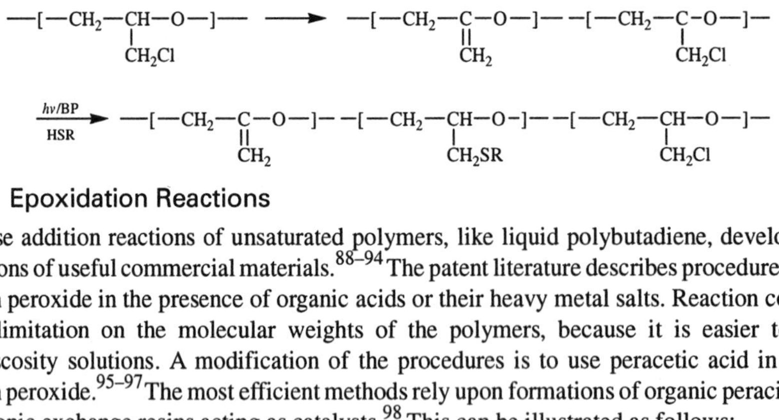

$$-[-CH_2-CH(CH_2Cl)-O-]- \longrightarrow -[-CH_2-C(=CH_2)-O-]--[-CH_2-CH(CH_2Cl)-O-]-$$

$$\xrightarrow[HSR]{h\nu/BP} -[-CH_2-C(=CH_2)-O-]--[-CH_2-CH(CH_2SR)-O-]--[-CH_2-CH(CH_2Cl)-O-]-$$

8.2.5.7. Epoxidation Reactions

These addition reactions of unsaturated polymers, like liquid polybutadiene, developed into preparations of useful commercial materials.[88–94] The patent literature describes procedures that use hydrogen peroxide in the presence of organic acids or their heavy metal salts. Reaction conditions place a limitation on the molecular weights of the polymers, because it is easier to handle lower-viscosity solutions. A modification of the procedures is to use peracetic acid in place of hydrogen peroxide.[95–97] The most efficient methods rely upon formations of organic peracids *in situ* with cationic exchange resins acting as catalysts.[98] This can be illustrated as follows:

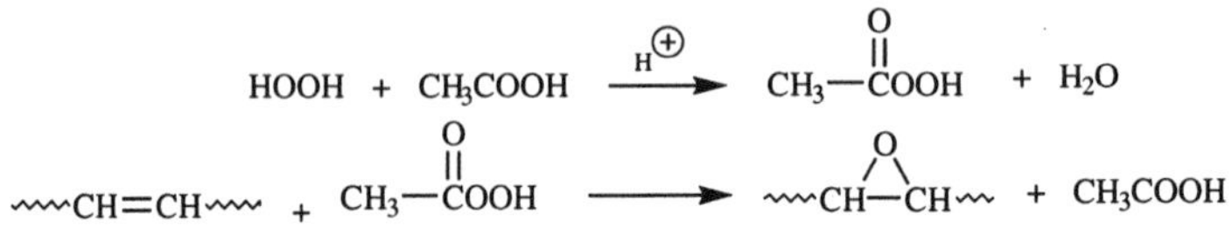

$$HOOH + CH_3COOH \xrightarrow{H^{\oplus}} CH_3-C(=O)OOH + H_2O$$

$$\sim\sim CH=CH \sim\sim + CH_3-C(=O)OOH \longrightarrow \sim\sim CH\overset{O}{-}CH \sim\sim + CH_3COOH$$

The reaction is accompanied by formations of byproducts:

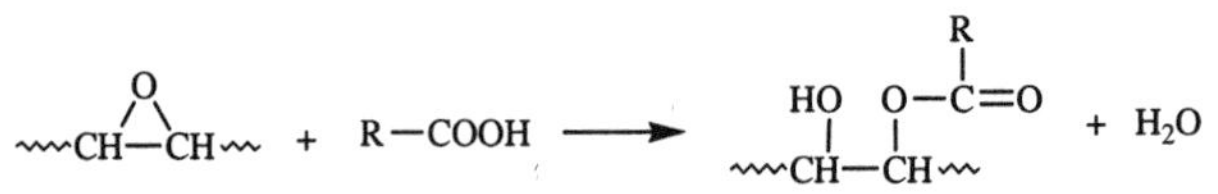

$$\sim\sim CH\overset{O}{-}CH \sim\sim + R-COOH \longrightarrow \sim\sim CH(OH)-CH(O-C(=O)R) \sim\sim + H_2O$$

Polybutadienes that are high in 1,4-structures tend to epoxidize more readily and yield less viscous products.[100,101] The epoxidation reaction can also be carried out on poly(1,4-cyclopentadiene)[99]:

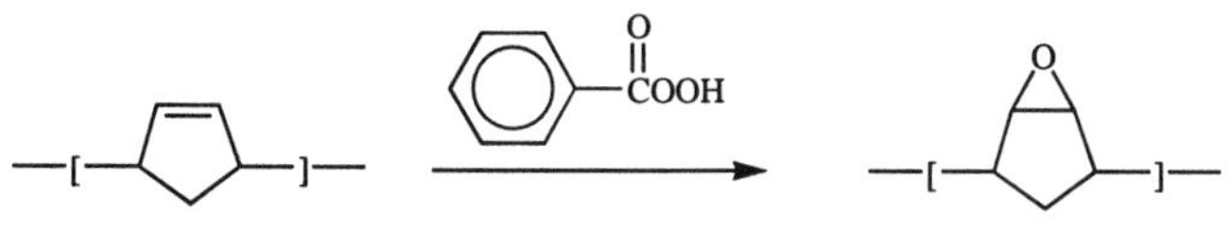

Perbenzoic acid is an effective reagent in chloroform and in methylene chloride solutions at 0–20 °C. The conversions are high, yielding brittle materials soluble in many solvents. The products can be cast into transparent films.[99]

Monoperphthalic and *p*-nitroperbenzoic acids are also efficient epoxidizing agents. They can, however, cause crosslinking, as is the case in epoxidation of polycyclopentadiene.[99] The products react like typical epoxy compounds[99]:

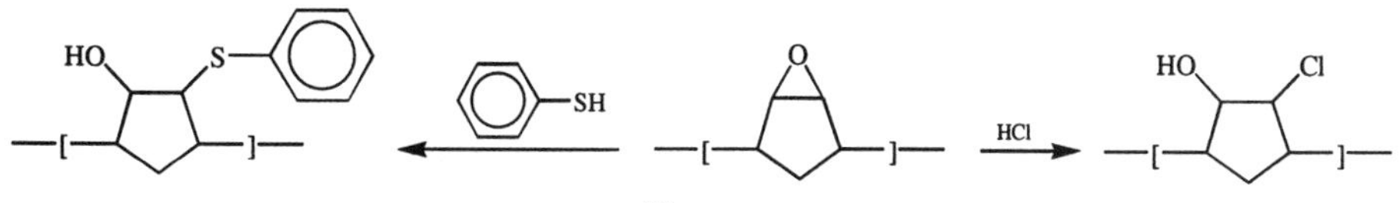

Some other reactions of the epoxy groups are[99]:

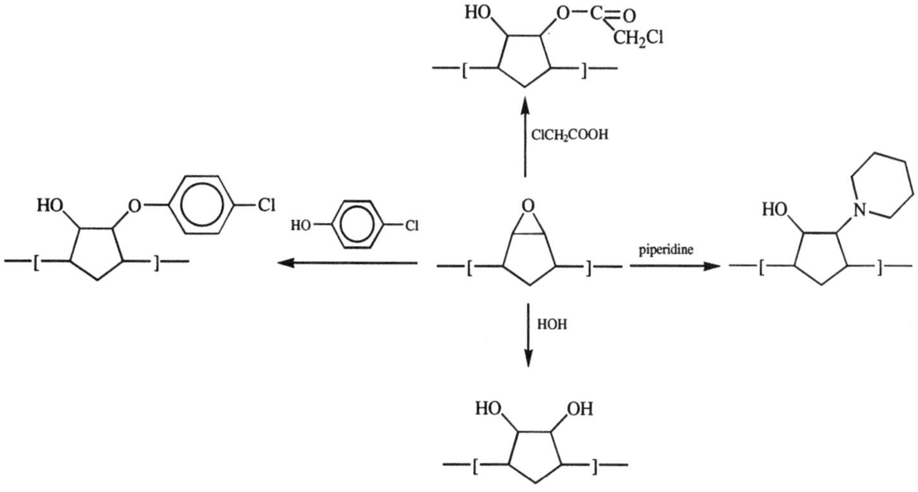

8.3. Rearrangement Reactions

There are different types of polymeric rearrangements. One of them is isomerizations of polymers with double bonds.

8.3.1. Isomerization Reactions

The isomerization of *cis*-polybutadiene can be carried out with the aid of ultraviolet light or with gamma radiation. When light is used, the free-radical reaction requires photoinitiators.[102,103] The mechanism involves freely rotating transitory free radicals on the polymer backbone. These form from additions of photoinitiator fragments to the double bonds. The adducts break up again, releasing the attached initiator fragments and reestablishing the double bonds. The new configurations are *trans* because they are more thermodynamically stable.[102,103] It can be shown as follows:

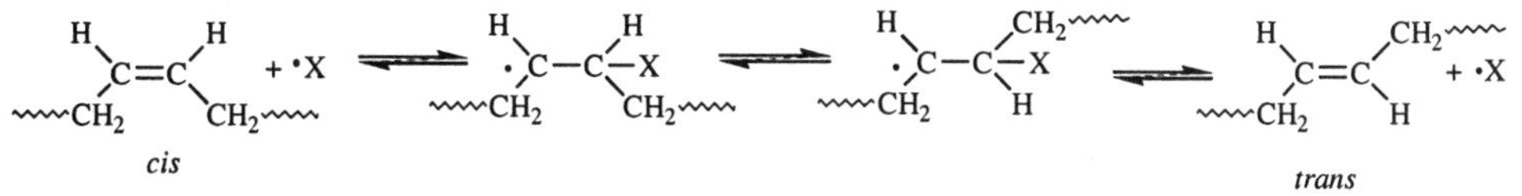

where X· represents a free-radical fragment from a photoinitiator.

With gamma radiation there is no need for any additives.[104,105] Here, the mechanism of isomerization is believed to involve direct excitation of the π-electrons of the double bonds to antibonding orbitals where free and geometric interconversions are possible. In benzene solutions, energy transfers take place from excited benzene molecules to the polymer double bonds.

When 1,4-polyisoprene films are irradiated with light,[122] *cis–trans* isomerizations occur. In the process 1,4-unsaturations decrease. Also, vinyl and vinylidene double bonds and cyclopropyl groups form. This can be illustrated as follows:

$$-[-CH_2(CH_3)C{=}CH-CH_2-]- \xrightarrow{h\nu} -[-CH_2(CH_3)\dot{C}-\dot{C}H-CH_2-]- \xrightarrow{\text{isomerization}} -[-CH_2(CH_3)C{=}CH(CH_2-]-)$$

hydrogen shift

$$-[-CH(CH_3)\dot{C}\;\cdot CH_2\!-\!\dot{C}H-]- \longrightarrow -[-CH_2(CH_3)\underset{\triangle}{C-CH_2-CH}-]-$$

and

$$\sim\!CH_2-C(CH_3){=}CH-CH_2-CH_2-C(CH_3){=}CH-CH_2\!\sim$$
$$\downarrow$$
$$\sim\!CH_2-C(CH_3){=}CH-CH_2\cdot + \cdot CH_2-C(CH_3){=}CH-CH_2\!\sim$$
$$\updownarrow \qquad\qquad \updownarrow$$
$$\sim\!CH_2-\dot{C}(CH_3)-CH{=}CH_2 + CH_2{=}C(CH_3)-\dot{C}H-CH_2\!\sim$$
$$\downarrow$$
$$\sim\!CH_2-C(CH_3)(CH{=}CH_2)-CH(C(CH_3){=}CH_2)-CH_2\!\sim$$

Polybutadiene can also isomerize *cis–trans* and lose unsaturation when irradiated in the solid state. This must be done in vacuum with ultraviolet light of 1236 Å or 2537 Å.[106,107] Both free-radical and ionic mechanisms are suspected to operate simultaneously.

8.3.2. Cyclizations and Intramolecular Rearrangements

Cyclization reactions of natural rubber and other polymers from conjugated dienes have been known for a long time. The reactions occur in the presence of Lewis and strong protonic acids. They result in loss of elastomeric properties and some unsaturation. Carbon cations form in the intermediate step and subsequent formation of polycyclic structures[108,109]:

$$\sim\!CH_2-C(CH_3){=}CH-HCH-HCH-C(CH_3){=}CH-HCH-HCH\!\sim \xrightarrow{H^{\oplus}} \sim\!CH_2-\overset{\oplus}{C}(CH_3)-CH_2-HCH-HCH-C(CH_3){=}CH-HCH-HCH\!\sim \longrightarrow$$

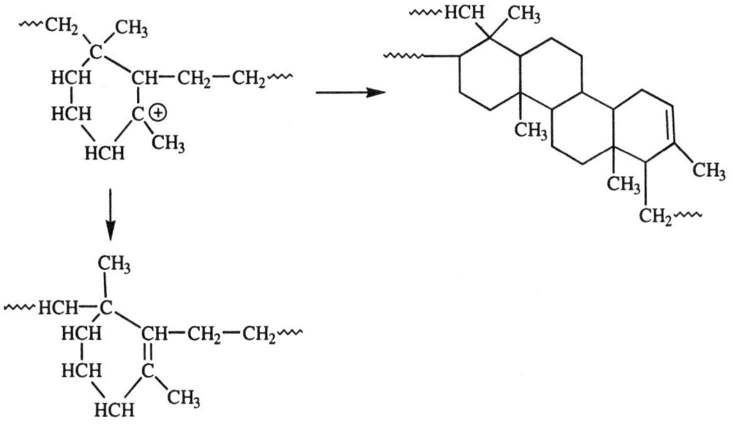

In a similar manner, polymers with pendant unsaturation undergo cyclization reactions in benzene in the presence of BF_3 or $POCl_3$ yielding ladder structures. The exact nature of the initiation process is not clear. Water may be needed for the initiation step[110,111]:

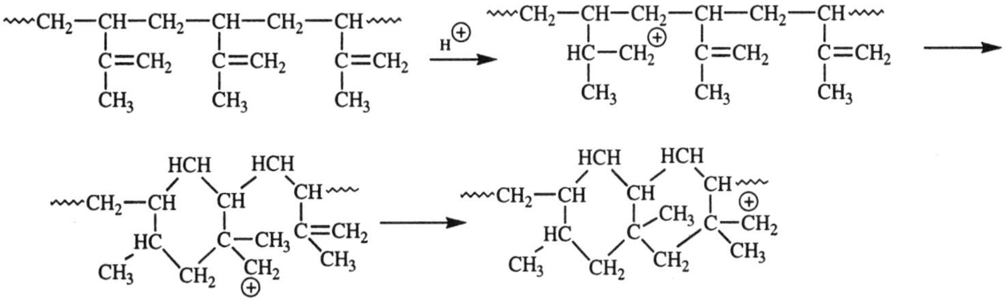

The reactions result in formations of six-membered monocyclic and fused polycyclic units. These reactions of carbon cations should also lead to molecular rearrangements, like 1,2-shifts of protons, resulting in formations of five-membered rings and spiro cyclopentane repeat units.[111]

Cyclization reactions of polyisoprene can be catalyzed by $TiCl_4$[112] and by sulfuric acid.[113–115,117] The products appear similar in the infrared spectra[110] with only a few minor differences. Also, there is only a small number of fused rings in the product.

Polyacrylonitrile converts to a red solid when heated above 200 °C.[116] Only a small amount of volatile material is given off. Further heating of the red residue to about 350 °C or higher converts it to a black brittle material. This black material has a ladder structure:

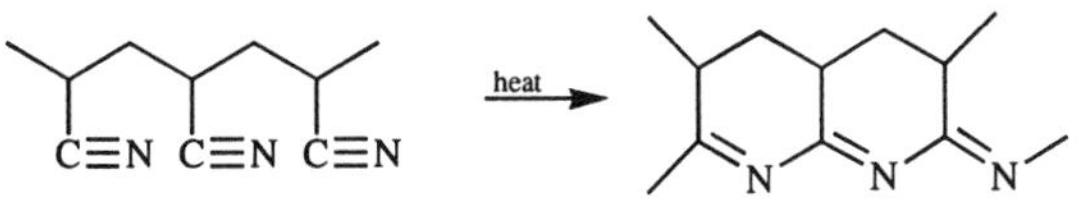

Heating of polyacrylonitrile in the presence of oxygen yields some quinone-type structures:

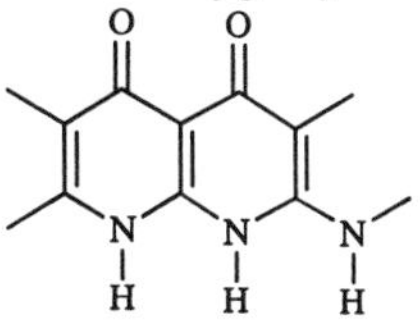

Further heating of the polymer at very high temperatures, in excess of 2000 °C, results in formation of graphite-like structures.[116] All, or almost all, nitrogen is lost, probably as HCN and N_2.

Migration of double bonds is a well established phenomenon in polymers from conjugated dienes. It occurs, for instance, during vulcanization of rubber (discussed in a later section). It also occurs upon simply heating some polymers.[117–119] Thus, the double bonds shift in natural rubber when it is heated to temperatures of 150 °C or above[120]:

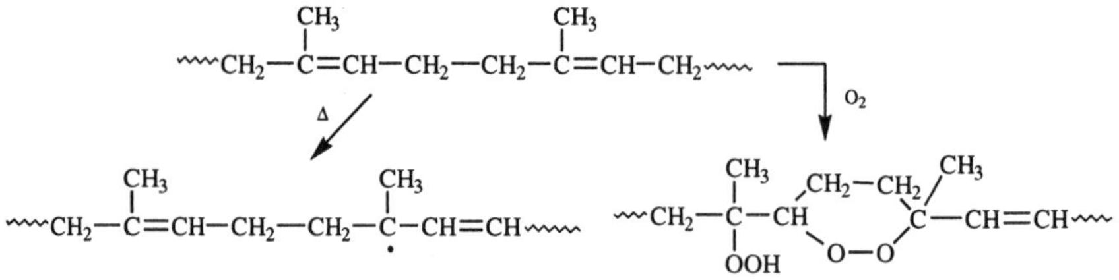

$$\sim CH_2-\overset{CH_3}{\overset{|}{C}}=CH-CH_2-CH_2-\overset{CH_3}{\overset{|}{C}}=CH-CH_2\sim \xrightarrow{\Delta} \sim CH_2-\overset{CH_3}{\overset{|}{C}}=CH-CH_2-CH_2-\overset{CH_3}{\overset{|}{\dot{C}}}-CH=CH\sim$$

$$\xrightarrow{O_2} \sim CH_2-\overset{CH_3}{\overset{|}{\underset{OOH}{\underset{|}{C}}}}-CH\overset{CH_2-CH_2}{\underset{O-O}{\quad}}\overset{CH_3}{C}-CH=CH\sim$$

Hydrochlorination of rubber in solution causes a different kind of double-bond shift[121]:

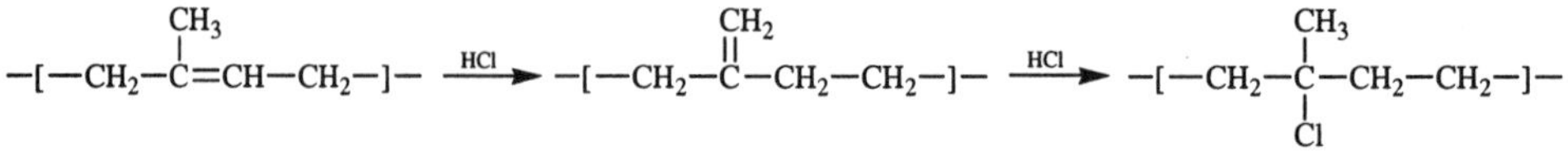

$$-[-CH_2-\overset{CH_3}{\overset{|}{C}}=CH-CH_2-]- \xrightarrow{HCl} -[-CH_2-\overset{CH_2}{\overset{||}{C}}-CH_2-CH_2-]- \xrightarrow{HCl} -[-CH_2-\overset{CH_3}{\overset{|}{\underset{Cl}{\underset{|}{C}}}}-CH_2-CH_2-]-$$

In polymers with pendant double bonds, facile cycloadditions occur upon heating[123–125]:

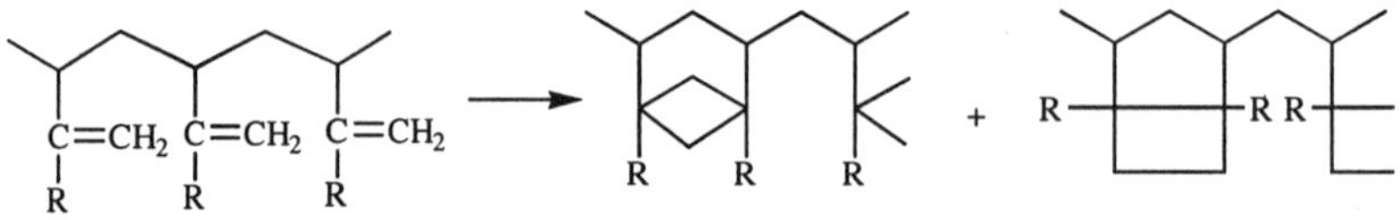

$$\text{polymer with pendant } \underset{R}{\underset{|}{C}}=CH_2 \text{ groups} \longrightarrow \text{cyclobutane-fused products} + \text{cyclobutane-fused products}$$

Many other polymeric structures can rearrange under proper conditions. For instance, poly(4,4′-diphenylpropane isophthalate) rearranges upon irradiation with UV light to a structure containing *o*-hydroxybenzophenones[126]:

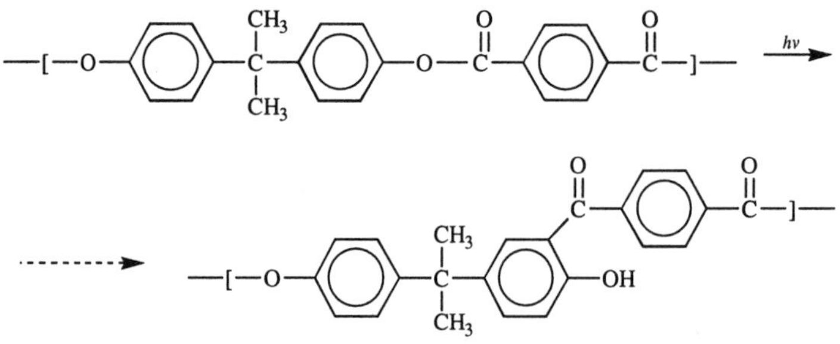

$$-[-O-C_6H_4-\overset{CH_3}{\overset{|}{\underset{CH_3}{\underset{|}{C}}}}-C_6H_4-O-\overset{O}{\overset{||}{C}}-C_6H_4-\overset{O}{\overset{||}{C}}-]- \xrightarrow{h\nu}$$

$$\dashrightarrow -[-O-C_6H_4-\overset{CH_3}{\overset{|}{\underset{CH_3}{\underset{|}{C}}}}-C_6H_3(OH)-\overset{O}{\overset{||}{C}}-C_6H_4-\overset{O}{\overset{||}{C}}-]-$$

The mechanism is believed to be that of a photoinduced *Fries rearrangement.*[127] It may be similar to one catalyzed by Lewis acids. Another example is a spontaneous rearrangement in the solid state of poly(α-phenylethylisonitrile)[128]:

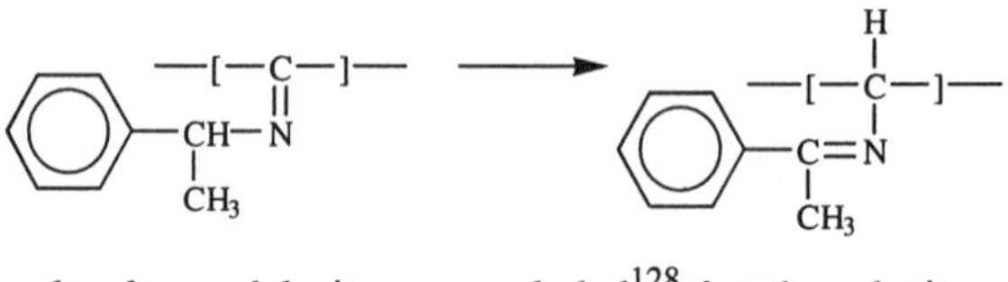

$$-[-\overset{}{C}-]- \;\text{with}\; \overset{||}{N}-\underset{CH_3}{\underset{|}{CH}}-C_6H_5 \longrightarrow -[-\overset{H}{\overset{|}{C}}-]- \;\text{with}\; \underset{|}{C}=N\text{, } C_6H_5-\underset{CH_3}{\underset{|}{C}}=N$$

Based on studies of molecular models, it was concluded[128] that the substituents force progressive, one-handed twisting of the helix shape of the molecules. This occurs to such an extent that the vicinal imino double bonds are out of conjunction with each other. As a result, each benzylic hydrogen atom becomes localized over the electrons of the imino groups. It proceeds in a constant screw direction around the axis of the helix. Such steric confinement causes tautomeric rearrangements, as shown above, which results in some relaxation of the compression. An additional driving force is a gain in conjugation between the imino double bonds and the aromatic rings of each substituent.[128]

Intramolecular rearrangements of polymers with ketone groups were subjects of several studies. The *Schmidt* and *Beckmann rearrangements* were carried out on copolymers of ethylene and carbon monoxide[129]:

$$-[-(-CH_2-CH_2-)_x-\overset{O}{\overset{\|}{C}}-]- \xrightarrow[H_2SO_4]{HN_3} -[-(-CH_2-CH_2-)_x-\overset{H}{\overset{|}{N}}-\overset{O}{\overset{\|}{C}}-]-$$

$$\downarrow NH_2OH$$

$$-[-(-CH_2-CH_2-)_x-\overset{N-OH}{\overset{\|}{C}}-]- \longrightarrow -[-(-CH_2-CH_2-)_x-\overset{H}{\overset{|}{N}}-\overset{O}{\overset{\|}{C}}-]-$$

The starting materials were low and medium molecular weight copolymers. Infrared spectra of the products from the Schmidt reaction show a high degree of conversion.[129] The yields of the oximes and subsequent Beckmann rearrangements are also high.

A Schmidt reaction on poly(*p*-acetylstyrene) yields a product containing acetylaminostyrene groups in high yields and the products are surprisingly pure.[130] The solvent is acetic acid and the converted material precipitates out as a sticky mass:

$$-[-CH_2-CH(C_6H_4-C(=O)-CH_3)-]- \xrightarrow[H_2SO_4]{HN_3} -[-CH_2-CH(C_6H_4-N(H)-C(=O)-CH_3)-]- \xrightarrow{H^{\oplus}} -[-CH_2-CH(C_6H_4-NH_2)-]-$$

Another rearrangement reaction is isomerization of unsaturated polyesters upon heating. Thus, for instance, maleate polyesters rearrange to the fumarate analogs at elevated temperatures[131]:

$$-[-CH_2\text{-}O-\overset{O}{\overset{\|}{C}}-CH{=}CH-\overset{O}{\overset{\|}{C}}-O-CH_2-]- \ (cis) \longrightarrow -[-CH_2\text{-}O-\overset{O}{\overset{\|}{C}}-CH{=}CH-\overset{O}{\overset{\|}{C}}-O-CH_2-]- \ (trans)$$

8.4. Substitution Reactions

Many substitution reactions are carried out on polymers in order to replace atoms in the backbones or in the pendant groups with other atoms or groups of atoms. These reactions do not differ much from those of the small molecules.

8.4.1. Substitution Reactions of Saturated Polymeric Hydrocarbons

It is often desirable to replace hydrogens with halogens. *Fluorination of polyethylene* can be carried out in the dark by simply exposing the polymer, either in sheet or in powder form, to fluorine gas. The reaction is exothermic and it is best to dilute the gas with nitrogen to allow a gradual introduction of the fluorine and avoid destruction of the polymer.[132,133] In this manner, however, only the surface layers are fluorinated and the substitutions occur only a few molecular layers deep.

In surface fluorination with vacuum ultraviolet glow discharge plasma, the photon component of the plasma enhances the reactivity of the polymer surfaces toward fluorine gas[134]:

Gas Phase: $F_2 + h\nu\ (210\text{–}360\ nm) \longrightarrow 2F\bullet$

Solid Surface $\sim CH_2-CH_2 \sim + h\nu\ (< 160\ nm)$

$$\longrightarrow \sim CH_2-\overset{\oplus}{C}H \sim + e^{\ominus} \xrightarrow{e^{\ominus}} \sim CH_2-\overset{\bullet}{C}H \sim$$

The surface free radicals can also cause elimination of hydrogen radicals and formation of double bonds. Whether as free radicals or through unsaturation, the units are now more reactive toward fluorine.

A film that is three mils thick can be completely fluorinated on a 100-mesh phosphor bronze gauze, if the reaction is allowed to proceed for several days.[135] Fluorination can also be carried out with mercuric or cupric fluorides in hydrofluoric acid. The reaction must be carried out at 110 °C for 50 hours. As much as 20% fluorine can be introduced.[136]

In direct fluorination of powdered high-density polyethylene with the gas, diluted with helium or nitrogen, the accompanying exotherm causes partial fusion. In addition, there is some destruction of the crystalline regions.[137] On the other hand, fluorination of single crystals of polyethylene can result in fluorine atoms being placed on the carbon skeleton without disruption of the crystal structure. The extent of crosslinking, however, is hard to assess.[138] The reaction has all the characteristics of free-radical mechanism[139]:

$$F_2 \longrightarrow 2F\cdot$$

$$RH + F\cdot \longrightarrow R\cdot + HF$$

$$R\cdot + F_2 \longrightarrow RF + F\cdot$$

Chlorinations of polyethylene can be carried out in the dark or in the presence of light. The two reactions, however, are different, though both take place by a free-radical mechanism. When carried out in the dark at 100 °C or higher, no catalyst is needed, probably because there are residual peroxides from oxidation of the starting material. Oxygen must be excluded because it inhibits the reaction and degrades the product.[140] Also, the reaction is catalyzed by traces of $TiCl_4$.[141] Such trace quantities may be residual titanium halide from a Ziegler–Natta catalyst left over in the polymer from the polymerization reaction. When it is carried out at 50 °C in chlorobenzene, –CHCl– groups form[142]:

$$\sim CH_2–CH_2–CH_2–CH_2\sim + Cl_2 \rightarrow \sim CH_2–CH_2–CHCl–CH_2\sim$$

This slows the chlorination of adjacent groups.

Trace amounts of oxygen catalyze chlorinations in the presence of visible light.[140] The same reaction in ultraviolet light is accompanied by crosslinking. The photochemical process can be illustrated as follows:

1. $Cl_2 + h\nu \longrightarrow 2Cl\cdot$

2. $Cl\cdot + \sim CH_2-CH_2-CH_2\sim \longrightarrow \sim CH_2-\dot{C}H-CH_2\sim + HCl$

3. $\sim CH_2-\dot{C}H-CH_2\sim + Cl_2 \longrightarrow \sim CH_2-CHCl-CH_2\sim + Cl\cdot$

etc.

Chlorination of polyethylene can result in varying amounts of hydrogen atoms being replaced by chlorine. It is possible to form a product that contains 70% by weight of chlorine. The amount of chlorination affects the properties of the product. At low levels of substitution the material still resembles the parent compound. When, however, the level of chlorine reaches 30–40%, the material becomes an elastomer. At levels exceeding 40% the polymer stiffens again and becomes hard.

Commercial chlorinations of polyethylene are usually conducted on high-density ($D > 0.96$) linear polymers. The molecular weights of the starting materials vary. High molecular weight polymers form tough elastomers. Low molecular weight materials, however, allow easier processing of the products. The reactions are carried out in carbon tetrachloride, methylene dichloride, or chloroform at reflux temperatures of the solvents and at pressures above atmospheric to overcome poor solubility. The solubility improves with the degree of chlorination. Two different procedures are used. In the first, the reactions are conducted at 95–130 °C. When the chlorinations reach a level of about 15%, the polymers become soluble and the temperatures are lowered considerably.[143] In the second procedure, the reactions are conducted on polymers suspended in the solvent. When the chlorine content reaches 40% and the polymers become soluble, chlorinations are continued in solution. By continuing the reaction a chlorine content of 60% can be reached. The products from the two processes differ. The first one yields a homogeneous product with the chlorine atoms distributed uniformly throughout the molecules. Chlorination in suspension, on the other hand, yields heterogeneous materials with only segments of the polymeric molecules chlorinated. Some commercial chlorinations are conducted in water suspensions. These reactions are carried out at 65 °C until approximately 40% levels of chlorine are achieved. The temperatures are then raised to 75 °C to drive the conversions further. In such procedures, agglomerations of the particles can be a problem. To overcome that, water is usually saturated with HCl or $CaCl_2$.[144] Problems with agglomeration are also encountered during suspension chlorinations in solvents, like CCl_4. Infrared spectra of chlorinated polyethylenes show the presence of various forms of substitutions. There are ~CHCl–CHCl~ as well as $-CCl_2-$ groups present in the materials.[145,146] Surface photochlorination of polyolefin films[146] considerably improves the barrier properties of the films to permeations of gases.

Chlorinations of polypropylene usually result in severe degradations of the polymer. When $TiCl_4$ is the chlorination catalyst, presumably, less degradation occurs.[140] Studies of *bromination of polypropylene* (atactic) show that when the reaction is carried out in the dark, in CCl_4 at 60 °C, the substitution reactions proceed at the rate of 0.5% per hour.[147]

Chlorosulfonation or polyethylene is a commercial process. The reaction resembles chlorination in the step of hydrogen abstraction by chlorine radicals. It is catalyzed by pyridine[148–150] and can be pictured as follows[147]:

$$Cl\cdot + \sim CH_2-CH_2-CH_2\sim \longrightarrow HCl + \sim CH_2-\dot{C}H-CH_2\sim \overset{SO_2}{\rightleftharpoons}$$

$$\sim CH_2-\underset{}{\overset{SO_2\cdot}{\overset{|}{C}H}}-CH_2\sim \xrightarrow{Cl_2} \sim CH_2-\overset{SO_2Cl}{\overset{|}{C}H}-CH_2\sim + Cl\cdot$$

$$\sim CH_2-\dot{C}H-CH_2\sim + Cl_2 \longrightarrow \sim CH_2-CHCl-CH_2\sim + Cl\cdot$$

The amount of SO_2 vs. chlorine in the reaction mixture affects the resultant ratios of chlorosulfonation vs. chlorination of the polymer. These ratios and the amounts of conversion vary with the temperature.[149]

Commercially produced chlorosulfonated polyethylene contains approximately 26–29% chlorine and 1.2–1.7% sulfur.[151] The material is an elastomer that remains flexible below –50 °C. It is commonly crosslinked (vulcanized) through the sulfonyl chloride groups. Heating it, either as a solid or in solution, to 150 °C results in loss of SO_2 and HCl. If the material is heated for two hours at 175 °C, all SO_2Cl groups are removed. In gamma radiation-induced chlorosulfonations and sulfoxidations of polyethylene powders[153] at room temperature, the ratios of SO_2Cl groups to Cl groups decrease with increases in radiation.

Polypropylene can be chlorosulfonated to the extent of containing 6% chlorine and 1.4% sulfur without embrittlement. The reaction can be conducted in CCl_4 at 55 °C. There is apparently less degradation than in a direct chlorination reaction.[148,152]

8.4.2. Substitution Reactions of Halogen-Bearing Polymers

Procedures for commercial *chlorinations of poly(vinyl chloride)* vary. Low-temperature chlorinations are conducted on aqueous dispersions of the polymers that are reacted with chlorine gas in the presence of swelling agents, like chloroform. These are light catalyzed reactions, usually carried out at about 50 °C. They result in substitutions of methylene hydrogens.[158,159]

$$-[-CH_2-\underset{Cl}{\underset{|}{C}H}-]- \xrightarrow{h\nu,\ Cl_2} -[-\underset{Cl}{\underset{|}{C}H}-\underset{Cl}{\underset{|}{C}H}-]- + HCl$$

Some breakdown of the polymers accompanies the reactions in suspension.[154] Irregularity in the structures and the release of HCl significantly contribute to this. It appears that the degradations are initiated by HCl, that is released as a result of the chlorination. In the process double bonds form. Immediately upon their formation they become saturated with chlorines.

A study was conducted on solution chlorination of poly(vinyl chloride) in the presence of free radicals[155] generated from azobisisobutyronitrile. The reaction appears to proceed in two stages. The first takes place until 60% of all the CH_2 groups have reacted. After that, in the second stage the original –CHCl groups are attacked. Some unreacted $-CH_2-$ groups, however, remain.[155] Photochlorination can achieve 90% conversion.[156,157] Typical commercially chlorinated poly(vinyl chloride), however, has a chlorine content of 66–67%. The material contains methylene groups and is in effect a copolymer of vinyl chloride and 1,2-dichloroethylene.

Chlorination raises the softening temperature of poly(vinyl chloride). The products exhibit poorer heat stability than the parent material and are higher in melt viscosity. High-temperature chlorinations in chlorinated solvents at 100 °C result in extensive substitutions of the methylenic hydrogens as well. The reactions, however, are accompanied by extensive chain scissions. The products are soluble in solvents like acetone and methylene chloride, have low softening points, low impact strength, and poor color stability.

Chlorination of poly(vinylidene chloride) can be carried out with sulfuryl chloride using azobisisobutyronitrile.[155] The reaction appears to proceed in three steps:

1. Formation of $SO_2Cl\cdot$ free radicals by abstraction.
2. Formation of free radicals on the polymeric backbones:

$$\sim CCl_2-CH_2-CCl_2\sim + SO_2Cl\cdot \longrightarrow \sim CCl_2-\dot{C}H-CCl_2\sim$$

3. Transfer reactions of the active sites:

$$\sim CCl_2-\dot{C}H-CCl_2\sim + SO_2Cl_2 \longrightarrow \sim CCl_2-CHCl-CCl_2\sim + SO_2Cl\cdot$$

Chlorination of poly(vinyl fluoride) yields a product with 40–50% chlorine content.[158] *Fluorination* of poly(vinylidene fluoride) was reported.[160] When mixtures of fluorine and nitrogen gases are used, the reactions are limited by the amount of diffusion of fluorine into the polymer network. X-ray photoelectron spectroscopy shows the presence of $-CF_2-$, –CHF–, and $-CH_2-$ groups in the product.[161]

Many attempts at other modifications of poly(vinyl chloride) were reported in the literature. Often, the reactions are based on expectations that the polymers will react like typical alkyl halides. Unfortunately, in place of nucleophilic substitutions, the polymers often undergo rapid and sequential eliminations of HCl along the chains. Nevertheless, many substitution reactions are still possible and can be successfully carried out. One example is a replacement of 43% of the chlorine atoms with azide groups[162]:

$$\sim CH_2-\underset{Cl}{CH}-CH_2-\underset{Cl}{CH}\sim \xrightarrow{NaN_3} \sim CH_2-\underset{Cl}{CH}-CH_2-\underset{N_3}{CH}\sim$$

The reaction can be carried out at 60 °C for 10 hours and the polymer does not crosslink. It proceeds faster in DMF than in other solvents. Once substituted, the polymer becomes photosensitive and crosslinks when irradiated with ultraviolet light. The presence of azide groups allows further modifications. One such modification is a reaction with an isocyanate[162]:

$$\sim CH_2-\underset{Cl}{CH}-CH_2-\underset{N_3}{C}\sim + C_6H_5-N=C=O \longrightarrow \sim CH_2-\underset{Cl}{CH}-CH_2-\underset{N=C=N-C_6H_5}{CH}\sim$$

By a similar reaction, phosphinimine groups can be formed on the polymer[162]:

$$\sim CH_2-\underset{Cl}{CH}-CH_2-\underset{N_3}{CH}\sim + (C_6H_5)_3-P \longrightarrow \sim CH_2-\underset{Cl}{CH}-CH_2-\underset{N=P(C_6H_5)_3}{CH}\sim$$

Some other modifications of the azide group containing polymers are[163,164]:

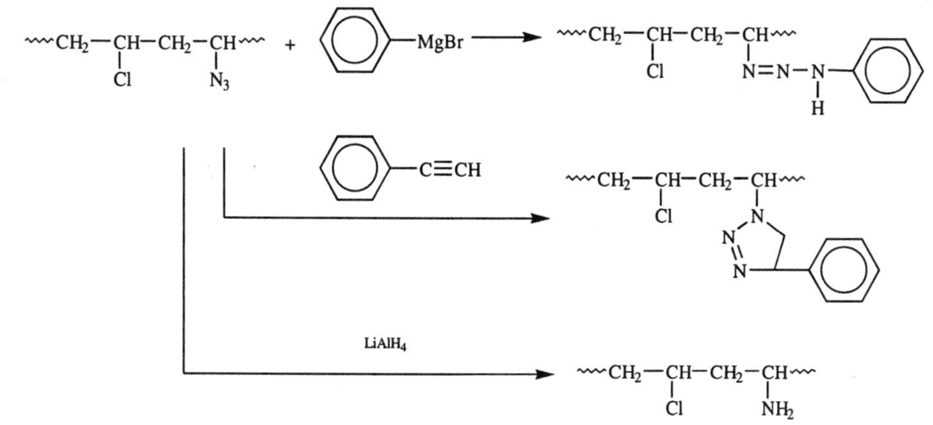

$$\sim CH_2-\underset{Cl}{CH}-CH_2-\underset{N_3}{CH}\sim + C_6H_5MgBr \longrightarrow \sim CH_2-\underset{Cl}{CH}-CH_2-\underset{N=N-NH-C_6H_5}{CH}\sim$$

$$\xrightarrow{C_6H_5C\equiv CH} \sim CH_2-\underset{Cl}{CH}-CH_2-\underset{\text{triazole ring with phenyl}}{CH}\sim$$

$$\xrightarrow{LiAlH_4} \sim CH_2-\underset{Cl}{CH}-CH_2-\underset{NH_2}{CH}\sim$$

Note: The illustration of the reactions with $LiAlH_4$, as shown above, implies that all chloride atoms remain intact on the polymer backbones. It appears likely, however, that some of them might get removed and double bonds might form instead in the backbone.

Poly(vinyl chloride) reacts with various thiols in ethylene diamine to produce monosulfide derivatives[165]:

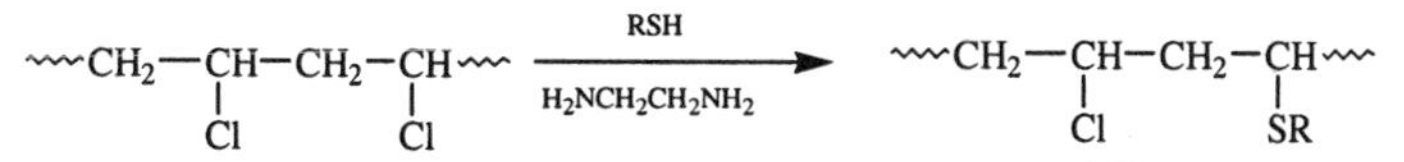

$$\sim CH_2-\underset{Cl}{CH}-CH_2-\underset{Cl}{CH}\sim \xrightarrow[H_2NCH_2CH_2NH_2]{RSH} \sim CH_2-\underset{Cl}{CH}-CH_2-\underset{SR}{CH}\sim$$

When sulfur is added to the reaction mixture, disulfides form instead[165]:

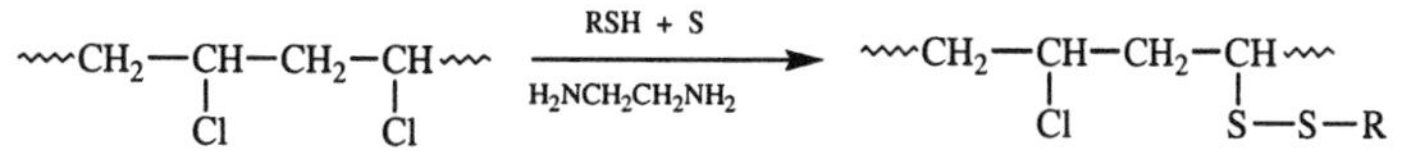

$$\sim CH_2-\underset{Cl}{CH}-CH_2-\underset{Cl}{CH}\sim \xrightarrow[H_2NCH_2CH_2NH_2]{RSH\ +\ S} \sim CH_2-\underset{Cl}{CH}-CH_2-\underset{S-S-R}{CH}\sim$$

The thiolate ions can serve as strong nucleophiles and also as weak bases.[166] When poly(vinyl chloride) is suspended in water in the presence of swelling agents and phase transfer catalysts, it reacts with mercaptans, like 3-[*N*-(2-pyridyl)carbamoylpropyl]thiol:

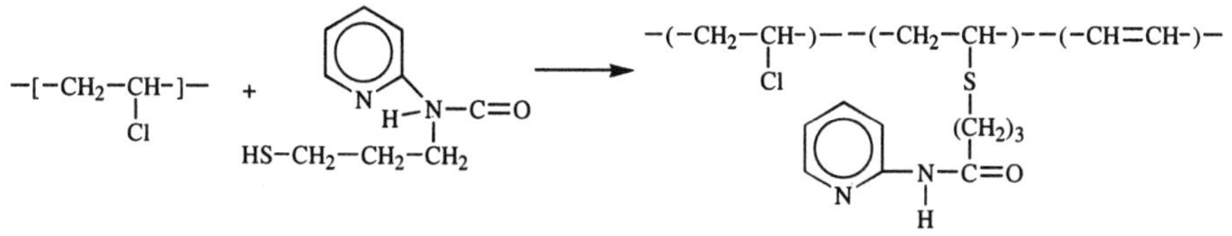

$$-[-CH_2-\underset{Cl}{CH}-]- + HS-CH_2-CH_2-CH_2-C(=O)-NH-(2\text{-pyridyl}) \longrightarrow -(-CH_2-\underset{Cl}{CH}-)--(-CH_2-\underset{S-(CH_2)_3-C(=O)-NH-(2\text{-pyridyl})}{CH}-)--(-CH=CH-)-$$

Acetoxylation of poly(vinyl chloride) can be carried out under homogeneous conditions.[167] Crown ethers, like 18-crown-6, solubilize potassium acetate in mixtures of benzene, tetrahydrofuran, and methyl alcohol to generate unsolvated, strongly nucleophilic “naked” acetate anions. These react readily with the polymer under mild conditions.[167] Substitutions of the chlorine atoms on the polymeric backbones by anionic species take place by a S_N2 mechanism. The reactions can also proceed by a S_N1 mechanism. That, however, requires formations of cationic centers on the backbones in the rate-determining step and substitutions are in competition with elimination reactions. It is conceivable that anionic species may (depending upon basicity) also facilitate elimination reactions without undergoing substitutions.[167]

Reactions of poly(vinyl chloride) with aromatic amines, amino alcohols, or aliphatic amines in DMF solution result in both substitutions and in eliminations.[168] Reactions with aniline yield the following structure[168]:

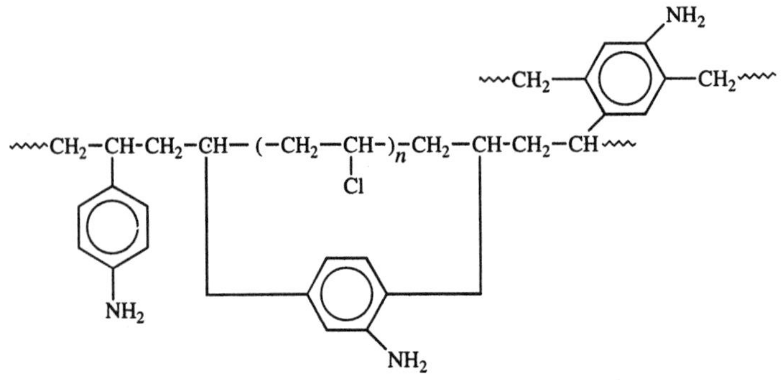

Carbanionic reaction sequences can be used to introduce various photosensitive groups.[169] The process consists of creating carbanionic centers on the backbone and then reacting them with various halogenated derivatives:

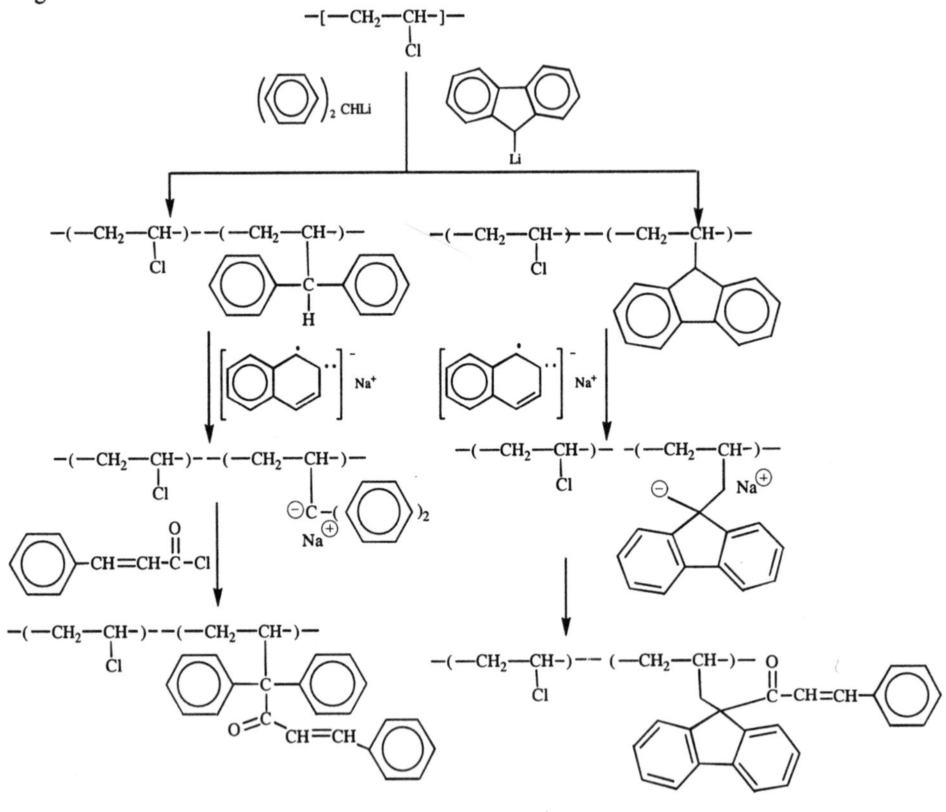

A redox cyclopentadienyl iron moiety can also be introduced into the poly(vinyl chloride) backbone by a similar technique.[170] Many other attempts were reported at replacing the halogens of poly(vinyl chloride), poly(vinyl bromide), and poly(vinyl iodide) with an alkali metal or with a hydrogen. For instance, in an effort to form poly(vinyl lithium), the polymers were reacted with organolithium compounds and with metallic lithium. The reactions with alkyllithium, however, resulted in substitutions by the alkyl groups, similarly to the reactions shown previously[171]:

$$-[-CH_2-CH(Br)-]- \; + \; C_4H_9Li \longrightarrow -[-CH_2-CH(C_4H_9)-]- \; + \; LiBr$$

Reactions with metallic lithium lead to formations of polyenes.[171] On the other hand, when poly(vinyl chloride) is reacted with metal hydrides, like lithium aluminum hydride in a mixture of tetrahydrofuran and decalin at 100 °C, macroalkanes form[172]:

$$-[-CH_2-CH(Cl)-]- + LiAlH_4 \longrightarrow -[-CH_2-CH_2-]- + LiCl$$

Replacement of the chlorine with *N,N*-dialkyldithiocarbamate was reported to occur at 50–60 °C in DMF solvent[173]:

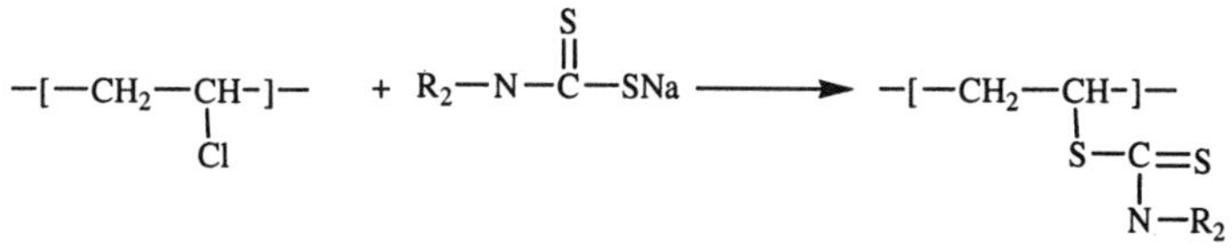

$$-[-CH_2-CH(Cl)-]- + R_2-N-C(=S)-SNa \longrightarrow -[-CH_2-CH(S-C(=S)-N-R_2)-]-$$

The reaction is catalyzed with ethylene diamine.[174]

8.4.3. Substitution Reactions of Polymers with Aromatic Rings

There are some interesting reports in the literature on reactions carried out on the backbones of polystyrenes. There are also many reports in the literature on aromatic substitution reactions of polystyrene. Only a few, however, are in industrial practice.

8.4.3.1. Halogenation Reactions of Polystyrene

Photochlorination of polystyrene involves replacement of hydrogens at the α and β positions.[175,176] It was believed in the past that the chlorine atoms react preferentially at the α position until the chlorine content of the product reaches 20% by weight. After that, it was thought that the chlorines are introduced into the other position. Later, however, this was contradicted.[177] In fact, when polystyrene is photochlorinated in carbon tetrachloride at low temperatures, like 13 °C, it is substituted equally at both positions. At higher temperatures, like 78 °C, substitutions at the β position actually predominate.[177]

Chlorinations of poly(*p*-methyl styrene) are somewhat more selective for the pendant methyl groups and result in di- and tri-substitutions at the *p*-methyl position. Only small amounts of chlorine are introduced into the polymer backbones.[178] Substitutions at the backbones, however, are possible with the use of SO_2Cl_2 as the chlorinating agent. In this case, half of the chlorines still replace the methyl hydrogens, but the other half replace hydrogens on the backbone.

Free-radical additions of maleic anhydride to polystyrene backbones can be carried out with the help of either peroxides or ultraviolet light.[164] Approximately 2% of the anhydride can be introduced. If, however, the additions are carried out on α-brominated polystyrene, the anhydride content of the polymer can be raised to 15%[164]:

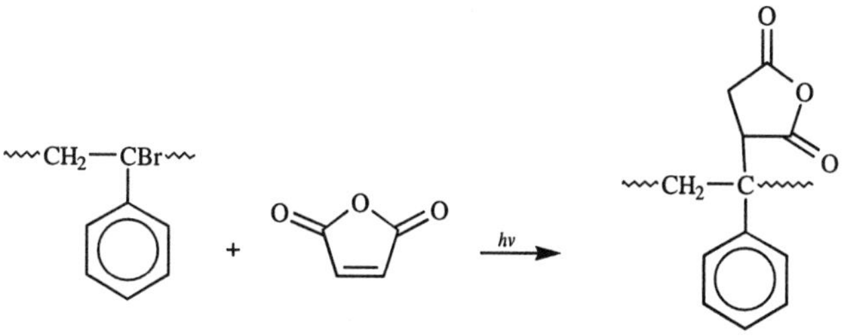

$$\sim CH_2-CBr(C_6H_5)\sim + \text{maleic anhydride} \xrightarrow{h\nu} \sim CH_2-C(C_6H_5)(\text{succinic anhydride})\sim$$

Note: The extra hydrogen shown above on the anhydride moiety of the product presumably comes from chain transferring.

Bromination of polystyrene with *N*-bromosuccinimide and benzoyl peroxide in CCl_4 at room temperature can achieve a 61% conversion in four hours. Considerable degradation, however, accompanies this reaction.[180]

8.4.3.2. Chloromethylation Reactions

Chloromethylation reactions of the aromatic rings of polystyrene and styrene copolymers are being carried out extensively. Chlorodimethyl ether (a carcinogen) is a good solvent for these polymers. It is therefore commonly employed as the reagent.[181,182] Laboratory preparations can be carried out in mixtures of carbon disulfide and ether, using zinc chloride as the catalyst. A 9-hour reaction at room temperature yields 10% substitution.[183] The chloromethylation process[184] occurs in two steps. Benzyl methyl ether forms as an intermediate. A crosslinking reaction between the aromatic nuclei and the formed CH_2Cl group occurs as a side reaction. There are strong indications that the chloromethylation takes place only at one position on the ring.[185] The same is true of bromomethylation.[185]

Stannic chloride is a very effective catalyst for this Friedel-Craft reaction.[186] Iodomethylation can also be carried out in the same manner with similar results.[179] When the reactions are carried out on crosslinked styrene copolymers with chlorodimethyl ether and stannic chloride catalyst, they are accompanied by strong morphological changes.[187] If these reactions are carried out with low levels of chloromethylating agents or catalysts, they occur more or less homogeneously. Larger levels of either of them, however, result not only in greater levels of chlormethylation, but also in higher degrees of secondary crosslinkings and in uneven distributions of the chloromethyl groups.[188]

Another technique of chloromethylating polystyrene is to react it with methylal and thionyl chloride in the presence of zinc chloride[189]:

$$-[-CH_2-\underset{\displaystyle C_6H_5}{CH}-]- + CH_3-O-CH_2-O-CH_3 + SOCl_2 \xrightarrow{ZnCl_2} -[-CH_2-\underset{\displaystyle C_6H_4CH_2Cl}{CH}-]- + (CH_3)_2-SO_3$$

Chloromethyl substituted polystyrenes can also be prepared from poly(*p*-methyl styrene)s by treating them with aqueous sodium hypochlorite in the presence of a phase transfer catalysts, like benzyltriethylammonium chloride.[190] The conversions of methyl to chloromethyl groups can be as high as 20% without any detectable formation of dichloromethyl groups. When, however, the reactions are pushed to conversions as high as 61%, some dichloromethyl groups form.

It should be noted that vinyl benzyl chloride monomer is available commercially. It is therefore possible to simply prepare the chloromethylated polystyrene or copolymers from the monomer without the chloromethylation reactions.

Techniques for chloromethylating polyarylether sulfones, polyphenylene oxide, phenolic resins, and model compounds were described recently.[191] When the subsequent products are converted to quaternary amines, there is a decrease in the quaternization rate with increase in degree of substitution. This may be due to steric effects imposed by restricted rotation of the polymeric chains.[191] This phenomenon was not observed in quaternization of poly(chloromethyl styrene). The chloromethylation reaction of a polysulfone with chloromethyl ether, catalyzed by stannic chloride, can be illustrated as follows:

$$\sim C_6H_4-SO_2-C_6H_4-O-C_6H_4-C_6H_4-O-C_6H_4-SO_2-C_6H_4\sim$$

$$\longrightarrow \sim C_6H_4-SO_2-C_6H_4-O-C_6H_3(CH_2Cl)-C_6H_3(CH_2Cl)-O-C_6H_4-SO_2-C_6H_4\sim$$

8.4.3.3. Reactions of Halomethylated Polymers

Many known reactions of the halomethyl groups on polymers are possible.[206,207] One can, for instance, convert poly(chloromethyl styrene) to poly(hydroxymethyl styrene).[183] Iodomethylated polystyrene can be treated with triethylphosphite in order to carry out an *Arbuzov* reaction[192]:

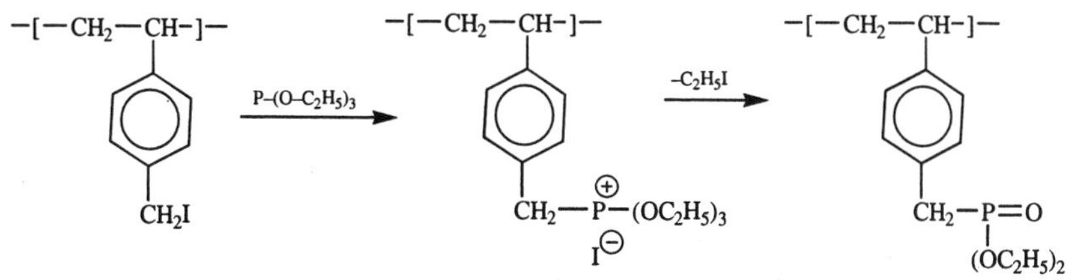

Chloromethylated polystyrene can also be converted to a phosphonium salt for use in the *Wittig* reaction[193]:

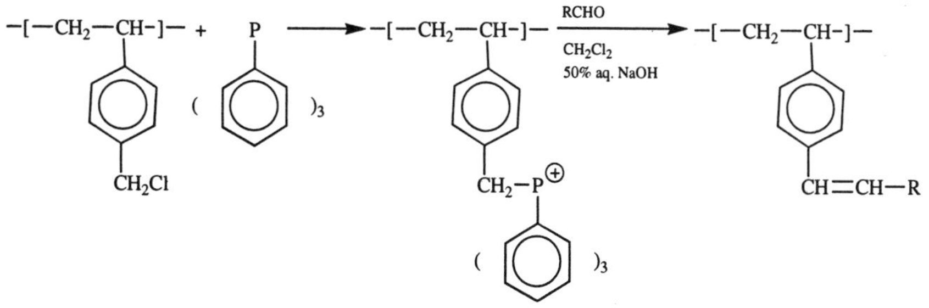

The product of a reaction of chloromethylated polystyrene and triphenylphosphine can also convert to nucleophiles.[194] In addition, use of a phase transfer catalyst converts soluble chloromethylated polystyrenes to phosphine oxides. Reactions with dioctylphosphine can serve as an example.[195] Sometimes, phase transfer reactions are easier to carry out than conventional ones. This is the case with a Witting reaction. Both linear and crosslinked chloromethylated polystyrenes react smoothly with triphenylphosphine to give derivatives that react with various aldehydes.[196,197] Phase transfer catalysts can also be used in carrying out nucleophilic substitutions with the aid of sulfides, like tetrahydrothiophine[198]:

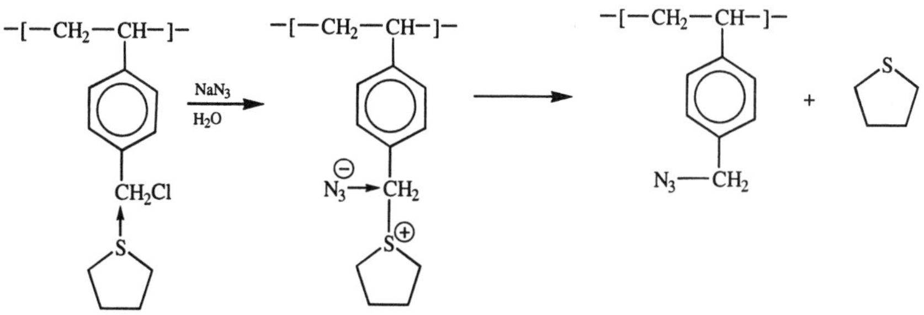

When chloromethylated crosslinked polystyrene is reacted with potassium superoxide, the yield depends upon the type of solvent used. In dimethylsulfoxide, in the presence of 18-crown-6 ether, the conversion to hydroxymethyl groups is 45%. In benzene, however, it is only 25%. High conversions are obtained by catalyzing the reaction with tetrabutylammonium iodide in a mixture of solvents. This results in 85% conversions to hydroxymethyl groups, while the rest become iodide groups.[199]

Quaternary salts are more effective than crown ethers in reactions with salts of oxygen anions, such as carboxylate and phenolate.[200] On the other hand, lipophilic crown ethers, like dicyclohexyl-18-crown-6, exhibit higher catalytic activity than the quaternary salts in reactions with salts of the sulfur anions. Also, the catalytic activity of the phase-transfer catalysts toward nitrogen anions is intermediate between that toward oxygen and that toward sulfur anions. Solid–liquid two-phase systems generally give higher degrees of conversion than do liquid–liquid systems. When, however, lipophilic phase-transfer catalysts are used with lipophilic reagents, high degrees of substitutions are achieved in liquid–liquid two-phase systems.[200]

Conversion of chloromethylated styrene to anionic exchange resins is done commercially by amination reactions to form quaternary ammonium groups.[201,202] These reactions can be illustrated as follows[203]:

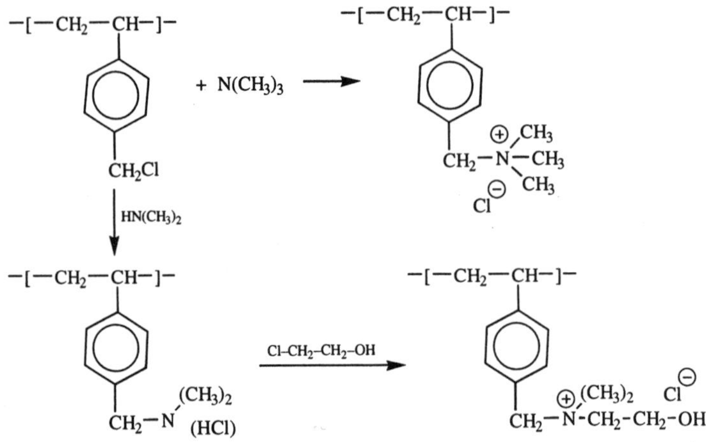

The kinetics of amination of chloromethylated polystyrene with monohydroxydialkyl tertiary amines show that the reactions proceed in two steps, at two different rates. The rate changes take place at conversions of 45–50%.[205] These rates are favorably influenced by increases in the dielectric constants of the solvents.[204] Two different rate constants also exist in reactions with 3-alkylamino-propionitrile.

8.4.3.4. Friedel-Craft Alkylation Reactions

Friedel-Craft acylations of polystyrene can be carried out in CS_2 or in CCl_4 at reflux temperatures of the solvents. The yields are high[209]:

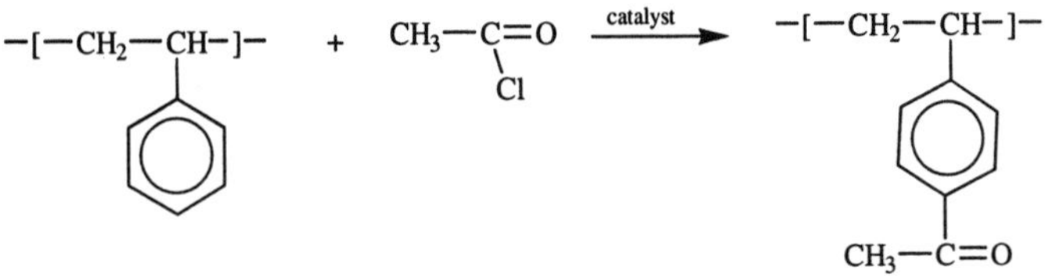

Chloromethylated copolymers of styrene with divinyl benzene undergo Friedel-Craft-type reactions in condensations with 1-phenyl-3-methylpyrazolone or with 1-phenyl-2,3-dimethylpyrazolone-5 in the presence of either $ZnCl_2$, BF_3, or $SnCl_4$[208]:

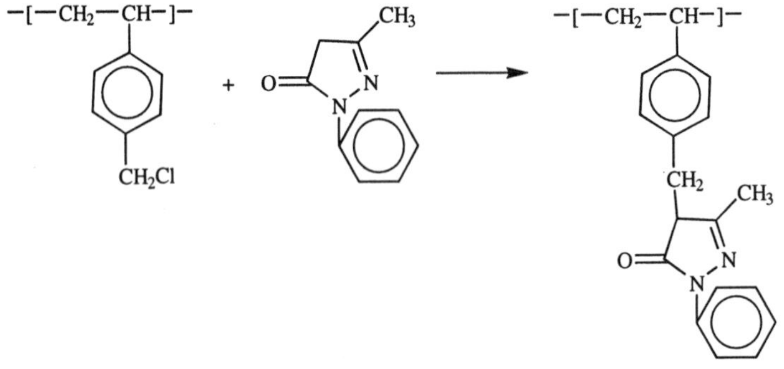

Polystyrene can also crosslink by a Friedel-Craft reaction[210]:

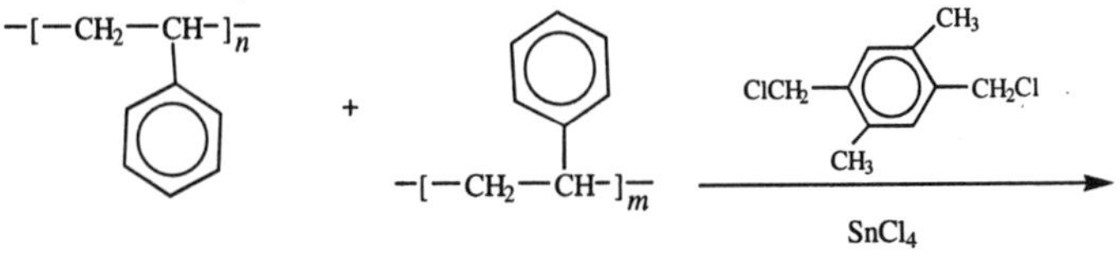

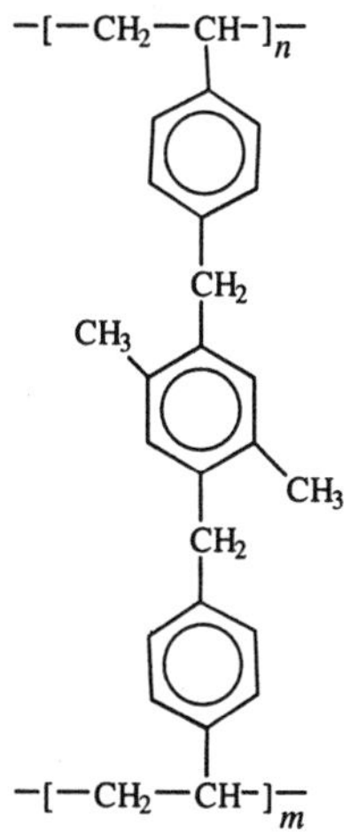

$-[-CH_2-CH-]_n-$

CH_2

CH_3

CH_3

CH_2

$-[-CH_2-CH-]_m-$

The above condensation takes place in dichloroethane, with stannic chloride catalyst at 50 °C.[210] The maximum reaction rate varies with both the initial concentration of 1,4-dimethyl-2,5-dichloromethylbenzene, shown above, and the initial concentration of $SnCl_4$. Crosslinked polystyrene particles or beads also form by Friedel-Craft suspension crosslinking of polystyrene with 1,4-dichloromethyl-2,5 dimethylbenzene.[211] The polymer is dissolved in nitrobenzene and a two-phase reaction occurs in 70% by weight of an aqueous suspension of $ZnCl_2$. Poly(vinyl alcohol) can be used as the suspending agent.

8.4.3.5. Sulfonation Reactions

Sulfonation reactions of polystyrene and its copolymers with divinyl benzene are carried out commercially to prepare ion exchange resins. Partial sulfonations of polystyrenes are achieved in the presence of ethers. When more than 50% of the aromatic rings are sulfonated, the polymers become water soluble. At lesser amounts of sulfonation (25–50%), the polymers are solvent soluble.[212,213]

When polystyrene is sulfonated in chlorinated hydrocarbons with a complex of dioxane-SO_3, the polymer precipitates from solution at low concentrations.[214,215] Complexes of ketones with SO_3 can also be used to sulfonate polystyrene in halogenated solvents.[216] The ratio of sulfonation is more favorable for poly(vinyl toluene) than it is for polystyrene at the same conditions.[217] Also, sulfur dioxide swells polystyrene. The polymer can be sulfonated in this medium with sulfur trioxide or with chlorosulfonic acid.[218] Polystyrene, sulfonated in CS_2 with aluminum chloride catalyst, is water insoluble in a free acid form.[219]

8.4.3.6. Nitration, Reduction, and Diazotization

Nitration of polystyrene was originally carried out a long time ago.[220] A nitrating mixture of nitric and sulfuric acids dissolves the polymer and a nitro derivative forms at 50 °C within three hours[221]:

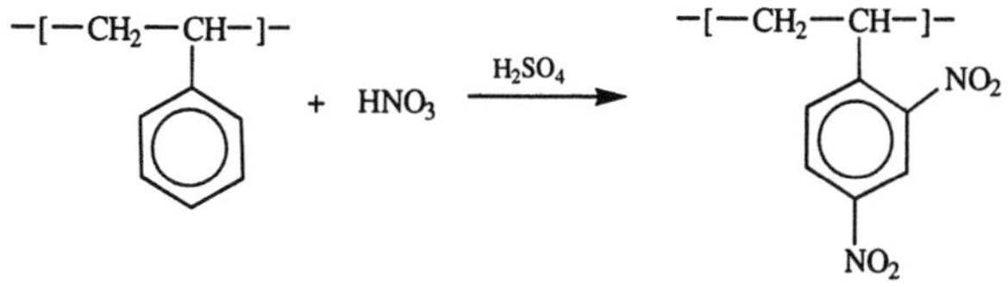

$-[-CH_2-CH-]-$ + HNO_3 $\xrightarrow{H_2SO_4}$ $-[-CH_2-CH-]-$ (NO_2, NO_2)

The reaction is accompanied by a loss of molecular weight. Nitration of isotactic polystyrene yields a more crystalline product (about 1.6 NO_2/ring) than the parent compound.[222] Here too, however, a loss in molecular weight accompanies the reaction.[223] Polystyrene can be nitrated under mild conditions using acetyl nitrate. The product contains approximately 0.6 nitro groups per each benzene ring.[224]

The nitro groups of polynitrostyrene are reduced by phenyl hydrazine that acts as a hydrogen donor[225]:

$$-[-CH_2-CH(C_6H_4NO_2)-]- + 3\,C_6H_5-NH-NH_2 \longrightarrow -[-CH_2-CH(C_6H_4NH_2)-]- + 3\,C_6H_6 + 3N_2 + H_2O$$

Polyaminostyrene can undergo typical reactions of aromatic amines, such as diazotization.[226] The diazonium salt decomposes with ferrous ions to yield polymeric free-radicals:

$$-[-CH_2-CH(C_6H_4NH_2)-]- \xrightarrow[H^{\oplus}]{HNO_3} -[-CH_2-CH(C_6H_4N_2^{\oplus}\,Cl^{\ominus})-]- \xrightarrow{Fe^{\oplus\oplus}} -[-CH_2-CH(C_6H_4^{\bullet})-]-$$

8.4.3.7. Metalation Reactions

Functional polystyrene derivatives are starting materials for further reactions in many multistep syntheses. An example is a metalation of polystyrenes for use as intermediates[227]:

$$-[-CH_2-CH(C_6H_4I)-]- \xrightarrow{C_4H_9Li} -[-CH_2-CH(C_6H_4Li)-]- \xrightarrow{R(R')C=O} -[-CH_2-CH(C_6H_4-C(R)(R')-OH)-]-$$

Some other reactions of lithiated polystyrene are[227]:

$$-[-CH_2-CH(C_6H_4-C(R)=O)-]- \xleftarrow{RCN} -[-CH_2-CH(C_6H_4Li)-]- \xrightarrow{(C_6H_5)_2-PCl} -[-CH_2-CH(C_6H_4-P(C_6H_5)_2)-]-$$

When polystyrenelithium is aminated by a reagent prepared from methoxyamine and methyllithium, two reaction mechanisms are possible. One may proceed via nitrene intermediates and the other via electophilic nitrenium ions.[228] Many other reactions of polystyrenelithium can be found in the literature.[229–232] Sodium metalated polystyrene reacts in a similar manner[228]:

$$-[-CH_2-CH(C_6H_4Cl)-]- + 2\,[C_{10}H_8]^{\bullet -}\,Na^{+} \longrightarrow -[-CH_2-CH([C_6H_4]^{\ominus}\,Na^{\oplus})-]- \xrightarrow{Cl-Si(CH_3)_2-CH_2Cl} -[-CH_2-CH(C_6H_4-Si(CH_3)_2-CH_2Cl)-]-$$

Polyiodostyrene is a good starting material for many other reactions. Some of them are[233,234]:

$$-[-CH_2-CH(C_6H_4I)-]- + 2\,CH_3-COOH \longrightarrow -[-CH_2-CH(C_6H_4-\overset{\oplus}{I}(O-CO-CH_3))-]-\ \ CH_3-CO-O^{\ominus}$$

$$-[-CH_2-CH(C_6H_4I)-]- \xrightarrow{Cl_2} -[-CH_2-CH(C_6H_4ICl_2)-]- + K_2CO_3 \longrightarrow -[-CH_2-CH(C_6H_4IO)-]-$$

Poly(2,6-dimethyl-1,4-phenylene oxide) can be brominated with *N*-bromosuccinimide.[235] The product can also subsequently be used for further reactions,[235] such as phosphorylation with triethylphosphite:

$$-[-C_6H_2(CH_3)_2-O-]- \xrightarrow[CCl_4]{NBS} -[-C_6H_2(CH_2Br)(CH_3)-O-]- \longrightarrow -[-C_6H_2(CH_2-P(=O)(OC_2H_5)_2)(CH_3)-O-]-$$

8.4.4. Reactions of Acrylic, Methacrylic, and Related Polymers

The functional groups of polymers from acrylic and methacrylic esters can undergo all the typical reactions of such groups. There are therefore numerous reports in the literature on such reactions.

8.4.4.1. Reduction of the Ester Groups

Perhaps the most reported reactions of these polymers are *reductions* of the functional groups. Among them is the reaction with lithium aluminum hydride to reduce the ester groups. The success, however, depends upon the reaction medium. Poly(methyl methacrylate) is reduced to poly(methallyl alcohol) in ether solvents[236]:

$$-[-CH_2-C(CH_3)(C{=}O-O-CH_3)-]- + LiAlH_4 \longrightarrow -[-CH_2-C(CH_3)(CH_2-OH)-]-$$

The results, however, are inconclusive, because combustion analyses fail to match the theoretical composition for poly(methallyl alcohol). It is impossible to tell to what extent the reduction takes place.[236] Inconclusive results are also obtained in similar reductions of poly(methyl acrylate) in mixtures of tetrahydrofuran and benzene. When a product of such reduction is acetylated with acetic anhydride in pyridine[237] as follows:

$$-[-CH_2-CH(C{=}O-O-CH_3)-]- + LiAlH_4 \longrightarrow -[-CH_2-CH(CH_2-OH)-]-$$

$$-[-CH_2-CH(CH_2-OH)-]- + (CH_3-CO)_2O \underset{}{\overset{\text{pyridine}}{\rightleftharpoons}} -[-CH_2-CH(CH_2-O-C(=O)-CH_3)-]-$$

Even after hydrolysis in water, treatment with hot *m*-cresol, and extracted with hydrochloric acid to remove the suspended inorganic materials,[237] the product is still only soluble in pyridine and *m*-cresol.

Somewhat similar results are obtained in reductions of high molecular weight poly(methyl acrylate) with lithium aluminum hydride[238] in tetrahydrofuran. The reaction yields a product that is only soluble in mixtures of hydrochloric acid with either methyl alcohol, dioxane, or tetrahydrofuran. The problem is apparently due to some residual aluminum that is hard to remove.[239] If, however, the reduction is carried out in a *N*-methylmorpholine solution, followed by addition of potassium tartrate, a pure product can be isolated.[240] *N*-Methylmorpholine is a good solvent for reductions of various macromolecules with metal hydrides.[236] In addition, the solvent permits the use of strong NaOH solutions to hydrolyze the addition complexes that form. Other polymers that can be reduced in it are those bearing nitrile, amide, imide, lactam, and oxime pendant groups. Reduction of polymethacrylonitrile, however, yields a product with only 70% of primary amine groups.[241]

Complete reductions of pendant carbonyl groups with $LiAlH_4$ in solvents other than *N*-methylmorpholine, however, were reported. Thus, a copolymer of methyl vinyl ketone with styrene was fully reduced in tetrahydrofuran.[242]

Reductions with metal hydrides are often preliminary steps for additional reactions. For instance, a product of $LiAlH_4$ reduction of syndiotactic poly(methyl methacrylate) can be reacted with succinic anhydride and then converted to an amide[243]:

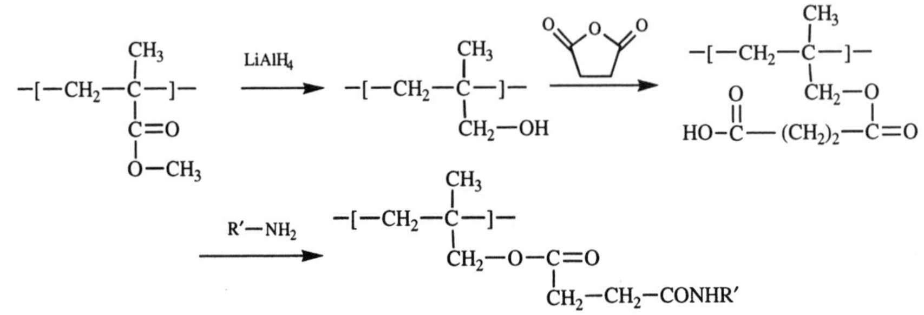

$$-[-CH_2-\underset{\underset{O-CH_3}{|}}{\overset{CH_3}{|}}{C}(C{=}O)-]- \xrightarrow{LiAlH_4} -[-CH_2-\overset{CH_3}{\underset{CH_2-OH}{C}}-]- \xrightarrow{\text{succinic anhydride}} -[-CH_2-\overset{CH_3}{\underset{CH_2-O-C(=O)-(CH_2)_2-C(=O)-OH}{C}}-]-$$

$$\xrightarrow{R'-NH_2} -[-CH_2-\overset{CH_3}{\underset{CH_2-O-C(=O)-CH_2-CH_2-CONHR'}{C}}-]-$$

Substituted succinic anhydride can be used as well. When poly(dimethyl itaconate) is reduced with $LiAlH_4$ in THF, the product contains some ash, but 93% of the functional groups are reduced[239]:

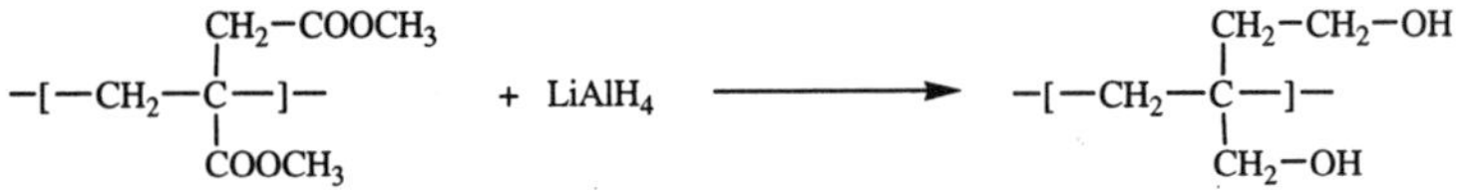

$$-[-CH_2-\overset{CH_2-COOCH_3}{\underset{COOCH_3}{C}}-]- + LiAlH_4 \longrightarrow -[-CH_2-\overset{CH_2-CH_2-OH}{\underset{CH_2-OH}{C}}-]-$$

8.4.4.2. Nucleophilic and Electrophilic Substitutions

Many other conversions of functional groups of acrylic and methacrylic resins were reported. One of them is a conversion of methyl acrylate to a hydrazide by a direct reaction with hydrazine[244]:

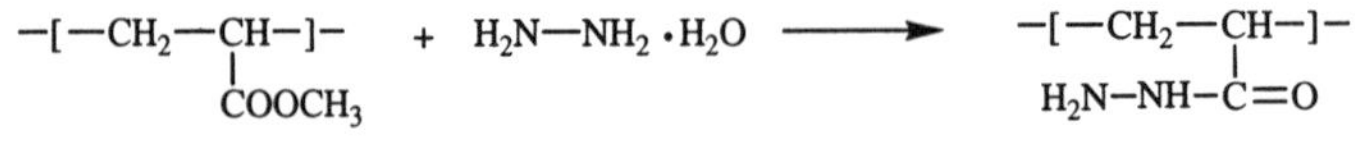

$$-[-CH_2-\underset{COOCH_3}{CH}-]- + H_2N-NH_2\cdot H_2O \longrightarrow -[-CH_2-\underset{H_2N-NH-C{=}O}{CH}-]-$$

The above reaction requires a 10:1 ratio of hydrazine to the ester groups. In the laboratory it can be carried out on a steam bath over a period of 2–3 hours. Approximately 60–80% of the ester groups convert.[244] The hydrazides can form various hydrazones through reactions with aldehydes and ketones:

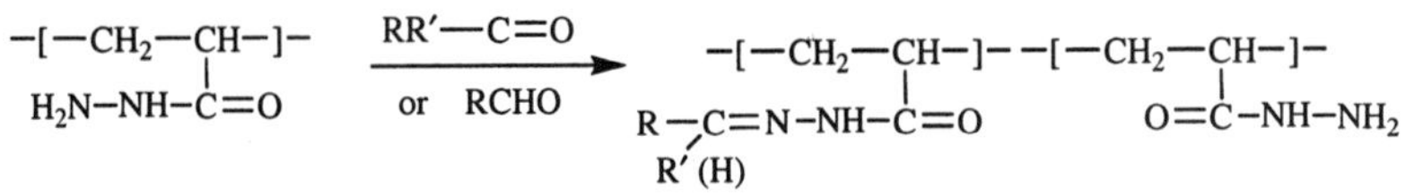

$$-[-CH_2-\underset{H_2N-NH-C{=}O}{CH}-]- \xrightarrow[\text{or RCHO}]{RR'-C{=}O} -[-CH_2-\underset{R(R'(H))C{=}N-NH-C{=}O}{CH}-]--[-CH_2-\underset{O{=}C-NH-NH_2}{CH}-]-$$

The hydrazide can also be converted to an azide[244]:

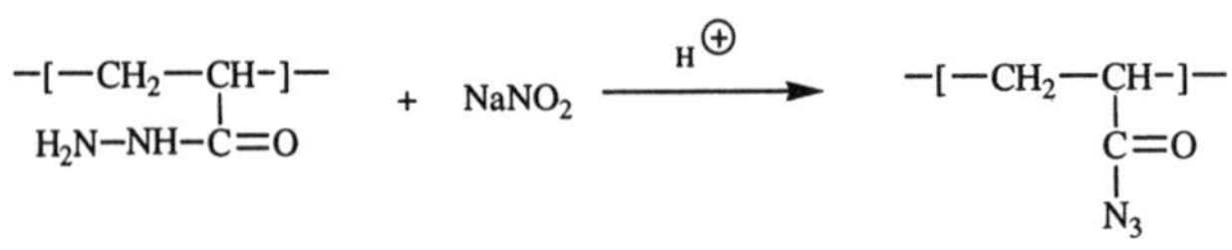

$$-[-CH_2-\underset{H_2N-NH-C{=}O}{CH}-]- + NaNO_2 \xrightarrow{H^{\oplus}} -[-CH_2-\underset{\underset{N_3}{|}}{CH}(C{=}O)-]-$$

A *Hoffman reaction* of the azide yields a crosslinked polymer.[244] Syndiotactic poly(methyl methacrylate) converts to a hydrazide in a similar manner.[245]

Nucleophilic substitution reactions can be carried out on poly(methyl methacrylate) with heterocyclic organolithium reagents.[246] The reactions are conducted in homogeneous solutions in tetrahydrofuran or in benzene combined with hexamethylphosphoramide. Copolymers will form with tautomeric keto-β-heterocyclic structures. The following heterocyclic reagents are useful[246]: 2-picolinyllithium, [(4,4-dimethyl-2-oxazole-2-yl)methyl]lithium, quinaldinyllithium, and [2-thiazole-2-yl-methyl]lithium.

In attempts to carry out *Arndt–Eister* reactions on poly(methacryloyl chloride), the polymer was reacted with diazomethane in various molar ratios and at different temperatures.[247] Initially, acid chloride groups do react with diazomethane as expected. The products, however, undergo subsequent reactions with neighboring acid chloride groups and form cyclic structures[247]:

$$-[-CH_2-C(CH_3)(C{=}O\,Cl)-CH_2-C(CH_3)(C{=}O\,Cl)-]- + CH_2N_2 \longrightarrow -[-CH_2-C(CH_3)(C{=}O\,CHN_2)-CH_2-C(CH_3)(C{=}O\,Cl)-]- \longrightarrow -[-CH_2-(\text{cyclic diketone ring with } CH_3, CH_3, =O, =O, =N_2)-]-$$

When *Curtius* and *Lossen* rearrangement reactions are attempted on poly(acryloyl chloride),[248] the products are fairly regular polyampholytes:

$$-[-CH_2-CH(C{=}O\,Cl)-CH_2-CH(C{=}O\,Cl)-]- \xrightarrow[2.\ HNO_2]{1.\ NH_2-NH_2} -[-CH_2-CH(C{=}O\,N{=}\overset{\oplus}{N}{=}\overset{\ominus}{N})-CH_2-CH(C{=}O\,OH)-]-$$

$$\longrightarrow -[-CH_2-CH(\overset{\oplus}{N}H_3)-CH_2-CH(C{=}O\,O^{\ominus})-]-$$

Somewhat similar results are obtained with a *Hoffman reaction* on polyacrylamide.[249] A *Schmidt reaction* on poly(acrylic acid) also yields mixed results.[250] When it is run in acetic acid the intermolecular reactions appear to predominate over the intramolecular ones. Also, the products formed in acetic acid have a higher nitrogen content that those formed in dioxane.[250] The NMR spectra show the presence of some acid anhydride groups. This has an additional effect of lowering the yield.

Diels-Alder reactions can be carried out on poly(furfuryl methacrylate) with dienophiles like maleic anhydride or a maleimide.[252] Dilute solutions (10%) of the polymers in benzene can be used, requiring up to 30% molar excess of the dienophiles:

$$-[-CH_2-C(CH_3)(C{=}O-O-CH_2-\text{furyl})-]- + \text{maleic anhydride} \longrightarrow -[-CH_2-C(CH_3)(O{=}C-O-CH_2-\text{oxabicyclic anhydride adduct})-]-$$

The additions take place at room temperature and the reactions take from 7–30 days[252] to complete.

Polyacrylonitrile reacts with hydroxylamine and the product can be metalated by elements from Group IV (Sn, Ge, Si). This is a convenient route to formation of polymers with pendant organometallic compounds[253]:

$$-[-CH_2-CH(C{\equiv}N)-]- + H_2NOH \longrightarrow -[-CH_2-CH(H_2N-C{=}NOH)-]- \xrightarrow{ClSnR_3} -[-CH_2-CH(H_2N-C{=}N-O-SnR_3)-]-$$

The yields are high when the reactions are carried out on dilute solutions of polyacrylonitrile in dimethylformamide at 75 °C. The solutions must contain 1.5 moles each of hydroxylamine hydrochloride and sodium carbonate per mole of acrylonitrile groups.[253]

A *Ritter* reaction can be carried out on atactic polyacrylonitrile with *N*-hydroxymethylamides of acetic, benzoic, and benzene-sulfonic acids.[254] When the same reactions are carried out with *N*-hydroxymethylimides of succinic or phthalic acids in tetramethylene sulfone, there is a stronger tendency toward crosslinking.

Copolymers of methacroyl chloride will undergo an *Arbuzov rearrangement* in reactions with triethyl phosphite in dimethylformamide, dioxane, or benzene at 75 °C.[255] The conversions are high, ranging between 96–98%.

The aldehyde groups of polyacrolein can be reduced by the *Meerwein–Ponndorf reaction.* There is a limit, however, to the amount of alcoholate that can be used and to the concentrations of free aldehyde groups in the staring material.[256] Also, ester condensations take place (*Tischenko* reaction) at the same time as the reductions occur.[256]

$$-[-CH_2-\underset{\underset{H}{\diagdown}\!\!\!\!\!\underset{|}{C}=O}{CH}-]- \; + \; (H-\underset{CH_3}{\overset{CH_3}{\underset{|}{\overset{|}{C}}}}-O)_3-Al \longrightarrow -[-CH_2-\underset{CH_2OH}{\underset{|}{CH}}-]- \; + \; CH_3-\overset{O}{\overset{||}{C}}-CH_3$$

8.4.5. Substitution Reactions of Poly(vinyl alcohol)

There are many practical uses for products from reactions of poly(vinyl alcohol). Among them are commercial preparations of poly(vinyl acetal)s formed through condensations with aldehydes. Two materials that are currently being marketed are poly(vinyl formal) and poly(vinyl butyral). The first is formed from partially hydrolyzed poly(vinyl acetate) that is dissolved in aqueous acetic acid and excess formaldehyde. The mixture is heated, sulfuric acid is added, and the reaction is allowed to proceed at 70–90 °C for 6 hours. Sulfuric acid is then neutralized and the formal precipitates out (see also Chapter 5).

Two different industrial processes are used for preparations of the butyral. In both of them acetate-free poly(vinyl alcohol) is used. In the first one, 10% solutions of the starting material are treated with butyraldehyde and sulfuric acid. The mixtures are heated to 90 °C for one and a half hours and the products precipitate. They are neutralized, washed, and dried. In the second one poly(vinyl alcohol) is suspended in ethanol/ethyl acetate, and butyraldehyde together with a strong mineral acid are added. The solutions are then neutralized. The butyrals separate out. They are neutralized and the resin washed and dried.

If poly(vinyl alcohol) films are reacted with formaldehyde in water containing salt and an acid catalyst (heterogeneous formalization), crosslinking occurs. The number of the crosslinks increases with decreasing acid concentration and fixed amounts of formaldehyde and salt.[257]

Direct reactions of poly(vinyl alcohol) with aldehydes in the *Kornblum reaction* result in formations of acetals that also contain residual hydroxyl group and often acetate groups. The acetate groups can be there from incomplete hydrolysis of the parent poly(vinyl acetate) that was used to form the poly(vinyl alcohol). Reactions of poly(vinyl alcohol) with ketones yield similar ketals. At present no ketals are offered commercially.

Alkyl etherifications of poly(vinyl alcohol) occur when the polymer is combined with *n*-alkyl halides in dimethylsulfoxide combined with pyridine.[258,259] It was suggested that the alkyl halides convert to aldehydes and acids and then act as intermediates in the dimethylsulfoxide-pyridine solution[258,259]:

$$RCH_2Br \; + \; CH_3-\overset{O}{\overset{||}{S}}-CH_3 \xrightarrow{\text{pyridine}} RCHO \; + \; CH_3-\overset{O}{\overset{||}{S}}-CH_3 \; + \; HBr$$

Many modifications of poly(vinyl alcohol) were carried out to form photosensitive materials. Thus, unsaturation was introduced into the pendant groups for photocrosslinking. One example is a condensation with pyridinium and quinolinium salts[260]:

$$-[-CH_2-\underset{OH}{CH}-CH_2-\underset{OH}{CH}-]- + H-C(OR)_2-CH_2-C_6H_4-CH{=}CH-C_5H_4N^{\oplus}-CH_3 \longrightarrow -[-CH_2-\text{(1,3-dioxane ring, O, O)}-]- \text{ with } -CH_2-C_6H_4-CH{=}CH-C_5H_4N^{\oplus}-CH_3$$

The material cyclodimerizes on exposure to light.[261] (See Section 8.5 for additional discussion of this subject.)

Schotten–Baumann esterifications of poly(vinyl alcohol) are used extensively in preparations of various derivative. The reactions appear to proceed well when acid chlorides are employed in two-phase systems.[262] The polymers are dissolved in water and the solutions are blended in 1:1:1 equal volumes with NaOH solutions and cyclohexanone. They are then mixed thoroughly with solutions of the acid chlorides in mixtures of cyclohexanone and toluene. The reaction mixture is stirred vigorously for about 90 minutes at –5 to 5 °C to obtain the desired product.

Metalation of poly(vinyl alcohol) is used to form ether derivatives[263]:

$$-[-CH_2-\underset{OH}{CH}-]- + [CH_3-\overset{O}{\overset{\|}{S}}-CH_2]^{\ominus} Na^{\oplus} \longrightarrow -[-CH_2-\underset{ONa}{CH}-]- + (CH_3)_2-S{=}O$$

$$-[-CH_2-\underset{ONa}{CH}-]- + Br-(CH_2)_x-CH_3 \longrightarrow -[-CH_2-\underset{O-(CH_2)_x-CH_3}{CH}-]- + NaBr$$

The *Ritter reaction* on poly(vinyl alcohol) yields soluble products. Only some of the hydroxy groups, however, are converted to amide structures[264]:

$$-[-CH_2-\underset{OH}{CH}-CH_2-\underset{OH}{CH}-]- \xrightarrow[H_2SO_4]{RCN} -[-CH_2-\underset{O=C-N(R)-H}{CH}-CH_2-\underset{OH}{NH}-]-$$

Also, it is possible that neighboring group interactions may lead to cyclizations and formations of 1,3-oxazines.[264]

8.4.6. Miscellaneous Exchange Reactions

Many miscellaneous exchange reactions are reported in the literature.[249–265] A few are presented here. One such reaction is reduction of pendant carbonyl groups of poly(vinyl methyl ketone) with metal hydrides[242,265]:

$$-[-CH_2-\underset{\underset{R}{C=O}}{CH}-]- \xrightarrow[\text{or } KBH_4]{LiAlH_4} -[-CH_2-\underset{\underset{R}{H-C-OH}}{CH}-]-$$

Another is introduction of unsaturation into pendant groups by a Wittig reaction on pendant carbonyls[266]:

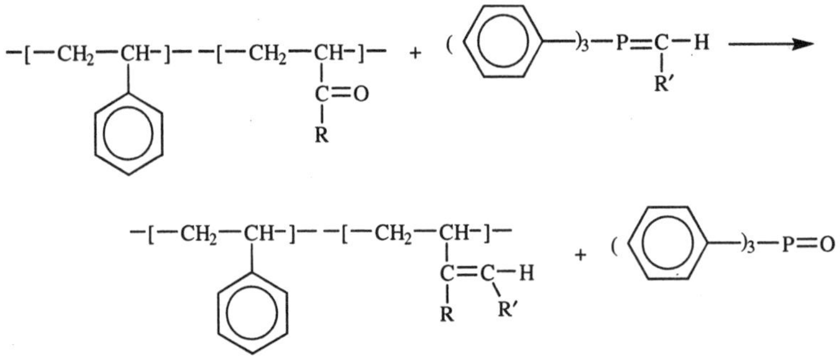

$$-[-CH_2-CH(C_6H_5)-]-[-CH_2-CH(C(=O)R)-]- + (C_6H_5-)_3-P=CH-R' \longrightarrow$$

$$-[-CH_2-CH(C_6H_5)-]-[-CH_2-CH(C(R)=CH-R')-]- + (C_6H_5-)_3-P=O$$

The same reaction can also be carried out on a copolymer of ethylene and carbon monoxide.[266]

Polycaprolactam can be treated with either SO_2Cl_2, $POCl_3$, or PCl_5 at 70 °C to introduce ionic chlorine groups.[267] The main product is poly(α,α-dichlorocaprolactam):

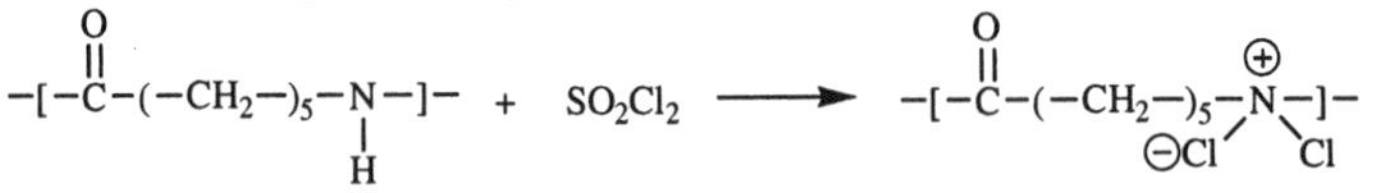

$$-[-C(=O)-(-CH_2-)_5-N(H)-]- + SO_2Cl_2 \longrightarrow -[-C(=O)-(-CH_2-)_5-N^{\oplus}(Cl)-]- \ \ominus Cl$$

Also, when Nylon 6,6 is reacted with trifluoroacetic anhydride, trifluoroacetyl nylon forms[268]:

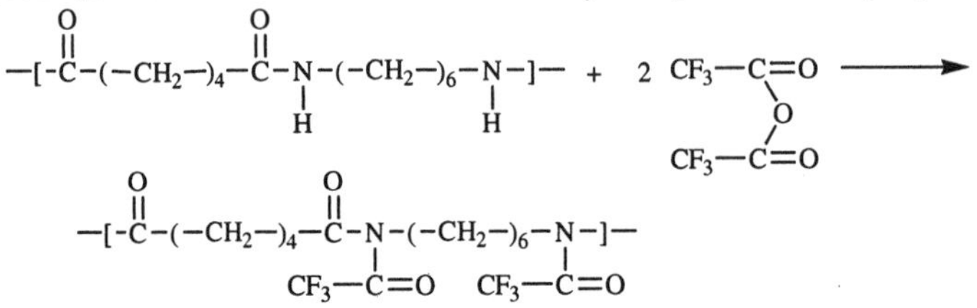

$$-[-C(=O)-(-CH_2-)_4-C(=O)-N(H)-(-CH_2-)_6-N(H)-]- + 2\ (CF_3-C=O)_2O \longrightarrow$$

$$-[-C(=O)-(-CH_2-)_4-C(=O)-N(C(=O)-CF_3)-(-CH_2-)_6-N(C(=O)-CF_3)-]-$$

Syndiotactic poly(2-methallyl hydrogen phthalate) can be amidated according to the following scheme[269]:

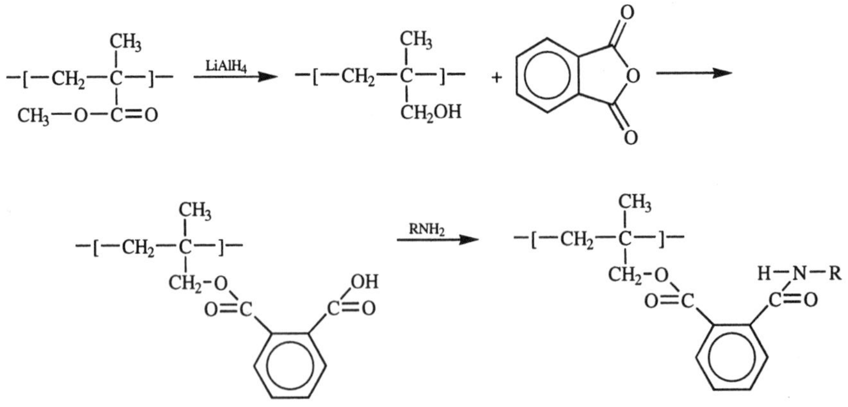

$$-[-CH_2-C(CH_3)(C(=O)-O-CH_3)-]- \xrightarrow{LiAlH_4} -[-CH_2-C(CH_3)(CH_2OH)-]- + \text{phthalic anhydride} \longrightarrow$$

$$-[-CH_2-C(CH_3)(CH_2-O-C(=O)-C_6H_4-C(=O)OH)-]- \xrightarrow{RNH_2} -[-CH_2-C(CH_3)(CH_2-O-C(=O)-C_6H_4-C(=O)-NH-R)-]-$$

Dimethyl sulfate alkylates almost quantitatively sterically hindered aromatic poly(pyridine ether)s and poly(pyridine ether sulfone)s in nitrobenzene.[270] The reaction can be illustrated as follows:

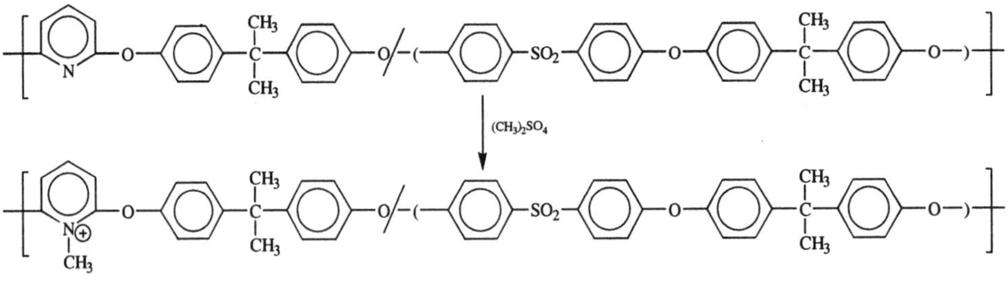

$$\left[-C_5H_3N-O-C_6H_4-C(CH_3)_2-C_6H_4-O-/-(-C_6H_4-SO_2-C_6H_4-O-C_6H_4-C(CH_3)_2-C_6H_4-O-)-\right]$$

$$\downarrow (CH_3)_2SO_4$$

$$\left[-C_5H_3N^{\oplus}(CH_3)-O-C_6H_4-C(CH_3)_2-C_6H_4-O-/-(-C_6H_4-SO_2-C_6H_4-O-C_6H_4-C(CH_3)_2-C_6H_4-O-)-\right]$$

A 20% mole excess of the alkylating reagent is required and the reaction must be conducted for 6 hours at 80 °C.

Dichloroketene, generated by the ultrasound-promoted dechlorination reaction of thrichloroacetyl chloride with zinc, adds to the carbon–carbon double bonds of poly(methyl-1-phenyl-1-silane-*cis*-pent-3-ene).[271] This can be illustrated as follows:

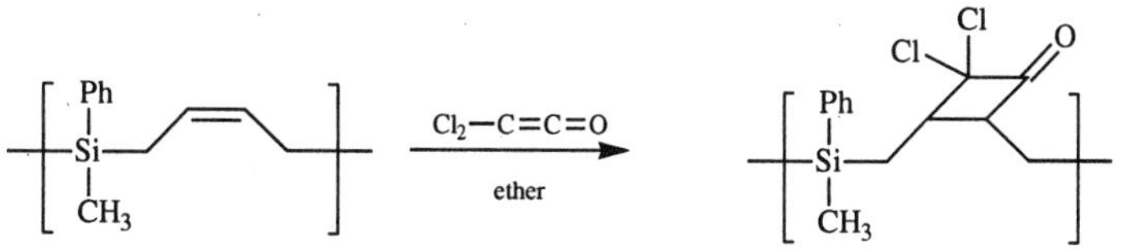

$$\left[-Si(Ph)(CH_3)-CH_2-CH=CH-CH_2-\right] \xrightarrow[\text{ether}]{Cl_2-C=C=O} \left[-Si(Ph)(CH_3)-CH_2-\underset{\text{cyclobutanone, }CCl_2,\ C=O}{CH-CH}-CH_2-\right]$$

8.5. Crosslinking Reactions of Polymers

These reactions are quite numerous and have been utilized for a long time. They also include all thermosetting processes of polymers. Many are discussed in previous chapters.

8.5.1. Vulcanization of Elastomers

Crosslinking of natural rubber was discovered by Goodyear back in 1839. Sulfur, which was the original crosslinking agent, is still utilized today in many processes. Early studies demonstrated that the crosslinks are mainly polysulfides:

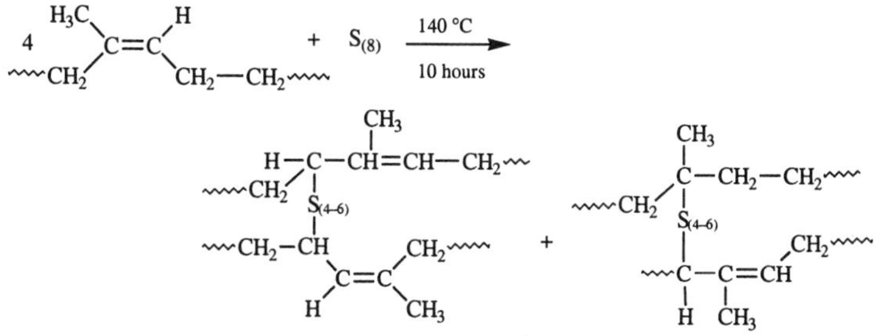

$$4\ \sim CH_2(H_3C)C=CH-CH_2-CH_2\sim \ + \ S_{(8)} \xrightarrow[\text{10 hours}]{140\ ^\circ C}$$

$$\sim CH_2-\underset{S_{(4-6)}}{CH(H)}(CH_3)-CH=CH-CH_2\sim \ \text{linked via } S_{(4-6)} \text{ to } \sim CH_2-CH-C(H)=C(CH_3)-CH_2\sim \ + \ \sim CH_2-C(CH_3)(S_{(4-6)})-CH_2-CH_2\sim \ \text{linked to } \sim C(H)-C(CH_3)=CH-CH_2\sim$$

The reactions take place at all temperatures, but industrially they are carried out from 50–75 °C and above. At lower temperatures, however, the process may take days to complete. At temperatures of 135–155 °C approximately 8% of sulfur (by weight of rubber) reacts.[272] Also, sulfur dissolves in unvulcanized rubber even at room temperature. The overall mechanism of the reaction is still being studied. Most evidence points to an ionic mechanism and a sulfonium ion intermediate.[272] It was shown[273] that a straightforward reaction of sulfur with rubber is insufficient. Somehow, between 40–100 atoms of sulfur must be combined in order to obtain one crosslink. Out of 40–100 atoms only 6–10 are actually engaged in the formation of the crosslinks. The rest of the atoms form cyclic sulfide units that become spread along the main chain.[273]

To improve the efficiency of the vulcanization reaction, various accelerators were developed. Among them are zinc oxide combined with fatty acids, and/or amines. Zinc oxide forms zinc mercaptides like $(XS)_2ZnL_2$, where X is an electron-withdrawing substituent and L is a ligand from a carboxyl or an amine group. The function of the ligand is to render the complex soluble. The mercaptide complexes are believed to react with additional sulfur to form zinc perthiomeraptides.

The accelerators that are most commonly used are derivatives of 2-mercaptobenzothiazole. They are very effective when used in combinations of metal oxides with fatty acids (referred to as *activators*). The favorite activators are zinc oxide combined with stearic acid. The combinations permit rapid vulcanizations that take minutes compared to hours when sulfur is used alone. In the process of vulcanization 2,2′-dithiobisbenzthiazole forms initially and then reacts with sulfur to form polysulfides[273]:

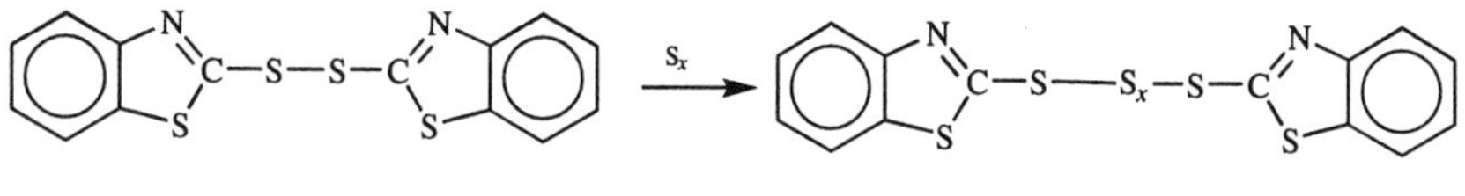

$$\text{BT}-C-S-S-C-\text{BT} \xrightarrow{S_x} \text{BT}-C-S-S_x-S-C-\text{BT}$$

The products from reactions with sulfur in turn react with natural or synthetic rubber at any allylic hydrogen. This is a concerted reaction that results in the formation of sulfur-containing adducts of the polymers:

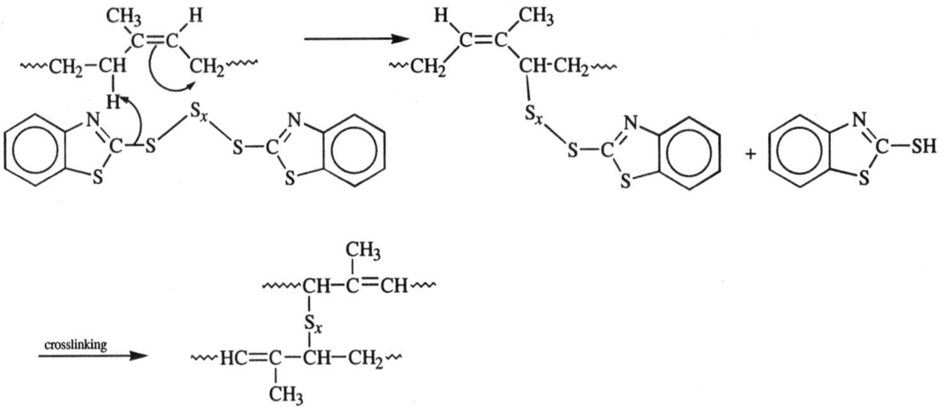

crosslinking

Once the crosslinks are formed, further transformations takes place. Some of them consist of reactions that shorten the polysulfide links:

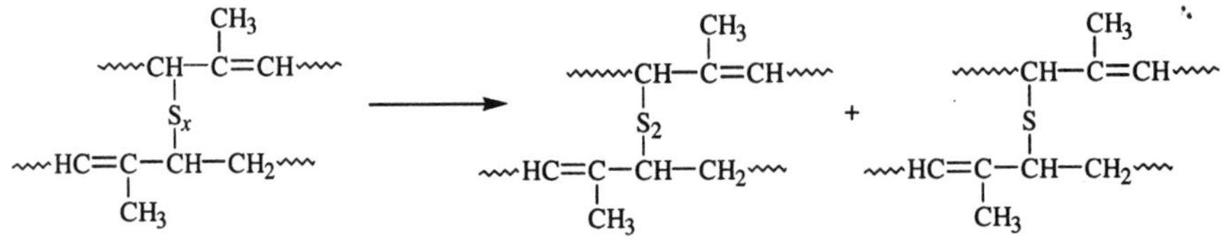

Also, some crosslinks are lost through elimination reactions:

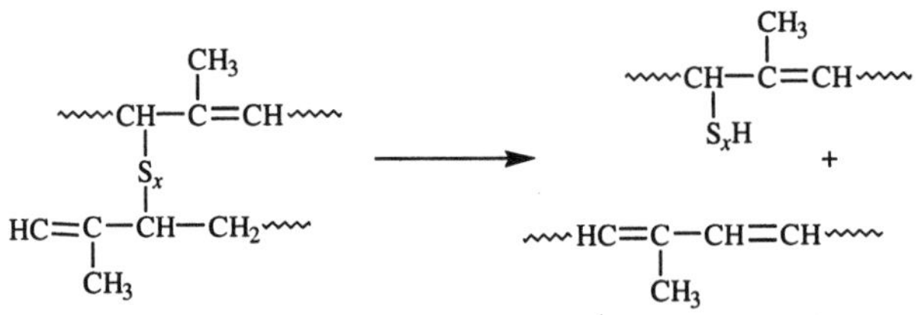

In addition, some cyclic sulfur compounds form in the process[273]:

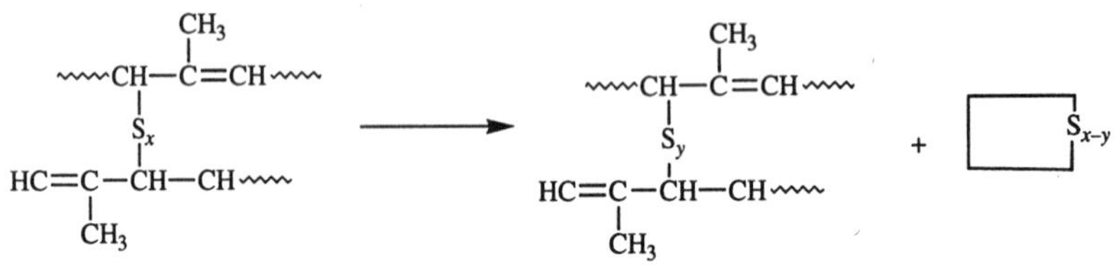

8.5.2. Crosslinking With the Aid of Peroxides

This can be used with many polymers such as polyolefins, polymers of dienes, and others. The reactions with natural rubber can be illustrated as follows[274]:

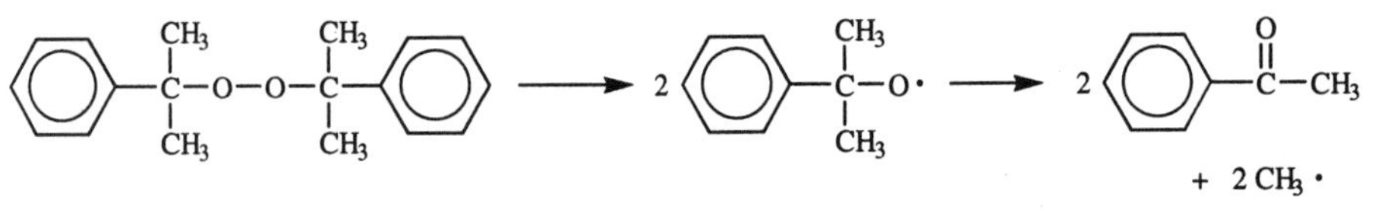

$$+\ 2\,CH_3\cdot$$

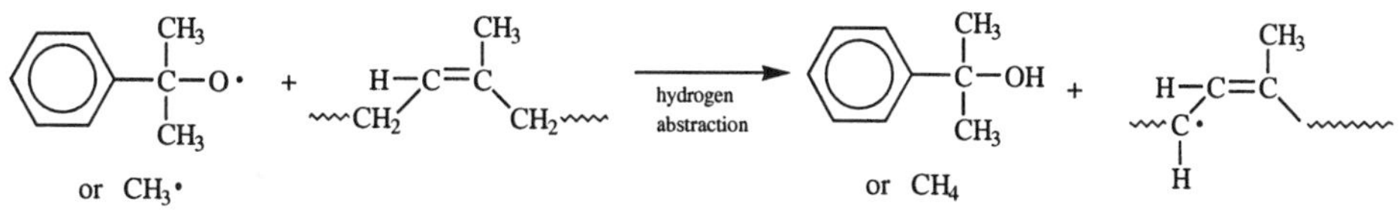

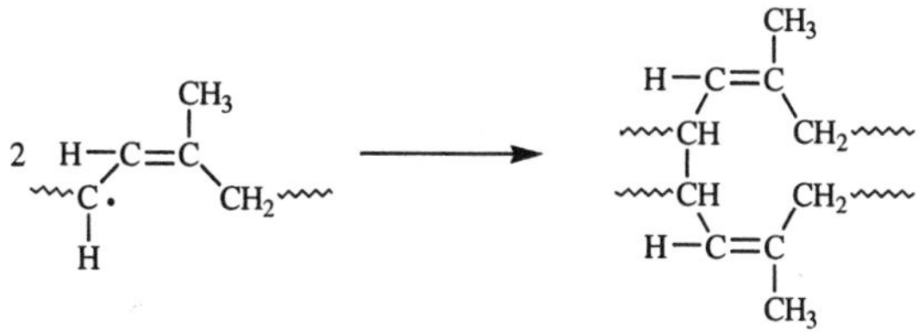

$$C_6H_5{-}C(CH_3)_2{-}O\cdot \text{ (or } CH_3\cdot) + \sim CH_2{-}C(CH_3){=}CH{-}CH_2\sim \xrightarrow[\text{abstraction}]{\text{hydrogen}} C_6H_5{-}C(CH_3)_2{-}OH \text{ (or } CH_4) + \sim \dot{C}H{-}CH{=}C(CH_3)\sim$$

$$2\ \sim \dot{C}H{-}CH{=}C(CH_3){-}CH_2\sim \longrightarrow \begin{matrix} H{-}C{=}C(CH_3){-}CH_2\sim \\ \sim CH \\ | \\ \sim CH \\ H{-}C{=}C(CH_3){-}CH_2\sim \end{matrix}$$

The extent of the crosslinking, as shown above, is not clear. It is known, however, that cleavage reactions, that are followed by free-radical recombinations, can take place[274]:

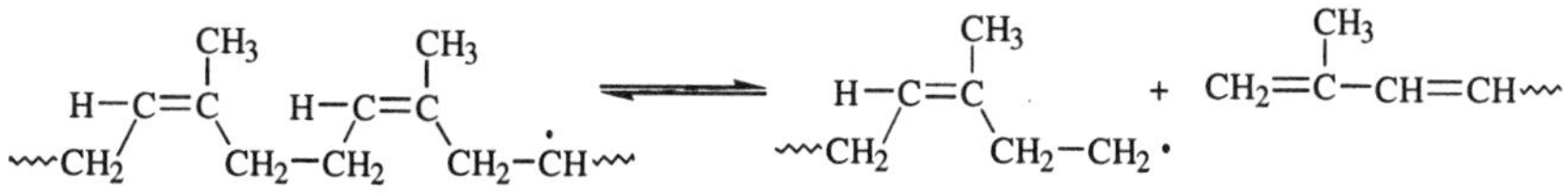

$$\sim CH_2{-}CH{=}C(CH_3){-}CH_2{-}CH_2{-}CH{=}C(CH_3){-}CH_2{-}\dot{C}H\sim \rightleftharpoons \sim CH_2{-}CH{=}C(CH_3){-}CH_2{-}CH_2\cdot + CH_2{=}C(CH_3){-}CH{=}CH\sim$$

Polymeric chains, bearing free radicals, combine with each other to give branched structures. Additions of chains with a free radical to double bonds result in formations of crosslinks.[274]

8.5.3. Miscellaneous Crosslinking Reactions of Polymers

Many miscellaneous crosslinking reactions of polymeric materials are reported in the literature. For instance, poly(acryloyl chloride) can be crosslinked with diamines[275]:

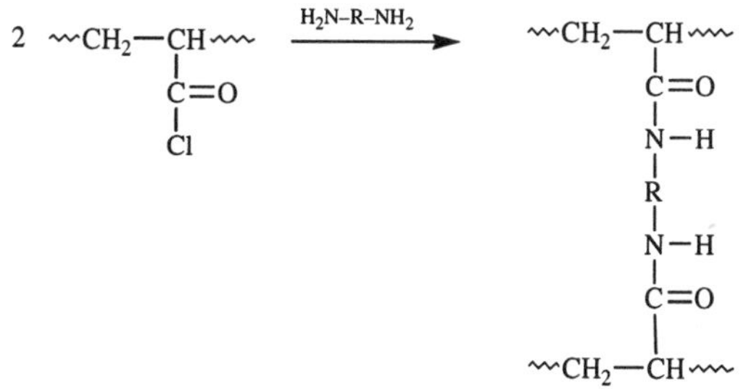

$$2\ \sim CH_2{-}CH(C{=}O{-}Cl)\sim \xrightarrow{H_2N{-}R{-}NH_2} \sim CH_2{-}CH\sim{-}C(=O){-}NH{-}R{-}NH{-}C(=O){-}\sim CH_2{-}CH\sim$$

In a similar manner, polymers with pendant chlorosulfonate groups crosslink when reacted with diamines or with glycols.[275]

8.5.4. Radiation Crosslinking

This industrial technology includes crosslinking of polymeric materials with the aid of ultraviolet light, ionizing radiation, or electron beams. All three processes are used industrially in many applications.

Both electron beams and ultraviolet light initiate free-radical polymerizations with very low activation energy. This allows high polymerization rates at room temperature because the rates are not temperature dependent. Once initiated, free-radical polymerizations follow typical paths. It is, however, peculiar to radiation curing of coating materials that the gel states form at very early stages of the reactions. This is due to extensive use of polyfunctional monomers and prepolymers. In fact, it was demonstrated that the gel points occur at around 5% conversion of the prepolymers in typical commercial formulation, yet conversions to about 63% of the prepolymers to polymers are required to obtain dry films.[299]

Ultraviolet light initiated polymerizations are precise reactions that require a match between the wavelength of the light source and the light absorbing and reacting molecule. A direct relationship exists between the chemical structure of a particular molecule and the light energy that it is capable of absorbing. The effect of accelerated electrons from electron beams upon matter, however, is not that precise and varies considerably with the energy of the beam. If it is too high, the electrons tend to pass through without causing any reactions. Yet, if the energy is low enough, the electrons may be completely absorbed and high concentrations of initiating radicals may form. If they are absorbed and the electrons impinge upon matter, ionized tracks form along the trajectories. This is a result of formation of ion pairs. These ion pairs consist of heavy organic positive ions and less energetic secondary electrons that form when the electrons are displaced from the molecular orbitals. The ion pairs have strong interaction cross-sections with the materials and cause formations of secondary ionized tracks in which disturbed systems try to restore their equilibrium. Attempts to restore the equilibrium result in ruptures of chemical bonds, promotions of electrons into higher excited levels, and formations of many "hot radicals." These initiate polymerizations of vinyl monomers.[299] After the free-radical polymerizations have been initiated, whether photochemically or by electron beams, the reactions of propagation and termination follow very similar paths.

8.5.5. Photocrosslinking of Polymers

This is actually an ancient art that dates back to the days of the Babylonians who photocrosslinked pitch for decorative purposes. History aside, however, Minsk *et al.*[276] may be the first chemists who synthesized a photocrosslinkable polymer, poly(vinylcinnamate). The crosslinking reactions can be illustrated as follows:

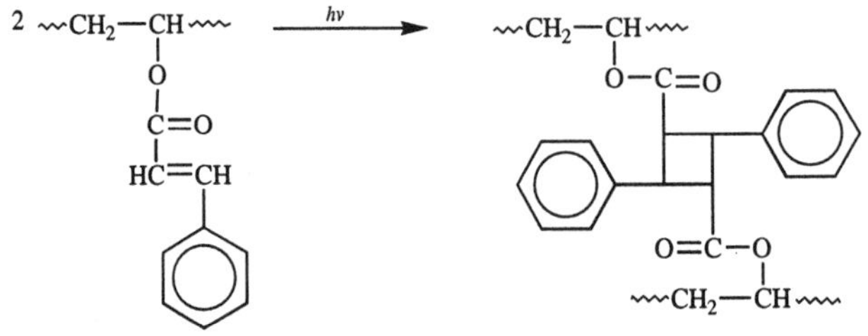

$$2\ \sim CH_2-CH(-O-C(=O)-HC=CH-C_6H_5)\sim \xrightarrow{h\nu} \text{cyclobutane dimer: } \sim CH_2-CH(-O-C=O)\sim \ \ / \ \ O=C-O-\ \sim CH_2-CH\sim$$

Today, photosensitive polymers are utilized in many technologies from photoresists to microelectronic components, miniature and integrated circuits, photoengraving, precision chemical milling, and formation of protective and decorative coatings. Many photosensitive polymers are simply formed by attaching photosensitive groups to existing known polymeric materials. Other approaches involve special syntheses of macromolecules from photosensitive monomers. It also includes formations of photosensitive polyfunctional oligomers for photocrosslinking reactions.

A detailed discussion of various photoreactions belongs in a book on photochemistry. The following are merely a few brief comments on the background of the reactions that occur in photocrosslinkings to aid those not familiar with the subject.

When a quantum of light interacts with a molecule in the ground state, the energy is absorbed by that molecule and an electron is promoted to a higher energy level. The molecule in the process attains an excited state. This phenomenon obeys four laws of organic photochemistry, as was stated by Turro[394]:

1. Photochemical changes take place only because light is absorbed by the system.
2. One photon or one quantum of light activates only one molecule.
3. Each quantum or photon that is absorbed by a molecule has a given probability of populating either the singlet state or the lowest triplet state.
4. In solution, the lowest excited singlet and triplet states are the starting points for the photochemical processes.

The relationship between the amount of light or the number of photons absorbed, and the number of molecules that as a result undergo a reaction, is defined as the *quantum yield.* After a molecule (designated below as A) absorbs a photon and reaches an excited state, it may dissipate the extra energy and return to the ground state by one of three processes[394]:

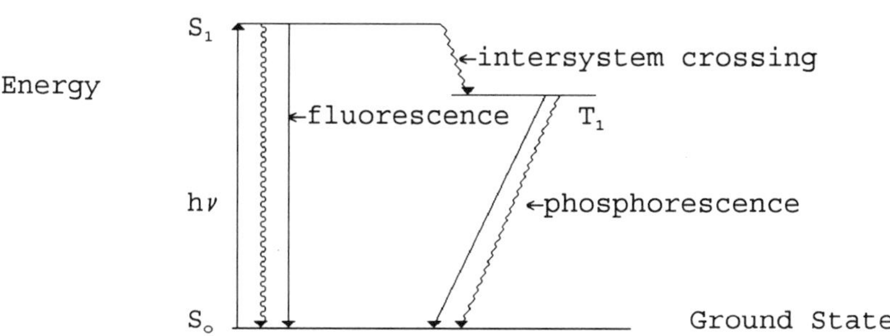

FIGURE 8.1. Energy diagram: So represents the ground state while S_1 and T_1 represent the singlet and triplet states, respectively.

1. $A + h\nu \rightarrow A^* \rightarrow A + h\nu'$
2. $A + h\nu \rightarrow A^* \rightarrow A + \text{heat}$
3. $A + h\nu \rightarrow A^* \rightarrow \text{Chemical Products}$

In process 1 the molecules return from the excited singlet to the ground state through light emission in the form of *fluorescence*. In the ground state an electron is usually paired off with another one that has an opposite spin. During the excitation, one of the paired electrons is promoted to a higher energy level while still maintaining the direction of its spin. This is an excited *singlet state*. Instead of returning to the ground state via fluorescence, however, the molecule might lose only part of the energy of excitation. In the process it undergoes an intersystem crossing, reverses the spin of the electron, and becomes an excited *triplet*. Two electrons in the triplet state have the same spin and the molecule can act as a diradical. The lifetime of the excited triplet state is very much longer than that of the excited singlet state. This intersystem crossing is shown in process 2. No absorption or emission of light radiation accompanies the conversion. Instead, the excess vibrational energy is dissipated rapidly as heat. The molecule can also return to the ground state from the triplet state through light emission in the form of phosphorescence. A very simplified energy diagram may illustrate the process, as shown in Fig. 8.1. Process 3 can lead to many reactions, such as isomerizations.

There are at least five different known ways that photoexcited molecules can generate free radicals:

1. Through transferring the energy to monomer molecules. This can, however, lead to an easy return to the ground state without initiating any reactions.
2. By cleaving into radicals.
3. Through additions to olefinic bonds.
4. By acting as donor molecules and transferring electrons to neighboring molecules. This leads to formations of pairs of ion radicals.
5. By abstracting hydrogen from solvent, monomer, or any other molecule present in the reaction mixture. This is accompanied by formation of two radicals.

8.5.5.1. Photocrosslinking of Polymers with Light-Sensitive Groups

The photocrosslinking reactions of poly(vinyl cinnamate), shown earlier, are reversible when irradiated with light of proper wavelength. Some intramolecular cyclizations take place as well.[277] Nevertheless, these reactions found many commercial applications and led to the development of other materials with similar photosensitive groups. All of them photocrosslink in a similar manner. Some of these groups are shown below[278]:

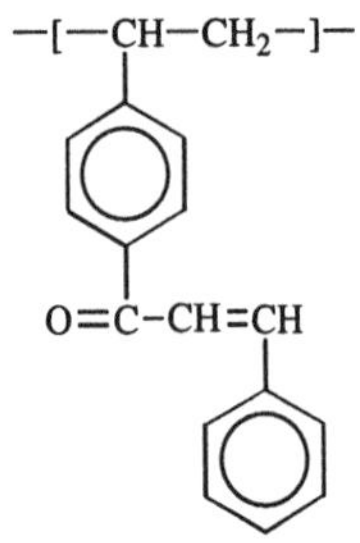

chalcone 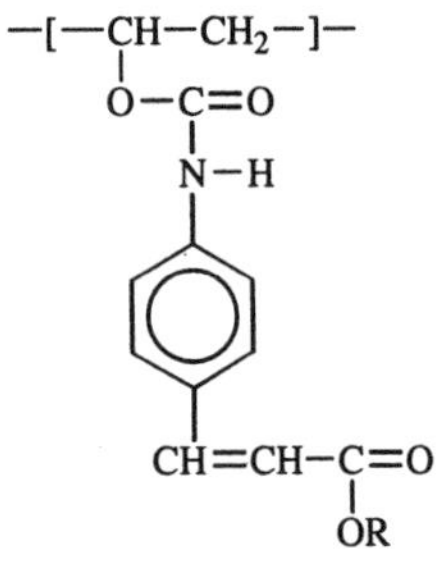   cinnamate urethane 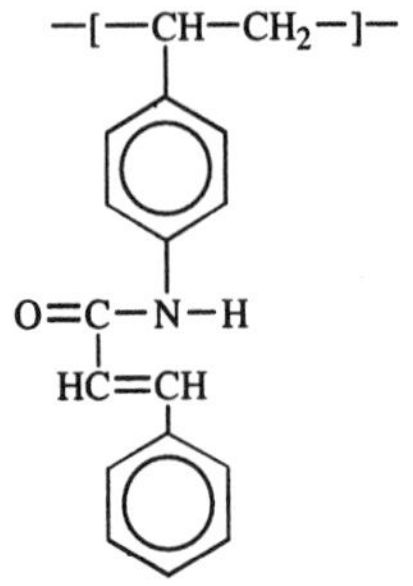   cinnamide

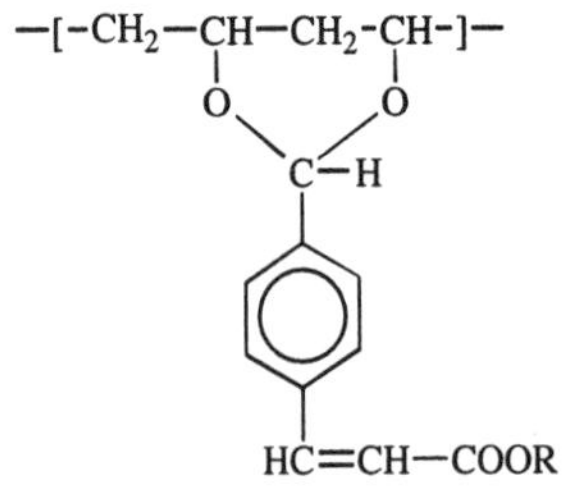

acetal 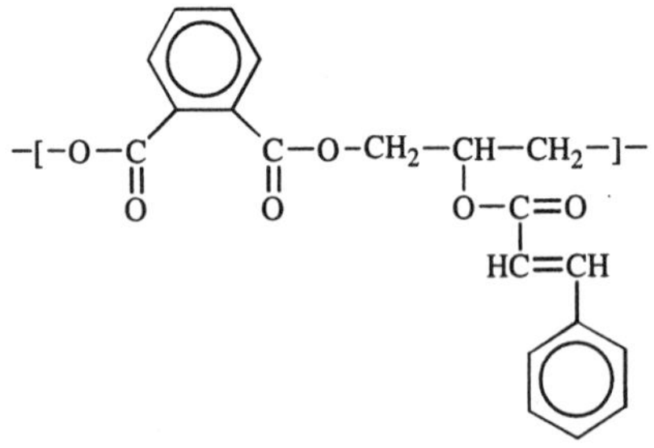   polyester

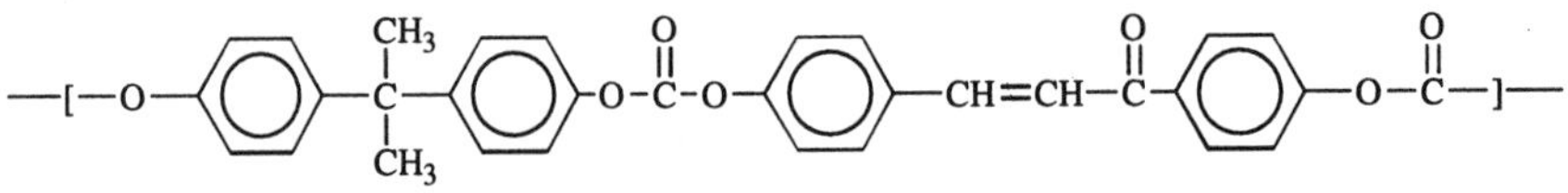

polycarbonate

The dimerizations take place upon irradiation with light of a wavelength longer than 300 nm.[279–282] The dimers, however, dissociate back to the monomeric forms upon irradiation with light of 254 nm.[283]

The rates of the crosslinking reactions can be greatly accelerated by addition of photosensitizers. They are molecules that, after absorption of light and attaining the triplet state, transfer the energy to other molecules. The effectiveness of sensitizers in accelerating the reactions of poly(vinyl cinnamate) is illustrated in Table 8.1.

Another photosensitive group is the azide. It photodecomposes into a very reactive nitrene group and nitrogen.[284–286] When it is irradiated with light of appropriate wavelength, a very efficient intersystem crossing occurs. The dissociation of the molecule follows almost every transition from an excited ($n \rightarrow \pi^*$) state to a high vibrational level of the ground state.[284-286]

Introduction of a pendant azide group into the polymer is possible in a variety of ways and many papers were published describing different approaches. One example is the introduction of the group into novolacs[287]:

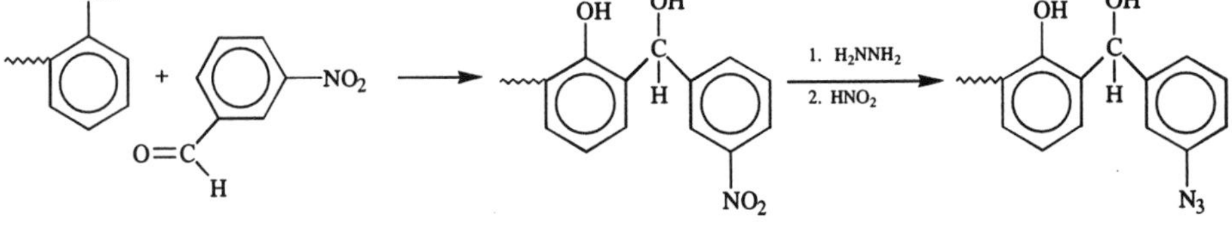

1. H_2NNH_2
2. HNO_2

Another example is formation of acetals by reacting formyl-1-naphthyl azide with poly(vinyl alcohol)[287]:

Table 8.1. Effect of Sensitizers on Relative Speed of Crosslinking of Poly(vinyl cinnamate)[395]

Sensitizer	Relative speed
—	1
Naphthalene	3
Benzanthrone	7
Phenanthrene	14
Chrysene	18
Benzophenone	20
Anthrone	31
5-Nitroacenaphthene	84
4-Nitroaniline	100
2-Nitrofluorene	113
4-Nitromethylaniline	137
4-Nitrobiphenyl	200
Picramide	400
4-Nitro-2,6-dichloro-dimethylaniline	561
Michler's ketone	640
4-Nitro-2,6-dibromo-dimethylaniline	797
N-Acyl-4-nitro-1-naphthylamine	1100

$$H-\overset{O}{\overset{\|}{C}}-C_{10}H_6-NH_2 \longrightarrow H-\overset{O}{\overset{\|}{C}}-C_{10}H_6-N_3 \xrightarrow{\sim CH_2-CHOH\sim} \sim CH_2-CH-CH_2-CH\sim \text{ (O, O linked via } C-H \text{ to } C_{10}H_6-N_3)$$

Preparations of azide derivatives from styrene-maleic anhydride copolymers, cellulose, and gelatin by attaching aromatic azide compounds[288] are described in the literature. Most of the resultant polymers crosslink rapidly when exposed to light of 260 mμ wavelength. Also, as much as 90% of the hydroxy groups of poly(vinyl alcohol) can be esterified with *p*-azido-benzoyl chloride[289]. These reactions must be carried out in mixtures of chloroform and aqueous sodium hydroxide. Based on infrared spectroscopy, the following crosslinking mechanism was proposed[289]:

$$2\ \sim CH_2-CH(-O-C(=O)-C_6H_4-N_3)\sim \xrightarrow{h\nu} 2\ \sim CH_2-CH(-O-C(=O)-C_6H_4-\dot{N}\cdot)\sim + N_2 \longrightarrow \sim CH_2-CH(-O-C(=O)-C_6H_4-N{=}N-C_6H_4-C(=O)-O-)CH-CH_2\sim$$

The preparations and properties of many polymers containing pendant aryl azide groups were described by Delzenne and London.[290] There were attempts made to combine azide groups with cinnamoyl groups in one pendant substituent on polymers[291]:

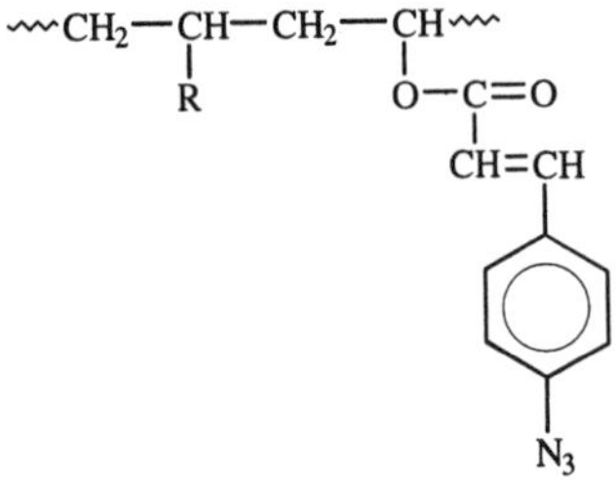

Addition of the azide groups, however, does little to increase bond activity. Also, marked wavelength dependence on $\pi \rightarrow \pi^*$ and $n \rightarrow \pi^*$ transitions occurs in both functional groups, the azide and the cinnamoyl.[291]

Polymers bearing pendant furan groups are also photosensitive and some were synthesized, for instance[292]:

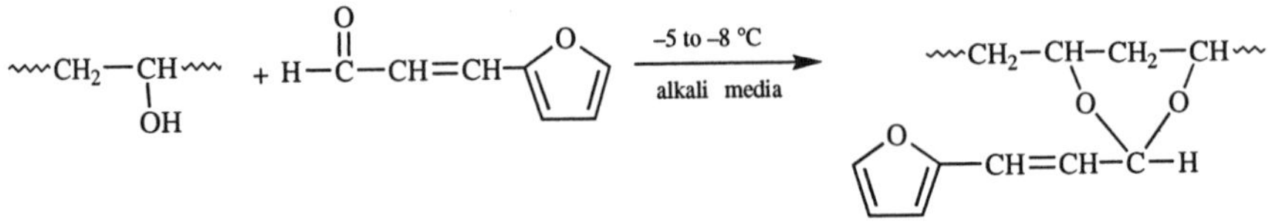

–5 to –8 °C

alkali media

The above polymer is more photosensitive than poly(vinyl cinnamate). The proposed mechanism of photocrosslinking, however, is similar.[292] Other attempts at forming photosensitive materials include attaching photoinitiators to macromolecules, such as benzoin or furoin[293]:

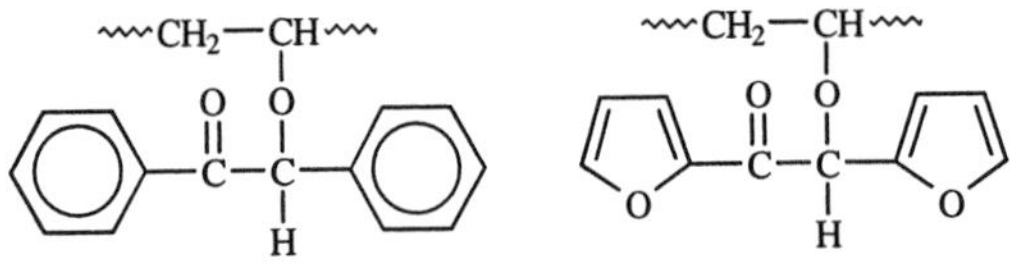

poly(benzoin acrylate) poly(furoin acrylate)

A few other, similar polymers are shown below.[294] One is formed by reacting benzoin with a styrene-maleic anhydride copolymer:

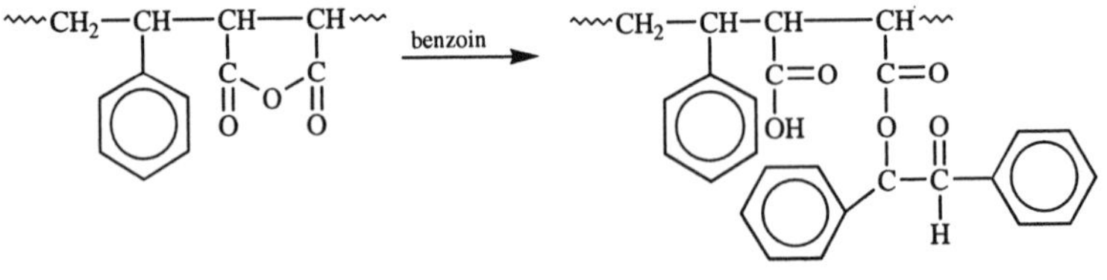

benzoin

Another is formed by reacting poly(vinyl alcohol) with α-chloro-deoxybenzoin:

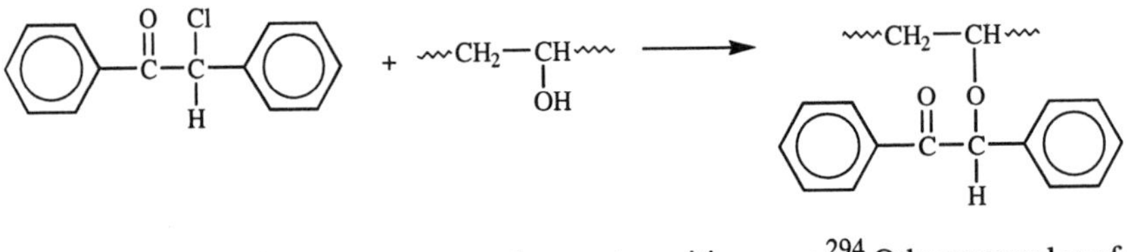

Furoin substituted polymers are among the most sensitive ones.[294] Other examples of photosensitive polymers are macromolecules with pendant episulfide groups.[295] They also crosslink rapidly upon irradiation:

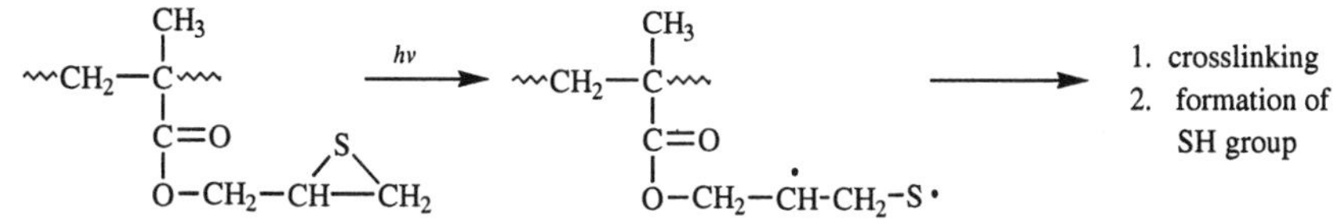

1. crosslinking
2. formation of SH group

Currently, industry uses many polymers that crosslink mostly by $2\pi + 2\pi$ type dimerizations with formations of cyclobutane rings. This is like the crosslinking reaction of the cinnamate group.[301,302] Such materials are used in photoresists, photolithography, and in many ultraviolet light curable compositions intended for other uses.[300] One example is poly(naphthyl vinyl acrylate). The crosslinking can be illustrated as follows:

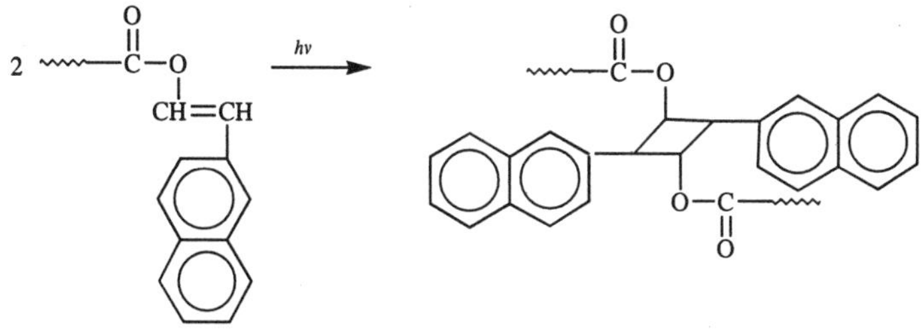

The naphthalenes become bonded to cyclobutane rings in 1,2 and 1,3 positions, on the same side of the cyclobutane ring. Some other functional groups that also photocrosslink by $2\pi + 2\pi$ addition when attached to polymers are[296–299]:

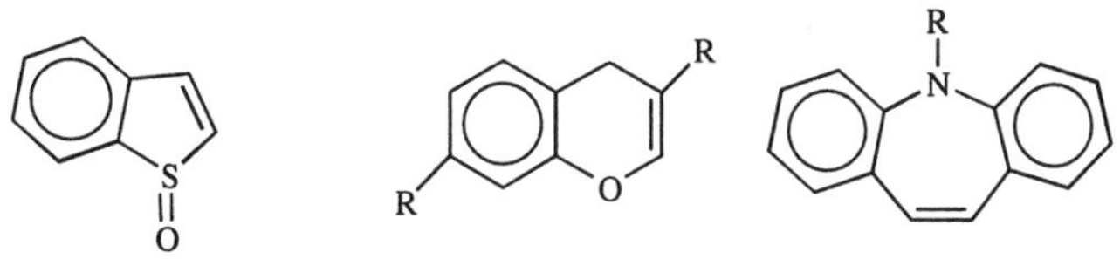

benzothiophene oxide **coumarin** **diben[*b,f*]azepine**

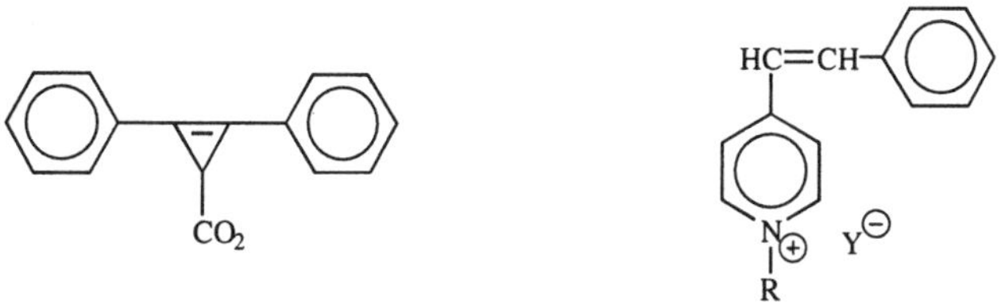

diphenylcyclopropenecarboxylate **stilbazole**

Other functional groups that can be used to form photocrosslinkable polymers are alkyne, maleimides, stilbene, styrene, 1,2,3-thiadiazole, thymine, and anthracene. Pendant groups with anthracene moiety, however, are believed to crosslink by a $4\pi + 4\pi$ type of cycloaddition:

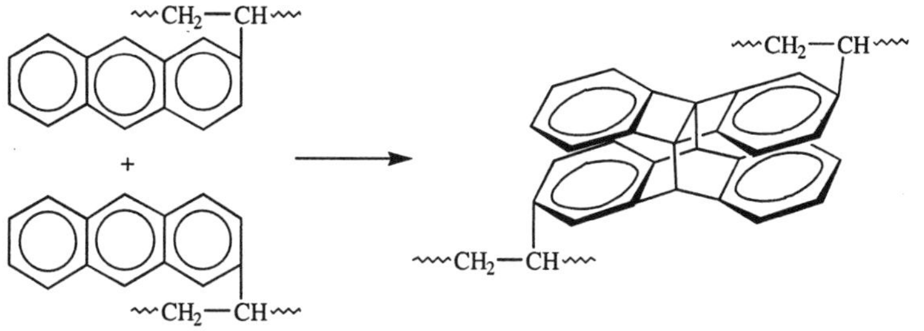

8.5.5.2. Ultraviolet Light Curing Technology of Coatings

Many photocrosslinkable ink and coating compositions in commercial use today consist of prepolymers or oligomers that are combined with polyfunctional monomers and photoinitiators.[298,299] There are two types: those that crosslink by free-radical mechanisms and those that crosslink by ionic mechanisms. In the first the monomers are often mixtures of di- and triacrylate esters. They are combined with photosensitive prepolymers, but lack adequate photosensitivity to crosslink by themselves. Instead, they rely upon photoinitiators as the sources of free radicals for the cure. The diacrylate esters usually serve as diluents to lower the viscosity of the mixtures. These can be various glycol diacrylates, such as butanediol diacrylate:

$$H_2C{=}CH{-}\overset{O}{\overset{\|}{C}}{-}O{-}CH_2{-}CH_2{-}CH_2{-}CH_2{-}O{-}\overset{O}{\overset{\|}{C}}{-}CH_2{-}CH{=}CH_2$$

Sometimes, other diluents that are not acrylic esters, like *N*-vinyl pyrrolidone, are also used. The triacrylate esters may be compounds, like pentaerythritol triacrylate or trimethylolpropane triacrylate:

$$CH_3{-}CH_2{-}C{-}({-}O{-}\overset{O}{\overset{\|}{C}}{-}CH{=}CH_2)_3$$

The prepolymers can be different types of materials. They must, however, contain residual unsaturation in order to react and crosslink with the monomers. Examples of such materials may be polyurethane acrylates that are prepared from urethane prepolymers. The excess isocyanate groups are treated with hydroxyethyl or hydroxypropyl acrylates. Other prepolymers with terminal and/or pendant hydroxy groups are also often esterified with acrylic acid. The oligomers might also be bisphenol A diglycidyl ethers prereacted with acrylic acid to form terminal acrylate groups. An example of these would be:

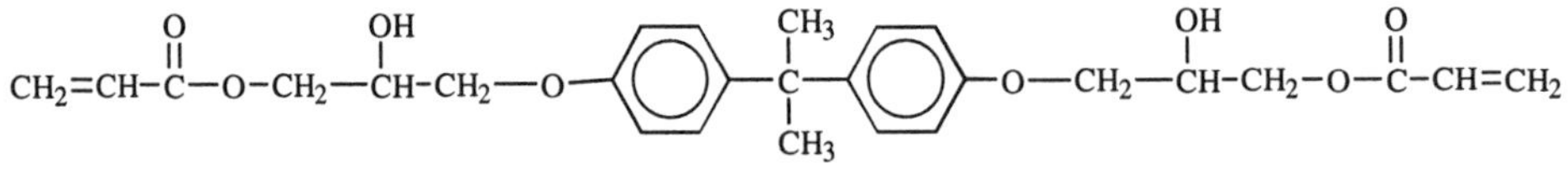

$$CH_2{=}CH{-}\overset{O}{\overset{\|}{C}}{-}O{-}CH_2{-}\overset{OH}{\overset{|}{C}H}{-}CH_2{-}O{-}C_6H_4{-}\underset{CH_3}{\underset{|}{\overset{CH_3}{\overset{|}{C}}}}{-}C_6H_4{-}O{-}CH_2{-}\overset{OH}{\overset{|}{C}H}{-}CH_2{-}O{-}\overset{O}{\overset{\|}{C}}{-}CH{=}CH_2$$

Unsaturated polyesters are also used as prepolymers. Such polyesters might, in addition, have the unreacted hydroxy groups esterified with acrylic acid. Photocrosslinkable unsaturated polyesters can also be prepared by many other means. One example is introduction of unsaturation into the backbones or into the pendant groups via a Knoevenagel reaction[303]:

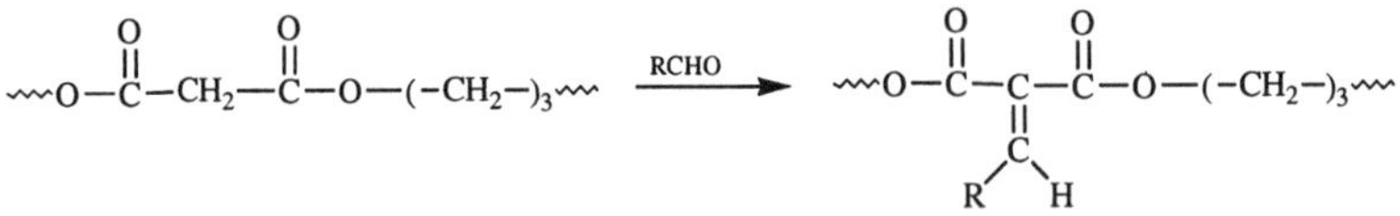

$$\sim O{-}\overset{O}{\overset{\|}{C}}{-}CH_2{-}\overset{O}{\overset{\|}{C}}{-}O{-}({-}CH_2{-})_3\sim \xrightarrow{RCHO} \sim O{-}\overset{O}{\overset{\|}{C}}{-}\underset{\underset{R\quad H}{C}}{\underset{\|}{C}}{-}\overset{O}{\overset{\|}{C}}{-}O{-}({-}CH_2{-})_3\sim$$

Use of photosensitizers is often limited commercially to photocrosslinking some special coatings and a small number of special lithographic inks. Ultraviolet light curing of most coatings and inks usually requires greater crosslinking speeds than most photosensitizers can deliver. *Photoinitiators* are therefore used instead. These materials, upon irradiation with light, either fragment into initiating radicals, or react with other molecules in such a way as to form initiating species. Many compounds fit the definition. They can be peroxides, disulfides, azo compounds, ketones, aldehydes, and others. As an example one might cite photocleavage of diphenyldisulfide. It is used in some coating formulations that are based on styrene-unsaturated polyesters:

$$C_6H_5{-}S{-}S{-}C_6H_5 \xrightarrow{h\nu} 2\ C_6H_5{-}S\cdot$$

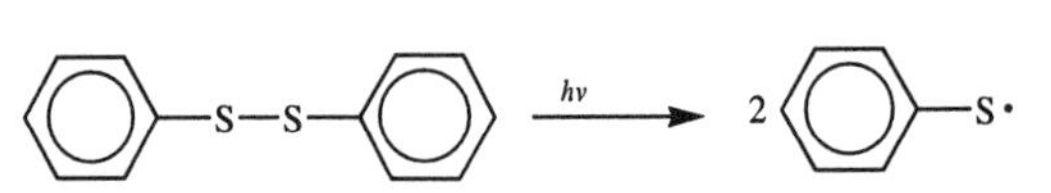

Most commercial formulations, however, contain aromatic carbonyl compounds as initiators. They are preferred due to the higher speed with which they form free radicals. Many undergo Norrish Type I cleavage. Prominent among them are various benzoin ethers and benzil dialkyl ketals. Benzoin itself is believed to fragment from the triplet state while the benzoin ethers are generally thought to fragment from singlet excited states[390]:

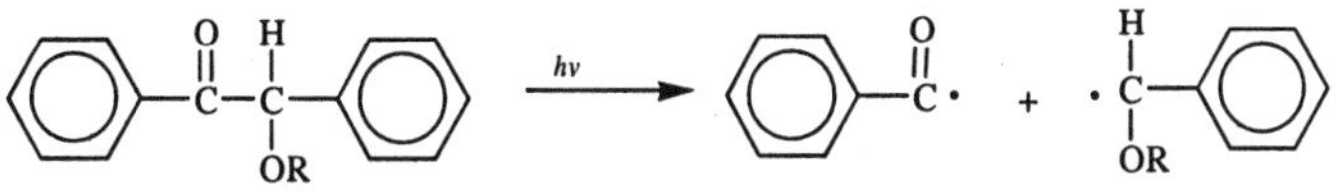

The unpaired electron of the benzoyl radical is not delocalized over the whole aromatic ring and is highly reactive and very efficient. The benzyl ether radical, however, is more stable and while it may initiate some polymerizations, it primarily contributes to chain terminations.

The relative efficiencies of different benzoin derivatives vary considerably in photoinitiation.[390,396] The nature of the monomer and the environment exert an effect as well. Also, the efficiency decreases when radical recombination is favored by the "cage" effect as a result of an increase in the viscosity of the medium. It is also possible that hydrogen abstractions in a viscous medium contribute to photoinitiations by benzoin and its derivatives.

Benzil dialkyl ketals are also very efficient photoinitiators. They too are believed to decompose by the Norrish Type I cleavage:

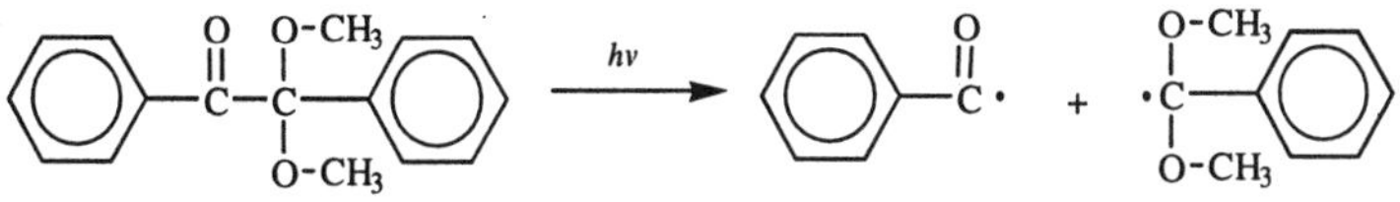

The efficiency of the dimethyl ketal, however, is attributed to the formation of very reactive methyl radicals[397]:

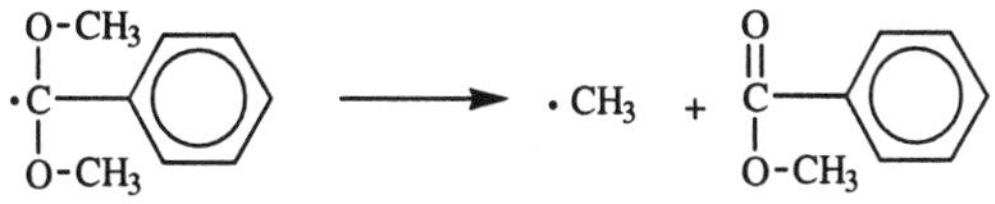

$$\cdot CH_3 \quad +$$

Many aromatic ketones are even more efficient in initiating photopolymerizations as a result of photoreductions in the triplet state. The mechanism of photoreduction is now commonly accepted to be one of electron transfer.[398,399] Upon irradiation, one of the n, nonbonding, electrons of the oxygen atom is promoted into either a σ^* or π^* antibonding orbital. These are $n \rightarrow \sigma^*$ and $n \rightarrow \pi^*$ transitions. They are the most readily observed transitions of carbonyl compounds, requiring less energy than $\pi \rightarrow \pi^*$ transitions, and can be illustrated as follows[400]:

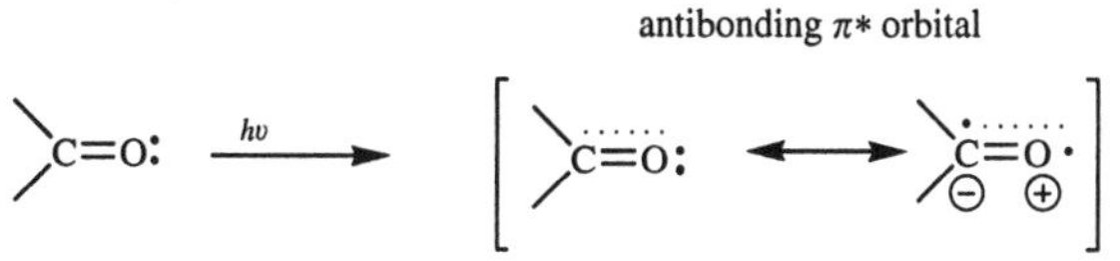

antibonding π^* orbital

In the above illustration one of the unshared electrons of oxygen is promoted to the antibonding π^* orbital. It also shows that the carbon atom has a higher electron density in the excited state than in the ground state. This results in a greater localized site for photochemical activity at the n orbital of oxygen. Because the carbonyl oxygen in the excited state is electron deficient, it reacts similarly to electrophilic alkoxy radicals:

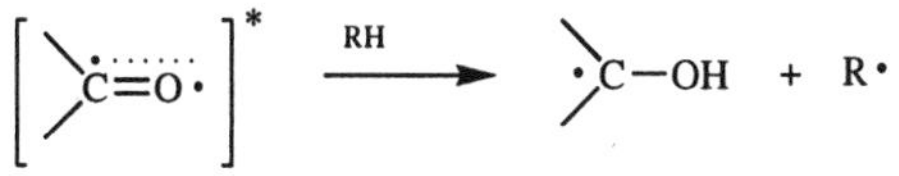

The above can be illustrated on a photoreduction reaction of benzophenone by isopropyl alcohol[401]:

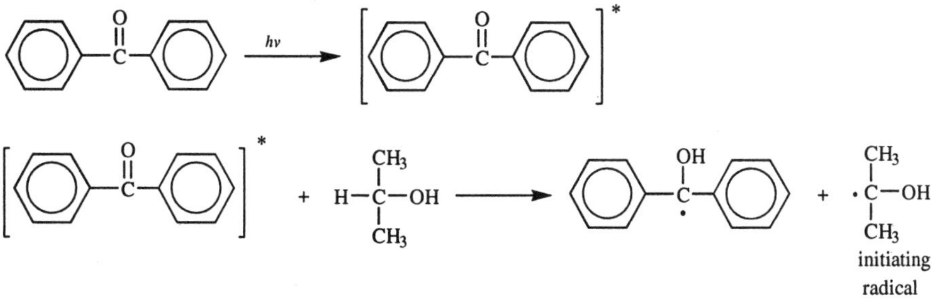

initiating radical

The ketyl radical is too stable and mainly dimerizes to benzpinacol or participates in chain terminations. It is not capable of initiating polymerizations.[398]

Higher conjugated aromatic carbonyl compounds, such as *p*-phenylbenzophenone, fluorenone, or xanthone, exhibit lowest π to π^* excited triplet states and do not abstract labile hydrogen atoms from compounds like ethers of alcohols. They can, however, be readily photoreduced by electron donors.[399] Such donors may be ground state amines or sulfur compounds. Electron transfer takes place from the donor molecules to the acceptors after the excited triplet carbonyl compounds form exiplexes, intermediate encounter complexes with the donors. The collapse of the complexes result in formations of pairs of radical ions. The anions then abstract protons from the amine cation radicals with a subsequent formation of two radicals.[399] The photoinitiation efficiency depends upon the relative rates of all reactions involved. The electron transfer reaction and the subsequent formation of initiating radicals can be illustrated as follows:

$$C_6H_5{-}C_6H_4{-}C(=O){-}C_6H_5 + :N(CH_3)(CH_2CH_2OH)_2 \xrightarrow{h\nu} \left[C_6H_5{-}C_6H_4{-}\dot{C}(O^{\ominus}){-}C_6H_5 \quad \cdot\overset{\oplus}{N}(CH_3)(CH_2{-}CH_2{-}OH)_2 \right]$$

$$\longrightarrow C_6H_5{-}C_6H_4{-}\dot{C}(OH){-}C_6H_5 + (HO{-}CH_2{-}CH_2)(CH_3)N{-}\dot{C}H{-}CH_2{-}OH$$

initiating radical

Originally, the success of the UV curing coating systems has been in clear materials. Pigmented materials remained a challenge sometimes, because many photoinitiators lack sufficient absorption in the UV/visible part of the spectrum to cure them. A major improvement came with the introduction of α-amino ketone compounds for photoinitiation.[470] The effectiveness of these compounds is even further enhanced by addition of sensitizers, like thioxanthone or its derivatives.[471] The improvement is particularly noticeable in relatively thick, highly pigmented films. This enhancement is claimed to be due to longer-wavelength absorptions ($\lambda_{max} \approx 380$ nm) of the thioxanthones.[472] A recent publication compares curing speeds of coating compositions containing combinations of α-morpholine ketones with isopropyl thioxanthone to those containing the α-morpholine ketones alone. Formulations of a white pigmented lacquer and a blue silk screen ink containing the sensitizer cured two and a half times faster than those without the sensitizer.[472] An α-morpholine ketone initiator can be illustrated as follows:

$$R{-}C_6H_4{-}C(=O){-}C(CH_3)_2{-}N(CH_2CH_2)_2O$$

where R is H, O–CH_3, S–CH_3, or N–$(CH_3)_2$.

The enhancement of the curing rate by additions of thioxanthone was studied with time-resolved laser spectroscopy and ^{1}H-NMR CIDNP. It led to the conclusion that two competitive processes occur from the excited triplet-state sensitizers. One is electron transfer and the other is energy transfer.[472] Whether it is the former or the latter depends upon the structure of the initiator used.[472]

The compositions of materials photocrosslinkable by *cationic mechanism* consist of mixtures of various vinyl ethers, or epoxides, or both. Difunctional cycloaliphatic epoxides have been used extensively in some UV curable systems, often as diluents for the various epoxy resins described in Chapter 6. Use of various divinyl ethers is also extensive. Because some cationic photoinitiators also generate free radicals, some compositions may contain mixtures of both types of materials, those that cure by cationic and those that cure by free-radical mechanisms.

The photoinitiators for cationic polymerizations are mainly of two types. One is based on aryldiazonium salts and the other on diaryliodonium, triarylsulfonium, or triarylselenonium salts. The diazonium salts photodecompose to yield both Lewis acids and strong protonic acids:

$$C_6H_5\text{-}N_2^{\oplus}\,PF_6^{\ominus} \xrightarrow{h\nu} C_6H_5\text{-}F + PF_5 + N_2$$

$$PF_5 + H_2O \longrightarrow H^{\oplus}\,PF_5OH^{\ominus}$$

The anions that are preferred are mostly of low nucleophilicity, such as $SbF_6^{\ominus}$, $AsF_6^{\ominus}$, and $PF_6^{\ominus}$. These systems yield low termination rates, cure at room temperature, and exhibit considerable amounts of postcuring. There are some disadvantages to the use of initiators based on diazonium salts. They tend to gel, unless stored in the dark. The films also tend to be colored, further yellowing on ageing. Also, nitrogen evolution causes bubbles or pinholes in thicker films.

The photodecomposition mechanisms of both iodonium and sulfonium salts are similar.[299] They can be illustrated as follows:

$$(C_6H_5)_2I^{\oplus}X^{\ominus} \xrightarrow{h\nu} \left[(C_6H_5)_2I^{\oplus}X^{\ominus}\right]^* \longrightarrow C_6H_5\text{-}I^{\oplus}\!\cdot\, X^{\ominus} + \cdot C_6H_5$$

$$C_6H_5\text{-}I^{\oplus}\!\cdot\, X^{\ominus} + RH \longrightarrow C_6H_5\text{-}IH^{\oplus}X^{\ominus} + R\cdot \longrightarrow C_6H_5\text{-}I + H^{\oplus}X^{\ominus}$$

where $H^{\oplus}X^{\ominus}$ is the initiating species. These systems generate acids of the type HBF_4, $HAsF_6$, HPF_6, and $HSbF_6$. The decomposition mechanism of the triarylsulfonium salt is believed to be:

$$(C_6H_5)_3S^{\oplus}X^{\ominus} \xrightarrow{h\nu} \left[(C_6H_5)_3S^{\oplus}X^{\ominus}\right]^* \xrightarrow{RH} (C_6H_5)_2S + H^{\oplus}X^{\ominus} + \cdot C_6H_5$$

A somewhat different mechanism, however, is visualized for the photodecompositions of dialkylpehenacylsulfonium salts[299]:

$$C_6H_5\text{-}\overset{O}{\overset{\|}{C}}\text{-}CH_2\text{-}\overset{\oplus}{S}(R')(R)\;X^{\ominus} \underset{}{\overset{h\nu}{\rightleftharpoons}} C_6H_5\text{-}\overset{O}{\overset{\|}{C}}\text{-}CH{=}S(R')(R) + H^{\oplus}X^{\ominus}$$

ylid

The cure rates by the triarylsulfonium salts are also influenced by the size of the anions. Larger anions give faster cures.[299] The rates are also influenced by the temperatures. However, they reach an optimum. The mechanisms of photoinitiations by triarylselenonium salts are believed to be generally similar to those by triarylsulfonium and diaryliodonium salts.

8.5.6. Crosslinking of Polymers With Electron Beams

The techniques of crosslinking polymeric material with the aid of electron beams have been studied since 1930.[299] In the 1950s serious attempts were made to utilize the technology to crosslink organic coatings. One such application was in the preparation of panels for automobiles (Ford Electrocure Process). Today, electron beam curing is used to crosslink wire and cable coatings and insulations. It is also used on heat shrinkable packaging films and to cure coatings on wood panels in such diverse places as Japan, Brazil, and The Netherlands. Electron beams are also used extensively in curing resists for electronic circuitry.

The mechanism of electron beam crosslinking is mainly free radical in nature. It is believed to initially involve ejections of hydrogen radicals from materials like polyethylene as a result of bombardment with high energy electrons:

$$\sim CH_2{-}CH_2\sim \longrightarrow \sim CH_2{-}\dot{C}H\sim + H\cdot$$

The hydrogen radical then abstracts another hydrogen from a neighboring polymer molecule and the two radical polymers combine or crosslink:

$$\sim CH_2{-}CH_2\sim + H\cdot \longrightarrow \sim CH_2{-}\dot{C}H\sim + H_2$$

$$\sim CH_2{-}\dot{C}H\sim + \sim CH_2{-}\dot{C}H\sim \longrightarrow \begin{array}{c}\sim CH_2{-}CH\sim \\ | \\ \sim CH_2{-}CH\sim\end{array}$$

Wire coatings are often based on polyolefins or on poly(vinyl chloride). Many other electron-beam curable materials for surface coatings, however, are very similar in composition to the ultraviolet-light curable ones. No photoinitiators, however, are needed.

8.6. Polymeric Reagents

These are macromolecules with functional groups that can be used in various reactions. The polymeric reagents differ from any other reagents only in the fact that the functional moiety is immobilized on a macromolecule. Such materials find applications in organic syntheses, biochemical reactions, special separations, and analyses. A big advantage of some polymeric reagents is that they can be separated, often easily by filtration, from the products that are small molecules, particularly when they are crosslinked. Crosslinked polymeric reagents have an additional advantage in that several different polymeric reagents can be used simultaneously without the functional groups being accessible to each other for interaction. Polymeric reagents are also used as special immobilizing media. Reactions of some compounds in solution require high dilutions. Immobilization, however, may permit the same reactions to be carried out at relatively high concentrations. Immobilization can also be very useful in syntheses that consist of many steps, where the byproducts after each step can simply be washed away. This avoids lengthy isolation and purification procedures. Also, an immobilizing polymer may provide microenvironmental effects for the attached species for the reactions. These may include special electronic and steric conditions that could be different from those existing in bulk or in solution.

8.6.1. Immobilized Reagents

Crosslinked polystyrene (copolymer with divinyl benzene) is now a favorite support material. Perhaps the main reason for choosing crosslinked polystyrene is that it can be functionalized in many ways. It can be nitrated, chloromethylated, sulfonated, lithiated, carboxylated, and acylated. The greatest use has been made of the chloromethylated and lithiated derivatives. This is because these two derivatives can react with nucleophilic and electrophilic reagents, respectively, resulting in a wide range of functionalized polymers. See Section 8.4.3 for an illustration.

Two types of crosslinked polystyrene are used. One type is a polymer that is crosslinked by only 1–2% of divinyl benzene. The material, though fairly strong mechanically, swells and expands significantly in volume when dispersed in a solvent. It is *microporous*. A copolymer with up to 20% divinyl benzene is the second type of support. It is prepared in the presence of large quantities of diluents to retain the products in expanded form. As a result the structures are *macroporous* or *macroreticular*. The advantages of macroporous over microporous structures are faster reactions, less fragility, and easier handling. Macroreticular supports, though less often used, have the advantage of being useful in almost any solvent. Some other polymers that have been employed as support materials are cellulose, starch, polyamides, and poly(glycidyl methacrylate). They are more useful with hydrophilic or polar reactants and solvents.

The chemical uses of polymeric reagents were classified according to the general type of reaction.[304] Some applications of polymeric reagents follow:

1. Polymer-attached reagents are used in *special separations* to selectively bind one or a few species out of complex mixtures:

$$\text{(P)-A} + \text{B,C,D,E} \longrightarrow \text{(P)-A...B} + \text{C,D,E} \xrightarrow{\text{filter}} \text{(P)-A...B} \xrightarrow{\text{recover}} \text{(P)-A} + \text{B}$$

   The polymer-bound compound is separated from the mixture and then released.
2. The polymer may have a catalyst attached. Such catalysts can be enzymes, inorganic compounds, or organometallic compounds:

$$\text{(P)-A} + \text{substrate} \longrightarrow \text{(P)-A} + \text{product}$$

   where A is the catalyst site.
3. The polymeric reagent may be used as a transfer agent. Low molecular weight reactants transfer the functional moiety with the aid of the polymeric agent. This leaves the products in pure form after filtration and solvent removal.
4. The polymers are used as *carriers* or as *blocking groups* in syntheses[302]:

$$\text{(P)-} + \text{A} \longrightarrow \text{(P)-A} \xrightarrow{\text{B}} \text{(P)-A-B} \xrightarrow{\text{C}} \text{(P)-A-B-C} \longrightarrow \text{(P)-} + \text{A-B-C}$$

It is claimed[305] that many types of reactions can be performed more easily with multiple polymers. To avoid undesirable side reactions two reacting species can be used with each attached to a different polymer. Such polymer-bound reactants can coexist in the same reaction vessel without interacting. An example is the preparation of benzoyl acetonitrile by Patchornik and coworkers.[304] Molecules of triphenylmethane lithium, attached to polystyrene supports, were combined with also immobilized *o*-nitrophenol. The *o*-nitrophenols were prereacted with benzoyl chloride. The two species were combined and acetonitrile molecules containing acidic hydrogens were introduced into the reaction mixture. This resulted in hydrogens being abstracted from the introduced molecules and formation of short-lived carbanions:

$$\text{(P)-}C_6H_4\text{-}C^{\ominus}(C_6H_5)_2\ Li^{\oplus} + CH_3-C\equiv N \longrightarrow Li^{\oplus}\ {}^{\ominus}CH_2-C\equiv N + \text{(P)-}C_6H_4\text{-}CH(C_6H_5)_2$$

The carbanions in turn reacted with the ester groups and yielded the desired product:

$$\text{(P)-}C_6H_3(NO_2)\text{-O-}\underset{\|}{\overset{}{C}}(=O)\text{-}C_6H_5 + Li^{\oplus}\ {}^{\ominus}CH_2-C\equiv N \longrightarrow C_6H_5-C(=O)-CH_2-C\equiv N + \text{(P)-}C_6H_3(NO_2)\text{-OLi}$$

The yield from the above reaction was found to be about 90%. This compares very favorably with a 27% yield obtainable without polymeric supports.[306]

Another example is benzoylation of γ-butyrolactone. When an acylation reaction is being carried out on an ester in solution, the ester enolate must be completely formed before the acylating agent can be introduced. Otherwise, the acylating agent reacts with the base instead. During this period, however, the ester enolate can undergo self-condensation. Using polymer-attached reagents, however, benzoylation of γ-butyrolactone or similar compounds can be achieved in 95% yields[307]:

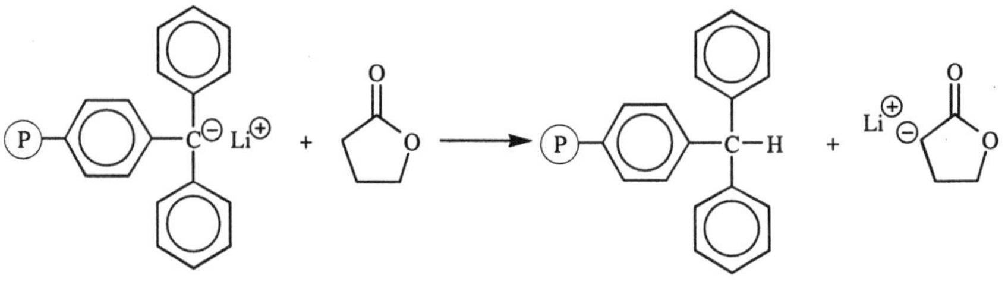

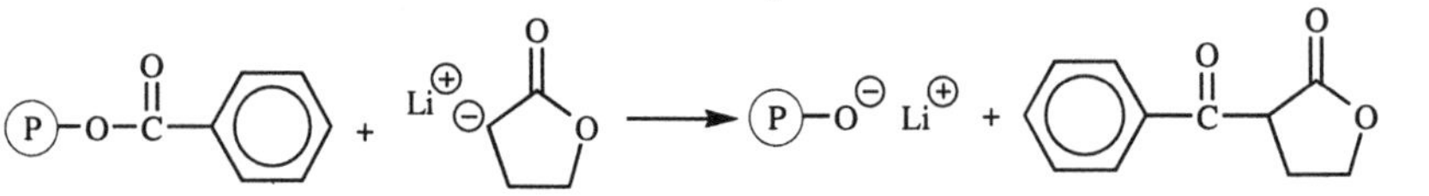

By comparison, when the above reaction is performed in solution, without any support, the yield is only about 31%.[307] Patchornik termed the use of supports in this manner "wolf and lamb chemistry."[304]

It appears unlikely that all organic reactions can benefit from the use of polymeric reagents or catalysts. Nevertheless, such reagents and catalyst do appear to be promising in many applications. The fact that polymer-bound reagents and catalyst are more expensive, however, precludes wide industrial applications, except in special situations.

8.6.2. Polymeric Catalysts

It may be more accurate to refer to many of them as catalysts immobilized on polymeric supports. Such catalysts can be inorganic compounds, such as Lewis acids, attached to organic polymers. They can also be organic or biochemical catalysts. Perhaps the biggest group among the polymeric catalysts attached to supports are the *immobilized enzymes*. They are used in industrial processes as well as in research laboratories. Immobilization often tends to improve stability and, in some cases, activity over a broader range of pH and temperatures. Another advantage is elimination of enzyme contamination of waste streams. On the other hand, immobilized enzymes are often less active after immobilization.

Several major techniques of enzyme immobilization are used. An important one is covalent bonding of the enzyme to the support material. Such attachment usually consists of reacting some functional group of the enzyme, not active in the enzymatic process, with a functional group on another polymer that is the carrier. Hydrophilic groups are preferred for reactions with enzymes in aqueous media.

An immobilization of an enzyme on cellulose with azide groups[308] attached can serve as an illustration. Carboxymethyl cellulose is the starting material:

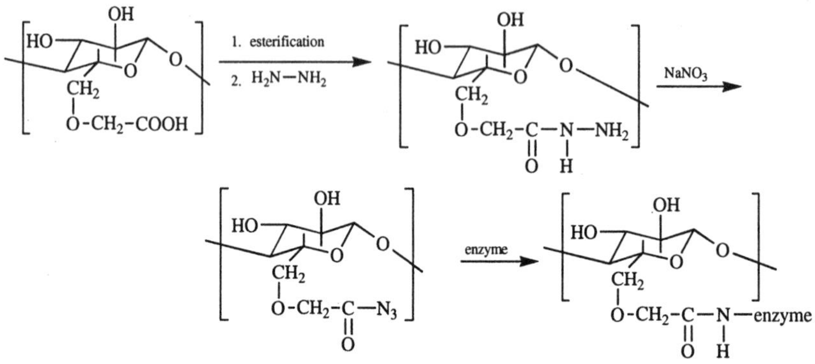

In other techniques, the protein may be bound by some copolymer of maleic anhydride, where the anhydride groups react with some available amine groups on the enzyme.[308] Yet other techniques may utilize cyanuric chloride attached to polysaccharides for immobilization[308]:

Polyaminostyrene can be diazotized or treated with thiophosgene and then used in enzyme immobilization[309]:

Epoxy groups on carrier molecules are capable of reacting with amine or carboxylic acid groups of the enzymes. This is also used in enzyme immobilizations. A variation on the technique is to react a vinyl monomer that contains an epoxy groups, like glycidyl methacrylate or glycidyl acrylate, with the enzyme first. The product is then polymerized or copolymerized through the vinyl portion.[310] As stated earlier, in many cases immobilization of enzymes is accompanied by some loss of activity. In some instance, the loss of activity is severe. A new technique, however, was reported recently.[311] Enzymes called protease are immobilized on a polymer with aminoglucose units to form covalently bonded carbohydrate–protein conjugates. In aqueous solution, the conjugated enzymes show about the same catalytic activity as native enzymes. At elevated temperatures, however, they exhibit enhanced stability. In addition, they are capable of catalyzing reactions in organic solvents that denature and inactivate the native enzymes.[311]

There are also many uses for nonenzymatic *polymeric catalysts.* For instance, polymer-bound crown ethers, cryptates, and channel compounds behave as polymeric phase-transfer catalysts. The catalytic activity is based on selective complex formation. An example is the use of polystyrene-attached oxygen heterocycles [18]-crown-6 or a cryptand[222] to catalyze replacements of bromine in *n*-octyl bromide by an iodine or by a cyanide group[312]:

A 95% yield is achieved. The catalytic activity, as a result of the complexation of the cations, results in an increased nucleophilicity of the anions.

Interactions of ions and ion pairs with vinyl polymers of crown ethers were shown to be considerably more efficient than such interactions with unattached crown ethers.[313] Also, studies of diazo-4,7,13,16-tetraoxacyclooctadecane bound to polyacrylamide gel show an enhancement of cationic complexation when compared to ligands that are not bound to polymers.[314] On the other hand, polymer-bound crown ethers do not offer any advantage over unbound ligands in the Koening–Knorr reaction.[315]

The catalytic properties and solute binding capabilities of the pendant crown ethers and glyme ligands apparently depend on the spacing between the ligands. They also depend upon the structure and length of the chains connecting the ligand bound ions, and the solvent.[316] In low-polarity solvents, the ligands activate anionic reactants through modification of their ion-pair structures.

Among other polymer-bound catalysts, onium compounds can be used successfully in halogen exchange reactions between activated and nonactivated halides. This is the case, for instance, in additions to the double bonds of dichlorocarbene to form substituted cyclopropanes[317] and in C-alkylating nitriles.[318] When optically active polymeric ammonium compounds are used as catalysts in carbene addition reaction, chiral products form.[319,320]

Ion exchange resins have been used for a long time now to catalyze some reactions. This is mentioned, for instance, in Section 8.2.5.7 on epoxidation reactions. Basic ion exchange resins can also be used to catalyze condensations of furfural with aliphatic aldehydes.[321]

A still different kind of polymeric catalyst is one that has pendant photosensitizers attached. To be effective, the sensitizer portion must absorb light and undergo a transition from a singlet to a longer-lived triplet state. It must then, without emitting radiation, activate a substrate molecule and return to the ground state. Some dyes function in this manner. An example is Rose Bengal. When it is attached to crosslinked polystyrene[322] it can be used to produce singlet oxygen. The excited oxygen in turn hydroperoxidizes olefins. The structure of polymer-bound Rose Bengal can be illustrated as follows:

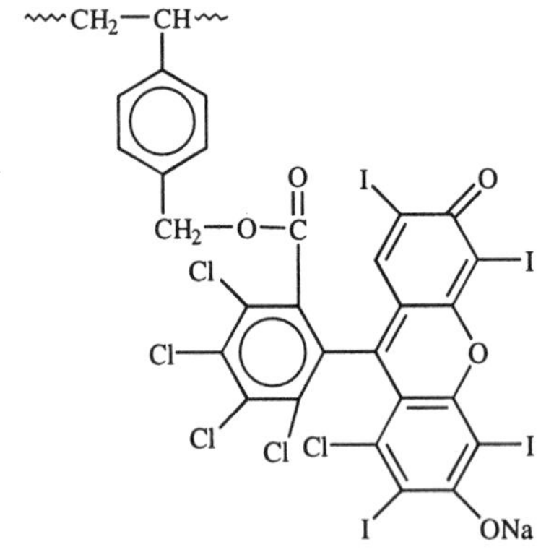

polymer bound Rose Bengal

The photosensitized hydroperoxidation reaction of olefins[322] can be shown as follows:

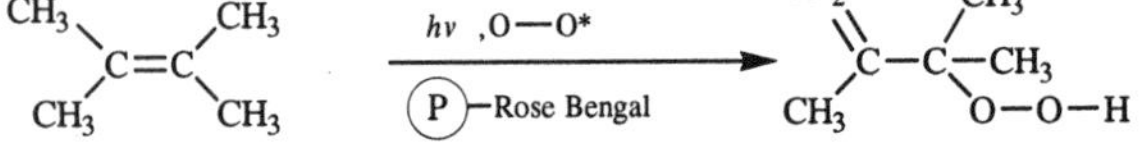

$$(CH_3)_2C{=}C(CH_3)_2 \xrightarrow[\text{(P)–Rose Bengal}]{h\nu,\ O{-}O^*} CH_2{=}C(CH_3){-}C(CH_3)_2{-}O{-}O{-}H$$

8.7. Formation of Graft Copolymers

This is an important part of polymer syntheses that is used in many industrial processes. In 1967 Battaerd and Tregear[323] published a book on the subject that contains 1000 references to journal publications and 1200 references to patents. In addition, there are several monographs and many review papers.[324] The synthetic methods developed to date range from using free-radical attacks on polymeric backbones to highly refined ionic reactions. There are examples where these ionic reactions attach the side chains at well-designated locations.

8.7.1. Free-Radical Grafting by the Chain-Transferring Technique

This technique is probably one of the simplest ways to form graft copolymers. It consists of carrying out free-radical polymerizations of monomers in the presence of polymers preformed from different monomers. A prerequisite for this synthesis is that the active sites must form on the polymeric backbones during the course of the reactions. Ideally, this occurs if the steps of initiations consist only of attacks by the initiating radicals on the backbones:

$$\text{Initiation:} \quad I{-}I \longrightarrow 2\,I\cdot$$

$$I\cdot + \sim\!A{-}A{-}A\!\sim \longrightarrow I{-}H + \sim\!A{-}\dot{A}{-}A\!\sim$$

Propagations then proceed from the backbone sites:

$$\text{Propagation:} \quad \sim\!A{-}\dot{A}{-}A\!\sim + n\,M \longrightarrow \sim\!A{-}\underset{}{\overset{(M)_{n-1}M\cdot}{\overset{|}{A}}}{-}A\!\sim$$

Terminations can take place in many ways. Of course, termination by combination will lead to crosslinked insoluble polymers and that is undesirable. An ideal termination takes place by chain transferring to another site on a polymer backbone to initiate another chain growth.

The above, however, is an ideal picture. In reality, the efficiency of grafting by this technique depends upon the following:

1. Competitions between the different materials present in the reaction mixture, such as monomer, solvent, and polymer backbone for the radical species. This includes competition between chain growth and chain transferring to any other species present.
2. Competition between the terminating processes, such as disproportionation and chain transferring.

The conditions can vary considerably and it is possible to carry out the reactions in bulk, solution, or in emulsion. When the reaction take place in emulsion, success depends much on the experimental techniques. The rate of diffusion is a factor and anything that affects this rate must be considered. Because grafting efficiency depends upon chain transferring to the backbone, knowledge of the chain-transferring constants can help predict the outcome of the reactions. Sometimes, the information on the chain-transferring constants is not available from the literature. It may, however, be possible to obtain the information from reactions of low molecular weight compounds with similar structures.[325–327] One assumes equal reactivity toward attacking radicals. The validity of such an assumption was demonstrated on oligomers.[328–330]

The reactivity of the initiating radicals toward the backbones can vary and this can also change the efficiency of grafting. Benzoyl peroxide initiated polymerizations of methyl methacrylate monomer, for instance, in the presence of polystyrene[331] yield appreciable quantities of graft copolymers. Very little graft copolymers, however, form when di-*t*-butyl peroxide initiates the same reactions. Azobisisobutyronitrile also fails to yield appreciable quantities of graft copolymers. This is due to very inefficient chain transferring to the polymer backbones by *t*-butoxy and isobutyronitrile radicals.

Not all chain transferring to the backbones results in formations of graft copolymers. An example is polymerization of vinyl acetate in the presence of poly(methyl methacrylate). No graft copolymers form and this is independent of the reactivity of the initiators.[332] In fact, grafts of poly(vinyl acetate) to poly(methyl methacrylate) and to polystyrene cannot be prepared by this technique.[333]

Grafting efficiency may increase with temperature.[333,334] This could be due to higher activation energy of the transfer reaction than that of the propagation reaction.[335] Meaningful effects of temperature, however, are not always observed. In grafting polystyrene to poly(butyl acrylate) in emulsion, for instance, there is no noticeable difference between 60–90 °C by this technique.[336]

The presence of sites with high transfer constants on the polymeric backbone enhances the efficiency of grafting. Such sites can be introduced deliberately. These can, for instance, be mercaptan groups[337,338] that can be formed by reacting H_2S with a polymer containing epoxy groups:

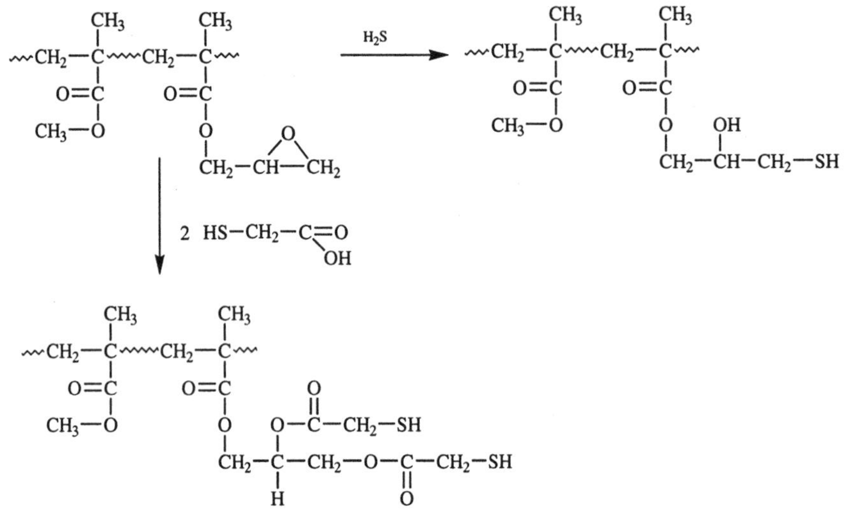

Free-radical polymerizations of acrylic and methacrylic esters in the presence of the above backbones result in high yield of graft copolymers.[337,338]

Another example is formation of mercaptan groups on cellulose in order to form graft copolymers[339]:

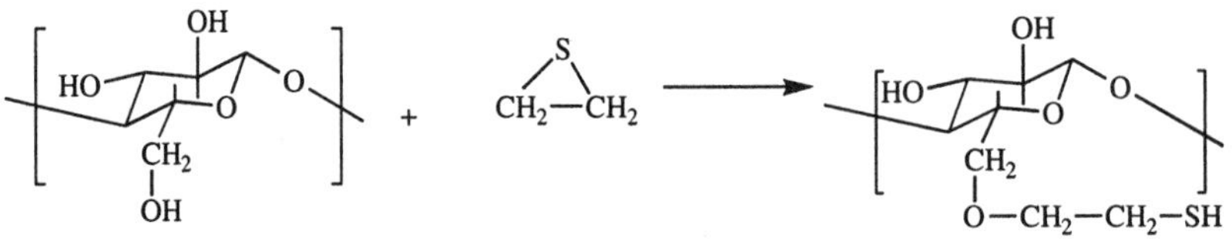

Pendant nitro groups are also effective in chain-transfer grafting reactions. Thus, graft copolymers of polystyrene with cellulose acetate *p*-nitrobenzoate[340] and with poly(vinyl *p*-nitrobenzoate)[341] form readily. Nitro groups appear to be more effective in formations of graft copolymers by a radical mechanism than are double bonds located as pendant groups.[340]

8.7.2. Free-Radical Grafting Reactions to Polymers with Double Bonds

Carbon-to-carbon double bonds, either in the backbone or in the pendant groups, are potential sites for free-radical attacks. In addition to the double bonds, the hydrogen atoms on the neighboring carbons are allylic and potential sites for chain transferring. Because rubbers, natural and synthetic, possess such unsaturations, they are used extensively as backbones for various grafting reactions. Whether the reactivity of the initiating radical is important in determining grafting efficiency has not been completely established. Graft copolymers of poly(methyl methacrylate) on gutta-percha, however, form in good yields when the initiator is benzoyl peroxide.[342,343] Yet, when azobisisobutyronitrile is used, only a mixture of homopolymers forms. Work with ^{14}C-labeled initiators shows that the primary radicals react both by addition to the double bonds and by transfer to the methylene group.[342] Grafting reactions to polybutadiene, however, proceed via chain transferring from the growing-chain radical to the backbone.[344] Nevertheless, strong evidence also shows that the initiator radicals can interact directly with polymeric backbones.[345–347]

When graft copolymers of polystyrene to natural rubber form, the chain length of the attached branches equals the chain lengths of the unattached polystyrene homopolymer that forms simultaneously.[348] This led to the following conclusions[368]:

1. As the concentration of rubber increases, the length of the grafted branches diminishes, while their number remains the same.
2. When the concentration of the initiator increases, the length of the branches diminishes, but the number of branches increases.
3. When the concentration of monomers increases, the length of the branches also increases, but their number diminishes.
4. When the polymerization time increases, the length of the branches remains the same, but their number increases.

8.7.3. Preparation of Graft Copolymers With the Aid of Macromonomers

The chain-transferring methods for preparing graft copolymers suffer from the disadvantages of low efficiency and contamination by homopolymers. The efficiency in forming graft copolymers, however, increases with the use of *macromonomers*. A macromonomer is a macromolecular monomer, an oligomer, or a polymer with a polymerizable end group. When macromonomers are copolymerized with other monomers, comb-shaped polymers form.[367,368] The copolymerizations can be free-radical or ionic in mechanism. Some examples of macromonomers are presented in Table 8.2.

A preparation of graft copolymers with the aid of macromonomers can be illustrated as follows[379]:

$$n\ CH_2{=}\underset{Y}{\overset{R'}{C}} + m\ CH_2{=}CH{-}\overset{O}{\overset{\|}{C}}{-}O{-}CH_2{-}\underset{OH}{CH}{-}CH_2{-}O{-}\overset{O}{\overset{\|}{C}}{-}R{-}S{-}({-}\underset{X}{\overset{R''}{C}}{-}CH_2{-})_o{-}H$$

$$\xrightarrow{\text{initiator}} -[-CH_2-\underset{Y}{\overset{R'}{C}}-]_n-[-CH_2-\underset{C(=O)-O-CH_2-CH(OH)-CH_2-O-C(=O)-R-S-(-C(R'')(X)-CH_2-)_o-H}{CH}-]_m-$$

Many variations on the above technique are possible.

8.7.4. Initiations of Polymerizations from the Backbone Polymers

High degrees of grafting by the free-radical mechanism can be attained when polymerizations are initiated from the backbones of the polymer. One way this can be done is to form peroxides on the backbone structures. Decompositions of such peroxides can yield initiating radicals. Half of them will be attached to the backbones. An example is the preparation of graft copolymers of polystyrene[380,381]:

Table 8.2. Some Macromonomers Reported in the Literature

Macromonomer	Reference
$CH_2{=}C(CH_3){-}COOCH_2CH_2{-}({-}CH(C_6H_5){-}CH_2{-})_n{-}C_4H_9$	369
$CH_2{=}C(CH_3){-}COO{-}CH_2{-}CH(OH){-}CH_2{-}OOC{-}CH_2{-}S{-}({-}CH_2{-}C(CH_3)(COOCH_3){-})_n{-}H$	370
$CH_2{=}C(CH_3){-}COOCH_2{-}CH_2OOC{-}NH{-}C_6H_3(CH_3){-}NH{-}COO{-}CH_2{-}CH_2{-}S{-}({-}CH_2{-}C(CH_3)(COOCH_3){-})_n{-}H$	371
$CH_2{=}C(CH_3){-}CH_2{-}({-}C(CH_3)_2{-}CH_2{-})_n{-}C(CH_3)_2{-}CH_3$	372
$CH_2{=}CH{-}C_6H_4{-}(CH_2{-}C(CH_3)_2{-})_n{-}Cl$	373
$CH_2{=}CH{-}C_6H_4{-}CH_2{-}[{-}O{-}({-}CH_2{-})_4{-}]_n{-}O{-}CH_2CH_3$	374
$CH_2{=}C(CH_3){-}C({=}O){-}[{-}O{-}({-}CH_2{-})_4{-}]_n{-}O{-}C_6H_5$	375
$CH_2{=}C(CH_3){-}C({=}O){-}({-}O{-}CH_2{-}CH_2{-})_n{-}CH{-}[C_6H_5]_2$	376
$CH_2{=}C(CH_3){-}C({=}O){-}O{-}({-}CH_2{-}CH_2{-}O{-})_n{-}H$	377
$CH_2{=}CH{-}C_6H_4{-}Si(CH_3)_2{-}({-}O{-}Si(CH_3)_2{-})_n{-}Cl$	378
$CH_2{=}C(CH_3){-}COO{-}({-}CH_2{-})_3{-}Si(CH_3)_2{-}({-}O{-}Si(CH_3)_2{-})_n{-}R$	378
$CH_2{=}CH{-}C_6H_4{-}CH_2{-}({-}O{-}CH_2{-}CH_2{-})_n{-}O{-}CH_3$	405

$$\sim\sim CH_2-CH(C_6H_4)\sim\sim \xrightarrow[\text{catalyst}]{CH_2=CH-CH_3} \sim\sim CH_2-CH(C_6H_4-C(CH_3)_2-H)\sim\sim \xrightarrow{O_2} \sim\sim CH_2-CH(C_6H_4-C(CH_3)_2-O-O-H)\sim\sim$$

Thermal cleavage of the above peroxides results in macromolecules with free-radicals sites. Hydroxy radicals also form and initiate formations of homopolymers. Decompositions of the peroxides by redox mechanisms, however, increase the yields of graft copolymers, but do not stop all formations of hydroxy radicals[380]:

$$\sim\sim CH_2-CH(C_6H_4-C(CH_3)_2-O-O-H)\sim\sim \xrightarrow{Fe^{\oplus\oplus}} \sim\sim CH_2-CH(C_6H_4-C(CH_3)_2-O\cdot)\sim\sim + Fe^{\oplus\oplus\oplus} + OH^{\ominus}$$

Some homopolymers still form.[382]

Air oxidation of polypropylene can result in the formation of hydroperoxide units at the sites of the tertiary hydrogens.[383] The polymer can also be oxidized when dissolved in cumene that contains some cumene hydroperoxide at 70–80 °C. A product containing 0.8% oxygen by weight as a hydroperoxide[383] can be formed and can subsequently be reacted to form graft copolymers. Various monomers[384–386] can be used, such as vinyl acetate or 2-vinyl pyrrolidone.

Many other hydroperoxidations of polymers were reported in the literature. The materials are used in formations of graft copolymers. One example is hydroperoxidation of poly(ethylene terephthalate) and poly(ε-caproamide). The products yield graft copolymers with various acrylic and methacrylic esters and acrylic and methacrylic acids.[387–389]

Ozone treatment of polymers can also cause hydroperoxidation of labile hydrogens. It can, however, also cause extensive degradation of the backbone polymers, because attacks by ozone on double bonds in the backbones converts them to unstable ozonides. Starch can be ozonized to form graft copolymers.[340,343] The same is true of cellulose,[341] poly(vinyl chloride),[342,344] and polyethylene.[345] Hydroperoxides form without noticeable degradation. This allows subsequent preparations of graft copolymers.

In a similar manner it is possible to start with copolymers of acryloyl or methacryloyl chloride and react them with hydroperoxides.[346] This can be illustrated as follows:

$$-[-CH_2-C(CH_3)(C(=O)-O-CH_3)-]--[-CH_2-CH(C(=O)-Cl)-]- \xrightarrow{(CH_3)_3COOH} -[-CH_2-C(CH_3)(C(=O)-O-CH_3)-]--[-CH_2-CH(C(=O)-O-O-C(CH_3)_3)-]-$$

The decomposition of the pendant peroxide in the presence of vinyl monomers yields mixtures of graft copolymers and homopolymers.

Preparation and subsequent decomposition of polymers with diazonium salts can also be used to form graft copolymers. An example is a nitrated polystyrene, reduced to the amine derivative and then diazotized.[347] The decomposition of the diazonium salt results in the formation of radicals:

$$\sim CH_2-CH\sim (C_6H_5) \xrightarrow{HNO_3} \sim CH_2-CH\sim (C_6H_4NO_2) \longrightarrow \sim CH_2-CH\sim (C_6H_4NH_2)$$

$$\xrightarrow{HNO_2} \sim CH_2-CH\sim (C_6H_4N_2Cl) \xrightarrow{Fe^{++}} \sim CH_2-CH\sim (C_6H_4^{\bullet}) + Fe^{+++}$$

The radical sites are capable of initiating polymerizations of monomers. A similar approach can be taken with cellulose.[348] Mercerized cotton and sodium salt of carboxymethyl cellulose will react with *p*-aminophenacyl chloride:

$$[\text{cellulose unit}-CH_2-O-Na]_n + Cl-\overset{O}{\overset{\|}{C}}-C_6H_4-NH_2 \longrightarrow [\text{cellulose unit}-CH_2-O-\overset{O}{\overset{\|}{C}}-C_6H_4-NH_2]_n$$

The material can be converted to diazonium salts and then decomposed with ferrous ions in the presence of some vinyl monomers to form graft copolymers. Acrylonitrile forms graft copolymer readily without formation of any homopolymers. Styrene and vinyl acetate, however, do not. A modification of this technique is to conduct the diazotization reaction in the presence of emulsifiers.[349] The amounts of graft copolymers that form with acrylic and methacrylic monomers and *N*-vinylpyrrolidone depend upon the nature and pH of the emulsifiers, the reaction time, and the temperature.

Ceric ions form graft copolymers with various macromolecules by a redox mechanism. The reactions can be illustrated as follows:

$$R-CH_2OH + Ce^{4+} \longrightarrow [\text{ R-CH}_2\text{-OH-Ceric complex }]$$

$$\longrightarrow R-\dot{C}H-OH \text{ (or } R-CH_2-O\bullet \text{)} + Ce^{3+} + H^{\oplus}$$

The almost exclusive formation of free radicals on the polymeric backbones results in formations of many products that are close to being free from homopolymers.[350] The reactions are widely used in formations of graft copolymers of poly(vinyl alcohol) and particularly of cellulose and starch. The grafting reaction fails, however, when attempted on polysaccharides that lack free hydroxy groups on the second and third carbons. This led to speculation[351] that the bond between these carbons cleaves. In the process, free radicals presumably form on the second carbons and aldehyde structures on the third carbons of the glucose units. This point of view, however, is not generally accepted. Instead, it was proposed that more likely positions for attacks by the ceric ions are at the C_1 carbons of the glucoses at the end of the polysaccharide chains.[352] This is supported by the observation that oxidation of cellulose is an important prerequisite for the formation of graft copolymers.[353]

Graft copolymerizations by a redox mechanism can also take place with the aid of other ions. This includes grafting on cellulose backbones with ferrous ions and hydrogen peroxide.[354] Redox grafting reactions can also take place on nylon and on polyester. For instance, graft copolymers of methyl methacrylate on nylon 6 can be prepared with manganic, cobaltic, and ferric ions.[355] Another example is grafting poly(glycidyl methacrylate) on poly(ethylene terephthalate) fibers with the aid ferrous ion–hydrogen peroxide. The reaction depends on the concentration of the monomer, hydrogen peroxide, time, and temperature.[356]

8.7.5. Photochemical Syntheses of Graft Copolymers

Photolabile groups on polymers can serve as sites for photoinitiated graft copolymerizations. For instance, when polymers and copolymers of vinyl ketone decompose in ultraviolet light in the presence of acrylonitrile, methyl methacrylate, or vinyl acetate, graft copolymers form[357]:

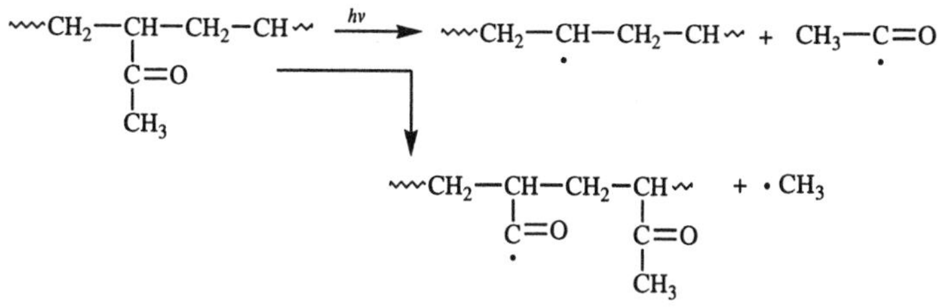

The free radicals that are unattached to the backbone polymers, like the methyl radical shown above, also initiate polymerizations and considerable amounts of homopolymers form as well.

In some instances, graft copolymers form as a result of chain transferring that takes place after photodecomposition of the photolabile materials. An example is formation of graft copolymers of polyacrylamide on natural rubber, poly(vinyl pyrrolidone), or dextrin with the aid of benzophenone and ultraviolet light.[358] The free radicals from photodecomposition of benzophenone react with the polymers by chain transferring. The growth of acryalamide is subsequently initiated from the polymer backbones. Phototendering dyes can be used in this manner with cellulose.[359] Thus, anthraquinone dyes can be adsorbed to cellulose. Upon irradiation, proton abstractions take place, creating initiating radicals on the backbone polysaccharide:

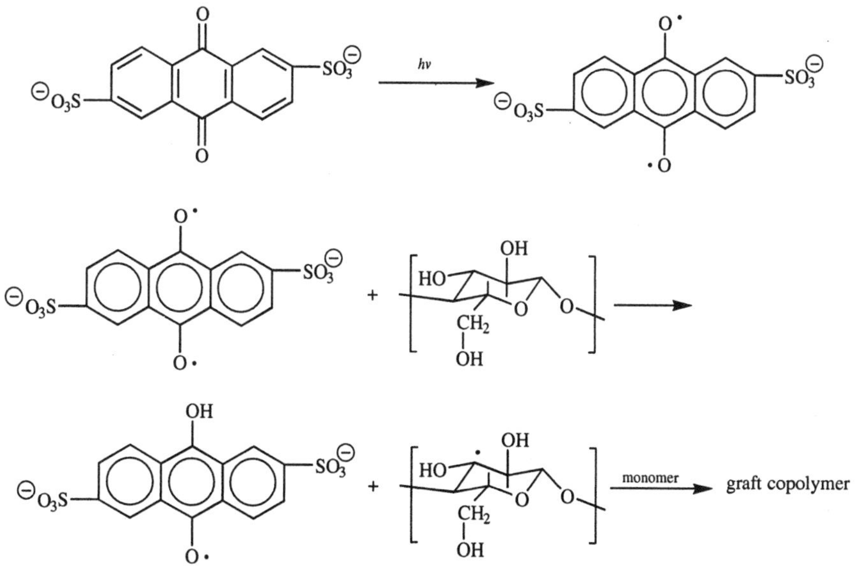

It is believed that the above dye monoradicals disproportionate to hydroquinones and quinones. Transfer reactions to solvent lead to formations of homopolymers. This gives high yields of graft copolymers of methyl methacrylate with cellulose. The same is true of acrylonitrile.[360] On the other hand, only small quantities of graft copolymers form with styrene or vinyl acetate monomers.[360]

It is also possible to form graft copolymers on the surface of fibers by coating them with photoinitiators, like benzophenone, together with a monomer and then irradiating them with ultraviolet light.[414] Similar to the action of the anthraquinone dyes shown above, benzophenone in the excited triplet state mainly abstracts hydrogens and forms radicals on the surface.[415]

8.7.6. Graft-Copolymer Formation With the Aid of High-Energy Radiation

High-energy radiation sources include gamma rays from radioactive elements, electron beams from accelerators, and gamma rays from nuclear reactors. The energy radiated by these sources is sufficiently high to rupture covalent bonds. This results in formations of free radicals. Several different methods are used to form graft copolymers. One of them is *irradiation in open air*. The free radicals that form scavenge oxygen and form peroxides and hydroperoxides on the polymeric chains. These are subsequently decomposed in the presence of monomers to form graft copolymers. When they are decomposed thermally,[361] the hydroperoxides yield much greater quantities of homopolymers than do the peroxides. However, when the decompositions are conducted at room temperature by redox mechanisms, formations of homopolymers are reduced.[361,362] Another method is *irradiation in vacuum*. This results in formations of trapped radicals on the polymer backbones. After irradiation, the polymers are heated in the absence of oxygen and in the presence of vinyl monomers. The best results are obtained when irradiations are done at low temperatures, below T_g of the polymers. A high degree of crystallinity is also beneficial, because mobility of the chains results in loss of trapped radicals. When the monomers are added, however, heat must be applied, but this can result in loss of some of the radicals. The third method is *mutual irradiation* in an inert atmosphere of polymers and monomers together. The polymers are either swelled or dissolved by the monomers. The relative sensitivities of the two species, the monomer and the polymer, to radiation can be important factors in this third procedure. Efficiency of grafting depends upon formations of free radicals on the polymer backbones. If only a small number of free radicals form, the irradiations produce mainly homopolymers. Also, if the polymers tend to degrade from the irradiation, block copolymers form instead. The presence of solvents and chain-transferring agents tends to lower the amount of the grafting.[363]

Recently, it was shown[409] that energetic heavy ions can also produce graft copolymers. The results appear similar to those obtained by electron beams. Also, many papers reported the use of plasma for surface modification of films. The process can result in the formation of graft copolymers when it is accompanied by the introduction of a monomer or monomers. One such example is the use of an argon plasma to graft polyacrylamide to polyaniline films.[410] The near-ultraviolet-light plasma induces the reaction. Other monomers that can be grafted by this reaction are 4-styrenesulfonic acid and acrylic acid.[410]

8.7.7. Preparation of Graft Copolymers With Ionic Chain-Growth and Step-Growth Polymerization Reactions

Both anionic and cationic mechanisms can be used to form graft copolymers. A typical example of graft-copolymer formation by an anionic mechanism is grafting polyacrylonitrile to polystyrene[364]:

$$\sim CH_2-CH\sim (C_6H_4Cl) + 2\,[C_{10}H_8]^{\bullet -}Na^+ \longrightarrow \sim CH_2-CH\sim (C_6H_4^{\ominus}\,Na^{\oplus}) + NaCl + 2\,C_{10}H_8$$

$$\sim CH_2-CH\sim (C_6H_4^{\ominus}\,Na^{\oplus}) + CH_2{=}CH{-}C{\equiv}N \longrightarrow \sim CH_2-CH\sim (C_6H_4-CH_2-\underset{C\equiv N}{CH}-CH_2-\underset{C\equiv N}{CH}\sim)$$

Another example of ionic graft copolymerization is a reaction carried out on pendant olefinic groups using Ziegler–Natta catalysts in a coordinated anionic-type polymerization.[365] The procedure consists of two steps. In the first, diethylaluminum hydride is added across the double bonds. In the second the product is treated with a transition metal halide. This yields an active catalyst for polymerizations of α-olefins. By this method polyethylene and polypropylene can be grafted to butadiene styrene copolymers. Propylene monomer polymerization results in formations of isotactic polymeric branches:

$$\sim CH_2-\underset{CH=CH_2}{\underset{|}{CH}}\sim \xrightarrow{Al(C_2H_5)_2H} \sim CH_2-\underset{CH_2-CH_2-Al(C_2H_5)_2}{\underset{|}{CH}}\sim$$

$$\xrightarrow[CH_2=CH-CH_3]{TiCl_4\ (TiCl_3,\ VoCl_3)} \sim CH_2-\underset{CH_2-CH_2-(-CH_2-\overset{CH_3}{\overset{|}{CH}}-)_n-}{\underset{|}{CH}}\sim$$

Another example is formation of graft copolymers of formaldehyde with starch, dextrin, and poly(vinyl alcohol).[366] This procedure is also carried out in two steps. Potassium naphthalene is first reacted with the backbone polymer in dimethylsulfoxide. The formaldehyde is then introduced in gaseous form to the alkoxide solution.

A similar reaction can be used to form graft copolymers of poly(ethylene oxide) on cellulose acetate.[391] Poly(ethylene oxide) can also be grafted to starch. For instance, a preformed polymer[392] terminated by chloroformate end groups can be used with potassium starch alkoxide:

$$\text{starch}-O^{\ominus}K^{\oplus} + Cl-\overset{O}{\overset{\|}{C}}-O-(-CH_2-CH_2-O-)_n-CH_2-CH_2-O-\overset{O}{\overset{\|}{C}}-Cl \longrightarrow$$

$$\text{starch}-O-\overset{O}{\overset{\|}{C}}-O-(-CH_2-CH_2-O-)_n-CH_2-CH_2-O-\overset{O}{\overset{\|}{C}}-Cl$$

The products are water soluble. The efficiency of the coupling process, however, decreases with an increase in the DP of poly(ethylene oxide).

Lithiated polystyrene reacts readily with halogen-bearing polymers like polychlorotrifluoroethylene.[411] This can be utilized in the formation of graft copolymers. The reactions can be conducted in solutions as well as in preparations of surface grafts on films.[411]

An example of a cationic grafting reaction is the formation of a graft copolymer of polyisobutylene on polystyrene backbones.[393] Polystyrene is chloromethylated and then reacted with aluminum bromide in carbon disulfide solution. This is followed by the introduction of isobutylene:

$$\sim CH_2-\underset{C_6H_5}{CH}-CH_2-\underset{C_6H_5}{CH}\sim + Cl-CH_2-O-CH_3 \longrightarrow \sim CH_2-\underset{C_6H_5}{CH}-CH_2-\underset{C_6H_4-CH_2-Cl}{CH}\sim$$

$$\xrightarrow{AlBr_3} \sim CH_2-\underset{C_6H_5}{CH}-CH_2-\underset{C_6H_4-CH_2^{\oplus}\ AlBr_3Cl^{\ominus}}{CH}\sim \xrightarrow{CH_2=C(CH_3)-CH_3} \sim CH_2-\underset{C_6H_5}{CH}-CH_2-\underset{C_6H_4-(CH_2)_2-\overset{\oplus}{C}-(CH_3)_2\ \overset{\ominus}{AlBr_3Cl}}{CH}\sim \xrightarrow{etc.}$$

The above, however, yields only 5–18% of a graft copolymer, even at –60 °C. It is possible that considerable amounts of crosslinking occur at the reaction conditions and may perhaps be the reason for the low yield.[393]

Another example is grafting to cellulose. BF_3 can be adsorbed to the surface of the polymer. It then reacts with hydroxy groups and yields reactive sites for cationic polymerization of α-methyl styrene and isobutylene.[402] These reactions are carried out at –80 °C.

Cationic graft copolymerizations of trioxane can be carried out with the help of reactive C–O–C links in a number of polymers, like poly(vinyl acetate), poly(ethylene terephthalate), and poly(vinyl butyral).[403] Many graft copolymers can also be formed by ring opening polymerizations.[404] The reactions with active hydrogens on the pendant groups, like hydroxy, carboxyl, amine, amide, thiol, and others, can initiate some ring-opening polymerizations. An example is preparation of graft copolymers of ethylene oxide with styrene.[405] Copolymers of styrene with 1–2% of hydrolyzed vinyl acetate (vinyl alcohol after hydrolysis) can initiate polymerizations of ethylene oxide and graft copolymers form.

Recently, solutions of polysilanes were treated with controlled amounts of triflic acid (CF_3SO_2OH) in CH_2Cl_2 and afterward with tetrahydrofuran. This yielded a graft copolymer of poly(tetramethylene oxide) on polysilane backbones.[412]

An interesting series of papers was published by Kennedy and coworkers on the use of alkylaluminum compounds as initiators of graft copolymerizations.[407] Allylic chlorines form very active carbon cations in the presence of this initiator. This is also true of macromolecular carbon cation sources.[402] As a result, very high grafting efficiency is achieved in many different polymerizations using macromolecular cationogens and alkylaluminum compounds. In some instances the formation of graft copolymers is greater than 90%. The grafting reaction can be illustrated as follows[407]:

~~CH₂—CH~~ (p-CH_2Cl phenyl) + 2-methyl-2-oxazoline (H_3C, N, O ring) $\xrightarrow{Al(C_2H_5)Cl}$ ~~CH₂—CH~~ (phenyl) CH_2–(–N(–C(=O)–CH_3)–CH_2–CH–)$_n$

The temperatures of the reactions and the nature of the aluminum compounds are the most important synthetic variables.[407]

On the other hand, many other graft copolymerizations by a cationic mechanism suffer from low grafting efficiencies. They are also often accompanied by large formations of homopolymers. The use, however, of living cationic processes appears to overcome this drawback. An illustration of this can be another preparation of a graft copolymer of polyisobutylene on a polystyrene backbone[413]:

~~CH₂—CH~~ ~CH₂—CH~~ (phenyl; phenyl–$C(CH_3)_2$–O–C(=O)–CH_3) $\xrightarrow[CH_2Cl_2]{BCl_3}$ ~~CH₂—CH~~ ~CH₂—CH~~ (phenyl; phenyl–$C(CH_3)_2$–O(- -BCl_3)–C(=O)–CH_3)

$\xrightarrow{\text{isobutylene}}$ ~~CH₂—CH~~ ~CH₂—CH~~ (phenyl; phenyl–$C(CH_3)_2$–(CH_2–$C(CH_3)_2$–)$_x$)

8.7.8. Miscellaneous Graft Copolymerizations

In a rather interesting reaction, ethylene oxide can be graft copolymerized with nylon-6,6.[406] Formation of the graft copolymer greatly enhances flexibility of the material, while the high melting point of the nylon is still maintained. Thus, nylon 6,6 that contains as much as 50% by weight of grafted poly(ethylene oxide) still melts at 221 °C and has an apparent T_g below –40 °C. It also maintains flexibility and other useful properties over a wide range of temperatures[406]:

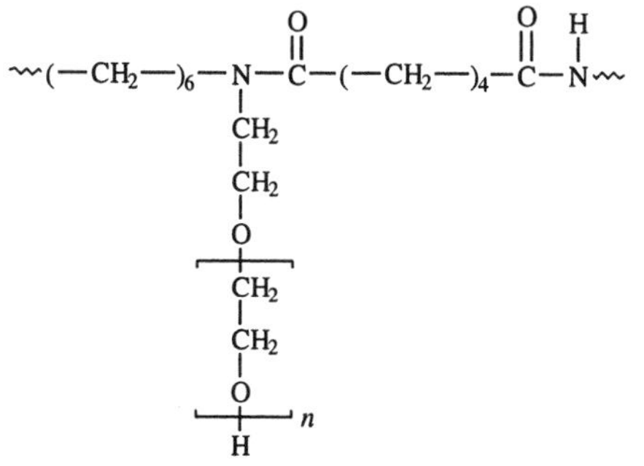

$$\sim\!\!(-CH_2-)_6-\underset{\underset{\underset{\underset{\left[\underset{\underset{O}{|}}{\overset{CH_2}{\underset{|}{CH_2}}}\right]_n}{|}}{O}}{\underset{|}{CH_2}}}{\underset{|}{CH_2}}}{N}-\overset{O}{\overset{\|}{C}}-(-CH_2-)_4-\overset{O}{\overset{\|}{C}}-\overset{H}{\overset{|}{N}}\!\!\sim$$

(terminal H on the grafted poly(ethylene oxide) chain)

An entirely different procedure can be used to form graft copolymers by a step-growth polymerization.[408] Formaldehyde is condensed with either phenol, *p*-cresol, or *p*-nonyl phenol and the resin is attached to either nylon 6, nylon 6,6, nylon 6,10, or nylon 11 backbones. Initially, the formaldehyde is prereacted with an excess of phenol in the presence of the nylon, but without any catalyst, at temperatures high enough to cause condensation. This is followed by addition of toluenesulfonic acid at a lower temperature. At that point, when free formaldehyde is no longer present in the reaction mixture to cause gelation, the novolac molecules attach themselves to the nylon backbones. The excess phenol is washed away, leaving pure graft copolymers.

8.8. Block Copolymers

These polymers consist of two or more strands of different polymers attached to each other. There does not appear to be any general stipulation to the minimum size of each block. There does appear to be, however, general agreement that each sequence should be larger than just a few units. In describing a block copolymer it is helpful if the following structural parameters are available to characterize the material:

1. Copolymer sequence distribution as well as the length and the number of blocks.
2. The chemical nature of the blocks.
3. The average molecular weight and the molecular weight distribution of the blocks and of the copolymer.

Block copolymers, particularly of the A–B–A type, can exhibit properties that are quite different from those of random copolymers and even from mixtures of homopolymers. The physical behavior of block copolymers is related to their solid state morphology. Phase separation occurs often in such copolymers. This can result in dispersed phases consisting of one block dispersed in a continuous matrix from a second block. Such dispersed phases can be hard domains, either crystalline or glassy, while the matrices are soft and rubber-like.

8.8.1. Block Copolyesters

Two polyester homopolymers can react and form block copolymers in a molten state at temperatures high enough for ester interchange.[414] As the reaction mixtures are stirred and heated, the interchanges initially involve large segments. With time, however, smaller and smaller segments form as the transesterifications continue. To prevent eventual formation of a random copolymer, the reactions should be limited in time.

Ester interchange reactions can be retarded, particularly when esterification catalysts like zinc of calcium acetate are present, by adding phosphorous acid or triphenyl phosphite.[415] This improves the chances of forming block copolymers. It can be applied to preparations of block copolymers of poly(ethylene terephthalate) with poly(ethylene maleate), poly(ethylene citraconate), and poly(ethylene itaconate).[416] With ester interchange catalysts, like titanium alkoxides or their complexes, melt randomization may be inhibited by adding arsenic pentoxide that deactivates them.[417]

Block copolyesters also form in reactions between hydroxy and acid chloride-terminated prepolymers.[419] This can take place in the melt or in solution in such solvents as chlorobenzene or *o*-dichlorobenzene.[418] For relatively inactive acid chlorides, like terephthaloyl chloride, high reaction temperatures are required. Phosgene also reacts with hydroxy-terminated polyesters to form block copolymers. The reactions must be carried out in an inert solvent. Block copolyethers also form readily by ester interchange reactions with low molecular weight diesters[420]:

$$(n+1)\ CH_3O-\overset{O}{\overset{\|}{C}}-C_6H_4-\overset{O}{\overset{\|}{C}}-OCH_3 + n\ HOCH_2CH_2OH + H\text{-}[O(CH_2)_4]_m\text{-}OH \longrightarrow$$

$$-[-\overset{O}{\overset{\|}{C}}-C_6H_4-\overset{O}{\overset{\|}{C}}-(-OCH_2CH_2O-\overset{O}{\overset{\|}{C}}-C_6H_4-\overset{O}{\overset{\|}{C}}-)_{\bar{n}}-(-OCH_2CH_2CH_2CH_2-)_{\bar{m}}-O-]_{\bar{o}}- \quad + (2n+2)\ CH_3OH$$

Acetates of tin, lead, manganese, antimony, and zinc as well as esters of orthotitanates catalyze the reactions.[421] Optimum temperatures for these reactions are between 230–260 °C at 0.03–1 mm Hg pressure.[421] Block copolymers can also form by ring-opening polymerizations of lactones, when carboxy-terminated macromolecular initiators are used[422]:

$$\sim CH_2-\underset{C_6H_5}{CH}-CH_2-\underset{C_6H_5}{\overset{\ominus}{CH}}\ \overset{\oplus}{Na} + CO_2 \longrightarrow \sim CH_2-\underset{C_6H_5}{CH}-CH_2-\underset{C_6H_5}{CH}-\overset{O}{\overset{\|}{C}}-O\ Na$$

$$\xrightarrow{x\ \text{(}H_2C-C(CH_3)_2-C(=O)-O\text{, ring)}} \sim CH_2-\underset{C_6H_5}{CH}-CH_2-\underset{C_6H_5}{CH}-\overset{O}{\overset{\|}{C}}-(-O-CH_2-\underset{H_3C}{\overset{H_3C}{C}}-\overset{O}{\overset{\|}{C}}-)_x-O^{\ominus}$$

8.8.2. Block Copolyamides

Simple melt blending reactions can also be applied to preparations of block copolyamides, similarly to the process for polyesters. With time, total equilibrium conditions also are gradually achieved in the melt.[423] Interfacial polycondensation is also useful in preparation of block copolymers. When mixed diacid chlorides and/or mixed diamines are reacted, the more active diacid chlorides and/or diamines react preferentially and blocks form. In addition, it is possible to carry out the growth of one of the segments first, to a fairly large size, and follow it by addition of the other comonomers.[424]

8.8.3. Polyurethane–Polyamide Block Copolymers

These block copolymers can be formed in many ways. One technique is to prereact a diamine with a diacid chloride. The polyamide that forms is then treated with bischloroformate to attach to polyurethane blocks.[425] The process can be reversed, and the polyurethane can be formed first and then attached to polyamide blocks.[425]

8.8.4. Polyamide–Polyester Block Copolymers

Block copolymers consisting of polyamide and polyester blocks can form through melt blending.[426] The reactions probably involve aminolyses of the terminal ester groups of the polyesters by the terminal amine groups of the polyamides. Ester interchange catalysts accelerate the reaction.[427]

Block polyester–polyamides also form through initiation of ring-opening polymerizations of lactones by the terminal amine groups of the polyamides[428]:

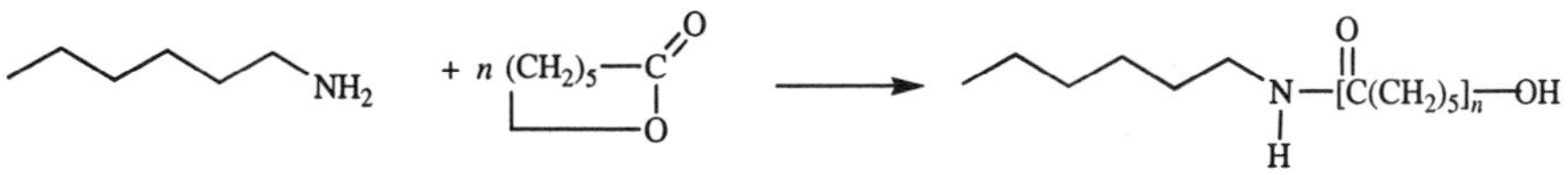

$$\sim\sim NH_2 + n\ (CH_2)_5\text{—}C(=O)\text{—}O\ (\text{ring}) \longrightarrow \sim\sim N(H)\text{—}[\overset{O}{\overset{\|}{C}}(CH_2)_5]_n\text{—}OH$$

If the polyamide has terminal amine groups at both ends, then triblock copolymers form.

8.8.5. Polyurethane Ionomers

These materials were reviewed as a special class of block copolymers.[433] They are linear polyaddition products of diisocyanates containing nonrandom distributions of ionic centers. The preparations are similar to those of polyurethane elastomers that are described in Chapter 6. One example is a material prepared from a high molecular weight polyester that is free from ionic centers and that is terminated by isocyanate groups at each end. The prepolymer is coupled with *N*-methylamino-2,2′-diethanol to form a segmented polymer:

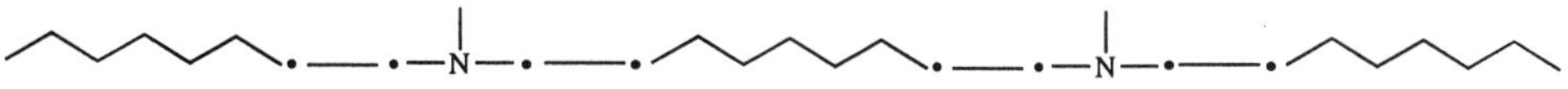

$$\sim\sim\bullet\text{——}\bullet\text{—}\underset{|}{\overset{|}{N}}\text{—}\bullet\text{——}\bullet\sim\sim\bullet\text{——}\bullet\text{—}\overset{|}{N}\text{—}\bullet\text{——}\bullet\sim\sim$$

Similar products form from isocyanate-terminated polyethers. These materials can be crosslinked with difunctional quaternizing agents, such as 1,4-bis(chloromethyl)benzene[434]:

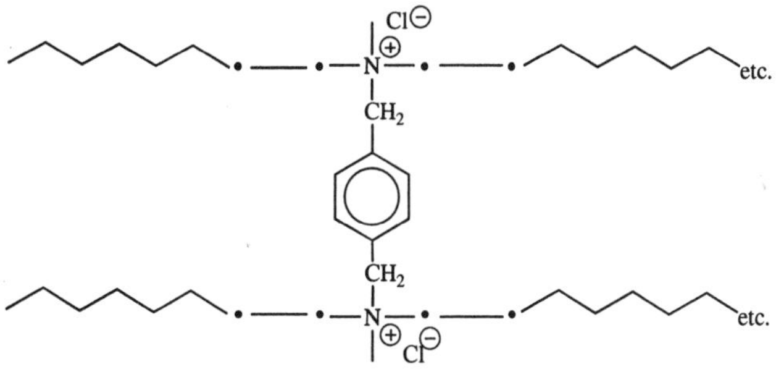

$$\sim\sim\bullet\text{——}\bullet\text{—}\overset{|\ Cl^{\ominus}}{\underset{CH_2}{\overset{\oplus}{N}}}\text{—}\bullet\text{——}\bullet\sim\sim \text{etc.}$$
$$|$$
$$C_6H_4$$
$$|$$
$$\sim\sim\bullet\text{——}\bullet\text{—}\overset{CH_2}{\underset{|\ Cl^{\ominus}}{\overset{\oplus}{N}}}\text{—}\bullet\text{——}\bullet\sim\sim \text{etc.}$$

The products are cationic ionomers. Anionic ionomers form very similarly through coupling of chains with bifunctional anionic "chain lengtheners"[435]:

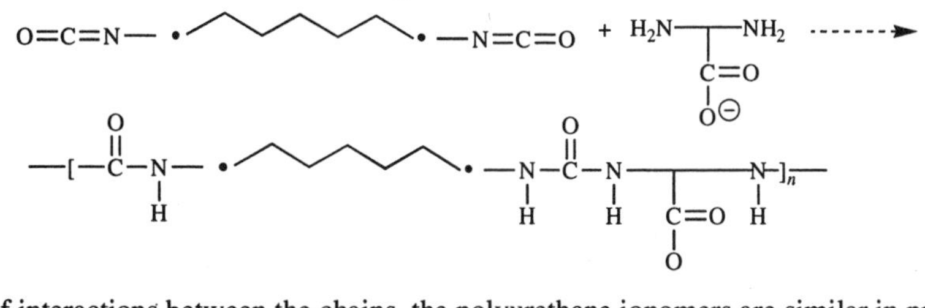

$$O{=}C{=}N\text{—}\bullet\sim\sim\bullet\text{—}N{=}C{=}O + H_2N\text{—}\underset{\underset{O^{\ominus}}{|}}{\underset{C=O}{|}}\text{—}NH_2 \dashrightarrow$$

$$\text{—}[\text{—}\overset{O}{\overset{\|}{C}}\text{—}\underset{H}{N}\text{—}\bullet\sim\sim\bullet\text{—}\underset{H}{N}\text{—}\overset{O}{\overset{\|}{C}}\text{—}\underset{H}{N}\text{——}\underset{\underset{O}{|}}{\underset{C=O}{|}}\text{——}\underset{H}{N}\text{—}]_n\text{—}$$

Because of interactions between the chains, the polyurethane ionomers are similar in properties to crosslinked elastomers. In solution they are strongly associated.

8.8.6. Block Copolymers of Poly(α-olefin)s

These block copolymers form readily when appropriate Ziegler–Natta catalysts are used.[436] This is discussed in Chapters 3 and 5. In addition, there is a special technique for preparations of such block copolymers. At the outset of the reaction, only one monomer is used in the feed. A typical catalyst might be α-$TiCl_3$/$(C_2H_5)_3Al$. After the first monomer has been bubbled in for a short period, perhaps 5 minutes, the addition is stopped and the unreacted monomer removed by evacuation or by flushing. The second monomer is then introduced and may also be bubbled in for the same period. The addition of the second monomer may then be stopped, the second monomer evacuated, and the whole process might be repeated. If an equal length of each block is desired, the addition times of each monomer may be varied to adjust for different rates of polymerization.[436]

8.8.7. Simultaneous Use of Free-Radical and Ionic Chain-Growth Polymerizations

This technique allows the formation of many different types of block copolymers.[437] Lithium metal can be used to initiate polymerizations in solvents of varying polarity. Monomers, like styrene, α-methylstyrene, methyl methacrylate, butyl methacrylate, 2-vinylpyridine, 4 vinylpyridine, acrylonitrile, or methyl acrylate, can be used. The mechanism of initiation depends upon the formation of ion radicals through reactions of lithium with the double bonds:

$$Li + CH_2{=}CH(C_6H_5) \longrightarrow H_2\dot{C}{-}\overset{\ominus}{C}H(C_6H_5)\ Li^{\oplus}$$

Propagation reactions proceed from both active sites, the radical and the carbanion. When two different monomers are present, free-radical propagation favors formation of copolymers, while propagation at the other end favors formation of homopolymers. There is therefore a tendency to form AB——B type block copolymers.

8.8.8. Preparation of Block Copolymers by Homogeneous Ionic Copolymerization

Formation of block copolymers by this method depends upon the ability to form "living" chain ends. Among anionic systems, the following polymerizations fit this requirement:

1. Polymerizations of nonpolar monomers with alkali metal–aromatic electron transfer initiators in ethers.[438]
2. Polymerizations of nonpolar monomers with organolithium compounds in hydrocarbon solvents.[439]
3. Acrylonitrile polymerizations in dimethyl formamide initiated by sodium triethylthioisopropoxyaluminate at –78 °C.[440]
4. Copolymerizations of hexafluoroacetone and cyclic oxides initiated by CsF.[441]
5. Polymerization of alkyl isocyanates initiated by organoalkali species in hydrocarbons at –78 °C.[442]

Among the cationic "living" polymerizations that can be used for block copolymer formation are:

1. Polymerizations of isobutylene[461] and/or vinyl ethers[463] with appropriate catalysts. This includes formation of block copolymers from the two types of monomers.[465]
2. Polymerizations of tetrahydrofuran with the aid of chlorobenzenediazonium hexafluorophosphate,[443] triphenylmethyl hexachloroantimonate,[444] or phosphorus pentafluoride.[445]
3. Polymerization of *p*-methyl styrene, *N*-vinylcarbazole, and indene with appropriate catalysts.

The preparations by anionic mechanism of A——B——A type block copolymers of styrene and butadiene can be carried out with the styrene being polymerized first. Use of alkyl lithium initiators in hydrocarbon solvents is usually a good choice, if one seeks to form the greatest amount of *cis*-1,4 microstructure.[446] This is discussed in Chapter 3. It is more difficult, however, to form block copolymers from methyl methacrylate and styrene, because "living" methyl methacrylate polymers fail to initiate polymerizations of styrene.[447] The poly(methyl methacrylate) anions may not be sufficiently basic to initiate styrene polymerizations.[445]

A "living" cationic polymerization of tetrahydrofuran, using BH_3 as the initiator in the presence of epichlorohydrin and 3,3-bis(chloromethyl)oxacyclobutane,[448] results in the formation of block copolymers. Two types form. One is an A——B type. It consists of polytetrahydrofuran blocks attached to blocks of poly(3,3-bis(chloromethyl)-oxacyclobutane. The other one is an A——AB——B type.[448]

The preparation of well-defined sequential copolymers by an anionic mechanism has been explored and utilized commercially for some time now. Initially, the cationic methods received less attention until it was demonstrated by Kennedy[450] that a large variety of block copolymers can be formed. The key to Kennedy's early work is tight control over the polymerization reaction. The initiation and propagation events must be fundamentally similar, although not identical[450]:

Ion generation:

$$RX + MeX_n \rightleftharpoons R^{\oplus}\,{}^{\ominus}MeX_{x+1}$$

Cationization and propagation:

$$R^{\oplus} + CH_2{=}C\!\!<\; \longrightarrow R{-}CH_2{-}C^{\oplus} \xrightarrow{+\;CH_2=C<} R{-}CH_2{-}C\!\sim\!\sim\!\sim\!\sim CH_2{-}C^{\oplus}$$

In this scheme, chain transfer to monomer must be absent and the termination is well defined:

Termination:

$$R{-}CH_2{-}C\!\sim\!\sim\!\sim\!\sim CH_2{-}C^{\oplus}\;{}^{\ominus}MeX_{n+1} \longrightarrow R{-}CH_2{-}C\!\sim\!\sim\!\sim\!\sim CH_2{-}CX + MeX_n$$

where X is a halogen and Me is a metal.

This allows formation of macromolecules with terminal halogens. They can be used to initiate new and different polymerizations.

Three methods were developed to overcome transfer to monomer.[450] These are: (1) use of *inifers*; (2) use of proton traps; and (3) establishing conditions under which the rate of termination is much faster than the rate of transfer to monomer. The first one, the inifer method, is particularly useful in formation of block copolymers. It allows the preparation of *head* and *end* (α and ω) functionalized telechelic polymers. Bifunctional initiators and transfer agents (*inifers*) are used. The following illustrates the concept[450]:

Ion generation:

$$XRX + MeX_n \rightleftharpoons XR^{\oplus} + MeX_{n+1}^{\ominus}$$

Cationization and propagation:

$$XR^{\oplus} + CH_2{=}C\!\!<\; \longrightarrow XR{-}CH_2{-}C^{\oplus} \xrightarrow{n\;CH_2=C<} XR\!\sim\!\sim\!\sim\!\sim CH_2{-}C^{\oplus}$$

Chain transfer with inifers:

$$XR\!\sim\!\sim\!\sim\!\sim CH_2{-}C^{\oplus} + XRX \longrightarrow XR\!\sim\!\sim\!\sim\!\sim CH_2{-}C{-}X + XR^{\oplus}$$

$$\downarrow$$

$$X{-}C{-}CH_2\!\sim\!\sim\!\sim\!\sim R \sim\!\sim\!\sim\!\sim CH_2{-}C{-}X$$

In the above scheme, the inifer, XRX, is usually an organic dihalide. If chain transferring to the inifer is faster than chain transferring to the monomer, the polymer end groups become exclusively terminated with halogens.

It is also possible to carry out "living" cationic polymerization of isobutylene, initiated by a difunctional initiator.[461] This results in a formation of bifunctional "living" segments of polyisobutylene that are soft and rubbery. Upon completion of the polymerization, another monomer, one that yields stiff segments and has a high T_g value, like indene, is introduced into the living charge. Polymerization of the second monomer is initiated from both ends of the formed polyisobutylene. When the reaction is complete, the polymerization is quenched. Preparations of a variety of such triblock and star block polymers have been described.[461]

A new technique was developed recently, by introducing cationic to anionic transformation.[464] A "living" carbocationic polymerization of isobutylene is carried out first. After it is complete, the ends of the chains are transformed quantitatively to polymerization-active anions. The additional blocks are then built by an anionic polymerization. A triblock polymer of poly(methyl methacrylate)–polyisobutylene–poly(methyl methacrylate) can thus be formed. The transformation involves several steps. In the first, a compound like toluene is Friedel-Craft alkylated by α,ω-di-*tert*-chloropolyisobutylene. The ditolylpolyisobutylene which forms is lithiated in step two to form α,ω-dibenzyllithium polyisobutylene. It is then reacted with 1,1-diphenylethylene to give the corresponding dianion. After cooling to –78 °C and dilution, methyl methacrylate monomer is introduced for the second polymerization[464] in step three.

Formation of block copolymers from polymers with functional end groups has been used in many ways. In anionic polymerization, various technique were developed for terminating chain growth with reactive end groups. These end groups allow subsequent formations of many different block copolymers. One such active terminal group can be toluinediisocyanate.[449] The isocyanate group located *ortho* to the methyl group is considerably less reactive toward the lithium species due to steric hindrance.[448] The unreacted isocyanate group can be used for attachment of various polymers that are terminated by hydroxy, carboxy, or amine groups. Other functional compounds that can be used in such reactions are alkyl or aryl halides, succinic anhydride, *n*-bromophthalimide,[448] and chlorosilanes.[449]

Block copolymers can often offer properties that are unattainable with simple blends or random copolymers.[364] This led to many efforts to combine dissimilar materials, like hydrophilic with hydrophobic, or hard with soft segments, as was shown earlier. A recent paper[462] describes the formation of block copolymers containing helical polyisocyanide and an elastomeric polybutadiene. Compound $[(\eta^3\text{-}C_3H_5)\text{–}Ni(OC(O)CF_3)]_2$ was used to carry out "living" polymerization of butadiene and then followed by polymerization of *tert*-butyl isocyanide to a helical polymer.

8.8.9. Special Reactions for Preparation of Block Copolymers

A special case is the use of the *Witting* reaction. Poly(*p*-phenylene pentadienylene)[455] is prepared by this reaction first. This is utilized in a preparation of a block copolymer[456] according to the following scheme:

$$Br^{\ominus}\,(C_6H_5)_3\text{–}\overset{\oplus}{P}\text{–}(\text{–}CH_2\text{–})_{n+2}\text{–}\overset{\oplus}{P}\text{–}(C_6H_5)_3\,Br^{\ominus} \xrightarrow[\text{THF}]{n\text{ BuLi}} (C_6H_5)_3\text{–}P{=}CH\text{–}(\text{–}CH_2\text{–})_n\text{–}CH{=}P\text{–}(C_6H_5)_3$$

polymer 1

$$H\text{–}\overset{O}{\overset{\|}{C}}\text{–}[\text{–}C_6H_4\text{–}CH{=}CH\text{–}CH{=}CH\text{–}CH{=}CH\text{–}]_m\text{–}C_6H_4\text{–}\overset{O}{\overset{\|}{C}}\text{–}H + \text{polymer 1} \longrightarrow$$

$$\left[{=}\underset{H}{\underset{|}{C}}\text{–}[\text{–}C_6H_4\text{–}CH{=}CH\text{–}CH{=}CH\text{–}CH{=}CH\text{–}]_m\text{–}C_6H_4\text{–}\underset{H}{\underset{|}{C}}{=}CH\text{–}(\text{–}CH_2\text{–})\text{–}CH{=}\right]_z$$

Recently, the preparation of block copolymers by "living" ring-opening olefin metathesis polymerization was reported.[457] Initially norbornene or *exo*-dicyclopentadiene polymerize by bis(η^5-cyclopentadienyl)titanacylbutane. The resulting living polymers react with terephthaldehyde to form polymers with terminal aldehyde groups. The aldehyde group in turn initiates polymerizations of *t*-butyldimethylsilyl vinyl ether by aldol-group transfer polymerizations.[458] The following is an illustration of the process:

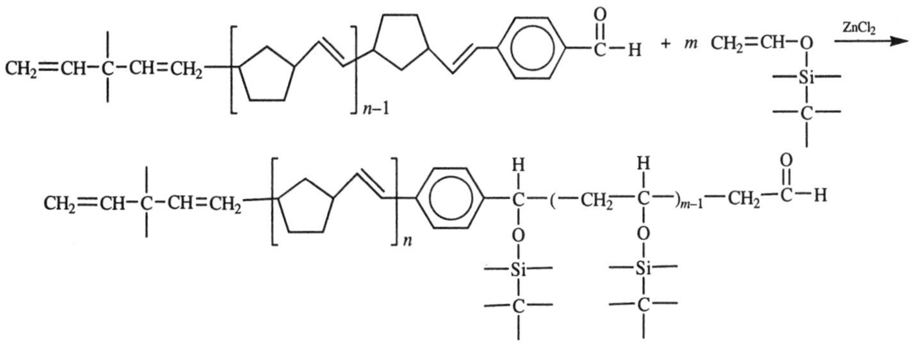

Subsequently, the terminal aldehyde groups are reduced with $NaBH_4$ and the silyl groups cleaved off by treatment with tetrabutylammonium fluoride to produce a hydrophobic–hydrophilic A–B diblock copolymer.

8.8.10. Miscellaneous Block Copolymers

Polyamide–polyether block copolymers can be formed by a variety of techniques. One of them consists of initial preparation of amine-terminated polyethers. This can be done by reacting hydroxy-terminated polyethers with acrylonitrile and then reducing the nitrile groups to amines[429]:

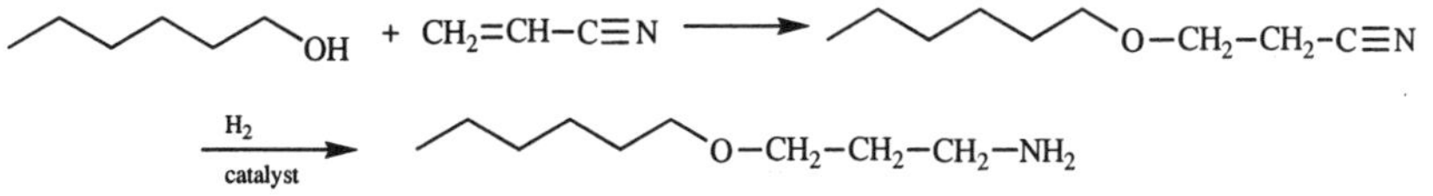

$$\sim\!\sim\!OH + CH_2{=}CH{-}C{\equiv}N \longrightarrow \sim\!\sim\!O{-}CH_2{-}CH_2{-}C{\equiv}N \xrightarrow[\text{catalyst}]{H_2} \sim\!\sim\!O{-}CH_2{-}CH_2{-}CH_2{-}NH_2$$

The products of the reduction condense with carboxylic acid-terminated polyamides to form block copolymers.

Another technique is to form the polyethers with terminal chloride groups.[430] Hydroxy-group-terminated polyethers, for instance, can be converted to halogen-terminated polyethers. The products will react with ammonia and the amine-terminated polymer will react with carboxylic acid-terminated polyamides.[430]

A British patent describes preparations of block copolymers in two steps. In the first, two different salts of hexamethylenediamine are formed: one with carboxylic acid terminated polyoxyethylene and the other with adipic acid (nylon-6,6 salt). In the second step the two salts are reacted in the melt. Caprolactam can be used in place of the second salt.[431] Also, a Japanese paper describes formations of block copolymers by reacting polyoxyethylene in melt condensation reactions with caprolactam in the presence of dicarboxylic acids.[432]

Polyamide-6 (nylon-6) can form block copolymers with rubber[459] and with poly(dimethylsiloxane).[460] In the latter case, the polysiloxane forms first by "living" polymerization and is terminated by an acylated caprolactam. The caprolactam portion of the molecule is then polymerized with the aid of lithium caprolactamate:

$$R-(\underset{CH_3}{\overset{CH_3}{Si}}-O)_x-(\underset{CH=CH_2}{\overset{CH_3}{Si}}-O)_y-\underset{CH_3}{\overset{CH_3}{Si}}-O\sim\sim\sim\overset{O}{\overset{\|}{C}}-\underset{\llcorner(CH_2)_5\lrcorner}{N-C=O} + Li^{\oplus}\,{}^{\ominus}\underset{\llcorner(CH_2)_5\lrcorner}{N-C=O}$$

$$\downarrow z\,H-\underset{\llcorner(CH_2)_5\lrcorner}{N-C=O}$$

$$R-(\underset{CH_3}{\overset{CH_3}{Si}}-O)_x-(\underset{CH=CH_2}{\overset{CH_3}{Si}}-O)_y-\underset{CH_3}{\overset{CH_3}{Si}}-O\sim\sim\sim\overset{O}{\overset{\|}{C}}-(\underset{H}{N}-(CH_2)_5-\overset{O}{\overset{\|}{C}})_z-\underset{\llcorner(CH_2)_5\lrcorner}{N-C=O}$$

This diblock copolymer can be melt-annealed at *ca* 250 °C. It exhibits superior mechanical properties to nylon-6 homopolymer.[460]

8.8.11. Mechanochemical Techniques for Formation of Block Copolymers

These techniques rely upon high shear to cause bond scissions. Ruptured bonds result in formations of free-radical and ionic species.[451] When this application of shear is carried out in the presence of monomers, block copolymers can form. This approach is exploited fairly extensively. Such cleavages of macromolecules can take place during cold mastication, milling, and extrusion of the polymers in the viscoelastic state. Both homolytic and heterolytic scissions are possible. The first yields free-radical and the second ionic species. Heterolytic scissions require more energy but should not be written off as completely unlikely.[451] Early work was done with natural rubber.[452] It swells when exposed to many monomers and forms a viscoelastic mass. When this swollen mass is subjected to shear and mechanical scission, the resultant radicals initiate polymerizations. The mastication reaction was shown to be accompanied by formation of homopolymers.[453] Later, the technique was applied to many different polymers with many different monomers.[454]

8.9. Conductive Polymers

Most polymeric materials, as we usually know them, are insulators. During the past 15–20 years, however, a new class of organic polymers evolved with the ability to conduct electric current.[466] At the present time, it is not completely understood by what mechanism the electric current passes through them. We do know, however, that all conductive polymers are similar in one respect. They all consist of extended π-conjugated systems, namely, alternating single and double bonds along the chain.

One of the earliest known conductive polymers is polysulfurnitride $(SN)_x$, an inorganic material that tends to be explosive, but becomes superconducting at 0.3 K.[466] Since then, many conductive polymers have evolved. The most investigated ones appear to be polyacetylene, polyaniline, polypyrrole, polythiophene, poly(phenylene sulfide), and poly(phenylene vinylene).

Polyacetylene, which was originally prepared by a polymerization of acetylene by Shirakawa with the aid of a Ziegler–Natta catalyst, is an insulator. It can be shaped into a silvery-looking film. Partial oxidation of the film, however, with iodine or other materials transforms it and increases its conductivity 10^9-fold. The process of transforming a polymer to its conductive form through chemical oxidation or reduction is called *doping*. Polyacetylene, which can be *cis* or *trans*, is more thermodynamically stable in the *trans* form and converts from *cis* to *trans* when heated above 150 °C.

Two types of polyacetylene doping are possible:

$$(CH)_n + 1/2\,I_2 \longrightarrow (CH)_n^{\oplus}\;(I_3)_{0.33}^{\ominus} \quad \text{oxidative doping (p-type)}$$

$$(CH)_n + x\text{Na} \longrightarrow (\text{Na}^{\oplus})_x\,[(CH)_n]^{x\ominus} \quad \text{reductive doping (n-type)}$$

The doping process can be reversed and conductive polymers can be undoped again by applying an electrical potential. It causes the dopant ions to diffuse in and out of the structure.

Additional improvements in preparations of polyacetylene came from several developments. One is the use of metathesis polymerization of cyclooctatetraene, catalyzed by a titanium alkylidene complex. The product has improved conductivity, though it is still intractable and unstable. By attaching substituents it is possible to form soluble and more stable materials that can be deposited from solution on various substrates. Substitution, however, lowers the conductivity. This is attributed to steric factors introduced by the substituents that force the double bonds in the polymeric chains to twist out of coplanarity.[467] Recently, a new family of substituted polyacetylenes was described.[468] These polymers form from ethynylpyridines as well as from ethynyldipyridines. The polymerization reaction takes place spontaneously by a quaternization process:

C≡C—H (4-ethynylpyridine) + R—X ⟶ polymer with pyridinium rings (N+ —R, X^-)

where X is a bromine or an iodine.

Like other substituted polyacetylenes, these materials are fairly stable in air and are soluble in polar solvents, also in water. The conductivity, however, is improved over previously reported substituted polyacetylenes to within the range of semiconductors.

Preparation of a highly conductive polyacetylene is achieved when the original Ziegler–Natta catalyst used by Shirakawa is aged in silicone oil at 150 °C. It is believed that this results in the formation of polymers with less defects in the structures. The conductivity of these materials, when doped, approaches that of copper.

In 1979, it was demonstrated that polypyrrole can be formed as a film by electrochemical oxidative polymerization of the pyrrole monomer in acetonitrile. The polymers that forms on the surface of the electrode can be peeled off as flexible, shiny blue-black films. Subsequently, in 1982 it was shown that thiophene can also be electropolymerized oxidatively at the anode. The method allows control over the oxidative potential during the polymerization, yielding doped films with optimized polymer properties. Both polypyrrole and polythiophene differ from polyacetylene in that both form during the polymerization in the doped form and both are stable in air. They are, however, less conductive than the doped polyacetylene. The exact structures of polypyrrole and polythiophene are not known at this time. The process of oxidative polymerization involves very reactive cation radical intermediates. Much of the coupling of the heterocyclic rings together is at the 2 and 5 positions. X-ray photoelectron spectroscopy shows that the polypyrrole formed in this manner has about 30% of the linkages at other than 2 and 5 positions. They might be in the 2 and 3 positions. This introduces "defects" into the hypothetically ideal chain and reduces the conjugation length and, with it, the conductivity.

The flexible films of polypyrrole that form upon electrochemical oxidation are not only stable in air and water, but may also be heated to 200 °C without much change in electrical properties. The oxidative polymerization of pyrrole can be illustrated as follows:

$$\text{pyrrole} \xrightarrow[R_4NClO_4]{\text{oxidation}} [\text{—}C_4H_2NH\text{—}]_n\,(ClO_4)_{0.33} + 2H^{\oplus} + 2e^{\ominus}$$

A regioselective synthesis of highly conductive poly(3-alkylthiophene)s was reported recently.[469] The following synthetic procedure was used:

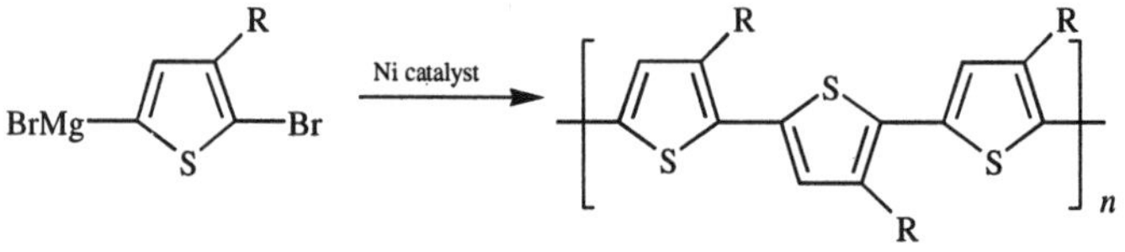

The iodine-doped, unoriented poly(3-dodecyl thiophene)s exhibit an average conductivity of 600 S cm^{-1} and a maximum conductivity of 1000 S cm^{-1}.[469]

Polyaniline was first prepared at the turn of the century. Several oxidation states are known. The conductivity and the color of the material vary progressively with oxidation. Only one form, however, known as the emeraldine salt, is truly conducting. The material can be prepared readily by either electrochemical or chemical oxidation of aniline in aqueous acid media. Common oxidants, such as ammonium peroxydisulfate, can be used. Flexible emeraldine films can be cast from solutions of *N*-methylpyrrolidone and made conductive by protonic doping. This is done by dipping the films in acid or exposing them to acid vapors. The process results in protonation of the imine nitrogen atoms:

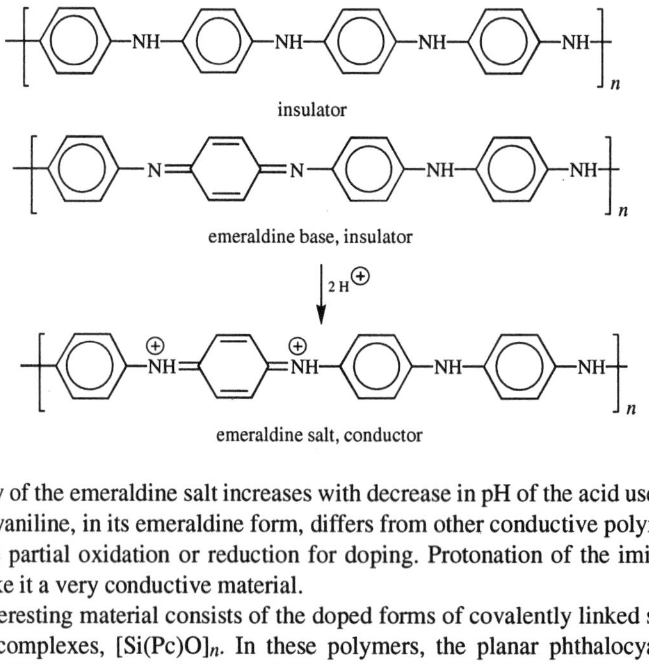

The conductivity of the emeraldine salt increases with decrease in pH of the acid used to dope it. In this respect, polyaniline, in its emeraldine form, differs from other conductive polymers because it does not require partial oxidation or reduction for doping. Protonation of the imine nitrogens is sufficient to make it a very conductive material.

Another interesting material consists of the doped forms of covalently linked siloxane-phthalocyanine (Pc) complexes, $[Si(Pc)O]_n$. In these polymers, the planar phthalocyanine units are apparently stacked face-to-face and form columns, due to the silicon–oxygen–silicon bonds. The polymers appear to be intrinsically metallic systems. The principal pathways of conductivity are perpendicular to the phthalocyanine planes. The extended π–π systems that form result from face-to-face overlaps of the phthalocyanine units. This enables the electrons or holes to travel in a perpendicular direction.

Orientation of conductive polymeric films yields large additional increases in conductivity. Thus, for instance, films of doped, oriented poly(phenylene vinylene)[467] not only have the strength of high-performance polymers, but also their conductivities measure as high as 10^4 S per centimeter. This is approximately one-thousand times greater than that of the unoriented films.

Review Questions

Section 8.1

1. What are the necessary conditions for a fair comparison of the reactivity of functional groups on macromolecules with those on small molecules?
2. When is unequal reactivity observed between large and small molecules?
3. Discuss the effect of diffusion-controlled reactions on the reactivity of macromolecules.
4. Discuss how paired group and neighboring group effects influence random irreversible reactions.
5. Discuss reactions that favor large molecules.

Section 8.2

1. Explain, showing chemical equations, how hydrochlorination of natural rubber is often accompanied by cyclization.
2. Discuss chlorination of natural rubber. How can natural rubber and polybutadiene be brominated?
3. Discuss, with the aid of chemical equations, hydrogenation of 1,4-polybutadiene.
4. How can polyisoprene and polybutadiene be modified by additions of carbenes? Explain and discuss, showing the structures of the starting materials and the products.
5. Illustrate the Prins reactions of rubber with formaldehyde and with glyoxal. How can this reaction be carried out in the solid phase?
6. Discuss polar additions to unsaturated polymers. Give examples. Include the *ene* reaction.
7. How can the Ritter reactions be carried out on isoprenes?
8. How can the Diels-Alder reactions be carried out with unsaturated polymers? Use chemical equations to illustrate.
9. Explain how polybutadiene and isoprene can be epoxidized, giving reagents and showing all byproducts.

Section 8.3

1. Show how, and explain why, *cis*-polybutadiene rearranges to *trans* as a result of irradiation by gamma rays or ultraviolet light. Also, show what happens to polyisoprene when it is irradiated with the ultraviolet light.
2. Discuss, with the aid of chemical equations, the cyclization reactions of rubber.
3. How are polyacrylonitrile fibers converted to graphite-like fibers? Explain, showing all the steps in the thermal cyclization of polyacrylonitrile.
4. Discuss migration of double bonds in polymers.
5. How does poly(4,4′-diphenylpropane isophthalate) rearrange upon irradiation with ultraviolet light?
6. Give examples of the Schmidt and Beckmann rearrangements.

Section 8.4

1. Discuss, with the help of chemical equations, the fluorination reactions of polyethylene.
2. Discuss chlorination of polyethylene and polypropylene.
3. How is chlorosulfonation of polyethylene carried out industrially? Explain and write the chemical equations for the reactions. How is the product used?
4. Discuss chlorination of poly(vinyl chloride).
5. Discuss chlorination of poly(vinylidine chloride) and poly(vinyl fluoride).
6. Describe the reactions of sodium azide with poly(vinyl chloride) and the subsequent reactions of the azide group. Use chemical equations.
7. Describe reactions of poly(vinyl chloride) with sulfur compounds.

8. What are the products of reactions of poly(vinyl chloride) with aniline? Show and explain.
9. Explain how carbanionic centers can be formed on the backbones of poly(vinyl chloride) molecules. Show what subsequent reactions can take place.
10. What happens when poly(vinyl chloride) is treated with organolithium compounds? Explain with the help of chemical equations.
11. Discuss photochlorination of polystyrene.
12. Discuss chloromethylation reactions of polystyrene and its copolymers, showing all chemical reactions.
13. Discuss various reactions of chloromethylated polystyrene with the help of chemical equations.
14. Show examples of the Friedel-Craft reactions of polystyrene and chloromethylated polystyrene.
15. Discuss sulfonation of polystyrene.
16. Discuss nitration of polystyrene, reduction of nitropolystyrene, and the subsequent diazotization reaction.
17. Discuss metalation of polystyrene and subsequent reactions.
18. Discuss reduction reactions of the polymers of methacrylic and acrylic esters with metal hydrides. What solvents give optimum conditions?
19. Discuss nucleophilic substitution reactions of poly(methyl methacrylate), the Arndt–Eister reactions of poly(methacryloyl chloride), and the Curtius and Lossen rearrangements of poly(acryloyl chloride).
20. Show a Diels-Alder reaction of poly(fulrfuryl methacrylate) with maleic anhydride.
21. What are the industrial processes for preparation of poly(vinyl butyral)? Describe the process, showing all chemical reactions.
22. How can the Schotten–Baumann esterifications of poly(vinyl alcohol) be carried out? Explain.

Section 8.5

1. Discuss the vulcanization reactions of rubbers. Show all the chemical reactions.
2. Discuss ionizing radiation crosslinking of polymers.
3. Show the reaction of photocrosslinking of poly(vinyl cinnamate) and explain the mechanism.
4. Show the reactions of photocrosslinking of polymers bearing azide groups.
5. Draw examples of polymers with pendant benzoin and furoin groups.
6. Explain the mechanism of photocrosslinking by $2\pi + 2\pi$ type dimerization. Give examples.
7. Discuss typical ultraviolet-light curable coating compositions by free-radical and by cationic mechanisms.
8. Why are photoinitiators preferred to photosensitizers in UV curable materials?
9. Discuss, with the help of chemical equations, the mechanism of photoreduction of aromatic ketone photoinitiators by hydrogen donors and by electron donors.
10. Discuss photoinitiators for photocrosslinking by cationic mechanisms, showing all reactions and products.
11. Discuss crosslinking reactions by electron beams.

Section 8.6

1. What is meant by microporous and macroreticular supports?
2. Discuss some applications of polymeric reagents. Show examples.
3. How are enzymes immobilized? Discuss, showing chemical reactions.
4. Discuss nonenzymatic catalysts.

Section 8.7

1. What is the mechanism of free-radical graft copolymer formation by the chain-transferring technique? Explain by examples and discuss advantages and disadvantages.
2. Discuss free-radical grafting reactions to polymers with double bonds. Give examples and show reactions.

3. How can macromonomers be used to form graft copolymers? Give several examples.
4. How can polymerizations be initiated from the backbones of the polymers? Explain and give examples.
5. Discuss photochemical preparation of graft copolymers.
6. How is high-energy radiation used to form graft copolymers?
7. Discuss formations of graft copolymers by ionic chain-growth and step-growth polymerizations.

Section 8.8

1. Discuss the advantages of block copolymers over polymer blends.
2. How can block copolyesters be formed? Explain with chemical equations.
3. How can block copolyamides be formed? Explain as in question 2.
4. How can polyurethane–polyamide and polyamide–polyester block copolymers be formed? Explain and show chemical reactions.
5. What are polyurethane ionomers? How are they prepared?
6. Describe the technique for the preparation of block copolymers of poly(α-olefins).
7. How do block copolymers form in simultaneous free-radical and ionic chain-growth polymerizations? Explain and give examples.
8. Discuss formations of block copolymers by homogeneous ionic copolymerization.
9. How can block copolymers form by mechanochemical techniques?

Section 8.9

1. How is polyacetylene formed and made conductive?
2. How are polypyrrole and polythiophene prepared?
3. What is emeraldine salt and how does it form?

References

1. P.J. Flory, *Principles of Polymer Chemistry*, Chapter II, Cornell University Press, Ithaca, New York, 1953.
2. T. Alfrey Jr., Chapt. IA in *Chemical Reactions of Polymers* (E.M. Fetters, ed.), Wiley-Interscience, New York, 1964.
3. D.T.L. Chen and H. Morawetz, *Macromolecules*, **9**, 463 (1976).
4. S. Malkin and E. Fisher, *J. Phys. Chem.*, **66**, 2482 (1962).
5. H. Morawetz, *J. Macromol. Sci.-Chem.*, **A13**, 311 (1979).
6. M. Kamachi, *Makromol. Chem., Suppl.*, **14**, 17 (1985).
7. A. Okamoto, Y. Shimanuki, and I. Mita, *Eur. Polym. J.*, **18**, 545 (1982); *ibid.*, **19**, 341 (1983).
8. K.F. O'Driscoll and H.K. Mahabadi, *J. Polym. Sci., Chem. Ed.*, **11**, 869 (1976).
9. S. Benson and A.M. North, *J. Am. Chem. Soc.*, **84**, 935 (1962).
10. M. Kamachi, *Adv. Polym. Sci.*, **38**, 56 (1981).
11. P.J. Flory, *J. Am. Chem. Soc.*, **61**, 1518 (1939); T.H.K. Barron and E.A. Boucher, *Trans. Faraday Soc.*, **65** (12), 3301 (1969).
12. C.S. Marvel and C.L. Levvesque, *J. Am. Chem. Soc.*, **61**, 3234 (1939).
13. H. Morawetz and P.E. Zimmerling, *J. Phys. Chem.*, **58**, 753 (1954).
14. H. Morawetz and E. Gaetjens, *J. Polym. Sci.*, **32**, 526 (1958).
15. P.E. Zimmerling, E.W. Westhead Jr., and H. Morawetz, *Biochem. Biophys. Acta*, **25**, 376 (1957).
16. V.O. Cherkezyan and A.D. Litmanovich, *Eur. Polym. J.*, **21,** 623 (1985).
17. A.D. Litmanovich, *Eur. Polym. J.*, **16**, 269 (1980).
18. S.Sawant and H. Morawetz, *Macromolecules*, **17**, 2427 (1984).
19. I. Sakurada, *Off. J. IUPAC*, **16** (2–3), 263 (1968).
20. D. Pini, R. Settambolo, A. Raffaelli, and P. Salvadori, *Macromolecules*, **20**, 58 (1987).
21. E. Gaetjens and H. Morawetz, *J. Am. Chem. Soc.*, **83**, 1738 (1961).
22. N. Plate, *Vysokomol. Soyed.*, **A10** (12), 2650 (1968).
23. W.H. Daly, *J. Macromol. Sci.-Chem.*, **A22**, 713 (1985).
24. A.R. Mathieson and J.V. McLaren, *J. Polym. Sci.*, **A,3**, 2555 (1965).
25. J.C. Layte and M. Mandel, *J. Polym. Sci., A,2*, 1879 (1964).

26. F. Fleming and G.R. Marshal, *Tetrahedron Lett.*, 2403 (1970).
27. A. Patchornik and M.A. Kraus, *J. Am. Chem. Soc.*, **92**, 7587 (1970).
28. E. Bondi, M. Fidkin, and A. Patchornik, *Israel J. Chem.*, **6**, 22 (1968).
29. O. Fuchs, *Angew. Chem., Int. Ed. Engl.*, **7**(5), 394 (1968).
30. M. Tsuda, *Makromol. Chem.*, **72**, 174, 183 (1964).
31. D.M. Jones, *Adv. Carbohydr. Chem.*, **19**, 229 (1964).
32. E.M. Fetters, in *Crystalline Olefin Polymers*, Part II (R.A. Raff and K.W. Doaks, eds.), Interscience, New York, 1964.
33. S. Tazuke, *Makromol. Chem., Suppl.*, **14**, 145 (1985).
34. M.R. Gomez-Anton, J.G. Rodriguez, and I.F. Pierola, *Macromolecules*, **19**, 2932 (1986).
35. A.D. Roberts (ed.), *Natural Rubber Science and Technology*, Oxford University Press, New York, 1988.
36. H. Staudinger and H. Staidinger, *J. Prakt. Chem.*, **162**, 148 (1943); *Rubber Chem. Technol.*, **17**, 15 (1944).
37. C.S. Ramakrishnan, D. Raghunath, and J.B. Ponde, *Trans. Inst. Rubber Ind.*, **29**, 190 (1953); *ibid.*, **30**, 129 (1954).
38. M. Troussier, *Rubber Chem. Technol.*, **29**, 302 (1956).
39. C.F. Broomfield, *J. Chem. Soc.*, 114 (1944); *Rubber Chem. Technol.*, **17**, 759 (1944).
40. C.S. Ramakrishnan, D. Raghunath, and J.B. Ponde, *Trans. Inst. Rubber Ind.*, **30**, 129 (1954); *Rubber Chem. Technol.*, **28**, 598 (1955).
41. I.A.Tutorskii, L.V. Sokolova, and B.A. Dogadkin, *Vysokomol. Soyed.*, **A13** (4), 952 (1971).
42. J. Royo, L. Gonzalez, L. Ibarra, and M. Barbero, *Makromol. Chem.*, **168**, 41 (1973).
43. G. Dall'Asta, P. Meneghini, and U. Gennaro, *Makromol. Chem.*, **154**, 279 (1972).
44. G. Dall'Asta, P. Meneghini, I.W. Bassi, and U. Gennaro, *Makromol. Chem.*, **165**, 83 (1973).
45. I.W. Bassi and R. Scordamaglia, *Makromol Chem.*, **166**, 283 (1973).
46. F. Stelzer, K. Hummel, C. Graimann, J. Hobisch, and M.G. Martl, *Makromol. Chem.*, **188**, 1795 (1987).
47. K. Hummel, *J. Mol. Catal.*, **28**, 381 (1987); K. Hummel, *Pure Appl. Chem.*, **54**, 351 (1982).
48. Y.G. Garbachev, K.A. Garbatova, O.N. Belyatskaya, and V.E. Gul, *Vysokomol. Soed.*, **7**, 1645 (1965).
49. P.J. Canterino, *Ind. Eng. Chem.*, **49**, 712 (1957).
50. E. Ceausescu, R. Bordeianu, E. Buzdugan, I. Cerchez, P. Ghioca and R. Stancu, *J. Macromol. Sci.-Chem.*, **A22**, 803 (1985).
51. M. Bruzzone and A. Carbonaro, *J. Polym. Sci., Chem. Ed.*, **23**, 139 (1985).
52. Y. Doi, A. Yano, K. Soga, and D.R. Burfield, *Macromolecules*, **19**, 2409 (1986).
53. N.A. Mohammadi and G.L. Rempel, *Macromolecules*, **20**, 2362 (1987).
54. H. Bradbury and M.C. Semale Perera, *Brit. Polym. J.*, **18**, 127 (1986); A. Konietzny and U. Bietham, *Angew. Makromol. Chem.*, **74**, 61 (1978).
55. S. Siddiqui and R.E. Cais, *Macromolecules*, **19**, 595 (1986).
56. S. Siddiqui and R.E. Cais, *Macromolecules*, **19**, 998 (1986).
57. R.E.Cais and S. Siddiqui, *Macromolecules*, **20**, 1004 (1987).
58. H. Stuger and R. West, *Macromolecules*, **18**, 2349 (1985).
59. H.J. Prins, *Chem. Weekblad*, **17**, 932 (1917); *ibid.*, **16**, 1510 (1919).
60. F. Kirchhoff, *Chem. Ztg.*, **47**, 513 (1923).
61. J. McGavack, U.S. Pat. # 1,640,363 (Aug. 30, 1927).
62. D.F. Twiss and F.A. Jones, British Pat. # 348 303 (March 24, 1930).
63. D.F. Twiss and F.A. Jones, British Pat. # 523 734 (July 22, 1940).
64. G.H. Latham, U.S. Pat. # 2,417,424 (March 18, 1947).
65. S. Hirano and R. Oda, *J. Soc. Chem. Ind. (Japan)*, **47**, 833 (1944); *Chem. Abstr.*, **42**, 7982 (1948).
66. D.F. Twiss and F.A. Jones, U.S. Pat. # 1,915,808 (June 27, 1933).
67. C. Harris, *Ber.*, **37**, 2708 (1904).
68. C. Pinazzi and R. Pautrat, *Compt. Rend.*, **254**, 1997 (1962); *Chem. Abstr.*, **56**, 15647 (1962).
69. C. Pinazzi and R. Pautrat, *Rev. Gen. Coutchuc*, **39**, 799 (1962).
70. C. Pinazzi, R. Pautrat, and R. Cheritat, *Compt. Rend.*, **256**, 2390, 2607 (1963).
71. S.K. Lee. *Polym. Prepr., Am. Chem. Soc.*, **8**, 700 (1967).
72. Y. Minoura and H. Ikeda, *J. Appl. Polym. Sci.*, **15**, 2219 (1971).
73. G. Caraculacu and I. Zugravescu, *J. Polym. Sci., Polym. Lett.*, **6**, 451 (1968).
74. C.G. Gubelein, *J. Macromol. Sci.-Chem.*, **A5**, 433 (1971).
75. K. Toda, Y. Numata, and T. Katsumura, *J. Appl. Polym. Sci.*, **15**, 2219 (1971).
76. K.W. Leong and G.B. Butler, *J. Macromol. Sci.-Chem.*, **A14**, 287 (1980); C.S.L. Baker and D. Bernard, *Polym. Prepr., Am. Chem. Soc.*, **26** (2), 29 (1985).
77. D.A. Bansleben and F. Jachimowicz, *Polym. Prepr., Am. Chem. Soc.*, **26** (1), 106 (1985).
78. M. Anavi and A. Zilkha, *Eur. Polym. J.*, **5**, 21 (1969).
79. J.W. Mench and B. Fulkerson, *Ind. Eng. Chem.*, **7**, 2 (1968).
80. J.P. Kennedy and R.A. Smith *J. Polym. Sci., Chem. Ed.*, **18**, 1523 (1980).

81. B. Ivan, J.P. Kennedy and V.S.C. Chang, *J. Polym. Sci., Chem. Ed.* **18**, 3177 (1980).
82. J.P. Kennedy, V.S.C. Chang, and W.P. Francik, *J. Polym. Sci., Chem. Ed.*, **20**, 2809 (1982).
83. W.A. Thaler, S.J. Brois, and F.W. Ferrara, *Macromolecules*, **20**, 254 (1987).
84. E. Ceausescu, S. Bittman, V. Fieroiu, E. Badea, E. Gruber, A. Ciupitoiu, and V. Apostol, *J. Macroml. Sci.-Chem.*, **A22** (5–7), 525 (1985).
85. T. Nishikubo, T. Shimokawa, and H. Arita, *Makromol. Chem.*, **188**, 2105 (1987).
86. T. Katayama and H.K. Hill Jr., *Macromolecules*, **20**, 1451 (1987).
87. H.L. Cohen, *J. Polym. Sci., Chem. Ed.*, **22**, 2293 (1984).
88. G.J. Carlson, J.R. Skinner, C.W. Smith, and C.H. Wilcoxen Jr., U.S. Pat. # 2,870,171 (May 6, 1958).
89. C.W. Smith and G.P. Payne, U.S. Pat. # 2,786,854 (March 26, 1957).
90. C.M. Gable, U.S. Pat. # 2,870,171 (Jan. 20, 1959).
91. W.C. Smith, *Recl. Trav. Chim. Pays-Bas*, **29**, 686 (1930); *Chem. Abstr.*, **24**, 4261 (1930).
92. F.P. Greenspan and R.E. Light Jr., U.S. Pat. # 2,829,130 (April 1958).
93. S. Tocker, U.S. Pat. # 3,043,818 (July 1962).
94. R.W. Rees, U.S. Pat. # 3,050,507 (August 1962).
95. H. Lee and K. Neville, *Handbook of Epoxy Resins*, McGraw-Hill, New York, 1967.
96. W.G. Potter, *Epoxy Resins*, Springer-Verlag, New York, 1970.
97. C.A. May (ed.), *Epoxy Resins: Chemistry and Technology,* 2nd ed., Dekker, New York, 1988.
98. A.F. Chadwick, D.O. Barlow, A.A. D'Adieco, and J.W. Wallace, *J. Am. Oil Chem. Soc.*, **35**, 355 (1958).
99. K. Meyersen and J.Y.C. Wang, *J. Polym. Sci.*, **A,5**, 725 (1967).
100. H.S. Makowski, M. Lynn, and D.H. Rotenberg, *J. Macromol. Sci.-Chem.*, **A4**, 1563 (1980).
101. D. Zuchowska, *Polymer*, **21**, 514 (1980).
102. M.A. Golub, *J. Am. Chem. Soc.*, **80**, 1794 (1958).
103. M.a. Golub, *J. Am. Chem. Soc.*, **81**, 54 (1959).
104. M.A. Golub, *J. Am. Chem. Soc.*, **82**, 5093 (1960).
105. M.A. Golub, *J. Phys. Chem.*, **66**, 1202 (1962).
106. M.A. Golub and C.L. Stephens, *J. Polym. Sci.*, **C-6** (16), 765 (1967).
107. R.W. Lenz, K. Ohata, and J. Funt, *J. Polym. Sci.-Chem. Ed.*, **11**, 2273 (1973).
108. M.A. Golub and J. Hiller, *Can. J. Chem.*, **41**, 937 (1963).
109. M.A. Golub and J. Hiller, *Tetrahedron Lett.*, 2137 (1963).
110. R.J. Angelo, M.L. Wallach, and R.M. Ikeda, *Polym. Prepr., Am. Chem. Soc.*, **8** (1), 221 (1967).
111. M.A. Golub and J. Heller, *J. Polym. Sci., Polym. Lett.*, **4**, 469 (1966).
112. N.G. Gaylord, I. Kossler, M. Stolka, and J. Vodehnal, *J. Am. Chem. Soc.*, **85**, 641 (1963).
113. N.G. Gaylord, I. Kossler, M. Stolka, and J. Vodehnal, *J. Polym. Sci.*, **A-1,2**, 3969 (1964).
114. M. Stolka, J. Vodehnal, and I. Kossler, *J. Polym. Sci.*, **A-1,2**, 3987 (1964).
115. I. Kossler, J. Vodehnal, and M. Stolka, *J. Polym. Sci.*, **A-1,3**, 2081 (1965).
116. P. Ehrburger and J.B. Donnet, in *High Technology Rubbers*, Part A (M. Lewin and J. Preston, eds.), Dekker, New York, 1985; G. Henrici-Olive and S. Olive, *Adv. Polym. Sci.*, **51**, 1 (1983); E. Fitzer, *Angew. Chem., Int. Ed. Engl.*, **19**, 375 (1980).
117. E.M. Bevilacqua, *Rubber Chem. Technol.*, **30**, 667 (1957).
118. E.M. Bevilacqua, *J. Am. Chem. Soc.*, **79**, 2915 (1957).
119. E.M. Bevilacqua, *Science*, **126**, 396 (1957).
120. E.M. Bevilacqua, *J. Am. Chem. Soc.*, **81**, 5071 (1959).
121. G. Salmon and Chr. van der Schee, *J. Polym. Sci.*, **14**, 287 (1954).
122. M.A. Golub and C.L. Stephens, *J. Polym. Sci.*, **A-1,6**, 763 (1968).
123. M.A. Golub, *J. Polym. Sci., Polym. Lett.*, **12**, 615 (1974); *ibid.*, **12**, 295 (1974).
124. M.A. Golub and J.L. Rosenberg, *J. Polym. Sci., Chem. Ed.*, **18**, 2548 (1980).
125. M.A. Golub, *J. Polym. Sci., Chem. Ed.*, **19**, 1073 (1981).
126. R.W. Lenz, K. Ohata, and J. Funt, *J. Polym. Sci., Chem. Ed.*, **11**, 2273 (1973).
127. S.B. Maerov, *J. Polym. Sci.*, **A-1,3**, 487 (1963).
128. F. Millich and R. Sinclair, *Chem. Eng. News*, p. 30 (June 26, 1967).
129. R.H. Michel and W.A. Murphy, *J. Polym. Sci.*, **55**, 741 (1961).
130. S. van Paeschen, *Makromol. Chem.*, **63**, 123 (1963).
131. J.A. Banchette and J.D. Cotman Jr., *J. Org. Chem.*, **23**, 117 (1958).
132. H. Morawetz and B. Vogel, *J. Am. Chem. Soc.*, **91**, 563 (1969).
133. H. Morawetz and G. Gordimer, *J. Am. Chem. Soc.*, **92**, 7532 (1970).
134. G.A. Gordon, R.E. Cohen, R.F.Baddour, *Macromolecules*, **18**, 98 (1985).
135. A.J. Rudge, British Pat. # 710,523 (June 16, 1954).
136. I. Vogt and W. Krings, German Pat. # 1,086,891 (Aug. 11, 1960).
137. M. Okada and K. Makuuchi, *Kagyo Kagaku Zashi*, **73**, 1211 (1970) (from a private English translation).

138. H. Schonhorn, P.K. Gallagher, J.P. Luongo, and F.J. Padden Jr., *Macromolecules*, **3**, 800 (1970).
139. W.T. Miller and A.L. Dittman, *J. Am. Chem. Soc.*, **78**, 2793 (1956).
140. J.R. Myles and P.J. Garner, U.S. Pat. # 2,422,919 (June 24, 1947).
141. W.N. Baxter, U.S. Pat. # 2,849,431 (Aug. 26, 1958).
142. L.B. Krentsel, A.D. Litmanovich, I.V. Patsukhova, and V.A. Agasandyan, *Vysokomol. Soyed.*, **A13**, 2489 (1971).
143. M. Goren, *Polim. Vehomerim Plast.*, #1, 8 (1971).
144. R.S. Taylor, U.S. Pat. # 2,592,763 (April 20, 1952).
145. H.W. Thompson and P. Torkington, *Trans. Faraday Soc.*, **41**, 254 (1945).
146. K. Nambu, *J. Appl. Polym. Sci.*, **4**, 69 (1960).
147. T. Nakagawa and S. Yamada, *J. Appl. Polym. Sci.*, **16**, 1997 (1972).
148. G.D. Jones, in *Chemical Reactions of Polymers* (E.M. Fetters, ed.), Wiley-Interscience, New York, 1964.
149. British Pat. # 811,848 (April 15, 1959); *Chem. Abstr.*, **53**, 23100f (1951).
150. H. Noeske and O. Roeleau, U.S. Pat. # 2, 889,259 (June 2, 1959).
151. A. McAlevy, U.S. Pat. # 2,405,971 (Aug. 20, 1946).
152. R.E. Brooks, E.D. Straom, and A. McAlevy, *Rubber World*, **127**, 791 (1953).
153. G. Natta, G. Mazzanti, and M. Buzzoni, Italian Pat. # 563,508 (April 24, 1956).
154. K. Konishi, K. Yamaguchi, and M. Takahisha, *J. Appl. Polym. Sci.*, **15**, 257 (1971).
155. P. Bertican, J.J. Bejat, and G. Vallet, *J. Chim. Phys, Physicochim. Biol.*, **67**, 164 (1979).
156. R.T. Sikorski and E. Czerwinska, *Polymer*, **25**, 1371 (1984).
157. R.T. Sikorski and E. Czerwinska, *Eur. Polym. J.*, **22**, 179 (1986).
158. C. Decker, *J. Polym. Sci., Polym. Lett.*, **25**, 5 (1987).
159. G. Svegliado and F.Z. Grandi, *J. Appl. Polym. Sci.*, **13**, 1113 (1969).
160. M. Kolinsky, D. Doskocilova, B. Schneider, and J. Stokr, *J. Polym. Sci.*, **A-1,9**, 791 (1971).
161. Backskai, L.P. Lindeman, and J.W. Adams, *J. Polym. Sci.*, **A-1,9**, 991 (1971).
162. G.G. Scherer, P. Pfluger, H. Braun, J. Klein, and H. Weddecke, *Makromol. Chem., Rapid Commun.*, **5**, 611 (1984).
163. M. Takeishi and M. Okawara, *J. Polym. Sci., Polym. Lett.*, **7**, 201 (1969).
164. M. Takeishi and M. Okawara, *J. Polym. Sci., Polym. Lett.*, **8**, 829 (1970).
165. M. Okawara, T. Endo, and Y. Kurusu, in *Progress of Polym. Science, Japan* (K. Imahori, ed.), Vol. 4, p. 105, 1972.
166. K. Mori and Y. Nakamura, *J. Polym. Sci., Polym. Lett.*, **9**, 547 (1971).
167. G. Levin, *Makromol Chem., Rapid Commun.*, **5**, 513 (1984).
168. M.K. Naqvi and P. Josph, *Polym. Commun.*, **27**, 8 (1986).
169. Z. Wolkober and I. Varga, *J. Polym. Sci.*, *C* (16), 3059 (1967).
170. H. Calvayrac and J. Gole, *Bull. Soc. Chim. France*, 1076 (1986).
171. E. Roman, G. Valenzuela, L. Gargallo, D. Radic, *J. Polym. Sci.*, **21**, 2057 (1983).
172. D. Braun and E. Seeling, *Makromol. Chem.*, **86**, 124 (1965).
173. W. Hahn and W. Muller, *Makromol. Chem.*, **16**, 71 (1955).
174. M. Okawara, *Coatings and Plastics Chemistry Preprints, Am. Chem. Soc.*, **40**, 39 (1979).
175. Y. Nakamura, K. Mori, and M. Kaneda, *Nippon Kagaku Kaishi*, 1620 (1976); from Ref. # 164.
176. R.K. Jenkings, N.R. Byrd, and J.L. Lister, *J. Appl. Polym. Sci.*, **12**, 2059 (1968).
177. R.A. Haldon and J.N. Hay, *J. Polym. Sci.*, **A-1,5**, 2295 (1967).
178. J.C. Bevington and L. Rotti, *Eur. Polym. J.*, **8**, 1105 (1972).
179. R. Tarascon, M. Hartney, M.J. Bowden, *Polym. Prepr., Am. Chem. Soc.*, **25** (1), 289 (1984).
180. T. Saigusa and R. Oda, *Bull. Inst. Chem. Res., Kyoto Univ.*, **33,** 126 (1955); *Chem. Abstr.*, **50**, 1357 (1956).
181. F. Helfferich, *Ion Exchange*, McGraw-Hill, New York, 1962; G.A. Olah and W.S. Tolgyesi, in *Friedel-Craft and Related Reactions* (G.A. Olah, ed.), Wiley-Interscience, New York, 1964; W.E. Wright, E.G. Topikar, and S.A. Svejda, *Macromolecules*, **24**, 5879 (1991).
182. G. Pozniak and W. Trochimszuk, *J. Appl. Polym. Sci.*, **27**, 1833 (1982).
183. C.H. Bamford and H. Linsay, *Polym.*, **14**, 330 (1973).
184. R. Hauptman, F. Wolf, and D. Warnecke, *Plaste Kautsch.*, **18**, 330 (1971).
185. E.E. Ergojin, S.P. Rafikov, and B.A. Muhidtina, *Izv. Akad. Nauk Kaz. SSR*, **19** (16), 49 (1969).
186. G.Z. Esipov, L.A. Derevyanki, V.P. Markidin, R.G. Rakhuba, and K.L. Poplavskii, *Vysokomol. Soyed.*, **B12**, 274 (1970).
187. S. Belfer, R. Glozman, A. Deshe, and A. Warshawsky, *J. Appl. Polym. Sci.*, **25**, 2241 (1980).
188. S.R. Rafikov, G.N. Chelnokova, and G.M. Dzhilkibayeva, *Vysokomol. Soyed.*, **B10**, 329 (1968).
189. L. Galeazzi, German Pat. # 2,455,946 (June 1975).
190. W.H. Daly, *J. Macromol. Sci., Chem.*, **A22**, 713 (1985).
191. S. Mohanraj and W.T. Ford, *Macromolecules*, **19**, 2470 (1986).
192. S.R. Rafikov, G.M. Dzhilkibayeva, G.N. Chelkova, and G.B. Shaltuper, *Vysokomol. Soyed.*, **A12** (7) 1608 (1970).
193. P. Hodge and J. Waterhouse, *Polymer*, **22**, 1153 (1981).

194. M. Hassanein, A. Akelah, *Polym. Prepr., Am. Chem. Soc.*, **26** (1), 88 (1985).
195. T.D. N'guyen, J.C. Gautier, and S. Boileau, *Polym. Prepr., Am. Chem. Soc.*, **23**, 143 (1982).
196. P. Hodge, B.J. Hunt, E. Khoshdel, and J. Waterhouse, *Polym. Prepr., Am. Chem. Soc.*, **23**, 147 (1982).
197. P. Hodge, B.J. Hunt, J. Waterhouse, and A. Wightman, *Polym. Commun.*, **24**, 70 (1983).
198. M. Takeishi and N. Umeta, *Makromol. Chem., Rapid Commun.*, **3**, 875 (1982) .
199. B.N. Kolarz and A. Rapak, *Makromol. Chem.*, **185**, 2511 (1984).
200. T. Iizawa, T. Nishikubo, Y. Masuda, and M. Okawara, *Macromolecules*, **17**, 992 (1984).
201. G.D. Jones and S.J. Goetz, *J. Polym. Sci.*, **25**, 201 (1957).
202. G.D. Jones, *Ind. Eng. Chem.*, **44**, 2686 (1952).
203. F. Marinola and G. Naumann, *Angew. Makromol. Chem.*, **4/5**,185 (1968).
204. S. Dragan, I. Petrariu, and M. Dima, *J. Polym. Sci., Chem. Ed.*, **10**, 3077 (1986).
205. S. Dragan, V. Barboiu, D. Csergo, I. Petrariu, and M. Dima, *Makromol. Chem.*, **187**, 157 (1986).
206. R.E. Kesting, *Synthetic Polymeric Membranes*, McGraw-Hill, New York, 1971.
207. W. Chen, R. Mesrobian, and D. Ballantine, *J. Polym. Sci.*, **23**, 903 (1957).
208. G.V. Kharitonov and L.P. Belikova, *Vysokomol. Soyed.*, **A11** (11), 2424 (1969) .
209. W.O. Kenyon and G.P. Waugh, *J. Polym. Sci.*, **32**, 83 (1958); C.C. Unruh, *J. Appl. Polym. Sci.*, **2**, 358 (1959).
210. W.R. Bussing and N.A. Peppas, *Polymer*, **24**, 209 (1983).
211. D.G. Barar, K.P. Staller, and N.A. Peppas, *J. Polym. Sci., Chem. Ed.*, **21**, 1013 (1983).
212. R. Singer, A. Demagistri, and C. Miller, *Makromol. Chem.*, **18/19**, 139 (1956).
213. R. Singer and A. Demagistri, *J. Chim. Phys.*, **47**, 704 (1950).
214. M. Baer, U.S. Pat. # 2,533,211 (Dec. 12, 1950).
215. R. Singer, U.S. Pat. # 2,604, 456 (July 22, 1952).
216. B. Blaser, M. Rugenstein, and T. Tischbirek, U.S. Pat. # 2,764,576 (Sept. 25, 1956).
217. H.H. Roth, *Ind. Eng. Chem.*, **46**, 2435 (1954).
218. J. Eichhorn, U.S. Pat. # 2,877,213 (Mar. 10, 1959).
219. C.F.H. Aleen and L.M. Minsk, U.S. Pat. #2,735,841 (Feb. 21, 1956) .
220. J. Blyth and A.W. Hofman, *Ann. Chem.*, **53**, 316 (1984).
221. H. Zenftman, *J. Chem. Soc.*, 820 (1950).
222. A.S. Matlack and D.S. Breslow, *J. Polym. Sci.*, **45**, 265 (1960).
223. J.A. Blanchette and J.D. Cotman Jr., *J. Org. Chem.*, **23**, 117 (1958).
224. F. Cramer, H. Helbig, H. Hettler, K.H. Scheit, and H. Seliger, *Angew. Chem.*, **78**, 64 (1966); *ibid., Int. Ed. Engl.*, **5**, 601 (1966).
225. B. Philipp, G. Reinisch, and G. Rafler, *Plaste Kautsch.*, **3**, 190 (1972).
226. W. Kern, R.C. Schulz, and D. Braun, *Chem. Ztg./Chem. Apparatur*, **84** (12), 385 (1960).
227. G. Greber and J. Toelle, *Makromol. Chem.*, **52**, 208 (1962).
228. R.P. Quirk and P.L. Chemg, *Macromolecules*, **19**, 1291 (1986).
229. D. Braun, *Makromol. Chem.*, **30**, 85 (1959).
230. D. Braun, *Makromol. Chem.*, **33**, 181 (1959).
231. D. Braun, H. Daimond, and G. Becker, *Makromol. Chem.*, **62**, 183 (1963).
232. D. Braun, *Chimia*, **14**, 24 (1960).
233. Y. Yamada and M. Okawara, *Makromol. Chem.*, **152**, 153 (1972).
234. Y. Yamada and M. Okawara, *Makromol. Chem.*, **152**, 163 (1972).
235. K. Chander, R.C. Anand, and I.K. Varma, *J. Macromol. Sci.-Chem.*, **A20**, 697 (1983).
236. J. Petit and B. Houel, *Comp. Rend.*, **246**, 1427 (1958).
237. B. Houel, *Comp. Rend.*, **246**, 2488 (1958).
238. R.C. Schulz, P. Elzer, and W. Kern, *Chimia*, **13**, 237 (1959).
239. C.S. Marvel and J.H. Shpherd, *J. Org. Chem.*, **24**, 599 (1961).
240. H.L. Cohen, D.G. Borden, and L.M. Minst, *J. Org. Chem.*, **26**, 1274 (1959).
241. W. Frey and E. Klesper, *Eur. Polym. J.*, **22**, 735 (1986).
242. H.L. Cohen and L.M. Minsk, *J. Org. Chem.*, **24**, 1404 (1959).
243. J.A. Banchette and J.D. Cotman, *J. Org. Chem.*, **23**, 1117 (1958).
244. W. Kern and R.C. Schulz, *Angew. Chem.*, **69**, 153 (1957).
245. R. Vartan-Boghossian, B. Dederichs, and E. Klesper, *Eur. Polym. J.*, **22**, 23 (1986).
246. R. Roussel, M.O. De Guerrero, and J.C. Gallin, *Macromolecules*, **19**, 291 (1986).
247. S. Rondou, G. Smets, and M.C. De Wilde-Delvaux, *J. Polym. Sci.*, **24**, 261 (1957).
248. M. Vrancken and G. Smets, *J. Polym. Sci.*, **14**, 521 (1954).
249. M. Mullier and G. Smets, *J. Polym. Sci.*, **23**, 915 (1957).
250. A. Ravve, *J. Polym. Sci.*, **A-1,6**, 2889 (1968).
251. G. Van Paeschen, *Makromol. Chem.*, **63**, 123 (1963).
252. H. Boudevska, *Izv. Bulgar. Acad. Sci.*, **3**, 303 (1970).

253. C.E. Carraher Jr. and L.S. Wang, *Makromol. Chem.*, **152**, 43 (1972).
254. V. Janout and P. Cafelin, *Makromol. Chem.*, **182**, 2989 (1981).
255. N.S. Bondareva and E.N. Rastovskii, *Zh. Prikl. Khim. (Leningrad)*, **43** (1), 215 (1970).
256. I.V. Andreyeva, MM. Kotin, L.Ya. Modreskay, Ye.I. Pokrovskii, and G.V. Lyubimova, *Vysokomol. Soyed.*, **A14** (7), 1565 (1972).
257. K. Ogasawara, N. Yamagami, and S. Matuzawa, *Angew. Makromol. Chem.*, **25**, 15 (1972).
258. K. Imai, T. Shiomi, T. Tsuchida, C. Watanabe, and S. Nishioka, *J. Polym. Sci.*, **A-1,21**, 305 (1983).
259. K. Imai, T. Shiomi, Y. Tezuka, and M. Miya, *J. Polym. Sci., Chem. Ed.*, **22**, 841 (1984).
260. K. Ichimura, *J. Polym. Sci., Chem. Ed.*, **A-1,20**, 1411 (1982).
261. K. Ichimura and S. Watanabe, *J. Polym. Sci., Chem. Ed.*, **A-1,20**, 1419 (1982).
262. M. Tsuda, H. Tanaka, H. Tagami, and F. Hori, *Makromol. Chem.*, **167**, 183 (1973).
263. P. Gramain and J. Le Moigue, *Eur. Polym. J.*, **8**, 703 (1972).
264. M. Anavi and A. Zilkha, *Eur. Polym. J.*, **5**, 21 (1969).
265. L. Merle-Aubry, Y. Merle, and E. Selegny, *Makromol. Chem.*, **172**, 115 (1973).
266. L.X. Mallavarapu and A. Ravve, *J. Polym. Sci.*, **A,3**, 593 (1965) .
267. G. Reinisch and Dietrich, *Eur. Polym. J.*, **6**, 1269 (1970).
268. E.J. Gunster and R.S. Schulz, *Makromol. Chem.*, **180**, 1891 (1979).
269. J. Klein and E. Klesper, *Makromol. Chem., Rapid Commun.*, **5**, 701 (1984).
270. H.R. Kricheldorf, P. Jahnke, and N. Scharnag, *Macromolecules*, **25**, 1382 (1992).
271. H.S.J. Lee and W.P. Weber, *Macromolecules*, **24**, 4749 (1991).
272. H.L. Fisher, *Chemistry of Natural and Synthetic Rubber*, Reinhold, New York, 1957.
273. P.W. Allen, D. Barnard, and B. Saville, *Chem. Brit.*, **6**, 382 (1970).
274. C.G. Moore and J. Scanlan, *J. Polym. Sci.*, **43**, 23 (1960).
275. N.G. Gaylord and F.S. Ang, in *Chemical Reactions of Polymers* (E.M. Fetters, ed.), Wiley-Interscience, New York, 1964.
276. L.M. Minsk, J.G. Smith, W.P. van Deusen, and J.F. Wright, *J. Appl. Polym. Sci.*, **2**, 302 (1959).
277. F.I. Sonntag and R. Srinivasan, S.P.E. Photopolymers Meeting, Nov. 6–7, Elenville, New York, 1967.
278. J.L.R. Williams, S.P.E. Photopolymers Meeting, Nov. 6–7, Elenville, New York, 1967.
279. G.M.J. Schmidt, *J. Chem. Soc.*, 2014 (1964).
280. J. Bergman, G.M.J. Schmidt, and F.I. Sonntag, *J. Chem. Soc.*, , 2021 (1964).
281. M.D. Cohen, G.M.J. Schmidt, and F.I. Sonntag, *J. Chem. Soc.*, 2000 (1964).
282. H.G. Curnie, C.C. Natale, and D.J. Kelly, *J. Phys. Chem.*, **71**, 767 (1967).
283. H. Tanaka and E. Otomegawa, *J. Polym. Sci., Chem. Ed.*, **12**, 1125 (1974).
284. A. Reiser, H.M. Wagner, R. Marley, and G. Bowes, *Trans. Faraday Soc.*, **63**, 2403 (1967).
285. A. Reiser and R. Marley, *Trans. Fraday Soc.*, **64**, 1806 (1968).
286. A. Reiser, F.W. Willets, G.C. Terry, V. Williams, and R. Marley, *Trans. Faraday Soc.*, **65**, 3265 (1989).
287. T. Tsunoda and T. Yamaoka, S.P.E. Photopolymer Meeting, Nov. 6–7, Elenville, New York, 1967.
288. S.H. Merrill and C.C. Unruh, *J. Appl. Polym. Sci.*, **7**, 273 (1963).
289. T. Tsunoda, T. Yamaoka, G. Nagamatsu, and M. Hiroheshi, S.P.E. Photopolymer Meeting, Oct. 15–16, Elenville, New York, 1970.
290. G.A. Delzenne and U.L. London, *J. Polym. Sci.*, **C,22**, 1149 (1969) .
291. A. Yabe, M. Tsuda, K. Honda, and H. Tanaka, *J. Polym. Sci.*, **A-1,10**, 2376 (1972).
292. M. Tsuda, *J. Polym. Sci.*, **A-1,7**, 259 (1969).
293. Y. Kurusu, H. Nishiyama, and M. Okawa, *Makromol Chem.*, **138**, 49 (1979).
294. A.Ya. Dringerg, B.M. Fundlyer, G.N. Gorelik, and A.M. Frost, *Zh. Prikl. Khim.*, **32**, 1348 (1959).
295. M. Tsunooka, T. Ueda, and M. Tanaka, *J. Polym. Sci., Polym. Lett.*, **19**, 201 (1981); *ibid., Chem. Ed.*, **22**, 2217 (1984) .
296. M.J. Farrall, M. Alexis, and M. Trecarton, *Polymer*, **24**, 114, (1983); A.O. Patil, D.D. Deshpande, and S.S. Talwar, *Polymer*, **22**, 434 (1981); O. Zimmmer and H. Meier, *J. Chem. Soc., Chem. Commun.*, 481 (1982); M.P. Stevens and A.D. Jenkins, *J. Polym. Sci., Polym. Chem. Ed.*, **17**, 3675 (1979); M. Kata, M. Hasegawa, and T. Ichijo, *J. Polym. Sci.*, **B-8**, 263 (1970).
297. G.E. Green, B.P. Stark, and S.A. Zahir, *J. Macromol. Sci., Rev. Macromol. Chem.*, **C21**, 187 (1981–1982).
298. G. Oster and N. Yand, *Chem. Rev.*, **68**, 125 (1968).
299. D.R. Randell (ed.), *Radiation Curing of Polymers*, CRC Press, Boca Raton, 1987.
300. Z. Wang, F.R.M. McCourt, and D.A. Holden, *Macromolecules*, **25**, 1576 (1992) .
301. R.C. Daly and R.H. Engebrecht, Canadian Pat. # 1,106,544 (1981).
302. W.G. Herkstroeter and S. Farid, *J. Photochem.*, **35**, 71 (1986).
303. A. Ravve, G. Pasternack, K.H. Brown, and S.B. Radlove, *J. Polym. Sci., Chem. Ed.*, **11**, 1733 (1973).
304. M.A. Kraus and A. Patchornik, *J. Polym. Sci., Macromol. Rev.*, **15**, 55 (1980).
305. M.A. Kraus and A. Patchornik, *Israel J. Chem.*, **9**, 269 (1971).

306. B.J. Cohen, M.A. Kraus, and A. Patchornik, *J. Am. Chem. Soc.*, **103**, 7620 (1981).
307. T.H. Mough II, *Science*, **217**, 719 (1982).
308. H.D. Orth and W. Brummer, *Angew. Chem., Int. Ed. Engl.*, **11** (4), 249 (1972).
309. G. Manecke and W. Stock, *Angew. Chem., Int. Ed. Engl.*, **17**, 657 (1978).
310. L. D'Anguiro, G. Mazzola, P. Cremonesi, and B. Focher, *Angew. Makromol. Chem.*, **72**, 31 (1978).
311. M.D. Bednarski and M.R. Callstrom, *J. Am. Chem. Soc.*, **114**, 378 (1992).
312. N. Cinouini, S. Colojna, H. Molinari, and P. Rundo, *J. Polym. Sci., Chem. Commun.*, 394 (1976).
313. S. Kopolow, Z. Machacek, U. Takaki, and J. Smid, *J. Macromol. Sci.-Chem.*, **A7** (5), 1015 (1973).
314. P. Kutchukov, A. Ricard, and C. Quivron, *Eur. Polym. J.*, **16**, 753 (1980).
315. J. Capillon, A. Richard, and C. Quivron, *Polym. Prepr., Am. Chem. Soc.*, **23** (1), 168 (1980).
316. J. Smid, *Polym. Prepr., Am. Chem. Soc.*, **23** (1), 168 (1982).
317. S.L. Regen, *J. Org. Chem.*, **42**, 875 (1977).
318. H. Komeili-Zadeh, H.J.-M. Don, and J. Metzger, *J. Org. Chem.*, **43**, 156 (1978).
319. S. Colonna, R. Fornasier, and U. Pfeiffer, *J. Chem. Soc., Perkin Trans. 1*, 8 (1978).
320. E. Chiellini and R. Solaro, *J. Chem. Soc., Chem. Commun.*, 231 (1977).
321. A. Mastagli, A. Froch, and G. Durr, *C.R. Acad. Sci., Paris*, **235**, 1402 (1952).
322. E.C. Blossey, D.C. Neckers, C. Douglas, A.L. Thayer, and A.P. Schaap, *J. Am. Chem. Soc.*, **95**, 5820 (1973).
323. H.A.J. Battaerd and G.W. Tregear, *Graft Copolymers*, Wiley-Interscience, New York, 1967.
324. P. Dreyfus and R.P. Quirk, "Graft Copolymers," in *Encyclopedia of Polymer Science and Engineering*, Vol 7, 2nd ed. (H.F. Mark, N.M. Bikales, C.G. Overgerger, and G. Menges, eds.), Wiley-Interscience, New York, 1986.
325. S. Basu, J.N. Sen, and S.R. Palit, *Proc. Roy. Soc. (London)*, **A202**, 485 (1950).
326. R.A. Gregg and F.R. Mayo, *J. Am. Chem. Soc.*, **75**, 3530 (1953).
327. D. Lim and O. Wichterle, *J. Polym. Sci.*, **5**, 606 (1961).
328. M. Morton and I. Purma, *J. Am. Chem. Soc.*, **80**, 5596 (1958).
329. S. Okamura and K. Katagiri, *Makromol. Chem.*, **28**, 177 (1958).
330. G.V. Schulz, G. Henrici, and S. Olivé, *J. Polym. Sci.*, **17**, 45 (1955).
331. G. Smets, J. Roovers, and W. Van Humbeek, *J. Appl. Polym. Sci.*, **5**, 149 (1961).
332. G. Smets and M. Claesen, *J. Polym. Sci.*, **8**, 289 (1952).
333. R.A. Hays, *J. Polym. Sci.*, **11**, 531 (1953).
334. R.A. Gregg and F.R. Mayo, *J. Am. Chem. Soc.*, **75**, 3530 (1953) .
335. A. Gsperowicz, M. Kolendowicz, and T. Skowronski, *Polymer*, **23**, 839 (1982).
336. M.S. Gluckman, M.J. Kampf, J.L. O'Brien, T.G. Fox, and R.K. Graham, *J. Polym. Sci.*, **37**, 411 (1959).
337. T.G. Fox, M.S. Guckman, F,B, Gornick, R.K. Graham, and G. Gratch, *J. Polym. Sci.*, **37**, 397 (1959).
338. D.K. Ray Chandhuri and J.J. Hermans, *J. Polym. Sci.*, **48**, 159 (1960).
339. S. Nakamura, E. Yoshikawa, and K. Matsuzaki, *J. Appl. Polym. Sci.*, **25**, 1833 (1988).
340. S. Nakamura, H. Sato, and K. Matsuzaki, *J. Polym. Sci., Polym. Lett.*, **11**, 221 (1973).
341. P.W. Allen and F.M. Merrett, *J. Polym. Sci.*, **22**, 193 (1956).
342. P.W. Allen, G. Ayrey, C.G. Moore, and J. Scanlan, *J. Polym. Sci.*, **36**, 55 (1959).
343. R.J. Ceresa (ed.), *Block and Graft Copolymers*, Wiley-Interscience, New York, 1973.
344. B.D. Gesner, *Rubber Chem. Technol.*, **38**, 655 (1965).
345. P.G. Ghosh and P.K. Sengupta, *J. Appl. Polym. Sci.*, **11**, 1603 (1967).
346. R.E. Wetton, J.D. Moore, and B.E. Fox, *Makromol. Chem.*, **132**, 135 (1970).
347. L.V. Valentine and C.B. Chapman, *Ric. Sci. Suppl.*, **25**, 278 (1955).
348. G.N. Richards, *J. Appl. Polym. Sci.*, **5** (17), 553 (1961).
349. G.I. Simionescu and S. Dumitriu, *J. Polym. Sci., C* **37**, 187 (1972).
350. G. Mino and S. Kaizerman, *J. Polym. Sci.*, **31**, 242 (1958).
351. R.M. Livshits, V.P. Alachev, M.V. Prokofeva, and Z.A. Rogovin, *Vysokomol. Soyed.*, **6** (4), 655 (1964).
352. Y. Iwakura, T. Kurosaki, and Y. Imai, *J. Polym. Sci.*, **A-1, 3**, 1185 (1965); Y. Iwakura, Y. Imai, and K. Yagi, *J. Polym. Sci.*, **A-1, 6**, 801 (1968); *ibid.*, **A-1, 6**, 1625 (1968); Y. Imai, E. Masuhara, and Y. Iwakura, *J. Polym. Sci., Polym. Lett.*, **8**, 75 (1970).
353. Y. Ogiwara, Y. Ogiwara, and H. Kubota, *J. Polym. Sci.*, **A-1, 6**, 1489 (1968).
354. A.A. Gulina, R.M. Livshits, and Z.A. Rogovin, *Vysokomol. Soyed.*, **7** (9) 1693 (1965).
355. S. Lenka and P.L. Nayak, *J. Appl. Polym. Sci.*, **27**, 1959 (1962).
356. A. Hebeish, S. Shalaby, A. Waly, and A. Bayazeed, *J. Polym. Sci.*, **28**, 303 (1983).
357. J.E. Guillet and R.G.W. Norrish, *Nature*, **173**, 625 (1954); *Proc. Roy. Soc. (London)*, **A233**, 172 (1956).
358. G.C. Menon and S.L. Kapur, *J. Appl. Polym. Sci.*, **3**, 54 (1960).
359. N. Geacintov, V. Stannett, and E.W. Abrhamson, *Makromol. Chem.*, **36**, 52 (1959); A. Chapiro, *J. Polym. Sci.*, **29**, 321 (1958); *ibid.*, **34**, 439 (1959).
360. G. Smets, W. De Winter, and G. Delzenne, *J. Polym. Sci.*, **55**, 767 (1961).
361. A. Chapiro, *J. Polym. Sci.*, **48**, 109 (1960).

362. Y. Hachihama and S. Takamura, *Technol. Rep. Osaka Univ.*, **11** (485) 431 (1961); *Chem. Abstr.*, **57**, 6179 (1962).
363. A.J. Restaino and W.N. Reed, *J. Polym. Sci.*, **36**, 499 (1959).
364. A. Noshay and J.E. McGrath, *Block Copolymers*, Academic Press, New York, 1977.
365. G. Gruber and G. Egle, *Makromol. Chem.*, **53**, 206 (1962).
366. G. Sasson and A. Zilkha, *Eur. Polym. J.*, **5**, 315 (1969).
367. R. Waak and M.A. Doran, *Polymer*, **2**, 365 (1961).
368. R. Milkovich, *Polym. Prepr., Am. Chem. Soc.*, **21**, 40 (1980).
369. R. Milkovich and M.T. Chiang, U.S. Pat. # 3,786,116 (1974).
370. M. W. Thompson and F.A. Waite, British Pat. # 1,096,912 (1967).
371. B.W. Jackson, U.S. Pat. # 3,689,593 (1972).
372. J.P. Kennedy, V.S.P. Chang, R.A. Smith, and B. Ivan, *Polym. Bull.*, **1**, 575 (1978).
373. J.P. Kennedy, Preprints, 5th International Symposium on Cationic and Other Ionic Polymerizations, Vol. 6, 1980.
374. R. Asami, M. Takaki, K. Kita, and E. Asakura, *Polym. Bull.*, **3**, 83 (1980).
375. J.S. Vrgas, J.G. Zillox, P. Rempp, and E. Frauta, *Polym. Bull.*, **3**, 83 (1980).
376. P. Rempp, *Polym. Prepr. Jpn.*, **29**, 1397 (1980).
377. Y. Tanizaki, K. Minagawa, S. Takase, and K. Watanabe, Abstracts of the A.C.S.–C.S.J. Chemical Congress, INDE, p. 136 (1979).
378. G. Gerber and E. Reese, *Makromol. Chem.*, **55**, 96 (1962).
379. Y. Yamashita, *J. Appl. Polym. Sci., Appl. Polym. Symp.*, **36**, 193 (1981).
380. R.B. Mesrobian and H. Mark, *Textile J.*, **23**, 294 (1953).
381. D.J. Metz and R.B. Mesrobian, *J. Polym. Sci.*, **16**, 345 (1955).
382. J.D. Matlack, S.N. Chinai, R.A. Guzzi, and D.W. Levi, *J. Polym. Sci.*, **49**, 533 (1961).
383. G. Natta, E. Beati, and F. Severini, *J. Polym. Sci.*, **34**, 685 (1959).
384. D. Makulova and D. Berek, *Sb. Prac. Chem. Fak. Sloven Vysokej. Skily Tech.*, **141** (1961); *Chem. Abstr.*, **57**, 8729 (1962).
385. J.J. Wu, Z.A. Rogovin, and A.A. Konkin, *Khim Volokna*, **6**, 11 (1962); *Chem. Abstr.*, **58**, 11509 (1963).
386. E. Beati, S. Toffano, and F. Severini, *Chem. Ind. (Milan)*, **45**, 690 (1963); *Chem. Abstr.*, 7669 (1963).
387. V.V. Korshak, K.K. Mozgova, and M. Shkolina, *Vysokomol. Soyed.*, **4**, 1469 (1962).
388. V.V. Korshak, K.K. Mozgova, and S.P. Krukovskii, *Sov. Plast. (Engl. Transl.)*, **7**, 6 (1963).
389. V.V. Korshak, K.K. Mozgova, M. Schkolina, I.P. Nagdasova, and V.A. Berestnev, *Vysokomol. Soyed.*, **5**, 171 (1963).
390. J. Hutchison and A. Ledwith, *Polymer*, **14**, 405 (1973) .
391. A. Bar-Ilan and A. Zilkha, *Eur. Polym. J.*, **6**, 403 (1970).
392. G. Ezra and A. Zilkha, *J. Appl. Polym. Sci.*, **13**, 1493 (1969).
393. G. Kockelbergh and G. Smets, *J. Polym. Sci.*, **33**, 227 (1958).
394. N.J. Turro, *Molecular Photochemistry*, W.A. Benjamin, New York, 1967.
395. M. Tsuda, *J. Polym. Sci.*, **7**, 259 (1969); K. Nakamura and S.K. Kuchi, *Bull. Chem. Soc. Japan*, **41**, 1977 (1968).
396. H.G. Heine, H.J. Resenkranz, and H. Rudolph, *Angew. Chem., Int. Ed. Engl.*, **11**, 974 (1974);.
397. C.J. Groeneboom, H.J. Hageman, T. Overeem, and A.J.M. Weber, *Makromol. Chem.*, **183**, 281 (1982).
398. J. Hutchison, M.C. Lambert, and A. Ledwith, *Polymer*, **14**, 250 (1973).
399. A. Ledwith, *Pure Appl. Chem.*, **49**, 431 (1977).
400. W.A. Pryor, *Free Radicals*. McGraw-Hill, New York, 1966.
401. M.J. Gibian, *Tetrahedron Lett.*, 5331 (1967).
402. G. Rausing and S. Sunner, *Tappi*, **45** (1), 203A (1962).
403. V. Jacks and W. Kern, *Makromol. Chem.*, **83**, 71 (1965).
404. P. Weiss, J. F. Gerecht, and I.J. Krems, *J. Polym. Sci.*, **35**, 343 (1959).
405. K. Ito, Y. Tomi, and S. Kawaguchi, *Macromolecules*, **25**, 1534 (1992).
406. S.G. Cohen, H.S. Haas, and H. Slotnick, *J. Polym. Sci.*, **11**, 913 (1953); H.S. Haas, S.G. Cohen, A.C. Oglesby, and E.R. Karlin, *J. Polym. Sci.*, **15**, 427 (1955); S.T. Rafikov, G.N. Chelnokova, I.V. Zhuraleva, and P.N. Gribkova, *J. Polym. Sci.*, **53**, 75 (1961) .
407. J.P. Kennedy (ed.), *J. Appl. Polym. Sci., Appl. Polym. Symp.*, **30** (1977); J.P. Kennedy, U.S. Pat. # 3,349,065 (Oct. 24, 1967); J.P. Kennedy, in *Polymer Chemistry of Synthetic Elastomers*, (J.P. Kennedy and E. Tornquist, eds.), Wiley-Interscience, New York, 1968; J.P. Kennedy and J.K. Gillham, *Adv. Polym. Sci.*, **10**, 1 (1972); J.P. Kennedy and M. Nakao, *J. Appl. Polym. Sci., Appl. Polym. Symp.*, **30**, 73 (1977).
408. A. Ravve and C.W. Fitko, *J. Polym. Sci.*, **A-1,4**, 2533 (1966) .
409. N.Betz, A. Le Moel, J.P. Duraud, and C. Darnez, *Macromolecules*, **25**, 213 (1992).
410. E.T. Kang, K.G. Neoh, K.L. Tan, Y. Uyama, N. Morikawa, and Y. Ikada, *Macromolecules*, **25**, 1959 (1992) .
411. B.U. Kolb, P.A. Patton, and T.J. McCarthy, *Polym. Prepr., Am. Chem. Soc.*, **28** (2), 248 (1987).
412. J. Hirkach, K. Ruehl, and K. Matyjaszewski, *Polym. Prepr., Am. Chem. Soc.*, **29** (2), 112 (1988).
413. Y. Jiang and J.M.J. Frechet, *Polym. Prepr., Am. Chem. Soc.*, **30** (1), 127 (1989).

414. J.F. Kenney, *Polym. Eng. Sci.*, **8** (3), 216 (1968); W.H. Charch and J.C. Shivers, *Text. Res. J.*, **29**, 536 (1959) .
415. Imperial Chemical Industries Ltd. British Pat. # 802,921 (Nov. 23, 1955).
416. A. Noshay and J.E. McGrath, *Block Copolymers: Overview and Critical Survey*, Academic Press, New York, 1977.
417. D.H. Solomon, Chapt. 1 in *Step Growth Poymerization* (D.H. Solomon, ed.), Dekker, New York, 1972.
418. J.A. Brydson, *Plastic Materials*, 4th ed., Butterworth Scientific, London, 1982.
419. General Electric Co., British Pat. # 984,522 (June 1, 1960).
420. H. Schollenberger, in *Polymer Technology* (P.F. Bruins, ed.), Interscience, New York, 1969.
421. R. Ukielski and H. Wojcikiewicz, *Int. Polym. Sci. Technol.*, **3** (11), T65 (1976); G.K. Hoexchele and W.K. Witsiepe, *Angew. Makromol. Chem.*, **29–30**, 267 (1973).
422. Y. Yamashita and T. Hane, *J. Polym. Sci., Polym. Chem. Ed.*, **11**, 425 (1973); Y. Yamashita, Y. Murase, and K. Ito, *J. Polym. Sci., Polym. Chem. Ed.*, **11**, 435 (1973).
423. V.V. Korshak and T.M. Frunze, *Synthetic Hetero-Chain Polyamides*, S. Manson, Jerusalem, 1987.
424. P.W. Morgan, *J. Polym. Sci.*, **C4**, 1075 (1964); D.J. Lyman and S.L. Yung, *J. Polym. Sci.*, **40**, 407 (1959).
425. D.J. Lyman and S.L. Yung, *J. Polym. Sci.*, **40**, 407 (1959).
426. V.V. Korshak, S.V. Vinogradova, M.M. Teplyakov, R.D. Fedorova, and G.Sh. Papara, *Vysokomol. Soyed.*, **8** (12), 2155 (1966); V.V. Korshak, S.V. Vinogradova, M.M. Teplyakov, and A.D. Maksomov, *Izv. Vysshk. Uchebn. Zaved. Khim. Khim. Tekhnol.*, **10** (6), 688 (1967) .
427. V.V. Korshak, S.V. Vinogradova, and M.M. Teplyakov, *Vysokomol. Soyed.*, **7** (8), 1406 (1965).
428. Y. Goto and S. Miwa, *Chemistry of High Polymers, Japan*, **25**, 595 (1968).
429. K.L. Mittal, *Polyamides: Syntheses, Characterizations and Applications*, Vols. 1 and 2, Plenum Press, New York, 1984.
430. Imperial Chemical Industries, British Pat. # 807,666 (Jan. 4, 1957).
431. Imperial Chemical Industries, British Pat. # 1,148,068 (Aug. 9, 1965).
432. M. Tsuruta, F. Kabayashi, and K. Matsuyata, *Kogyo Kagaku Zasshi*, **62**, 1084, 1087 (1959); *Chem. Abstr.*, **57**, 13977 (1962).
433. D. Dieterich, W. Keberle, and H. Witt, *Angew. Chem., Int. Ed. Engl.*, **9** (1), 40 (1970) .
434. D. Dieterich, O. Bayer, J. Peter, and E. Muller, German Pat. # 1,156,977 (June 5, 1962).
435. D. Dieterich and O. Bayer, British Pat. # 1,078,202 (1964); W. Keberle and D. Dieterich, British Pat. # 1,078,688 (1964).
436. C.A. Lukach and H.M. Spurlin, in *Copolymerization* (G.E. Ham, ed.), Interscience, New York, 1964; H.F. Mark and S.M. Atlas, *Pet. Refiner*, **39**, 149 (1960); G. Natta, *J. Polym. Sci.*, **34**, 21, 531 (1959).
437. A.V. Tobolsky and K.F. O'Driscoll, DAS 1 114 323, June 24, 1959.
438. M. Szwarc and J. Smid, *Progress in Reaction Kinetics*, Vol. II, p. 219, Pergamon Press, Oxford, 1964; M. Szwarc, *Nature*, **178**, 1168 (1956).
439. S. Bywater, *Fortsch. Hochpolym.-Forsch.*, **4**, 66 (1965).
440. R. Chiang, J.H. Rhodes, and R.A. Evans, *J. Polym. Sci.*, **A-1,4**, 3089 (1966).
441. N.L. Madison, *Polym. Prepr., Am. Chem. Soc.*, **7**, 1099 (1966).
442. H.Yu, A.J. Bur, and L.J. Fetters, *J. Chem. Phys.*, **44**, 2568 (1966).
443. M.P. Dreyfuss and P. Dreyfuss, *Polymer*, **6**, 93 (1965).
444. C.E.H. Bawn, R.M. Bell, and A. Ledwith, *Polymer*, **6**, 95 (1965).
445. D. Sims, *Makromol. Chem.*, **64**, 151 (1963).
446. S. Bywater, *Macromolecular Chemistry*, Butterworths, London, 1962; M. Morton, *Anionic Polymerization: Principles and Practice*, Academic Press, New York, 1983.
447. R.K. Graham, D.L. Dunkelburger, and W.E. Goode, *J. Am. Chem. Soc.*, **82**, 400 (1960).
448. T. Saegusa, S. Matsumoto, and Y. Hashimoto, *Macromolecules*, **3**, 377 (1970).
449. L.J. Fetters, *J. Polym. Sci., C*, **26**, 1 (1969).
450. J.P. Kennedy, *Makromol. Chem., Suppl.*, **3**, 1 (1979).
451. R.J. Ceresa, *Block and Graft Copolymers*, Butterworths, Washington D.C., 1962.
452. H. Staudinger, *Ber.*, **57**, 1203 (1924); *Kautchuk*, **5**, 128 (1929).
453. D.J. Angier and W.F. Watson, *J. Polym. Sci.*, **20**, 235 (1956).
454. R.J. Ceresa, *Polymer*, **1**, 477 (1960).
455. B. Gordon III and L.F. Hancock, *Polym. Prepr., Am. Chem. Soc.*, **26** (1), 98 (1985).
456. P.J. Hans and B. Gordon III, *Polym. Prepr., Am. Chem. Soc.*, **28** (2), 310 (1987).
457. W. Risse and R.H. Grubbs, *Polym. Prepr., Am. Chem. Soc.*, **30**, 193 (1989).
458. D.Y. Sogah and O.W. Webster, *Macromolecules*, **19**, 1775 (1986).
459. D. Patit, R. Jerome, and Ph. Teyssie, *J. Polym. Sci., Polym. Chem. Ed.*, **17**, 2903 (1979); W.T. Allen and D.E. Eaves, *Angew. Makromol. Chem.*, **58/59**, 321 (1977); B.H. Wondraczek and J.P. Kennedy, *J. Polym. Sci., Polym. Chem. Ed.*, **20**, 173 (1982).
460. C.A. Veith and R.E. Cohen, *Polym. Prepr., Am. Chem. Soc.*, **31** (1), 42 (1990).

461. J.P. Kennedy, B. Keszler, Y. Tsunogae, and S. Midha, *Polym. Prepr., Am. Chem. Soc.*, **32** (1), 310 (1991).
462. T.J. Deming and B.M. Novak, *Macromolecules*, **24**, 5478 (1991).
463. T. Higashimura, S. Aoshima, and M. Sawamoto, *Makromol. Chem. Macromol Symp.*, **13/14**, 457 (1988).
464. J.P Kennedy and J.L. Price, *Polym. Mater. Sci.*, 64 (1991); J.P. Kennedy, J.L. Price, and K. Koshimura, *Macromolecules*, **24**, 6567 (1991).
465. T. Pernecker, J.P. Kennedy, and B. Ivan, *Macromolecules*, **25**, 1642 (1992) .
466. M.G. Kanatzidis, *Chem. Eng. News*, p. 38 (Dec. 3, 1990); L.Y. Chiang, P.M. Chaikin, and D.O. Cowan (eds.), "Advanced, Organic Solid State Materials," *Mater. Res. Soc. Symp. Proc.*, **173** (1990); J.R. Reynolds, *Chemtech*, **18**, 440 (1988).
467. X.-F. Sun, S.B. Clough, S. Subrmanyam, A. Blumstein, and S.K. Tipathy, *Polym. Prepr., Am. Chem. Soc.*, **33** (1), 576 (1992).
468. S. Subramanyam and A. Blumstein, *Macromolecules*, **24**, 2668 (1991).
469. R.D. McCullough and R.D. Lowe, *Polym. Prepr., Am. Chem. Soc.*, **33** (1), 195 (1992).
470. W. Rutsch, G. Berner, R. Kirchmayr, R. Husler, G. Rist, and N. Buhler, in *Organic Coatings—Science and Technology* (G.D. Parfitt and A.V. Patsis, eds.), Vol. 8, Dekker, New York, 1986.
471. K. Dietliker, M. Rembold, G. Rist, W. Rutsch, and F. Sitek, Rad-Cure Europe 87, Proceedings of the 3rd Conference Association of Finishing Processes SME, Dearborn, MI, 1987.
472. G. Rist, A. Borer, K. Dietliker, V. Desobry, J.P. Fouassier, and D. Ruhlmenn, *Macromolecules*, **25**, 4182 (1992).

Appendix

The following data are for use with the programs on the diskette included in this volume.

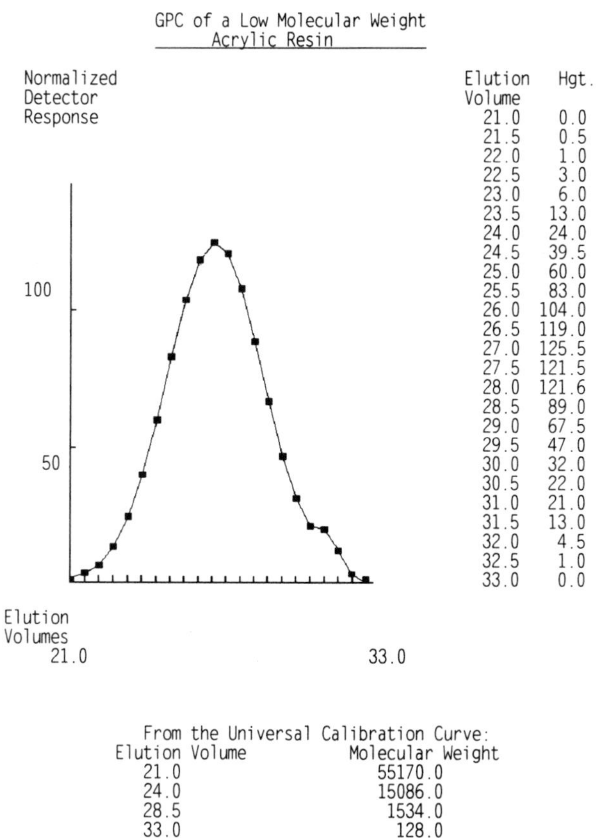

Elution Volume	Hgt.
21.0	0.0
21.5	0.5
22.0	1.0
22.5	3.0
23.0	6.0
23.5	13.0
24.0	24.0
24.5	39.5
25.0	60.0
25.5	83.0
26.0	104.0
26.5	119.0
27.0	125.5
27.5	121.5
28.0	121.6
28.5	89.0
29.0	67.5
29.5	47.0
30.0	32.0
30.5	22.0
31.0	21.0
31.5	13.0
32.0	4.5
32.5	1.0
33.0	0.0

From the Universal Calibration Curve:

Elution Volume	Molecular Weight
21.0	55170.0
24.0	15086.0
28.5	1534.0
33.0	128.0

90 MHz ^{13}C NMR Spectra of the Methyl Region of Polypropylene:

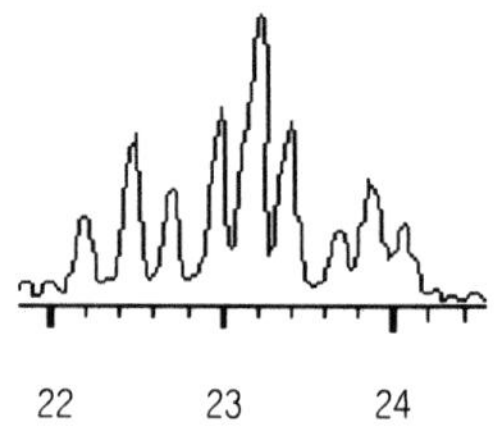

From NMR spectra of atactic polypropylene the observed pentads are:

mmmm	0.186	α = 0.795, σ = 0.231, 8.51
mmmr	0.119	ω = 0.574, β = 0.163
rmmr	0.037	
mmrr	0.107	
mmrm + rmrr	0.168	
rmrm	0.024	
rrrr	0.159	
mrrr	0.120	
mrrm	0.061	

Index